OXFORD MONOGRAPHS ON METEOROLOGY

Editor

P. A. SHEPPARD

THE PHYSICS OF CLOUDS

BY

B. J. MASON, D.Sc., F.R.S.

DIRECTOR-GENERAL METEOROLOGICAL OFFICE
FORMERLY PROFESSOR OF CLOUD PHYSICS
IMPERIAL COLLEGE OF SCIENCE AND TECHNOLOGY, LONDON

SECOND EDITION

CLARENDON PRESS · OXFORD

1971

Oxford University Press, Ely House, London W. 1

GLASGOW NEW YORK TORONTO MELBOURNE WELLINGTON
CAPE TOWN SALISBURY IBADAN NAIROBI DAR ES SALAM LUSAKA ADDIS ABABA
BOMBAY CALCUTTA MADRAS KARACHI LAHORE DACCA
KUALA LUMPUR SINGAPORE HONG KONG TOKYO

© OXFORD UNIVERSITY PRESS 1957, 1971

FIRST EDITION 1957
SECOND EDITION 1971

PRINTED IN NORTHERN IRELAND
AT THE UNIVERSITIES PRESS, BELFAST

Preface to the Second Edition

This book has the same structure and scope as the first edition, but the text has been completely re-written, revised, and enlarged to accommodate the significant advances that have taken place during the intervening decade. It attempts to provide a fairly complete and critical account of all the important developments up to the end of 1969 and includes some important work published in 1970.

Again the emphasis is on the microphysical processes of condensation of water vapour to form droplets, the supercooling, nucleation, and freezing of water droplets, the growth and aggregation of snow crystals, the mechanisms of raindrop, snowflake, and hailstone formation, the radar detection of precipitation elements, and the various processes of cloud electrification, with only sufficient background information on the structure and dynamics of cloud systems to allow realistic discussion of the particle physics. However, in the preceding preface, written thirteen years ago, reference was made to the strong interactions that exist between these microphysical processes and the air motions in and around clouds, and to the importance of acquiring a much deeper understanding of the cloud dynamics for the development of the subject as a whole. The hopes expressed then for much greater effort and progress in this direction, and for the appearance of a major text on this aspect of the subject, have yet to be realized, but there is now a much greater awareness of the need. For, while considerable progress has been made in establishing the physical laws that govern the nucleation, growth, and aggregation of particles, extrapolation of the knowledge obtained under controlled laboratory conditions to the evolution of populations of particles in the vastly more complex environment of a natural cloud requires great caution while our understanding of cloud dynamics is so rudimentary. Indeed it may be difficult to identify the important gaps in our current understanding of the microphysical events until this has been formulated in a dynamical context and tested against the results of observation, measurement, and prediction. The use, in recent years, of new, especially Doppler, radar techniques, of modern methods of data processing, and of numerical models, has given the subject a new impetus in this direction. I hope that this may lead, in the next decade, to advances comparable to those made in the laboratory during the 1950s and 1960s and which form the basis of this book.

Again, I am grateful to the learned societies, publishers, and authors, acknowledged in the text, for permission to reproduce many of the diagrams.

PREFACE TO THE SECOND EDITION

I am indebted to Miss Heather May and Mrs Janet Bolton for their help in preparing diagrams and an earlier typescript of the text, and especially to my present secretary, Miss Eileen Forde, for her invaluable assistance in producing the final manuscript, the index, and the bibliography. Lastly, I wish to acknowledge the skill and care which the Clarendon Press have exercised in the production of the book.

Bracknell
January 1971

B. J. MASON

Preface to the First Edition

CLOUD physics is concerned with those processes which are responsible for the formation of clouds and the release of precipitation. The subject has expanded enormously during the last decade and is now one of the most flourishing branches of meteorological physics. During this short period it has developed from the pursuit of a few individual scientists with modest resources into a group activity requiring elaborately instrumented aircraft, radar, and a wide range of laboratory facilities. This rapid development along a broad front has produced a large and diverse literature, the growth of which has made life increasingly difficult for the research worker and impossible for the student. Accordingly, it appeared that an account of the present state of the subject might serve a useful purpose although the disadvantages of writing a book at a time when the field is rapidly growing have become only too apparent. In this volume I have attempted to give a fairly comprehensive account of recent researches, both experimental and theoretical, on the micro-physical processes of nucleation, condensation, droplet growth, the initiation and growth of ice crystals, and the mechanisms of precipitation release. There follow also discussions on the present status of rain-making experiments, which have stimulated much of the fundamental research; on radar studies of precipitating clouds; and on the electrification of clouds which, I feel, should be regarded as an integral part of cloud physics.

Although the emphasis here is upon the *micro-physical* processes, it is important to recognize that these are largely controlled by the atmospheric motions which are manifest in clouds. These *macro-physical* features of cloud formation and growth, which might more properly be called a *dynamics*, provide a framework of environmental conditions confining the rates and duration of the microphysical events. For example, the growth or freezing of cloud droplets is accompanied by the release of great quantities of latent heat, profoundly influencing the motion of cloudy air masses, while the motions which ultimately cause evaporation of the cloud determine its duration, and will set a limit to the size which its particles can attain. Progress in cloud physics has been hindered by a poor appreciation of these interrelations between processes ranging from nucleation phenomena on the molecular scale to the dynamics of extensive cloud systems on the scale of hundreds or thousands of kilometres.

Unfortunately, our present understanding of the large-scale physics of clouds is rudimentary. This is partly because this aspect of the subject has not received the attention it deserves, but mainly because of the difficulty of obtaining observational information about air motions on a scale too large to be simulated in the laboratory and yet too small to be defined by the observational network used in weather forecasting. Because I am convinced that future progress will be largely governed by our improved understanding of cloud dynamics, I hope that the next few years will see a greatly increased effort in this direction. At this stage it seems advisable to stress only those aspects which are firmly based upon observation and which are necessary to provide an adequate background for discussion of the micro-physical processes. Cloud dynamics will, I hope, eventually form the subject of a separate volume.

In attempting to write a coherent and integrated account of the subject as I see it at present, I have aimed at being critical rather than exhaustive, and consequently the treatment reflects, to some extent, my personal views. In a new and rapidly growing field there is, of course, room for differences of opinion and of interpretation; it is difficult and perhaps undesirable to remain dispassionate in the heat of the battle. However, I have done may best to provide an up-to-date review of the subject and to point out those gaps in our knowledge which appear to merit urgent attention. I shall be pleased if it proves to be of some use to my fellow research workers, and if it draws the attention of the student to a field which offers enormous scope for research into a wide range of phenomena, both in the free atmosphere and in the laboratory, and by theoretical methods. I hope also that I have managed to convey a little of the continual pleasure and excitement which I have experienced while working in this field during the last few years.

In the preparation of this book I have received help from a number of people to whom I should like to express my grateful thanks. To my colleague, Mr. F. H. Ludlam, I am greatly indebted not only for writing the Introduction on 'The Large-scale Physics of Clouds' but also for many stimulating discussions on all aspects of the subject during the last seven years. I am very grateful to those friends who, by sending me their papers in advance of publication, have ensured that this book will appear only a few months out of date. For permission to reproduce many of the diagrams I am indebted to those learned societies, publishers, and authors acknowledged in the text. I am particularly indebted to Mrs. S. M. Devers, Miss E. M. Lea, Miss M. L. J. Pinhard, Mrs. M. Brookfield, and Miss S. A. Latta for their invaluable assistance in preparing the typescript and diagrams, and to the Clarendon Press for the skill and care which they have exercised in the production of the book. Lastly, I owe much to my wife who typed a good deal of the manuscript, and without whose continual help and encouragement this book would not have been written.

B. J. M.

Imperial College, London
February 1957

Contents

1. THE NUCLEATION OF WATER-VAPOUR CONDENSATION

   1.1. Homogeneous condensation — 1
      1.1.1. Thermodynamics of phase changes — 2
      1.1.2. Thermodynamic derivation of Kelvin's formula — 3
      1.1.3. Embryo droplets in statistical equilibrium with the vapour — 5
      1.1.4. Statistical equilibrium between embryos of different sizes — 6
      1.1.5. The unbalanced steady state in a supersaturated system — 7
      1.1.6. The current in the steady state—the rate of formation of droplets — 9
      1.1.7. Time to reach to the steady state — 11
      1.1.8. Further extensions of the theory — 12
      1.1.9. Comparison of theory with experiment — 13
   1.2. Condensation on ions — 17
   1.3. Nucleation by insoluble particles — 20
   1.4. Condensation on soluble particles — 24
   1.5. Condensation on nuclei of mixed constitution — 29
   1.6. Concluding remarks — 30

2. THE NUCLEI OF ATMOSPHERIC CONDENSATION

   2.1. Collection, measurement, and identification of atmospheric nuclei — 32
      2.1.1. Techniques for counting and examining Aitken nuclei (5×10^{-7} cm $< r <$ 0·1 μm) — 32
      2.1.2. Sampling and examination of large and giant nuclei (0·1 μm $< r <$ 10 μm) — 39
   2.2. The concentration, size, and size distribution of atmospheric condensation nuclei — 50
      2.2.1. The concentrations of Aitken nuclei and their variation with place, time, and meteorological factors — 52
      2.2.2. Concentrations and size distributions of large and giant nuclei — 56
   2.3. The chemical composition of atmospheric aerosols — 63
   2.4. The production of natural aerosols — 66
      2.4.1. Production of nuclei by combustion and by chemical reactions — 67
      2.4.2. Production of sea-salt nuclei — 75
      2.4.3. Production of nuclei over the continents — 79
      2.4.4. The growth of nuclei by coagulation — 80

CONTENTS

- 2.5. Nuclei involved in cloud formation ... 83
- 2.6. The removal of aerosols from the troposphere ... 87

3. THE GROWTH OF DROPLETS IN CLOUD AND FOG

- 3.1. The size distribution of cloud droplets ... 92
  - 3.1.1. Experimental techniques ... 92
  - 3.1.2. Results ... 98
  - 3.1.3. Summary of results ... 111
- 3.2. The liquid-water content of clouds ... 113
  - 3.2.1. Techniques ... 113
  - 3.2.2. Results ... 119
- 3.3. Theoretical studies of cloud droplet growth ... 121
  - 3.3.1. Growth of a single droplet by condensation ... 123
  - 3.3.2. Growth of a population of droplets by condensation in cumulus ... 125
  - 3.3.3. The formation of large droplets in small cumulus ... 139
  - 3.3.4. Droplet growth by condensation in layer clouds ... 140
  - 3.3.5. The effect of random fluctuations in supersaturation on the broadening of the droplet spectrum ... 142
  - 3.3.6. Droplet growth by collision and coalescence ... 145
  - 3.3.7. The effect of small-scale turbulence on collisions between droplets ... 153
- 3.4. Concluding remarks ... 154

4. INITIATION OF THE ICE PHASE IN CLOUDS

- 4.1. The supercooling of water containing foreign particles ... 156
- 4.2. The homogeneous nucleation of supercooled water ... 164
- 4.3. Evidence for the supercooling of water below $-40°$ C ... 172
- 4.4. Ice nuclei in the atmosphere ... 174
  - 4.4.1. Experimental techniques ... 174
  - 4.4.2. Concentrations of ice nuclei as a function of temperature ... 183
  - 4.4.3. Variations of ice-nucleus concentration in space and time ... 187
  - 4.4.4. Mode of action of ice nuclei ... 189
  - 4.4.5. Nature and origin of ice nuclei ... 194
- 4.5. Secondary processes of ice-nucleus production ... 207
  - 4.5.1. Fragmentation of snow crystals ... 207
  - 4.5.2. Shattering and splintering of freezing drops ... 207
- 4.6. Artificial ice nuclei ... 212
  - 4.6.1. Ice-nucleating properties of inorganic compounds ... 213
  - 4.6.2. Ice-nucleating properties of organic compounds ... 223
  - 4.6.3. Relation between nucleating properties and crystalline structure ... 226
  - 4.6.4. The production and behaviour of silver iodide smokes ... 227

5. THE FORMATION OF SNOW CRYSTALS

5.1. The classification of solid precipitation	236
5.2. The mass, dimensions, and fall velocities of snow crystals	237
5.3. Fixation and photography of snow crystals	242
5.4. The occurrence of ice crystals in natural clouds	243
5.4.1. The forms of individual snow crystals	243
5.4.2. Aggregation of ice crystals to form snowflakes	248
5.5. Studies of ice crystal growth in the laboratory	251
5.5.1. Variation of crystal habit with temperature and super-saturation	251
5.5.2. The influence of impurities on ice crystal habit	264
5.5.3. The growth of ice crystals in an electric field	265
5.6. The mechanism of habit change	267
5.7. The surface structure of ice crystals	270
5.8. The growth rates of ice crystals	274

6. THE PHYSICS OF NATURAL PRECIPITATION PROCESSES

6.1. Forms of precipitation	282
6.2. Physical processes responsible for release of precipitation	283
6.3. Observational data on the characteristics of precipitating clouds	288
6.3.1. Observations on clouds in middle latitudes	289
6.3.2. Observations on clouds in tropical and subtropical latitudes	294
6.4. The release of precipitation from layer clouds	297
6.4.1. Characteristics and structure of precipitating layer clouds	297
6.4.2. The growth of precipitation elements in layer clouds	306
6.5. The release of precipitation from shower clouds	314
6.5.1. Characteristics and structure of shower clouds	314
6.5.2. Production of showers by coalescence of cloud droplets	317
6.5.3. Raindrop multiplication by break-up of large drops	327
6.5.4. The release of showers by the growth of ice particles	329
6.6. Hail	332
6.6.1. Occurrence of hail	332
6.6.2. Classification of hail	332
6.6.3. The size and shape of hailstones	333
6.6.4. The structure of hail	335
6.6.4.1. Types of ice structure	335
6.6.4.2. The density, crystal structure, and air content of accreted ice	336
6.6.4.3. Structure of soft hail	340
6.6.4.4. Structure of small hail pellets	341
6.6.4.5. Structure of hailstones	341
6.6.5. The density of hailstones	346

6.6.6. The aerodynamics of hailstones — 347
6.6.7. Theories of hailstone growth — 348
6.6.8. The melting of hailstones — 366

7. ARTIFICIAL MODIFICATION OF CLOUDS AND PRECIPITATION

7.1. Historical introduction — 369
7.2. The experimental seeding of cumuliform clouds — 371
    7.2.1. Experiments with dry ice — 371
    7.2.2. Seeding of cumulus with silver iodide — 374
    7.2.3. Seeding of cumulus with water drops and hygroscopic nuclei — 377
7.3. The experimental seeding of layer clouds — 380
7.4. Large-scale cloud-seeding operations — 382
    7.4.1. Evaluation procedures — 383
    7.4.2. Seeding with silver iodide from aircraft — 385
    7.4.3. Seeding with silver iodide from ground generators—Project 'Grossversuch III' — 390
7.5. Suppression of large hail — 391
7.6. Discussion — 393

8. RADAR STUDIES OF CLOUDS AND PRECIPITATION

8.1. Basic radar theory — 400
    8.1.1. Calculation of back-scattered power from a target — 400
    8.1.2. Choice of radar parameters — 403
8.2. The scattering and attenuation of radar waves by meteorological particles — 404
    8.2.1. Scattering of spherical particles of $D \ll \lambda$ — 404
    8.2.2. Scattering by non-spherical particles of $D \ll \lambda$ — 405
    8.2.3. Experimental verification of scattering theory — 413
    8.2.4. Scattering by large hydrometeors of $D > \lambda/20$ — 417
    8.2.5. Scattering by non-precipitating clouds — 426
    8.2.6. Attenuation of radar waves by clouds and precipitation — 426
8.3. The presentation of radar information — 429
8.4. Radar echoes from different cloud systems — 432
8.5. Analysis of the radar signal — 434
    8.5.1. Evaluation of the echo intensity — 434
    8.5.2. Signal fluctuations and Doppler radar — 437
    8.5.3. The Doppler spectrum and spectrum of intensity fluctuations — 441
8.6. Meteorological information from Doppler radar — 441
    8.6.1. Determination of drop-size distributions from the Doppler spectrum — 441
    8.6.2. Measurements of air motions and precipitation growth in showers and thunderstorms — 446

CONTENTS

8.6.3. Determination of horizontal winds, wind shear, and convergence	454
8.7. The structure of precipitating layer clouds as revealed by radar	458
8.7.1. General features	458
8.7.2. The region above the $0°$ C level	459
8.7.3. The melting region	462
8.7.4. The region below the melting band	463
8.7.5. Polarization of the radiation scattered by hydrometeors	463
8.7.6. The echo intensity as a function of height	464
8.7.7. Theory of the melting band	465
8.7.8. Radar upper bands	468
8.8. Detection of non-precipitating clouds with millimetric radar	471
8.9. The cumulonimbus and thunderstorm as revealed by radar	471
8.10. Measurement of rainfall by radar	477

9. THE ELECTRIFICATION OF CLOUDS

9.1. The vertical electric field and current in fine weather	483
9.2. The electric fields produced at the earth's surface by thunderstorms and lightning discharges	487
9.2.1. Field changes due to lightning flashes	487
9.2.2. Recovery of the field after a discharge—generation of electric moment	494
9.3. The structure of the lightning flash	496
9.3.1. Photography of the lightning flash	496
9.3.2. Electrical field-changes during a lightning flash	500
9.3.3. The mechanisms of electrical breakdown	509
9.4. The electrical structure of the thunderstorm	511
9.5. Correlation between lightning and precipitation	517
9.6. Mechanisms of charge generation and separation in thunderstorms	520
9.6.1. Basic requirements of a satisfactory theory	520
9.6.2. Basic mechanisms of cloud electrification	521
9.6.2.1. Influence mechanisms	521
9.6.2.2. Electrification produced by the rupture of large drops	523
9.6.2.3. Convective theories of thunderstorm electrification	525
9.6.2.4. Electrification associated with the freezing and melting of water	526
9.6.2.5. Thermoelectric effects in ice	531
9.6.2.6. Electrification associated with the collision and fracture of ice crystals	539
9.6.2.7. Electrification associated with the freezing and splintering of water drops and the formation of rime	543
9.6.3. The generation of electric charges and fields in precipitating clouds	550

CONTENTS

   9.6.3.1. The Wilson mechanism of selective ion capture   550
   9.6.3.2. Electrification produced by the rebound of cloud particles from hydrometeors in polarizing electric fields   551
   9.6.3.3. Electrification associated with the rupture of droplets impacting and freezing on hail pellets   555
 9.7. The transfer of electricity between the atmosphere and earth—the maintenance of the earth's charge   557
  9.7.1. The charge transferred by lightning discharges   557
  9.7.2. Charge transfer by point-discharge currents   558
  9.7.3. Charge transported by precipitation   559
  9.7.4. Electrical balance-sheet for the earth's surface   566

APPENDIX A THE COLLISION AND COALESCENCE OF WATER DROPS FALLING IN AIR   569

APPENDIX B THE PHYSICAL PROPERTIES OF FREELY FALLING RAINDROPS   592

SOME USEFUL PHYSICAL CONSTANTS   614

BIBLIOGRAPHY AND AUTHOR INDEX   616

SUBJECT INDEX   661

PLATES

Fig. 2.2	facing page	34	Fig. 6.22b, c	facing page	342
Fig. 2.15	,,	76	Fig. 6.23b, c	,,	344
Fig. 2.16	,,	77	Fig. 6.24	,,	345
Fig. 4.6	,,	176	Fig. 6.25	,,	346
Fig. 4.12	,,	207	Fig. 6.32	,,	363
Fig. 4.16	,,	226	Fig. 7.1	,,	370
Fig. 4.17	,,	227	Fig. 7.2	,,	372
Table 5.1	between pp.	232–3	Fig. 7.4	,,	381
Figs. 5.5–5.12	,,	248–9	Figs. 8.13–4	,,	429
Fig. 5.19	facing page	260	Figs. 8.15–6	,,	430
Figs. 5.20 & 21	,,	261	Figs. 8.17–8	,,	431
Fig. 5.23	,,	266	Figs. 8.20–1	,,	432
Fig. 5.24	,,	267	Fig. 8.22	,,	433
Fig. 5.25	,,	268	Fig. 8.30	,,	468
Fig. 5.27	,,	270	Fig. 8.32	,,	471
Fig. 5.28	,,	271	Fig. 9.12	,,	523
Figs 5.30–2	,,	274	Fig. A.9a	,,	590
Figs. 6.20–1	,,	340	Fig. A.9b	,,	591

1
The Nucleation of Water-Vapour Condensation

CLOUDS are formed by the lifting of damp air which then cools adiabatically by expansion as it encounters continuously falling pressures at higher levels in the atmosphere. The relative humidity consequently rises and eventually the air becomes saturated with water vapour. Further cooling produces a supersaturated vapour, but the excess vapour condenses on to some of the multitude of tiny particles suspended in the air to form a cloud composed of minute water droplets. The growth of these droplets tends to oppose further increase in the supersaturation, which reaches a peak value of usually less than 1 per cent and then decreases. In the absence of foreign particles and ions, much higher supersaturations are required for droplet condensation. Although we are never concerned with this process in natural clouds, we begin with a discussion of homogeneous nucleation of water vapour because it represents the simplest form of condensation process. Moreover, it provides the most straightforward illustration of the theoretical approach to nucleation problems in general, some examples of which will be discussed later.

1.1. Homogeneous condensation

It was found by C. T. R. Wilson (1897) that when air initially saturated with water vapour and freed as far as possible from foreign particles was subjected to a sufficiently large, rapid expansion, a cloud of very small water droplets appeared spontaneously from the vapour. He reported that with an initial temperature of 20° C a cloud, as distinct from a small number of larger droplets, appeared only if the expansion ratio exceeded 1·37, thus producing momentarily almost eightfold supersaturation.

In the absence of foreign particles, aggregates of the condensed phase can arise only as the result of chance collisions of molecules in the supersaturated vapour. These small aggregates (or embryos) are continually formed and disrupted because of microscopic thermal and density fluctuations in the vapour, but only if they surpass a certain critical size determined by the prevailing supersaturation and temperature, can they survive and continue to grow—otherwise they evaporate and disappear. The supersaturated vapour is a metastable phase, and the embryos can grow to become *nuclei* for the development of the liquid phase only if they attain a size at which they are more stable than the vapour phase. The probability of formation of an aggregate of critical size increases as the supersaturation is increased. The critical radius r^*, which the aggregate, assumed spherical, must attain in order to be in (unstable) equilibrium with the vapour, was first deduced by Kelvin (1870) and later by Gibbs (1875) from thermodynamical arguments, as

$$r^* = 2M\sigma_{\text{LV}}/\rho_{\text{L}}\mathbf{R}T \ln p/p_\infty, \qquad (1.1)$$

where p is the pressure of the supersaturated vapour and p_∞ the equilibrium vapour pressure at temperature T over a plane surface of the liquid, σ_{LV} the specific surface energy of the liquid-vapour interface, ρ_{L} the density of the liquid (strictly the difference between the densities of the liquid and vapour), $\mathbf{R}$ the universal gas constant, and M the molecular weight of the liquid.

1.1.1. *Thermodynamics of phase changes*

Phase-equilibria and changes of phase may be conveniently discussed with the aid of two thermodynamic free-energy functions: the Helmholtz free energy

$$F = U - T\mathscr{S} \qquad (1.2)$$

and the Gibbs free energy

$$G = U - T\mathscr{S} + pV, \qquad (1.3)$$

where U is the internal energy, T the absolute temperature, $\mathscr{S}$ the entropy, p the pressure, and V the volume of the system under consideration. Substitution for dU from the first law of thermodynamics,

$$T\,d\mathscr{S} = dU + p\,dV, \qquad (1.4)$$

in the differential forms of (1.2) and (1.3) yields the following relations for a system of constant composition *in equilibrium*:

$$dF = -\mathscr{S}\,dT - p\,dV \qquad (1.5)$$

and

$$dG = -\mathscr{S}\,dT + V\,dp. \qquad (1.6)$$

Thus if a system is in equilibrium at constant temperature and volume, F is stationary, and can be shown to be a minimum. Similarly, for a system in equilibrium at constant temperature and pressure, G is a minimum. We shall be concerned mainly with systems in which the pressure is held constant and therefore with the Gibbs free energy G.

§1.1 NUCLEATION OF WATER-VAPOUR CONDENSATION

If the system is not in equilibrium and its composition n_i (quantity of component i) is subject to change, we introduce the chemical potential $\mu_i = \left(\frac{\partial G}{\partial n_i}\right)_{T,p}$, and the change in free energy becomes

$$dG = V\,dp - \mathscr{S}\,dT + \mu_i\,dn_i. \tag{1.7}$$

We now consider a small change involving two phases, a, b, of a chemically homogeneous system, for example the condensation of an infinitesimal quantity of vapour. This involves a change $(\mu_b - \mu_a)\,dn$ in the chemical potential of the system and also an infinitesimal increase dA in the surface area of the liquid phase and a quantity of work $\sigma_{LV}\,dA$. The total change in the Gibbs free energy of the system is then

$$\Delta G = (\mathscr{S}_a - \mathscr{S}_b)\,dT + (V_b - V_a)\,dp + (\mu_b - \mu_a) + \sigma_{LV}\,dA, \tag{1.8}$$

where all the quantities may be taken as referring to one molecule. For an isothermal, isobaric change, $dT = 0$, $dp = 0$, and

$$\Delta G = (\mu_b - \mu_a) + \sigma_{LV}\,dA. \tag{1.9}$$

If the two phases are in equilibrium, then $dG = 0$, $dA = 0$, and $\mu_a = \mu_b$, so that the chemical potential of a molecule in the saturated vapour over a plane liquid surface is the same as that of a molecule in the liquid.

1.1.2. Thermodynamic derivation of Kelvin's formula

The thermodynamic theory of phase transitions, which rests on the assumption that the two phases are in stable equilibrium, cannot deal with the *rate* at which the transition proceeds; this is assumed to be infinitesimal, which is not strictly the case, particularly during the initial stages of development of the new phase. Nevertheless, we consider a closed system at absolute temperature T, containing vapour at pressure p, and one droplet of liquid, radius r, consisting of g molecules. Let the chemical potential per molecule in the vapour phase be μ_a and in the liquid phase be μ_b. Imagine the droplet to have been formed by the condensation of g molecules of vapour. The total change in the free energy of the system is then

$$\Delta G = (\mu_b - \mu_a)g + 4\pi r^2 \sigma_{LV} \tag{1.10}$$

$$= (\mu_b - \mu_a)g + \alpha g^{\frac{2}{3}}, \tag{1.10a}$$

where α is a constant such that $\alpha g^{\frac{2}{3}} = 4\pi r^2 \sigma_{LV}$. The second term on the right of (1.10a) is the surface free-energy term which makes an important contribution to ΔG because of the large surface-to-volume ratio of the small droplet.†

The way in which the free energy ΔG varies with the size of an embryo is shown in Fig. 1.1. Curve (a) is for a saturated vapour where $\mu_a = \mu_b$, while curve (b) is for a supersaturated vapour for which $\mu_a > \mu_b$. In the latter case, there is a maximum of the free energy of the system when g has a certain critical value g^*. When the embryo has this critical size it is in equilibrium with the vapour, but the equilibrium is unstable. On surpassing the critical size, the droplet will grow with a

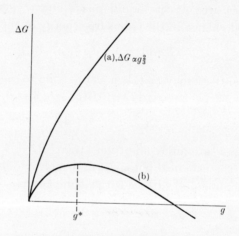

FIG. 1.1. The free energy ΔG required to form a droplet containing g molecules; curve (a) in saturated vapour; curve (b) in supersaturated vapour. g^* is the critical size.

decrease of free energy and so tend to become larger still. If, however, the vapour phase in thermodynamically stable ($\mu_a < \mu_b$), embryos of the new phase reach only a relatively small size and then decay.

The condition for equilibrium between the droplet and the vapour can be obtained by differentiating (1.10a). It is found that $d(\Delta G)/dg = 0$ when

$$\mu_b - \mu_a = -\tfrac{2}{3}\alpha g^{-\frac{1}{3}} = -8\pi\sigma_{\mathrm{LV}}\, r^2/3g, \tag{1.11}$$

but

$$g = \tfrac{4}{3}\pi r^3 \frac{\mathscr{N} \rho_{\mathrm{L}}}{M}, \tag{1.12}$$

† It is assumed that the surface tension of such a small droplet is that appropriate to a plane surface of the liquid at the same temperature. The effect of the curvature of small droplets on their surface tension has been the subject of several theoretical treatments, e.g. Tolman (1949), Kirkwood and Buff (1949), but such indirect experimental evidence as exists does not support their conclusions (see Mason 1951).

§1.1 NUCLEATION OF WATER-VAPOUR CONDENSATION

where $\mathcal{N}$ is Avogadro's number, and therefore combining (1.11) and (1.12) we obtain for the equilibrium condition,

$$\mu_a - \mu_b = \frac{2M\sigma_{LV}}{\mathcal{N}\rho_L r^*}, \qquad (1.13)$$

where r^* is the radius of the critical nucleus containing g^* molecules.

We may obtain an alternative expression for $(\mu_a - \mu_b)$ by considering the transfer of one molecule from the vapour to a plane surface of the liquid, the vapour pressure being p, the equilibrium vapour pressure p_∞, with the temperature T remaining constant. Then, treating the vapour as an ideal gas and remembering that $\mu_b = \mu_a(p_\infty)$, we may write

$$\mu_a - \mu_b = \mu_a(p) - \mu_a(p_\infty)$$
$$= \int_p^{p_\infty} d\mu_a = \int_p^{p_\infty} v_a\, dp = kT \int_p^{p_\infty} \frac{dp}{p}, \qquad (1.14)$$

or $\qquad \mu_a - \mu_b = kT \ln p/p_\infty = kT \ln S, \qquad (1.15)$

where μ_a is the free energy and v_a the volume occupied per molecule at pressure p and temperature T, k is Boltzmann's constant, and $S = p/p_\infty$ is the saturation ratio.

Substituting for $\mu_a - \mu_b$ from (1.15) into (1.13) gives

$$\ln p/p_\infty = 2M\sigma_{LV}/\rho_L \mathbf{R} T r^* \qquad (1.16)$$

which is Kelvin's formula (1.1).

1.1.3. *Embryo droplets in statistical equilibrium with the vapour*

According to (1.10) and (1.11),

$$\Delta G = \tfrac{1}{3}\alpha g^{\frac{2}{3}} = \tfrac{4}{3}\pi r^2 \sigma_{LV}, \qquad (1.17)$$

and so the system is in true equilibrium (minimum value of ΔG), only when $g = 0$. For small finite values of ΔG, there is always a small probability, $e^{-\Delta G/kT}$, that the system will be found in a non-equilibrium state because of microscopic density and temperature fluctuations. If, in a system containing n_0 molecules, there are n_g embryos each consisting of g molecules, then the most probable distribution n_g is shown by Frenkel (1946) to be given by

$$n_g = n_0 \exp(-\Delta G/kT) = n_0 \exp\!\left(g \ln S - \frac{\alpha g^{\frac{2}{3}}}{kT}\right). \qquad (1.18)$$

The way in which the number of embryos of given size varies with the number of molecules g in each is shown in Fig. 1.2, in which $\ln n_g$ is plotted against g. Curve (a) corresponds to a system in which the vapour is just saturated; n_g decreases continuously as g is increased. But, for a supersaturated system, represented by curve (b), n_g has a minimum value when the number of molecules in the embryo has a critical value g^* corresponding to the critical radius r^* in (1.1). For values of $g > g^*$, n_g increases and becomes greater than n_0 when g is very large. Clearly, this part of curve (b) cannot correspond to reality, so that for a super saturated vapour, (1.18) breaks down at large values of g. In order to determine the true distribution curve, it is necessary to examine the equilibrium conditions between embryos of various sizes in more detail The treatment now to be given closely follows that of Farley (1952).

1.1.4. *Statistical equilibrium between embryos of different sizes*

We now suppose that the equilibrium distribution given by (1.18) has been set up, and examine the microscopic balancing conditions for dynamic equilibrium between embryos of size g, embryos of size $g-1$, and the vapour. It is assumed that g is a moderately large number so that the equations can be expressed in differential form.

The number of vapour molecules β_g striking an embryo of size g, radius r, per second is

$$\beta_g = 4\pi r^2 \mathcal{N} p (2\pi MRT)^{-\frac{1}{2}}. \tag{1.19}$$

It is assumed that all the vapour molecules that hit the embryo are captured by it, i.e. the condensation coefficient is unity. Therefore, the number of embryos of size $g-1$ becoming embryos of size g per second is $\beta_{g-1} n_{g-1}$. Similarly, if γ_g is the number of molecules evaporating from an embryo of size g per second, then the number of embryos of size g becoming embryos of size $g-1$ per second is $\gamma_g n_g$.

For dynamic equilibrium we have

$$\beta_{g-1} n_{g-1} = \gamma_g n_g \tag{1.20}$$

which, with $\xi = \gamma_g/\beta_{g-1}$, can be written

$$\ln \xi = \ln n_{g-1} - \ln n_g = -\frac{d}{dg} \ln n_g. \tag{1.21}$$

Substitution from (1.18) now gives

$$\xi = \exp(-\ln S + 2\alpha g^{-\frac{1}{3}}/3kT). \tag{1.22}$$

We have used the thermodynamically-derived distribution (1.18) to deduce the ratio ξ between the number of molecules evaporating from

§ 1.1   NUCLEATION OF WATER-VAPOUR CONDENSATION

an embryo and the number hitting an embryo one size smaller. For very small embryos, (1.22) shows that ξ is large (as $\alpha g^{-\frac{1}{3}} \propto 1/r$), i.e. the tendency is for the aggregates to evaporate rather than grow. In this region of the equilibrium distribution, dynamic equilibrium is maintained because, according to (1.18), there are more small embryos than large ones. According to (1.22), as the embryo size increases, ξ falls steadily to the limiting value $1/S$. For a supersaturated vapour ($S > 1$), ξ is less than unity for large embryos. In this region the dynamic equilibrium can be maintained only if n_g increases as g increases, as shown in Fig. 1.2, an impossible requirement which must be recognized below. For an embryo of a certain size, ξ has the value 1. In this case (1.21) shows that $n_{g-1} = n_g$, a situation which therefore corresponds to a minimum value of n_g and to embryos of critical size g^*. Putting $\xi = 1$ in (1.22) we obtain

$$\ln S = 2\alpha g^{-\frac{1}{3}}/3kT,$$

which is the condition for the embryos to have critical size.

The mechanism of detailed balancing just considered is more fundamental than the purely thermodynamic approach and will enable the analysis to be extended to include non-equilibrium states, for example to modify (1.18) so that it corresponds more closely to reality for large values of g. We next discuss the distribution of embryos in a real supersaturated system.

1.1.5. *The unbalanced steady state in a supersaturated system*

In practice, the distribution given by n_g in (1.18) and represented by curve (b) of Fig. 1.2 cannot be set up because it requires an infinite

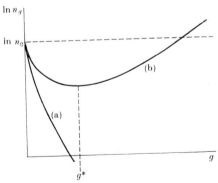

FIG. 1.2. The equilibrium distribution of droplets: curve (a) in a saturated system; curve (b) in a supersaturated system. n_g is the number of droplets that contain g molecules, n_0 is the total number of molecules in the system, g^* is the critical size. (From Farley (1952).)

number of very large embryos and these are not available. When the vapour first becomes supersaturated there are very many small embryos and no large ones. The detailed balance condition (1.20) is therefore not satisfied in the supersaturated vapour; there is a net growth in the population of embryos of *all sizes* and the approach to the equilibrium distribution n_g never succeeds because of the physical impossibility at large values of g. As the embryos grow the number of molecules in the vapour phase must decrease, but in order to carry out the analysis, it will be assumed that the vapour is maintained supersaturated by a continuous influx of new molecules from outside.

Suppose then, that at any time, the number of embryos of size g is f_g instead of n_g. Eqn (1.20) is no longer valid for the new distribution and we have instead

$$I_g = \beta_{g-1} f_{g-1} - \gamma_g f_g, \qquad (1.23)$$

where I_g is the net number of embryos passing from size $g-1$ to size g per second. At first f_g changes with time, $\partial f_g/\partial t = I_g - I_{g+1}$, but eventually a steady distribution of embryos is reached such that $I_g = I$, a constant independent of g. In this distribution the population f_g is steady, but there is a net transfer of embryos from small to large sizes. In effect, I embryos are formed per second from the vapour and pass through the distribution to emerge as droplets at the rate of I per second. I is called the *current* of droplets. The present problem is to calculate the steady distribution and in this way to arrive at a value for I.

For small values of g the steady distribution f_g need depart only slightly from the equilibrium distribution n_g to give the current I, because here the value of n_g is large. But as g increases, n_g falls and f_g

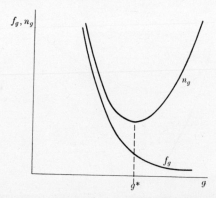

FIG. 1.3. The equilibrium distribution n_g in a supersaturated system, and the steady distribution f_g. n_g and f_g are the number of droplets which contain g molecules. (From Farley (1952).)

§ 1.1   NUCLEATION OF WATER-VAPOUR CONDENSATION

must deviate progressively more and more from n_g to give the same current. If, in its growth process, an embryo exceeds the critical size g^*, then ξ becomes less than unity, and further growth becomes more probable than evaporation. The droplet thereafter grows unchecked to larger and larger sizes, and in this region the number of drops of a given size is inversely proportional to the mean life in this size, i.e., $f_g \propto 1/\beta_g$. As β_g increases as $g^{\frac{2}{3}}$, the number of droplets f_g of any size g decreases as $g^{-\frac{2}{3}}$, and this is represented in Fig. 1.3.

We now proceed to set up a differential equation for f_g and then to evaluate I, the number of visible droplets formed per second in a supersaturated vapour.

1.1.6. *The current in the steady state—the rate of formation of droplets*

The equation for I_g, the net number of embryos passing from size $g-1$ to size g per second in the steady state is,

$$I_g = \beta_{g-1} f_{g-1} - \gamma_g f_g, \tag{1.23}$$

which, from (1.20),

$$= n_{g-1} \beta_{g-1} \left(\frac{f_{g-1}}{n_{g-1}} - \frac{f_g}{n_g} \right)$$

$$= -\beta_g n_g \frac{d}{dg}(f_g/n_g). \tag{1.24}$$

When the distribution f_g is steady, I_g has the constant value I. It is also assumed that β_g is constant $(=\beta)$, which will be approximately true for a small range of embryonic sizes around the critical size. Under these conditions (1.24) may be integrated to give

$$\frac{f_g}{n_g} = -\frac{I}{\beta} \int \frac{dg}{n_g} + A, \tag{1.25}$$

where A is a constant. To evaluate A, consider the validity of (1.25) when g is very large. Substituting in (1.25) from (1.18) gives

$$\frac{f_g}{n_g} = A - \frac{I}{\beta n_0} \int \exp\left(\frac{\alpha g^{\frac{2}{3}}}{kT}\right) \exp(-g \ln S) \, dg. \tag{1.26}$$

For large values of g, the second exponential term in the integral is the more important, so that

$$\frac{f_g}{n_g} \simeq A - \frac{I}{\beta n_0} \exp\left(\frac{\alpha g^{\frac{2}{3}}}{kT}\right) \int \exp(-g \ln S) \, dg = A + I/\beta n_g \ln S, \tag{1.27}$$

and as for large values of g, $f_g \propto 1/\beta$, $A = 0$, and (1.25) may be written

$$f_g/n_g = \frac{I}{\beta}\int_g^\infty \mathrm{d}g/n_g. \qquad (1.28)$$

Eqn (1.28) satisfies (1.24) and agrees with (1.27) when $g \to \infty$.

Let us now consider what happens when $g \to 0$. The theory is not valid for very small embryos, but when g is small, $f_g \to n_g$. Assuming $f_g/n_g \to 1$ as $g \to 0$, and that the extrapolation is valid, we obtain from (1.28)

$$I = \beta \Big/ \int_0^\infty \mathrm{d}g/n_g. \qquad (1.29)$$

To calculate I, the number of droplets formed per second, it is necessary to evaluate the integral in (1.29) in which we can substitute for n_g from (1.18) to give

$$\frac{1}{n_g} = \frac{1}{n_0}\exp\!\left(\frac{\alpha g^{\frac{2}{3}}}{kT} - g\ln S\right) = \frac{1}{n_0}\exp\!\left(\frac{\Delta G}{kT}\right). \qquad (1.30)$$

Now $1/n_g$ has a sharp maximum value $1/n^*$ when $g = g^*$, so in this region we can replace ΔG by its expansion with respect to the difference $g-g^*$, viz.:

$$\Delta G = \Delta G_{\max} + \frac{1}{2}\!\left(\frac{\partial^2 \Delta G}{\partial g^2}\right)_{g=g^*}\!(g-g^*)^2 = \Delta G_{\max} - \frac{4\pi}{9g^{*2}}\sigma_{LV} r^{*2}(g-g^*)^2$$

giving
$$\frac{1}{n_g} = \frac{1}{n_0}\exp(\Delta G_{\max}/kT)\exp\!\left\{-\frac{4\pi\sigma_{LV} r^{*2}}{9kT g^{*2}}(g-g^*)^2\right\}. \qquad (1.31)$$

Since from (1.18), (1.10a), and (1.11)

$$n^* = n_0\exp(-\Delta G_{\max}/kT) = n_0\exp(-4\pi r^{*2}\sigma_{LV}/3kT), \qquad (1.32)$$

$$\int_0^\infty \frac{\mathrm{d}g}{n_g} = \frac{1}{n^*}\int_0^\infty \exp\{-Y(g-g^*)^2\}\,\mathrm{d}g = \frac{1}{n^*}\!\left(\frac{\pi}{Y}\right)^{\!\frac{1}{2}}, \qquad (1.33)$$

and
$$I = \beta\Big/\int_0^\infty \mathrm{d}g/n_g = 4\pi r^{*2}\mathcal{N} p(2\pi MRT)^{-\frac{1}{2}}\frac{2}{3}\frac{r^*}{g^*}\!\left(\frac{\sigma_{LV}}{kT}\right)^{\!\frac{1}{2}}\!n^*$$

$$= \frac{2p}{(2\pi MRT)^{\frac{1}{2}}}\frac{M}{\rho_L}\!\left(\frac{\sigma_{LV}}{kT}\right)^{\!\frac{1}{2}}\!n^*. \qquad (1.34)$$

Substituting from (1.32) for n^*, and from (1.1) for r^*, we have that I, the number of embryos transformed into droplets per second, is given by

$$\ln I = \ln \frac{1}{\mathbf{R}^2\rho_L}\!\left(\frac{2\mathcal{N}^3 M\sigma_{LV}}{\pi}\right)^{\!\frac{1}{2}} + 2\ln\frac{p_\infty}{T} + 2\ln\frac{p}{p_\infty} - \frac{16\pi M^2\sigma_{LV}^3}{3\rho_L^2 k\mathbf{R}^2 T^3 \ln^2(p/p_\infty)}, \qquad (1.35)$$

§ 1.1   NUCLEATION OF WATER-VAPOUR CONDENSATION

where $\mathcal{N}$ is Avogadro's number, p the prevailing vapour pressure, p_∞ the equilibrium vapour pressure at temperature T, and $p/p_\infty = S$ is the saturation ratio.

The derivation of (1.35) follows closely that of Zeldovich (1942). Earlier treatments were given by Volmer and Weber (1926), Farkas (1927), and Becker and Döring (1935). Those of Volmer and Weber and of Farkas were based on thermodynamical and statistical arguments in which the embryos were treated as isolated entities, the mutual interaction between them being ignored; they did not consider the detailed balancing between growing and evaporating embryos. Becker and Döring rejected this thermodynamical treatment and considered the kinetics of the condensation process, taking into account not only the condensation of the vapour on to the surfaces of embryos but also the reverse process of re-evaporation. Their treatment of the problem is very similar to that given above. The differences lead to slightly different versions of eqn (1.35), but the last two terms, which are the most important, remain the same. The two treatments predict very much the same variation of nucleation rate with change of supersaturation, but rather different values for the absolute nucleation rate.

1.1.7. *Time to reach to the steady state*

In deriving the expressions (1.34) or (1.35) for I, we have assumed that the steady distribution of embryos is indeed established. It is therefore of interest to calculate the relaxation time, i.e. the time required to establish this steady distribution once the vapour is brought into a supersaturated state.

Initially the number of embryos f_g of size g varies with time, and we have from (1.24) and assuming for simplicity that β is constant,

$$\partial f_g/\partial t = I_g - I_{g+1} = -\frac{\partial I}{\partial g} = \beta \frac{\partial}{\partial g}\left\{n_g \frac{\partial}{\partial g}(f_g/n_g)\right\}. \tag{1.36}$$

Putting $F(g, t) = f_g/n_g$, (1.36) becomes

$$n_g \partial F/\partial t = \beta \frac{\partial}{\partial g}\left(n_g \frac{\partial F}{\partial g}\right). \tag{1.37}$$

This equation, with n_g given by (1.18), determines the relaxation time, but there is no formal solution in this case. Farley obtains an estimate for the relaxation time by solving (1.37) with n_g constant, assuming initially $F = 0$ for $g > 0$, $F = 1$ for $g = 0$, and in the final state, $F = 1$ for all values of g. The solution is then

$$F = 1 - \text{erf}\{g(4\beta t)^{-\frac{1}{2}}\}, \tag{1.38}$$

indicating that 90 per cent of the population of embryos reach size g after time $t = 25g^2/\beta$. For water droplets attaining a critical size in a typical expansion chamber, this time is about 10^{-5} s†, i.e. about one-thousandth of the sensitive time of the expansion chamber, so that the time lag is unimportant.

This will not be the case, however, in very rapid expansions such as occur in supersonic wind tunnels where the steady-state theory just discussed fails to predict the onset of condensation. In air containing 0·1 per cent by weight of water vapour and expanded at a Mach number of 1·385 from a stagnation pressure of 1·163 atmospheres and temperature 300 K, Wegener and Smelt (1950) found that condensation occurred after delay times of about 4×10^{-5} s, whereas the theory predicts that an equilibrium distribution of embryos would occur after 4×10^{-6} s. Probstein (1951), following earlier work by Kantrowitz (1951), obtained an approximate solution for the non-steady state eqn (1.36) in an attempt to determine whether the time lag associated with the approach to a steady-state rate of droplet formation could account for the discrepancy. He took into account the heating of the embryo by the bombardment of molecules of the condensing phase and the consequent inability of the embryos to absorb all the incident molecules (this implies a condensation coefficient less than unity) as additional factors contributing to the time lag. Probstein's analysis shows, however, that the calculated magnitude of the time lag is insufficient to account for the above-mentioned discrepancy between theory and experiment, unless the condensation coefficient is assumed to be about 0·02, much lower than the value of 0·4 recently calculated by Lothe and Pound (1962) for an embryo of sixty molecules.

1.1.8. *Further extensions of the theory*

There have been a number of recent attempts to extend and refine the theory of homogeneous nucleation, notably by Lothe and Pound (1962) and by Oriani and Sundquist (1963), who argue that quantum statistical translational and rotational states contribute greatly to the entropy of the droplet embryo. The inclusion of these terms causes the

† More detailed treatments by Probstein (1951), Wakeshima (1954), and Courtney (1963) give values between 10^{-6} and 10^{-5} s.

theoretical value of the nucleation rate I to be increased by a factor of about 10^{17} and makes it impossible to reconcile this with experimental observation. Lothe and Pound think that the situation may be partly retrieved by treating the nucleation as a transient rather than as a steady-state process but Oriani and Sundquist argue that the Lothe–Pound correction can be largely offset by allowing for the fact that the surface free energy of the small embryo is a function of its size. Having shown that the Kirkwood–Buff treatment would give impossible answers, Oriani and Sundquist put forward a new bond-breaking model for the surface of the embryo and calculated how the surface free energy might vary with the embryo size.

These discussions all recognize that there are logical difficulties in regarding the embryos as well-defined spherical 'droplets' having the thermodynamic and physical properties of macroscopic drops. The macroscopic concepts become very vague when the aggregates contain as few as twenty molecules, and are probably meaningless when applied to embryos of sub-critical size that are not in equilibrium with the vapour phase. The early stages of embryo formation, especially, must be thought of in terms of successive attachments of vapour molecules to a small aggregate, where the probability of capture probably needs to be assessed in terms of binding forces, the relative orientation of the molecules, etc., rather than in terms of macroscopic concepts.

However, attempts to formulate a quantitative theory in these terms have so far met with little success and although we have little cause to be satisfied with eqn (1.35) it is probably the best we have and its value must be judged by comparison with experiments.

1.1.9. *Comparison of theory with experiment*

Substitution of numerical values in (1.35) allows calculation of the number I of droplets per second in each cm^3 of vapour. Fig. 1.4 shows a plot of log I against the saturation ratio $S = p/p_\infty$ for water vapour, with $T = 260$ K and σ_{LV} taking the bulk value of 77·5 erg cm^{-2}. The nucleation rate increases rapidly with increasing values of S and attains a value of 1 cm^{-3} s^{-1} only when $S \to 5$.

Experiments have been made by several different workers in which very clean, water-saturated air has been subjected to rapid expansion in a cloud chamber and the minimum supersaturation required to produce a detectable droplet concentration determined. The results are summarized in Table 1.1. In the literature there has been a tendency to underline an apparent good agreement between the experimental

NUCLEATION OF WATER-VAPOUR CONDENSATION

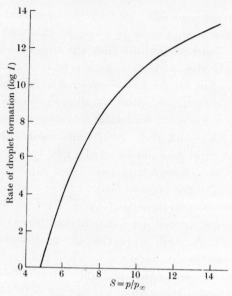

FIG. 1.4. The rate of nucleation I as a function of the saturation ratio S calculated from eqn (1.35).

results and the predictions of eqn (1.35), but these claims merit a more critical examination.

It is rather difficult to compare the experiments of Wilson (1899) and of Powell (1928) with the theory since the rates of droplet formation corresponding to their reported appearance of a 'cloud', as distinct from

TABLE 1.1

Experimental and observed rates of homogeneous nucleation in water vapour

	T_i(K)	T_f(K)	$S_{\text{expt.}}$	Obs. droplet concn. (cm^{-3})	$I_{\text{obs.}}$	$S_{\text{theor.}}$
Wilson (1899)	293	257	7·90	'cloud limit'	$\geqslant 10^6$	$\geqslant 7·0$
Powell (1928)	291	256·6	7·80	probably $>10^3$		
Volmer and Flood (1934)		261	5·03	~ 1	$\sim 10^2$	5·42
Frey (1941)		263	5·0	10	10^5	5·90
Sander and Damköhler (1943)		261	4·36	1	10^2	5·42
Barnard (1954)	293·7	261	6·60	10^3	10^6	6·40
Madonna et al. (1961)		261	5·70	10	10^3	5·68
		238	6·40	10	10^3	9·50

sparse 'rain-like' condensation, are uncertain. If, however, we assume that the droplet concentration must have exceeded 10^3 cm^{-3} in order to be detected above the background condensation on ions (which were not removed) and, furthermore, that the effective sensitive time of the chamber was about 10^{-3} s, we arrive at a minimum value of $I = 10^6$ cm^{-3} s^{-1}. Substitution of this value in eqn (1.35) gives a lower limit for the theoretical saturation ratio of $S = 7.0$ to be compared with the measured value of 7·8 or 7·9. Part of this discrepancy may be accounted for by the fact that, with such high concentrations of droplets, the actual saturation ratio achieved in the cloud chamber must have been less than that calculated on the basis of a perfectly adiabatic expansion due to the abstraction of water vapour and release of latent heat by the growing droplets (see Mason 1951).

Volmer and Flood (1934), who removed foreign particles from the air by repeated expansions, and ions by a strong electric field, determined a critical expansion ratio which just produced a noticeable increase in the number of droplets above the background. They estimated the minimum observable increase in droplet density to correspond to $I \simeq 1$ cm^{-3} s^{-1}, for which the corresponding theoretical value of S, at a final temperature of 261 K, was 4·96 according to Volmer and Flood and 5·05 according to eqn (1.35); these are to be compared with their measured value of 5·03. Similar good agreement was obtained for a number of other pure vapours and consequently this work became accepted as a verification of the theory and as a basis for much subsequent work in the general field of nucleation phenomena. However, it is very doubtful whether Volmer and Flood could have detected such a small increase in the number of droplets in a chamber whose sensitive time was only about 10^{-2} s. An increase of one droplet/cm^3 is just about detectable by careful visual observation; this would correspond to $I = 10^2$ cm^{-3} s^{-1}, for which eqn (1.35) gives $S = 5.42$. It is accordingly doubtful whether Volmer and Flood obtained such good agreement between experiment and theory as they believed.

Frey (1941) attempted to determine the droplet concentration more accurately by photographing the transient cloud. The minimum detectable concentration was 10 droplets/cm^3 and the effective sensitive time of the chamber probably not much in excess of 10^{-4} s. This minimum concentration of droplets was observed at a final temperature of 261 K and saturation ratio of 5·0; the corresponding theoretical value of S is 5·90. But Frey's method of cleaning his chamber before each experiment appears inadequate and it seems highly probable that the initial condensation that he observed occurred on foreign nuclei or on ions. A similar criticism may be made of Sander and Damköhler (1943).

In some more recent and very careful experiments by Barnard (1954), who was well aware of the difficulty of removing all foreign nuclei and of preventing their formation by rubber and metal surfaces under the influence of ultraviolet light, the concentration of droplets was photographed as the expansion ratio was increased in very small steps. It was found that the increase of droplet concentration with increasing supersaturation was systematically less than that predicted by the theory. Thus if a figure for the sensitive time of the chamber (and hence for the nucleation rate I) was chosen to make theory and observation agree for an observed droplet concentration of 1 cm^{-3} and $S \equiv 5 \cdot 20$, a droplet concentration of 10^3 cm^{-3} occurred at $S = 6 \cdot 60$ instead of at the theoretical value of 6·40. However, the agreement here is quite good, and probably part of the discrepancy may be attributed to the actual supersaturation being rather less than the nominal adiabatic value.

The most recent work is that by Madonna, Scuilli, Canjar, and Pound (1961), who used purified and filtered nitrogen as a carrier gas and an electric field to remove the ions. Working with a minimum detectable droplet concentration of 10 cm^{-3}, these authors obtained good apparent agreement between the experimental and theoretical values of the critical supersaturation when the final temperatures at the end of the expansion were only a few degrees below 0°C (see Table 1.1). However, much larger discrepancies occurred at lower temperatures, droplets now appearing at much lower supersaturations than are indicated by the theory. This led Madonna et al. to conclude that, unless the surface tension of water decreases, rather than increases, as the temperature is lowered below about −25°C, (measurements show it to increase smoothly down to −23°C), these serious discrepancies at low temperatures suggest that the agreement obtained at higher temperatures may have been fortuitous; indeed the condensation may not have been homogeneous but have occurred on ions not removed by the field or upon nuclei produced, perhaps, by chemical reactions between the water vapour, the carrier gas, and small irremovable traces of impurities under the influence of ultraviolet light.

In conclusion, it may be said that a critical analysis of the various experimental results, summarized in Table 1.1, indicates that the Becker–Döring theory of homogeneous condensation has not been

§1.2 NUCLEATION OF WATER-VAPOUR CONDENSATION

satisfactorily confirmed by experiment over a wide temperature range. There are certainly unsatisfactory and unconvincing features of the theory, but the discrepancies between the theory and the experimental results of different authors are not systematic and cannot be reconciled by inserting different values for such parameters as the condensation coefficient and surface tension in eqn (1.35). Much of the trouble lies in the experiments where there are great difficulties in excluding entirely small traces of contamination and sources of foreign nuclei, in determining accurately the droplet concentration, the sensitive time of the chamber, and hence the nucleation rate. There remains a need for an *experimentum crucis* in which one can be certain of observing homogeneous nucleation in the absence of foreign particles and in which the droplet concentration will be sufficiently low to ensure that the peak supersaturation can be accurately computed in terms of the expansion ratio and initial temperature for a truly adiabatic system. The experiment should probably be carried out in all-glass apparatus, with a number of highly-purified carrier gases, and over a fairly wide range of temperature.

1.2. Condensation on ions

We have seen that homogeneous condensation of water vapour occurs only if the supersaturation reaches several hundred per cent, but the process is greatly facilitated in the presence of foreign particles that serve as ready-prepared aggregates (nuclei) for condensation.

In a classic investigation, Wilson (1899) discovered that, after removing foreign particles from the air of his cloud chamber by making repeated expansions and allowing the droplets to settle out, further expansions produced no droplets until the expansion ratio exceeded 1·25, corresponding to a saturation ratio S of about 4. The nuclei responsible for these droplets could not be removed by successive expansions nor by filtration through cotton wool. The fact that condensation occurred at similar supersaturations when the air was irradiated with X-rays suggested that the responsible nuclei were small ions. By ionizing the air in his chamber by a very short exposure to X-rays and by application of electric fields of the appropriate polarity, Wilson was able to produce an excess of ions of either sign. He then found that expansion from an initial temperature of 293 K produced condensation on some negative ions with an expansion ratio of 1·25, for which $S = 4$, that practically all the negative ions were involved when the ratio exceeded 1·28, but that to effect condensation on positive ions, the expansion ratio had to exceed 1·31 ($S \simeq 6$). Wilson's results have been confirmed by several later workers, a representative sample of results being given in Table 1.2. The conditions under which condensation occurs upon ions seem to be more sharply defined and more reproducible than is the case for homogeneous nucleation; the results obtained by different workers are in good agreement.

TABLE 1.2

Saturation ratios at which condensation occurs on small ions

	Ion sign	T_1 (K)	T_2	V_2/V_1	S
Wilson (1899)	−	293	267·8	1·252	4·2
	+	293	..	1·31	6·0
Przibram (1906)	−	293	..	1·236	3·7
	+	293	..	1·31	6·0
Laby (1908)	−	..	267·6	1·256	4·2
Andrén (1917)	−	..	267·8	1·253	4·1
Powell (1928)	−	291	266·5	1·245	3·98
Flood (1933)	−	..	265	1·252	4·1
Loeb, Kip, and Einarsson (1938)	−	295	..	1·25	..
	+	295	..	1·31	..
Scharrer (1939)	−	292	..	1·25	4·14
	+	292	..	1·28	4·87
Sander and Damköhler (1943)	−	..	265	..	3·9

Note: T_1 and V_1 are respectively the initial temperature and volume of the air, while T_2, V_2 refer to the values at the end of the expansion.

If a droplet of radius r and dielectric constant ϵ, condenses on an ion of charge q and radius r_0 in an environment of saturation ratio S and dielectric constant ϵ', the change in the free energy of the system is

$$\Delta G = -\tfrac{4}{3}\pi r^3 \rho_\mathrm{L} \frac{N}{M} kT \ln S + 4\pi r^2 \sigma_\mathrm{LV} + \frac{q^2}{2}\left(\frac{1}{\epsilon'} - \frac{1}{\epsilon}\right)\left(\frac{1}{r} - \frac{1}{r_0}\right). \quad (1.39)$$

The condition for equilibrium between the droplet and the vapour is given by $\partial \Delta G/\partial r = 0$ which, if we assume $\partial \sigma_\mathrm{LV}/\partial r = 0$ and $\epsilon' = 1$, gives

$$\frac{RT\rho_\mathrm{L}}{M} \ln S = \frac{2\sigma_\mathrm{LV}}{r} - \frac{q^2}{8\pi r^4}\left(1 - \frac{1}{\epsilon}\right). \quad (1.40)$$

Comparing this with the Kelvin equation (1.1), it is seen that the saturation ratio S at equilibrium is smaller for a charged than an uncharged droplet. Eqn (1.40), expressing the variation of the equilibrium supersaturation with the droplet radius, is represented by the full curve of Fig. 1.5. The value of $r = r_1$ at which S attains a maximum value is given by $r_1^3 = q^2(1-1/\epsilon)/4\pi\sigma_{LV}$; when the droplet exceeds a radius r_1, it will continue to grow with a decrease of free energy as long as the supersaturation is maintained, and the nucleus droplet becomes a stable aggregate of the condensed phase.

In order to calculate the rate at which droplets will, under the influence of statistical fluctuations, grow from an initial radius $r_a(<r_1)$ to a radius $r_c(>r_1)$ that will be in equilibrium with the prevailing saturation ratio S, Tohmfor and Volmer (1938) modified the

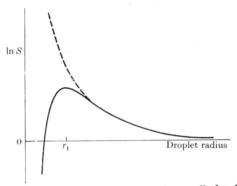

FIG. 1.5. The equilibrium vapour pressure for small droplets (full curve: charged droplets; broken line: uncharged droplets).

Becker–Döring theory and used eqn (1.39) to obtain the change in free energy ΔG^* involved in the formation of a droplet of critical size. They derived the following expression for the rate of production of droplets at 265 K;
$$I \simeq 10^\gamma N_i \exp(-\Delta G^*/kT), \tag{1.41}$$
where N_i is the number of ions per cm³ in the vapour. On substituting experimental values of $S = 4\cdot1$, $T = 265$ K, $\sigma_{LV} = 77$ erg cm⁻², $q = 4\cdot8 \times 10^{-10}$ e.s.u., $I = 1$ cm⁻³ s⁻¹, and $N_i = 10^3$ cm⁻³ in (1.39), (1.40), and (1.41), Tohmfor and Volmer found that the equations could be satisfied simultaneously only if $r_a = 4\cdot8 \times 10^{-8}$ cm, $r_c = 10\cdot2 \times 10^{-8}$ cm and $\epsilon = 1\cdot85$. This is not an entirely convincing result because, although the effective dielectric constant ϵ in the strong field of an ion is likely to be considerably less than the static value of 80 for bulk water, it is unlikely to be less than the high-frequency value of 3. This apart, it is clear that the above theory is incomplete in that it fails to account for the fact that negative ions promote condensation at lower supersaturations than do positive ions.

This simple treatment ignores polarization of the droplet by the ion. In water, where the strongly polar molecules form an oriented surface layer, the surface energy of the droplet will probably be modified to an extent depending not only on the magnitude of the charge as indicated by eqn (1.39), but also on its polarity. Therein may lie the explanation of negative ions being able to promote condensation at lower supersaturations than can positive ions.

Loeb, Kip, and Einarsson (1938), who obtained experimental confirmation of this sign preference, argue that small ions cannot grow into droplets by successive attachments of water vapour molecules, but rather that an ion is captured by a molecular aggregate or embryo, which is not yet large enough to possess a regular surface layer and a statistically conditioned electrical double layer that are characteristic of a true droplet. The molecules in the surface layer of the embryo are imagined to be oriented with the oxygen atoms directed outwards; the surface force field will then tend to orient approaching vapour molecules with their protons towards the surface, so favouring H-bond linkages and propagation of the surface structure. The capture of a negative ion by the aggregate will enhance the surface force field and so favour capture of vapour molecules in the correct orientation, while molecules striking a positively charged embryo would have to rearrange themselves before becoming bound. Thus, condensation may be expected to occur more readily on negative ions than on positive ions.

Again, it appears that eqn (1.40), which is derived in terms of macroscopic concepts, cannot apply to the pseudo-crystalline embryo droplets just discussed; indeed, the existence of sign preference, which is not predicted by (1.40), is evidence of the breakdown of this equation in the very early stages of the condensation process.

1.3. Nucleation by insoluble particles

The atmosphere is never sufficiently clean for condensation to occur homogeneously, or upon ions. Particles of dust and smoke, hygroscopic crystals, and small droplets of solution may also serve as centres of condensation. We shall now discuss the properties of these condensation nuclei beginning with insoluble particles.

Volmer (1939) extended the original Volmer–Weber nucleation theory to treat the case of condensation on a plane, solid surface. A condensing droplet is assumed to take the shape of a spherical cap characterized by its radius of curvature r, and ϕ, the angle of contact of the liquid on the substrate—see Fig. 1.6. Using the subscripts L, C, and V to

§1.3 NUCLEATION OF WATER-VAPOUR CONDENSATION

denote the liquid, catalyzing substrate, and vapour, Fletcher (1962a) writes

$$m = \cos\phi = (\sigma_{CV} - \sigma_{CL})/\sigma_{LV} \tag{1.42}$$

and the change in free energy involved in forming the droplet is

$$\Delta G = \rho_L \frac{\mathcal{N}}{M}(\mu_L - \mu_V)V_L + \sigma_{LV}A_{LV} + (\sigma_{CL} - \sigma_{CV})A_{CL}, \tag{1.43}$$

where

$$A_{LV} = 2\pi r^2(1-m), \quad A_{CL} = \pi r^2(1-m^2), \quad V_L = \frac{\pi r^2}{3}(2+m)(1-m)^2.$$

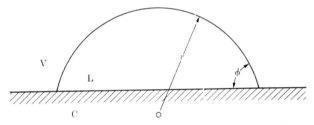

FIG. 1.6. A spherical-cap embryo of liquid (L) in contact with its vapour (V) and a nucleating surface (C).

The radius of the critical embryo is obtained by putting $\partial/\partial r(\Delta G) = 0$ to give

$$r^* = 2M\sigma_{LV}/\rho_L RT \ln(p/p_\infty) \tag{1.44}$$

and, after substitution in (1.43),

$$\Delta G^* = \frac{16\pi M^2 \sigma_{LV}^3}{3\{\rho_L RT \ln(p/p_\infty)\}^2} f(m), \tag{1.45}$$

where

$$f(m) = (2+m)(1-m)^2/4. \tag{1.46}$$

These results differ from those for homogeneous nucleation only by the factor $f(m)$ in the expression for ΔG^*. Since $-1 \leqslant m \leqslant 1$, $0 \leqslant f(m) \leqslant 1$, so that the presence of a foreign surface reduces the free energy necessary to form a critical embryo except in the special case of the surface being completely hydrophobic ($\phi = 180°$).

Calculation of the nucleation rate follows the same lines as for homogeneous nucleation and yields the following expression for the rate of production of critical embryos on unit surface area:

$$I \simeq \{p/(2\pi MRT)^{\frac{1}{2}}\}\pi r^{*2}n'\exp(-\Delta G^*/kT), \tag{1.47}$$

in which ΔG^* is given by (1.45), $p/(2\pi MRT)^{\frac{1}{2}}$ is the rate of impact of molecules on unit area of surface, πr^{*2} is the surface area of the critical embryo, and n' the concentration of adsorbed molecules on the substrate. The pre-exponential factor in (1.47) clearly depends upon the vapour pressure p and upon the nature of the adsorption on the nucleating surface. If the adsorbed molecules form an appreciable fraction of a monolayer at the prevailing supersaturation, then near 0°C the kinetic factor is of order 10^{24} to 10^{27} cm^{-2} s^{-1}. Inserting this and other appropriate numerical values into (1.47) allows the saturation ratio for a

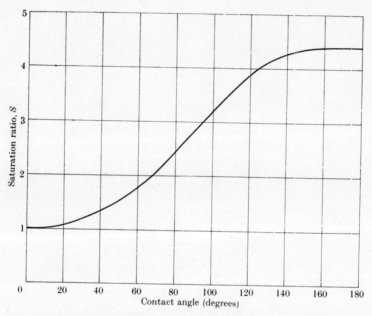

FIG. 1.7. Critical saturation ratio S for $I = 1$ cm^{-3} s^{-1} on a plane substrate of given contact angle at 0° C. (From Fletcher 1962, p. 55.)

given nucleation rate to be determined as a function of contact angle; the results of a specimen calculation are plotted in Fig. 1.7.

A further extension of the theory to include the effects of particle size has been made by Fletcher (1958). If the nucleating particle is considered to be a sphere of radius r, then ΔG^* can be evaluated as before except that the geometry is rather more complicated. The result is

$$\Delta G^* = \frac{16\pi M^2 \sigma_{\text{LV}}^3}{3\{\rho_\text{L} \mathbf{R} T \ln(p/p_\infty)\}^2} f(m, x), \qquad (1.48)$$

where
$$x = \frac{r}{r^*} = \frac{r\rho_\text{L} \mathbf{R} T \ln p/p_\infty}{2M\sigma_{\text{LV}}}$$

§ 1.3 NUCLEATION OF WATER-VAPOUR CONDENSATION

and $f(m, x) = \frac{1}{2}\left[1 + \left(\frac{1-mx}{g}\right)^3 + x^3\left\{2 - 3\left(\frac{x-m}{g}\right) + \left(\frac{x-m}{g}\right)^3\right\} + 3mx^2\left(\frac{x-m}{g} - 1\right)\right]$

with $\qquad g = (1 + x^2 - 2mx)^{\frac{1}{2}}.$

The nucleation rate per particle is proportional to the particle area and is

$$I \simeq \{p/(2\pi MRT)^{\frac{1}{2}}\}4\pi r^2 r^{*2} n' \exp(-\Delta G^*/kT), \qquad (1.49)$$

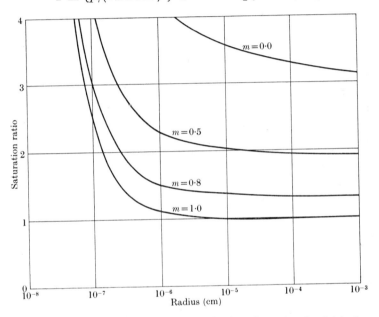

FIG. 1.8. Critical saturation ratio for nucleation of a water droplet in 1 s on a particle of given radius and surface properties defined by $m = \cos \phi$. (From Fletcher (1958).)

the pre-exponential factor having the approximate value of $10^{25}\, 4\pi r^2$. Fletcher has used these expressions to calculate the nucleation behaviour of insoluble particles as a function of radius and contact angle, a typical set of curves being plotted in Fig. 1.8. No detailed experimental confirmation of these curves has yet been produced although the confirmation by LaMer and Gruen (1952) of Kelvin's equation for water droplets provides indirect confirmation of the curve for $m = 1$.

In practice, the problem may be further complicated by the presence on the substrate of impurities, and of imperfections in the form of steps and pits that may promote condensation at supersaturations lower than those required for nucleation on a plane or convex surface.

1.4. Condensation on soluble particles

The equilibrium vapour pressure p_r over the surface of a droplet of pure water of radius r exceeds that over a plane water surface at the same temperature p_∞ as expressed by (1.16), viz.

$$\ln p_r/p_\infty = 2\sigma_{\mathrm{LV}} M/\rho_\mathrm{L} \mathbf{R} T r. \tag{1.50}$$

Thus a drop introduced in to a just saturated atmosphere will evaporate; a droplet cannot persist, much less grow, unless the environment is supersaturated by an amount required by (1.50). If, however, a droplet is formed on a wholly or partially soluble nucleus, the equilibrium vapour pressure at its surface is reduced by an amount depending on the nature and concentration of the solute, which means that condensation will be able to set in at a lower supersaturation than on an insoluble particle of the same size.

An expression for the equilibrium vapour pressure at the surface of a droplet of solution was first derived by Köhler (1921, 1926). A modified version was given by Wright (1936) and has since been used rather uncritically in the literature. A new derivation, leading to an equation of rather different form, will now be given.

We shall consider a solution droplet, radius r, at the surface of which the vapour pressure is p'_r, to be in equilibrium with an atmosphere in contact with a plane water surface whose equilibrium vapour pressure is p_∞. If an elemental mass dm of water is transferred from the droplet to the plane water surface, the resulting decrease in the free energy of the solution will be

$$\Delta G = \sigma' \, dA - P \, dV = \frac{2\sigma' \, dm}{r \rho'_L} - \frac{P \, dm}{\rho'_L}, \tag{1.51}$$

where dA ($= 8\pi r \, dr$) and dV are the changes in the surface area and the volume of the droplet, σ', ρ'_L, and P are respectively the surface tension, density, and osmotic pressure of the *solution*. Alternatively, the decrease of free energy is given by the work gained in evaporating the mass dm of water at pressure p'_r, expanding the vapour to the lower pressure p_∞ and condensing it at pressure p_∞, the whole process being carried out reversibly and isothermally. Assuming the water vapour to behave as an ideal gas, we now have

$$\Delta G = dm \, \frac{\mathbf{R} T}{M} \ln \frac{p'_r}{p_\infty}, \tag{1.52}$$

where $\mathbf{R}$ is the universal gas constant, T the temperature, and M the

§1.4 NUCLEATION OF WATER-VAPOUR CONDENSATION

molecular weight of water. Equating the two expressions for ΔG gives

$$\ln \frac{p'_r}{p_\infty} = \frac{2\sigma' M}{\rho'_L RTr} - \frac{PM}{\rho'_L RT}. \tag{1.53}$$

For solutions whose densities vary linearly with concentration (this is very nearly true for NaCl, MgCl$_2$, etc.), $P = (\mathbf{R}T\rho_L/M)\ln(1+in_1/n_2)$, where n_1 is the number of moles of solute dissolved in n_2 moles of water, ρ_L is the density of water, and i is van't Hoff's factor which depends upon the chemical nature and the degree of dissociation (i.e. on the concentration) of the solute. Thus (1.53) may be written as

$$\ln \frac{p'_r}{p_\infty} = \frac{2\sigma' M}{\rho'_L RTr} - \frac{\rho_L}{\rho'_L} \ln \left\{ 1 + \frac{imM}{W(\tfrac{4}{3}\pi r^3 \rho'_L - m)} \right\} \tag{1.54}$$

or

$$\frac{p'_r}{p_\infty} = \left(\exp \frac{2\sigma' M}{\rho'_L RTr} \right) \left\{ 1 + \frac{imM}{W(\tfrac{4}{3}\pi r^3 \rho'_L - m)} \right\}^{-(\rho_L/\rho'_L)}, \tag{1.54a}$$

where m is the mass of the solute in grammes.

This expression may be compared with that of Wright (1936), viz.

$$p'_r/p_\infty = \exp(2\sigma M/\rho_L \mathbf{R}Tr) - imM/\tfrac{4}{3}\pi r^3 \rho_L W,$$

where the unprimed symbols refer to pure water. Wright assumed i to be constant ($= 2\cdot 22$) for all concentrations; this is not permissible as pointed out by McDonald (1953) who gives experimentally determined values of i varying from $2\cdot 0$ for infinitely dilute solutions, to $2\cdot 91$ for saturated solutions of sodium chloride.† The numerical difference between (1.54) and Wright's equation are largest for small, highly concentrated droplets, but normally do not exceed a few per cent. For example, putting $m = 10^{-14}$ g of NaCl, $r = 2\times 10^{-5}$ cm, $T = 273$ K, $\sigma' = 83\cdot 5$ erg cm^{-2}, $\rho'_L = 1\cdot 23$ g cm^{-3}, and $i = 2\cdot 7$ in (1.54) yields $p'_r/p_\infty = 0\cdot 832$, while the corresponding value according to Wright is $0\cdot 802$. At higher humidities, the differences become smaller, and when $p'_r/p_\infty \rightarrow 1$, they are negligible. Nevertheless, it is not permissible, particularly when considering the early stages of growth of a hygroscopic nucleus, to ignore the dependence of σ', ρ'_L, and i on the concentration of the solute.

When the solution droplet is in equilibrium with the surrounding atmosphere, with its surface temperature equal to that of the air, p'_r must equal the partial pressure of the water vapour, so $H/100 = p'_r/p_\infty$, where H is the relative humidity of the air. Thus the condition for

† Tables of values of i for eight electrolytes have been calculated by Low (1969).

equilibrium becomes

$$\frac{H}{100} = \left(\exp\frac{2\sigma' M}{\rho'_\text{L} \mathbf{R} T r}\right)\left\{1 + \frac{imM}{W(\frac{4}{3}\pi r^3 \rho'_\text{L} - m)}\right\}^{-(\rho_\text{L}/\rho'_\text{L})} \tag{1.55}$$

an equation which can be used to calculate the radius of drops in equilibrium with an atmosphere of a given relative humidity. Specimen equilibrium curves for NaCl nuclei of different masses are plotted in Fig. 1.9. When a nucleus droplet reaches equilibrium with an atmosphere of 100 per cent relative humidity, the depression of the equilibrium vapour pressure due to the presence of the dissolved solute is equal and opposite to the elevation caused by the curvature of the droplet surface. At higher values of the relative humidity, the equilibrium radius of the nucleus droplet increases, until a critical value of the supersaturation is

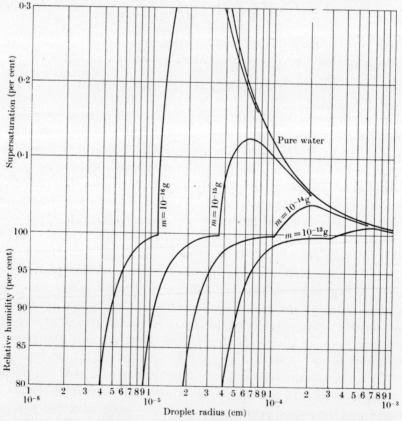

Fig. 1.9. The equilibrium relative humidity (or supersaturation) as a function of droplet radius for solution droplets containing the indicated masses of sodium chloride.

§1.4   NUCLEATION OF WATER-VAPOUR CONDENSATION

reached corresponding to the maximum of the relevant curve in Fig. 1.9. At this stage, the solution is so dilute that, to a high degree of accuracy, (1.55) may be written

$$\frac{H}{100} = e^{B/r}\left(1 - \frac{C}{r^3}\right) \simeq 1 + \frac{B}{r} - \frac{C}{r^3}, \qquad (1.55a)$$

where $B = 3 \cdot 2 \times 10^{-5}/T$ and $C = 8 \cdot 6 \, m/W$ are constants for a given temperature and a particular nucleus. The condition for the curve to attain a maximum, corresponding to critical values of the supersaturation (H_c) and droplet radius (r_c), is $dH/dr = 0$ whence, to a high degree of accuracy, $r_c^2 = 3C/B$ and $H_c/100 = 1 + (4B^3/27C)^{\frac{1}{2}}$. The supersaturations required for NaCl nuclei of various masses to reach the critical radius are shown in Table 1.3.

If a nucleus droplet exceeds the critical radius and the supersaturation is maintained, it will continue to grow with a decrease of free energy to form a water droplet. Under these conditions, the transition from a *nucleus droplet* ($r < r_c$) to a fully developed *cloud droplet* ($r > r_c$) would occur very rapidly, and in theory, the drop would grow without limit. In practice, one is not concerned with a droplet growing in isolation under steady-supersaturation conditions, but rather, with a whole population of droplets competing for the water vapour being released in a cooling air mass. In this case, the supersaturation will not, in general, remain steady; when vapour is being extracted at a faster rate than it is

TABLE 1.3

Critical radii and supersaturations for nuclei of various sizes ($T = 273$ K)

(a) *Hygroscopic nuclei of* NaCl

log m (g)	−16	−15	−14	−13	−12	−11	−10	−9	−8
$r(\mu m)$ at $H = 78\%$	0·039	0·084	0·185	0·39	0·88	1·85	4·1	8·8	18·5
$r_c(\mu m)$†	0·20	0·62	2·0	6·2	20	62	200	620	2 000
$H_c - 100$ (= supersat. %)‡	0·42	0·13	$4 \cdot 2 \times 10^{-2}$	$1 \cdot 3 \times 10^{-2}$	$4 \cdot 2 \times 10^{-3}$	$1 \cdot 3 \times 10^{-3}$	$4 \cdot 2 \times 10^{-4}$	$1 \cdot 3 \times 10^{-4}$	$4 \cdot 2 \times 10^{-5}$
r (of crystal) (μm)	0·022	0·048	0·103	0·22	0·48	1·03	2·2	4·8	10·3

r at $H = 100\%$ is approx. $r_c/\sqrt{3}$.
† For other nuclear substances of molecular weight W, multiply by $(58 \cdot 5/W)^{\frac{1}{2}}$.
‡ For other nuclear substances of molecular weight W, multiply by $(W/58 \cdot 5)^{\frac{1}{2}}$.

(b) *Insoluble wettable nuclei*

log r (cm)	−7	−6	−5	−4	−3
$100 p_r/p_\infty$	323	112·5	101·2	100·12	100·01

being released, the supersaturation is forced to retreat and so growth of the droplets is restricted.

The uppermost curve in Fig. 1.9 shows the supersaturations necessary for continued condensation to occur on drops of pure water as calculated from (1.50) and applies very nearly to insoluble wettable nuclei that do not possess capillary-active surfaces. It is evident from the diagram that a higher supersaturation is required to activate such a particle than a hygroscopic nucleus having an equal radius at say, 99 per cent relative humidity. It will be seen from Table 1.3 that all sodium chloride nuclei of $m > 10^{-15}$ g, and all insoluble non-hydrophobic nuclei of $r > 1$ μm, require supersaturations of less than 0·1 per cent for them to act as centres of continued condensation.

In deriving the curves of Fig. 1.9 we have neglected to mention that solution droplets do not remain liquid at very low humidities. As the droplet shrinks the solution becomes increasingly concentrated until, eventually, salt crystals begin to separate out. Presumably the droplet remains supersaturated until it becomes infected by a nucleus on which the salt crystals can form, or until it becomes so supersaturated that it crystallizes spontaneously. As soon as crystallization begins, the vapour pressure of the drop becomes higher than that of the surrounding air, evaporation proceeds rapidly, and may appear instantaneous. Several separate crystals may appear as a result.

Experimental confirmation of these events was obtained by Owens (1926), and later by Dessens (1949), who studied the growth of salt nuclei suspended from very fine spiders' threads in unsaturated atmospheres of controlled humidity and thereby verified the form of the curves in Fig. 1.9 for NaCl and $ZnCl_2$. Dessens also found that, instead of crystallizing, these solution droplets could remain liquid at relative humidities as low as 40 per cent, whereas, in the case of NaCl, the phase-change is expected at 78 per cent humidity. A similar result was obtained by Junge (1952a) and is portrayed in Fig. 1.10. Here the relatively sharp onset of solution contrasts with the rather wide range of humidity over which crystallization takes place. Twomey (1953a, 1954) also noted that the phase-change from solid to liquid always occurred at a sharply defined and reproducible humidity and therefore proposed this phase-change as a convenient method of identification for large hygroscopic particles. More recently, Orr, Hurd, Hendrix, and Junge (1958) have found curves similar to Fig. 1.10 for particles in the range 10^{-6} to 10^{-5} cm, and also that phase transitions occurred at slightly lower humidities for very small particles.

§ 1.5   NUCLEATION OF WATER-VAPOUR CONDENSATION

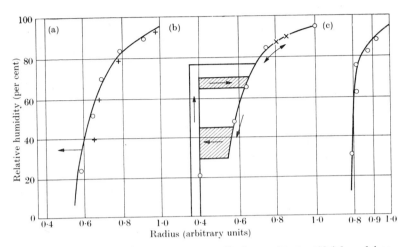

Fig. 1.10. Experimental values of the equilibrium radii of artificial nuclei as a function of relative humidity, and the corresponding theoretical curves: (a) a nucleus of pure $CaCl_2$ solution became supersaturated at $H < 35$ per cent. (b) A nucleus of pure NaCl—crystallization did not occur until humidity fell below 30 per cent. The theoretical curve predicts that crystallization should have occurred at 78 per cent relative humidity. (c) Artificial nucleus composed of mixture of $CaCl_2$ and $CaSO_4$. (From Junge (1952a).)

1.5. Condensation on nuclei of mixed constitution

So far, we have discussed the behavior of completely insoluble particles and of pure hygroscopic particles as potential condensation nuclei. That natural aerosols are unlikely to fall into just these two categories was suggested by some early experiments of Junge (1936) in which he determined the size of nuclei produced by burning coal gas at various relative humidities, by measuring their mobilities as large ions. His curves agreed fairly well with the solution-nucleus theory at humidities above about 75 per cent, but at lower humidities, the nuclear radius remained practically constant, a result that was not to be expected if the nuclei were wholly hygroscopic.

In a later discussion of these measurements and of observations on atmospheric visibility, Junge (1950) found it possible to obtain good agreement between the observations and calculations based on the assumption that the particles consisted of an insoluble core surrounded with a thin coat of hygroscopic substance. He called such particles 'mixed nuclei'. At humidities above 70 per cent, the calculations indicated that these mixed nuclei would react to changes in humidity in much the same way as wholly soluble nuclei of equivalent size, but at humidities below about 70 per cent, the solution coat would shrink, the particle becoming almost wholly solid. In a later investigation, Junge (1952a) used the spider-thread technique to determine the relation between the equilibrium radii and the relative humidity for artificially produced mixed droplets of $CaCl_2$ and $CaSO_4$; the results, which are shown in Fig. 1.10(c), may be compared with those for a pure solution droplet in Fig. 1.10(a).

1.6. Concluding remarks

It appears, then, that atmospheric aerosols consist of solid insoluble particles, of soluble matter giving rise to pure solution droplets, and of mixed nuclei that are partly soluble, partly insoluble. The last two groups of nuclei, being hygroscopic, will obviously be preferred as centres of condensation, their efficiency being determined by both their size and chemical nature. Whether or not non-hygroscopic particles will play a significant role in cloud formation will depend largely on the numbers and sizes of the more favoured hygroscopic nuclei present. In any case, because insoluble particles may obtain a hygroscopic film, either by capturing small solution droplets or by acting as condensation platforms for trace impurities, it may be that truly non-hygroscopic particles are rather rare in the atmosphere.

2

The Nuclei of Atmospheric Condensation

WE have seen in Chapter 1 that in air freed from aerosols and from ions, condensation of water vapour occurs only if the supersaturation reaches several hundred per cent. In the atmosphere, abundant foreign particles serve as ready-prepared aggregates (nuclei) for condensation and prevent these large supersaturations from being achieved. The peak supersaturation that will be attained in an air mass containing a population of condensation nuclei will depend upon the temperature and the rate of cooling of the air, and upon the concentration, size distribution, and nature of the particles—subjects that provide the material for this chapter.

The first experiments to demonstrate condensation of water vapour on nuclei were carried out by Coulier (1875) and by Aitken† (1880–1), who showed that a cloud formed in slightly supersaturated air was made more dense on introducing combustion products, but that thinner clouds were obtained by either filtering the air through cotton-wool, allowing it to stand for some days, or by repeated cloud formation in an enclosed volume of air. It was found that practically all the nuclei could be removed by these methods and that the resulting clean air could sustain an appreciable supersaturation without droplets appearing in the body of the gas. Coulier enclosed air and water in a flask and produced supersaturation by compressing a hollow rubber ball connected to the flask and suddenly releasing it. In his early experiments, Aitken produced supersaturation by blowing steam into a large vessel containing the air under test. Later, he developed independently an expansion method similar in principle to that of Coulier, but used a pump for partial evacuation of the apparatus.

Aitken concluded that when water vapour condensed in the atmosphere it always did so upon the dust particles present, using the term 'dust' in a wider sense to include hygroscopic and non-hygroscopic particles, however produced. Indeed, he felt that there were two types of nuclei, those with an affinity for water vapour on which condensation begins below water saturation, and non-hygroscopic nuclei that require

† Careful study of Aitken's original papers on condensation (*Collected Scientific Papers*, 1923) is strongly recommended.

an appreciable degree of supersaturation to act as centres of condensation. He believed that those of the first class were mainly responsible for fog formation, while the others produced haze.

In more recent years, Aitken's work has been extended particularly to study the larger airborne particles and it is now clear that atmospheric aerosols cover a wide range of particle sizes, from about 10^{-7} cm radius for the small ions consisting of a few neutral air molecules clustered around a charged molecule, to more than 10 μm (10^{-3} cm) for the largest salt and dust particles. Their concentrations, expressed as the number of particles per cm^3 of air, also cover an enormous range, from <100 cm^{-3} over the oceans, to perhaps 10^6 cm^{-3} in the highly polluted air of large industrial cities. The small ubiquitous ions almost certainly play no part in atmospheric condensation because of the very high supersaturations (about 300 per cent) required for their activation, while particles of $r > 10$ μm, having fall speeds of several cm s^{-1}, are able to remain airborne for only a limited time.

2.1. Collection, measurement, and identification of atmospheric nuclei

The wide range of particle sizes and concentrations necessitates the use of several different experimental techniques for the collection, sizing, and examination of atmospheric aerosols. For the purposes of further discussion, it is convenient to divide the spectrum of potential condensation nuclei into three parts: (a) those nuclei that can be detected in an Aitken counter, having radii between 5×10^{-7} cm and 10^{-5} cm (0·1 μm), which we shall call 'Aitken nuclei'; (b) particles of radii between 0·1 and 1 μm, which we shall call 'large nuclei'; and (c) particles of radius larger than 1 μm, which we shall call 'giant nuclei'. Division according to these particular size groups has the advantage that it coincides quite well with divisions based either on the techniques of measurement, or on the processes by which they are produced.

2.1.1. *Techniques for counting and examining Aitken nuclei*
(5×10^{-7} cm $< r < 0\cdot1$ μm)

(*a*) Expansion-type counters. By definition, these nuclei are those that are detected by the Aitken nucleus counter, three forms of which were developed by Aitken himself in the years following 1887.

The most frequently used condensation-nucleus counter is a modified form of Aitken's portable counter, in which a fixed volume of saturated air is expanded quickly by a predetermined amount. Condensation

§ 2.1   THE NUCLEI OF ATMOSPHERIC CONDENSATION

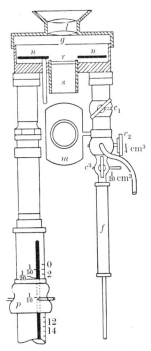

FIG. 2.1. The portable Aitken nucleus counter.

occurs on the airborne particles, and the resulting droplets fall on to a stage where they may be counted through a magnifying glass. The instrument shown in Fig. 2.1 has a test chamber 1 cm deep and volume 35 cm³, the walls of which are lined with wet blotting paper to keep the air saturated with water vapour. The chamber is first flushed with filtered air which is checked for being free of nuclei by rapid expansion following the sharp withdrawal of the piston. A known sample of atmospheric air is then introduced into the chamber which is then sealed, an expansion is made, and the resulting droplets falling on the stage are counted against a scale ruled in millimetre squares. It is the practice to repeat the expansion until no more droplets fall, and to use the sum of all the counts on one square as the representative figure for the evaluation. The appearance of five droplets on one square after 1 cm³ of atmospheric air is admitted, would correspond to a nucleus concentration in the air of 25 000 cm^{-3}. If the air should contain much higher concentrations of nuclei, it can be diluted by introducing smaller samples through the stopcocks c_2 and c_3, which have cavities of 0·25 cm³ and 0·05 cm³ respectively. This instrument has several unsatisfactory features; errors are introduced through turbulence producing a non-uniform distribution of droplets on the stage, and it is susceptible to leaks. A rather more satisfactory version was introduced by Scholz in 1932, in which the cloud chamber volume was 100 cm^3, but this also was apt to develop leaks. A modified form of the Aitken counter in which the graticule can be observed simultaneously by either two persons, or by one person and a 35-mm camera, has been described by Pollak (1952).

Many of the defects of the Aitken counter, particularly the errors that arise because of leakage, and the difficulties of counting, have been overcome in a very convenient and satisfactory instrument developed by Nolan and Pollak (1946), Pollak and O'Connor (1955), and Pollak and Metnieks (1957). It is shown in Fig. 2.2. It consists of a vertical tube 60 cm long and 2·5 cm in diameter fitted with a wet porous ceramic lining and sealed at both ends by electrically-heated glass plates. On the top there is an illuminator, the light source of which produces a parallel beam of light. At the bottom is a photocell. Air containing the nuclei under investigation is drawn through the tube until the previous contents are removed. Filtered air is then pumped into the tube until a selected overpressure is reached. After about 1 min, when the air has become saturated with water vapour and lost its heat of compression, the photocell current (I_0) is read and the pressure is released. The sudden expansion cools the moist air, a fog is formed on the nuclei and the current falls to a new value (I). The relationship between the extinction $(I_0-I)/I_0$ and the concentration of nuclei has been determined experimentally by reference to an Aitken counter. The instrument holds its calibration very well and gives very reproducible results.

A well-designed instrument allowing of size discrimination is described by Rich (1955). The air sample starts at atmospheric pressure and expands into two evacuated chambers. The expansion into the first chamber is as rapid as possible and thus determines the maximum supersaturation. The expansion in the second is slow enough to permit the growth of water drops and bring the supersaturation below the initial value, and yet fast enough to be practically adiabatic. Only those nuclei smaller than the size determined by the initial supersaturation are activated in the second expansion and contribute to the extinction of light, which is measured photo-electrically.

Another version of the photo-electric instrument has been described by Verzar (1953) in which all the operations are carried out automatically by electromagnetic valves and a measurement can be taken

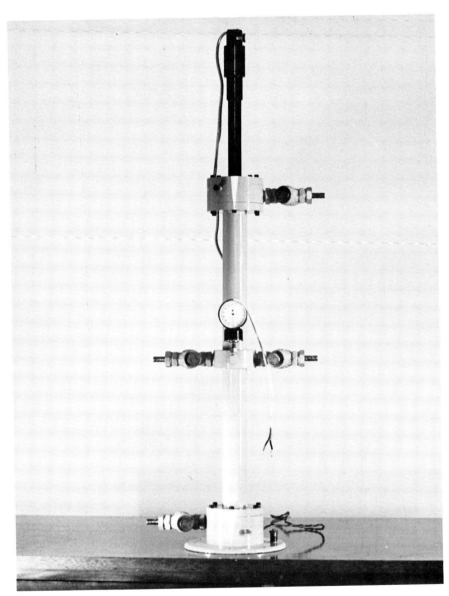

Fig. 2.2. The Nolan–Pollak counter.

every 4 min. The readings of the galvanometer are recorded photographically.

The maximum supersaturation achieved, and hence the size of the smallest nuclei detected in an Aitken counter, is not well established. The peak supersaturation cannot be calculated from the expansion ratio because the exchanges of heat and water vapour between the air and the walls of the small chamber introduce important non-adiabatic effects. That counts of less than 100 cm^{-3} have been measured over the oceans suggests that the ubiquitous small ions are not detected. Using an expansion chamber much larger than a portable Aitken counter, and an electron microscope to examine the deposition of condensation products, Hosler (1950) found that an expansion ratio of 1·15 was sufficient to activate nuclei of radius 5×10^{-7} cm. The conventional Aitken counter, though having a rather larger maximum expansion ratio, gave considerably smaller counts, indicating that not all the smaller nuclei were being detected. The upper limit for the size of particles detected in the counter can be estimated by assuming that larger nuclei will grow so rapidly in the saturated air of the chamber that they will fall out during the interval between admitting the sample and making the expansion. A delay of 5 s would ensure settling out of all droplets with $r > 3·8$ μm, the corresponding nuclear radius, in the case of NaCl, being $2·2 \times 10^{-5}$ cm. It appears reasonable, then, to take the lower size limit of Aitken nuclei as 5×10^{-7} cm and the upper limit as 10^{-5} cm, although these values may vary a little with the particular instrument.

(b) *Diffusion cloud chambers.* The diffusion cloud chamber was originally designed by Langsdorf (1936) to produce a high continuous supersaturation in clean air for the study of the ionization tracks left by cosmic rays. In the thermal-gradient version of this instrument, the experimental chamber consists of a cylinder, in the bottom of which is water at a temperature T_1. The top of the cylinder is closed with a porous plate which is kept wet with water at a higher temperature T_2. Water vapour is distributed through the chamber by mixing and diffusion, and if the side walls are well lagged, both the temperature and vapour density gradients will be linear. Since, however, the saturation vapour pressure of water is not linearly related to its temperature, this mixing and diffusion leads to a supersaturated condition as shown in Fig. 2.3. The supersaturation is given by $(p-p')/p'$ and has a maximum near the temperature $(T_2-T_1)/2$. If there are slight stirring motions within the chamber, all particles may be expected to experience this maximum supersaturation unless they grow and fall

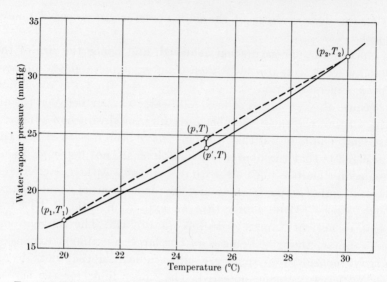

Fig. 2.3. The principle of the thermal gradient cloud chamber. (After Wieland (1956).)

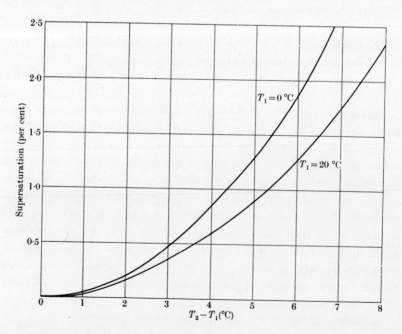

Fig. 2.4. Supersaturation produced in a thermal gradient chamber. (After Wieland (1956).)

§ 2.1   THE NUCLEI OF ATMOSPHERIC CONDENSATION   37

out in the meantime. Fig. 2.4 shows this supersaturation as a function of the temperature difference between the top and bottom of the chamber. A chamber of this type was used by Wieland (1956) and is most useful in the range of supersaturations from 0·1 to a few per cent that is not easily achieved by an expansion chamber.

An automatically operating and recording thermal diffusion chamber, incorporating several improvements, has been developed by Radke and Hobbs (1969a). Instead of photographing a known small volume of the chamber and then counting by eye the number of bright spots produced by the droplets on the film, a very time-consuming process, the nuclei activated at a pre-determined supersaturation are allowed to grow to a pre-determined, very nearly uniform size, and their concentration is determined by measuring the light-scattering coefficient of about 20 cm³ of the cloud with an integrating nephelometer. The temperatures of the base and lid of the chamber at any time can be recorded directly. The maximum supersaturations employed are generally 0·2, 0·5, and 1·0 per cent. The test sample of air is drawn into the chamber, left to settle, and the droplets allowed to grow to a given size, their light-scattering coefficient is measured, the sample is removed by suction and replaced by a fresh sample, this sequence of operations being carried out automatically within a few seconds. The authors claim that nucleus counts obtained by this method are in good agreement with those obtained by direct visual counting. This type of counter has, however, the disadvantage that if the walls are anywhere cooler than the vapour that comes into contact with them, quite high local supersaturations may be produced and upset the measurement.

An attempt to overcome this difficulty was made by Twomey (1959a, b) in his chemical gradient chamber, which has water at the top and concentrated HCl in the bottom, both at the same temperature. In this case, water vapour diffuses downwards and HCl gas diffuses upwards. In a given air parcel the pressure of water vapour is a linear function of the concentration of HCl, but the equilibrium vapour pressure over HCl solution is not and so supersaturation can result. The maximum supersaturation achieved depends only upon the concentration of the acid solution, and values ranging from about 0·05 per cent to over 100 per cent can be produced and accurately determined. Fig. 2.5 shows the maximum supersaturation as a function of HCl concentration over the range of interest.

In Twomey's counter, a set of glass beakers is used with a range of acid concentrations. A sample of the aerosol under test is introduced

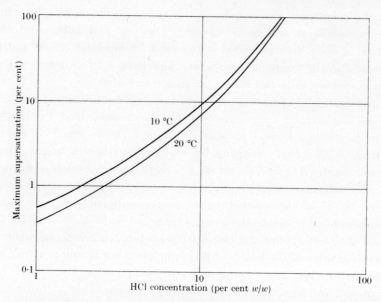

Fig. 2.5. Supersaturation produced in a chemical gradient chamber using dilute HCl. (After Twomey (1959a).)

into each beaker and the droplet density is measured by photographing them in a bright, well-defined beam from a mercury arc.

Difficulties in this method arise from continuous dilution of the acid by condensing water vapour, the fall-out of the large nuclei before they enter the region of highest supersaturation, and the possibility of chemical reaction altering the efficiency of the nuclei. It has, however, given some interesting information on the efficiency of natural condensation nuclei, as discussed on p. 83.

(c) *Diffusion and mobility measurements.* The sizes of nuclei with radii less than 0·1 μm may be determined from measurements of either the diffusion coefficient (e.g. Nolan and Doherty 1950) or of ionic mobility (Israel 1931, 1932). Many of the nuclei capture a single positive ion or an electron, so that if their mobility (velocity in an electric field of 1 V cm^{-1}) is measured, their size can be computed from Stokes's law. Such an ion spectrum is not quite equivalent to a nucleus spectrum, because some ions have multiple charges, particularly when their number is small and their size large.

(d) *Identification of Aitken nuclei.* Positive identification of the Aitken nuclei, which are too small to resolve under the optical microscope, is very difficult even with the electron microscope, not only because of their small size but because they often partially evaporate in the high vacuum or are affected by electron bombardment in the microscope. Although present in large *numbers*, they occur in only very small *mass* concentrations, rarely more than 10 μg in 1 m³ of air, which is very small for chemical analysis and also difficult to separate from the larger quantities of material contributed by the large and giant nuclei. There is, therefore, little direct information on the composition of Aitken nuclei, but an alternative approach is to study the most likely and most efficient sources of these particles. These matters are discussed in § 2.4.1.

2.1.2. *Sampling and examination of large and giant nuclei*
  (0·1 μm $< r <$ 10 μm)

The mean size of atmosphere nuclei has been deduced by Wright (1939) from observations of visibility. The visibility will depend upon the nucleus concentration and the relative humidity; it will decrease if either of these factors increases, the other remaining constant. Wright analysed the observations of visibility made over a number of years both in clear and foggy conditions at various stations in the British Isles and calculated mean values of the atmospheric opacity over different ranges of relative humidity. From the available data, he estimated the respective contributions to the opacity of molecular scattering, absorption and scattering by solid particles, and scattering of light by water drops. Assuming that, over a long period, the average number of nuclei was independent of humidity, he ascribed the changes of visibility to the change of nuclear size with humidity. At all stations there was a definite reduction of visibility with rise of humidity, indicating that some of the responsible nuclei were hygroscopic. Having found the relative change of mean nucleus size with relative humidity, it was only necessary to find which of the family of curves shown in Fig. 1.9 best fitted the observations to obtain the mean mass of the nuclei; the concentration was then determined from the relation between the visibility and the size and concentration of the nuclei. In a later paper, Wright (1940) applied the method in detail to visibility observations made over a 5-year period at Valentia in south-west Ireland. He assumed that the responsible nuclei consisted mainly of sea salt and found that the curve of closest fit was obtained for nuclei of mean mass $1·1 \times 10^{-13}$ g. Their average concentration was then calculated to be 63 cm^{-3}.

A number of methods have been devised for collecting nuclei from the atmosphere so that they can be examined under either the optical or the electron microscope. If the particle-laden air is drawn past a narrow obstacle the particles will tend to impact on it because of their inertia. The efficiency of collection may be improved by subjecting the particles to centrifugal, thermal, or electrostatic forces.

(a) Impaction techniques. In order to capture a true, representative sample of the aerosol, the 'collection efficiency' of the collector (i.e. the fraction of the particles lying in its projected cross-section that actually hit the collector) should be unity for particles of all sizes. In practice, the smaller particles tend to follow the flow lines of the air and to be deflected round the obstacle and only the larger particles have sufficient momentum to strike it. The collection efficiency is governed by the particle size, the velocity of the air-stream, and the geometry of the obstacle, in such a way that the collection of small particles requires a fast air-stream and a collector of high surface curvature. The trajectories of particles relative to cylinders, ribbons and spheres, and the corresponding collection efficiencies, have been calculated by Langmuir and Blodgett (1946) and Ranz and Wong (1952). Fig. 2.6 shows the

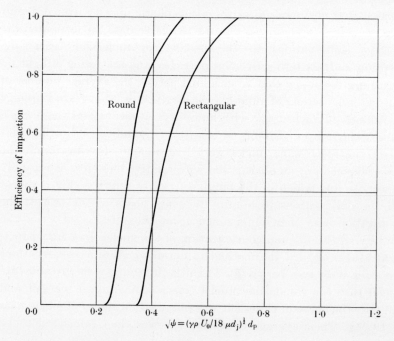

FIG. 2.6. Theoretical impaction efficiencies of aerosol jets. (From Ranz and Wong (1952).)

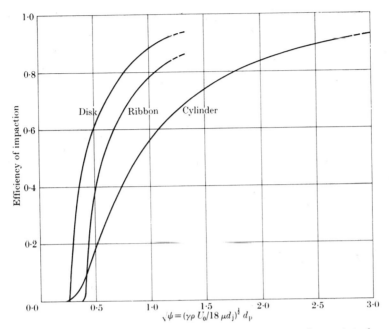

Fig. 2.7. Theoretical impaction efficiencies of an aerosol on obstacles of various shapes. (After Ranz and Wong (1952).)

collection efficiency of a system consisting of an infinitely long, round or rectangular jet of diameter or width d_j impinging on a flat plate of infinite extent, and Fig. 2.7 shows the collection efficiencies of various obstacles of diameter or width d_j placed in an air-stream of infinite extent. The curves, in all cases, are plotted in terms of a dimensionless parameter;

$$\sqrt{\psi} = (\gamma \rho U_0/18\mu d_j)^{1/2} d_p, \qquad (2.1)$$

where d_p is the diameter of the aerosol particle, ρ its density, U_0 the free air-stream velocity and μ the viscosity of air, all in c.g.s. units. γ is a constant defined by

$$\gamma = 1 + \frac{2\lambda}{d_p}\{1 \cdot 23 + 0 \cdot 41 \exp(-0 \cdot 44\, d_p/\lambda)\}, \qquad (2.2)$$

where λ is the mean free path of gas molecules. The collection efficiency approaches unity for large particles but falls off for small particles, so it is usually necessary to apply weighting factors in favour of small particles to correct the measured size distribution.

Several forms of impactor device are available. Two of these, the Owens dust counter (Owens 1922) and the Zeiss konimeter (Junge

1952a), are similar in principle. By means of a pump, the air under test is sucked through a narrow opening, the resulting air jet then strikes a glass plate on which the particles are deposited. The plates can be removed and the particles counted under the microscope. In the konimeter, the aerosol is arrested by a sticky layer on the glass plate. In the dust counter, the air is humidified so that the particles acquire a water skin which allows them to adhere to the glass. In both cases, the size range of the particles that adhere to the plate is governed by the air velocity in the jet and the width of the aperture; smaller particles are deposited with higher air velocities and narrower apertures. Junge (1953) describes a two-stage konimeter that divides the particles into two different size groups determined by the widths of the two slits. Using an air flow of 400 cm³ s⁻¹, Junge found that the deposit at the first slit, 0·85 mm wide, consisted of particles of $r > 0.9$ μm, the smaller ones being deflected towards the second slit of width 0·2 mm where those deposited were predominantly in the size range 0.2 μm $< r < 0.9$ μm. In this way, he separated the large from the giant nuclei in atmospheric aerosol.

A rather similar instrument, which consists essentially of a system of four jets and sampling slides in series, is the cascade impactor described by May (1945), and illustrated in Fig. 2.8. Any form of aerosol

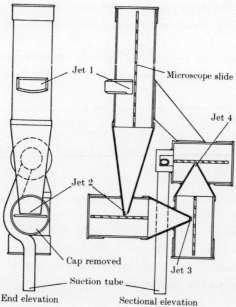

Fig. 2.8. Diagrammatic section of the cascade impactor. (From May (1945).)

in which the particle diameter is greater than 1 μm can be sampled by means of this instrument. When air is drawn through the current commercial instrument at the normal rate of 17·5 litres/min, the velocities at the four jets are 2·2, 10·2, 27·5, and 77 m s^{-1} respectively. Under these conditions, the largest droplets of unit density that impact on the second, third, and fourth slides are 20, 7, and 2 μm in diameter respectively, these being the minimum sizes impacted with 100 per cent efficiency by the previous slide. Because of the high rate of aspiration, the sampling time is limited by the accumulation of deposit on the slides and can only be very brief when the concentration of aerosol is high. With sampling periods of less than 1 s, there is a serious risk of starting, stopping, and dead-space errors. In all such devices, errors may arise in the sampling of large particles if the free air-stream velocity is substantially different from the air velocity in the intake of the instrument i.e. if the sampling is not isokinetic. The intake orifice of the cascade impactor was originally designed to give minimum loss of large drops when the instrument was facing directly into a wind of 3·5 m s^{-1}, and this design gave a high intake efficiency when the surrounding air was stagnant. When sampling fog, however, the air is rarely stagnant but, on the other hand, rarely achieves 3·5 m s^{-1}; consequently large drops are lost. Under these conditions, May (1961) found it necessary to place the impactor in a small portable wind tunnel and to draw air past it at a steady velocity of 5 m s^{-1} when the sampling rate was 17·5 litres/min. In order to reduce the loss of small fog droplets by evaporation in the instrument, May removed the second and fourth stages and gave the remaining two slides a uniform reciprocating motion of short period across the line of the jet. The sample was then spread over a larger area of the slide and allowed the sampling period to be greatly increased. This version of the cascade impactor allows droplets of diameter greater than 1 μm to be collected with high efficiency.

In sizing aerosol particles, it is of great convenience if they are sorted according to size by the sampling apparatus. An instrument that fulfils this purpose is the conifuge described by Sawyer and Walton (1950). The airborne particles are drawn into the apparatus, which is essentially a conical centrifuge, and winnowed by an internally circulated stream of clean air in such a way that the particles are classified according to their settling velocities and deposited on a glass slide in a continuously graded sample. When the particles are of similar shape and density, the settling velocities are a unique function of the size, and the deposit is in the form of a spectrum of particle sizes with a high degree

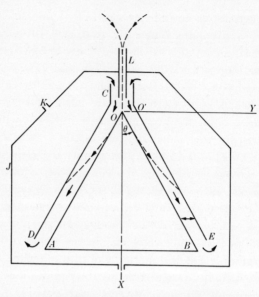

Fig. 2.9. Diagrammatic section of conifuge: →, clean air; --→, particulate cloud. (From Sawyer and Walton (1950).)

of purity. For spherical particles of unit density, the spectrum extends from about 30 to 0·5 μm diameter. The principle of operation is shown in Fig. 2.9 in which OAB represents an inner solid cone, and CDE an outer conical shell of the same vertical angle, the two being rotated about their common axis OX and enclosed in a stationary outer casing J. Rotation of the cones causes a circulation of air through the annular space between them, air entering at C and being expelled through the annulus DA–BE at the base. A controlled amount of the circulating air is allowed to leave the outer container through a small orifice K, causing an equal amount (the sample) to enter through the tube L and impinge on to the apex O of the inner cone. Here it streams over the surface of the latter in a symmetrical film separated from the outer cone by a larger quantity of clean air circulating through the system. Suspended particles in the sampled air, on arriving in the annular space between the cones, are subjected to a centrifugal force due to the rotation which causes them to move radially outwards at speeds proportional to (and greatly in excess of) their settling velocities under gravity. At the same time, the particles are carried forward by the air flow so that each strikes the outer cone at a distance from the apex characterized by its settling velocity, the larger ones being deposited near O' and the smaller ones towards E. The apparatus is normally driven by an electric motor at 3000 rev/min and samples at the rate of 25 cm³ min⁻¹; it can then handle particles of settling velocities up to 2 cm s⁻¹ ($d \approx 30$ µm) without loss by impaction on the inner cone.

A sophisticated instrument, similar in principle to the conifuge, is the aerosol spectrometer devised by Goetz and Preining (1960), which separates aerosol particles in the diameter range 0·03–3 µm. A laminar air-stream carrying the particles is subjected to a large centrifugal force (up to 26 000 g) by leading it through a helical channel in a rapidly spinning rotor. The periphery of the rotor forms a groove and is covered with a flexible foil that seals the groove to an airtight channel. The particles fall out of the air-stream according to Stokes's law and impinge on the foil. After sampling, the foil is removed for analysis of the continuous, spiral, band-shaped deposit, either by a microscope, or by photometric recording over areas of 10^{-4} cm² of the light scattered by the particles under reflected dark-field illumination. If the time of residence in the channel is sufficient for all suspended particles to reach the outer wall, i.e. the inner surface of the foil, the deposit thereon represents a complete size spectrum of the aerosol, effected by differences of radial velocity between the large and small particles. The former are deposited near the entrance, the latter towards the end of the channel. The locus of a deposited particle is thus determined by the geometry of the channel, the flow rate, the angular velocity of the rotor, and the Stokes diameter of the particle. This instrument has excellent resolving power, can handle flow rates of up to 7·5 litres/min, and has the great advantage that measurement of the deposit yields the size (and mass) distribution of the aerosol, independent of its fate after deposition.

As an alternative to employing an air-stream of high velocity to ensure the efficient deposition of particles on a large collector, one may use very narrow collectors and correspondingly low air-speeds. This technique was applied in a very elegant manner by Dessens (1946a, b; 1947a, b; 1949), who caught aerosol particles on very fine spiders' threads of about 10^{-6} cm diameter and examined them under a microscope in an atmosphere of controlled humidity. A spider is encouraged to spin a mesh on a small metal frame that can either be exposed across the intake of a small pump, or rotated at a known constant speed on a gramophone turn-table. Aerosol captured in the open country was found to consist of some small solid particles but, for the most part, of solution droplets that remained liquid even at relative humidities as low as 40 or 50 per cent. The size distribution of the particles with

$r > 0.2$ μm could be determined by measurement under the microscope with a calibrated eyepiece, and their concentrations from the number adhering to unit length of the thread, the time of exposure, and the speed of the air past the thread. The only difficulty lies in choosing the time of exposure so that the risk of coagulation between captured and incident droplets, and between neighbouring droplets on the threads, is very small. This method was also used by Fournier d'Albe (1951) to study the nuclei produced over the Bay of Monaco, and by Junge (1952a) to determine the size distribution of mixed nuclei over land and their variation of radius with relative humidity.

Woodcock and Gifford (1949) used thin glass strips cut from microscope slides and silver rods of 1 mm² cross-sectional area as collectors for nuclei of sea salt. They were unable to collect nuclei smaller than 0.5 μm radius and 10^{-13} g in mass because they were swept round the collectors, the smallest of which were 1 mm wide. These were coated with a hydrophobic film on which droplets assumed a hemispherical shape, thus allowing their volumes to be determined from a simple measurement of their diameters. The collectors were mounted on a wind vane, and having been exposed to the nucleus-laden air for a suitable period, were placed on a microscope stage in a thermostatically controlled box containing dilute sulphuric acid of known concentration. The droplets formed on the nuclei were brought into equilibrium with the air over the acid solution, the vapour pressure of which was known, and the numbers and diameters of the droplets were measured. Assuming these droplets to consist of sea-water, and having previously determined its vapour pressure and density as functions of temperature and chlorinity, the weight of chloride in the individual nuclei could be found. This is known as the 'isopiestic' or equal-pressure method. The concentration of nuclei of a particular size is given by n/UEt, where n is the number of these nuclei on unit area of the slide, U the mean velocity of the air relative to the collector during the time of exposure t, and E is the collection efficiency of the slide for these nuclei at wind speed U. In a critical examination of errors that can arise in this method, Toba (1966) stresses the need for careful temperature control of the humidity chamber.

(b) *Thermal precipitation.* An apparatus that is more efficient than impaction devices for the collection of sub-micron particles is the thermal precipitator described by Watson (1936). The sampling rate is rather slow—7 cm³ min⁻¹, but samples of highly polluted air can be taken over periods of several minutes without serious overlapping of the deposit. The design of the instrument centres upon the fact that around a hot body there exists a dust-free space, the thickness of which depends on the temperature difference between the hot surface and its surroundings. An electrically heated wire is placed between two glass cover-slips, so that the dust-free space is larger than the separation of the latter. Air containing the particles is drawn past the wire at a known controlled rate by an aspirator, at a sufficiently slow speed that the particles, while following the convection currents directed towards the cold cover-slips, are unable to penetrate the dust-free barrier and are therefore deposited outside it as line deposits. A modified version of the instrument, in which the deposit is distributed over a wider region to facilitate microscopic examination and counting and to reduce the risk of aggregation of particles, has been described by Walton (1950). The glass cover-slip is replaced by a collodion-coated grid which is oscillated with respect to the heated wire; this results in an areal deposit suitable for electron microscopy.

When working with very dilute aerosols, it is a great advantage to sample the air much more quickly than is possible with the thermal precipitator. For this purpose, Millipore filters† made from a very porous membrane of a cellulose ester, have been developed. The filters, in which about 80 per cent of the volume consists of conical holes, (the diameters of which vary from 0·1 to 1·5 μm according to the type), can be used to remove with 100 per cent efficiency all particles of radius greater than 0·2 μm from the air drawn through them. The filter, which acts as a sieve, rapidly acquires a high electrostatic charge and ensures that the particles are retained near its surface. For microscopic examination this has an obvious advantage over the usual filters in which the particles are embedded throughout the depth. The particles on the surface are kept apart so that it is easy to prevent overlapping and agglomeration. The filters are supported against a porous carbon plate to prevent buckling, and the air is drawn through the whole assembly by a pump; with a filter of 2 inches diameter, a sampling rate of up to 30 l/min may be used. The filters have a refractive index of 1·54, so that when immersed in cedar-wood oil or other suitable liquids of the same refractive index, they become transparent to transmitted light, enabling the aerosol to be microscopically examined without background interference; also the filters can be readily dissolved in acetone to leave the particle deposit on a glass slide. The use of

† Obtainable from Lowell Chemical Company, Watertown, Mass., U.S.A.

these filters for aerosol sampling and analysis is described in a paper by First and Silverman (1953).

(c) Optical methods. The counting and measurement of particles while they are suspended in the air is of special value when examining unstable systems such as rapidly coagulating smokes and evaporating aerosols. This is conveniently done by illuminating the cloud as it streams through a small glass cell with a shallow ribbon of light and viewing the Tyndall beam of scattered light with a low-power microscope. Ultramicroscopes designed on this principle are capable of detecting particles of $r > 0.1$ μm. Special measures are necessary to reduce the errors that arise from secondary scattering by particles outside the beam and from the loss of particles by diffusion to the walls of the apparatus.

An automatic instrument for counting particles of $r > 0.3$ μm at rates of up to 20 s^{-1} was developed by Gucker, O'Konski, Pickard, and Pitts (1947). An improved version, which sorts the particles into size groups, is described by Gucker and Rose (1954). The particle-laden stream, sharply bounded by a sheath of clean air, flows through a dark tube in which extreme precautions are taken to eliminate stray light. The light scattered by the particles between 1° and 20° in the forward direction is collected by a photomultiplier and each particle eventually produces a voltage pulse of amplitude proportional to the particle diameter. The pulses are sorted electronically into various size groups that may be calibrated by using monodisperse aerosols of known particle size. This instrument can detect particles of $r > 0.2$ μm; for particles of $r < 0.5$ μm the accuracy is limited by the signal-to-noise ratio. A device working on similar principles has been used on an aircraft in Russia—see p. 97. An automatic apparatus that counts the aerosol particles containing sodium chloride by the pulses of light they produce when they enter a non-luminous flame, has been developed by Soudain (1951) and also by Vonnegut and Neubauer (1953). The two instruments are very similar in principle and construction; in the latter model (Fig. 2.10), the air to be analysed is passed through a pure hydrogen flame at 5 litres/s when low concentrations of nuclei are being measured, and at about 5 litres/min for smaller, more numerous nuclei. The flame is enclosed in a vertical steel tube to protect it from draughts that would cause spurious responses. The image of the flame is focused by a lens on to a photo-multiplier cell, an optical filter having a transmission peak at the wavelength of the sodium D lines being placed in front of the photo-cell to exclude light produced by elements

§ 2.1   THE NUCLEI OF ATMOSPHERIC CONDENSATION   49

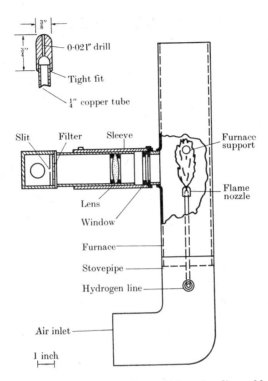

FIG. 2.10. Automatic flame counter for particles of sodium chloride. (From Vonnegut and Neubauer (1953).)

other than sodium. The voltage pulses produced across the load of the photo-cell are amplified and registered on a mechanical recorder which prints the number of particles recorded in a 15-min period. The counting circuit is designed so that pulses having less than a certain amplitude fail to trigger the counter. The sensitivity of the apparatus can be controlled to regulate the amplitude of the pulse produced by a particle of given size. The instruments at present in use are not calibrated in terms of particle size, but are generally set to record particles of radius greater than, say, 1 μm.

(d) *Identification of large and giant nuclei.* Large and giant nuclei are large enough to be resolved under the optical microscope. Sometimes it is possible to identify particles from their appearance and the hygroscopic particles can be detected by allowing them to grow in a humid atmosphere. We have seen on p. 46 that the latter technique can be used, in fairly clean air, to determine the masses of salt-particles, while the chemical nature of a salt particle may be decided by noting the relative humidity at which it begins to dissolve. In very polluted air, the very mixed nature of the aerosol may make it difficult to identify even the largest particles.

Electron microscopy enables very small particles to be resolved, but liquid droplets and some of the more volatile solid particles evaporate in the high vacuum of the microscope and in the heat generated by the electron beam. Moreover, as only very small samples can be studied, examination of a representative portion of an aerosol is a very laborious undertaking. Electron diffraction suffers from similar disadvantages, but it can identify the chemical constitution of very small quantities of material.

Modern microchemical techniques allow very small quantities of material to be analyzed but give only the anion and cation contents separately and not in combination. Very promising and relatively simple methods of identifying individual airborne particles, in the micron and sub-micron range, have been described by Seely (1952) and Lodge (1954). The particle is brought into contact, by impaction or otherwise, with a gelatine film sensitized with a specific chemical reagent that reacts with the particle to produce a persistent spot or halo that can be recognized under the microscope. For the detection of chloride particles, Seely used mercurous fluosilicate as the reagent, the reaction producing a halo of insoluble mercurous chloride that grows uniformly out from the centre by diffusion until the chloride is completely dissolved. The halo is then about nine times the diameter of the original particle, which allows the size of particles of $m > 10^{-14}$ g ($d > 0.2$ μm) to be determined from measurement of the rings. Vittori (1955) used a similar method with silver nitrate as the reagent, the precipitate being silver chloride. He was able to detect chloride particles of mass 10^{-16} g but, in order to obtain sufficient sensitivity for such small particles, the gelatine has to be very thin and calibration becomes difficult. Lodge (1954), who described other reagents to detect calcium, magnesium, sulphates, nitrates, and ammonium salts, collected the particles with a Millipore filter (see p. 47), which was then floated on the reagent solution and, after development, washed and dried, placed on a microscope slide coated with immersion oil (to make the filter invisible), and examined under dark-field illumination for the characteristic reaction sites.

2.2. The concentration, size, and size distribution of atmospheric condensation nuclei

There appears to be a certain regularity in the size distribution of natural aerosols which is the result of dynamic equilibrium between

§ 2.2   THE NUCLEI OF ATMOSPHERIC CONDENSATION

the production of nuclei from different sources and their simultaneous loss by coagulation, sedimentation, etc. Both the number and mass concentrations of aerosol particles vary a good deal in space and time, but typical values for an industrial city and mid-ocean are as follows:

		Aitken	Large	Giant
Industrial city	number/cm³	10^4	10^2	1
	mass (μg/m³)	10	20	20
Mid-ocean (force 4 winds)	number/cm³	10^2–10^3	1–10	1
	mass (μg/m³)	0·5	0·5	5

A survey of the size ranges of various types of aerosol particles is shown in Fig. 2.11.

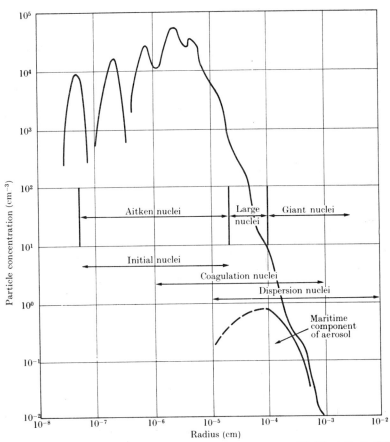

FIG. 2.11. A generalized representation of the size distribution of natural aerosols in heavily polluted air over land. (From Junge (1952b).)

2.2.1. *The concentrations of Aitken nuclei and their variation with place, time, and meteorological factors*

A vast number of determinations of the nucleus content of the air has been made with the Aitken counter and its successors in many different geographical and topographical positions, on land and sea, at ground level, on mountain tops, and in the upper air. For a comprehensive review of these measurements the reader is referred to an excellent monograph by Landsberg (1938). A very condensed summary of the numerous observations is given in Table 2.1 which shows the enormous

TABLE 2.1
Nucleus contents of the atmosphere in different types of localities (after Landsberg)

Locality	No. of places	No. of observations	Average concentrations	Average maximum	Average minimum	Absolute maximum	Absolute minimum
City	28	2500	147 000	379 000	49 100	4 000 000	$3l/00 cm^3$
Town	15	4700	34 300	114 000	5 900	400 000	$620/cm^3$
Country inland	25	3500	9 500	66 500	1 050	336 000	$180/cm^3$
Country seashore	21	2700	9 500	33 400	1 560	150 000	0
Mountain:							
500–1000 m	13	870	6 000	36 000	1 390	155 000	30
1000–2000 m	16	1000	2 130	9 830	450	37 000	0
2000 m	25	190	950	5 300	160	27 000	6
Islands	7	480	9 200	43 600	460	109 000	80
Ocean	21	600	940	4 860	840	39 800	2

range of particle concentrations—from 0 to $4 \times 10^6/cm^3$—which have been encountered. The average concentration is smallest over the oceans and in the upper air, and is greater by several orders of magnitude in air subject to industrial pollution.

Interpretation of the observed variations in nucleus concentration at any one place involves taking a large number of factors into consideration, for example the geographical location, the local topography and the nature of the terrain, the proximity of human settlements with their domestic and industrial pollution, the source and track of the existing air mass in relation to possible sources and sinks of nuclei, besides the influence of meteorological factors such as wind speed and direction, the intensity of convective and turbulent mixing and transport, the intensity of solar heating, humidity, and the duration and intensity of precipitation. Landsberg (1938) has attempted to correlate the rather inadequate data with a number of these factors but, because at any one station the situation is often dominated by local influences, several of which may operate simultaneously, it is difficult from the scattered

§ 2.2 THE NUCLEI OF ATMOSPHERIC CONDENSATION

and often short-term observations, to arrive at firm conclusions which may have general validity.

Observations made at land stations indicate that the nucleus concentration has a diurnal variation. In populous areas it shows a maximum in the early morning when human activities start, a minimum around noon due probably to enhanced upward vertical transport, an increase again in the afternoon as convection dies down and a fall again at night when human activities are diminished. At land stations where there is little disturbance from local pollution, there is only a slight diurnal variation, with a slight increase in the afternoon (maximum about 1400 hours) which has usually been attributed to the advection of pollution from other regions. Stations located near human settlements show an annual variation, with a maximum in the winter months that can largely be accounted for by the greater production of nuclei by fires, although in some places there is evidence that the reduced convective activity in winter may also be a contributory factor.

It appears that there is a fairly strong correlation between the nucleus concentration and the relative humidity, high humidities being associated with low counts. This may perhaps be associated with the growth of the larger hygroscopic nuclei at high humidities and their increased efficiency in removing the smaller Aitken nuclei by coagulation. The well-known sweeping action of precipitation may also be relevant in this context. The number of nuclei must be profoundly influenced by the direction and speed of the wind, which may not only bring nuclei from sources upwind, but produce them locally as dust or spray. Many observers have noted increased concentrations when the wind direction has changed to bring nuclei from a highly polluted region. While there are several reports from land stations of the nucleus concentration decreasing with increasing wind speed, the reverse is true on other occasions. A stronger wind is associated with increased turbulence which tends to distribute the nuclei through a greater vertical depth of atmosphere, but, on the other hand, over a suitable terrain, it may increase the numbers by raising dust. Indeed, a high correlation has been found between the nucleus concentration as measured by an Aitken counter and the dust content of the air as measured by an Owens dust counter—see Boylan (1926), Wright (1932). As an example of the influence of solar radiation, one may recall Aitken's (1912) very interesting, but not yet fully elucidated observation, that the nucleus concentration on a beach at low tide increased twentyfold when the sun came out.

TABLE 2.2
Average vertical distribution of nuclei from balloon ascents (after Landsberg)

Altitude (m)	0–500	500–1000	1000–2000	2000–3000	3000–4000	4000–5000	>5000
Concentration (cm^{-3})	22 300	11 000	2500	780	340	170	80

Measurements made at various altitudes on mountains show that the nucleus concentration decreases quite rapidly with height (Table 2.1), and much less so when the air is unstable than in the presence of a ground inversion. In fact Landsberg (1934), from measurements made at Mt. Taunus Observatory, found that the count fell from 2000 cm^{-3} just beneath a sharp temperature inversion to only 100 cm^{-3} just above it. Israël (1930), during a balloon ascent over an industrial region, found a concentration of 28 400 cm^{-3} just below an inversion with a value of 800 cm^{-3} only 30 m higher up. The average values for nucleus concentration as a function of height, determined from twenty-eight balloon ascents, are given in Table 2.2. A very similar vertical distribution was reported by Weickmann (1957a) on the basis of measurements made from an aeroplane.

Over the oceans, remote from human activity, and not complicated by local variations in topography and terrain. the situation is less complex, but unfortunately the observations are rather few and sporadic. Most of our information comes from observations made at 221 localities on board the *Carnegie* (see Shiratori 1934), which show the concentrations to be generally much lower than on land. The relative frequencies of various concentrations are given in Table 2.3; on more than two-thirds of occasions the concentrations were less than 400 cm^{-3}. Hess

TABLE 2.3
Relative frequencies of nucleus concentrations over the oceans as measured aboard the Carnegie

No. cm^{-3}	0–100	100–200	200–300	300–400	400–500	500–600	600–700
% frequency	10·5	29·7	15·2	12·6	4·6	6·4	1·4

No. cm^{-3}	700–800	800–900	900–1000	1000–1500	1500–2000	>2000
% frequency	3·2	2·3	0·9	4·1	1·8	7·5

(1948) made some interesting measurements while travelling from New York to France and back. The maximum concentration measured was 4480 cm^{-3} in the English Channel and the minimum was 78 cm^{-3}. The average value for the eastbound voyage was 527 cm^{-3} and for the westbound voyage 659 cm^{-3}. In each case the mean counts were greater over the western part of the Atlantic than over the eastern part, a result which Hess attributed to the air masses travelling predominantly from west to east. The nucleus counts on foggy days and during a rough sea were not appreciably different from the general average value. On 79 per cent of occasions, the concentration was less than 600 cm^{-3} and exceeded 1000 cm^{-3} on only 14 per cent of occasions. During a similar voyage in 1951, Hess (1951) obtained a maximum count of 14 000 cm^{-3} in the English Channel. In mid-Atlantic the minimum concentration was 182 cm^{-3} and the maximum concentration was 2500 cm^{-3}. He confirmed his previous observations of higher counts in the western Atlantic but, on this occasion, reported appreciably higher concentrations with very rough seas.

The results of some measurements made during a week's cruise in the Pacific Ocean, over 400 miles from land, were reported by Ohta (1951), who obtained an average count of 290 cm^{-3}, a maximum of 690 cm^{-3} and a minimum of 70 cm^{-3}. No diurnal change, nor changes in nucleus concentration with wind speed, were found over the open ocean. Measurements made on an island in the Japan Sea, however, showed increased counts around midday. The ratio of charged to uncharged nuclei was 1·9 on the Pacific Ocean and 1·5 on the Japan Sea.

Parkinson (1952) made some observations during a voyage from New York to Rio de Janeiro with prevailing easterly winds. The mean of thirty sets of observations gave a value of 676 nuclei/cm^3, 7 per cent of the counts being under 300 cm^{-3}, 40 per cent between 300 and 600 cm^{-3}, 40 per cent between 600 and 1000 cm^{-3}, and 13 per cent above 1000 cm^{-3}.

An interesting series of measurements with a portable Aitken counter was made by Moore (1952, 1955) on board an ocean weather-ship in the North Atlantic, for periods of one month each during 1951 and 1952. He obtained concentrations varying from 77 to 2500 cm^{-3} with mean values of 703 in 1951 and 445 in 1952. The smallest populations existed under convective conditions (cumulus or cumulonimbus), highest counts with stratus clouds, and intermediate concentrations with stratocumulus or a mixture of cumulus and stratocumulus. Moore was also able to correlate the nucleus concentrations with average lapse rate in the lowest 2 km as determined from radio-sonde ascents from the weather ships, the average concentration being 1080 cm^{-3} on those occasions when the lapse rate was less than 3° C/km, and only 374 cm^{-3} when the lapse rate exceeded 6·5° C/km. He found little overall variation of nucleus content with wind speed, apart from a slight tendency for slightly higher counts to occur at very low wind speeds. There appeared to be very little correlation between the nucleus concentration and the height of the waves.

2.2.2. *Concentrations and size distributions of large and giant nuclei*

In summarizing the results obtained with a variety of techniques in many different places, it is convenient to separate the data obtained over the land from those obtained over the sea.

Using the spiders'-thread technique, Dessens (1946a, b; 1947a, b; 1949) found that, in open country in France, the number concentrations of haze drops varied from day to day, but were typically about 100 cm^{-3} for drops of $r > 0·1$ μm and 1 cm^{-3} for drops of $r > 1$ μm measured at 78 per cent relative humidity. Similar results were obtained by Dessens, Lafargue, and Stahl (1952) in Greenland.

Junge (1953) used a two-stage konimeter to determine the size distribution of aerosol at Frankfurt, on the Zugspitze, and on Mt. Taunus, the results being shown in Fig. 2.12. He found that particles of $r > 0·1$ μm obeyed a size-distribution law of the form

$$\left.\begin{aligned} n(r) &= \frac{dN}{d(\log r)} = \frac{A}{r^3} \dagger \\ N' &= \frac{A'}{r^3} = \frac{B}{m}, \end{aligned}\right\} \quad (2.3)$$

or

where $dN = n(r)\,d(\log r)$ is the number of particles per cm^3 in the radius interval $d(\log r)$, and N' is the total number of particles per cm^3 of radius greater than r or mass greater than m; A, A' and B are constants. Equation (2.3) implies that the particles in each logarithmic interval of radius contribute equally to the mass concentration of the aerosol, and therefore that the large and giant nuclei contribute equal amounts of material. The Aitken nuclei, despite their large numbers, do not account for more than about 20 per cent of the aerosol substance. Junge found that, in cloud-free air, the large nuclei, which varied in

† If $n'(r)\,dr$ were defined as the number of particles per cm^3 of radii between r and $r+dr$, then $n'(r) = 2·30\,A/r^4$.

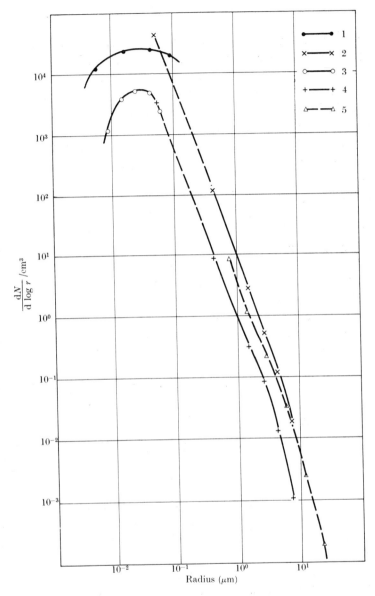

FIG. 2.12. Complete size distributions of natural aerosols, average data. Frankfurt/Main, curves 1, 2, and 5. Curve 1: ion counts converted to nuclei numbers. Curve 2: data from impactors. The point below 0·1 μm radius was obtained from the total Aitken-nuclei number under the assumption that the radius interval of the Aitken nuclei is d log $r = 1·0$. Curve 5; average sedimentation data over a period of 11 days. Curves 1, 2, and 5 are *not* simultaneous. Zugspitze, 3000 metres above m.s.l.: curves 3 and 4 correspond to curves 1 and 2 for Frankfurt and were obtained at approximately the same time. The dashed curves between 8×10^{-2} μm and 4×10^{-1} μm are interpolated. (From Junge (1958).)

concentration between 45 and 360 cm^{-3} with a mean value of 130 cm^{-3}, appeared to represent a rather constant component of the aerosol in contradistinction to the large fluctuations of Aitken nuclei. The largest nuclei, with radii 15–20 μm, were present in concentrations of only about 10 m^{-3}, the concentration of all giant nuclei varying between 0·5 and 10 cm^{-3}. Again their size distributions obeyed, on average, an r^{-3} law. All particles of $r > 5$ μm consisted almost entirely of solid matter, but many smaller particles contained a proportion of hygroscopic substance.

Cartwright, Nagelschmidt, and Skidmore (1956), who collected particles by thermal precipitation and examined them under the electron microscope, were able to make measurements in the size range 0·03 μm $< r <$ 0·9 μm, well beyond that accessible to light microscopy, and found that the general shape of the size distribution function was similar to that found by Junge. Further confirmation came from Twomey and Severynse (1963, 1964), who determined the size distribution of atmospheric aerosols with respect to their diffusion coefficient using diffusion batteries, and a Nolan–Pollak condensation nucleus counter to determine the total concentrations of particles in the aerosol at the various stages of decay. These authors found a apparent second peak in the distribution at $r = 5 \times 10^{-6}$ cm. Friedlander and Pasceri (1965a, b) removed particles of 0·4 μm $< r <$ 20 μm from the air with a four-stage cascade impactor, sized and counted these under the optical microscope, and measured the mass concentration of aerosol by filtering a known volume of air through a Millipore filter. Smaller particles of 7×10^{-7} cm $< r <$ 0·1 μm were collected by Brownian diffusion to an electron microscope grid placed at the centre of a rotating disc; with laminar flow this resulted in uniform deposition of particles over the surface of the disc. The particles were shadowed with platinum and then sized and counted under the electron microscope. Although individual samples showed appreciable differences in particle size distribution, especially for particles of $r > 10$ μm, the size spectra for particles of $r > 0·4$ μm obeyed quite well, on average, the r^{-3} law of Junge, the closest empirical relation being $n(r)\,\mathrm{d}(\log r) = 0·017\,\phi r^{-3}$ or $n'(r) = 0·04\,\phi r^{-4}$, where ϕ is the total volume of the particles in unit volume of air. Particles of $r < 0·05$ μm deviated appreciably from this relation in much the same manner as reported by Junge in that the particle concentration reached a maximum at $r \simeq 0·01$ μm, the smaller nuclei being present in lower concentrations then would be expected from an extrapolation of the r^{-3} law.

§ 2.2   THE NUCLEI OF ATMOSPHERIC CONDENSATION

The fact that size distributions of atmospheric aerosol measured in several different parts of the world all show the r^{-3} dependence, especially in the size range $0{\cdot}05\ \mu\text{m} < r < 1\ \mu\text{m}$, led Friedlander (1960, 1961)—see also Swift and Friedlander (1964)—to suggest that, in the absence of sources of particulates, quasi-stationary states exist for this middle range of the size spectrum in which the rate of production of particulate matter by coagulation is balanced by the rate of loss by sedimentation. On this assumption it was possible to obtain a particular solution of the equation describing the time rate-of-change of the size distribution function according to which the latter assumes a self-preserving form

$$n'(r) = \frac{N_t^{\frac{4}{3}}}{\phi^{\frac{1}{3}}}\, \psi_r(\eta_r) = \frac{N_t^{\frac{4}{3}}}{\phi^{\frac{1}{3}}}\, \psi_r\{r(N_t/\phi)^{\frac{1}{3}}\}, \qquad (2.4)$$

where N_t is the total number of particles per unit volume of air. Thus $n'(r)$ is a function only of N_t and the mass-mean radius of the particles. In order to test this self-preservation theory, Clark and Whitby (1967) measured the concentrations and size distributions of particles of $10^{-7}\ \text{cm} < r < 3\ \mu\text{m}$ in urban atmospheres using a continuous sampling system consisting of a nucleus counter, an electrical particle counter, and an optical counter, and plotted fifty-eight different measured size distributions in the non-dimensional coordinates ψ_r vs. $\eta_r = r(N_t/\phi)^{\frac{1}{3}}$. As eqn (2.4) would lead one to expect, the distributions all fell close to a single curve when $\eta_r > 1{\cdot}0$ corresponding to $r > 0{\cdot}05\ \mu\text{m}$, and then ψ_r obeyed the relation $\psi_r = 0{\cdot}05\{r(N_t/\phi)^{\frac{1}{3}}\}^{-4}$ which, on substitution in eqn (2.4), gives $n'(r) = 0{\cdot}05\ \phi r^{-4}$ in close agreement with the result obtained by Friedlander and Pasceri (1965). This study represented a severe test of the theory in that the fifty-eight size distributions were recorded under a variety of weather and source conditions and included a broad range of particle sizes.

The first detailed information on the size distribution of nuclei in marine air was obtained by Woodcock and Gifford (1949), who exposed their hydrophobic collectors on an aircraft to sample these nuclei at different levels up to 300 m over the sea near Woods Hole (Mass.) and up to 1150 m over Bermuda. With surface wind speeds up to $10\ \text{m s}^{-1}$, salt nuclei of $m > 4 \times 10^{-13}\ \text{g}$ were found in concentrations of up to $2\ \text{cm}^{-3}$, while nuclei as large as $5 \times 10^{-9}\ \text{g}$ were present to the extent of 1/litre. In the thermally stable conditions prevailing at Woods Hole, the numbers of nuclei fell off rapidly at higher altitudes, but in the well-mixed air encountered over Bermuda, they were much the same at all

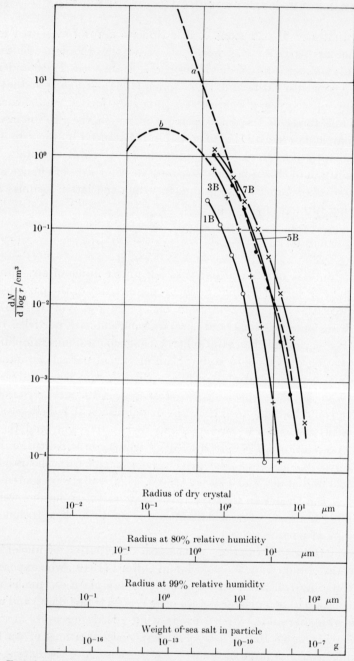

Fig. 2.13. Average size distribution of sea-salt nuclei measured over the oceans for wind forces 1, 3, 5, 7 on the Beaufort scale according to Woodcock (1953). The dashed line, a, gives the size distribution of continental aerosol for comparison. The dashed line, b, is an extrapolated size distribution of marine aerosol. (From Junge (1958).)

levels investigated. Woodcock (1950a) made some interesting measurements from the top of a lighthouse during a tropical storm off the coast of Florida. In winds of more than 30 m s^{-1}, nuclei of $m > 10^{-11}$ g were present in concentrations of about 1 cm^{-3}, and the largest nuclei, of $m = 4 \times 10^{-8}$ g, in concentrations of 2/litre. In these conditions, the average concentration of sea salt in the air was 6×10^{-10} g cm^{-3}, which may be compared with values of about 10^{-11} g cm^{-3} in moderate winds. Woodcock (1953) also made a series of measurements on the size distributions of giant sea-salt nuclei over the sea near Hawaii, particular attention being directed to their variation with surface wind speed and altitude. Fig. 2.13 shows the salt-nucleus content of the air at cloud-base level as a function of the surface wind speed. It appeared that increasing winds were associated with a rather consistent pattern of increase in both the numbers and sizes of the particles. Woodcock suggested that these increases were due primarily to the increased numbers of bubbles trapped in white-caps which, on bursting, projected small droplets of sea water into the air.

The dashed extension, b, of the curve for a force 3 wind in Fig. 2.13 is based on measurements made by Moore and Mason (1954) on a weather-ship in the eastern North Atlantic, 250 miles from land, in which hygroscopic particles of mass down to 2×10^{-14} g were detected, and on measurements over Ireland by Metnieks (1958), who collected chloride particles of mass down to 2×10^{-15} g with a cascade impactor. On the basis of these data and a chemical analysis of particles of $0.08 < r < 0.8$ μm caught in the clean maritime air of Hawaii, Junge (1963) deduces that the most frequent particle size is about 10^{-14} g and that the number concentration falls off for smaller sizes. Junge estimates the total concentration of salt nuclei to be 2 cm^{-3} in a wind of Beaufort force 3 (5 m s^{-1}), and not to exceed about 20 cm^{-3} in stronger winds.

The curves of Fig. 2.13 depart considerably from Junge's r^{-3} law for continental aerosol, which is plotted for comparison. However, very similar curves to those of Woodcock were obtained by Moore and Mason when the aerosol was collected in winds stronger than 7 m s^{-1} that produced 'white-caps' on the ocean. They also found that the concentration of salt nuclei increased with increasing wind speed to reach values of 10 cm^{-3} in winds of 15 m s^{-1}. Surprisingly, even higher concentrations were sometimes measured when the wind was less than 7 m s^{-1} and the sea surface was fairly calm, but then the size distributions were much more like those obtained by Junge for continental aerosol. This suggests that, although some salt particles were identified among these nuclei, they may have been reinforced by hygroscopic nuclei of land origin. On the other hand, air arriving at Hawaii, having had a long uninterrupted track over the ocean, may be almost free of continental aerosol. An interchange of nuclei between land and sea is also indicated by the fact that Moore (1952) and Ohta (1951) found the concentration of Aitken nuclei over the ocean to be independent of the wind speed and the height of the waves, and were therefore probably of land origin.

A marked change in the size distribution of the aerosol at Frankfurt during an influx of maritime air was noted by Junge (1952b). In particular, the slope of the mass distribution curve over the size range 0·4 to 2·0 μm decreased abruptly from a value of about 1·0 to 0·7. Many of the larger nuclei were found to contain cubic crystals of sodium chloride. That large and giant sea-salt nuclei may be carried considerable distances inland was also established by Woodcock (1952), who found that 100 km inland the concentration and size distribution of particles

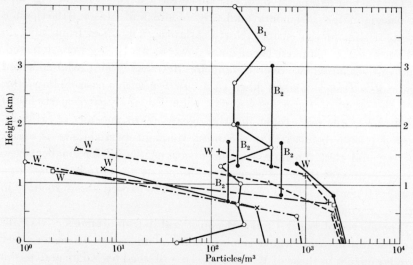

FIG. 2.14. Vertical distribution of the number of sea-salt particles having a dry radius $\geqslant 3$ μm (equivalent to a radius of 7 μm at 80 per cent relative humidity and a mass of 5×10^{-10} g). The curves marked W are measured by Woodcock (1953) in regions of the trade winds; the portion of the curve between the ocean surface and the measured values around 0·5 km is assumed. The values show the rapid decrease at the level of the trade-wind inversion. The curves marked B are given by Byers et al. (1957). Curve B_1 is an average of three soundings made in Illinois. Curves B_2 are average concentrations for four overland flights southward from Chicago. The length of the bar indicates the altitude range during each flight. (From Junge 1963.)

were much the same at all levels from 150 to 1500 m and were not appreciably different from similar measurements made over the sea. During a series of flights over south-eastern Australia, Twomey (1955) found that, in air undisturbed by precipitation or convective activity, the total concentration of salt nuclei of $m > 10^{-10}$ g was not greatly reduced after being transported 100 km inland, but the particles became much more uniformly distributed in the vertical. Their numbers fell off very rapidly above cloud-base level in convective situations, and above post-frontal subsidence inversions, and were appreciably reduced by precipitation. Similar results were obtained by Byers, Sievers, and Tufts (1957), on several flights from Illinois to the Gulf of Mexico. The particles were collected by an impactor and deposited on a gelatine strip impregnated with mercurous fluosilicate. A chloride particle produced a bluish halo about 10 times the size of the particle. Fig. 2.14 represents a summary of the results for particles of $m > 5 \times 10^{-10}$ g. The vertical profiles over land and those over the sub-tropical ocean are quite different. The particle concentration over land is fairly constant with altitude apart from a sharp drop in the lowest 200 m which may be caused by fallout and impaction of particles on obstacles on the earth's surface.

2.3. The chemical composition of atmospheric aerosols

Atmospheric particulates consist of solid matter, droplets of solution, or a mixture of both. Insoluble particles, of smokes and mineral dusts, are widespread, but a considerable proportion of natural aerosol consists of solution droplets of hygroscopic salts which grow with increasing humidity of the air. That a high proportion of particles collected overland are of a mixed nature may be seen by observation under the electron microscope when the soluble and more volatile components disappear. At relative humidities above 70 per cent, the majority of these mixed nuclei act as droplets. In pure maritime air, on the other hand, the sea-salt particles contain practically no insoluble components.

It is very difficult to determine the composition of the Aitken nuclei and most of the large nuclei because they are too small to see under the optical microscope, and in the electron microscope they partially evaporate or become damaged by the electron beam. Also, they may be masked by higher mass concentrations of large and giant nuclei. Accordingly there is virtually no direct information on the constitution of Aitken nuclei but microchemical methods have yielded some interesting data on the larger particles. Junge (1953, 1954, 1957a) collected these in a cascade impactor, separating them into their large ($0.08 < r < 0.8$ μm) and giant ($0.8 < r < 8$ μm) components, and analyzed the samples dissolved in distilled water by microchemical spot tests.

In highly industrialized Western Germany, he found that the large particles contained much NH_4 and SO_4, the ratio of the components corresponding to those of $(NH_4)_2SO_4$. The presence of ammonium sulphate was also confirmed by electron diffraction. The giant nuclei, on the other hand, contained little NH_4, so presumably the abundant SO_4 was associated with another cation; most probably it existed as H_2SO_4 formed by the conversion of SO_2. Chlorine was present in both large and giant particles, but the fact that the chlorine content of the giant particles increased and that of the large particles decreased with the advection of maritime air, suggests that, while the giant particles were mainly of maritime origin, the large nuclei originated mainly from land sources. Very similar results were found in the more rural surroundings of Round Hill on the east coast of the United States. Here the nitrate was limited almost exclusively to the giant nuclei, this being closely correlated with the chlorine. The results of these and similar investigations in Florida and Hawaii have shown that with increasing maritime influence, the $SO_4^=$, NH_4^+, NO_3^-, and Cl^- contents of the large particles progressively fall with NH_4^+ and $SO_4^=$ following parallel trends, while for the giant particles, only the first three decline and the Cl^- increases because of the increasing contribution from sea salt. Over the continents, the NH_4 and SO_4 components were much more prominent than the marine aerosols in the large particle range, while in the giant particles, there was a marked surplus of SO_4 over NH_4 at all the sampling sites.

The origin of the nitrate poses an interesting problem. The fact that its average concentration decreases nearer the sea would appear to suggest a continental origin. Junge (1956) found that advection of maritime air in Boston was accompanied by a low NO_3 content comparable to that measured in Hawaii, but parallel measurements in the centre and suburbs of Boston indicated that the urban atmosphere was not a major source of NO_3. The evidence points to the smallest concentrations occurring in the centres of oceans and in clean continental air masses, and the largest concentrations occurring in the marine air masses of the north-eastern coast of the United States. This implies that NO_3 is most easily formed in coastal areas where the maritime air comes into contact with continental air, and some additional evidence for this is provided by the fact that high concentrations of NO_3 are found in San Francisco and Los Angeles.

Direct identification of the nuclei actually contained in cloud and fog droplets has been attempted mainly by Japanese workers. Kuroiwa (1951, 1953, 1956) studied the residues of 225 sea-fog droplets under the electron microscope and found that about 50 per cent were composed mainly of insoluble combustion products sometimes coated with a film of hygroscopic substance, about 40 per cent consisted of sea-salt particles, and 10 per cent of soil particles. Most of the combustion and soil particles were smaller than 1 μm, but most of the sea-salt particles were between 1 and 2 μm in radius. A rather similar study, carried out by Ogiwara and Okita (1952) at Sendai, 12 km from the Pacific Ocean, revealed that most of the sea-fog droplet residues were initially hygroscopic but appeared to lose this property after irradiation by the electron beam of the microscope. In this respect they behaved very much like the combustion products of coal, lignite, pine-wood, grass, etc., and unlike the crystals produced by spraying a salt solution. The conclusion was that combustion products were the main source of condensation nuclei in these sea-fog droplets, sea salt playing a subsidiary role. A further investigation in the same region by Yamamoto and Ohtake (1953), while confirming the predominance of combustion nuclei (34 per cent), found a considerably higher proportion of sea salt nuclei (20 per cent). In a later paper, Yamamoto and Ohtake (1955) distinguish between fog and mist nuclei. From an electron microscope examination of 340 nuclei collected at three mountain stations and also over the Pacific Ocean, they found that 57 per cent of the nuclei of fog droplets, with predominant diameter about 0·3 μm, were combustion nuclei, whereas 61 per cent of the much larger mist particles (predominant diameter about 4 μm) consisted of sea salt. Isono (1957) collected cloud droplets on a mountain 40 km from, and 1500 m above, the sea, again in a region relatively free of industrial pollution. Using electron microscopy and micro-electron diffraction, he identified 30 per cent of the nuclei to consist mainly of NaCl with masses between 10^{-13} and 10^{-12} g, 20 per cent of the droplets contained soluble nuclei that could not be identified, 35 per cent contained insoluble particles, and the remainder had no visible nuclei. Ammonium sulphate could not be detected in any of the nuclei.

All these investigations were made in fairly clean country air not far from the sea. Even so, combustion nuclei were usually predominant and sea salt was identified in only about 25 per cent of the droplet residues. On the other hand, direct observation of giant nuclei over mid-ocean and in remote places such as Greenland shows them to be composed very largely of sea salt, the ratios of the various components being very similar to those of sea-water with no evidence of decomposition. In industrial regions, the soluble component of the aerosol amounts to only 10–20 per cent, the remainder being mainly soot, tar, ashes, and mineral dusts.

One has, of course, to be very cautious in making generalizations about the nature of the nuclei involved in cloud formation on the basis of the direct identification of only very small numbers of droplet residues that may not be representative of the bulk of condensation nuclei. Furthermore, one has to be careful not to discriminate in favour of large particles (or droplets), which are more easily captured, and of prominent crystals such as NaCl, which are easily identified.

2.4. The production of natural aerosols

The main sources of raw material for the manufacture of atmospheric aerosols are: the gases released into the atmosphere from land and ocean surfaces, and by plants; trace gases produced by reactions involving the atmospheric constituents including water vapour; particles of soil, mineral dust, and sea salts carried up from the earth's surface by the wind; and a wide variety of particulate and gaseous products of combustion. The materials so produced are present in small concentrations of order 1 to 100 $\mu g/m^3$, which usually show large variations in space and time because they spend only a relatively short time in the atmosphere before being removed by rain. Chemical analyses of the large and giant nuclei, and of rain water, show the major constituents to be sulphate, chloride, nitrate, and ammonia. The fact that sulphur, ammonia, and nitrogen oxides are usually present in much higher concentrations in the gaseous then in the particulate phase suggests that these gases, which have been identified in the form of SO_2 and H_2S, NH_3, N_2O, and NO_2, are partially converted into particulate matter and are then attached to clouds and raindrops. Atmospheric aerosol is believed to originate in four main ways: (1) by condensation and sublimation of vapours during the formation of smokes by heat and combustion; (2) by chemical reactions between trace gases produced either from natural sources or by combustion, and involving, perhaps, water vapour, ultraviolet radiation, and solid particles; (3) by the mechanical disintegration and dispersal of matter at the earth's

§ 2.4   THE NUCLEI OF ATMOSPHERIC CONDENSATION   67

surface, either as droplets of sea-water over the oceans, or as soil and mineral-dust particles over the continents; (4) by coagulation of nuclei leading to the formation of larger particles of mixed constitution.

2.4.1. *Production of nuclei by combustion and by chemical reactions*

In combustion processes the substances are molecularly dispersed; on mixing with the cool atmosphere, the vapour becomes strongly supersaturated and either droplets or solid particles are formed according to the nature of the particular substance. Typical substances formed in large quantities in this way are ashes, soot, tar products, and oils, as well as sulphuric acid and sulphates in cases where the fuel contains sulphur. A great variety of particles is formed in this way by industrial operations, by natural and man-made fires. The sizes of the particles cover a wide range, the great majority lying within that of the Aitken nuclei.

Some data on average concentrations of sulphur, ammonia, and nitrogen compounds, and sea salts found in the air, in aerosols, and in rain-water are given in Table 2.4 together with their likely sources and sinks. In many cases the information is based on the results of measurements made at only a few places and must therefore be regarded as tentative and not necessarily representative of the whole atmosphere.

Sulphur. In the gaseous form sulphur exists mainly as SO_2 and H_2S. Most of the natural sulphur is believed to exist as H_2S, produced by bacterial decomposition of organic matter and reduction of sulphate in the soil and in sea-water, but there have been very few measurements of the H_2S (as opposed to the total sulphur) content, especially in unpolluted atmospheres. Junge (1960) has made some measurements near Boston and found the concentration of H_2S to be fairly constant at about 9 $\mu g/m^3$ with the SO_2 content more variable but, on the average, about 2·5 times larger. Jacobs, Braverman, and Hochheiser (1957) found the mean concentration of H_2S in New York City to be 3 $\mu g/m^3$ and the concentration of SO_2 to vary between 280 and 1500 $\mu g/m^3$. Since H_2S cannot be removed rapidly in the gaseous phase by solution in cloud and rain drops, and because the estimated average residence time of local sulphur in the atmosphere is only about 20 days, it seems likely that H_2S must be partially converted to sulphate, probably in cloud droplets or on the surfaces of particles. Oxidation of H_2S to SO_4 is indicated by the fact that the SO_4 content of rain in rural areas is higher in the summer, whereas in polluted regions it is higher in the winter because of the higher SO_2 production.

TABLE 2.4

Sources and concentrations of major nucleogenic agents

| | | Over land | | | | Over oceans | | | |
| | | Concentrations | | | | | Concentrations | | |
Substance	Sources	In air ($\mu g/m^3$)	In aerosols ($\mu g/m^3$)	In rain (mg/l)	Sources	In air ($\mu g/m^3$)	In aerosols ($\mu g/m^3$)	In rain (mg/l)	Residence time in atmosphere
SO_2	Combustion of fuels Forest fires Volcanoes	~10 C 100–1000 P			None	0–1			~5 days
SO_4	Oxidation of SO_2 to SO_3 and $SO_3 + H_2O \rightarrow H_2SO_4$ $2NH_3 + H_2SO_4 \rightarrow$ $(NH_4)_2SO_4$ Similar formation of $CaSO_4$ and $MgSO_4$ Airborne soil particles		~1 C ~10 P	~1 C ~10 P	Sea salt		0·1–1·0	~1	
H_2S	Decay of organic matter in soil and coastal banks Reduction of SO_4 in soil	~10			Reduction of SO_4 in sea?				~40 days
NH_3	Decay of organic matter in soil	1–5 C			Decay of organic matter on sea surface	~1			
NH_4	Combustion of fossil fuels Forest fires $2NH_3 + H_2SO_4 \rightarrow$ $(NH_4)_2SO_4$	10 P	0·1–1·0 C 10 P	0·1–0·5 C 1 P			~0·01		

§2.4 THE NUCLEI OF ATMOSPHERIC CONDENSATION

	Source / Process					
NO_2	Industrial processes and automobiles $2NO + O_2 \rightarrow 2NO_2$? $NO + O_3 \rightarrow NO_2 + O_2$?	~1 C 10–100 P		~1		
NO_3	$3NO_2 + H_2O = 2HNO_3 + NO$ $HNO_3 + NaCl = NaNO_3 + HCl$	<1 C ~1 P	~0.5 C		~0.1	0.1–0.5
Cl (gas)	Industrial processes Formation of HCl, e.g. $HNO_3 + NaCl = Na_2NO_3 + HCl$ $H_2SO_4 + 2NaCl = Na_2SO_4 + 2HCl$	1–5		~1		
Chloride Na	Advection of sea salts Airborne soil particles	<1	0.1–0.5	Production of small droplets of sea water by bursting air bubbles	1–10	1–5	~5–10 days

C = country air; P = polluted air.

Large quantities of SO_2 are produced by the combustion of fuels, with lesser quantities supplied by forest fires and volcanoes. It is estimated that about 30 per cent of the total sulphur content of the atmosphere is contributed by man-made processes. The SO_2 content of country air is of the order of 10 $\mu g/m^3$ but concentrations 10 or 100 times larger are found in industrial cities. Over mid-ocean the concentrations are usually <1 $\mu g/m^3$, and in very remote areas is sometimes almost completely absent. The average residence time of SO_2 has been estimated to be about 5 days. Overland the concentration of SO_2 is usually ten or more times greater than that of SO_4, but even over the oceans, where appreciable quantities of SO_4 are present in sea salt particles and the SO_2 content may be very low, the situation is often reversed. SO_2 and H_2S can also be absorbed by plants and by the ocean surface, which therefore act as sinks as well as sources.

The conversion of SO_2 to SO_4 occurs by photochemical oxidation in the gas phase and catalytic oxidation in cloud droplets. According to Gerhard and Johnstone (1955), gaseous SO_2 is converted into SO_3 at the rate of 0·1–0·2 per cent per hour in strong sunlight and this, in turn, is readily transformed into H_2SO_4 in the presence of water vapour.

The oxidation of SO_2 in cloud droplets is an efficient process. According to the laboratory experiments of Junge and Ryan (1958), minute traces of metal ions in concentrations of the order of 1 p.p.m. are essential as catalysts. For a given concentration of SO_2 in the air and of catalyst in the water, conversion to SO_4 increases with time at first, but the process stops after about 1 h. Using 1 p.p.m. of $FeCl_2$ as a catalyst, for example, the final concentration of SO_4 in the water was found to be given by

$$\text{concentration of } SO_4 \ (\mu g/cm^3) = 3\cdot 3 \times 10^{-3} \times \text{concentration of } SO_2 \ (\mu g/m^3).$$

Thus for an SO_2 concentration of 10 $\mu g/m^3$, the maximum concentration of SO_4 in the water would be $3\cdot 3 \times 10^{-2}$ mg/litre. Even so, this is two orders of magnitude smaller than the SO_4 content of rain-water, and this suggests that there is an even more efficient conversion mechanism at work.

A clue to this mechanism was provided by Junge (1953, 1954) when he discovered that the large nuclei of continental aerosols contained a good deal of both ammonia and sulphate and in the proportions that suggested the compound to be $(NH_4)_2SO_4$. He also confirmed the presence of this salt by electron diffraction and showed that its concentration could be markedly increased by adding SO_2 and NH_3 gases to the aerosol. Junge and Ryan (1958) have demonstrated that the presence of dissolved ammonia in the form of NH_4OH acts as a catalyst for the conversion of SO_2 to SO_4 in oxygenated water and compute that the maximum concentrations of NH_4^+ and $SO_4^=$ to be expected are

$$NH_4^+(\mu g/cm^3) = 0\cdot 33 \times NH_3(\mu g/m^3) \times SO_2(\mu g/m^3)$$

and $$SO_4^=(\mu g/cm^3) = 0\cdot 87 \times NH_3(\mu g/m^3) \times SO_2(\mu g/m^3),$$

where the concentrations of NH_3 and SO_2 refer to those present in the air after completion of the oxidation. If we take the concentrations of SO_2 and NH_3 in country air to be 10 $\mu g/m^3$ and 3 $\mu g/m^3$, respectively, the maximum concentration of $SO_4^=$ to be expected in rain-water is 15 $\mu g/cm^3$ or 15 mg/litre, which is to be compared with average measured values of 1–3 mg/litre. It seems, therefore, that this mechanism could well account for the sulphate content of rain-water but there is some evidence to suggest that $SO_4^=$ may also be neutralized by Ca^{++} and Mg^{++} ions besides those of NH_4^+.

Cadle and Robbins (1960) have found that the reaction between droplets of dilute (10%) H_2SO_4 of diameter 0·9 μm and ammonia proceed extremely rapidly in the concentrations to be found in polluted atmospheres. The rate of the whole process is governed by the rate of diffusion of NH_3 to the droplets and it appears that practically every NH_3 molecule striking a droplet is converted into sulphate. Van den Heuvel and Mason (1963) have studied the formation of $(NH_4)_2SO_4$ in larger water drops, of 0·1 to 1 mm diameter, by exposing them to an air-stream containing controlled concentrations of NH_3, SO_2, and water vapour. Chemical analysis of the salt produced by evaporation of the drop showed it to be composed almost entirely of sulphate, the mass of salt being proportional to the product of the surface area of the drop and the time of exposure. According to Scott and Hobbs (1967), the rate-limiting step in the formation of the sulphate is probably an oxidation involving the $SO_3^=$ ion in the reaction $SO_2 + H_2O \rightarrow 2H^+ + SO_3^=$, the ammonia reaction $NH_3 + H_2O \rightarrow NH_4^+ + OH^-$ effectively catalyzing the oxidation process by releasing the OH^- ions required to neutralize the excess H^+ ions. The concentrations of gases used in these experiments were much higher than those that occur in the atmosphere but, if the results are scaled down linearly according to the concentration, they suggest that a cloud droplet of radius 10 μm, in an industrially polluted atmosphere containing 100 $\mu g/m^3$ of SO_2, could produce 10^{-11} g of ammonium sulphate in 1 h. Accordingly, it seems most likely that the large sulphate nuclei found in polluted air are formed by the absorption of SO_2 and NH_3 in cloud and fog droplets.

Moreover, since most cloud droplets evaporate, an individual particle may act several times as a condensation nucleus before being precipitated in rain. It is therefore possible that the irreversible conversion of SO_2 into particulate sulphate may be repeated in several cycles of condensation and result in the progressive accumulation of SO_4. The marked increase observed by Radke and Hobbs (1969b) in the concentration of condensation nuclei active at 1 per cent supersaturation in a thermal diffusion chamber following the dissipation of clouds upwind of the sampling station could be explained in terms of such a process.

A striking proof of the importance of SO_4 formation, even in clean air, is provided by the fact that ice and snow in central Greenland has an $SO_4^=$ content about ten times higher than that of any other soluble constituent including Na and Cl from sea-water, while the large aerosol particles ($r > 0.2$ μm) also consist largely of sulphate. Thus, Fenn, Gerber, and Wasshausen (1963) estimated that 40 per cent of the total aerosol mass, which amounted to about 0.4 μg/m³ with particle concentrations of about 1 cm⁻³, was sulphate. On some occasions ammonium was present in comparable concentrations but, on others, it was much less abundant than sulphate suggesting that, on some occasions, sulphate particles were present as compounds other than $(NH_4)_2SO_4$. In the Antarctic, Cadle, Fisher, Frank, and Lodge (1968) found that particles of $r = 0.1$–1.0 μm, present in concentrations of 0.1–1.0 cm⁻³, contained much higher concentrations of sulphur, mainly sulphate, than similar samples collected in other parts of the world. Electron-diffraction analysis indicated the presence of $(NH_4)_2SO_4$ and also other ammonium-sulphur compounds. Cadle *et al.* suggest that these particles were largely of stratospheric origin and were brought down from the layer of sulphate aerosols that exists as about 20 km altitude.

Ammonia. Ammonia exists predominantly in the gaseous phase in concentrations often lying in the range 1–5 μg/m³. The main sources are probably decaying organic matter in the soil and on the surface of the oceans, although water and soil having pH values less than 7 absorb ammonia from the air. Eriksson (1952) suggests that fossil fuels are also an important source. An average value for the NH_4^+ content of rain-water is 0.25 mg/litre, which might well be accounted for by dissolved NH_3 reacting with SO_2 in the manner just described. Rain-water analyses over the United States show low concentrations of NH_4 over the ocean ($<10^{-2}$ mg/litre), maximum values in the summer, which suggests cultivated soil as a source, but low values in the southeastern states where the soils have a low pH and probably prevent the escape of NH_3. The higher NH_4 content of rain-water in Europe has been attributed to the intense cultivation of land (Junge), and to the high consumption of coal (Eriksson), but there are no firm estimates for the rate of release of NH_3 by either source.

Ammonia is removed from the atmosphere by being absorbed in the oceans and in low-pH soils, and by being converted into ammonium sulphate and precipitated by rain as described above.

Oxides of nitrogen. We have seen that nitrates are present in the giant condensation nuclei and in rain-water; their formation, no doubt, involves the oxides of nitrogen in the atmosphere of which nitrous oxide (N_2O) and nitrogen dioxide (NO_2) have been detected, and the presence of NO is generally assumed.

N_2O is very stable and inert in the troposphere; its mixing ratio is fairly constant with height at about $3 \cdot 5 \times 10^{-7}$ corresponding to a concentration of 690 $\mu g/m^3$ at s.t.p. It is most likely produced by soil bacteria causing the decomposition of nitrogen compounds.

NO_2 is found in concentrations of the order of 1 $\mu g/m^3$ in clean, country air, but values of 10–100 $\mu g/m^3$ are produced in polluted urban atmospheres by industrial processes and automobiles. The following reactions for the production of NO_2,

$$2NO + O_2 \rightarrow 2NO_2; \quad NO + O_3 \rightarrow NO_2 + O_2$$

have been suggested, but these have not been thoroughly studied.

The nitrate (NO_3) content of aerosols is usually about 0·1 $\mu g/m^3$ in maritime and clean country air and perhaps 1 $\mu g/m^3$ in polluted air. On the basis of some laboratory experiments by Robbins, Cadle, and Eckhardt (1959), it seems probable that the NO_3 is produced by the reaction of NO_2 with sea-salt particles. Robbins *et al.* suggest that the first step of the reaction involves the hydrolysis of NO_2 to form nitric acid vapour:

$$3NO_2 + H_2O = 2HNO_3 + NO.$$

This vapour is then either adsorbed by dry NaCl particles or, at high relative humidities, goes into solution in the droplets. The nitric acid then reacts with the sodium chloride to form sodium nitrate and hydrochloric acid:

$$HNO_3 + NaCl \rightarrow NaNO_3 + HCl,$$

the latter being desorbed either immediately or during subsequent evaporation of the droplets.

Using NaCl and NO_2 in concentrations some three orders of magnitude greater than usually occur in the atmosphere, Robbins *et al.* (1959)

found that the reaction attained equilibrium in less than 10 min at room temperature, and that at humidities of about 80 per cent, the majority of the NaCl was converted to $NaNO_3$.

Chlorine. In the gaseous form, chlorine is produced mainly by industrial processes and by escape from the sea, and probably exists mainly in the form of HCl gas. Its concentration overland varies from about 1–5 $\mu g/m^3$, depending upon the degree of pollution, and is $\sim$1 $\mu g/m^3$ over the oceans. We have just discussed the reaction by which HCl might be released from sea-spray particles by the oxidizing action of NO_2. It has also been suggested that a similar reaction involving SO_3 or H_2SO_4 might lead to the formation of HCl and Na_2SO_4, i.e.

$$H_2SO_4 + 2NaCl \rightarrow Na_2SO_4 + 2HCl.$$

Chloride particles of marine origin are found far inland but they may also arise from salt deposits and industrial operations. The high concentrations of small chloride nuclei found by Rau (1955) in Bavaria, and by Podzimek (1959) and Podzimek and Černock (1961) during winter anticyclonic conditions in Czechoslovakia, were probably of industrial origin.

Sodium. Sodium exists mainly as sodium chloride, by far the most important source of which is sea salt. The mass ratio of Cl/Na in sea water is 1·8, which is also found in aerosols and rain-water collected over the sea and in coastal regions, but measurements made particularly in Sweden show that the Cl/Na ratio in rain-water tends to fall as the air moves inland and, on some occasions, may have almost zero values. This result may be interpreted either as evidence for a loss of chlorine or for the addition of sodium from land sources.

We have already seen how reactions between NaCl aerosol and H_2SO_4 and HNO_3 in polluted atmospheres may lead to the release of HCl and a corresponding reduction in the Cl/Na ratio of the aerosol. The fact that these reactions will proceed further on the smaller salt particles, which remain airborne for longer periods before being washed out, suggests at least a partial explanation for the progressively lower Cl/Na ratios with increasing distance from the coast. Although the conversion of NaCl to $NaNO_3$ probably proceeds more slowly in the low concentrations of the atmosphere than in the laboratory experiments of Robbins *et al.* (1959), it seems possible that the Cl/Na ratio of cloud water might well be reduced by a factor of 2 over a period of several hours, with a corresponding increase in the NO_3/Na ratio. Unfortunately the occurrence of larger NO_3/Na or SO_4/Na ratios at inland stations does

§ 2.4   THE NUCLEI OF ATMOSPHERIC CONDENSATION

not constitute evidence for these processes because of the presence of additional land sources of NO_3 and SO_4. A much more detailed analysis, by X-ray or electron diffraction, of the aerosols to detect the progressive formation of $NaNO_3$ and Na_2SO_4 and the presence of Na-bearing dusts will be necessary to settle this issue. On the evidence so far available, it seems that the very low Cl/Na ratios are most likely caused by airborne dusts supplying additional sodium.

The fact that the K/Na and Mg/Na ratios of both aerosol and precipitation increase with increasing distance from the coast has been explained by Wilson (1959) and Oddie (1960) in terms of the smaller sea-salt droplets being enriched in K and Mg relative to the coarser spray. Oddie points out that this fits well with Mason's view that the smaller salt nuclei are produced by the bursting of bubble films which, according to Wilson, will be rich in organic matter and therefore in potassium. The smaller nuclei, enriched in potassium, will be carried considerable distances inland before being washed out, whereas the larger droplets, produced by the break-up of small jets and having the same constitution as bulk sea-water, will fall out much nearer the coast.

2.4.2. *Production of sea-salt nuclei*

It was once thought that sea spray, produced in the foam of breaking waves, was the main source of sea-salt nuclei. However, apart from those that originate in surf, such particles are not very numerous and are generally too large to remain airborne for very long. In recent years Woodcock and his co-workers (Woodcock, Kientzler, Arons, and Blanchard 1953, Kientzler, Arons, Blanchard, and Woodcock 1954, and Blanchard and Woodcock 1957), following earlier suggestions by Jacobs (1937), Owens (1940), and Boyce (1954), have demonstrated by high-speed photography that salt particles larger than 1 μm are produced mainly by the bursting of numerous small air bubbles in the foam of breaking waves. The photographs of Fig. 2.15, taken by Knelman, Dombrowski, and Newitt (1954), show that the spherical bubble cap bursts near the summit, where it is thinnest, and the air, in escaping through the hole, ruptures the film. After bursting of the cap, water rushes into the crater to form an unstable jet, and this breaks up to produce between one and five drops that are about one-tenth the diameter of the parent bubble. Bubbles of about 2-mm-diameter project to heights of nearly 20 cm but, for both larger and smaller bubbles, the maximum height of projection is less than this. Blanchard and Woodcock (1957) suggest that the lower size limit of the bursting bubbles is determined by their dissolution in sea-water. The smallest bubbles observed to reach the ocean surface in 'white-caps' were about 100 μm in diameter and these produce salt particles of $m \sim 10^{-11}$ g. Much smaller bubbles with a mean diameter of 40 μm were observed only when snowflakes melted in sea-water. Under normal conditions, the lower size limit of the bubbles is not small enough to explain the production of the smaller salt nuclei by the evaporation of drops from the breaking bubble-jets.

But, in addition to these drops, Mason (1954) found that bubbles larger than about 2 mm in diameter produced a few drops with diameters between 5 and 30 μm and containing between 2×10^{-12} g and 5×10^{-10} g of salt; these were found very close to the water surface and were apparently ejected at low angles to the horizontal. Also, by studying the bursting of bubbles in a cloud chamber, Mason discovered that the disruption of the bubble cap produced much larger numbers of particles too small to be detected by conventional methods. Clean filtered air was passed through identical jets into two chambers, one containing sea water, and the other, distilled water. After an equal number of bubbles had burst in each chamber, an expansion was made; a dense cloud of tiny droplets was then observed in the space above the sea-water but not above the distilled water. It appeared that the bursting bubbles produced very small salt particles that acted as condensation nuclei during the expansion. It was estimated that bubbles of between 0·5 and 3 mm in diameter produced 100–200 nuclei with salt contents between 10^{-15} and 10^{-14} g. In a later paper, Mason (1957) revised this figure to 300 ± 80 nuclei per bubble almost independent of bubble diameter in the range 0.25–2.15 mm. This result was confirmed by Twomey (1960) using a diffusion cloud chamber, but Blanchard (1963), using both a Scholz counter and a diffusion cloud chamber, found that the number of bubble-film droplets produced by *single* bubbles decreased with decreasing bubble diameter below 2 mm and that bubbles of $d < 200$ μm produced no droplets.

This question has been carefully re-examined by Day (1964), working in Mason's laboratory and photographing the bursting of individual bubbles in the stable supersaturated atmosphere of a diffusion cloud chamber. From photographs of the kind shown in Fig. 2.16, Day confirmed that the average number of bubble-film droplets is dependent on bubble size, and ranges from none for bubbles of $d < 100$ μm, to 300–400 for bubbles of $d = 4$ mm bursting in sea-water and saline solutions. His results are plotted in Fig. 2.17, and those of Mason

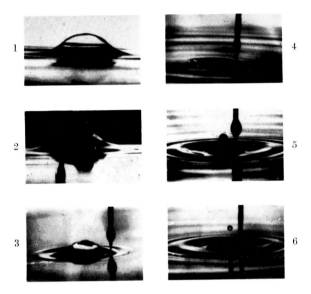

Fig. 2.15. Six stages in the collapse of an air bubble of 3 mm diameter in water. In the initial stage a protuberance develops on the hemispherical surface of the bubble (2), the collapse of which produces a cloud of fine droplets and leaves a large gap in the bubble envelope (3). The escape of gas through this gap disperses the cloud and imparts to the droplets a high velocity. Thereafter a crater is formed at the site of the bubble (4), and the incoming rush of liquid produces an unstable jet (5), which breaks up into one or more large drops (6). (From Knelman *et al.* (1954).)

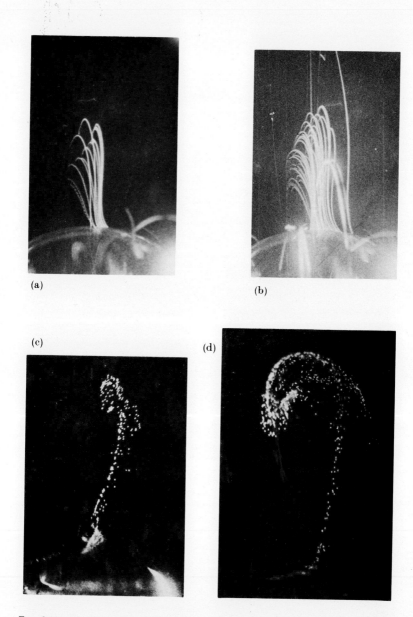

Fig. 2.16. (a), (b) tracks of film droplets produced by 1 mm diameter bubbles bursting in distilled water. A drop produced by the break-up of a jet is shown bouncing from the water surface in (b). (c) Family of film droplets from a 1·8-mm diameter bubble bursting in salt water. A second family of droplets is ejected at a low angle to the horizontal. (d) Inverted bowl pattern produced by subsidiary vortex ring of film droplets resulting from the burst of a 3-mm diameter bubble in salt water. (From Day (1964).)

§ 2.4   THE NUCLEI OF ATMOSPHERIC CONDENSATION   77

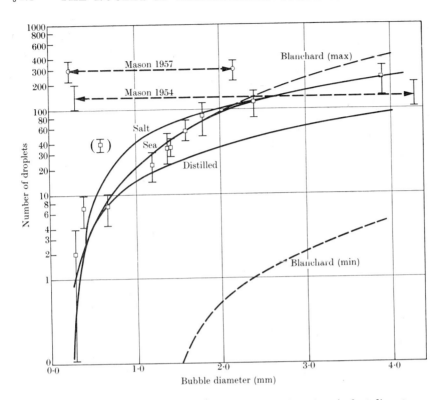

FIG. 2.17. Mean number of film droplets as a function of equivalent diameter of air bubbles. Data points showing the mean value and one standard deviation are presented only for sea water. Values reported by Mason and by Blanchard are included. (From Day (1964).)

and of Blanchard are shows for comparison. The average number of droplets produced was proportional to the square root of the area of the bubble cap. It is emphasized that these results were obtained with single bubbles bursting instantaneously on arriving at a clean water surface. Bubbles formed in clusters or having longer lifetimes on contaminated surfaces gave much less reproducible results.

In discussing the relative importance of the bubble-film and bubble-jet mechanisms of nucleus production, we may combine the observations of Blanchard and Woodcock (1957), that the number of bubbles of radius between R and $R+dR$ bursting per cm²/s in a foam patch is approximately $3 \times 10^{-6}\, dR/R^4$, with Day's data of Fig. 2.17 showing that the number of film droplets produced per bubble is roughly $10^4 R^2$. The total rate of production of film droplets by bubbles of

$R > 50\ \mu\text{m}$ is then

$$0 \cdot 03 \int_{5 \times 10^{-3}\text{cm}}^{\infty} \frac{\mathrm{d}R}{R^2} = 6\ \text{cm}^{-2}\ \text{s}^{-1}.$$

The rate of drop production by the breaking jets, assuming that each jet produces one drop, is

$$3 \times 10^{-6} \int_{5 \times 10^{-3}\text{cm}}^{\infty} \frac{\mathrm{d}R}{R^4} = 7 \cdot 5\ \text{cm}^{-2}\ \text{s}^{-1}.$$

On this basis, it seems that the two mechanisms are of about equal importance in the production of droplets, but those produced by the jets are much larger than the film droplets and are less likely to be carried up to cloud level. Taken together, these two estimates suggest a figure of about 10 cm^{-2} s^{-1} for the rate of production of salt nuclei in those regions where bubbling and wave-breaking are taking place.

However, these estimates are based on data obtained with single bubbles, whereas the experiments of Mason and of Twomey suggest that higher rates of nucleus production obtain when the bubbles burst in clusters so that fresh, clean surfaces are rapidly produced. Using a wind-wave tunnel, in which it was possible to simulate conditions over the ocean at a height of about 10m, Moore and Mason (1954) found that with an equivalent wind of 16 m s^{-1} at this level, the rate of production of salt nuclei of $m > 10^{-13}$ g at the water surface was 40 cm^{-2} s^{-1}, in excellent agreement with an independent estimate made from their size-distribution curves of large and giant nuclei collected over the ocean. Mason (1957), having found that bursting bubbles produced salt nuclei smaller than 10^{-13} g and probably as small as 10^{-15} g, suggested that the total concentration of salt nuclei over areas of breaking waves might be of order 100 cm^{-3}, the corresponding rate of production at the surface being 10^3 cm^{-2} s^{-1}. On the assumption that about 10 per cent of the ocean would be active in producing nuclei at any one time, he arrived at an overall average oceanic production rate of 100 cm^{-2} s^{-1}. Blanchard (1969), from measurements of the fluxes of nuclei emanating from breaking waves in the open ocean off Hawaii, estimated a nucleus production rate of 4000 cm^{-2} s^{-1} over these areas; using a figure of 3·5 per cent for the active fraction of the ocean surface, he also arrived at an average overall production rate of about 100 cm^{-2} s^{-1}, and argues that such a value is consistent with the average nucleus concentrations throughout the depth of the troposphere over the open oceans being 200 cm^{-3} and an average residence time of 20 days. Blanchard regards a production rate of 4000 $cm^{-2} s^{-1}$ in areas of breaking waves as reasonable in the light of some laboratory experiments that simulated the violent agitation of the sea surface, but there is a need for much better designed experiments capable of determining the total production rate and size spectrum of salt nuclei under realistic oceanic conditions. Meanwhile, an average production rate of 100 $cm^{-2} s^{-1}$ for the oceans as a whole, represents the best estimate that can be made at present. This may be compared with Squires's (1966) estimate of 500 $cm^{-2} s^{-1}$ for the average production rate of potential cloud-forming nuclei over the land mass of the Northern Hemisphere, and suggests that while sea-salt particles are probably a major source of condensation nuclei in maritime clouds, they make only a minor contribution to clouds forming well inland.

2.4.3. *Production of nuclei over the continents*

Of the natural processes involved in the primary production of aerosol particles over land, the more important are: combustion (e.g. forest fires, volcanoes); chemical reactions involving trace gases particularly in the presence of water, either in cloud and fog droplets or on the surfaces of solid particles; the raising of soil and dust particles from the earth's surface; the emission of pollen and volatile substances by plants. In addition, man-made activities, notably combustion, contribute large quantities of nucleogenic matter in both vapour and particulate form. Despite the increasing scale of atmospheric pollution in modern times, one would expect natural sources of cloud-forming nuclei to outweigh man-made sources unless the constitution of clouds has changed markedly over the last two hundred years. Squires (1966), from measurements made in SE. Australia and Colorado, U.S.A. of the concentrations of nuclei that become active at supersaturations of 0·5 per cent in a thermal diffusion chamber, and from estimates of the residence times of the air over these semi-arid regions, deduced the average production rate of such nuclei at the surface to be 500 $cm^{-2} s^{-1}$. Allowing for the fact that land covers only 40 per cent of the total area, he arrived at an overall rate of 200 $cm^2 s^{-1}$ for the production of cloud-forming nuclei of natural terrestrial origin over the Northern Hemisphere. Squires also attempted an assessment of the production of man-made nuclei over Denver, Colorado, by determining the transport of particles downwind of the city. Measurements made on four different days, on which the wind distribution with height was similar, indicated the average production rate of Aitken nuclei over the city to be

10^5 cm^{-2} s^{-1}, and that for larger nuclei active at 0·5 per cent supersaturation to be 10^4 cm^{-2} s^{-1}. On the assumption that man-made production of such nuclei is proportional to the total consumption of fuel of all kinds within any region, the results for Denver were extrapolated to larger regions—Colorado, the United States, and the Northern Hemisphere—for which the production rates of the larger 'cloud' nuclei were estimated to be about 20, 70, and 10 cm^{-2} s^{-1} respectively. In other words, the man-made contribution was estimated to be 14 per cent of the total for the United States and only 5 per cent for the Northern Hemisphere as a whole. Since the oceans probably make only a minor contribution to the total supply of cloud-forming nuclei, the implication is that the major sources are natural combustion processes and chemical reactions involving the absorption of such vapours as SO_2, H_2S, NH_3, and Cl by cloud droplets or on the surfaces of wetted solid particles.

There is considerable geological and other evidence that mineral dusts are transported and deposited over distances of hundreds, or even thousands, of miles from their source. Unfortunately, investigations have been limited to particles larger than 1 μm and almost nothing is known about the quantities of sub-micron dust particles in the atmosphere. In dust storms, the mass distribution of the particles near the ground has a maximum at about 25-μm radius but this value decreases with increasing height and increasing distance from the source. Quartz particles deposited in the Pacific Ocean are mostly between 0·5 and 15 μm in radius, the distribution having a maximum at a settling radius of 2–5 μm. Twomey (1960) provides evidence that dessication and heating of soil can generate 10^4–10^5 particles/cm^2 of surface. This is not a very efficient source unless the process is repeated many times under changing meteorological conditions. He thinks that the particles, which seem to be as small as 0·01 μm, are formed by crystallization of salt dissolved in the soil water or by thermally induced rupture of salt layers on the soil surface.

2.4.4. *The growth of nuclei by coagulation*

In a dense population of nuclei, chance collisions lead to coagulation of particles, an increase in their size, and a corresponding decrease in their concentration. It is an experimental fact (see Whytlaw-Gray and Patterson (1932)) that, if in a given volume of smoke the number of particles is plotted against the time, a linear relation of the form $\frac{1}{n} - \frac{1}{n_0} = Kt$ is obtained, where n is the concentration after time t, n_0

the concentration at time zero, and K is the coagulation constant for a given smoke. K varies with the nature and slightly with the weight concentration of the smoke; for an ammonium chloride smoke, its value is 7×10^{-10} cm³ s⁻¹. A similar relation holds for the recombination of small ions for which $K = 1.6 \times 10^{-6}$ cm³ s⁻¹, for coagulation of Aitken nuclei ($K = 1.4 \times 10^{-9}$ cm³ s⁻¹), and for small ions and Aitken nuclei ($K = 5$–10×10^{-6} cm³ s⁻¹). Thus it appears that larger particles coagulate more slowly than smaller ones, and that the factors chiefly affecting the rate of coagulation are the size of the particles and their degree of uniformity.

In an aerosol in which coagulations occur between two groups of spherical particles of radius r_1 and r_2, and concentrations n_1 and n_2, the rate of decrease of the particle concentration n_2 is

$$-dn_2/dt = \pi(D_1+D_2)(S_1+S_2)n_1 n_2, \qquad (2.5)$$

where D_1, D_2 and S_1, S_2 are respectively the diffusion constants and the radii of the spheres of influence of the particles.

$$D = \frac{RT}{\mathcal{N}} \frac{1+Al/r}{6\pi\mu r},$$

where R is the universal gas constant, T the absolute temperature, $\mathcal{N}$ Avogadro's number, l the mean free path of the air molecules, μ the viscosity of the air, and A is Cunningham's constant. ($Al = 9 \times 10^{-6}$ cm at ordinary temperatures.) Writing $S_1 = sr_1$ and $S_2 = sr_2$, we have

$$-\frac{dn_2}{dt} = \frac{RTs}{6\mu\mathcal{N}}\left(\frac{1+Al/r_1}{r_1}+\frac{1+Al/r_2}{r_2}\right)(r_1+r_2)n_1 n_2. \qquad (2.6)$$

For a homogeneous aerosol consisting of spherical particles, $r_1 = r_2$, $s = 1$, and (2.6) reduces to

$$-\frac{dn}{dt} = \frac{2}{3}\frac{RT}{\mu\mathcal{N}}\left(1+\frac{Al}{r}\right)n^2. \qquad (2.7)$$

In the early stages of coagulation in which r does not vary greatly, (2.7) may be integrated directly to give

$$\frac{1}{n} - \frac{1}{n_0} = \frac{2}{3}\frac{RT}{\mu\mathcal{N}}\left(1+\frac{Al}{r}\right)t, \qquad (2.8)$$

which has been verified experimentally. This equation predicts that smaller particles coagulate more quickly than larger ones. For example, particles of $r = 10^{-5}$ cm decay to half their initial concentrations of 10^4 cm⁻³ and 10^3 cm⁻³ in periods of $66\frac{2}{3}$ and 667 hours respectively, the corresponding times for particles of $r = 10^{-6}$ cm being 15 h and 150 h.

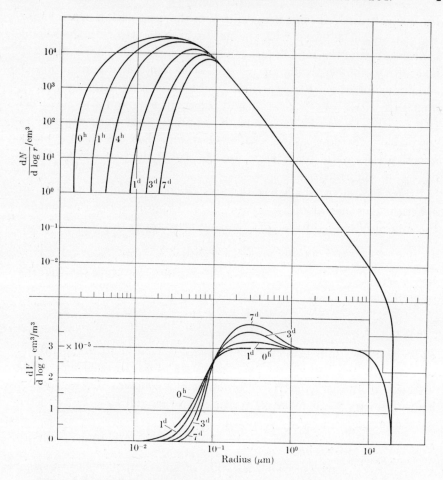

Fig. 2.18. Calculated change in the size and volume distribution of natural aerosols due to coagulation resulting from Brownian motion. h = hours, d = days. (From Junge (1958).)

For a two-component system in which $r_1 \gg r_2$, $n_1 \ll n_2$ so that r_1, r_2, and n_1 remain sensibly constant, $n_2(t)$ may be calculated by direct integration of (2.6). The more general case is treated by Junge (1963). Fig. 2.18 shows how a distribution typical of continental aerosol in industrially-polluted air changes with time by coagulation alone. The modification is almost completely confined to particles of $r < 0.1$ μm about which size the distribution tends to become rather sharply peaked after about 1 day.

2.5. Nuclei involved in cloud formation

The question now arises as to the nature of the nuclei actually involved in cloud formation. Because the concentrations of aerosol particles are usually much greater than those of cloud droplets, only a small fraction of the potential nuclei are utilized in cloudy condensation. We have already seen that the larger hygroscopic particles will be favoured, while small and medium ions will not, in general, serve as condensation nuclei because of the high supersaturations required for their activation.

In a cooling air mass, condensation occurs first on the largest hygroscopic nuclei, and whether or not particles of a certain size are activated depends upon the concentration of more efficient nuclei and the rate at which water vapour is made available for condensation by cooling of the air. Thus the concentrations and sizes of nuclei actually involved in droplet condensation will depend upon the maximum degree of supersaturation attained, and this will vary in different situations.

Measurements on the spectrum of active condensation nuclei as a function of supersaturation were first made by Wieland (1956) with his temperature-gradient diffusion cloud chamber. Fig. 2.19 shows some of Wieland's curves obtained in Switzerland in air masses of various origins. Curves 5, 13, and 15 are for Föhn air-masses that have risen high over mountains and been depleted of nuclei by precipitation before descending to the place of observation. The air-mass 19, on the other hand, is described as 'mixed' and presumably had a long continental trajectory.

Measurements made in various air masses of well-defined origin near Sydney, Australia, by Twomey (1959b) using a chemical-gradient diffusion chamber, are shown in Fig. 2.20. Continental air masses, especially in drought conditions, were found to be an order of magnitude richer in condensation nuclei active at 1 per cent supersaturation than were maritime air masses, and this accords with Squires's (1958a) finding that continental clouds in this region contained much higher concentrations of droplets than maritime clouds. Squires and Twomey (1966) used a thermal diffusion chamber to make comparative measurements of the concentrations of potential cloud-forming nuclei over Colorado, U.S.A. and the Caribbean ocean, as being typical of a continental and a maritime location in the Northern Hemisphere. The mean concentrations of nuclei activated at 0·35 per cent supersaturation

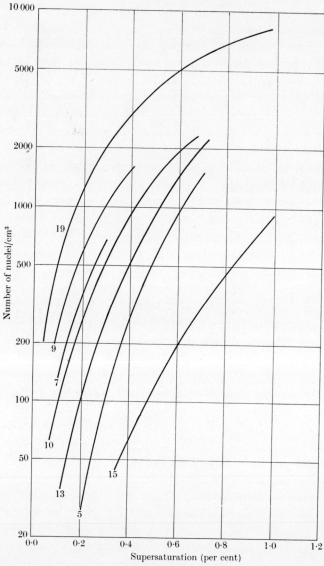

FIG. 2.19. Nucleus spectra in typical air masses. Curves 5, 13, are and 15 for Föhn conditions and curve 19 for a 'mixed' air mass. (After Wieland (1956).)

over Colorado varied from about 300 per mg of air at 2000 ft (600m) above the surface, to about 50/mg at 19 000 ft (5700m); the corresponding figures over the Caribbean were 100/mg at 2000 ft and 30/mg at 1300 ft (390m). Thus the mean concentration over Colorado was about three times that over the Caribbean in the lower atmosphere, but the contrast decreased with increasing height, and at about 5 km there was no appreciable difference.

Unfortunately only a few attempts have been made to measure size spectra of both condensation nuclei and cloud droplets on the same occasion. Twomey and Squires (1959), working 300 km inland from Sydney, Australia, obtained cloud-droplet samples from non-precipitating cumuli and collected air samples from beneath cloud base in both continental and maritime air-streams. The nucleus content of the air was measured in a chemical diffusion cloud chamber immediately after each flight. Having obtained the nucleus spectrum with respect to supersaturation, the authors calculated the total concentrations of droplets that such a spectrum should produce in a cloud sustained by an updraught of 1 m s^{-1}, and compared these with the actual measurements

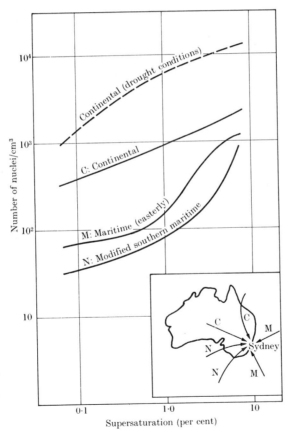

FIG. 2.20. Nucleus content of typical air masses. (From Twomey (1959b).)

of cloud droplet populations. The very good agreement obtained in twelve out of fifteen cases suggested that the systematic differences observed in cloud droplet spectra are primarily caused by variations in condensation nucleus populations. A very similar result was obtained by Twomey and Warner (1967) in a similar investigation over Queensland, but using an improved droplet sampler and a thermal diffusion chamber. In this case, good agreement was obtained if the updraught was assumed to be 3 m s^{-1}, but the result is not very sensitive to the updraught speed, the total number concentration of cloud droplets, as distinct from their size, being determined early in the ascent of the nuclei above cloud base. Jiusto (1966) made similar measurements both inland and over the sea from Hawaii. The concentration of nuclei active at 1 per cent supersaturation in a thermal diffusion chamber was about 50 cm^{-3} on the coast and doubled to 100 cm^{-3} at 3 km inland. This was reflected in the cloud droplet concentrations, which averaged 45 cm^{-3} with average radius 11 μm in maritime cumulus, and 100 cm^{-3} with average radius 7 μm in orographic clouds inland. The concentrations of nuclei active at 0·35 per cent supersaturation in the cloud chamber were about half the above values and very similar to the measured cloud droplet concentrations. The concentrations of Aitken nuclei over the ocean were usually <200 cm^{-3} below 2 km altitude; inland they sometimes reached 500 cm^{-3}, but were smaller by 1 to 2 orders of magnitude than the average values measured over the central United States.

These results all suggest that there is a close correspondence between concentrations of cloud droplets and the concentrations of condensation nuclei that can be activated by supersaturations of a few tenths of 1 per cent, and that both quantities are, on average, an order of magnitude lower in clean maritime air than over polluted continents. The question remains whether the effective nuclei over the oceans consist mainly of sea salt. Moore and Mason (1954) found salt nuclei of $m > 2 \times 10^{-14}$ g in concentrations of only 10 cm^3 over the ocean, even in winds exceeding 15 m s^{-1}. There is some evidence that even smaller nuclei are produced by bursting air bubbles in foam patches, so that the total concentrations of salt nuclei may be somewhat higher than this. Hence, in maritime areas such as Hawaii, where cloud droplet concentrations are only about 50 cm^{-3}, a significant fraction of the nuclei may consist of sea salt but, in continental clouds containing some hundreds of droplets per cm^3, sea-salt nuclei are likely to contribute only a few per cent of the nuclei involved.

2.6. The removal of aerosols from the troposphere

The trace gases involved in the formation of condensation nuclei may be removed from the troposphere by absorption and/or decomposition at the earth's surface, by decomposition in the atmosphere by reactions that result in the formation of aerosols or other gases, and by being absorbed by water droplets and finally washed out by precipitation. Ammonia is probably removed mainly by the first mechanism and SO_2 by the last, but little is known about the rates at which these processes act.

The removal of aerosols is possible by fallout under gravity, impaction of particles on obstacles on the earth's surface, and wash-out by precipitation.

We have seen that fall-out of aerosols is important only for particles larger than 10–20 μm radius; for smaller particles it is of minor importance compared with washout by precipitation. Junge (1963) reports that, in regions with normal amounts of rainfall, the dry fall-out of aerosols is between 5 and 40 per cent of the total, and that this percentage varies with the component.

That removal by impaction may be quite efficient, at least locally, is indicated by two lines of evidence. Eriksson (1955) found that the amount of chloride received at the ground in Sweden by rain-water was only about one-quarter that carried away by the rivers and explained this by the collection of sea-spray particles from the air on the foliage of trees. Byers *et al.* (1957) found the concentration of chloride particles at ground level to be only about one-fifth that at 300 m and attributed the loss to impaction on trees and other obstacles on the ground.

Wash-out, by which is meant the over-all mechanism by which trace substances are removed from the atmosphere by precipitation, includes the following processes.

(1) The consumption of condensation nuclei during the formation of cloud elements.

(2) The subsequent attachment of aerosol particles to the cloud elements as a result of Brownian motion.

(3) The attachment of aerosol particles to cloud elements by diffusiophoresis.

(4) The absorption and fixation of trace gases on cloud elements.

(5) The collection of aerosol particles by falling raindrops.

Process (4) has already been dealt with and we shall now consider the others in a little more detail.

In clean maritime air over mid-ocean, the toal concentration of salt nuclei of $m > 2 \times 10^{-14}$ g is about 10 cm^{-3} and the concentration of droplets in maritime clouds is about the same. This suggests that nearly all the salt particles available at the cloud-forming levels are involved in condensation, and since the average residence time of water vapour in the atmosphere is about 10 days, the mean residence time of the salt particles will be about the same. Overland, in Europe and America, cloud-droplet concentrations are usually a few hundreds per cm^3; if we take Junge's measurements on the size distribution of aerosols on the Zugspitze to be rather typical of those at cloud level over well-populated continents, we find that a cloud-droplet population of 300 cm^{-3} would involve all particles of $r > 0.1$ μm, i.e. nearly all the large and giant nuclei, which comprise the greater part of the mass of the aerosol present at these levels.

We next consider the removal of particles by their attachment to cloud droplets through Brownian motion. The fractional rate of removal of particles of radius r_p and concentration n_p by cloud droplets of radius r_c and concentration n_c is given by (2.6) as

$$-\frac{1}{n_p}\frac{dn_p}{dt} = \frac{RT}{3\mu\mathcal{N}}\left(\frac{1+Al/r_p}{r_p}+\frac{1+Al/r_c}{r_c}\right)(r_p+r_c)n_c = Kn_c. \quad (2.9)$$

Thus
$$n_p(t) = n_p(0)\exp(-Kn_c t), \quad (2.10)$$

and the time constant λ, i.e. the time taken for concentration of particles to fall to 1/e of the initial value, is

$$\lambda = 1/Kn_c.$$

If $r_c = 10$ μm, we have

$r_p = 0.01$ μm	0.1 μm	1 μm	
$K = 7.3 \times 10^{-7}$	1.4×10^{-8}	9.6×10^{-10}	cm^3 s^{-1}.

If $n_c = 100$ cm^{-3},

$\lambda = 3.8$ h	200 h	2900 h.

Thus only particles of $r < 0.1$ μm, i.e. the Aitken nuclei, are effectively removed by Brownian capture on cloud droplets within a few hours; the capture of the large and giant particles by this process is extremely slow.

Diffusiophoresis is a phenomenon whereby particles suspended in the vicinity of an evaporating or condensing surface experience a force that moves them in the direction of the diffusive flux of vapour. The forces on the particle arise from the combined effects of diffusive and

§ 2.6   THE NUCLEI OF ATMOSPHERIC CONDENSATION

hydrodynamic flow in the gas and vapour mixture. Goldsmith, Delafield, and Cox (1963) have shown by experiment that the velocity V_p (cm s^{-1}) imposed on sub-micron particles by water vapour diffusion through air at s.t.p. is independent of the particle size and is given by

$$V_p = -1 \cdot 9 \times 10^{-4} \frac{dp}{dx}, \qquad (2.11)$$

where dp/dx is the water vapour-pressure gradient in mb/cm. This result is in excellent agreement with the theory developed by Waldmann (1959).

Goldsmith et al. deduced that the capture of sub-micron particles by growing cloud droplets would be very slow and confirmed this by a cloud chamber experiment in which droplets of salt solution, of up to 10 μm in radius, were grown in a cloud of radioactive particles of $r \approx 0 \cdot 1$ μm. They further concluded that diffusiophoresis could account for less than 1 per cent of the total wash-out of nuclear-bomb debris during one condensation cycle in the atmosphere.

Finally, we consider the removal of aerosol particles by impaction on raindrops. The rate at which particles of radius r and concentration $n(r)$ in the subcloud layer are swept up by raindrops of radius R falling at velocity V may be written

$$\frac{dm}{dt} = \pi R^2 V \sum^r E(R,r) n(r) \tfrac{4}{3}\pi r^3 \rho, \qquad (2.12)$$

where dm/dt is the rate of mass accretion of aerosols, $E(R, r)$ is the collection cross-section of the raindrops for a particle, and ρ is the particle density. The mass concentration of aerosol material in the raindrops after they have fallen through a height $Z = \int V\, dt$ will then be

$$C = m/\tfrac{4}{3}\pi R^3 = \frac{\pi z \rho}{R}\left\{\sum^r E(R,r) n(r) r^3\right\}. \qquad (2.13)$$

If we take the size distribution of the aerosol to be that measured by Junge (1953) in Frankfurt, we can write

$$n(r) r^3 = a\, d(\log r) = 10^{-11}\, d(\log r),$$

and

$$C = \frac{\pi z \rho}{R} \times 10^{-11} \int_{\log r_{min}}^{\log r_{max}} E(R,r)\, d(\log r). \qquad (2.14)$$

Some values of $E(R, r)$ calculated for an air pressure of 900 mb and temperature 0° C and spherical particles of unit density, are given in

Appendix A. We see that particles of $r < 1$ μm have practically zero collision efficiency for all sizes of raindrops, and thus only the giant aerosol particles are removed by direct impaction on raindrops.† The concentration of material to be expected in raindrops of various sizes after falling through 1 km of Junge's 'Frankfurt' atmosphere are as follows:

R(mm)	0·1	0·6	1·0	1·8
C(mg/litre)	72	22	13	8

These values are of the same order as those measured and indicate the importance of this process for rain chemistry. Turner (1955) found, in agreement with this treatment, that the chloride concentration in raindrops was generally inversely proportional to their radii if the drops did not suffer appreciable evaporation during their fall.

The rate at which particles of radius r are removed by rain composed of a spectrum of drop sizes is

$$-\frac{1}{n(r)}\frac{dn(r)}{dt} = -\pi \sum^{R} E(R, r) N(R) R^2 V(R), \qquad (2.15)$$

For widespread, persistent, frontal rain it may be shown that

$$\pi \sum^{R} N(R) R^2 V(R) \simeq 4 \times 10^{-4} p^{\frac{4}{5}},$$

where p is the precipitation rate in mm/h. For particles of radius 8 μm and unit density, $\bar{E} \simeq 0.80$, so their concentration is reduced to 1/e of its original value in

$$\lambda = 1/(3.2 \times 10^{-4} p^{\frac{4}{5}}) \text{ s}.$$

If

$p =$	1	5	10	mm/h,
$\lambda =$	52	14·4	8·2	min.

It is evident that giant particles ($r > 1$ μm) are efficiently washed out by rain. A decrease in the concentration of trace substances with time during precipitation has been noted by many observers. For example, Anderson (1915) found that the concentration of nitrate in rain-water was inversely proportional to the amount of rain that had fallen. But, as Junge points out, this does not necessarily mean that the major part of the trace substance in rain is collected by falling raindrops because the substances existing inside the cloud are also depleted

† These calculations of collection efficiency apply only to the capture of particles by the undersurface of a falling drop. It is possible that particles of $r < 1$ μm may be captured in its turbulent wake. This possibility has yet to be investigated experimentally.

by continuing rainfall, and the two effects cannot be distinguished in rain samples at the ground.

In summary, we can say that the Aitken nuclei ($r < 0.1~\mu$m) are removed mainly by becoming attached to cloud droplets through Brownian motion, the majority of the large and giant nuclei act as centres of cloud-droplet formation, and the giant nuclei existing in the subcloud layer are removed by direct impaction on raindrops. Cloud droplets, formed upon large or giant nuclei, having acquired smaller nuclei by Brownian capture will, on evaporation, leave behind nuclei of mixed constitution; several cycles of evaporation and condensation around the same nucleus may be an important mechanism for concentrating the trace substances and for converting these from gaseous to particulate form.

3
The Growth of Droplets in Cloud and Fog

So far attention has been confined mainly to the nature and properties of the condensation nuclei and to the early stages of growth of an individual nucleus. We shall now consider the further growth of populations of nuclei leading to the formation of cloud or fog. The growth rate of individual droplets depends not only on the hygroscopic and surface tension forces and the ambient humidity, as already discussed, but also on the processes of diffusion and thermal conduction that determine the rate of transfer of water vapour to and heat from the droplet. When we have to deal with a population of droplets, the problem becomes more complicated: since all the droplets compete for the available water vapour, their growth rate will depend upon the concentration, size distribution, and nature of the nuclei, the rate of cooling of the air, its temperature, and the scale and intensity of the turbulence in the cloud. The influence of these various factors on the growth of cloud particles has been partially investigated by theoretical methods and will be discussed later in this chapter, but first, we shall describe the results of some measurements of droplet-size distribution and liquid-water content in clouds. Such observations are of great importance in that the droplet-size spectrum represents the result of several simultaneous physical processes and of the influence of the many factors just described, the relative importance of which can only be estimated by theoretical computations whose crucial test is the accurate prediction of the observed droplet-size spectrum.

3.1. The size distribution of cloud droplets

3.1.1. *Experimental techniques*

Much effort has been devoted to developing experimental techniques for measuring the size distribution of cloud droplets. Direct methods, which involve the capture of the droplets on suitably prepared and exposed surfaces and the later counting under the microscope of the droplets or their impressions with reference to their diameters, although tedious, appear to be the most reliable. The accuracy of these methods depends upon obtaining a true representative sample from the cloud

§ 3.1   THE GROWTH OF DROPLETS IN CLOUD AND FOG

or fog, preventing evaporation between sampling and counting, and avoiding shattering or coalescence of the captured droplets.

In order to capture a true sample, the 'collection efficiency' of the collector (i.e. the fraction of the droplets lying in its projected cross-section which actually hit the collector) must be unity for droplets of all sizes occurring in the cloud; on approaching the collector, the droplets are deflected from their original paths by an amount depending mainly on the radii of the droplets, the width of the collector, and the wind speed—see p. 40. The measured drop-size distributions have therefore to be corrected to allow for the tendency of the collector to discriminate against the smaller drops. The representativeness of the sample will depend on the volume of air swept out by the collector during its exposure in relation to the dimensions of the cloud and the rate of change of the droplet population. There is a danger that large droplets will shatter on hitting the collector at high speed. According to Rupe (1950), water drops impacting in kerosene at 100 m s^{-1} will splash if their radii exceed 10 μm. Coalescence of droplets in the sample may occur if they are very close together; this can be avoided by choosing the exposure time so that the covering fraction does not exceed, say, 0·1. Prevention of evaporation of the droplets is most important, and unless immediate photography of the sample is feasible, it must be an inherent feature in the design of any method.

In one method designed to prevent evaporation, a glass slide is coated with an oily mixture into which the droplets penetrate and become completely submerged. According to Mazur (1952), the matrix should have a density close to unity, should be smooth and highly viscous, and have a refractive index sufficiently different from that of water to ensure good contrast. The density of the oil should be high enough to prevent the larger drops from settling on to the slide where they will become flattened and make their measurement uncertain. Fuchs and Petrjanoff (1937) found that good results could be obtained with a freshly melted mixture of light mineral oil and vaseline. Further refinements of the technique are described by May (1945). It was used in an aircraft by Mazur (1943) who saturated the mineral oil with distilled water to prevent the droplets evaporating into the oil; he reported that the droplets could be preserved for many hours without showing any perceptible change in size or shape.

An alternative method, having some advantages, is to coat a clean glass slide with a thin film of magnesium oxide in which the droplets, on impaction, leave a pit of diameter proportional to their size. The oxide is applied to the required thickness by passing a slide to and fro immediately above the flame from burning magnesium ribbon. The layer, which should be rather thicker than the largest droplet expected, is very soft in texture and has a grain size of about 0·5 μm. When droplets hit this layer, they penetrate the surface and leave permanent round holes which may be seen most clearly under the microscope by using strong transmitted light. The technique was developed by May (1950) who carried out a complete calibration for droplets of radii between 6 and 120 μm, a spinning disk (Walton and Prewett 1949) being used to produce water droplets of very uniform size. Droplets of the same size were collected on slides coated with magnesium oxide and with the oil–vaseline mixture, the diameters of the pits in the oxide being compared with those of the droplets in the oil. For droplets of radius greater than 10 μm, the ratio of the droplet diameter to the pit diameter remained constant with droplet size at 0·86; for smaller droplets, the ratio decreased, until below 5 μm, the method was of little use because of the effect of the grain size. The calibration factor is not sensitive to impact velocity over a wide range of the latter. This method has the advantages that droplets cannot coalesce (overlapping is obvious), and that samples keep indefinitely without change. However, it is rather unsuitable for drops of radius less that 5 μm, the oxide layer is rather fragile, and its texture varies with the burning conditions of the magnesium ribbon. An alternative method, employing a thin carbon film, is reported to give better results by Squires (1958a). A third method uses a slide coated with gelatine containing a water-soluble dye such as naphthol green B (Liddell and Wootten 1957). Impacting droplets produce clear spots surrounded by intense green rings whose diameters are 2·5 times those of the original drops. Droplets of $d > 1$ μm can be readily detected.

The magnesium oxide method was used for collecting cloud droplets from aircraft by Mazur (1943) and by Frith (1951). A microscope slide coated with the oxide was exposed by a shutter for $\frac{1}{100}$ s, during which time some 250 cm³ of air were sampled. The largest droplets were practically undeviated by the air flow round the collector and impinged near the centre of the slide, while smaller droplets were deviated towards the edges. Frith claimed, however, that with air speeds of about 400 km/h, there was no significant loss of droplets of radius less than 5 μm.

A very convenient form of apparatus has been developed by Squires and Gillespie (1952) in which successive exposures of magnesium oxide-coated rods are made every 3 s. The glass rods are fixed into the magazine of a gun; the particular one in the firing position is connected at one end to a compressed spring which, when released by the trigger, executes one complete oscillation, thrusting the rod out lengthwise across the air stream and pulling it back again in about $\frac{1}{30}$ s. Another rod is then placed in position by rotating the magazine and the process is repeated. The ten slides in the magazine can all be fired in about 30 s; reloading a new magazine takes 40–50 s. The collection efficiency of the 3-mm diameter rods at their stagnation points, with an air speed of 150 knots, is 0·94 for droplets of radius 5 μm, and exceeds 0·98 for droplet radii greater than 10 μm, but falls sharply for smaller sizes. A similar instrument using three oil-coated slides is described by Brown and Willett (1955) and, after further development, Clague (1965) has been able to expose eighteen soot-coated slides in rapid succession.

A continuous sampler, in which cloud particles are captured and permanently replicated using the well-known Formvar technique, has been developed by MacCready and Todd (1964) and Spyers-Duran and Braham (1967). A continuous tape made of 16- or 35-mm film leader is coated with Formvar plastic solution before being moved at uniform speed across a 4-mm-wide sampling slit. Alternatively, the tape is pre-coated with Formvar which is softened just before exposure to the cloud with a chloroform spray. After exposure the ribbon passes into a drying compartment, the plastic sets and encapsulates the droplets, which evaporate to leave behind spherical cavities that can be counted and measured under the microscope. Having checked the collection efficiency of the sampling head in a wind tunnel, and by comparing droplet spectra with those obtained by the Clague sampler at the same time, Spyers-Duran and Braham report that, in the hands of a skilful operator, the device is capable of giving reliable samples of cloud droplets and ice crystals provided they become completely embedded in the plastic film which is about 70 μm thick.

A camera for the direct photography of cloud droplets from an aircraft was used by McCullough and Perkins (1951), but the sampling volume was only a few cubic millimeters and a large number of photographs were required to obtain a representative sample of droplets. The instrument gave considerable trouble in operation and did not absolve one from laborious measurement of the images on the film.

Another interesting instrument is described by Keily and Millen (1960). The cloud droplets are sucked at sonic speed through a pinhole in an aerodynamically-shaped housing and strike an insulated metal target held at a potential of 400 V. While being projected across the air gap between the aperture and the target, the droplets are subjected to an electric field of 5000 V cm^{-1} and, on striking the target, produce in the detector circuit voltage pulses of 60 μs duration and of amplitude proportional to the surface area of the droplet. The authors believe that droplets of $r = 1$–30 μm rebound from the target without shattering and that the pulses result from the neutralization at the target of the induced charges acquired by the droplets in the electric field. Since the pulses produced by 1-μm droplets are only a few microvolts, the noise in the electronic circuits is a limiting factor. The collection efficiency of the instrument is quoted as 1·3 for droplets of radius 1 μm and 1·0 for droplets of radius 50 μm. Keily and Millen present a few measurements of droplet-size distributions obtained with this instrument in small cumulus and low stratus having water contents of $<0\cdot005$ g m^{-3}. The droplet radii range from 1 to 7 μm with mode radii of about 1·5 μm.

Some information on droplet sizes in supercooled clouds has been obtained by a rotating multi-cylinder apparatus. The instrument consists of five cylinders of different diameters mounted coaxially and rotated slowly with their long axes normal to the wind. The differing catches of rime per unit length are measured, from which the median-volume radius of the droplets and some general indication of their size distribution can be determined. This technique was used on aircraft in Germany by Diem (1942, 1948), on Mt. Washington, U.S.A., and during an extensive series of measurements of aircraft icing conditions reported by Lewis and Hoecker (1949) and Kline and Walker (1951). In order to obtain measurable deposits, the cylinders must sweep out a very long path in the cloud, so that the method yields only average values and cannot indicate rapid spatial or temporal variations in droplet size.

All impaction methods have the disadvantage that the collection efficiency is a function of droplet size, so that they tend to discriminate against the smaller droplets. One looks forward, therefore, to the development of method that will not disturb the sample.

An interesting method of determining droplet-size distribution from the intensity distribution within a small diffraction angle of the Fraunhofer pattern produced by the cloud, has been described by Oura and Hori (1953). A parallel beam of red light (6500 Å), after transmission through a 50-cm path of the fog, is focused on a red-sensitive plate where the diffraction pattern is produced. The intensity distribution of the diffraction pattern, as a function of the diffraction angle, is determined by a microphotometer. Since the fog particles are distributed at random in space and time, the intensity in the Fraunhofer pattern can also be calculated in terms of the particle-size distribution and the wavelength of the light, by superposing the elementary patterns produced by each particle. Assuming that the droplet-size distribution is a linear combination of several Gaussian distributions, the coefficients are chosen so that the calculated and measured intensity distributions agree. The authors claim that the technique is valid for droplets of $r < 10$ μm.

An instrument developed at the Massachusetts Institute of Technology was designed to determine the droplet-size distribution from measurements on the transmission of infrared light through the cloud, as a function of wavelength. A similar instrument, working with visible light, was developed earlier in Russia by Driving, Mironov, Morozov, and Khvostiknov (1943). In both cases, the technique is to compare the measured transmission with transmissions calculated on the basis of various assumed drop-size distributions, and to find the nearest match. Unfortunately, this method does not yield a unique solution. Preliminary measurements, made by Eldridge (1957) with the M.I.T. instrument, in orographic cloud, indicated that droplets of $r < 1$ μm occurred in much greater proportions than have been detected by other methods. Until this result receives independent confirmation, the reliability of this technique must remain in doubt.

A light-scattered technique for the measurement of cloud droplets has been developed by Laktionov (1959), improved upon and used on an aircraft by Kazas (1963). It is similar, in principle, to Gucker's instrument for the detection of aerosols described on p. 48. The drop-laden air is drawn through a capillary tube and is surrounded by a sheath of clean saturated air that prevents the droplets from evaporating. It then intersects a light beam, 2×0.3 mm, at the focus of an ellipsoidal mirror, and the light scattered by each individual droplet is collected by a photomultiplier. The output pulses are amplified and sorted electronically into five size groups, the first recording droplets of $r \geqslant 2.2$ μm, and the last, droplets of $r \geqslant 6.3$ μm. The maximum sampling rate is 20 cm³ s⁻¹ when the air speed is 20 m s⁻¹, these rates being controlled so that there is only a very small probability of two droplets appearing in the sampling volume at the same time. Some results obtained by this instrument are listed in Table 3.1. The accuracy is probably limited by the fact that the flow is not isokinetic and by some loss of droplets through evaporation and deposition in the tubes.

On the whole, we have to record little progress during the past decade, and that the development of a reliable, accurate, and labour-saving technique for cloud-droplet measurement remains an urgent requirement for research in cloud physics.

3.1.2. *Results*

When the droplet samples have been collected and measured under the microscope, the size distribution may be represented in a number of different ways to accentuate different features. While the absolute concentrations of droplets of different sizes are most conveniently displayed on a plot of $n(r)$ against r, when $n(r)\,\mathrm{d}r$ is the number of droplets per unit volume of radii between r and $r+\mathrm{d}r$, the relative concentrations of different sizes are best shown as a percentage of the total. If, however, one wishes to show the contributions that the different droplet sizes make to the total liquid-water content, a plot of $n(r)r^3\,\mathrm{d}r$ against r is suitable. The characteristics of the droplet spectrum can be expressed numerically in terms of a number of parameters of which we may define the following:

r_m, the mean radius, i.e. the sum of all the droplet radii divided by the total number of droplets;

r_d, the mode or most frequent radius, i.e. the radius corresponding to the peak of the $n(r)\,\mathrm{d}r$ curve;

r_v, the mean-volume radius, i.e. the radius of a droplet whose volume is equal to the average droplet volume;

r_{50}, the median radius; half the droplets have radii smaller (or larger) than this value;

r_p, the predominant radius, i.e. the radius that contributes most to the volume of water and corresponds to the peak of the $n(r)r^3\,\mathrm{d}r$ curve;

r_n, the median-volume radius, i.e. the radius of the droplet such that half the water is comprised by larger droplets.

We shall now examine the experimental data to see how the droplet size distribution varies with cloud type and, for a particular cloud, how it varies in space and time. An analysis of sampling errors, based on Poisson statistics, is given by Cornford (1967).

A few measurements of droplet size in fair-weather and dense cumuli were made by Mazur (1943), Diem (1948), and Squires and Gillespie (1952), but the first extensive series of measurements of droplet size in cumuliform clouds was made by Weickmann and Aufm Kampe (1953) using slides coated with castor oil. Simultaneous measurements were made of the water content (and hence of the droplet concentration) using a transmissometer (see p. 118). The slides were made an integral

§ 3.1 THE GROWTH OF DROPLETS IN CLOUD AND FOG

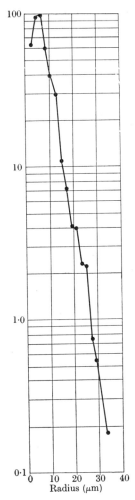

FIG. 3.1. The average droplet-size distribution in fair-weather cumulus. Average liquid-water content $1 \cdot 0$ g m^{-3}; average droplet concentration 302 cm^{-3}. (From Weickmann and Aufm Kampe (1953).)

part of the surface of a cylinder; at aircraft speeds of 150 miles/h, the smallest droplets that could be caught were of radius about 1·5 μm. To minimize the effects of droplet evaporation into the oil (amounting to about 0·05 μm s^{-1} for droplets of all sizes), the slides were photographed under the microscope in the aircraft. Figs. 3.1–3.3 show average droplet spectra in fair-weather cumulus, cumulus congestus and cumulonimbus. All available spectra were used in this presentation, irrespective of the position of the sampling point in the cloud. There is a very marked difference between the spectra of fair-weather cumulus and those of cumulus congestus and cumulonimbus. Whereas the recorded range of droplet radii in fair-weather cumulus extends from 3 to 33 μm, those of cumulus congestus and cumulonimbus extend from 3 or 4 μm to 100 μm, and sometimes beyond. In only one case were drops of radius greater than 100 μm found in non-precipitating cumulus congestus, but were often found within the precipitation areas of comulonimbus clouds. In fair-weather cumulus, the water content was small and the droplet numbers large; the reverse was true in the heavy types of cumulus. In general, near the base of the cumulus, the spectra were narrow and the

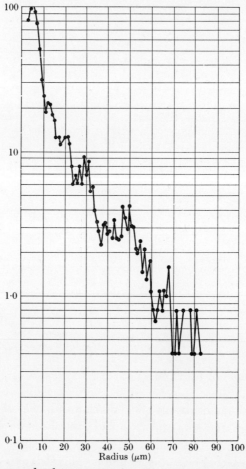

FIG. 3.2. Average droplet spectrum in upper part of cumulus congestus. Average liquid-water content 3·9 g m^{-3}; average droplet concentration 64 cm^{-3}. (From Weickmann and Aufm Kampe (1953).)

§ 3.1 THE GROWTH OF DROPLETS IN CLOUD AND FOG

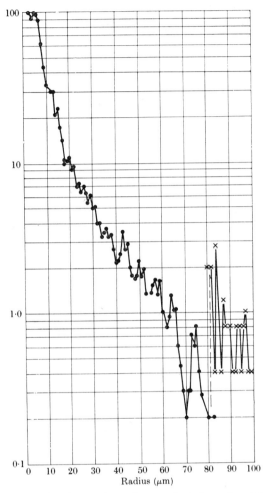

FIG. 3.3. Average droplet spectrum in cumulonimbus. Average liquid-water content 2·5 g m^{-3}; average droplet concentration 72 cm^{-3}. Note the shift in ordinate for radii >80 μm. (From Weickmann and Aufm Kampe (1953).)

droplet radii small, while about 2 km higher up the droplet size was a maximum and the spectra very broad. At still higher levels, the droplet size decreased and the spectra tended to become more narrow. The computed values of liquid-water content and droplet number per cm^3 are presented in Fig. 3.4, together with the average temperature of the observation point, the root-mean-square droplet radius and the minimum visibility. Values with a question mark indicate that the droplet sample was taken in a precipitation area and may have been influenced by the break-up of raindrops. The trend of the droplet concentration,

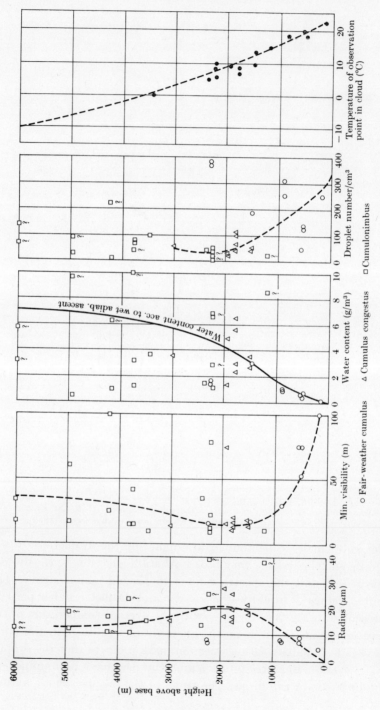

Fig. 3.4. The physical properties of cumuliform clouds as functions of height above cloud base. (From Weickmann and Aufm Kampe (1953).)

§ 3.1   THE GROWTH OF DROPLETS IN CLOUD AND FOG   103

from large numbers near the cloud base to small numbers near the 2-km level concurrent with an increase of liquid-water content with height, indicates that the droplets may have grown partly by coalescence. The changes that take place in the droplet spectra during the transition from fair-weather cumulus to cumulus congestus are illustrated in Figs. 3.1 and 3.2. Whereas the number of small droplets has decreased by an order of magnitude, the number of big droplets has increased. Moreover, the bulk of the liquid water has shifted to larger droplets (Fig. 3.5), only a few per litre of which are required to account for the

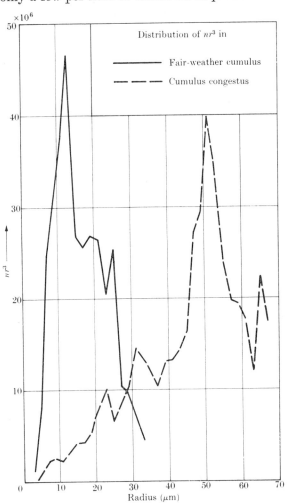

FIG. 3.5. The contribution made by droplets of various sizes to the liquid-water content of fair-weather cumulus and cumulus congestus. (From Weickmann and Aufm Kampe (1953).)

greater part of the liquid-water content. Though droplets of radius greater than 60 μm are present in concentrations of less than 100/litre, their significance in connection with the onset of precipitation is evident since they are large enough to start an efficient process of growth by accretion.

Results showing the same general trend as those just described were obtained by Zaitsev (1950) in cumulus humilis and cumulus congestus, one of his diagrams showing the variation of drop-size distribution, droplet concentration, and liquid-water content with height in a particular cumulus congestus being reproduced in Fig. 3.6. It must be pointed out that this diagram does not represent an instantaneous

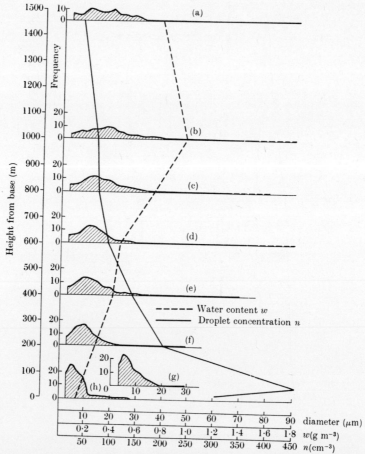

FIG. 3.6. The droplet-size distribution, liquid-water content, and droplet concentration as functions of height in a cumulus congestus. (From Zaitsev (1950).)

picture of the spatial distribution of the various quantities, since the aircraft traversed the cloud at successively higher levels over a period of several minutes; an instantaneous picture, allowing strict comparison of the micro-physical state of the cloud at all levels, would necessitate simultaneous traverses at different levels by several aircraft. Nevertheless, the data obtained by Zaitsev indicate that both the mode and maximum radii of the cloud droplets tend to increase with height, at any rate up to about 2 km above cloud base, while their concentrations decrease. He reports also that the droplets are smaller at the periphery of the cloud than in the centre.

Measurements made in recent years with multiple-exposure samplers have revealed that the droplet concentration may show rapid spatial variations on a scale of less than 300 m and that almost clear patches are often found within quite dense and compact clouds. Even so, these variations are not large enough to mask the characteristic differences between different types of cloud that have been revealed by Squires (1956, 1958a, b) and Squires and Warner (1957) in an extensive series of measurements in clouds that were classed as of maritime, continental, or transitional origin. These observations point to the importance of the nucleus population and the rate of ascent of the air in determining the number concentrations of droplets within the cloud. Fig. 3.7 summarizes measurements of droplet concentrations in three types of cloud forming over Hawaii all having summit temperatures above $-3°$ C and containing no ice crystals. The orographic cloud, which varied in thickness from 2000 ft (600 m) to 5000 ft (1500 m) and was sustained by upcurrents of only 10–25 cm s^{-1}, had low concentrations ($\sim$10 cm^{-3}) of fairly large drops. The dark stratus cloud, about 1000 ft thick, had a similar constitution. The cumulus clouds forming over the sea contained updraughts of about 1 m s^{-1} and, in consequence, the median droplet concentration was higher at 45 cm^{-3}.

Squires et al. found that clouds formed well inland in Australia were of markedly different constitution from those of maritime clouds of similar size. Table 3.2 shows that, compared with a median droplet concentration of 45 cm^{-3} for maritime cumuli, the corresponding figure for continental cumuli was 228 cm^{-3}, with clouds in transitional air masses having an intermediate concentration. Very similar results were reported by Battan and Reitan (1957); cumuli forming over the ocean near Puerto Rico had average droplet concentrations of 55 cm^{-3}, while for clouds over the central United States the figure was 200 cm^{-3}.

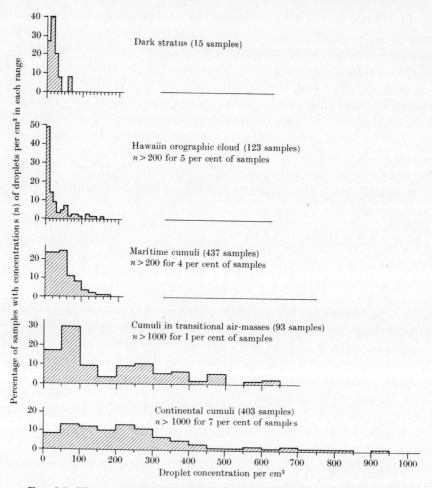

Fig. 3.7. Histograms of the percentages of samples taken in each of five cloud types for which the droplet concentrations fell in the ranges indicated on the horizontal axes. (From Squires (1958a).)

Since the liquid-water content of many small cumulus of all types averages about $0\cdot 5 \text{ g m}^{-3}$, it follows that clouds with large droplet concentrations consist of small droplets, while those with small droplet populations contain many large drops. There is indeed a correlation between the microstructure and the stability of the clouds. As Squires (1958a) points out, warm maritime cumuli more than 6000 ft deep usually produce a shower within half an hour, but such shallow clouds forming inland in Australia rarely rain unless they build to well above the 0° C level.

Some typical droplet size spectra obtained in Hawaii clouds by

§ 3.1   THE GROWTH OF DROPLETS IN CLOUD AND FOG   107

Squires are represented in Fig. 3.8. Going from orographic cloud and dark stratus to trade-wind cumuli, there is a progressive increase in droplet concentration, decrease in average droplet size, and a narrowing of the droplet-size spectrum, all with very little change in the liquid-water content. Since the nucleus population is probably the same for all

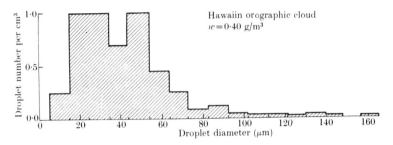

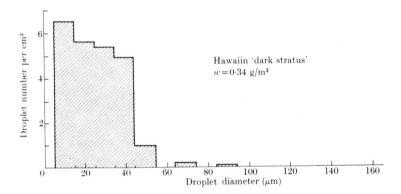

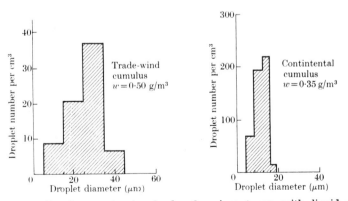

FIG. 3.8. Droplet spectra in clouds of various types, with liquid water contents w as shown. Cumulus samples were taken 2000 ft above cloud base, orographic and dark stratus values are averages. Note change in ordinate scale from figure to figure. (After Squires (1958a).)

cloud types, these differences probably reflect differences in cloud dynamics.

In continental cumuli, the droplet concentration is higher, the mean droplet size smaller, and the spectrum still narrower than in maritime clouds. Squires and Twomey (1958) demonstrated that these differences were not due to differences in the populations of giant nuclei and attributed them largely to differences in the *total* nucleus content of the air. Twomey and Warner (1967) found that the total concentrations of cloud droplets near the bases of modest, warm, nonprecipitating cumuli over Queensland were closely correlated with the concentrations of condensation nuclei sampled from the air below cloud base and activated at supersaturations of up to about 1 per cent in a thermal diffusion chamber. Furthermore, Warner (1968) reported that the *dispersion* of the droplet spectra in these clouds was independent of the droplet concentration and mean droplet diameter over a wide range of values, but was apparently correlated with the vertical velocity of the air.

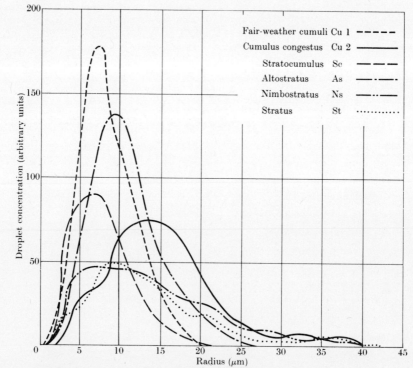

Fig. 3.9. The mean droplet-size distributions of various cloud types. (From Diem (1948).)

§ 3.1   THE GROWTH OF DROPLETS IN CLOUD AND FOG   109

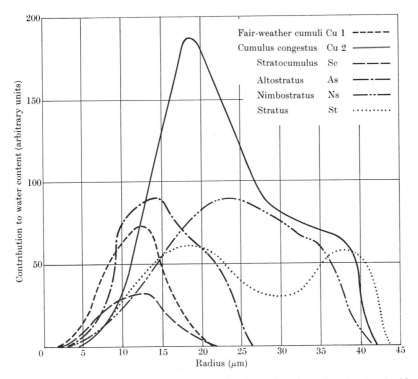

FIG. 3.10. The contribution made by droplets of various sizes to the liquid-water content of different types of cloud. (From Diem (1948).)

Squires's measurements in layer clouds may be supplemented by those of Diem (1942, 1948), who exposed small oil-coated slides each for about $\frac{1}{50}$ s. The slides were photo-micrographed within a minute of collection. Diem claimed that the collection was satisfactory for droplets of radius greater than 1·5 μm, but there is reason to believe that his calculations of the collection efficiency were incorrect, and that the instrument also discriminated against droplets somewhat larger than this. Average droplet spectra from clouds of various types are shown in Figs. 3.9 and 3.10 and their characteristic parameters in Table 3.1. Fair-weather cumulus, altostratus, and stratocumulus were found to have rather narrow spectra, but the cloud types that are normally associated with precipitation had much broader distributions.

A number of measurements using magnesium oxide slides were made in shallow layers of stratocumulus by Frith (1951), twelve samples being taken over a flight path of about 15 miles. Although in any one traverse, the samples showed wide variations in droplet concentration and in liquid-water content as determined from integration of the droplet-size distribution, on average, high droplet concentrations were associated with high water contents. The predominant droplet radii for different clouds varied from 10 to 20 μm.

Droplet sizes in California stratus were measured from an airship by Neiburger (1949). Small slides coated with a thin film of lampblack were exposed to the air-stream and the droplet diameters determined from the diameters of the rings left in the soot. Except near the cloud base, the droplet spectra had a well-pronounced mode at radius 7 μm, the radii ranging from 2 to 40 μm.

Extensive series of measurements of droplet sizes in stratiform clouds have been in various parts of Russia under a wide variety of climatic conditions. These measurements, made mainly with slide impactors, have been summarized by Borovikov et al. (1961). Bearing in mind that droplets of $r < 3$ μm were not collected, the mean radius and dispersion of the droplet spectra averaged over a large number of samples were remarkably similar for all types of stratiform cloud although individual samples showed wide variations. Indeed the variations at different heights and at different stages of development within the same cloud were often larger than the differences between the *average* characteristics of different cloud forms. In nearly all the clouds there was a tendency for the droplet size to increase upwards from the base and to reach a maximum value somewhere below the cloud top.

Measurements of droplet size in sea fog were made by Houghton and Radford (1938) on the north-east coast of the United States, by collecting them on a hydrophobic (vaseline) surface and photographing them under the microscope. Sampling errors occurred for droplets of radius less than about 10 μm, but may not have greatly influenced the results in view of the fact that the median-volume radii ranged from 13 to 38 μm, with an average of 23 μm. The largest drop measured was 60 μm in radius. The mean liquid-water content was found to be 0·13 g m^{-3}, with a range of from 0·01 to 0·30 g m^{-3}. The droplets were much larger than generally found in clouds; this, coupled with the relatively low water content, shows that the droplet concentration was very small. Chemical analysis of the fog water tended to confirm that the droplets had formed on a few relatively large sea-salt nuclei.

Hagemann (1935) obtained drop-size distributions in fog in Germany using oil-covered slides. He found that the most frequent size ranged from 4·5 to 17 μm radius, with an average of 7·8 μm. Kozima, Ono, and Yamaji (1953), using an oiled-slide method, report that in fogs on the

§ 3.1   THE GROWTH OF DROPLETS IN CLOUD AND FOG

south-eastern coast of Hokkaido (Japan) which are advected from the Pacific Ocean, the droplet-size distribution varies a good deal from one fog to another, depending upon the physical properties of the air mass. Some fogs were very rich in small droplets and had mode radii below 5 μm, while others contained a small number of very large drops ($r > 40$ μm) among those of ordinary size, and had mode radii greater than 10 μm. The occurrence of large droplets tended to be associated with high temperatures; the droplet concentrations varied between about 1 and 20 cm^{-3}. Measurements made at the same times with the optical method of Oura and Hori (1953) indicated the presence of many more smaller droplets than were caught on the oiled slides, the mode radius being generally less than 5 μm.

3.1.3. *Summary of results*

The direct methods of determining droplet size, involving the use of only simple equipment, appear most reliable at present, but, of course, are extremely laborious. Unless satisfactory electronic techniques are developed for scanning the samples and sorting the droplets automatically into size groups, a considerable effort will be required to obtain representative samples in all types of clouds in different stages of their life history, and under a variety of meteorological and geographical conditions. At present our data are too scanty to make comparisons between the characteristics of the droplet populations of different cloud types very profitable, but Table 3.1 indicates the tendency for clouds that are unlikely to precipitate (fair-weather cumulus and thin layers of stratocumulus and altostratus) to consist of relatively large concentrations of small droplets, and for the thicker, denser clouds (cumulus-congestus, cumulonimbus, and nimbostratus) to contain considerably larger droplets. Low stratus and fog also generally consist of rather low concentrations of relatively large droplets, but their constitution largely depends on their life history and age. Clouds formed over the middle of large oceans (e.g. near Hawaii) contain smaller concentrations of droplets than those of similar water content formed in continental air. This is, no doubt, a consequence of the higher nucleus content of continental air. Squires (1958b) gives, in Table 3.2, data on the frequency of values of droplet concentration in three different types of cloud having very similar concentrations of liquid water.

The experimental data of Mazur, Diem, Frith, and Hagemann were examined by Best (1951a) who found that, in many cases, the droplet-size

TABLE 3.1
Characteristics of cloud-droplet populations

Cloud type	Author	r_{min} (μm)	r_{50} (μm)	r_d (μm)	r_m (μm)	r_{max} (μm)	n(cm^{-3})	w(g m^{-3})
Small continental cumulus								
(<7000 ft) Australia	Squires	2·5	6	—	—	10	420	0·4
U.S.A.	Weickmann & aufm Kampe	3		6	9	33	300	1·0
England	Durbin			4	6	30	210	0·45
Russia	Kazas			4·3	4·7		310	0·15
U.S.A.	Draginis		6	7			300	0·40
Small maritime cumulus								
Hawaii	Squires	2·5	12	11	15	20	75	0·50
Caribbean	Draginis		9	11			45	0·4
Cumulus congestus								
U.S.A.	Weickmann & aufm Kampe	3		6	24	83	64	2·0
(600 m above base)								
Russia	Zaitsev	2		5·5		40	95	0·45
Cumulonimbus								
U.S.A.	Weickmann & aufm Kampe	2		5	20	100	72	2·5
Altostratus								
Germany	Diem	1		4·5	5	13	450	
Russia	Kazas			4·6	6·6		220	0·6
Nimbostratus								
Germany	Diem	1		4	6	20	330	
Stratus								
Germany	Diem	1		4	6	22	260	
Hawaii	Squires	2·5	13			45	24	0·35
Stratocumulus								
Germany	Diem	1		3·5	4	12	350	
England	Frith	3				25	500	
Orographic cloud								
Hawaii	Squires	5	13			34	45	0·30

distribution can be represented by the formula

$$1 - F = \exp\{-(x/a)^b\}, \tag{3.1}$$

where F is the fraction of liquid water in the air comprised by drops of diameter less than x, and a and b are numerical factors. There appears to be no significant difference between the values of a for different cloud types, but there does appear to be a correlation between the values of a and the liquid-water content w, such that $w = 1 \cdot 1 \times 10^{-3} a^{1 \cdot 79}$, where w is expressed in g m^{-3} and a in microns. There appears to

TABLE 3.2
Frequency of values of droplet concentration (cm^{-3}) in three types of cloud (Squires 1958b)

	Hawaiian orographic cloud	Maritime cumuli	Continental cumuli
First quartile concentration	4	22	119
Median concentration	10	45	228
Third quartile concentration	30	70	310
Maximum concentration observed	370	470	2800
Total number of observations	123	438	403

be no systematic variation of the distributive index b with cloud type or with liquid-water content, the mean value being 3·27. Most of the data examined by Best were obtained from either layers of stratocumulus or from fog, but it seems likely that with suitable adjustment of the parameters a and b, it will prove possible to fit his equation to most cloud-droplet spectra.

According to Levin (1954), the size distribution of cloud droplets can be represented quite well by a log-normal distribution, viz.

$$n(r) = \frac{N}{\sqrt{2\pi} \log \sigma_g} \exp\left\{-\frac{(\log r - \log r_g)^2}{2 \log^2 \sigma_g}\right\}, \qquad (3.2)$$

where $n(r)$ is the frequency of occurrence of radius r, N is the total number of droplets, σ_g is the standard deviation, and r_g the geometric mean radius.

It may also be represented quite well by a gamma (Pearson Type III) distribution of the form:

$$n(r) = \frac{N(\gamma\alpha)^{\gamma\alpha+1}}{\alpha \, e^{\gamma\alpha} \Gamma(\gamma\alpha+1)} \left\{1 - \frac{(r-r_d)}{\alpha}\right\}^{\gamma\alpha} \exp\{-\gamma(r-r_d)\}, \qquad (3.3)$$

where Γ is the Gamma function, $\alpha = r_d - r_{min}$, $\gamma = 1/r_m - r_d$, r_d and r_m are the mode and mean radii.

3.2. The liquid-water content of clouds

3.2.1. *Techniques*

The concentration of liquid water in a cloud, however dispersed, is of considerable meteorological importance. Its magnitude and spatial distribution in the cloud are important factors in the study of cloud dynamics, since they indicate the degree of mixing that has taken place between the rising cloudy air and its drier environment. Changes in the water content are important thermodynamically, since they are accompanied by large energy changes. More detailed knowledge of its magnitude and distribution than we have at present would facilitate more precise calculations on the rate of growth of precipitation elements and the evaluation of icing risks to aircraft. Unfortunately there is, as yet, no entirely satisfactory method of measuring this quantity which meets all the following requirements: (a) able to measure the concentration of liquid water both in the supercooled and non-supercooled state over the range 0·05 g m^{-3} to about 5 g m^{-3} with an accuracy of

0.05 g m^{-3}; (b) have a rapid response (i.e. a lag of not more than a few seconds), so that is can detect small-scale variations such as may occur in convective clouds; (c) cause as little disturbance of the environment as possible, so that the sample is truly representative of the cloud and not a function of the sampling characteristics of the instrument; (d) if not an absolute instrument, should lend itself to accurate calibration; (e) be robust and reliable when used in an aircraft; (f) give a continuous, easily interpreted record of the water content.

It is, of course, possible to evaluate the liquid-water content of a cloud by integration of the droplet-size distribution curves if the volume of air sampled is known. The method, first used by Diem (1942) and Mazur (1943), is liable to be inaccurate because a large contribution to the liquid water may be made by a very small number of very large droplets, and any error involved in counting these may lead to a serious error in evaluation of the water content.

The use of rotating cylinders for determining the water content, by weighing the amount of rime deposited while sweeping out a known volume of cloud, has already been mentioned in the discussion on droplet-size determination. The amount of icing is a function of the air speed, collection efficiency of the cylinders, and the liquid-water content of air, the latter being computed as a mean over a run of several miles. The instrument, which can be used only in supercooled clouds, though robust, has a very slow response and is cumbersome to use in an aircraft. It is useful for determining the average water content over a flight of some miles, to an accuracy of perhaps 5 per cent, but is not a suitable research tool for most cloud physics investigations.

A more convenient form of the rotating-cylinder instrument is the rotating-disk icing meter (Katz and Cunningham 1948). A thin disk rotates slowly with its edge exposed to the air stream. As ice accumulates, the diameter of the disk increases and moves a probe which is mechanically connected to a transducer. Deflection of the probe varies an air gap and thus an inductance is a previously balanced a.c. bridge; the out-of-balance current is rectified and gives a measure of the water content. Alternatively, movement of the probe can be made to deflect a mirror which moves the image of a lamp across a slit to give a photographic record (Lewis and Hoecker 1949). A scraper removes the ice as the disk rotates and the process is repeated, a change of deposit usually being achieved every 10 s, although this period may be decreased by increasing the rate of rotation when high water contents are anticipated. This instrument has many of the disadvantages of the riming cylinders since it works on the same principle, but it is easy to operate, reasonably rapid (lag about 10 s), suitable for continuous recording, robust, and quite small. It has been used in the United States, Canada, and this country.

For these instruments and others that depend on measuring the rate of accretion of ice on an unheated body, there are limiting values of the water content above which the measurements become unreliable. With high air speeds and high water contents, the instrument may collect supercooled water at a rate faster than it can be frozen by loss of the latent heat of fusion to the environment; the excess water may be flung off the instrument which will therefore underestimate the water content. The heat economy of a riming cylinder has been examined by Ludlam (1951b), who shows that, with an air speed of 400 km/h and air temperature $-10°$ C, the results obtained with a cylinder of $\frac{1}{8}$-in diameter will be unreliable if the true water content exceeds about 1 g m^{-3}; if, under similar conditions, the temperature were $-20°$ C, this limit would be raised to 2·4 g m^{-3}. His computations have been confirmed experimentally by Fraser, Rush, and Baxter (1952), who determined the working limits of both the rotating cylinder and rotating disk in an icing tunnel. For a disk of $2\frac{1}{8}$-in diameter and $\frac{1}{8}$-in thick, with air speed 400 km/h, and temperature $-12°$ C, the indicated water contents were less than the true values, if the latter were above 1·5 g m^{-3}. If these ice accretion instruments are used in supercooled clouds of high water content, they need to be refrigerated.

Promising thermal methods are described by Perkins (1951), Neel and Steinmetz (1952), and Neel (1955). In the latter, a wire or cylinder heated electrically to a high temperature is exposed to the air-stream. Droplets impinging on it are evaporated and cool its surface. The magnitude of this cooling, which is measured by either a resistance thermometer or a thermocouple, is a measure of the water content. Calibration is required beforehand and the air temperature, altitude, and air speed must be known, and the collection efficiency of the instrument computed. To ensure that the instrument responds only to changes in water content, a differential method can be used in which one cylinder is exposed to the cloud droplets and an exactly similar one is shielded from them in such a manner that both are under the same conditions of air flow. The cylinders form two arms of a Wheatstone bridge, previously balanced in clear air, the out-of-balance current giving a measure of the water content. Narrow cylinders can be used to cause little interference with the air-stream and a lag of less than 5 s can be achieved. The instrument is robust and can be adapted for continuous recording, but its power consumption is rather high. It also suffers from the disadvantage that large cloud drops tend to splash off the wires. In modified versions of the instrument described by Aufm Kampe, Weickmann, and Kelly (1956) and by Skatskii (1963), the rate at which water impinges on the surface is measured in terms of the electrical power necessary to evaporate the water as fast as it collects. The electric current used for heating is controlled by the resistance between two electrodes which varies with the quantity of water on the collector. Aufm Kampe, Weickmann, and Kelly fill the space between the electrodes with lithium chloride whose resistance decreases rapidly as it becomes wet; also, the boiling of the water creates a foam that prevents the impinging droplets from splashing. This instrument has a lag of about 0·5 s and an accuracy of ± 10 per cent for water contents up to 5 g m^{-3}. Skatskii claims an over-all accuracy of ± 20 per cent and a lag of 2 s. These thermal methods are applicable in both supercooled and non-supercooled clouds.

An instrument that has been much used in the United States is the capillary collector designed by Vonnegut (1949a). The collector head consists of a small cup made of sintered metal with pores of about 30-μm diameter; droplets colliding with its inner surface are sucked by the action of capillary forces into a measuring system, their total volume being measured by the displacement of water along a graduated capillary tube. The rate of change of the column is a measure of the instantaneous water content, and in a later modification, this is measured by the change in electrical capacity between two plates of a condenser as the air in the tube is replaced by water. This instrument has the following advantages. It is quick in response, reasonably accurate (although evaporation losses may occur at the collector head), provides an unambiguous method of determining the liquid-water content, and can give a continuous recording. Its effective use is limited to warm clouds as supercooled droplets tend to ice-up the collector head; heating the head to obviate this increases the evaporation errors.

Fig. 3.11 shows a very convenient instrument designed by Warner and Newnham (1952) for use in non-supercooled clouds. A paper tape 1 in wide, is moved at controllable speed past a narrow slit open to the air-stream, where it is wetted by the cloud droplets. After a short delay to allow the paper to absorb the droplets collected on its surface, the resistance across the tape is measured and exhibited on a recording milliammeter. The instrument has a collection efficiency exceeding 0·75

§ 3.2   THE GROWTH OF DROPLETS IN CLOUD AND FOG

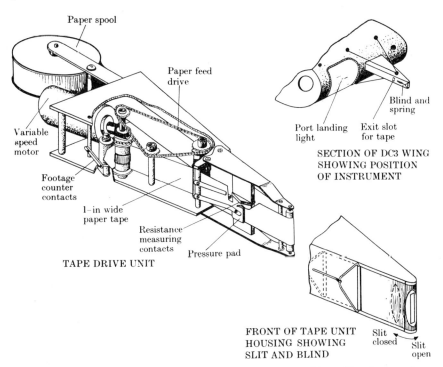

FIG. 3.11. The paper-tape liquid-water content meter. (From Newnham and Warner (1952).)

for droplets of radii greater than 5 μm in an air-stream of 300 km/h but this falls off rapidly for smaller droplets; the absolute accuracy is probably not better than 20 per cent, but the apparatus has a lag of <1 s, gives a continuous record, and provides useful information on the distribution of liquid water in a cloud. It is not suitable for use in supercooled clouds, nor in precipitation when the slit becomes covered with water and splashing occurs with the larger drops.

Zaitsev (1948) describes a method in which absorbent filter paper is exposed to the cloud droplets entering an intake, for a measured interval of 3 to 15 s. The exposure time and the speed of the aircraft are noted, allowing the volume of air entering the system to be determined from calibration curves. The weight of liquid water contained in this volume of air is found by measuring the diameter of the coloured spot it produces on the paper. This is an intermittent method, the accuracy of which obviously depends very much on the reliability of the calibrations involved, but it has the merit of simplicity. Again, it would be unsuitable for use in supercooled clouds and in heavy precipitation. In a recent variant of this method, described by Balabanova (1961), the cloudy air is sucked through two consecutive cotton-wool filters, the first of which retains both droplets and vapour, while the second absorbs vapour only. If the volume of air drawn through the filters is known, the liquid-water content of the air can be found by weighing both filters. The collection efficiency of the device is not clearly stated, but the method gave over-all averages that were in agreement with values calculated from droplet-size distributions, although in individual cases there were large differences.

As the visibility is obviously related to the liquid-water content of the cloud, it should be possible, in principle, to determine the latter from measurements of the former using a modified version of a formula derived by Trabert (1901), viz.

$$\mathscr{V} = \frac{C}{w} \sum n_r r^3 \Big/ \sum n_r r^2 = \frac{C}{w} k \bar{r},$$

where $\mathscr{V}$ is the visibility in metres, $\bar{r}$ the linear mean droplet radius in microns, w the liquid-water content in g m^{-3}, and C is a numerical factor which, according to scattering theory, takes the value 2·6. The method has recently been adapted for use on an aircraft by Aufm Kampe (1950). A searchlight giving a parallel beam is mounted on the wing of the aircraft and an optical system allows only parallel light to strike a photo-electric cell. The visibility, given by Koschmieder's formula, is then related to the light intensity recorded by the photo-cell by the equation $\mathscr{V} = 3\cdot91 d/\ln I_0/I_1$, where d is the distance between the light source and the receiver, I_0, I_1 are respectively the measured intensities in the absence, and in the presence of cloud. Using an optical path of $d = 16$ m, Aufm Kampe was able to measure visibilities between 20 and 1500 m. The water contents were calculated from Trabert's formula, the droplet-size distributions being obtained with oil-coated slides. If the droplets are all of the same size, k takes the value unity, but for a broad spectrum it takes higher values, exceeding, perhaps, 2.

The above review of experimental methods of measuring the liquid-water content of clouds is not, perhaps, complete, but mentions the more important ones known to the author. Several other techniques have been suggested from time to time and include: measurement of the change in dew point or refractive index of the air consequent on evaporation of the water droplets; measurement of the changes in the dielectric constant of the cloudy air brought about by variations in the liquid-water content; and measurement of the angular distribution of light

§ 3.2   THE GROWTH OF DROPLETS IN CLOUD AND FOG

from an artificial rainbow. But, so far as the author is aware, no instruments utilizing these principles have been made to work satisfactorily in the air.

3.2.2. *Results*

Although for the purpose of estimating the danger of ice accretion on aircraft, a measure of the average liquid-water content over a flight path of some miles may be adequate, for a detailed study of cloud structure it is necessary to have a continuously recording instrument with a lag of not more than a few seconds, so that spatial variations on the scale of 100 m can be detected. Ideally, we should like to measure the horizontal distribution of liquid water in a cumulus at several different levels simultaneously, in order to obtain an 'instantaneous' picture which could be correlated with the thermal and dynamical structure. Such comprehensive information has not yet been obtained.

A reasonably detailed picture of the distribution of liquid water in cumuliform clouds has been provided by Zaitsev (1950). He finds that in large cumulus, the water content increases with height above cloud base over the first kilometre, after which it decreases again. It is lower near the periphery of the cloud than in the centre, a result, no doubt, of mixing between the cloudy air and the drier environment at the edges of the cloud. In small cumulus, the water content rarely exceeds 1 g m^{-3} and is generally much smaller. In cumulus congestus, much higher values are found and, over small localized regions, may approach the maximum theoretical value calculated on the assumption of adiabatic ascent of the air from the cloud base. Values exceeding the adiabatic are sometimes recorded in precipitating clouds due to local concentrations of rainwater. Zaitsev made successive traverses of a cumulus congestus, taking three measurements during each traverse, from which he constructed the isopleths shown in Fig. 3.12. He reports that the distribution of liquid water is correlated with the distribution of vertical velocity and that, with strong updraughts, the maximum concentration of water is generally found higher up in the cloud.

Measurements of liquid-water concentrations made at various heights in fair-weather cumulus, cumulus congestus, and cumulonimbus by Weickmann and Aufm Kampe (1953) are shown in Fig. 3.4. These authors state that, in cumulus-congestus, the water content averaged over all the observations was $3 \cdot 9 \text{ g m}^{-3}$, but since the measurements were made mainly in the upper parts of the clouds, the average value for the whole clouds would probably be about 2 g m^{-3}. These average

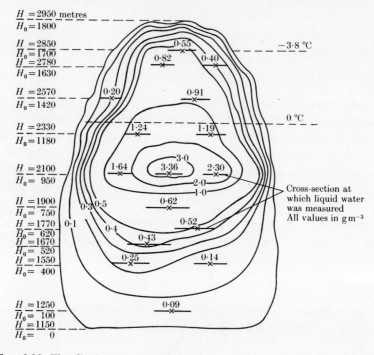

Fig. 3.12. The distribution of liquid-water in a cumulus congestus. (From Zaitsev (1950).)

values are not, however, of much significance; the physical properties of cumulus vary a good deal from cloud to cloud and it seems necessary at this stage to study individual clouds in detail.

That the water content and the mean droplet size first show a fairly steady increase with height above cloud base, followed by a sharp drop towards the cloud top, has been confirmed by Warner and Newnham (1952), Warner (1955), and Squires (1958c), in measurements on small cumulus usually less than 2 km deep. Using their continuously recording instrument, they encountered considerable fluctuations in the horizontal distribution of liquid water, the peak value measured at any one level being always considerably less than that calculated on the basis of adiabatic ascent from cloud base. The ratio of the peak to the adiabatic value was, on average, about 0·3 near cloud base, and decreased to about 0·1 at 5000–6000 ft. However, much higher values, occasionally approaching the adiabatic value, were sometimes found in localized regions of the clouds perhaps only 100 m across. Very similar results were obtained by Draginis (1958), who used the paper-tape instrument in the Caribbean and the central United States.

A similar vertical distribution of liquid water was found by Warner and Newnham in a layer of stratocumulus 8000 ft deep, where the maximum concentration was 0·3 g m^{-3}. A long series of measurements with rotating cylinders in stratocumulus layers has been summarized by Kline and Walker (1951) who quote values ranging from 0·06 to 1·30 g m^{-3}, with average values of 0·30 g m^{-3}. Frith (1951), from integration of droplet-size spectra, obtained values ranging from 0·07 to 0·83 g m^{-3}.

An extensive series of measurements of the liquid-water content of stratiform clouds has been made in Russia, the results being summarized by Borovikov et al. (1961). Measurements made in St, Sc, and Ac clouds at temperatures ranging from 0 to $-35°$ C showed that the average values of the liquid-water concentration ranged from 0·005 to 0·16 g m^{-3}, the higher values being usually associated with higher cloud temperatures. The temperature dependence of peak values was even more pronounced than for average values. The most frequent values of water content found in all types of stratiform cloud fell within the narrow range of 0·05–0·25 g m^{-3}. Low values were often found in deeply supercooled clouds, probably because the water was denuded by the growth of ice crystals. In deep nimbostratus clouds, the distribution of liquid water was very irregular due to presence of precipitation which, in any case, made the measurements very difficult.

Only a few measurements have been made in stratus. Neiburger (1949), working in California, found that in layers about 1600 ft thick, the vertical distribution of liquid water was very similar to that just described for stratocumulus, and gives maximum values of 0·67 g m^{-3}.

The water content of fogs was measured by Radford (1938) by drawing the air at a known rate through a series of wire screens and removing the central portion of the screen unit for weighing. He obtained values ranging from 0·1 to 0·22 g m^{-3}. With a balloon-borne apparatus, in which the droplets are caught on closely spaced vertical wires and then run down these to be recorded electrically as a series of large drops, Kuroiwa and Kinosita (1953) obtained records of the vertical distribution of liquid water in a fog. There was a general tendency for this to increase with height, values approaching 0·4 g m^{-3} at a height of 20 m being recorded in a dense advection fog with a strong inversion of temperature.

3.3. Theoretical studies of cloud droplet growth

We recognize two processes by which nucleus droplets may attain radii of several microns and so form a cloud or fog; the diffusion of water vapour to and its condensation upon their surfaces, and growth by the collision and coalescence of droplets moving relative to each other by virtue of Brownian motion, small-scale turbulence, electrical forces, and differential rates of fall under gravity. Both these processes will now be considered in some detail.

3.3.1. *Growth of a single droplet by condensation*

We imagine an isolated water drop of mass m, radius r, and density ρ_L to be stationary relative to an infinite atmosphere and growing very slowly by the diffusion of water vapour to its surface. If the temperature and vapour density of the remote environment remain constant, a steady-state diffusion field will be established round the droplet so that the mass of water vapour diffusing across any spherical surface of radius R centred on the droplet will be the independent of R and time. The flux of water vapour towards the drop will then be given by Fick's law of diffusion and is

$$F = \frac{dm}{dt} = 4\pi R^2 D \frac{d\rho}{dR} = \text{const.} = B, \text{ say}, \qquad (3.4)$$

where dm/dt is the rate-of-increase of droplet mass, $d\rho/dR$ is the radial gradient of vapour density, and D is the diffusion coefficient of water vapour in air.

Integrating (3.4) with respect to distance from the surface of the drop, where the vapour density is ρ_r, to infinity, where it is ρ, we have

$$4\pi D \int_\rho^{\rho_r} d\rho = \int_\infty^r \frac{B}{R^2} dR,$$

or
$$4\pi D(\rho_r - \rho) = -\frac{B}{r} = -\frac{1}{r}\frac{dm}{dt},$$

and therefore
$$\frac{dm}{dt} = 4\pi r D(\rho - \rho_r). \qquad (3.5)$$

Condensation of water vapour on the drop releases latent heat at a rate $L(dm/dt)$, where L is the latent heat of condensation, and this is dissipated mainly by conduction through the surrounding air. By analogy with (3.4), the equation for conduction of heat away from the droplet surface may be written

$$L\frac{dm}{dt} = -4\pi R^2 K \frac{dT}{dR}, \qquad (3.6)$$

§ 3.3 THE GROWTH OF DROPLETS IN CLOUD AND FOG

where the temperature gradient dT/dR is negative because the temperature decreases with increasing distance from the drop and K is the thermal conductivity of the air. Equation (3.6) may be integrated in a similar manner to (3.4) to give

$$L\frac{dm}{dt} = L4\pi r^2 \rho_L \frac{dr}{dt} = 4\pi Kr(T_r - T), \qquad (3.7)$$

where T_r is the surface temperature of the drop and T that of the distant environment.

Since the saturation vapour pressure and its temperature dependence are given by

$$p_s = \rho_s RT/M \quad \text{and} \quad 1/p_s \, dp_s/dT = LM/\mathbf{R}T^2,$$

$$\frac{d\rho_s}{\rho_s} = \frac{LM}{\mathbf{R}} \frac{dT}{T^2} - \frac{dT}{T}, \qquad (3.8)$$

M being the molecular weight of water and $\mathbf{R}$ the universal gas constant. Integrating (3.8) from the surface of the drop to infinity gives

$$\ln \frac{\rho_s(T_r)}{\rho_s(T)} = \frac{LM}{\mathbf{R}} \frac{(T_r - T)}{TT_r} - \ln \frac{T_r}{T} = \left(\frac{LM - \mathbf{R}T}{\mathbf{R}T^2}\right)(T_r - T) \qquad (3.9)$$

since $T_r \approx T$, and hence

$$\frac{\rho_s(T_r)}{\rho_s(T)} = 1 + \left(\frac{LM - \mathbf{R}T}{\mathbf{R}T^2}\right)(T_r - T) + \tfrac{1}{2}\left(\frac{LM - \mathbf{R}T}{\mathbf{R}T^2}\right)^2 (T_r - T)^2. \qquad (3.9a)$$

Because, in practice, $(T_r - T) \leqslant 1°$ C, the last term in (3.9a) may be neglected, and substitution from (3.7) gives

$$\frac{\rho_s(T_r) - \rho_s(T)}{\rho_s(T)} = \left(\frac{LM}{\mathbf{R}T} - 1\right)\left(\frac{T_r - T}{T}\right) = \frac{L}{4\pi KrT}\left(\frac{LM}{\mathbf{R}T} - 1\right)\frac{dm}{dt}. \qquad (3.9b)$$

Dividing (3.5) by $\rho_s(T)$ and adding to (3.9b), with the assumption that $\rho_r = \rho_s(T_r)$, yields

$$\frac{\rho - \rho_s(T)}{\rho_s(T)} = \left\{\frac{L}{4\pi KrT}\left(\frac{LM}{\mathbf{R}T} - 1\right) + \frac{1}{4\pi Dr\rho_s(T)}\right\}\frac{dm}{dt} \qquad (3.9c)$$

or

$$\frac{dm}{dt} = \frac{4\pi r\{\rho_s/\rho_s(T) - 1\}}{\frac{L}{KT}\left(\frac{LM}{\mathbf{R}T} - 1\right) + \frac{1}{D\rho_s(T)}}, \qquad (3.10)$$

or

$$r\frac{dr}{dt} = \frac{S - 1}{\left\{\frac{L\rho_L}{KT}\left(\frac{LM}{\mathbf{R}T} - 1\right) + \frac{\mathbf{R}T\rho_L}{DMp_s(T)}\right\}} \qquad (3.11)$$

$(S-1)$ being the supersaturation of the vapour.

Equation (3.11) describes the growth of a drop of pure water that is large enough for the curvature of the surface to have negligible influence upon the equilibrium vapour pressure. For a much smaller droplet growing on a salt nucleus of mass m and molecular weight W in a nearly saturated atmosphere, ρ_r in (3.5) must be replaced by

$$\rho'_r = \rho_s(T_r)\left(1 + \frac{2\sigma_{LV}M}{\rho_L R T_r} - \frac{imM}{\frac{4}{3}\pi r^3 \rho_L W}\right) = X\rho_s(T_r) \qquad (3.12)$$

so that

$$\frac{\rho - X\rho_s(T_r)}{\rho_s(T)} = \frac{1}{4\pi Dr}\frac{dm}{dt} = \frac{\rho_L}{D}r\frac{dr}{dt}. \qquad (3.13)$$

Adding (3.9b) and (3.13) we have

$$r\frac{dr}{dt}\left\{\frac{L\rho_L}{KT}\left(\frac{LM}{RT}-1\right) + \frac{\rho_L R T}{DMp_s(T)}\right\} = (S-1)-(X-1)\frac{\rho_s(T_r)}{\rho_s(T)},$$

and since $\rho_s(T_r)/\rho_s(T)$ is very close to unity for a drop growing at normal rates,

$$r\frac{dr}{dt} \simeq \frac{(S-1) - 2\sigma_{LV}M/\rho_L R T_r + imM/\frac{4}{3}\pi r^3 \rho_L W}{\left\{\frac{L\rho_L}{KT}\left(\frac{LM}{RT}-1\right) + \frac{\rho_L R T}{DMp_s(T)}\right\}} \qquad (3.14)$$

The growth of a droplet may be calculated from (3.14) if the properties, m and W, of the nucleus, the supersaturation, the temperature, and pressure of the air are specified. Some specimen results are given in Table 3.3. Very similar equations to those just derived have been

TABLE 3.3
Rate of growth of droplets by condensation on salt nuclei
(after Best 1951b)

Temperature $T = 273$ K		Pressure = 900 mb	
Nuclear mass m	10^{-14} g	10^{-13} g	10^{-12} g
Supersaturation = $100(S-1)\%$	0·05	0·05	0·05
Radius (μm)	Time (s) to grow from initial radius of 0·75 μm		
1	2·4	0·15	0·013
2	130	7·0	0·61
5	1 000	320	62
10	2 700	1 800	870
15	5 200	4 200	2 900
20	8 500	7 400	5 900
25	12 500	11 500	9 700
30	17 500	16 000	14 500
35	23 000	22 000	20 000
50	44 500	43 500	41 500

§ 3.3   THE GROWTH OF DROPLETS IN CLOUD AND FOG

verified experimentally by Keith and Arons (1954) from measurements on the growth of sea-salt particles of mass 10^{-8} to 10^{-9} g.

If the drop is falling through the air, its growth rate is increased because the diffusion and thermal fields do not extend to infinity but are confined to a boundary layer surrounding the drop. In consequence, the vapour and thermal gradients are enhanced, and the growth rate is increased by a factor that is a function of the Reynolds number $Re = 2\mu v r/\rho_a$ of the drop, the Schmidt number, $\mu/\rho_a D$, and the Prandtl number, $\mu c_p/K$, of the medium, μ, ρ_a, and c_p being respectively the dynamic viscosity, density, and specific heat of air. Measurements of the rates of evaporation of ventilated water drops by Frössling (1938), Kinzer and Gunn (1951), Ranz and Marshall (1952), and Dennis (1960), and measurements on ice spheres by Thorpe and Mason (1966), suggest that ventilation increases the rate of evaporation by a factor $(1+F\,Re^{\frac{1}{2}})$ in which F probably varies slowly with Re, but over the range $10 < Re < 100$ may, for all practical purposes, be taken as 0·23.

3.3.2. *Growth of a population of droplets by condensation in cumulus*

Having discussed the growth of a single droplet, we now consider the growth of a population of droplets leading to the formation of a cloud. The evolution of droplets growing on a population of condensation nuclei, while being lifted at an arbitrarily-assigned vertical speed in air that cools adiabatically *without* mixing with its surroundings, has been computed, in slightly different fashion, by several different workers, for example, Howell (1949), Squires (1952), Neiburger and Chien (1960), and Mordy (1959).

Having specified the initial temperature, pressure, humidity, and nucleus content of the air, the problem is defined by four differential equations expressing the rates-of-change of supersaturation, temperature, droplet radius (i.e. eqn (3.14)), and liquid-water content. The following treatment, taken from Mason and Ghosh (1957), is also very similar to that given by Squires (1952).

The supersaturation of an air mass can be expressed in terms of the difference between its dew point, T_d, and the actual temperature, T, of the air. If saturated air at temperature T and vapour pressure p_s is warmed to $T+dT$, then the incremental increase in vapour pressure, dp_s, required to keep it saturated is given by $dp_s/p_s = \epsilon L\,dT/\mathbf{R}_d T^2$, where L is the latent heat of condensation of water vapour, ϵ the specific gravity of water vapour relative to that of dry air, and

$$\mathbf{R}_d = 2\cdot 87 \times 10^6 \text{ erg g}^{-1}\,K^{-1}$$

is the gas constant for 1 gramme of dry air. If now the air is still kept saturated while being warmed to a slightly higher temperature, $T+\mathrm{d}T_\mathrm{d}$, and thereafter cooled to $T+\mathrm{d}T$ *without condensation occurring*, the vapour pressure will now be in excess of saturation by an amount $\mathrm{d}p$ such that the supersaturation $\mathrm{d}\sigma$ may be written

$$\mathrm{d}\sigma = \frac{\mathrm{d}p}{p_\mathrm{s}} = \frac{\epsilon L}{R_\mathrm{d} T^2}(\mathrm{d}T_\mathrm{d}-\mathrm{d}T), \qquad (3.15)$$

this being valid only for small changes in supersaturation.

Now the change in temperature of the air following a vertical displacement $\mathrm{d}z$ is given by

$$\mathrm{d}T = \left(\frac{-g\,\mathrm{d}z}{c_\mathrm{p}} - \frac{L\,\mathrm{d}x}{c_\mathrm{p}}\right), \qquad (3.16)$$

g/c_p being the dry adiabatic lapse rate, $\mathrm{d}x$ the change in the humidity mixing ratio and c_p the specific heat of air at constant pressure. Thus

$$\mathrm{d}\sigma = \frac{\epsilon L}{R_\mathrm{d} T^2}\left(\mathrm{d}T_\mathrm{d}+\frac{g}{c_\mathrm{p}}\mathrm{d}z+\frac{L\,\mathrm{d}x}{c_\mathrm{p}}\right). \qquad (3.17)$$

We may also use the following relations to represent changes occurring in slightly supersaturated air:

$$\mathrm{d}x+\mathrm{d}w = 0, \qquad (3.18)$$

where w is the liquid-water mixing ratio in grammes per gramme of dry air.

$$\mathrm{d}x = \epsilon\,\mathrm{d}\left(\frac{p_\mathrm{s}}{P}\right) = \frac{\epsilon}{P^2}(P\,\mathrm{d}p_\mathrm{s}-p_\mathrm{s}\,\mathrm{d}P), \qquad (3.19)$$

$$\frac{1}{p_\mathrm{s}}\frac{\mathrm{d}p_\mathrm{s}}{\mathrm{d}T} = \frac{\epsilon L}{R_\mathrm{d} T^2}, \qquad (3.20)$$

$$\frac{\mathrm{d}P}{\mathrm{d}z} = -\frac{gP}{R_\mathrm{d} T}, \qquad (3.21)$$

where P is the total pressure and g the acceleration due to gravity.

Substitution from (3.20) and (3.21) into (3.19) gives

$$\mathrm{d}x = -\mathrm{d}w = \frac{\epsilon}{P^2}\left(\frac{p_\mathrm{s} g P}{R_\mathrm{d} T}\mathrm{d}z + \frac{\epsilon P L p_\mathrm{s}}{R_\mathrm{d} T^2}\mathrm{d}T_\mathrm{d}\right), \qquad (3.22)$$

and finally, substitution for T_d from (3.22) into (3.17) yields

$$\mathrm{d}\sigma = \frac{\epsilon L}{R_\mathrm{d} T^2}\left\{g\left(\frac{1}{c_\mathrm{p}}-\frac{T}{L\epsilon}\right)\mathrm{d}z - \left(\frac{L}{c_\mathrm{p}}+\frac{PR_\mathrm{d} T^2}{L\epsilon^2 p_\mathrm{s}}\right)\mathrm{d}w\right\}. \qquad (3.23)$$

§ 3.3   THE GROWTH OF DROPLETS IN CLOUD AND FOG

If we consider the cloud droplets to be carried up in an updraught of velocity U, then (3.23) gives an expression for the time variation of the supersaturation, viz.

$$\frac{d\sigma}{dt} = AU - B\frac{dw}{dt}, \qquad (3.24)$$

where $A = \left(\frac{\epsilon Lg}{c_p R_d T^2} - \frac{g}{R_d T}\right) = \frac{3 \cdot 42 \times 10^{-4}}{T}\left(2 \cdot 6\frac{L}{T} - 1\right),$

L being measured in cal g^{-1}, and

$$B = \left(\frac{\epsilon L^2}{c_p R_d T^2} + \frac{P}{\epsilon p_s}\right) = \left(37 \cdot 8\frac{L^2}{T^2} + 1 \cdot 62\frac{P}{p_s}\right).$$

The time variation of temperature is given by substituting for dx from (3.22) into (3.16) to give

$$\frac{dT}{dt} = \frac{-gU}{c_p}\left(1 + \frac{L\epsilon p_s}{R_d PT}\right) \Big/ \left(1 + \frac{\epsilon^2 L^2 p_s}{c_p R_d PT^2}\right). \qquad (3.25)$$

Eqn (3.24) expresses the fact that the supersaturation is determined by the rate at which water vapour is released for condensation by lifting and cooling of the air minus the rate at which it is condensed on to the droplets.

The rate of growth of a droplet upon a nucleus of specified nature and mass is given by eqn. (3.14) and one such equation will be required for each separate class of condensation nuclei. Finally, the liquid-water mixing ratio is given by

$$\frac{dw}{dt} = \frac{4}{3}\pi \frac{\rho_L}{\rho_a}\frac{d}{dt}\sum nr^3, \qquad (3.26)$$

where n is the number of droplets per cm^3 of radius r and ρ_a is the air density. If the initial conditions, say, at cloud base are specified, eqns (3.14), (3.24), (3.25), and (3.26) may be integrated numerically, step-by-step, to give the drop-size, supersaturation, temperature, and liquid-water content in the cloud as functions of time and the corresponding heights above cloud base. Such a calculation was first made by Howell (1949), who simplified the problem by assuming the ambient temperature and pressure to remain constant during the cloud-forming process and omitted the Van't Hoff factor, i, contained in (3.14). His computations show that a group of droplets originating from nuclei of uniform size undergoes four distinct phases of growth as shown in Fig. 3.13. In the first phase, before the supersaturation reaches the value critical for activation of the nucleus, the droplets grow slowly and do

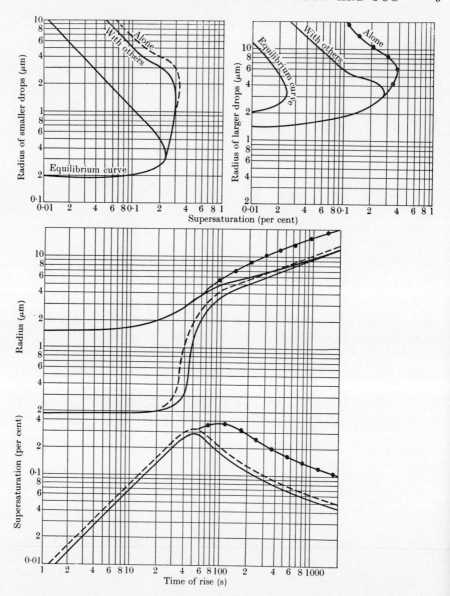

FIG. 3.13. The growth is steadily rising air of: – – – – – – a homogeneous group of nuclei containing 10^{-17} mol of NaCl; – · – · – · – · – a homogeneous group containing 10^{-15} mol; ——— both groups acting simultaneously. (From Howell (1949).)

§ 3.3   THE GROWTH OF DROPLETS IN CLOUD AND FOG

not lag far behind their equilibrium size except in the case of large nuclei with high rates of ascent. The second phase begins with rapid growth when the critical supersaturation is surpassed, and is terminated when increased condensation on the rapidly expanding liquid surfaces prevents further rise in the supersaturation. In the third phase, the rate of growth approaches equilibrium with the rate of supply of vapour; the supersaturation adjusts itself at a value just sufficient to cause the vapour to condense as fast as it is supplied above equilibrium pressure. The final phase is that of droplet growth after all transient effects of activation have become negligible.

Turning to more realistic conditions, Howell investigates the subsequent growth of a reasonable concentration and size distribution of salt nuclei under conditions of moderate, slow, and very slow, steady rates of ascent. His initial conditions and a summary of his results are shown in Table 3.4, the life history of the droplets and the course of the supersaturation being plotted in Figs. 3.14–3.16.

We see that higher rates of cooling produce larger peak supersaturations and cause a larger proportion of the smaller nuclei to be activated. Condensation sets in first on the larger nuclei; if they are present in only small numbers, and so are unable to assimilate the vapour as fast as it is released by the cooling air, the supersaturation rises and allows smaller nuclei to be effective. Moreover, the great majority of the nuclei that are activated grow to the critical size and become droplets; the maximum concentration of drops coincides with the peak supersaturation. Thereafter, the supersaturation falls because vapour diffuses to the many condensation centres more quickly than it can be replaced by cooling of the air, and now the smallest droplets start to evaporate. These droplets are generally less than 1 μm radius, however, and so contribute very little to the growth of the larger drops. The calculations also show that, in clouds containing a few hundred nuclei per cm^3, the supersaturation may attain a peak value of only about 0·1 per cent, and will surpass 1 per cent only under extreme circumstances. The most uniform droplet spectrum was produced in the case of relatively few nuclei growing in rapidly expanding air, but even with slow rates of cooling associated with activation of a large range of nuclear sizes, the final degree of heterogeneity in the drop-size spectrum was only small. Very similar calculations have been made by Mordy (1959) and Neiburger and Chien (1960) using large electronic computers. Mordy repeated two of Howell's computations and obtained very good agreement—a tribute to the accuracy of Howell's very laborious hand

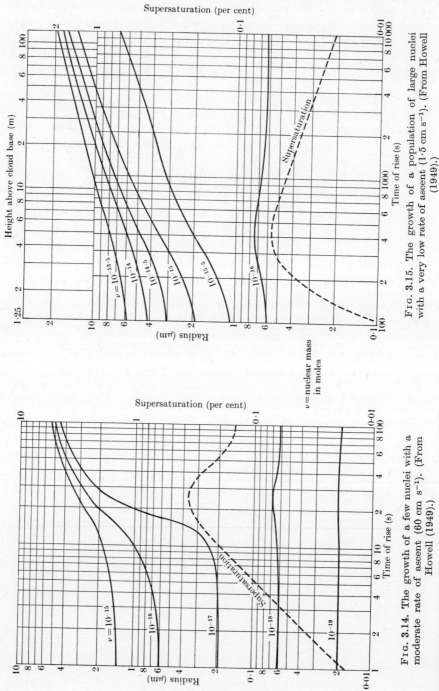

FIG. 3.15. The growth of a population of large nuclei with a very low rate of ascent (1·5 cm s⁻¹). (From Howell (1949).)

FIG. 3.14. The growth of a few nuclei with a moderate rate of ascent (60 cm s⁻¹). (From Howell (1949).)

ν = nuclear mass in moles

§ 3.3   THE GROWTH OF DROPLETS IN CLOUD AND FOG   131

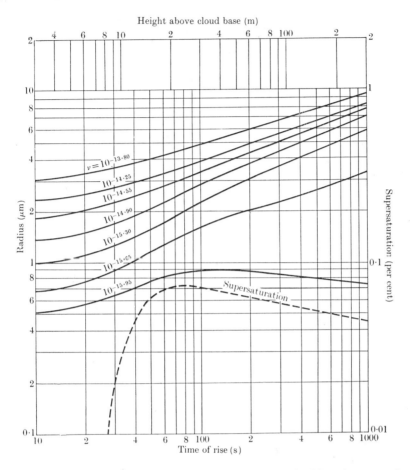

FIG. 3.16. The growth of a population of average nuclei with a slow rate of ascent (30 cm s^{-1}). (From Howell (1949).)

computations. Mordy's other computations were carried out for constant updraughts of 5, 15, 50, and 100 cm s^{-1}, and with broader nucleus spectra that included some giant hygroscopic nuclei. He investigated the influence of the ventilation factor $(1+F\ Re^{\frac{1}{2}})$ on the growth rate of the droplets and found it to be negligible for droplets of $r < 30\ \mu$m. Allowance was also made for the settling of the droplets under Stokes's law. In all cases, droplets formed on the smaller nuclei achieved nearly equal radii at distances of more than a few tens of metres above cloud base. Only the droplets formed on the very large nuclei continued to exhibit their initial differences of size for much longer growth periods.

TABLE 3.4
The growth of non-uniform groups of nuclei under various atmospheric conditions

Rate of ascent (cm s^{-1})	60	30	1·5
Total nucleus concentration (per cm^3)	500	2000	667
Range of nuclear masses (moles)	10^{-19} to 10^{-15}	10^{-16} to $10^{-13.8}$	10^{-16} to $10^{-13.5}$
Peak supersaturation (%)	0·36	0·072	0·056
Time to attain peak supersaturation (s)	25	70	
Mass range of activated nuclei (moles)	10^{-17} to 10^{-15}	$10^{-15.65}$ to $10^{-13.8}$	$10^{-15.5}$ to $10^{-13.5}$
Concentration of activated nuclei (per cm^3)	250	410	40
Radius at peak supersaturation (μm)	1·5 to 3	1·5 to 4·5	2 to 8
Time taken to grow to $r = 5$ μm (s)	150–100	2200–100	2450–30
Modulus of size distribution (m)† at 100 m above cloud base	0·06	0·33	0·23
Median-volume radius at 100 m above cloud base	~3 μm	~5·5 μm	~14 μm

† $m = \sqrt{(2\sigma/r_\text{n})}$ where r_n = median-volume radius and σ = standard deviation of drop volumes. A small value of m denotes a more nearly uniform spectrum.

These features were a consequence of the fact that the peak supersaturation was achieved only a few tens of metres above cloud base and that the main characteristics of the drop-size spectra were determined very early in the condensation process. The total concentration of cloud droplets is determined by the spectrum of condensation nuclei and the vertical velocity of the air, stronger updraughts tending to produce higher supersaturations and hence higher concentrations of droplets.

Neiburger and Chien (1960) allowed the updraught velocity to vary with height in the cloud but the general pattern of their results is very similar to those of Howell and Mordy. All these computations, in which a closed parcel of air containing a specified population of condensation nuclei is assumed to rise at an arbitrarily assigned vertical velocity and to cool adiabatically, agree in showing that the peak supersaturation is reached within a few seconds of the air becoming saturated and that, thereafter, the droplet spectrum narrows rather rapidly. Even if the initial nuclei show a wide dispersion, the resulting droplet-size spectra are much narrower than are found in natural clouds, and the model cannot be made to produce realistic spectra by varying the updraught speed or its variation with height. Only the few very large nuclei, that are unable to attain the equilibrium size appropriate to the prevailing supersaturation, are able to retain much of their initial size differential. It seems highly unlikely that the concentration and size distribution of the droplets are, in fact, largely determined during the first few tens of metres of ascent above cloud base, and that during the remainder of the ascent the droplets grow ever more slowly with the smaller ones catching up on the larger to narrow the spectrum.

The most unrealistic features of the closed parcel model are: the assumption that the cloudy air does not mix and become diluted with the surrounding drier air—the reality of the mixing is indicated by the measured liquid-water content of non-precipitating clouds being usually much less than the adiabatic value; the assumption that *all* droplets are retained and have equal lifetimes in the cloud; the assignment of an arbitrary vertical velocity since, as Mason and Emig (1961) have shown, a closed saturated parcel would undergo a strongly oscillatory vertical motion. The last-named authors developed equations for the time variations of vertical velocity, temperature, humidity-mixing ratio and liquid-water content of a saturated parcel as it ascends from the cloud base, rises through, and mixes with, its surroundings. This simplified treatment, which assumes that the rising cloudy thermal possesses horizontal homogeneity, arrives at rather critical values for a single parameter used to represent the rate of mixing. Mason and Chien (1962) have applied this very simple dynamical model to compute the growth of a population of cloud droplets in which, in addition to exchanging heat, momentum, and water vapour, the mixing processs is assumed to transfer droplets from the cloud to the drier environmental air where they are assumed to evaporate and be replaced by fresh condensation nuclei of the same mass. Some droplets now spend longer in the cloud than others and this obviously will lead to a broadening of the droplet-size spectrum.

The equation of vertical motion of the parcel equates its rate-of-change of momentum to the algebraic sum of the buoyancy and 'drag' forces. The buoyancy force per unit mass is $\{(T-T')/T' - w\}g$, where $(T-T')$ is the excess virtual temperature of the thermal over that, T', of its environment at the same level, and w is the liquid-water mixing ratio. Mason and Emig assume that, in addition to a net transfer of mass between the environment and the thermal due to entrainment, there is also a turbulent interchange of both heat and momentum which gives rise to a drag force in the manner described by Priestley (1953, 1954). Ignoring vertical motions in the environment, the equation of motion of the parcel reads

$$\frac{d}{dt}(MU) = M\left(\frac{T-T'}{T'}-\omega\right)g - MkU$$

or
$$\frac{dU}{dt} = \left(\frac{T-T'}{T'}-\omega\right)g - \left(k+\frac{1}{M}\frac{dM}{dt}\right)U, \qquad (3.27)$$

where U is the vertical velocity of the cloudy parcel, M is its mass, and k is Priestley's exchange coefficient. In order to simplify the calculations, we may write

$$\left(k+\frac{1}{M}\frac{dM}{dt}\right) = k' = \text{const.}$$

and assume that this damping factor is the same for the transfer of heat, momentum, and water.

The rate-of-change of temperature may now be written:

$$\frac{dT}{dt} = \frac{\dfrac{-gU}{c_p}\left\{1+\dfrac{L}{R_d T}\dfrac{\epsilon p_s}{(P-p_s)}\right\} - k'(T-T') - \dfrac{k'L}{c_p}\left\{(1+\sigma)\dfrac{\epsilon p_s}{P-p_s} - \dfrac{\epsilon p_s' S'}{P-p_s'}\right\} - \dfrac{k' c_w}{c_p}(\omega T - \omega' T')}{\left\{1+\dfrac{\epsilon L^2}{R_d c_p T^2}\left(\dfrac{\epsilon p_s}{P-p_s}\right)\right\}},$$

$$(3.28)$$

where S', p', and ω' are respectively the saturation ratio, saturation vapour pressure, and liquid-water mixing ratio of the environment, and c_w is the specific heat of liquid water.

The rate-of-change of supersaturation is

$$\frac{d\sigma}{dt} = -\left(\frac{P-p_s}{\epsilon p_s}\right)\left[\frac{d\omega}{dt}+k'(\omega-\omega')+k'\left\{(1+\sigma)\frac{\epsilon p_s}{P-p_s}-\frac{S'\epsilon p_s'}{P-p_s'}\right\}\right] -$$
$$-(1+\sigma)\left(\frac{\epsilon L}{R_d T^2}\frac{dT}{dt}+\frac{gU}{R_d T}\right)$$

$$(3.29)$$

and
$$\frac{d\omega}{dt} = \frac{4\pi}{3}\left\{\frac{d}{dt}\sum_r n_r r^3 - k'\left(\sum_r n_r r^3 - \sum_{r'} n_{r'}' r'^3\right)\right\}, \qquad (3.30)$$

where n_r is the number of droplets per gramme of air of radius r in the cloud and n_r' is the corresponding concentration of nucleus droplets of radius r' in the environment. Equations (3.27)–(3.30), together with

Table 3.5
Initial conditions of the computations represented in Figs. 3.17–3.19

Temperature and pressure at condensation level: 10° C, 900 mb						
Initial velocity of parcel $U_0 = 50$ cm s^{-1}						
Initial excess temperature $(T-T') = 0.7°$ C						
Lapse rate of environment $\alpha = 7°$ C km^{-1}						
Humidity of environment $H' = 90$ per cent						
Mixing parameter $k' = 4 \times 10^{-3}$ s^{-1}						
Mass of nucleus (g of NaCl)	10^{-15}	10^{-14}	10^{-13}	10^{-12}	10^{-11}	10^{-10}
Concentration of nuclei (g^{-1})	3×10^5	3×10^4	3×10^3	3×10^2	30	3
Initial droplet radius (μm)	0·34	1·0	2·2	3·7	6·0	10·0

(3.14) for the growth rate of a droplet upon a specified nucleus, describe the problem. Starting from the initial conditions specified in Table 3.5, these equations were integrated numerically in small time steps on a computer, to give values of U, T, $(T-T')$, σ, w, and r for each nucleus group, p, and the height, z, of the parcel. The initial values were chosen to give, according to the calculations of Mason and Emig, a small cumulus only about 1200 m (4000 ft) deep, with a maximum updraught of 3 m s^{-1}, and a maximum liquid-water content of 0·8 g/kg. Such a cloud is fairly representative of small non-precipitating cumulus in which drop sizes have been measured and are available for comparison.

The variation with time of vertical velocity, U, liquid-water mixing ratio, w, and supersaturation, σ, are plotted in Fig. 3.18, and their

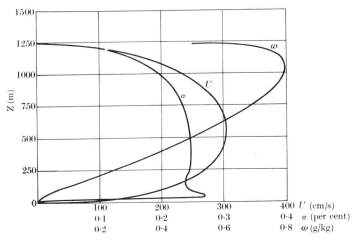

Fig. 3.17. The variation with height above cloud base of vertical velocity, U, liquid-water content, w, and supersaturation σ. (From Mason and Chien (1962).)

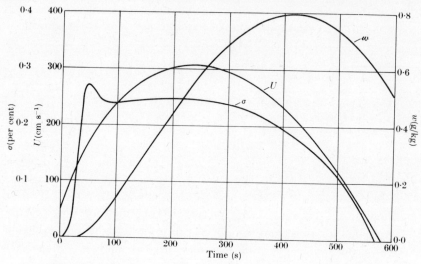

Fig. 3.18. The variation with time of the vertical velocity, U, liquid-water content, w, and supersaturation σ. (From Mason and Chien (1962).)

variation with height above cloud base in Fig. 3.17. The vertical velocity of the parcel reaches a maximum value of 306 cm s^{-1} (240 s and 556 m after it passes the condensation level), and reaches zero again after 580 s at a height of 1233 m. The liquid-water content reaches its maximum value of 0·8 g/kg (about 0·4 of the adiabatic value) after 420 s at 1050 m above cloud base. The supersaturation rises steeply at first and attains an artificial peak value after only 45 s, which arises because the assumed initial sizes of the more numerous nuclei were slightly too large. But this situation is soon adjusted and, after 90 s, the supersaturation rises slowly to another peak value of 0·25 per cent after 220 s (495 m), and thereafter continues to fall.

The cumulative droplet-size distributions attained at the instant when the vertical motion ceases are plotted for each nucleus-size group in Fig. 3.19. During the growth period of 580 s, the droplet spectrum, that extended initially from 0·34 to 10 μm radius, has continually spread and, finally, extends from 0·34 to 22·5 μm. But those droplets that have remained in the cloud for the whole 580 s range only from 12·1 to 22.5 μm. Salt nuclei of mass 10^{-9} g growing in these conditions would attain a radius of 33 μm in this period of nearly 10 min. As we shall see, once droplets of this size are formed, the spectrum becomes deformed rather rapidly by the coalescence of droplets, but up to this stage, we may compare these results with those obtained with the closed parcel

§ 3.3   THE GROWTH OF DROPLETS IN CLOUD AND FOG   137

model. For example, Mordy (1959) finds that, in a closed parcel rising at a constant velocity of 1 m s^{-1} and containing 450 salt nuclei per cm^3 ranging from 10^{-17} to $10^{-8 \cdot 5}$ g, a peak supersaturation of 1 per cent is reached after only 40 s; at this stage, droplets growing on nuclei of 10^{-15} g have increased in radius by 2 μm, and those on nuclei of 10^{-10} g, by only 1 μm. Thereafter, the spectrum begins to narrow as the supersaturation falls and the largest nuclei grow only very slowly. The

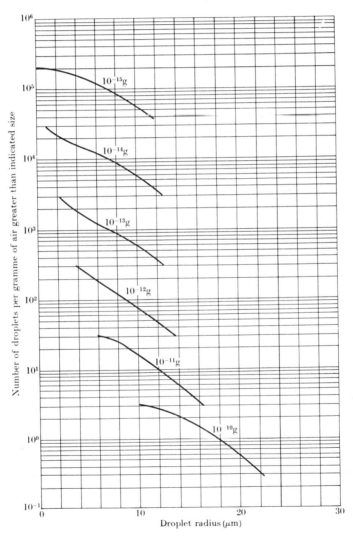

FIG. 3.19. The cumulative droplet-size distributions for each nucleus-size group reached after 580 s. (From Mason and Chien (1962).)

results of Mason and Chien seem much more realistic in that the peak supersaturation of 0·25 per cent is reached after 220 s and nearly half-way up the cloud; droplets growing on those nuclei that have remained in the cloud during this period have increased in radius by 8 μm, while those forming on nuclei of 10^{-10} g have grown by 8·7 μm. Thereafter the spectrum continues to broaden as shown in Fig. 3.19. Moreover, the magnitude and variation with height of the liquid-water content and the mean-volume droplet radius are rather similar to those reported, for example, by Durbin (1959) for clouds of this depth over southern England.

When the vertical motion ceases and no more water is released for condensation, the vapour pressure tends to come into equilibrium with the droplets of average size, the smaller ones tending to evaporate and the larger ones to grow, the average size increasing progressively with time. This aspect of the problem was treated by Langmuir (1944) and, in rather different fashion, by Elton, Mason, and Picknett (1958) as follows. The growth rate of a cloud droplet that is sufficiently large for the solute term to be ignored may be written (see eqn (3.14)) as

$$\frac{dm}{dt} = \frac{4\pi r}{Y}\left(S - 1 - \frac{G}{r}\right), \tag{3.31}$$

where $\quad Y = \left\{\frac{L\rho_L}{KT}\left(\frac{LM}{RT}-1\right) + \frac{\rho_L RT}{DMp_s(T)}\right\}, \quad G = \frac{2\sigma_{LV}M}{\rho_L RT},$

or

$$(S-1)r = \frac{Y}{4\pi}\frac{dm}{dt} + G. \tag{3.31a}$$

Summing over all the n droplets in unit volume of air,

$$(S-1)\sum_{1}^{n} r = \frac{Y}{4\pi}\frac{dw}{dt} + nG \tag{3.32}$$

or

$$(S-1) = \frac{Y}{4\pi\sum_{1}^{n} r}\cdot\frac{dw}{dt} + \frac{G}{\bar{r}}$$

where w is the liquid-water concentration in g cm^{-3} and $\bar{r}$ is the mean droplet radius.

On substituting for $(S-1)$ from (3.32) into (3.31) we have

$$\frac{dm}{dt} = \frac{4\pi}{Y}\left\{\frac{Y}{4\pi}\frac{r}{\sum r}\frac{dw}{dt} + G\left(\frac{r}{\bar{r}}-1\right)\right\}, \tag{3.33}$$

but since the updraught has ceased, $dw/dt = 0$, and

$$r^2 \frac{dr}{dt} = \frac{G}{Y\rho_L}\left(\frac{r}{\bar{r}}-1\right), \tag{3.34}$$

showing that all droplets of $r < \bar{r}$ will evaporate, and all droplets of $r > \bar{r}$ will grow. The mean radius $\bar{r}$ will, of course, vary as the evaporation–condensation proceeds but, if it changes very slowly, we may integrate (3.24) assuming $\bar{r}$ to remain constant and obtain the following expression for the time, t, required for droplets to grow from radius r to r':

$$\bar{r}^2 \ln\frac{r'-\bar{r}}{r-\bar{r}}+(r'-r)\{\bar{r}+\tfrac{1}{2}(r'+r)\} = \frac{Gt}{Y\rho_L}. \tag{3.35}$$

In a population having a mean droplet radius of $\bar{r} = 5$ μm, a droplet of $r = 10$ μm would grow to only 12 μm, and a droplet of $r = 20$ μm would grow to only 21·5 μm, in half an hour, even in a part of the cloud not diluted by mixing with the environment. We must conclude, then, that by the time the updraught has ceased, the majority of the droplets will usually have attained a size at which the vapour pressure differential over drops of different sizes is very small, and that this evaporation–condensation process will deform the spectrum only very slowly.

3.3.3. *The formation of large droplets in small cumulus*

We have traced the growth by condensation of cloud droplets up to the stage at which they attain radii of about 30 μm. While the main characteristics of cloud-droplet populations, for example mean droplet concentrations and sizes, may be accounted for in terms of condensation on typical observed spectra of condensation nuclei, the evolution of droplets up to a radius of 25 μm, beyond which they may continue to grow by coalescence and become incipient raindrops, is more difficult to explain.

In maritime cumulus, the appearance of droplets of $r = 25$–30 μm may be reasonably attributed to condensation on giant sea-salt nuclei of $m > 10^{-9}$ g, the time required being 5–10 min. However, in continental clouds, the growth of large droplets is likely to be impeded both by the absence of giant sea-salt nuclei and the presence of higher total concentrations of nuclei. It is particularly difficult to account for such observations as those of Durbin (1959) that droplets of $r > 30$ μm sometimes occur in small cumulus in concentrations as high as 1 cm^{-3} but, far from being typical, these observations may have been vitiated by splashing of large drops on the collectors, and one notes that

MacCready and Takeuchi (1968) found droplets of $r = 15-30$ μm in concentrations of only 20/litre in the unmixed cores of continental cumulus that contained concentrations of liquid water of 1 g m^{-3}. Even so, the release of showers from warm, continental clouds implies that droplets in concentrations of 10^2-10^3 m^{-3} grow to radii of 25 μm within 10–20 min of the air entering the cloud base, and this would appear to require a land source of hygroscopic or mixed nuclei comparable in efficiency to that provided by sea-salt particles over the oceans. A source of such nuclei has not yet been established, and much more careful observations, not only of droplet and nucleus populations, but also of cloud structure and evolution are required to determine the role that condensation on giant nuclei plays in extending droplet-spectra to sizes at which the coalescence process may take over and continue development towards the precipitation stage.

3.3.4. *Droplet growth by condensation in layer clouds*

In thin layer clouds it is difficult to specify the vertical motion and to calculate the progress of the supersaturation; on the other hand, the cloud may last much longer than a small cumulus. Best (1951b, 1952a) assumed the supersaturation to be uniform, and the lifetimes (and hence the sizes) of droplets to be governed by their turbulent transfer between the interior of the cloud and the drier air outside. Having shown the mean life of cloud droplets, as governed by turbulent diffusion, to be consistent with their observed mean size assuming the supersaturation to remain constant at 0·05 per cent, Best claims to have accounted for the observed distributions of cloud-droplet size by assuming that any increase in the concentration of droplets of a given radius, at a given level, is balanced by an equal loss by turbulent diffusion to the upper and lower boundaries of the cloud. His formulation of the problem in physical terms is not clear, but Best obtains a mathematical expression which indicates that the mean drop size should be directly proportional to the thickness of the cloud, to the square root of the supersaturation, and inversely proportional to the square root of the eddy coefficient of diffusion.

In a rather different approach, Mason (1952a, 1960a) treated the turbulent motion of a droplet inside the cloud as a random walk, the upper and lower cloud boundaries acting as absorbing barriers. Thus, any droplet carried to the boundary is assumed to evaporate and therefore lost to the cloud, but is replaced by a fresh nucleus of the same mass as that contained in the evaporating droplet. There is, therefore, a net loss of condensed water at the boundaries, but the concentration of

§3.3 THE GROWTH OF DROPLETS IN CLOUD AND FOG

condensation nuclei, and hence the total *number* of droplets in the cloud, remains constant.

If the eddies transporting the droplets are assumed to be small compared with the depth of the cloud, Mason shows that the probability that a droplet will remain in the cloud after a time t is

$$P(t) = \operatorname{erf} x + \sum (-1)^n \{\operatorname{erf}(2n+1)x - \operatorname{erf}(2n-1)x\}, \quad (3.36)$$

where
$$x = z_0/(2Nl^2 t)^{\frac{1}{2}}, \quad (3.37)$$

N being the number of random displacements of magnitude l suffered by a droplet in time t, and z_0 is the half-thickness of the cloud. If the half-life of the droplets in the cloud is λ,

$$P(\lambda) = 0{\cdot}5 \quad \text{and} \quad x = 0{\cdot}81.$$

The fraction of droplets that remain in the cloud for any other time, $m\lambda$, may then be calculated from (3.36) by putting $x = 0{\cdot}81/m^{\frac{1}{2}}$. These values of $P(m\lambda)$ enable one to calculate the fraction of droplets (the same fraction for all sizes) removed from the cloud in a given time interval and replaced by an equal number of inactivated nuclei. Mason took the half-life, λ, to be 1000 s on the assumption that a stratus layer, about 150 m deep, was forming at about 500 m above the ground where the horizontal wind speed was 5 m s^{-1}. Having specified the size distributions of the condensation nuclei, viz. 300 cm^{-3} of $m > 10^{-14}$ g with 3 cm^{-3} of $m > 10^{-12}$ g, and the supersaturation to remain constant at 0·05 per cent, eqn (3.14) was used to compute the growth of the drops in each time interval. The cumulative size distributions of the droplets after 6000 s is plotted in Fig. 3.20, when the median droplet radius is 4·8 μm, the mean-volume radius 7 μm, and the liquid-water content 0·4 g m^{-3}. The spectrum of droplet sizes is quite broad, extending to 20 μm in radius, these droplets arising on nuclei of only 10^{-12} g. The general shape of the spectrum was largely decided after a time 2λ or 2000 s, and changed only slowly after 3000 s. This computed spectrum is in good agreement with that actually measured by Neiburger in the middle of a layer of California stratus, 700 ft deep, when the total droplet concentration was 286 cm^{-3}. Of course the computation involves rather arbitrary assumptions concerning the nucleus distribution, supersaturation, and half-life of the droplets, but changes in these parameters result mainly in a shift in the spectrum to larger (or smaller) droplet sizes rather than in a radical change of shape. It seems to illustrate that the evolution of a broad spectrum, just as in the cumulus case, is largely determined by the turbulent diffusion of droplets between the cloud interior and its boundaries, or between regions of active condensation and regions of evaporation within the cloud.

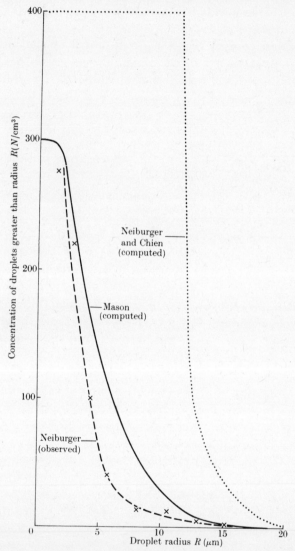

Fig. 3.20. Computed and observed cumulative size distributions of droplets in stratus cloud. (From Mason (1960a).)

3.3.5. *The effect of random fluctuations in supersaturation on the broadening of the droplet spectrum*

So far we have considered only situations in which the updraught velocity, temperature, and supersaturation vary continuously with height and time but, in reality, turbulent motions within the cloud will

§ 3.3   THE GROWTH OF DROPLETS IN CLOUD AND FOG

cause these quantities to fluctuate about mean values in a random manner. In effect, turbulent fluctuations in the vertical motion will produce corresponding fluctuations in supersaturation and these will cause additional dispersion of droplet sizes. Thus a droplet that experiences a series of positive fluctuations in updraught velocity will grow more quickly, and arrive at a given level with a particular size earlier, than an initially identical droplet that follows the mean steady updraught. The opposite will be true of a droplet that experiences a series of negative fluctuations, with the result that some droplets will grow faster, and some slower, than average.

This problem has been treated in formal mathematical terms in a number of papers by Russian workers, for example Belyaev (1961), Mazin (1965), Sedunov (1965), and Levin and Sedunov (1966, 1968), but they have not obtained quantitative solutions to the equations to calculate the evolution of droplet spectra. One can, however, demonstrate the magnitude of the dispersion of droplet size likely to be produced by the following simple calculation.

The growth rate of cloud droplets, large enough for the effects of the nucleus and curvature to be ignored, is given by eqn (3.11) as

$$r\frac{dr}{dt} = \zeta \bigg/ \left\{\frac{L\rho_L}{KT}\left(\frac{LM}{RT}-1\right) + \frac{RT\rho_L}{DMp_s(T)}\right\} = \frac{\zeta}{Y(T)}, \quad (3.38)$$

which, on integration, yields

$$\zeta t = \tfrac{1}{2}Y(T)(r^2 - r_0^2), \quad (3.39)$$

where ζ is the supersaturation and r_0 the initial radius. We now consider ζ to be a fluctuating quantity and define a random function

$$\psi(t) = \int^t \zeta(t')\,dt', \quad (3.40)$$

which describes the fluctuating values of ζ along the path of a drop. If ζ and hence ψ conform to a Gaussian distribution about a mean value of zero, it may be shown that for times much greater than the characteristic period τ_0 of the turbulence, the variances of the two related quantities after a time t are given by

$$\sigma_\psi^2 = 2\sigma_\zeta^2 \tau_0 t. \quad (3.41)$$

Fluctuations in supersaturation cannot be measured directly, but they can be related to fluctuations in vertical velocity as follows.

The time derivative of ζ in a steady updraught U is given by eqns

(3.24) and (3.38):

$$\frac{d\zeta}{dt} = AU - B\frac{d\omega}{dt} = AU - \frac{4\pi B \zeta N \bar{r}}{Y(T)}, \qquad (3.42)$$

where A and B are functions only of temperature and pressure, N is the number of droplets per gramme of air, and $\bar{r}$ is a 'mean' droplet radius. During a short-period fluctuation in updraught, ΔU, the mean supersaturation may be considered stationary, i.e. $d\bar{\zeta}/dt = 0$, so the corresponding fluctuation in supersaturation will be

$$\Delta \zeta = \frac{Y(T)}{4\pi} \frac{A}{B} \frac{1}{N\bar{r}} \Delta U. \qquad (3.43)$$

The standard deviation of the squares of droplet radii is

$$\sigma_r^2 = \frac{1}{N} \sum_{}^{n} (r^2 - \bar{r}^2) = \frac{2\sigma_\psi}{Y(T)} = \frac{2\sqrt{2}\,\sigma_\zeta(\tau_0 t)^{\frac{1}{2}}}{Y(T)} = \frac{1}{\sqrt{2}\,\pi} \frac{A}{B} \frac{1}{N\bar{r}} \sigma_u(\tau_0 t)^{\frac{1}{2}}$$

or $\qquad \sigma_r^2 = 4 \cdot 25 \times 10^{-9} \dfrac{1}{N\bar{r}} \sigma_u(\tau_0 t)^{\frac{1}{2}},$ (at 0°C, 900 mb). $\qquad (3.44)$

Taking the following values as representative of a swelling cumulus cloud; $N = 10^5\,\mathrm{g^{-1}}$, $\bar{r} = 10\,\mu\mathrm{m}$, $\sigma_u = 50\,\mathrm{cm\,s^{-1}}$, and $\tau_0 = 25\,\mathrm{s}$ (corresponding to a dissipation rate $\epsilon = \sigma_u^2/\tau_0 = 10^2\,\mathrm{cm^2\,s^{-3}}$), we have $\sigma_r^2 \simeq t^{\frac{1}{2}}$ (in $\mu\mathrm{m}^2$), i.e. after $t = 900\,\mathrm{s}$, $\sigma_r^2 = 30\,(\mu\mathrm{m})^2$, so that half of the droplets will have radii between 9 and 11 $\mu\mathrm{m}$. After 1 hr, half of the droplets will have radii ranging from 8 to 12 $\mu\mathrm{m}$.

Bartlett (1968) has used eqns (3.11) and (3.23)–(3.26), to follow the growth of droplets in a turbulent updraught in which the velocity fluctuations of an air parcel in time τ are simulated numerically by

$$\Delta v_r = n(\epsilon\tau)^{\frac{1}{2}}, \qquad (3.45)$$

where n is a random number selected from a normally-distributed population with a mean of zero and unit standard deviation, and the concentration of drops per unit mass of air is assumed to remain constant. He finds that although, because of the turbulent fluctuations, different drops take different times to reach a given height above the level of origin, they arrive there with almost the same size (very nearly the size that they would acquire by ascending along the saturated adiabatic in a steady updraught to that height). The reason for this is that droplets experiencing lower supersaturations are those that are in more slowly rising parcels of air and, therefore, have longer to grow,

§ 3.3   THE GROWTH OF DROPLETS IN CLOUD AND FOG

while those that grow more rapidly in the higher supersaturations associated with more rapidly rising air, have less time. More recent calculations suggest that in the case of a cloud of droplets that are initially monodispersed and in which there is no mixing with the surroundings, the standard deviation of the drop radii is less than about 0·05 μm when the mean droplet radius is about 15 μm, for any reasonable value of the turbulent energy.

Although the enhanced spread in the sizes of cloud droplets due to turbulence may be greater when a polydispersed initial spectrum is considered instead of the monodispersed case examined by Bartlett, it appears unlikely that internal turbulence can cause deviations greater than about 1 μm from the sizes predicted by the theory of condensation in a steady updraught.

3.3.6. Droplet growth by collision and coalescence

According to Hocking's (1959) calculations on collision efficiencies, cloud droplets must attain a radius of at least 19 μm by condensation before they are able to grow by collision with smaller droplets, but thereafter they may continue to grow at an ever-increasing rate by overtaking, colliding, and coalescing with smaller droplets. In the early treatments of this problem the small droplets, with constant size and number, were visualized as filling space with a uniform density of liquid water which the large drops sweep up continuously. On this *continuous model* all the large drops grow at the same rate given by

$$4\pi R^2 \frac{dR}{dt} = \pi R^2 \int_0^R \tfrac{4}{3}\pi n(r) r^3 E'(R, r)\{v(R) - v(r)\} \, dr \qquad (3.46)$$

or

$$\frac{dR}{dt} = \frac{\pi}{3} \int_0^R n(r) r^3 E'(R, r)\{v(R) - v(r)\} \, dr, \qquad (3.47)$$

where $n(r) \, dr$ is the number of droplets per unit volume of air with radii between r and $r+dr$, $E'(R, r)$† is the collection cross-section for two droplets of radii R and r and terminal velocities $v(R)$ and $v(r)$ respectively. In the simple case where all the small droplets have the same size and together constitute a liquid-water content of w grams per unit

---

† E' is related to the collection efficiency E as defined in Appendix A by

$$E' = E(R+r)^2/R^2.$$

volume of air, eqn (3.47) reduces to

$$\frac{dR}{dt} = \frac{E'(R, r)}{4\rho_L} w\{v(R) - v(r)\}, \tag{3.48}$$

where ρ_L is the density of liquid water.

This treatment calculates only the averaged and smoothed rate of growth of all droplets of a certain radius but, in fact, growth by coalescence is a discrete stepwise process in which some of the larger cloud droplets may undergo more than the average number of chance collisions (while others undergo less) and this small statistically-fortunate proportion will grow faster than the rest and produce a spread in the size distribution with a 'long tail' towards larger sizes. This 'stochastic' nature of the droplet growth process causes a spread in the time taken for originally identical droplets to reach a certain size and, in order to calculate how the size distribution changes with time, one is concerned with every possible combination of droplets that are able to coalesce, the probability of each coalescence, and the changes in these probabilities after each coalescence.

Telford (1955) was the first to point out the essential differences between the 'continuous' and 'stochastic' models particularly in relation to the rate of production of a few large drops. Using a simple cloud model containing initially only two discrete sizes of drop, and assuming a collision efficiency of unity for all sizes, he showed that the stochastic model produced a few large droplets of radius up to 45 μm about six times faster than the average rate given by the continuous model.

The general form of the stochastic equation for the growth of cloud droplets by coalescence, treated in various ways by Elton, Mason, and Picknett (1958), Golovin (1963), Twomey (1964, 1966), Berry (1967), Warshaw (1967), and Scott (1968), may be derived as follows. Defining $f(x) \, dx$ as the mean number of droplets of mass between x and $x+dx$ in unit volume of air, and a collection parameter $K(x, x')$ as the rate at which the space within which an x'-droplet will be captured is swept out by an x-droplet, i.e.

$$K(x, x') = \pi(r+r')^2 E(r, r')\{v(r) - v(r')\}, \tag{3.49}$$

the probability that a particular x-drop will collect an x'-droplet in time δt is

$$P(x, x') = K(x, x')f(x') \, dx' \, \delta t, \tag{3.50}$$

and the *mean* number of x-droplets that will collect x'-droplets in time δt in unit volume of air is

$$P(x, x')f(x) \, dx = K(x, x')f(x)f(x') \, dx \, dx' \, \delta t. \tag{3.51}$$

§3.3 THE GROWTH OF DROPLETS IN CLOUD AND FOG

The time rate-of-change of the number density of x-droplets is given by the difference between their rate of formation by the coalescence of pairs of droplets of masses x and x' and their rate of removal by combination with either larger or smaller droplets to form drops larger than x, viz.

$$\frac{\partial}{\partial t} f(x, t) = \frac{1}{2} \int_{x'=x_0}^{x'=x} K(x-x', x') f(x-x', t) f(x', t)\, \mathrm{d}x' - \int_{x'=x_0}^{x'=\infty} K(x, x') f(x, t) f(x', t)\, \mathrm{d}x', \quad (3.52)$$

where the factor $\tfrac{1}{2}$ cancels the effect of counting each collision twice. Eqn (3.52) is the *stochastic coalescence equation*.

Golovin (1963) was able to obtain an analytical solution of this equation if he assumed the collection parameter to be proportional to the sum of the droplet masses, i.e. $K(x, x') = b(x+x')$; this assumption is not too bad for drops of $r \gg 100\ \mu\mathrm{m}$, but is quite unrealistic for droplets of $r < 50\ \mu\mathrm{m}$ according to Hocking's calculations of E. Golovin's computations are accordingly not very relevant to the likely evolution of droplet spectra in clouds but enable one to obtain considerable insight into the effect of changing various parameters and to make useful checks of the more complex numerical integrations. The same is true of Scott's (1968) additional analytical solutions of the cases when $K(x, x') \propto xx'$ and $K(x, x') = $ constant.

The first attempts at detailed numerical computations using more realistic collection parameters were made by Twomey (1964, 1966) and Bartlett (1966). Twomey (1964), using values of collision efficiency based on the calculations of Hocking (1959) and Shafrir and Neiburger (1963), and assuming a coalescence efficiency of unity, showed that in a cloud containing $2 \cdot 56\ \mathrm{g\ m^{-3}}$ of liquid water and an initial droplet population of $135\ \mathrm{cm^{-3}}$ terminated at $25\text{-}\mu\mathrm{m}$ radius, the 100 largest drops in $1\ \mathrm{m^3}$ of air were growing ten times faster almost from the outset than calculated from the 'continuous growth' equation. In a later paper, Twomey (1966) integrated eqn (3.52) numerically in small time steps to compute the evolution of three initially Gaussian-type droplet spectra each of dispersion $\sigma/\bar{r} = 0 \cdot 15$ and divided into up to 100 radius intervals of increasing width at larger sizes, using values of $K(x, x')$ tabulated in up to 100×100 points. Numerical integration was terminated at the stage at which droplets of radius $50\ \mu\mathrm{m}$ appeared in concentrations of 100 m⁻³, and thereafter growth was assumed to follow the continuous equation. In a cloud containing 1 g m⁻³ of liquid water composed initially of 200 droplets/cm³ (mean-volume radius = 10 μm), and none greater than 21 μm in radius, the largest 100 droplets in 1 m³ of air reached 30 μm after 18 min and took 52 min to reach 100 μm. If, however, the droplet concentration was reduced to 50 cm⁻³ (mean-volume radius 16·5 μm), to produce a spectrum more typical of maritime cumulus, the largest 100 droplets/m³ attained a radius of 40 μm within 5 min and 100 μm within 14 min. This result demonstrated the importance of mean droplet size, but Twomey found that the growth rates of the larger drops were not very sensitive to the adopted values of collection efficiency because if, in the first-quoted calculation, all values of E were taken as unity instead of as given by Hocking, the time taken for the largest drops to reach 30 μm was increased only from 10 to 18 min. Twomey states, however, that the computed values of growth rate were critically dependent on Hocking's result that growth by coalescence cuts off sharply for drops of $r < 19$ μm.

These main conclusions were confirmed by Bartlett (1966) who, instead of solving eqn (3.52), divided his droplet spectrum into 1 μm-radius intervals, calculated the probabilities that during a small time interval the droplets in a given class will coalesce with droplets in each of the other classes, found the change in the number of drops in each class during the time step, and then repeated the whole procedure for a succession of 1-s time steps for a total period of 20 min. Calculations of droplet growth were extended only to a radius of 40 μm, drops larger than this being collected together in a cumulative 'overflow' regarded as incipient precipitation. In a cloud containing 1 g m⁻³ of liquid water composed of 205 droplets/cm³, none greater than 26 μm radius, droplets of 40 μm radius appeared in concentrations of 1/litre within $6\frac{1}{2}$ min and 7 per cent of the total liquid water was composed of drops larger than this after 20 min. In a cloud of the same total water content, but containing initially only 59 droplets/cm³ with the largest 30 μm in radius, the corresponding times were reduced to 2 min and 8 min respectively. Bartlett also demonstrated that neither increasing all Hocking's E values by 10 per cent, nor decreasing the minimum size or increasing the maximum size of the droplets that may be collected by a particular drop size, had little effect on the spread of droplet spectra, but that reducing the minimum collector drop radius from 19 to 17 μm, with corresponding shifts in all the other E-values, produced

§ 3.3   THE GROWTH OF DROPLETS IN CLOUD AND FOG

a marked effect on his 205 droplet/cm³ spectrum in that 40 μm droplets in concentrations of 1/litre now appeared 2 min earlier and the 'overflow' rate increased nearly threefold.

Bartlett's computations for his maritime cumulus model are confirmed by Warshaw (1967, 1968) who, using essentially the same procedures and data as Twomey, finds that in a cloud containing 1 g m⁻³ of liquid water composed of 50 droplets/cm³ with dispersion $\sigma/\bar{r} = 0.15$ and none greater than 28 μm in radius, the largest 100 droplets/m³ reach 40 μm within 2½ min—see Fig. 3.21. This is, however, twice as fast as computed by Twomey for the same cloud model; the reasons for this discrepancy are not clear.

The most satisfactory treatment of the problem is probably that of Berry (1967), who simplifies the numerical integration of eqn (3.52) by expressing the droplet radii on a logarithmic scale, which provides greater spacing at large radii, allows a reasonable number of droplets to be included in a counting interval, and allows the computation to be made with a much smaller number of numerical operations. The mass of droplets per gram of liquid water per cubic metre of air contained in the size range of unit $\ln r$ is defined as

$$g(\ln r) = x f(\ln r) = 3x^2 f(x), \qquad (3.53)$$

so that the area under the curve is

$$\int g(\ln r) \, \mathrm{d}(\ln r) = \int 3x^2 f(x) \frac{1}{3} \frac{\mathrm{d}x}{x} = \int x f(x) \, \mathrm{d}x, \qquad (3.54)$$

and is equal to the total liquid-water concentration, w, in the cloud; likewise the area between any two values of the radius gives the contribution of this size-range of drops to w. Berry starts with an initial drop-size distribution of the form

$$g(\ln r) = 3w \left(\frac{x}{\bar{x}_0}\right)^2 e^{-x/\bar{x}_0} \qquad (3.55)$$

or

$$f(\ln r) = 3n_0 \frac{x}{\bar{x}_0} e^{-x/\bar{x}_0},$$

where $\bar{x}_0$, the mean droplet mass, corresponds to a volume-mean radius of 10 μm, $w = 1$ g m⁻³, the initial droplet concentration, $n_0 = 240$ cm⁻³, and the dispersion in radius is 0·364. Using the Hocking and Shafrir–Neiburger (1963) values of collection efficiency, Berry finds that this spectrum broadens only slowly as shown in Fig. 3.22, where droplets of $r > 40$ μm constitute only about 7 per cent of the

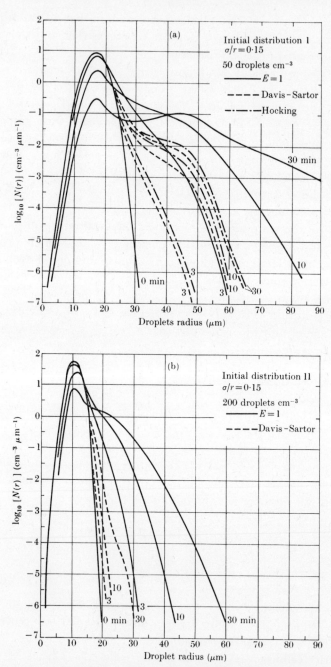

Fig. 3.21 (a), (b), Evolution of droplet-size distribution with time according to the stochastic theory of coalescence. (From Warshaw (1968).)

§ 3.3 THE GROWTH OF DROPLETS IN CLOUD AND FOG

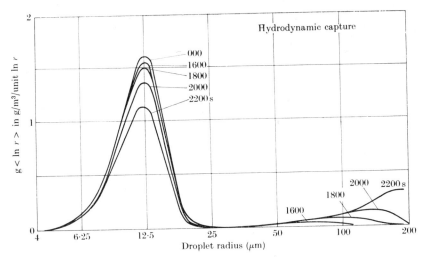

FIG. 3.22. The evolution of droplet-size distribution with time according to stochastic theory of coalescence for periods of up to 2200 s using Shafrir–Neiburger values of collision efficiency. (From Berry (1967).)

total liquid water after 30 min. Thereafter the larger droplets grow into precipitation elements quite rapidly, but a footnote in Berry's paper suggests that computations for this 'tail' of the spectrum are not too reliable.

There has been considerable argument between the various workers as to whether the stepwise methods of integrating the stochastic coalescence equation may not underestimate the broadening of the droplet spectrum because the changes are averaged over finite intervals of both time and droplet size and the history of each individual droplet in the initial distribution is not followed in detail. This point has been investigated, but probably not settled, by Kornfeld, Shafrir, and Davis (1968), who have used a great deal of time on a large computer to follow the coalescence events experienced by each individual droplet of one initial distribution during a period of 2 min; although they obtain a result that is practically identical with that obtained by Warshaw by stepwise integration of eqn (3.52), the calculations would probably have to be run for long periods to reveal significant differences.

Satisfactory mathematical techniques are therefore available for following the evolution of a cloud droplet population by the coalescence process, but further careful computations are needed to reconcile the discrepancies between the results of different authors. The results of all the calculations are critically dependent on the validity of Hocking's result that growth by coalescence is negligibly small for droplets of radius <20 μm. The importance of this result, which received experimental confirmation from Woods and Mason (1964), has stimulated some more accurate calculations by Davis and Sartor (1967), and by Hocking and Jonas (1970) who have improved Hocking's original treatment to calculate more accurately the forces resisting the approach of the two drops, and have included the effects of droplet rotation, both these corrections being particularly important at small separations. These new computations (see Appendix A), instead of predicting a sharp cut-off in collision efficiency for drops of radius <20 μm, show that droplets of radius as small as 10 μm have finite collision cross-sections, but these are <1 per cent. In fact, it turns out that Hocking (1959) *overestimated* the collision efficiencies between drops of radius 20 μm and droplets of $r = 9$–16 μm, the new values being only 2–3 per cent. However none of these changes is large enough to affect materially the results and conclusions of the computations of droplet growth just described. The conclusion stands that once droplets of radius >25 μm appear in concentrations of order 100 m^{-3} in the cloud, precipitation develops quite rapidly. Thus light rain may develop inside a typical small continental cumulus within 40–60 min, and within a typical small maritime cloud within 20 min. The rate-determining step may often be the long time taken for an adequate concentration of droplets to grow by coalescence from 20 to 25 μm radius; this may be typically 20 min. On some occasions this gap may be bridged by the presence of giant hygroscopic nuclei ($m > 10^{-9}$ g) that may grow to 25-μm radius by condensation in 5 min, but they are not usually present in sufficient concentrations to initiate shower production, and one is left with the impression that, in fact, showers do develop more easily and rapidly than can easily be accounted for by the current theories of droplet growth by condensation and coalescence.

There has been much discussion in the literature on the possibility of droplet growth being significantly accelerated in the presence of electric charges and fields. In natural clouds, droplets are usually charged, and may experience fields ranging from about 10 V cm^{-1} in swelling cumulus to more than 1000 V cm^{-1} in thunderstorms.

Accurate calculations of the forces between two charged conducting spheres, as functions of their radii and separation, have been published by Davis (1964). He has also calculated the force of attraction between two conducting spheres polarized in an applied electric field. In order to calculate the influence of electric fields on the relative trajectories, and hence on the collision efficiencies of droplet pairs, the equations of motion have been extended to include these electrostatic forces by Plumlee and Semonin (1964), and Hocking and Jonas (1970) using their improved formulation of the hydrodynamics. The results, summarized in Appendix A, show, for example, that for a pair of droplets of radii 20 and 10 μm the collision efficiency is not significantly increased unless the electric field exceeds 100 V cm^{-1}, when it attains a value of 0·08 compared with 0·02 in zero field. A comparable effect would be achieved by electrostatic attraction between a charged and an uncharged drop, if the former carried a charge of order 10^{-6} e.s.u. These conditions are likely to be met only in incipient thunderstorms where precipitation is already well under way. Accordingly we conclude that electrical forces are unlikely to accelerate the early growth of droplets by coalescence.

3.3.7. *The effect of small-scale turbulence on collisions between droplets*

Finally, we enquire whether the number of collisions between cloud droplets of different size might be increased in a turbulent air-stream because, if velocity fluctuations of the right frequency occur, the relative velocities between the large and small droplets may be increased owing to differences in their inertia. Clearly there will be an optimum frequency at which the induced relative velocity will be a maximum. At very low frequencies, all the droplets will tend to move with the air, and their relative velocities will be very small; as the frequency is increased, the larger drops will lag more and more behind the air motion relative to the smaller ones, but when the velocity fluctuations of the air become short in period compared with the time constant of the smaller droplets, these too will lag, and the relative velocities of large and small droplets will again be small. A theoretical examination of this problem has been made by East and Marshall (1954). For droplets of radii 15 and 10 μm, the optimum frequency is about 100 Hz. If the collision efficiency of large droplets for smaller ones is to be increased appreciably by the action of turbulence, accelerations of the order experienced when falling under gravity must be imparted to the droplets. This would imply that the air must undergo total root-mean-square accelerations of order g in the frequency band below about 100 Hz. It appears unlikely that turbulence of this intensity occurs in this frequency range in the free atmosphere, and it is accordingly doubtful whether this mechanism plays an important role in bringing about collisions between cloud droplets. But East and Marshall considered only the effect of turbulence on the motion of two drops relative to the air. The effect of spatial variations of the turbulent motion that might lead to collisions, even between droplets of equal size, due to the 'motion of drops with the air' has been investigated by Saffman and Turner (1956). For an initially homogeneous cloud, the latter show that the collision rates depend only on the sizes of the droplets, the rate of energy dissipated by the turbulence, ϵ, and the kinematic viscosity. Even assuming the collection efficiencies for cloud droplets of equal or near-equal size to be unity, their theory indicates that, for a cumulus cloud having liquid-water content $1\cdot5$ g m^{-3}, $\epsilon = 100$ cm^2 s^{-3}, only $2\cdot5$ per cent of the droplets will become involved once in collision after 10 min, and less than 1 in 10^9 droplets will achieve double the initial radius. This aspect of the turbulence therefore appears to be rather unimportant in determining the droplet-size distribution in cumulus clouds and even less important in layer clouds.

The possibility remains that the relative trajectories and hence the collision efficiencies of neighbouring droplets may be influenced by the velocity gradients existing in small eddies. It would therefore be worthwhile to extend the recent Hocking–Jonas calculations to include the effects of velocity shear.

3.4. Concluding remarks

The indications of these theoretical studies are that the size distribution of cloud droplets is controlled not only by the micro-physical processes of droplet growth by condensation and coalescence, but also by the cloud dynamics that determine the cloud dimensions, the degree of mixing with the environment, the distribution of vertical velocity, the scale and intensity of turbulence, and the period for which individual droplets remain in the cloud.

The basic characteristics of observed cloud droplet spectra, for example the mean droplet concentration and size, and the growth of droplets up to radius 20 μm, can be accounted for by condensation on typical observed nucleus spectra. Moreover, once droplets of $r > 25$ μm have appeared in concentrations of 10^2–10^3 m^{-3}, their further growth proceeds quite rapidly by coalescence and showers may be expected to develop within 20–30 min. There remains, however, the difficulty that our present theories of condensation and coalescence do not appear capable of explaining how droplets can transcend the critical size range 20–25 μm within a few minutes, except by appealing to condensation on giant hygroscopic nuclei that may be present in adequate concentrations over the oceans but are probably too scarce in the interiors of continents.

4

Initiation of the Ice Phase in Clouds

SINCE supercooled clouds are a common occurrence in the atmosphere, and because the co-existence of ice crystals and supercooled water droplets is germane to the release of snow, rain, and hail, the initiation and development of the ice phase in clouds assume fundamental importance.

Although large quantities of water such as lakes and ponds do not supercool appreciably, cloud droplets commonly exist in the supercooled state down to temperatures as low as $-20°$ C and, on occasion, down to $-35°$ C, while droplets of pure water, only a few microns in diameter, may be supercooled to $-40°$ C in the laboratory. At temperatures below $-40°$ C, such small droplets freeze automatically (or spontaneously) but, at higher temperatures, they can freeze only if they are infected with foreign particles which we shall call *ice nuclei*. The stability of natural supercooled clouds at temperatures above about $-15°$ C speaks for the rarity of *efficient* ice-forming nuclei in the atmosphere. Considerable interest therefore centres on the nature, origin, and mode of action of these particles but, before discussing these matters, it seems appropriate to review the experimental data on the supercooling and freezing of water.

In cloud physics one is concerned with the temperatures at which airborne drops, varying in diameter from a few microns, to about 5 mm for the largest raindrops, will freeze, and how the attainable degree of supercooling may depend upon the drop size, the rate of cooling and the purity of the water. In the absence of sufficiently detailed and careful experiments and of anything approaching a satisfactory theory, it was not possible, until recently, to gain some real insight into these problems. Although there was some indication in the extensive literature that the attainable degree of supercooling tends to increase with decreasing volume of the water sample, there was so much scatter in the results, with serious discrepancies between those of different workers, that no clear-cut relationships could be deduced. It appears that the earlier work may have failed to provide the required information for three main reasons. First, the water samples used by different investigators varied greatly in their origin and purity; secondly, they were usually contained in glass tubes or supported as drops on variously-treated metal surfaces so that nucleation may often have been induced by the solid boundaries; thirdly, in any one investigation, the volume of the sample was not usually varied sufficiently to establish clearly how this might be related to the attainable supercooling, particularly as there was usually a considerable spread in the freezing temperatures recorded for specimens of the same volume.

For these reasons, the whole subject has recently been examined afresh, and it is the results of this later work, both experimental and theoretical, that will now be reviewed.

The great majority of the experiments have been concerned with the study of heterogeneous nucleation in that the water used almost certainly contained foreign particles which initiated crystallization. The fact that the freezing temperature of the same water sample is not generally reproducible to within $0.5°$ C, or more, suggests that, to some extent, freezing is a random process, and that a given sample has only a statistical probability of freezing at a particular temperature. This is brought out even more clearly by the fact that, if a water sample is subdivided into a large number of equal volumes, the freezing temperatures of the latter show a simple probability distribution. This statistical character of the nucleation events, which has not always been clearly recognized, has the important implication that, in order to obtain significant and characteristic relationships, it is necessary to determine the freezing points of large numbers of samples.

4.1. The supercooling of water containing foreign particles

The results of some modern experiments in which the freezing points of large numbers of samples of widely different volumes have been determined, and which undoubtedly show a marked volume dependence, will now be summarized.

Vonnegut (1948a) placed sixty-four drops each weighing about 3 mg on a polished chromium-plated surface covered with a thin film of polystyrene. The plate was then held at a fixed temperature below 0° C, and the number of unfrozen drops was counted at successive intervals of time. The proportion of frozen to unfrozen drops, interpreted as the probability of freezing, increased more rapidly with time when the plate was held at a lower temperature. This result was interpreted by Vonnegut as indicating that the nucleation was time-dependent but, it would seem, only to the extent that there was a higher probability, at a lower temperature, of a drop becoming infected with an effective airborne freezing nucleus during the course of the experiment.

Heverly (1949) studied the freezing of drops of water and dilute aqueous solutions which were supported either on a small thermocouple junction or on waxed paper in a cryostat. The drop diameters varied from 50 μm to 1·1 mm and the cooling rates from 1 to 20° C/min. The freezing points of the drops fell roughly on the same diameter–temperature curve irrespective of the origin and contamination of the water. The freezing points of drops of 400 μm to 1·1 mm diameter were roughly constant at $-16°$ C, but there was a rapid fall of freezing point with decreasing diameter below 400 μm. The freezing points appeared to be independent of the cooling rate.

The results of an investigation by Hosler (1954), in which 490 droplets of distilled water were frozen on a platinum surface, show a variation of mean-freezing temperatures from $-19°$ C for a group of 200 μm diameter droplets, to $-30°$ C for 25-μm droplets, with an occasional droplet of the latter size being supercooled to $-34°$ C. Hosler also describes some experiments on the supercooling of water in Pyrex tubes of diameters ranging from 0·25 mm to 3 mm. Using ordinary laboratory-distilled water in tubes cleaned by prolonged washing with chromic acid and flushed with distilled water, he reported that the freezing points of the samples were independent of their length, volume or surface area, but appeared to depend only on the diameter of the containing tube. This challenge to the generally accepted view that the attainable degree of supercooling is dependent upon the volume of the sample seems to have been met by the much more careful experiments of Mossop (1955). He showed that Hosler's results were almost certainly a consequence of a thin film of water on the inner surface of a tube becoming nucleated by particles lodged on the walls, and that if the walls were coated with a hydrophobic film, volume-dependent freezing points were obtained.

Rau (1953a, b) studied the freezing of large number of drops, varying in diameter from 300 μm to 1 cm, on a highly polished metal surface in the presence of room air. Plots showing the relative frequencies with which drops of the same size froze at different temperatures revealed distinct maxima at about $-12°$ C and $-20°$ C, the former being more prominent for the larger drops, and the latter dominant for the smaller droplets. Very few drops froze at temperatures above $-4°$ C and the maximum supercooling attained in the presence of nuclei was $-33°$ C. Rau suggested that his observed bimodal distribution of freezing temperatures truly represented the activity–temperature relationship for airborne freezing nuclei. However, such a bimodal distribution was not observed by Bigg (1953) nor by Langham and Mason (1958) for the freezing temperatures of drops suspended at the interface of two liquids. It is possible that Rau's curves really represented the superposition of two probability distributions, the high-temperature peak being associated with the original nucleus population of the water, and the second with the activity of nucleation sites on the supporting surface. Such an interpretation would be consistent with Rau's findings that the amplitude of the $-12°$ C peak increased relative to that of the $-20°$ C peak for larger drops, and that there was a reverse trend if the water was progressively purified.

An interesting investigation on the nucleating effects of both atmospheric contamination and the containing walls on the supercooling of water in glass tubes was described by Wylie (1953). Vapour, produced from triply-distilled water and carried in a stream of particle-free oxygen, was condensed in glass tubes that had been flushed with chromic acid and steam, and heated to softening point to drive off nuclei from the walls. Despite these precautions, the experiments showed clearly that freezing was governed by the glass surface. Successive freezings of the same sample occurred at temperatures that lay in a narrow range governed by a probability distribution curve. Freezing was often observed to occur repeatedly at a particular nucleation site on the glass wall, and these sites were able to survive the heating of the glass to softening point. Volumes of about 2 cm³ usually froze at temperatures between -13 and $-20°$ C but, on two occasions, supercooling was prolonged to $-30 \pm 1°$ C.

The results of experiments with open tubes were much less reproducible, and showed that contamination by airborne dust sometimes caused the samples to freeze at temperatures as high as $-5°$ C.

A considerable improvement in the technique of investigating the supercooling of water was made by Bigg (1953), who eliminated the influence of solid supporting surfaces by suspending water drops at the interface of two liquids having different densities, where they were also protected from infection by airborne particles. He also investigated a wide range of drop diameters—from about 20 μm to 2 cm—and thus volumes differing by a factor of 10^9. The use of five pairs of supporting liquids, the members of a pair being practically immiscible with water and each other, established that the observed freezing temperatures of the drops were a property of the water and not of the surrounding media.

Bigg determined the freezing temperatures of large numbers of drops of various sizes, cooled at a constant rate, and divided them into groups according to drop diameter, each group containing not less than 100

§4.1 INITIATION OF THE ICE PHASE IN CLOUDS

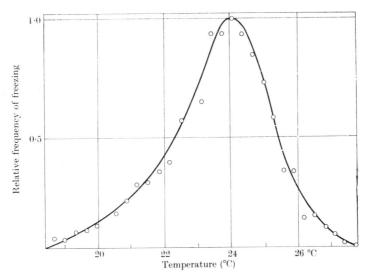

Fig. 4.1. The distribution of freezing temperatures of 1127 water drops of 1-mm diameter. (After Bigg (1953).)

individual observations. A curve showing a typical distribution of freezing temperatures of drops in a particular size group is shown in Fig. 4.1, which contains observations on 1127 drops of distilled water of 1-mm diameter. One significant feature of this curve, and of several similar curves obtained for other drop sizes, is that the scatter of the freezing points about the median value is much less than obtained with droplets supported on a solid surface which itself may provide a number of nucleation sites. The median freezing temperature for each size group, i.e. the temperature below which half the drops froze, is plotted against the drop diameter, on a logarithmic scale, in Fig. 4.2. Bigg thus discovered a linear relationship between the logarithm of the drop diameter (or volume) and the depth of supercooling $T_s = (273-T)°$ C, the line in Fig. 4.2 being represented by the relationship

$$\ln V = A - BT_s, \tag{4.1}$$

where A and B are constants for the particular sample of water used.

Bigg's work has been checked by Langham and Mason (1958) using a slightly improved version of his apparatus in which the drops were suspended between carbon tetrachloride and liquid paraffin (or a silicone fluid) in a shallow metal dish and cooled at the rate of 0·3° C/min. The results obtained with water varying in purity from that of rain-water to that produced by multiple distillation showed

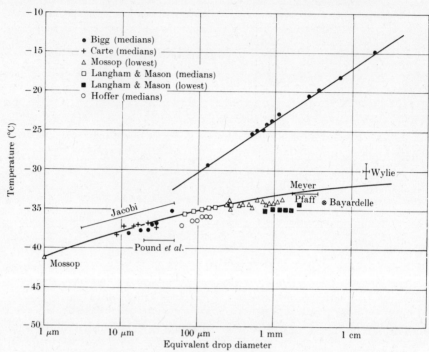

Fig. 4.2. The freezing temperatures of water samples as a function of their equivalent drop diameter.

the same general trend as those of Bigg; plots of the median-freezing temperatures of groups of drops against the logarithm of their diameters produced straight lines almost parallel to that of Bigg, but lying below or above it, depending upon whether the water used was more, or less, pure than that investigated by him. Further confirmation of the general validity of relation (4.1) was found by Mason (1956a) in re-analyzing the results of some earlier work by Dorsch and Hacker (1950) in which 5000 water droplets, with diameters between 8·75 μm and 1 mm, were cooled at rates varying from 6 to 15° C/min on a metal plate. Although the observed freezing temperatures of a group of drops of a particular size had a considerably greater spread than in Bigg's experiments, a plot of median temperatures versus drop diameter produced a line exactly parallel to his, but displaced by 7° C towards higher temperatures.

Bigg also investigated how the supercooling of water drops was affected by the rate at which they were cooled. With a cooling rate of 0·05° C/min, he found the median freezing temperature of 164 drops of

1 mm diameter to be 2° C higher than the corresponding value for a cooling rate of 0·5° C/min. A smaller effect has been reported by Carte (1956). For drops of about 20-μm diameter, a tenfold variation in cooling rate produced a shift of only 0·5° C in the freezing temperatures, but in the same sense as found by Bigg, and an effect of similar magnitude was found by Mossop (1955) for much larger volumes. It is not unlikely, however, that spurious shifts in temperature were introduced in these experiments, at high rates of cooling, by a lag in the thermometers, and a rather more careful investigation is required to establish the reality of the effect.

Although Bigg removed the grosser particles from his water samples by double distillation, his drops almost certainly contained large numbers of small particles which can be removed only by taking extreme measures (see p. 165). It therefore seems that his ln V–T_s relationship represents nucleation by foreign particles (heterogeneous nucleation). This is also suggested by the fact that other workers, who have taken greater precautions to purify their water, have been able to supercool specimens to temperatures well below those indicated by Bigg's line (see § 4.2 and Fig. 4.2).

Langham and Mason (1958) suggest that Bigg's relationship between droplet volume and depth of supercooling may be explained on the assumption that his water was contaminated by particles that were originally airborne, and whose efficiency as freezing nuclei might be expected to vary with temperature in much the same way as atmospheric aerosol.

The numbers of ice crystals appearing in a supercooled cloud, produced by condensation on atmospheric aerosols in cloud chambers, have been determined by several different workers; the results are summarized in § 4.4.2. In the temperature range −10 to −35° C, (above which natural ice nuclei are very rare, and below which the effects of homogeneous nucleation become apparent), the numbers of particles that initiate the freezing of supercooled droplets increases roughly logarithmically with decreasing temperature. Langham and Mason assume that the bulk water used in the supercooling experiments contained a representative sample of what were originally atmospheric particles, so that the activity of the contained freezing nuclei may also be represented by

$$n = n_0 \exp(aT_s), \qquad (4.2)$$

where n is the concentration of nuclei that become effective at temperatures between 0° C and $-T_s$° C, and n_0 and a are constants. For drops of volume V being cooled to a temperature $-T_s$ °C, and containing a randomly-distributed population of nuclei, the probability P of a drop containing at least one effective nucleus on reaching this latter temperature is given by

$$P = 1 - \exp(-Vn), \qquad (4.3)$$

or $\qquad \ln(1-P) = -Vn = -Vn_0 \exp(aT_s). \qquad (4.4)$

Now Bigg's empirical relationship (Fig. 4.2) shows the value of T_s for which drops of different volumes have a 50 per cent probability of freezing, i.e. $P = 0.5$, and for these, eqn (4.4) becomes

$$Vn_0 \exp(aT_s) = \text{const.}$$

or $\qquad \ln V = \ln C - aT_s, \qquad (4.5)$

which is Bigg's relationship. The value of the constant a fixes the shape of the freezing-nucleus distribution, while the value of C depends upon the absolute concentration of nuclei, i.e. upon the purity of the water.

The freezing point of water may, of course, be influenced by the presence of salts in solution. Bigg (see Mason (1953a)) determined the freezing temperatures of large numbers of 1-mm diameter drops of aqueous solutions of various alkali halides over a wide range of concentrations. In some cases, for example with NaCl, Bigg found that the freezing temperatures fell monotonically with increasing concentration of the solute, particularly large changes being produced as he proceeded from one-tenth saturated to fully-saturated solutions. Similar results have been obtained by Hoffer (1961) for smaller drops, about 100 μm in diameter, containing Mg_2SO_4. On the other hand, Bigg found that increasing the concentration of halides caused an apparent increase in the freezing temperatures of the drops although, at very high concentrations, they fell again to below the values obtained for pure water. In a very careful study, Pruppacher and Neiburger (1963) showed that these latter results were almost certainly due to the introduction of insoluble foreign particles with the salts and were not produced by the solutes themselves. When the insoluble particles were removed by filtration of the solutions through Millipore filters, the freezing temperatures of 2-mm diameter drops of a wide variety of solutions, supported between a fluorocarbon and refined liquid paraffin, fell progressively as the concentration of the solute was increased. Bigg's results can therefore be explained as the combined effect of added particles tending to raise the freezing temperature and of the dissolved salt tending to lower it.

§ 4.1 INITIATION OF THE ICE PHASE IN CLOUDS

One may imagine the freezing of a water sample to be initiated as follows. The structure of water can be described as a 'broken down' ice structure in which long-range order is lost but short-range order is preserved, and in which only a relatively few hydrogen bonds are broken but many are bent. As the temperature is lowered, the molecular arrangement in the supercooled liquid becomes progressively more and more ice-like. In the absence of foreign surfaces, nucleation of the ice phase may occur only by the chance orientation of local groups of water molecules into an ice-like configuration. A suitable solid particle, however, may cause water molecules to become 'locked' into the ice lattice under the influence of its surface force field. The molecular aggregate will not only be bound to the surface of the particle, but will have only one exposed surface; on both counts it will be less vulnerable to thermal bombardment than will a spontaneously formed aggregate, and will therefore have a higher probability of attaining the critical size at which it may nucleate the ice phase. Whether or not a stable ice nucleus is formed must be largely determined by the configuration of the surface force field of the substrate.

When electrolytes are dissolved in water, they are partly or fully dissociated into ions, and interaction between the electric fields of the ions and the water dipoles causes further distortion of the water structure. The breaking of additional bonds causes an increased number of vacant lattice sites and interstitial molecules and makes the formation of ice-like clusters of molecules less probable. Thus dissolved substances would be expected to hinder the nucleation and freezing processes and this is observed to be the case.

A recent, extensive, and apparently careful investigation by Vali (1968) has produced some surprising results which, if correct, would appear to challenge the interpretation of the measurements of natural ice nuclei described later. Vali has produced water drops of 1–2 mm diameter from samples of natural rain, snow, and hail, supported them on a silicone-coated aluminium cold stage, cooled them at a standard rate of 2° C min^{-1}, and determined their freezing temperatures by taking photographs at intervals of 0·2° C. Determination of the fraction $N(\theta)/N(0)$ of unfrozen drops at a given temperature θ allows the number, $K(\theta)$, of freezing nuclei per unit volume of water that become active between 0° C and $-\theta$° C to be calculated from

$$K(\theta) = -1/V \{\ln N(\theta)/N(0).\}$$

For example, Vali deduced that drops from a sample of melted hail contained 1 nucleus/cm³ active between 0° C and −6° C, and 500 cm⁻³ active between 0° C and −13° C. In another sample of melted hail, the number of active nuclei increased about one-thousandfold over the temperature range −14 to −16° C!

The most surprising results were obtained when the water samples were filtered through Millipore filters before making the drop-freezing tests. Filtration of water containing suspensions of humus-, loam-, or peat-rich soils, and of water from melted hailstones, through filters of pore diameter only 0·01 μm, caused the deduced concentrations of nuclei active at −7° C to be reduced, on average, by only a factor of 4, the largest reduction factor being 20 for one hail sample. Vali deduces from this that many of the most effective freezing nuclei in the water, and also in the atmosphere, are smaller than 0·01 μm in diameter, and are too small to be detected by cloud chambers and other conventional techniques that usually operate below −15° C. He suggests that a major source of natural freezing nuclei may be minute particles of organic matter, $<0\cdot01$ μm in diameter, that originate in the soil, but are carried aloft and captured by cloud droplets. It is difficult to understand how such small particles could be produced, carried aloft except by attachment to larger particles, or how they might act as very effective ice nuclei. This work will require careful checking and confirmation, particularly as we know very little about the efficiency of Millipore filters and the break-up of particle aggregates in water.

4.2. The homogeneous nucleation of supercooled water

A number of experimenters have reported on the possibility of supercooling small droplets of radius 5 μm to about −40° C before freezing occurs. Cwilong (1947a) and Fournier D'Albe (1949), by expanding atmospheric air in small cloud chambers, found that while at temperatures above −40° C a few ice crystals appeared among a much larger number of water droplets, below −41° C the fog was composed almost entirely of ice crystals. Mossop (1955), in a rather more careful study of the freezing of droplets of diameter about 1 μm, also found that the number of crystals increased rapidly as the temperature fell below −40° C and estimated that all the droplets had frozen after being maintained at −41·2±0·4° C for 0·6 s. Similar results were reported by Mason (1956a) from observations of droplets formed on small ions in the clean air of a diffusion cloud chamber as they fell at about ¼ cm s⁻¹ through a vertical temperature gradient.

As solid particles are excluded from these tiny droplets formed in clean air, it seems that nucleation of the ice phase must occur homogeneously or spontaneously by small groups of water molecules becoming locked by change into ice-like configurations. Such molecular aggregates will continually form and disappear as the result of microscopic thermal fluctuations. The probability of an aggregate reaching a given size increases as the temperature is lowered, until eventually it surpasses a critical size beyond which it can survive, continue to grow with a decrease of free energy, and form a nucleus for the ice phase.

Several workers have been able to supercool rather large droplets, usually supported in an immiscible liquid or on a metal surface, to temperatures well below $-30°$ C—probably to the threshold of homogeneous nucleation. The essential information on these experiments is summarized in Table 4.1. Langham and Mason (1958) went to some pains to produce, in appreciable quantity, water free from foreign particles and to study systematically the homogeneous nucleation of large numbers of drops of up to 2 mm in diameter. In Fig. 4.2 are plotted, for drops of $d < 0.5$ mm, the median freezing temperatures of groups containing about 100 drops of uniform size. The distribution of the individual freezing points of one hundred 90-μm diameter drops is shown in Fig. 4.3. The freezing temperatures of drops of $d > 0.5$ mm showed a wider spread; only about 10 per cent of them could be supercooled to the lowest temperatures indicated in Fig. 4.2 but, even so, drops of 2-mm diameter could be supercooled to $-35°$ C. Hoffer (1961) was able to supercool a group of fairly uniform-sized droplets ranging in diameter from 60 to 120 μm, and supported in silicone oil, to temperatures about 1° C below the corresponding values of Langham and Mason, but the significance of this cannot be assessed until the two sets of data are compared on the basis of equal cooling rates.

The results listed in Table 4.1 and plotted in Fig. 4.2 form a coherent pattern which Mason (1956a, 1960b) has shown to be consistent with the Turnbull–Fisher (1949) theory of homogeneous nucleation of a supercooled liquid in that the experimental points lie quite close to curve (b) of Fig. 4.3, which is calculated from eqn (4.16) on the basis of Mossop's observation that droplets of 1 μm diameter freeze at $-41°$ C within 0.6 s, to indicate the temperatures at which larger drops may be expected to freeze spontaneously within 1 s. However, doubts remain as to whether the freezing of droplets on a supporting liquid or solid surface may truly simulate the behaviour of freely-falling droplets in a natural cloud.

For this reason, Kuhns and Mason (1968) carried out an experiment

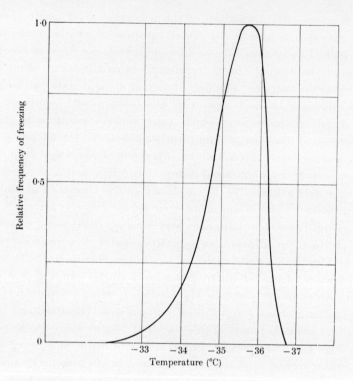

FIG. 4.3. The distribution of freezing temperatures of 100 droplets of 90 μm diameter of highly purified water. (From Langham and Mason (1958).)

in which the freezing of large numbers of individual droplets of very pure, particle-free, supercooled water, ranging in diameter from 5 to 120 μm, was photographed and measured as they fell through a vertical temperature gradient established in cold, purified helium, air, and other gases. Fig. 4.4 shows the median freezing temperature, as a function of droplet diameter, of droplet populations having diameters in the range 5–120 μm and falling freely in helium gas, which has the advantage of high thermal conductivity and low solubility in water. The directly observed freezing temperatures measured to within $\pm 0.1°$ C (curve a) are first corrected for thermal lag (curve b), and then standardized to a cooling rate of $0.1°$ C s^{-1} or $6°$ C min^{-1} to give the final determinations in curve c. It was very difficult to make measurements on droplets of diameter <5 μm, but when the curves of Fig. 4.4 are extrapolated over the diameter range 1 to 5 μm, they indicate that droplets of 1-μm diameter should freeze at about $-40°$ C in agreement with cloud-chamber observations. These results may be compared with those of

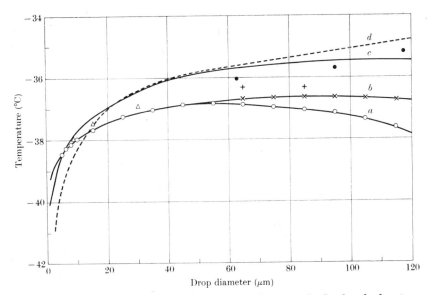

FIG. 4.4. The median freezing temperatures of groups of a few hundred water droplets falling in helium as a function of droplet diameter. *a*, Directly observed temperatures uncorrected; *b*, results corrected for thermal inertia of the droplets; *c*, results corrected for thermal inertia and standardized to a cooling rate of 6° C min^{-1}; *d*, freezing temperatures computed from the theory of homogeneous nucleation with $\sigma_{SL} = 19{\cdot}7$ erg cm^{-2}. Other experimental results: △, Carte; ●, Langham & Mason; + Hoffer, all adjusted to cooling rate of 6° C min^{-1}. (From Kuhns and Mason (1968).)

Carte (1956), Langham and Mason (1958), and Hoffer (1961) which, when corrected to the same cooling rate, lie below curve *c* by a few tenths of a degree. These slight differences may arise partly from experimental error and partly from the fact that the corrections of 0·5° C for a ten-fold change in the rate of cooling have been determined only under rather restricted conditions, but in view of the variety of experimental techniques and methods of preparing the water, it is gratifying that the results of the different workers agree as well as they do. They may also be compared with the theory of homogeneous nucleation of a liquid as formulated by Turnbull and Fisher (1949) and applied to supercooled water by Mason (1952*b*, 1960*b*).

The rate of formation of molecular aggregates of critical size per cubic centimetre per second is given by Turnbull and Fisher as

$$I \simeq \frac{nkT}{h}\exp\{-(U+\Delta G^*)/kT\}, \qquad (4.6)$$

where n is the number of molecules per cm^3 of the liquid, k is Boltzmann's constant, h Planck's constant, U the activation energy

TABLE 4.1

Summary of experimental data on the supercooling of water samples to very low temperatures—probably to the threshold of homogeneous nucleation

Author	Method of preparation of water	Equivalent drop diameters	Median freezing temp. (°C)	Lowest freezing temp. (°C)	Remarks
Meyer & Pfaff (1935)	multiple distillation and filtration through sintered glass and collodion filters. Test sample contained in quartz tube	between 1·8 and 3·8 mm (estimated)	—	−33	only a very few samples were supercooled to this temperature
Wylie (1953)	water condensed from a moist, nucleus-free stream of oxygen into fairly large glass tubes	between 1·4 and 1·7 cm (estimated)	—	−30 ± 1	only one sample supercooled to this low temperature on two occasions
Bigg (1953)	doubly-distilled water dispersed as small droplets at the interface of two immiscible liquids	15 to 50 μm	−38·0 to −35·4	—	six groups covered this diameter range with more than 100 droplets in each
Pound, Madonna, and Peake (1953)	water condensed as small droplets to form an emulsion in silicone oil	20 to 50 μm	−39	−42	no differentiation with respect to droplet size reported
Mossop (1955)	water condensed into glass, silica and silicone-treated glass capillaries	0·25 to 1·05 mm	—	−33·8 to −34·7	15 samples altogether supercooled to these low temperatures
	tiny droplets produced by condensation in an expansion chamber	1 μm	—	−41·2 ± 0·4	
Jacobi (1955)	droplets condensed on collodion films supported by highly-polished metal surfaces	3 to 50 μm	—	−37·4 to −34·2	—
Bayardelle (1954)	drops of doubly-distilled water suspended between mercury and a silicone liquid	4–6 mm	—	−34·1	—
Carte (1956)	droplets condensed on a metal surface and covered with silicone oil	9 to 30 μm	−38·2 to −37·3	−39 to −37·8	results of repeated freezings of 3 individual drops of 9, 15, 30 μm diam., and a single freezing of 100 droplets of 10-μm diam.
Langham and Mason (1958)	droplets of multiply-distilled water suspended in silicone oil	70 to 250 μm 1 to 2 mm	−35·6 to −34·4	−35·6	—
Hoffer (1961)	droplets of multiply-distilled water suspended in silicone oil	60 to 170 μm	−36·0 to −37·0	—	no systematic variation of median temperature with droplet size

§ 4.2   INITIATION OF THE ICE PHASE IN CLOUDS

for self-diffusion of a molecule in the liquid, and ΔG^* the work of nucleus formation.

Now
$$\Delta G^* = \tfrac{1}{3}\sigma_{SL} A^*, \tag{4.7}$$

where σ_{SL} is the specific surface free energy of the crystal/liquid interface and A^* is the total surface area of the critical nucleus. The free energy of formation of an ice embryo containing g molecules is

$$\Delta G = (\mu_S - \mu_L) g + A \sigma_{SL}, \tag{4.8}$$

where μ_S, μ_L are the chemical potentials per molecule in the solid and liquid phases. For (metastable) equilibrium between the embryo and supercooled liquid, ΔG must be a maximum and $d\,\Delta G/dg = 0$, when

$$\mu_L - \mu_S = \sigma_{SL}\frac{dA}{dg} = \sigma_{SL}\frac{dA}{dV}\frac{dV}{dg}, \tag{4.9}$$

where V is the volume of the nucleus.

But
$$\frac{d}{dT}(\mu_L - \mu_S) = -(s_L - s_S) = \frac{L}{T}, \tag{4.10}$$

where S_L, S_S are the entropies per molecule and L is the difference between the latent heats of sublimation and vaporization at temperature T.

Therefore
$$(\mu_L - \mu_S) = \int_T^{T_0} \frac{L\,dT}{T} \tag{4.11}$$

and
$$\sigma_{SL}\frac{M}{\mathcal{N}\rho_S}\frac{dA}{dV} = \int_T^{T_0} \frac{L\,dT}{T}, \tag{4.12}$$

where T_0 is the thermodynamic freezing point ($= 273$ K), M, ρ_S, the molecular weight and density of the solid phase, and $\mathcal{N}$ is Avogadro's number.

If the critical nucleus is assumed to be a hexagonal prism of height equal to the short diameter ($2r^*$),

$$A^* = 12\sqrt{3}\,r^{*2}, \quad V = 4\sqrt{3}\,r^{*3}, \quad \frac{dA}{dV} = \frac{2}{r^*},$$

and
$$r^* = \frac{2\sigma_{SL}}{\rho_S} \bigg/ \int_T^{T_0} \frac{L'\,dT}{T}, \tag{4.13}$$

where L' is now the latent heat per g. The number of molecules contained in the nucleus is

$$n^* = \frac{32}{\sqrt{3}} \times 10^{23} \frac{\sigma_{SL}^3}{\rho_s^2} \bigg/ \left\{ \int_T^{T_0} \frac{L'\,dT}{T} \right\}^3. \tag{4.14}$$

Therefore

$$\Delta G^* = \frac{1}{3}\sigma_{SL} A^* = \frac{16\sqrt{3}\sigma_{SL}^3}{\rho_s^2} \bigg/ \left(\int_T^{T_0} \frac{L'\,dT}{T} \right)^2, \tag{4.15}$$

and $\log_{10} I = 32 \cdot 84 + \log_{10} T - \dfrac{U}{2 \cdot 303 kT} - \dfrac{16\sqrt{3}\,\sigma_{SL}^3}{2 \cdot 303 \rho_s^2 kT \left(\int_T^{T_0} \dfrac{L'\,dT}{T} \right)^2}. \tag{4.16}$

The probability P that a droplet of volume V will freeze within a time t is given by

$$\ln(1-P) = IVt, \tag{4.17}$$

so that if $P = \frac{1}{2}$, $IVt = -\ln \frac{1}{2} = 0.7$, and if $t = 1$ s, $I = 0.7/V$. In principle eqn (4.16) can now be used to calculate the median nucleation temperature, T, for any value of V but, in practice, this cannot be done with confidence because I is very sensitive to the assumed value of σ_{SL} which cannot be accurately measured or calculated. Kuhns and Mason therefore substituted their experimental data for drops of diameter 20 μm freezing at $-37°$ C into (4.16) to calculate a value of $\sigma_{SL} = 19.7$ erg cm^{-2}, and then used this value (assumed to be independent of temperature) to predict the temperatures at which droplets of other sizes will freeze within 1 s. This yielded curve (d) of Fig. 4.3, which is of very similar shape to that of the experimental curve (c), and demonstrates that the theory closely matches the experimental results if σ_{SL} takes a value of about 20 erg cm^{-2}.

Incidentally, differentiation of eqn (4.17) gives

$$\frac{1}{(1-P)}\frac{dP}{dt} = IV$$

$$I = \frac{1}{(1-P)} \cdot \frac{1}{V}\left(\frac{dP}{dT}\right)\left(\frac{dT}{dt}\right), \tag{4.18}$$

showing that, since I increases with decreasing temperature and dP/dT is negative, a more rapidly cooling drop will tend to freeze at a lower temperature, and also that an x-fold increase in cooling rate should produce the same depression in the median freezing temperature as an x-fold reduction in droplet volume.

§ 4.2 INITIATION OF THE ICE PHASE IN CLOUDS

So far no assumptions have been made concerning the specific structure of liquid water other than that it contains some fluctuating aggregates that form and disappear as a consequence of local energy fluctuations. Frank and Wen (1957) described how these clusters can 'instantaneously' form and disappear by introducing an element of co-operativeness into the hydrogen bonding scheme (through the partially covalent character of the hydrogen bond), whereby bonds are not made and broken singly but several at a time, thus producing short-lived clusters of highly-bonded regions surrounded by non-hydrogen-bonded molecules. By postulating the existence of a 'chemical reaction' between monomeric water molecules and the hydrogen-bonded aggregates, Frank and Wen were able to account for several of the anomalous properties of water such as the maximum density at 4° C, the decrease of viscocity with increasing pressure, and the existence of a single dielectric relaxation time of order 10^{-11} s.

Némethy and Scheraga (1962) have applied the methods of statistical thermodynamics to this 'flickering cluster' model and successfully derived the thermodynamic parameters of liquid water such as the Helmholtz free energy, internal energy, specific entropy, and the specific heat at constant volume. Good agreement was obtained between the calculated and experimental values over the range 0–100° C, except in the case of C_v where the calculated value decreased too rapidly with increasing temperature. Némethy and Scheraga were also able to compute the concentration and size of the most probable cluster as functions of temperature, and by extrapolating their results to lower temperatures as shown in Fig. 4.5, Kuhns and Mason (1968) were able to show that the theoretical model not only predicts the phenomenon of supercooling, but gives a fairly precise estimate of the temperature at which homogeneous nucleation is likely to occur. In Fig. 4.5 Kuhns and Mason plot the number of molecules in a critical-sized ice-like nucleus, for various values of σ_{SL} and temperature, as given by eqn (4.16). We see that between 0 and $-30°$ C the most probable aggregate produced by thermal fluctuations is much smaller than the equilibrium-sized ice-like aggregate, and so the model predicts that pure water would exist in a metastable supercooled state. However, at temperatures below $-30°$ C, the most probable cluster size approaches the size of the critical ice nucleus, and if we take $\sigma_{SL} = 20$ erg cm^{-2}, the two curves intersect almost exactly at $-40°$ C! This excellent agreement with observation must be regarded as rather fortuitous in view of the uncertainty in the numerical value of σ_{SL}, and the uncertainties introduced into the calculation of the probable aggregate size by various simplifying assumptions in Némethy and Scheraga's treatment. Nevertheless, the fact that the predicted temperature region of homogeneous nucleation coincides so closely with

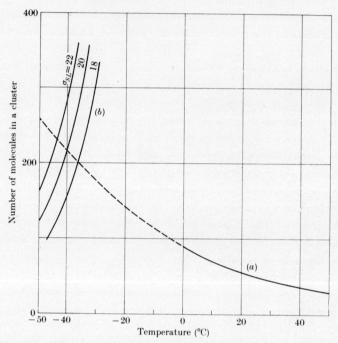

FIG. 4.5. Curve (a); the number of water molecules in the most probable cluster as a function of temperature between 0 and 50° C as calculated by Némethy and Scheraga. The curve has been extrapolated into the supercooled region as shown dashed. Curves (b); the number of molecules in a critical-sized ice-like nucleus as a function of temperature. Three curves are calculated for values of σ_{SL} = 18, 20, and 22 erg cm^{-2} respectively. (Kuhns and Mason (1968).)

the temperatures measured in our experiments leaves little doubt that the latter were concerned with the homogeneous nucleation of water drops and provides supporting evidence that the 'flickering cluster' model of liquid water has some validity. A more detailed discussion may be found in Fletcher (1970).

4.3. Evidence for the supercooling of water below −40° C

Although it has been established, by several different workers, that micron-size droplets of very pure water freeze within less than a second after being supercooled to −41° C, and that this temperature is generally accepted as the lower limit for their spontaneous nucleation, it should be mentioned that evidence has been offered to suggest that water may exist in the liquid state at considerably lower temperatures.

One such claim was made by Rau (1944), who stated that when all foreign nuclei were removed (or rendered impotent)!, it was possible to supercool drops supported on a metal plate to −72° C and to maintain them at that temperature for several hours. Below −72° C, the drops froze to produce crystals in the form of either octahedra, tetrahedra or cubes in contradistinction to the hexagonal crystals formed at higher temperatures. The crystals melted on being warmed above −72° C. Rau suggested that at −72° C spontaneous freezing occurred without the aid of nuclei, the result being a new crystalline ice structure. Attempts to repeat these observations have been made by Cwilong (1947b) and Brewer and Palmer (1951), who reported that they could not reproduce Rau's results unless they deliberately contaminated their water drops with fairly high concentrations of organic vapours such as acetone or alcohol, and criticized Rau's experimental arrangement on the grounds that alcohol contamination could have arisen from his cooling system. In a private communication, Rau says that he has repeated his experiments, eliminating any possibility of contamination of his water drops by alcohol, and that he has been able to reproduce his earlier results. Unfortunately, he gives so little information on his experimental arrangement, the methods he uses to remove impurities and to determine the physical state of his condensate (visual inspection is not good enough), that it is very difficult to assess his work, which will require independent confirmation before it is accepted. The fact that his crystals melted at −72° C makes it very difficult to believe that they could have been pure ice in any form. Evidence suggesting a phase change at about −62° C was obtained by Sander and Damköhler (1943), who measured the critical supersaturation for the precipitation from water vapour in air cooled, by rapid expansion in a cloud chamber, down to temperatures of −75° C. Until a terminal temperature of −62° C was reached, the precipitate appeared to consist of spherical particles that were reported as droplets. Below −62° C, the precipitate appeared to consist of angular ice crystals. These authors suggested that, either the crystals were formed by direct nucleation of ice from the vapour, or that −62° C represented the crystallization temperature of what, at higher temperatures, had been liquid droplets.

This work has been repeated rather more carefully by Madonna *et al.*

(1961) using clean nitrogen as the carrier gas. Their results are very similar to those of Sander and Damköhler, the appearance of spherical particles rapidly giving way to a cloud of glittering ice crystals at temperatures below −65° C. Madonna *et al.* suggest, though, that the non-glittering particles appearing at temperatures between −41° C and −65° C need not necessarily be liquid droplets but frozen spheres which have not developed crystalline faces. A similar suggestion was made earlier by Mason (1952*b*) to account for the appearance of iridescence in mother-of-pearl clouds that occur in the stratosphere at heights of about 23 km and temperatures of −80° C. No evidence for the sudden appearance of ice crystals below −65° C was obtained by Maybank and Mason (1959) who determined the numbers of ice crystals produced by the explosive expansion of a small volume of moist air for various values of the terminal temperature and supersaturation. In clean air, crystals appeared only when the supersaturation surpassed 400 per cent and the terminal temperature fell below −40° C, which suggests that they formed by condensation upon ions followed by homogeneous freezing. With increasing expansions, producing larger supersaturations and lower temperatures, the concentration of crystals increased continuously up to a maximum value of 4×10^6 cm^{-3}. There was no discontinuity in the curve at −65° C or any other temperature, and there was no evidence for the crystals being produced by direct sublimation of the vapour rather than by spontaneous condensation followed by instant freezing, even if the distinction were a profitable one under these conditions.

In conclusion then, there appears to be no convincing evidence for believing that small droplets of pure water can be supercooled below about −41° C for longer than about 1 s.

4.4. Ice nuclei in the atmosphere

4.4.1. *Experimental techniques*

The most commonly used methods of measuring the concentration of ice nuclei in the atmosphere involve cooling the air sample to form a supercooled cloud at a known temperature, and counting the ice crystals that form either by the freezing of droplets containing ice nuclei or by direct condensation on the nuclei themselves. A supercooled cloud may be formed by chilling an air sample below its dew point. The cooling may be achieved either by adiabatic expansion or by convectional cooling within a refrigerated chamber. Ideally, the air should be cooled at rates comparable to those experienced in natural clouds, but this is difficult to achieve in the laboratory with apparatus of convenient size, and so rates much faster than atmospheric are normally used.

An estimate of the concentration of ice crystals appearing in the cloud may be made either by counting their numbers or estimating their average spacing as they scintillate in a known illuminated volume of the cloud. Alternatively, the ice crystals may be allowed to settle out of the cloud on to a suitable detector and then counted. Cwilong (1947a) developed a sensitive method of detecting the threshold for ice-crystal production, the detector being a pool of supercooled water at the bottom of his cold chamber which froze rapidly when the first ice crystal fell into it. The method was improved by Schaefer (1949a), who used soap bubbles of polyvinyl alcohol solutions as detectors in which the rate of ice-crystal growth was slowed down sufficiently to allow individual crystals to be detected. The author has used a very stable soap film stretched across a metal ring. Ice crystals falling on the supercooled film soon grow to visible size and are easily counted. In a similar technique devised by Bigg (1957), the crystals are allowed to fall into a tray of sugar solution placed at the bottom of the chamber. Varying the concentration of the sugar solutions enables this technique to be used from about -5 to $-20°$ C, while the addition of glycerine or ethylene glycol allows the range to be extended to $-40°$ C.

Various combinations of these methods of cloud production and ice-crystal detection have been used, and we shall now describe a few of the experimental arrangements in more detail.

(a) *Expansion chambers.* The chamber is filled with atmospheric air, pre-cooled, and saturated with water vapour, and the cloud is then formed by rapid expansion of the air. Successive measurements of the ice crystals that appear are made at lower and lower temperatures achieved by larger and larger expansions.

A notable attempt to cool the air by expansion at rates comparable to those prevailing in the atmosphere was made by Findeisen (1942a), and Findeisen and Schulz (1944). They used a large chamber, of height 2·5 m, diameter 1 m, and volume 2 m³, and employed cooling rates corresponding to vertical air velocities of 5 to 22 m s^{-1}. Because the expansion lasted for long periods, of up to half an hour, special precautions were necessary to reduce heat transfer from the chamber walls to the cooling gas, and many of the original water droplets and ice must have fallen out before the end of the expansion.

The ice-nucleus content of out-door air was examined by Cwilong (1947a, 1947c) using a small expansion chamber of only a few hundred cm³ capacity and high rates of expansion but, because of the small volume, ice nuclei could be detected only at temperatures below $-32°$ C.

As an example of a modern expansion chamber, Fig. 4.6 shows the instrument designed by Warner (1957) that has been used extensively in Australia and the United States. It consists of a tank of 10-litres capacity, the walls of which are cooled to $-12°$ C by allowing a liquid refrigerant (Freon 12 or liquid CO_2) to evaporate in cooling coils surrounding the chamber walls and base. Evaporation takes place at a pressure which is held constant irrespective of flow rate, and at such a value that the boiling point of the liquid only differs by a few degrees centigrade from the desired temperature of the chamber as a whole. A thermostatic expansion valve is used to vary automatically the flow rate to the value appropriate to the heat load imposed by the surroundings. The chamber is filled with atmospheric air which is humidified by allowing it to come into equilibrium with the walls washed by a glycol–water mixture. The pressure in the chamber is raised by means of a small foot-pump and, after a period to allow equilibrium conditions to be reached, is suddenly released. A fog forms at a temperature depending upon the initial overpressure and consists of supercooled droplets, together with a relatively small number of ice crystals. During an expansion, the temperature, as measured by a resistance thermometer, falls to a minimum value in about 1/10 s, remains within a degree centigrade of this value for 5 s, and returns to its initial value with a time constant of about 15 s. The detector consists of a tray of sugar solution placed at the bottom of the cloud chamber. The water in the solution supercools, and when the tiny ice crystals fall into it, they quickly grow to visible size and are easily counted. In order to avoid the spurious production of ice crystals, it is particularly important to prevent frost formation on the walls of the chamber, and this is done by coating them with glycerol. Successive readings of ice nucleus concentration can be obtained at 4- to 5-min intervals, and the complete spectrum of nuclei, which covers a range of concentration from about 0·1/litre to 100/litre, may be examined in about 1 h.

(b) *Mixing cloud chambers.* In the 'mixing' method, the air sample is injected into a cold chamber, where it cools until thermal equilibrium is established after a period of some minutes. The air may be introduced into a chamber held at a fixed temperature and allowed to reach the wall temperature mainly by convective mixing, but this

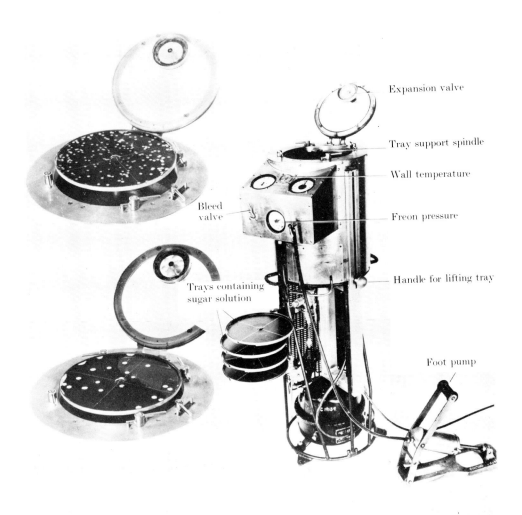

Fig. 4.6. Expansion cloud chamber for the measurement of ice-nucleus concentrations. (By courtesy of Mr. J. Warner.)

§ 4.4   INITIATION OF THE ICE PHASE IN CLOUDS   177

method allows only a single determination of the ice-nucleus concentration, and repeated measurements at different temperatures are necessary to obtain a temperature spectrum. Alternatively, the air may be placed in a warm container, and the wall temperature is then gradually reduced so that the nucleus concentration can be determined at various temperatures using the same air sample. But in this method, employed by Smith and Heffernan (1954) and by Georgii (1959), difficulties arise from the continuous fall-out of crystals, and time lags in temperature measurement and in the growth of ice crystals.

A well designed constant-temperature counter is described by Bigg (1957) and is shown in Fig. 4.7. The experimental space, of volume 10 litres, is enclosed within a multi-walled cylinder cooled by dry ice in a

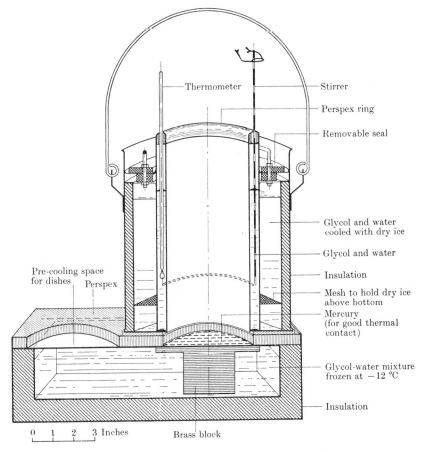

Fig. 4.7. Mixing-type cloud chamber for measurement of ice-nucleus concentrations. (From Bigg (1957).)

glycol–water mixture. When a humid air sample is drawn into the chamber, it cools rapidly towards the wall temperature, a cloud forms, and any ice crystals nucleated within it grow and fall out into a tray of sugar solution that covers the floor of the chamber and is held at a temperature of $-12°$ C. The inner surfaces of the chamber are carefully wiped over with glycerol before each measurement to prevent the formation of frost, the splintering of which causes spurious ice-crystal counts.

An interesting development is an automatic version of the chamber designed by Bigg and Meade (1959) to operate continuously throughout the 24 hours. It is found that measurements made only for short-periods each day may miss important changes in the ice nucleus count because some of the major peaks are short lived and may occur at night. The 10-litre chamber is held at $-20°$ C by a conventional refrigerating system and the sampled air, which enters in a steady stream, is brought to a standard humidity before being cooled. Sugar solution at $-10°$ C is carried on a moving belt underneath the chamber and the ice crystals that form in it are recorded by flash photography once per minute. Frost formation is prevented by a foam rubber roller saturated in glycol that circulates continuously around the cylindrical walls. In this arrangement the air is cooled fairly slowly and very high supersaturations are not attained because the incoming air mixes continuously with air already containing water droplets.

(c) *Precipitation and development techniques.* In an attempt to overcome some of the difficulties and limitations of cloud-chamber methods, Nathan and Hill (1957) and Fenn and Weickmann (1959) developed a thermal precipitation technique in which the particles in a measured volume of air are deposited on to a cooled surface of rhodium-plated copper. The collecting plate is further chilled and the numbers of ice crystals that appear at successively lower temperatures are counted. The method has the attraction that, with careful design and temperature control, the nuclei may be subjected to a constant temperature and a low supersaturation for quite long periods and thus enable one to distinguish between particles that act as sublimation and as freezing nuclei. Difficulties may arise from contamination of the ice nuclei and from the presence of nucleating sites on the metal surface.

The use of membrane filters, first described by Bigg, Miles, and Heffernan (1961), and also by Bigg, Mossop, Meade, and Thorndike (1963) has, in principle, many advantages over cloud-chamber methods in that large volumes of air can be sampled with simple apparatus and the number of collected ice nuclei can be determined by standardized processing in the laboratory that need not be near the sampling point. The nuclei are caught by sucking the air into the submicron pores of the filters mounted in sampling heads that can readily be adapted for use on aircraft and balloons. The originators develop the filters in the laboratory by allowing them to attain equilibrium at the required temperature, say $-15°$ C, exposing them for a standard time to a flux of small water droplets produced by a warm vapour source, and making the resulting ice crystals visible by pouring a supercooled solution of sodium silicate upon them. Unfortunately, with this technique, the number of crystals appearing on the filter depends rather critically upon the duration of the cooling and humidification procedure, and even unexposed filters sometimes give relatively high background counts. Also the number of crystals appearing on the filter is not proportional to the volume of the air sampled; larger volumes produce proportionately smaller counts, the relationship between the two varying from day to day and from place to place. Mossop and Thorndike (1966) attribute this apparent reduction in concentration of nuclei for larger sampled volumes to the capture of hygroscopic nuclei that denude the water vapour in their vicinity and prevent nearby ice nuclei from becoming activated. Further losses of nuclei in samples taken in highly polluted air are attributed to contamination of the nuclei and the filter by smoke particles, many of which are hygroscopic.

Many of these disadvantages have been removed by the important modifications recently introduced by Stevenson (1968). After exposure, the filter is impregnated with molten petroleum jelly that is allowed to block the pores of the filter but not flood the upper (collecting) surface. After allowing the jelly to set, the filter is placed on a cold metal stage in a small humidity chamber in which the air is saturated relative to the ice-coated walls maintained at $-16°$ C. The temperature of the cold stage and the filter is then lowered to, say, $-18°$ C, and the number of ice crystals appearing on the filter counted. Growing under these conditions, in an atmosphere barely supersaturated with respect to water, many hundreds of nuclei can be activated on a single filter without neighbouring crystals interfering. The concentration of nuclei is now found to be independent of the sampling volume except in very dusty atmospheres. The overall reproducibility of the method is now good and, apart from some doubts about the desirability of heating the filters to 70° C to melt the petroleum jelly, this now seems to be the most reliable and convenient method of determining the ice-nucleus content of the atmosphere. However, recent tests by Sax (1970) show that the Millipore filter technique as just described does not detect silver iodide nuclei of $r < 0.01$ μm which nevertheless may be capable of acting as freezing nuclei following their Brownian capture by supercooled droplets. Should the atmosphere contain significant concentrations of such small natural ice nuclei the Millipore filter method, in common with cloud chambers, is likely seriously to underestimate their contribution.

(d) *Assessment of techniques for determining ice-nucleus concentrations.* All the available techniques for counting ice nuclei involve activation of the nuclei, and the growth, detection, and counting of the resulting ice crystals. Every stage has its difficulties and sources of error and uncertainty.

In the first place, we have to recognize that, in a natural cloud, an ice nucleus may be activated in at least three ways. It may form an ice crystal by the direct deposition of water vapour into ice (i.e. act as a sublimation nucleus); it may act first as a condensation nucleus, (perhaps at humidities below water saturation if it has a soluble coat), and then cause the droplet to freeze; or it may be captured by a supercooled droplet and cause this to freeze. There is some laboratory evidence, (see p. 189), to indicate that these three mechanisms are unlikely to produce an ice crystal at identical temperatures. It is not known for certain which of the three is the most important in natural clouds; much must depend upon the rate of cooling of the air and the magnitude and duration of the supersaturation. Since it is almost impossible to reproduce in a cloud chamber the slow cooling and prolonged low supersaturations that prevail in the atmosphere, the ice nuclei may be activated in a different way and at different temperatures than would be the case in natural clouds. There are also serious discrepancies between the results obtained with different techniques. Thus Kline and Brier (1961), in extensive comparisons of measurements made with expansion and mixing chambers of identical geometry, and using the same (supercooled sugar solution) detector, found that, on average, the mixing-chamber counts were twenty times higher at $-15°$ C, and five times higher at $-20°$ C, than those recorded by the expansion chamber.

This discrepancy may arise largely because the minimum temperature achieved at the end of the expansion persists for only a few seconds and this may be insufficient for all the potential ice nuclei to be activated and grow into detectable ice crystals. Differences also arise from the quite different supersaturations achieved in the two instruments. Very high supersaturations, much higher than those attained in natural clouds, are achieved in large, rapid expansions. These promote condensation upon a larger fraction of the potential ice nuclei, hasten dissolution of any hygroscopic material on their surfaces and thereby improve their ice-nucleating properties. On the other hand, there is some evidence—by Mason and Maybank (1958), Mason and Van den Heuvel (1959), and Hoffer (1961)—that some partly soluble particles, some metallic oxides, and some clay minerals become less effective ice nucleators when immersed in liquid water than when they can adsorb a film of moisture from a slightly supersaturated atmosphere.

In a mixing chamber, cooling of the air takes place mainly near the glycerol-coated walls, which absorb much of the excess moisture and prevent water saturation being reached. In fact, with low ambient humidities, a water cloud will not form unless additional moisture is supplied, but then new problems arise because addition of moisture may produce high supersaturations in at least part of the cloud. But on the whole, the mixing chamber, though tedious to operate especially in warm humid climates, seems a more reliable method than the expansion chamber. The Bigg–Meade continuously-recording mixing chamber is attractive in principle in that cloud-like conditions are more nearly approached than in other types but, unfortunately, it has given a good deal of mechanical trouble and it cannot readily be used for making frequent measurements at more than one temperature.

Turning now to the growth of the ice crystals after nucleation, their failure to grow to detectable size will cause the concentration of ice nuclei to be underestimated. Thus if the crystals are to be detected in a beam of light, they must grow to a size at which they can be readily distinguished from water droplets, and the clouds must not be too dense for this reduces the contrast. If supercooled detecting solutions are used, care must be taken to see that ice saturation persists throughout the duration of the experiment so that the ice crystals do not evaporate before they reach the detector. This is a serious problem in very dry climates.

The detection methods themselves may give rise to serious errors. Because the ease of detection varies with the crystal form, crystal size, cloud density, illumination, and the observer's experience and vision, the direct visual counting of the crystals in a light-beam is highly subjective and very unreliable. Moreover, any glittering particles, such as specks of micaeous dusts, may be easily mistaken for ice crystals. Supercooled solutions, including soap films, provide a much more convenient and reliable measure of the total number of ice crystals falling from a cloud, but their temperature and concentration must be accurately controlled. If the solutions are too weak or too cold, ice crystals may be nucleated by contaminating particles in the liquid; if they are too warm or too concentrated, ice crystals will not grow at all. Sodium silicate solutions, whose preparation is described by Hoffer, Weber, and Fritzen (1964), are much less critical than sucrose solutions and operate over a greater range of temperature. If the solution is used repeatedly, being warmed up between experiments to destroy the ice crystals, its concentration may change with time, particularly if the environment is very warm and humid, or if it is very dry. If the cloud chamber walls are coated with glycerol to prevent frost formation, the equilibrium humidity of the air may be quite low, and evaporation of water from the detecting solution may make it too concentrated to grow ice crystals.

The formation of frost on the walls of the cloud chamber and the detachment of small splinters may give rise to spuriously high counts of ice nuclei. For example, Bigg and Meade (1959) noticed that, in their continuous chamber, the apparent concentration of ice nuclei built up over one thousandfold when glycerol-free patches developed on the walls.

The continuous acoustic nucleus detector recently described by Langer, Rosinki, and Edwards (1967) has some attractive features, notably that, carried on an aircraft, it can provide a continuous record of nucleus concentrations ranging from 0·1 to 500/litre at levels below 15 000 ft. But it appears to count only a fraction of the activated nuclei, and until it has been thoroughly tested and compared in flight with other methods, one cannot assess its ability to provide reliable measurements of ice-nucleus concentrations in the atmosphere as distinct from being a valuable detector in, say, the tracking of nucleus plumes, where the absolute values may not be too important.

The difficulties and uncertainties that attend the measurement of ice-nucleus concentrations are many and not sufficiently appreciated. Many of the earlier measurements, in which insufficient care was taken to prevent frost formation on the walls of the chambers, and which employed visual counting of the crystals in a beam of light, must be regarded as highly suspect. Even the best of the current techniques are

§ 4.4   INITIATION OF THE ICE PHASE IN CLOUDS   183

not capable of accurately determining the absolute concentrations of ice nuclei active at different temperatures, though relative values obtained by the same technique may be more reliable. Close and detailed comparison of measurements made with different instruments and observers, at different places and at different times, are hardly worth while. In any case, the observed concentrations of ice nuclei in clear air seem to bear very little relation to the numbers of ice crystals appearing in clouds at the same temperature. Apart from the fact that the sampling instruments cannot reproduce the conditions under which the ice nuclei may operate in a natural cloud, we shall argue in § 4.4.6 that, in a cloud, the primary ice nuclei may be multiplied a hundredfold or more by the fragmentation of ice crystals and freezing water droplets.

4.4.2. *Concentrations of ice nuclei as a function of temperature*

Many investigations have been made of the activity of atmospheric ice nuclei in relation to the air temperature. In most sampling methods the ice crystals are counted as they appear among a cloud of supercooled droplets, but the results obtained appear to depend upon the techniques used for producing the cloud and counting the crystals and, accordingly, it is convenient to divide the observations into two groups.

In the first group, we include those investigations that employed a relatively large chamber in which the temperature was reduced fairly slowly and the consequent gradual increase in the numbers of ice crystals was estimated visually.

Findeisen and Schulz (1944) used a large chamber, of capacity 2 m³, to study the ice-nucleus content of the surface air in Prague, and employed slow rates of expansion that corresponded to vertical air velocities of 5–22 m s^{-1}. On average, the number of ice particles increased steadily from about 1/m³ at $-12°$ C, to about 100/m³ at about $-20°$ C, and 1/litre at $-30°$ C. In this temperature range, the threshold temperature for the appearance of ice crystals in the lowest observable concentration of a few per m³ decreased with increasing rates of expansion, and was correlated with the rate prevailing at the time of cloud formation rather than with the subsequent expansion rate. The ice nuclei in this regime were called 'type 1' nuclei. When the temperature fell below about $-32°$ C, the concentration of ice particles increased almost one thousandfold to about 1/cm³. The temperature at which this rather sudden increase occurred was higher for increased rates of cooling, the responsible particles now being termed 'type 2' nuclei. At still lower temperatures, the numbers of crystals increased still further but, even near −40° C, they were considerably less than those of the water droplets.

The use of a large mixing chamber on an aircraft has been described by Smith and Heffernan (1954). A chamber of 76-litres capacity was cooled at a steady rate, and the number of crystals appearing in an illuminated volume counted visually at different temperatures. Precautions were taken to guard against the results being vitiated by frost splinters leaving the walls of the apparatus. Cognizance was also taken of the fact that there is a time lag in the activation of ice nuclei, that those crystals that appear first may settle out before others have appeared, and that losses by sedimentation will depend upon the rate at which the chamber is cooled. The observed crystal concentrations were therefore corrected by factors that were experimentally determined by using silver iodide smokes as test substances. The lower limit of nucleus detection varied a little with temperature, but was of the order 1 particle/litre.

The authors report the occurrence of three types of ice nucleus—see Fig. 4.8. One of these appeared only below −30° C, in reproducible

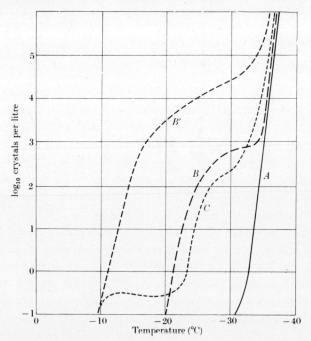

FIG. 4.8. The concentrations of ice crystals observed at various temperatures in a mixing chamber on an aircraft. (From Smith and Heffernan (1954).)

§ 4.4 INITIATION OF THE ICE PHASE IN CLOUDS

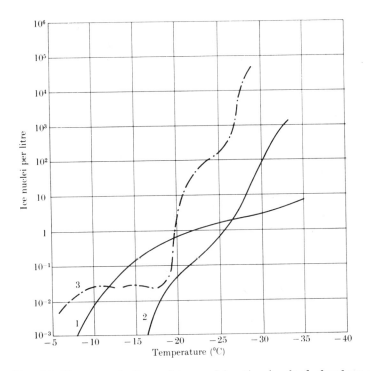

FIG. 4.9. The concentrations of ice nuclei active in cloud chambers as a function of temperature. 1, Findeisen and Schulz, 5 m s^{-1}; 2, Findeisen and Schulz, 20 m s^{-1}; 3, Murgatroyd and Garrod.

concentrations of about 1/litre at about $-32°$ C, increasing to about 10^6/litre (complete glaciation of the cloud) at $-37°$ C, and apparently correspond to Findeisen's 'type 2' nuclei. The second type was usually present in variable quantities and had rather variable characteristics, detectable numbers appearing at threshold temperatures between -10 and $-20°$ C. The third class, which was observed only occasionally, was characterized by the appearance of a low concentration (about 1/3 litres) of comparatively efficient nuclei active at about $-10°$ C. Unfortunately, subsequent experience has shown that this chamber with visual counting gives unreliable results and so the data on the second and third types or nuclei are probably not very significant although the consistent increase in concentration of crystals below $-30°$ C is probably real.

The results of all these investigations, shown in Fig. 4.9, show similar trends in that the concentration of ice crystals rises as the temperature falls below $0°$ C, reaches a plateau, and then rises very sharply as the temperature falls below about −30° C. Mason (1955a) has shown that this latter sudden rise, observed below −32° C in the large chambers used by Findeisen and Schulz and by Smith and Heffernan, may be attributed to spontaneous crystallization of the cloud droplets, but the crystals appearing at higher temperatures must have formed on freezing nuclei.

Experiments in our second group are characterized by the making of single measurements at a fixed temperature achieved either by a

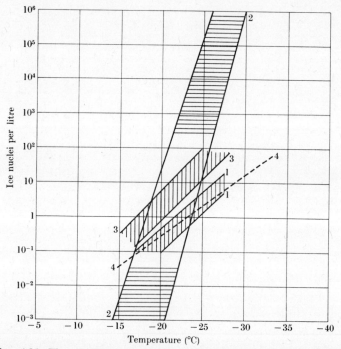

FIG. 4.10. The concentrations of ice nuclei active in cloud chambers as a function of temperature. 1, Palmer—small expansion chamber; 2, Workman and Reynolds—mixing chamber; 3, Fenn and Weickmann—thermal precipitator; 4, Warner—expansion chamber.

fast expansion, or by sudden chilling in a mixing chamber held at constant temperature. Fig. 4.10 summarizes the measurements of Palmer (1949) in England, Workman, and Reynolds (1950a) in New Mexico, and Warner (1957) in Australia. They all show a steady exponential increase in ice-nucleus concentration with decreasing temperature, although the slopes of the curves vary, and the concentrations at a given temperature differ by as much as two orders of magnitude. Although there may be large variations from day to day and from place to place, Fletcher (1962) deduces that, on average, the spectrum of atmospheric ice nuclei is probably best represented at temperatures between -10 and $-30°$ C by an exponential relation of the form

$$n(\Delta T) = n_0 \exp(\beta \Delta T), \tag{4.18}$$

where ΔT is the degree of supercooling, $n(\Delta T)$ is the number of nuclei per litre active at supercoolings less than ΔT, and n_0 and β are numerical factors. A reasonable average value for β is 0·6 with values between 0·4 and 0·8 being common; n_0 is typically about 10^{-5}/litre but may vary by several orders of magnitude. This gives typically a concentration of one active nucleus per litre at $-20°$ C, the concentration changing by a factor of ten for a temperature change of $4°$ C.

4.4.3. *Variations of ice-nucleus concentration in space and time*

Measurements made on land, at sea, and in aircraft show that the ice-nucleus content of the atmosphere may vary considerably with time and place. Characteristically the count at any one place may remain at a low level for several days or even weeks at a time, and then rise suddenly to a very high value—thousandfold increases are not unusual. Schaefer (1954), in a series of 3-hourly readings taken over a period of 6 years on the summit of Mt. Washington, found a tendency towards low counts in the summer and high counts in the winter. A similar result was found by Rau (1954) in a shorter series of measurements near Ravensburg. Schaefer reported individual counts varying between 1 and $10^7/m^3$. Some of the very high counts in winter may have been spurious, and due to frost splinters produced by the strong winds on the snow-covered mountain or from the walls of the chamber, but he also claimed that very high counts were associated with the arrival of air that had passed over the arid regions of the United States during periods of dust-storm activity. Observations made for only a short period each day may fail to reveal short-period fluctuations. But Bigg and Meade (1959), using a continuously-recording counter, discovered nucleus 'storms' that had typical lifetimes of less than 15 h during which concentrations sometimes increased one hundredfold in less than 1 hour.

In many cases, the fluctuations in ice-nucleus concentrations are probably due to local or regional sources of nuclei, for example the effluent from industrial plants such as steelworks, or dust storms, and then the counts may depend on the wind direction and the vertical stability of the atmosphere. Bigg (1961) suggested that vertical mixing is responsible for the onset of the nucleus storms. On very calm, clear nights, he found that the nucleus concentration beneath the ground temperature inversion steadily increased until soon after sunrise, and then suddenly diminished as the nuclei became carried to higher levels by turbulent mixing. On other occasions, a sudden influx of nuclei was observed some hours after sunrise, but the count diminished slowly thereafter. Bigg suggests that nuclei were trapped by an inversion above the measuring point and that the onset of turbulence after sunrise conveyed them rapidly downwards.

TABLE 4.2

Average concentrations of ice nuclei in surface air in various parts of the world (after Bowen 1961)

Location	Number of nuclei per litre at $-20°$ C	Source of data
(a) Mixing chambers		
Northern Hemisphere	2·6	France U.S.A. Japan
Southern Hemisphere	0·2	Australia
(b) Expansion chambers		
Northern Hemisphere	0·3	U.S.A. Great Britain France Sweden Germany
Southern Hemisphere	0·08	Australia South Africa New Zealand
Florida Hawaii Caribbean	0·1	
Antarctic Ocean	0·1	

Despite relatively large short-period variations, there is fair agreement between the *average* concentrations of active nuclei measured at widely separated places with the same type of instrument. A number of fairly long-term series of measurements taken in Australia, Europe, the United States, Japan, South Africa, and New Zealand provide some data on the average background levels of ice nuclei. Table 4.2 shows the mean of the lowest quarter of all readings for the Northern and Southern Hemispheres as obtained by both mixing and expansion chambers.

One notes that the ice-nucleus concentrations appear to be more than ten times higher in the Northern than in the Southern Hemisphere, except at oceanic locations such as Florida, Hawaii, and Puerto Rico. On the other hand, Bigg (1965), using only data from similar mixing chambers, found rather small differences between the two hemispheres. Table 4.2 also confirms that expansion chambers give consistently lower counts than mixing chambers. The available data suggest that, on about one-quarter of occasions, the nucleus concentrations in relatively calm air are about $10/m^3$ active at $-10°$ C, 1/litre at $-20°$ C, and 10/litre at $-25°$ C but that, on some occasions, and in special regions influenced by local sources, the concentrations may exceed these figures by more than a hundredfold.

In order to separate and assess the relative contributions from natural and man-made, global and local sources, it may be necessary to establish a global network of sampling stations, but this must await the perfection of more simple, convenient, and reliable devices than are presently available. However, proper interpretation of the data will probably require detailed analyses of air motions on both the synoptic and global scales and it is not at all clear that the effort would be worthwhile.

4.4.4. *Mode of action of ice nuclei*

(a) *Sublimation and freezing nuclei.* When the cloud chamber experiments of Fournier d'Albe (1949) and others showed that ice crystals formed only when the humidity of the atmospheric air well exceeded saturation with respect to ice and approached water saturation, Wegener's (1911) suggestion that ice crystals might form by the direct 'sublimation' of water vapour on suitable nuclei was largely discounted and it was supposed that the atmospheric process is one of condensation followed by droplet freezing. The appearance of some ice crystals slightly below water saturation, as reported by Findeisen (1942a), is not necessarily proof of them acting as sublimation nuclei because they may have acquired liquid water either by being hygroscopic 'mixed' nuclei or by capillary condensation. There is now, however, convincing evidence that the surfaces of many natural and artificial crystalline solids can nucleate ice crystals while remaining apparently dry; they probably act either by the adsorption of multimolecular films or patches of water on the surface, or by capillary condensation into steps, cracks and interstices. We shall follow Koenig (1962a) in distinguishing between these *sorption* nuclei and the *condensation-freezing* nuclei, and reserve the term *sublimation nucleus* for the special case involving condensation on to a (partly) ice-covered nucleus.

The size of the particles may also be an important factor in their mode of action. A very small hydrophobic particle, for example a silver-iodide particle of $r < 0.01$ μm, will be unable to act as a condensation-freezing nucleus unless the air becomes strongly supersaturated relative to a plane liquid water surface because of the curvature effect. Such particles are therefore not likely to become effective as ice nuclei in natural clouds unless they collide with supercooled cloud droplets or collect hygroscopic matter. Thus Edwards and Evans (1960) found that <0.5 per cent of AgI particles of $r < 0.01$ μm were effective below water saturation, even at $-18°$ C, and that the majority of the particles became active only when the relative humidity exceeded 110 per cent.

Using an aerosol of micron-sized silver-iodide particles, for which the curvature effect is negligible, Mason and Van den Heuvel (1959) observed that ice crystals appeared at temperatures between -4 and $-12°$ C only if the humidity surpassed water saturation but, below $-12°$ C, some ice crystals appeared if the air was sub-saturated relative to water but supersaturated relative to ice by at least 12 per cent. Recently, Roberts and Hallett (1968) found that about 1 in 10^4 particles of kaolinite, gypsum, glacial debris, and calcite behaved in a similar manner with the transition taking place at $-19°$ C and 20 per cent supersaturation relative to ice. After preactivation (see p. 191), both the threshold and transition temperatures were several degrees higher and ice crystals appeared at lower supersaturations than before.

From the practical point of view, it is important to know the number of nuclei that will act below or at water saturation relative to the number that will act by the freezing of bulk water at the same temperature. Working with nearly monodisperse AgI particles of $r \sim 2$ μm, Edwards and Evans (1968) found that the number of sorption nuclei effective at water saturation and at $-8°$ C was only about 1 per cent of those acting at freezing nuclei, while at $-16°$ C the fraction was about 10 per cent.

(b) *Behaviour of ice nuclei immersed in aqueous solutions.* There is no evidence to suggest that atmospheric particles act as sublimation nuclei; indeed, considering their mixed nature, the presence of hygroscopic matter will aid condensation and make sublimation impossible. The effect of the solute will be to lower the freezing temperature of the droplet, and Hoffer (1961) finds that, when salt is added to droplets containing particles of the kaolin clays, the depression in the freezing point is greater than the depression achieved by adding the salt to pure water. It may be that favourable nucleating sites on the surfaces of the nuclei are spoiled by the adsorption of solute ions.

Even the immersion of nuclei in fairly pure water may have a deleterious effect because, although the material may be very insoluble in bulk, very small particles and especially favourable nucleation sites may be dissolved. For example, Mason and Van den Heuvel (1959) found that, while the behaviour of a number of the most effective types of inorganic nuclei (including silver iodide) was the same whether the particles were ejected into a cloud chamber or immersed in bulk water, metallic oxides were much less effective when immersed in a water drop then when allowed to fall on a droplet surface, or when used as nuclei for cloud formation. Vaterite and a number of organic nuclei such as cholesterol also appear to be less effective when immersed in water, but Evans (1966) found the opposite for l-leucine. Hoffer (1961) reported that the freezing points of water drops containing clay minerals were appreciably lower than the corresponding threshold temperatures quoted by Mason for these substances in Table 4.3, but there is no guarantee that the sizes and purity of the particles were the same in the two sets of experiments.

(c) *Preactivation of nuclei.* The possibility that some ice nuclei may become more effective having once been involved in ice-crystal formation was first suggested by an observation of Fournier D'Albe (1949). He observed that, although cadmium iodide particles first produced ice crystals only at water saturation and below $-41°$ C, when the ice particles were allowed to evaporate, provided they were not warmed above $-9°$ C, they left behind nuclei on which ice would grow again at humidities only slightly above ice saturation. This behaviour of CdI_2 was confirmed by Mossop (1956a), who also found that a small proportion of the particles from sprayed suspensions of Iceland spar, gypsum, and sodium-bentonite clay behaved in a very similar manner. Mason and Maybank (1958) discovered that ten naturally-occurring minerals including kaolinite, volcanic ash, montmorillonite, and stony meteorite, were effective at higher temperatures after preactivation. Thus ice crystals grown on kaolinite nuclei at $-9°$ C, when evaporated and warmed to near, but not above, $0°$ C in a dry atmosphere, left behind nuclei which were thereafter active at $-4°$ C. Particles of montmorillonite, initially inactive even at $-20°C$, could be 'trained' to act a second time at $-10°$ C. Serpolay (1959) has reported similar behaviour for a number of metallic oxides including CuO, CrO_3, and Fe_3O_4. Day (1958) reported that the preactivation of calcite and pulverized fragments of stony meteorite was destroyed when the particles were exposed to a humidity of 80–90 per cent relative to ice at $-90°$ C for several minutes, and that clays and gypsum were deactivated in this humidity after only one minute. However, Higuchi and Fukuta (1966) found that samples of kaolinite, illite, and montmorillonite, kept at 50 per cent humidity and cooled to $-78°$ C, could subsequently be activated at temperatures as high as $-3°$ C when subjected to supersaturations relative to ice.

Mason (1950) first explained preactivation in terms of warming and drying causing only partial evaporation of the ice and the retention of a sub-microscopic film or patches of ice on the surfaces of the particle which would subsequently become activated when the humidity again exceeded ice saturation. Alternatively the phenomenon might be explained by the retention of embryos of ice in small cavities and capillaries that survive evaporation of the macroscopic ice because of the curvature effect.

The situation has recently become rather clearer as the result of a careful investigation by Roberts and Hallett (1968) in which initial activation, drying and warming, storage, and re-activation of the nuclei are all carried out in carefully controlled and slowly changing conditions of temperature and supersaturation that approximate to those prevailing in the atmosphere. They find that for 1–2 μm particles of kaolinite, montmorillonite, bentonite, gypsum, calcite, vaterite, albite, and glacial debris, there exists a threshold temperature for ice nucleation at water saturation and, below this, a 'critical' temperature below which nucleation can occur in any environment sub-saturated relative to water but supersaturated relative to ice by a constant value peculiar to the substance under test. Thus 1 in 10^4 of kaolinite particles had an initial nucleation threshold of $-10.5°$ C, a critical temperature of $-19°$ C, and a critical ice supersaturation of 20 ± 2 per cent. Corresponding figures for gypsum and calcite were $-17°$ C, $-20°$ C, and 20 per cent, while montmorillonite and bentonite had threshold temperatures of $-25°$ C and critical temperatures below $-28°$ C. When the macroscopic ice was evaporated and the nuclei held at humidities below ice saturation for long periods of up to one week, they were subsequently more effective as indicated by the new figures for kaolinite of $-4°$ C, $-12°$ C, and 12 per cent, and for montmorillonite of $-4°$ C, $-14°$ C, and 14 per cent. If the particles were subjected to temperatures above the threshold values, the enhanced activity was destroyed in a time dependent on temperature and humidity and which, for kaolinite at $-4°$ C and a subsaturation of 35 per cent relative to ice, was only 30 s.

These results are in broad agreement with those of Mason and Maybank, but differ from those of Day in that clays were found to retain preactivation for long periods when exposed to humidities of 80 per cent over ice. Regarding the mechanism of preactivation, Roberts and Hallett favour an explanation in terms of an adsorbed ice-like surface film rather than the retention of ice embryos in capillaries on two grounds: that the former explanation seems more consistent with the fact that the form of the supersaturation-temperature criteria for nucleation is similar for preactivated and non-preactivated nuclei; and that the observation that ice can spread on preactivated nuclei at quite low ice supersaturations would require capillaries of much larger diameter than are thought to exist. This last argument lacks conviction and the retention of embryos in capillaries cannot be ruled out.

Mossop suggested that the retention of ice embryos by particles involved in the formation of high ice clouds might enable them to nucleate lower clouds at temperatures only just below $0°$ C, while Mason and Maybank suggested that some of the more abundant soil particles such as montmorillonite, which are initially ineffective as ice nuclei, might be carried up and 'trained' at cirrus levels and thereafter nucleate lower clouds at much higher temperatures. Large quantities of ice crystals evaporate as they fall from Cb. anvils, cirrus, and other high-level clouds that give virga, and from frontal altostratus. They may leave behind preactivated nuclei that may remain suspended for a considerable time and remain active until they are either removed in precipitation, subjected to temperatures above about $-5°$ C, or to humidities of less than about 50 per cent relative to ice. Outside these limits, preactivation will probably be destroyed within a few minutes. The effective lifetime of these 'trained' nuclei will therefore depend very much on the persistence and distribution of humidity in the troposphere. In this connection, we recall the interesting report of Braham and Spyers-Duran (1967) who collected from an aircraft pyramidal, bullet-shaped, and hollow prismatic ice crystals in concentrations of up to $10^6/m^3$ that had apparently originated in cirrus trails and survived a fall of 20 000 ft in clear air having an average humidity of only about 25 per cent.

(d) *De-activation of ice nuclei by atmospheric pollutants.* The presence of a small concentration of ammonia in the air was reported by Reynolds, Hume, and McWhirter (1952) to increase the number of ice crystals that form on AgI particles in a supercooled cloud. On the other hand, the de-activation or 'poisoning' of active ice nuclei by certain organic vapours is well known. For instance, Birstein (1957, 1960) found that ethyl amine and methyl amine effectively inhibited the nucleation of ice by AgI and PbI_2, ethyl amine in concentrations as low as 1 p.p.m. depressing the threshold nucleation temperatures of PbI_2 from -6 to $-15°$ C. Methyl amine, at a partial pressure of about 2 mb, prevented ice crystal formation in outside air at temperatures down to $-52°$ C. Laboratory tests by Georgii (1963) showed that although the threshold temperatures of covellite, vaterite, and kaolinite were depressed by about $10°$ C by massive concentrations of ammonia the depression was less than $1°$ C for the highest concentrations that normally exist in industrially-polluted air. The effect of SO_2 was even smaller, and the combined effect of SO_2 and NH_3 was intermediate between that of the two gases separately. The conclusion is that these two major components of atmospheric pollution have negligible inhibiting effect on atmospheric ice nuclei. The Brownian capture of small smoke particles may be more important in this respect.

4.4.5. *Nature and origin of ice nuclei*

Since a cubic metre of atmospheric air may contain as few as ten ice nuclei active at $-10°$ C among, perhaps, 10^{11} other particles, the ice nuclei are very difficult to identify. Their nature and origin are, however, of considerable interest. The most natural assumption to make is that they originate mainly from the earth's surface as dust particles and are carried aloft by the wind, and Mason (1950) suggested that they consist largely of clay-silicate particles. Alternatively, Bowen (1953) has suggested that they may enter the top of the atmosphere as meteoritic dust.

Evidence as to the origin of ice nuclei may arise from three main sources. One may test, in the laboratory, particles of soils, mineral dusts, and smokes for their ice-nucleating ability and then assess their abundance in the atmosphere. A second approach is to collect snow crystals from the atmosphere and try to identify the nature of the nuclei on which they are formed. Thirdly, one may look for clues in the spatial and temporal variations of the ice-nucleus content of the atmosphere and how these are affected by geographical and meteorological factors.

(a) *Activity of natural mineral dusts.* Sprayed suspensions of various soil particles were introduced into a mixing-cloud chamber by Schaefer (1949*b*), who found that some clays produced ice crystals at temperatures as high as $-12°$ C and had a high level of activity at $-24°$ C. Pruppacher and Sänger (1955) found quartz, porcelain, clays, olivine, and activated charcoal to be effective in a supercooled cloud at temperatures below $-12°$ C, while Hama and Itoo (1956) found that the silicate minerals kaolinite, amphibole, and augite induced the freezing of supercooled water droplets suspended in silicone oil at temperatures below $-10°$ C.

A more extensive and detailed investigation on substances of higher purity was carried out by Mason and Maybank (1958) and Mason (1960c). They found that of thirty-five minerals tested in a mixing-cloud chamber, twenty-one, mainly silicate minerals of the clay and mica groups, were active nucleators to the extent of 1 particle in 10^4 producing an ice crystal at temperatures of $-15°$ C or above and, of these, ten were active above $-10°$ C—see Table 4.3. These substances are all minor constituents of the earth's crust and it is significant that common substances such as sea sand were not effective. The most abundant of these active substances is kaolinite; on the score of both natural abundance and effectiveness, Mason considered the kaolin minerals together with illites and halloysite to be the most important sources of natural terrestrial ice nuclei.

Supporting evidence for this view is provided by Isono, Komabayasi, and Ono (1959) and Isono and Ikebe (1960). Having discovered that high concentrations of ice nuclei in Tokyo were produced by the influx of yellow loess dust from China, they tested this material in the laboratory with a mixing chamber and a sugar-solution detector, and found that its threshold temperature, as defined in Table 4.3, was between -12 and $-15°$ C. They also tested several other mineral dusts in finely powdered form and, as shown in Table 4.3, obtained results very similar to those of Mason and Maybank for magnetite, albite, orthoclase, quartz, calcite, stony meteorites, and volcanic ash, but rather lower threshold temperatures for specularite, muscovite, and kaolinite. The activity of kaolinite varied a little according to the source and it seems that rather small differences in surface structure and composition may be important. There is also some evidence that the ice-nucleating ability of the clay minerals is reduced when they become immersed in bulk water and dilute aqueous solutions and they may therefore be less effective in the atmosphere than the laboratory tests suggest.

(b) *Industrial sources of ice nuclei.* While smokes from domestic and most industrial fires appear to provide very few ice nuclei, there are a number of specific sources, such as steel works, that produce large

TABLE 4.3
Ice nucleating ability of naturally occurring particles

Substance	Chemical composition	Symmetry	Threshold temperature (°C)	References
Covellite	CuS	Hex	-5	M & M
Vaterite-	$CaCO_3$	Hex	-7	M & M
β-Tridymite	SiO_2	Hex	-7	M & M
Magnetite	Fe_3O_4	Cubic	-8	M & M
			-9	I & I
Kaolinite	$Al_2(OH)_4Si_2O_5$	Triclinic	-9	M & M
			-13	I & I
Anauxite	$Al_2(OH)_4 \cdot Si_2O_5$	Monoclinic	-9	M
Illite		Monoclinic	-9	M
Metabentonite			-9	M
Microcline			-9	M & M
Hypersthene	$(Mg, Fe)_2(Si_2O_6)$	Rhombic	-10	I & I
Haematite	Fe_2O_3	Hex	-10	M & M
(Specularite)			-13	I & I
Pyrophyllite	$Al_2(OH)_2 \cdot Si_4O_{10}$	Monoclinic	-10	M
Gibbsite	$Al_2(OH)_6$		-11	M & M
Halloysite	$Al_2(OH)_4 \cdot Si_2O_5 \cdot 2H_2O$	Monoclinic	-12	M
			-13	M & M
Dickite	$Al_2(OH)_4 \cdot Si_2O_5$	Monoclinic	-12	M
Olivine	$(Mg, Fe)_2 \cdot SiO_4$	Rhombic	-12	P & S
			-18	I & I
Aquadag	C		-12	M & M
Dolomite	$CaMg(CO_3)_2$	Hex (Rhomb)	-14	M & M
Biotite		Monoclinic	-14	M & M
Attapulgite	$4H_2O \cdot (OH)_2Mg_5 \cdot Si_8O_{20} \cdot 4H_2O$	Monoclinic	-14	M
Muscovite		Monoclinic	-14	I & I
			<-18	M & M
Vermiculite		Monoclinic	-15	M & M
Nontronite		Monoclinic	-15	M
Montmorillonite		Monoclinic	-16	M
			<-18	M & M
Gypsum	$CaSO_4 \cdot 2H_2O$	Monoclinic	-16	M & M
Graphite	C		-16	M & M
Cinnabar	HgS	Hex	-16	M & M
Orthoclase	$KAlSi_3O_8$	Monoclinic	-17	I & I
			<-18	M & M
Anorthoclase			-17	M & M
Quartz	SiO_2	Hex	<-18	M & M
			-19	I & I
			<-20	I, K & O

TABLE 4.3 *Continued*

Substance	Chemical composition	Symmetry	Threshold temperature (°C)	References
Stony meteorite			−17	M & M
			−17	I & I
		(2 specimens)	< −18	M & M
		(3 specimens)	< −17	M
Volcanic ash:				
10 Japanese volcanoes			−12 to −16	I, K & O
Mt. Etna			−13	M & M
Crater Lake, Oregon			−16	S
Paricuten, Mexico			−23	S
Soils:				
American-loam, clay,			−8 to −25	S
Clay			−11	P & S
Loess, N. China			−12	I & I
			−15	I, K & O
Loess, Hanford, U.S.A.			−11	S

The following substances were found by Mason *et al.* to be inactive at temperatures above −18° C: Montmorillonite, sepiolite, albite, talc, sand, quartz, α-tridymite, and several samples of stony meteorite.

M & M = Mason and Maybank (1958), M = Mason (1960c), I & I = Isono and Ikebe (1960), I, K & O = Isono, Komabayasi, and Ono (1959), P & S = Pruppacher and Sanger (1955), S = Schaefer (1949b).

numbers of active particles. For example, Soulage (1958) estimated that a 25-ton electric-arc steel furnace produces, in one day, up to 3×10^{15} nuclei active at −20° C, and believes that they consist of oxides of iron and other metals. Telford (1960a) found electric furnaces producing manganese and silicon steels to be sources of effective nuclei.

The laboratory investigations of Serpolay (1958, 1959) and of Mason and Van den Heuvel (1959) confirm that many metallic oxides, notably those of iron, copper, aluminium, manganese, and nickel, are capable of acting as ice nuclei, some with threshold temperatures as high as −6° C. However, the daily output of Soulage's furnace is equivalent to the burning of only about 1 gramme of silver iodide and, since the efficiency of the nuclei is likely to be further reduced by coagulation and contamination in the atmosphere, such sources are unlikely to make a major contribution to the total supply of atmospheric nuclei, although locally they may mask the natural population.

(c) *The direct identification of ice nuclei.* Kumai (1951) and Aufm Kampe, Weickmann, and Kedesdy (1952) used the electron microscope to examine solid particles found at the centres of natural snow crystals when they were allowed to sublime. Identification was based on the similarity of their appearance to particles of known composition, but subsequent developments have made it possible to identify these particles more certainly from their electron-diffraction patterns. The application of this technique by Kumai (1961), Kumai and Francis (1962), and Rucklidge (1965) to particles in the centres of snow crystals collected in remote areas in Japan, the United States, and Greenland yielded results that show a remarkable measure of agreement as can be seen in Table 4.4. All the Japanese investigators agree that about 90 per cent of the central nuclei can be identified, that the particle diameters range from <0.1 to 8 μm, and that the vast majority of these are particles of clay, probably kaolinite.

These results appear to confirm the work of Mason and Maybank, but Kumai's technique has been questioned by Mossop (1963a) on the grounds that one cannot be certain that the large central particle, rather than one of the numerous smaller particles scattered throughout the crystal, acted as the ice nucleus, that clay particles may be favoured because they produce easily identifiable diffraction patterns, and that, in any case, they could be collected by the snowflake after its formation. Moreover, it is very difficult to reconcile the appearance of a clay

TABLE 4.4

Identification of central particles in snow crystals (percentage of various compositions)

Locality	Hokkaido, Japan	Honshu, Japan	Michigan, U.S.A.	Greenland	Missouri U.S.A.
Period	1948–56	1955–59	8 days in 1959	1960	1964
Authority	Kumai (quoted by Kumai (1961))	Isono (quoted by Kumai (1961))	Kumai	Kumai and Francis	Rucklidge
Clay minerals:					
Kaolin group			51	52	
Montmorillonite			14	16	
Illite			12	7	
Related minerals			10	10	
Total clay minerals	57	88	87	85	28
Hygroscopic particles	19	0	1	1	2
Combustion particles	8	4	2	0	3
Micro-organisms	1	0	0	0	1
Unidentified	10	8	9	11	40
(includes spheres)			2	2	—
Not observed	5	0	1	3	26
Total cases	307	52	271	356	250
Growth temperature of snowflakes	not known	not known	−5 to −18° C	mostly −4° to −18° C	

particle in nearly every snow crystal with Koenig's (1963) finding that such particles are almost entirely absent from ice crystals in cumulus clouds, and the arguments advanced in § 4.5 that the majority of snow crystals probably arise from ice splinters. It is also surprising that Kumai and Francis found as many as 5 out of 11 snow crystals collected from a stratus cloud seeded with dry ice contained large insoluble particles, probably of clay, unless these were captured by the snow crystals. On the other hand, Rucklidge was unable to find a nucleus in about one-quarter of his crystals, but since 70 of the 85 identified nuclei were clay particles, it is difficult not to conclude that these must be a very important source of at least the larger natural ice nuclei.

(d) *Correlation of ice-nucleus concentrations with meteorological and geographical factors.* A relatively long and extensive series of measurements with standardized equipment would be required to find statistically significant correlations between ice nucleus concentrations and other meteorological variables. Only very few such data are available. We have already mentioned that Schaefer (1954) found that high nucleus counts on Mt. Washington were associated with the arrival of air from dust-storm areas in the continental interior. Also, Isono, Komabayasi, and Ono (1959) observed that one-hundredfold increases in the ice nucleus concentration in Tokyo coincided with high concentrations of yellow loess dust in the air that were traced to dust storms in North China, about 2000 km away. Neither the ground nor the industry around Tokyo appeared to be major sources of nuclei. Maritime air usually contained low concentrations of nuclei, but there were occasional high counts that were associated with volcanic eruptions in southern Japan. Georgii and Metnieks (1958) measured sea-salt nuclei, Aitken nuclei, and ice nuclei at Valentia off the coast of Ireland and found a strong positive correlation between the Aitken- and the ice-nucleus counts. Since the Aitken nuclei were mainly of continental origin and the ice-nucleus counts were low in winds blowing from the sea, it was concluded that the ice nuclei were also primarily of continental origin.

Bigg (1956), on the other hand, found no significant correlation between ice-nucleus counts and wind direction at Carnavon in Western Australia, a consequence, perhaps, of the low nucleating power of Australian soils (Paterson and Spillane 1967), and the comparative lack of heavy industry. Bowen (1961), in summarizing world-wide measurements of ice-nucleus concentrations made in the period 1956–60, found that the reported average concentrations were about ten times higher in the Northern Hemisphere than in the Southern Hemisphere. The only exceptions to comparatively high counts in the Northern Hemisphere came from the oceanic stations Florida, Puerto Rico, and Hawaii. These results could be explained in terms of the Northern Hemisphere having a much larger land surface for the production of ice nuclei by both natural dusts and industrial processes, but Bigg (1965) thinks that these differences may arise, at least in part, from differences in experimental conditions.

Fenn and Weickmann (1959), using a thermal precipitation counter at Thule, Greenland, found ice-nucleus concentrations that were one-third to one-quarter of those found at the oceanic stations mentioned above. Counts made at latitude 78° S in Antarctica by Bigg and Hopwood (1963) were rather constant; they fall well within the spread of results for the Southern Hemisphere, and are slightly higher than the mean. The concentrations of large aerosol particles were, however, much *lower* than in Australia which suggests, perhaps, that the primary ice nuclei are small ubiquitous particles.

In order to establish firmly how the concentration of ice nuclei varies with meteorological and geographical factors, there is an obvious need for further measurements in both hemispheres with a reliable, standardized technique, in places that are unlikely to be greatly affected by *local* sources.

(e) *The variation of ice-nucleus concentration with height.* It might seem that the question of a terrestrial or an extra-terrestrial origin for the ice nuclei might be resolved by measuring the concentrations at various heights in the atmosphere. Cwilong (1947c), from measurements made on the Jungfraujoch, and Palmer (1949) using an expansion chamber in an aircraft, found that the nuclei that became active at $-32°$ C in surface air were absent above inversions, stratus cloud, and haze tops. On the other hand, mixing-chamber measurements in aircraft by Smith and Heffernan (1954) and Murgatroyd and Garrod (1957) failed to find a systematic tendency for the counts to decrease with increasing height. Indeed, on some occasions, the ice-nucleus count in the clean air above a low-level inversion was greater than in the smoke-laden air below, but perhaps this may be attributed to poisoning of the nuclei by the smoke.

Using a mixing chamber in the United States, Kassander, Sims, and McDonald (1957) found that air sampled at 15 000 ft had to be cooled by about 2° C lower than surface air to produce the same concentration of ice crystals, while Mossop, Carte, and Heffernan (1956), using the same type of instrument in South Africa, found little difference between the temperatures at which one crystal per litre appeared at levels between 500 and 15 000 ft above the ground.

Telford (1960b) used a Warner expansion chamber in an aircraft to make measurements up to 40 000 ft (12 km) above Australia, and found that the average counts in the lower stratosphere were not significantly different from those in the upper troposphere. Bigg, Miles, and Heffernan (1961), using the Millipore filter technique on balloons and U2 aircraft, found that the concentrations of nuclei active at $-10°$ C varied widely in the lower stratosphere—from 1 to 100 m^{-3} according to Mossop (1963a). Using the same technique on balloons up to heights of 27 km, Bigg and Miles (1963) report mean concentrations of about 40 m^{-3} active at $-10°$ C, and about 300 m^{-3} active at $-15°$ C, and point out that, per unit mass of air, these concentrations are comparable with maximum values measured at the ground.

However, these relatively high counts of ice nuclei at the higher levels do not prove that the particles are of extra-terrestrial origin. If they arise from distant sources on the earth, they may well be carried aloft by both large-scale and localized vertical motions and so become mixed throughout the depth of the troposphere within a few days; a sharp fall in concentration at greater heights is therefore not inevitable except, perhaps, over the source regions.

If the high nucleus concentrations in the troposphere come from the stratosphere they should be accompanied by radioactive debris, which has a mean residence time of months or years in the lower stratosphere. On the contrary, however, Bigg and Miles (1964) find no correlation between the ice-nucleus content and the β-ray activity of the air. Moreover, as the work of Junge and Manson (1961) and of Mossop (1963b) has shown, small insoluble particles collected from the stratosphere up to heights of about 20 km are embedded in particles of soluble sulphate, probably ammonium sulphate, which are completely inactive as ice nuclei at $-20°$ C, and are likely to depress the freezing activity of the insoluble particles.

The mechanism by which particles from the troposphere may penetrate into the lower stratosphere is at present in dispute, but they may be transported in the tops of large thunderclouds and through other breaks in the tropopause. If, as Junge believes, the sulphate particles in the stratosphere are manufactured from SO_2 and NH_3 of terrestrial origin, there is no reason why the upward eddy transport of these gases into the stratosphere should not be accompanied by small particles.

(f) *The meteor-dust hypothesis.* In order to explain the observation that ice nuclei are sometimes more abundant aloft than nearer the ground, and that long-term rainfall records show peaks on certain calendar dates, Bowen (1953) made the novel suggestion that the most effective ice nuclei in the atmosphere are particles of meteoritic dust entering the top of the atmosphere from outer space.

He found that the daily rainfall of Sydney, Australia, during January, summed over the period 1859–1952, shows distinct peaks on 13 and 22 January and 1 February. When similar peaks were found by Bowen (1956a) in long-period rainfall records of groups of stations in the State of New South Wales, New Zealand, South Africa, the United

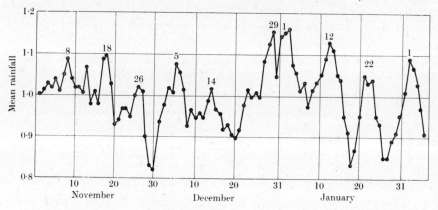

FIG. 4.11. The mean rainfall of November, December, and January for 300 stations distributed over the globe. (From Bowen (1956a).)

States, Japan, the Netherlands, and the British Isles, the evidence pointed to a world-wide rather than a local phenomenon. Combining all these data, Bowen (1956b) produced Fig. 4.11, a world rainfall curve incorporating data from approximately 300 stations distributed over the whole globe that had records for about 50 years. This diagram shows deviations from the mean of ±15 per cent and the three peaks on 13 and 22 January and 1 February are well marked. In similar analyses, Bowen (1956c, 1957) finds that the rainfall for November and December, August, September, and October also shows departures from the mean that are similar to those for January. Bowen also states that, in general, a peak in the rainfall curve is produced not by one or two very heavy falls on that day, nor by an extraordinarily large number of rainy days on that date, but by a greater than normal number of rainy days on which the rainfall was consistently higher than average. There has been much argument about the statistical significance of these rainfall peaks and the validity of the usual tests, because rainfall data tend to be serially correlated and do not obey a normal distribution. Thus the statistical tests used by Hannan (1955) did not disprove the hypothesis that the Sydney rainfall was homogeneous. On the other hand, O'Mahony (1962) concluded that the January rainfall at Sydney is not random, that significant peaks occur on 12 and 22 January, but that the December and February rainfall indicates a random distribution in time.

The wide geographical and persistent character of these rainfall peaks led Bowen to look for an extra-terrestrial cause that would have a yearly periodicity and be active simultaneously over the whole earth. The injection of meteoritic particles into the atmosphere to serve as ice-forming nuclei seemed the most plausible. In particular, Bowen associated the rainfall peaks of 13 and 22 January and 1 February with the occurrence of meteor showers on 13–14 December (Geminids), 22 December (Ursids) and 3 January (Quadrantids); another rainfall peak on 1–2 January was associated with the Bielids II shower on 2 December. The time lag of about 30 days between the meteor showers and the rainfall peaks is represented as the time taken for the particles to fall from an altitude of about 100 km into the lower atmosphere, where they may nucleate rain-clouds.

There are many aspects of the meteor-dust hypothesis that are difficult to accept. It implies that meteoritic dust can act as efficient ice nuclei, that the concentration of such particles at cloud level is comparable with the measured concentrations of ice nuclei, and that the incidence, persistence, and intensity of rainfall are largely controlled by the available concentrations of ice nuclei. There is no good evidence for any of these propositions.

The particles in a meteor shower cover the diameter range 1–1000 μm, and when these encounter the resistance of the upper atmosphere, particles of $d > 100$ μm burn up, those of $d = 10$–100 μm melt, and only those of $d < 10$ μm survive and enter the troposphere. Although calculations indicate that particles of diameter 10 μm should take about the 30 days required by Bowen's theory to fall into the troposphere, the much more numerous particles of $d \simeq 1$ μm would take about 1000 days. Indeed, it is because the particles have this wide range of size and fall speed, and because the lower atmosphere is effectively stirred over a period of several days, that it is difficult to imagine how a periodic input of dust into the top of the atmosphere could be preserved in the lower atmosphere and arrive at well-defined intervals of about 9 days.

Turning now to the concentration of meteoritic particles that may exist in the troposphere, the question has been raised of whether the rate of entry of small particles is significantly greater during meteor showers than at other times. Whipple and Hawkins (1956) demonstrate that the accretion rate for shower meteorites detected by radar or photographic means is less than twice the normal sporadic rate and therefore can have little effect on terrestrial phenomena. However, micrometeorite detectors on satellites and rockets indicate that the erosion of meteorite fragments in space may produce large numbers of particles that are too small to be detected by optical or radar techniques. Estimates of the concentration of particles in the atmosphere from studies of the accretion rate of meteoritic material have been made by Bowen (1953), who gives a figure of 1 m^{-3} in the stratosphere, by Junge (1957b), who uses various estimates of accretion to give values of between 10^{-4} and 1·5 particles per m^3 in the troposphere, and by Pettersson (1958), who gives 300 m^{-3} at the earth's surface. Even if these particles are effective ice nuclei, in concentrations of order 1 m^{-3} they are very unlikely to affect the rainfall, and even the higher estimates of 300 m^{-3} are inadequate to explain the ice-nucleus concentrations of at least 10^4 m^{-3} recorded by Bigg (1956) on the January days cited by Bowen.

In order to check whether there is a close correlation between the concentration of ice nuclei and the rainfall peaks, Bowen and his colleagues have organized daily measurements, during January and February, of nuclei at a number of stations in different parts of the world. Many of the results have been summarized by Bowen (1956d) and Bowen (1957). Measurements made with an airborne cloud chamber indicated the existence of rather broad peaks within a few days of 13, 22, and 30 January in Sydney in 1954 and 1955, but similar measurements made in Arizona in 1955 failed to reveal significant peaks, while ground-level observations in Hawaii and Panama did not fit the Bowen dates too well. Dates for January 1956 showed no peaks for Sydney, but relatively high concentrations of nuclei were reported from Western Australia on 13 and 22 January and 1 February, from California on 13 and 21 January, and from Pretoria, South Africa, on 13 and 29 January. In 1957, relatively high concentrations of nuclei were detected on the 16–17 January (dates of low rainfall according to

Fig. 4.11) in Sydney, Florida, and California, but when Bowen plots the mean temperature for the appearance of ice nuclei in concentrations of 0·1/litre for eight stations in Australia, United States, and South Africa taken together, peaks appear on 15, 25, and 31 January. However, the amplitude of these peaks is only about 1° C and, in view of the errors and the difficulties involved in obtaining reliable and reproducible results with cloud chambers, especially at these low concentrations, these results cannot be regarded as very significant. This is also underlined by recent measurements with continuously recording counters, which have revealed that high concentrations of ice nuclei may last for only short periods, and which stress the danger of regarding nucleus counts taken at infrequent intervals as being representative for the whole day.

The hypothesis that world-wide rainfall may be influenced by the influx of ice nuclei of meteoritic origin has recently become complicated by the discovery by Bradley, Woodbury, and Brier (1962) that the rainfall of the United States during the period 1900–49 shows a lunar periodicity, with two peaks occurring in the middle of the first and third weeks of the lunar synodic month. Following the discovery by Adderley and Bowen (1962) of a similar effect for New Zealand rainfall, Bigg (1963) and Bigg and Miles (1964) analysed long-term series of measurements of ice-nucleus concentrations made in Australia and South Africa, and reported a similar double-peak periodicity. Bowen has suggested a link with the meteor-dust hypothesis by postulating that the influx of meteoritic ice nuclei may be modulated by the moon. Although challenged by O'Mahony (1965), the periodicity in the rainfall data seems rather convincing, but no great significance can be attached to the ice-nucleus data since the amplitude of the peaks correspond to a variation of less than 1° C in cloud-chamber temperature.

There is no direct evidence for the existence of nuclei of meteorite origin in snow crystals and laboratory evidence as to their effectiveness is inconclusive. Mason (1960c) tested the fine dust produced by the grinding and vaporizing of a number of stony meteorites and found none of them to be effective at temperatures above $-17°$ C. Schaefer (1957) obtained similar results with metallic meteorites. However, Bigg and Giutronich (1967) argue, with some justification, that the methods of preparation of the aerosol in these experiments probably did not simulate the processes at work in the high atmosphere. They heated small stony and iron meteorites to boiling point in air at only 2 mmHg pressure, condensed the vapour on a slide, and used this as a substrate on which to grow ice crystals in a cold humidified chamber. The metallic meteorites produced black spheres of 5–25 μm diameter, 1 in 10^4 of which nucleated an ice crystal at $-8°$ C. The stony meteorite produced aggregates of sub-micron particles, 1 in 10^5 of which were active at $-8°$ C.

(g) *Summarizing remarks.* On general meteorological grounds, the incidence, intensity, and duration of rainfall is largely controlled by the frequency, distribution, and movement of weather systems and, in particular, by the vertical lifting of the air. It therefore seems more realistic to seek a dynamical explanation of the rainfall periodicities rather than attribute them to variations in the concentration of ice nuclei, especially as the supply of nuclei is unlikely to be critical in the deep cloud systems that produce much of the rain.

The direct identification of soil, particularly clay, particles in snow crystals, their high activity in laboratory tests, and the occurrence of higher concentrations of nuclei over land, particularly with the arrival of dust from specific sources, would all seem to point to the importance and efficacy of terrestrial dusts as nuclei. The fact that the average concentrations of nuclei in the Northern Hemisphere are higher than in the Southern Hemisphere also speaks for the earth's surface being the main source. On the other hand, it is difficult to believe that the appearance of large clay particles in about 90 per cent of snow crystals is truly representative of all snow crystals and much further careful work is required to test this. Also, the activity of the fairly pure samples of mineral particles used in the laboratory tests may have been higher than that of aged and contaminated particles residing in the atmosphere. There is strong evidence to suggest that the primary ice nuclei are much smaller than the 10-μm particles required by Bowen's 30-day hypothesis. They are probably of $d = 0 \cdot 1$–1 μm, having a lifetime in the troposphere of several days—sufficient to achieve global distribution. There is little direct evidence on the nature of these small particles. Kaolinite could well be the main source, but one cannot yet rule out meteoritic dust. On the other hand, there is little *direct* evidence that meteoritic dust does or can act as efficient ice nuclei. Evidence for the meteor-dust hypothesis, based on correlations between the occurrence of meteor showers and periodicities in rainfall, is indirect and circumstantial, and is unsupported by reasonable physical arguments on the possible magnitude and mechanism of the effect.

The origin and nature of ice nuclei must therefore be regarded as unsettled.

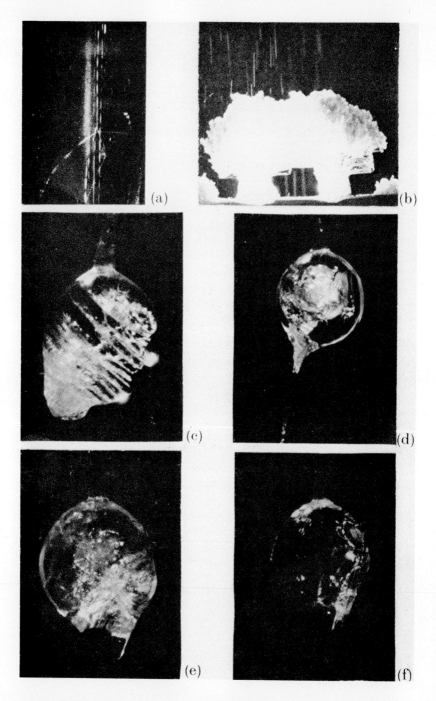

Fig. 4.12. The ejection of ice splinters by freezing droplets. (a) Tracks of splinters ejected from a droplet of $d \sim 100~\mu\mathrm{m}$ freezing during its fall through a column of air with a steep temperature gradient. (From Mason and Langham (1958).) (b) The white streaks mark the paths of splinters ejected by droplets of diamter $80~\mu\mathrm{m}$ freezing on a rime deposit. (c)–(f) Photographs of water drops about 1 mm in diameter nucleated near 0°C and freezing in air at −20°C. (c) shows air bubbles arranged in planar sheets; (d), (e), (f) show drops developing large spikes; the drop in (f) subsequently shattered violently. (Photographs by courtesy of Dr. D. A. Johnson.)

4.5. Secondary processes of ice-nucleus production

As already pointed out, measurements of ice nuclei in clear air may not serve as a reliable guide to the ice-crystal population of a cloud system because additional nuclei may be produced inside the cloud by fragmentation of ice crystals and freezing drops.

4.5.1. *Fragmentation of snow crystals*

The apparent necessity for the secondary production of ice nuclei during the steady release of snow from layer-cloud systems was pointed out by Mason (1955b), who showed that, in order to supply one nucleus per crystal, the air entering the cloud base would have to contain nucleus concentrations 10^2–10^3 times higher than are normally measured at $-15°$ C, and suggested that the additional nuclei are produced by small splinters being torn off the fragile dendritic crystals and that these can then serve as nuclei for new crystals. It may be shown that an original crystal would have to shed a splinter, on average, every 30 s, about thirty splinters in all, in order to sustain the numbers concerned and to multiply the original population of nuclei one hundredfold. This does not seem unreasonable, but the conditions under which snow crystals may splinter have not yet been investigated even in the laboratory.

4.5.2. *Shattering and splintering of freezing drops*

Ice splinters are also produced in certain circumstances when supercooled water drops freeze. Langham and Mason (1958) observed that when drops of radius 50–500 μm froze at the interface of two immiscible liquids they often split in half, that droplets of radius 50 μm shattered as they fell freely through a steep vertical temperature gradient in a cold column of air, sometimes ejecting visible ice fragments as shown in Fig. 4.12(a). Mason and Maybank (1960) carried out a detailed investigation with water drops of radius 30 μm to 1 mm suspended on fine fibres or thermocouples. Drops, initially at room temperature, were lowered in to a cold cell and nucleated as they cooled to temperatures between $-5°$ C and $-25°$ C, depending on the nature of the nuclei in the water. Alternatively, drops were nucleated in a second cell filled with tiny ice crystals produced by a fragment of solid CO_2 and then transferred to the main chamber where freezing proceeded at a lower temperature. Of the 1-mm drops nucleated at $-1°$ C and frozen at $-10°$ C, 46 per cent shattered and 90 per cent produced spikes (see Figs. 4.12(c)–(f)), an average of about fifty splinters per drop being detected in a dish of supercooled sugar solution placed beneath the drops. Drops nucleated below $-9°$ C failed to shatter and few produced spikes. The number of splinters produced was not very dependent upon the drop radius over the range 0·1–1 mm, but no splinters were observed from droplets of $r < 30$ μm. Drops of de-aerated water, frozen at $-13°$ C, shattered and splintered much more readily than similar drops containing an equilibrium concentration of air, and this was also true of drops freezing in an atmosphere of hydrogen, this being attributed to the lower solubility of hydrogen in water.

Mason and Maybank explained their observations as follows: Nucleation of the drop is followed by the formation of dendritic ice crystals spreading rapidly through the drop and around its surface; release of the latent heat of fusion momentarily raises the drop temperature everywhere to 0° C, and further freezing proceeds radially inwards at a rate controlled by the dissipation of latent heat to the environment. Expansion follows this second stage of freezing during which air is released from solution, many of the bubbles becoming trapped in the ice structure. When a slightly supercooled drop is nucleated by an ice crystal, freezing occurs initially only in the surface layers, where the very small quantity of air released can escape to produce a very thin, transparent shell of ice that is mechanically quite strong. Rupture of the shell, under pressure from the solidifying interior, frequently occurs at a weak spot where there first appears a bulge that may later develop a spike as liquid from the interior is extruded. If expansion of the interior cannot be accommodated by extrusion through spikes and cracks in the shell, large stresses develop, which may eventually cause the drop to explode into a few large, and perhaps many smaller, fragments. However, when a drop freezes at a much lower temperature, large quantities of air are released (because the solubility of air in water increases quite rapidly with decreasing temperature) and becomes trapped in the ice to give it a spongy texture. This yields much more readily to the expansion, so that although cracks and fissures may appear in the ice shell, the occurrence of spikes and violent shattering is rare, and splinter production is much reduced.

Muchnik and Rudko (1961) observed shattering of 1-mm-radius drops suspended on a thermocouple in a chamber cooled by solid carbon dioxide, and found that 50 per cent of the drops formed spikes and 20–50 per cent shattered when frozen at temperatures down to $-40°$ C. Evans and Hutchinson (1963) and Stott and Hutchinson (1965) suspended drops of diameter 0·9–1·7 mm on amylacetate fibres and nucleated them at about $-2°$ C by ice crystals produced in the wake of a pellet of solid carbon dioxide. The drops were subsequently frozen at $-15°$ C when about 20 per cent broke into two or more fragments. Dye and Hobbs (1966) froze 1-mm-diameter drops on a thermocouple in CO_2–air mixtures. The drops were first allowed to attain a temperature of $+2°$ C in the mixture and then lowered into a cell at $-10°$ C, where they were nucleated by ice crystals. Drops freezing in 1 per cent of CO_2 did not shatter, but the frequency of shattering increased as the concentration of CO_2 was raised, so that half of the drops shattered when this was 80 per cent. Dye and Hobbs suggested that fragmentation was caused by the pressure built up inside the drop by the large volume of CO_2 released during freezing, and that the frequent shattering observed by previous workers might have been caused by the presence of CO_2 in their apparatus.

Recognizing that the above experiments, in which the drops were not in thermal or solution equilibrium with the environment, did not simulate the freezing of freely-falling drops in the atmosphere, Johnson and Hallett (1968) carried out a rather detailed investigation to elucidate the mechanism of shattering under more nearly atmospheric conditions. They suspended 1-mm-diameter drops of distilled or deionized water from 15-μm shellac-coated polyester fibres and placed them in the centre of a cold cell having a vertical temperature gradient of $\frac{1}{2}°$ C/cm. Great care was taken to exclude CO_2. If drops, initially at about $+1°$ C, were suddenly plunged into the cold cell and nucleated by tiny ice crystals produced by a very cold needle, less than 5 per cent shattered if the air temperature was $-15°$ C, the frequency increasing to 10 per cent at $-20°$ C and 30 per cent at $-25°$ C, but was always lower than observed by Mason and Maybank. Moreover, if the drops were allowed to achieve thermal equilibrium with their surroundings before nucleation, shattering occurred only rarely at temperatures between -4 and $-20°$ C. These results, together with the fact that as many as 80 per cent of drops shattered when the air was replaced by hydrogen, and that 30 per cent shattered in helium, both these gases having a much higher thermal conductivity than air, suggest that rapid heat transfer and the formation of a complete ice shell around the drop during the early stages of freezing is conducive to shattering. For the same reason, the ventilation of the drop may be an important factor; thus Johnson and Hallett observed that when a drop was suspended in a horizontal air stream, freezing started at the stagnation point, spread progressively through the drop, did not form a complete shell, and did not shatter, but when a drop was rotated in the airstream, it froze symmetrically and subsequently disintegrated. They also noted that drops freezing in very soluble gases, such as CO_2 and N_2O, effervesced and threw off large numbers of small ice splinters.

In a very similar investigation, Dye and Hobbs (1968) found that when 1-mm-diameter drops that were allowed to achieve thermal and solution equilibrium before being nucleated in air at atmospheric pressure, 30 per cent produced spikes at temperatures above $-5°$ C, the frequency falling to 20 per cent at $-8°$ C, and to practically zero below $-15°$ C. No drop was observed to shatter, only three out of forty-eight produced splinters, but several ejected small droplets. Again, if warm drops were plunged into the cold chamber and cooled quickly, spike formation and fragmentation was much more frequent, while shattering occurred much more frequently in hydrogen and helium at $-10°$ C than in air. For drops nucleated at equilibrium in a mixture of air and carbon dioxide, the probability of shattering increased with increasing concentration of CO_2, reached a maximum of about 40 per cent for equal volumes of the two gases and fell again to about 25 per cent in pure carbon dioxide.

It seems, then, that the frequent fragmentation and splintering reported by earlier workers was due largely to their drops not being in thermal and solution equilibrium with the environment and may also have been effected by the presence of CO_2 in their apparatus. The experiments of Johnson and Hallett and of Dye and Hobbs are probably more representative of drops in a cloud, which rarely differ by more than a few tenths of a degree from the ambient temperature, except that the effect of free fall on the symmetry of the heat transfer and freezing was not properly simulated. This deficiency was partly remedied by Kuhns (1966) and Hobbs and Alkezweeney (1968), who studied the freezing of uniform drops falling freely through a cold column of air with a vertical temperature gradient of only about $\frac{1}{2}°$ C/cm. The latter observed that drops of diameter between 50 and 150 μm, freezing between -20 and $-32°$ C, or at $-8°$ C if containing silver iodide, shattered with a frequency independent of the nucleation temperature. Although their photographs, which are very similar to those of Langham and Mason, revealed only a few large fragments, a tray of supercooled sugar solution placed beneath the drops revealed many more smaller particles which became visible in the air when the humidity was raised to water saturation. But, drops of diameter

20–50 μm did not shatter, this being consistent with the original result of Mason and Maybank and also that of Kuhns, who found that drops of diameter 10–80 μm did not fragment at $-36°$ C in air though they did in carbon dioxide, helium, and hydrogen. Furthermore, Brownscombe and Thorndike (1968) observed that about 10 per cent of freely-falling drops of $d = 100$–250 μm shattered when nucleated by ice crystals at temperatures between -5 and $-15°$ C but smaller droplets were rarely seen to shatter.

Unfortunately the above experiments are not sufficiently consistent to provide reliable statistics on droplet shattering and splinter production and so it is not yet possible to make firm estimates of the likely rates of ice-crystal multiplication in clouds. However, as Mason and Maybank (1960) pointed out, the fragmentation of individual rain, drizzle, and cloud drops is unlikely to be important compared with the release of splinters during the impaction and freezing of large cloud droplets on pellets of soft, dry hail. This process was studied experimentally by Latham and Mason (1961b) who found that droplets of $d > 40$ μm freezing on a rimed surface ejected an average of ten splinters, but there is now reason to believe that many of the splinters they collected had other origins and that the splintering mechanism is much less powerful than their experiments indicated. Nevertheless certain features of the observations by Murgatroyd and Garrod (1960), Koenig (1963, 1968), Braham (1964), Mossop (1968), Mossop, Ruskin, and Heffernan (1968), Mossop and Ono (1969), and Mossop, Ono, and Wishart (1970), who report the appearance of ice crystals in the tops of cumulus clouds at temperatures as high as -5 to $-10°$ C, suggest that splintering during the growth of rimed particles may be responsible. Koenig (1963) found small, irregular crystals in concentrations of order 1 cm^{-3} together with snow and hail pellets of $d = 1$–2 mm when drops of $d > 30$ μm were present in concentrations of about 100/litre but, in a cloud of similar height containing very few such large drops, there were very few ice crystals. In a later paper, Koenig (1968) reported regular ice columns and needles and also many small irregular ice fragments in stratocumulus cloud with summit temperature above $-9°$ C; altogether ice particles were present in concentrations of about 40/litre.

Mossop et al. (1968) found regular columnar crystals and needles in concentrations of up to 40/litre in a long-lived cumulus whose summit temperature was only $-4°$ C, these being 10^3 times higher than the concentrations of ice nuclei measured below cloud base. This cloud contained frozen and unfrozen drops of $d > 100\ \mu$ in concentrations of 1/litre and also some pellets of soft hail. These observations and those of Mossop et al. (1970) of particularly high crystal concentrations ($\sim$100/litre) in maritime clouds warmer than $-10°$ C and containing large drops and rimed pellets in concentrations of 0·1/litre, together with the fact that very few of the ice particles collected by Koenig contained identifiable solid nuclei, suggest that they may have been produced by the freezing and fragmentation of the larger drops, probably on the surfaces of ice pellets. A simple calculation indicates that, starting from an initial concentration of 10 m^{-3} of rimed particles of $d = 0·5$ mm, the build up to a crystal concentration of 10/litre in 10 min would require the production of one ice splinter for every ten impacting droplets. Although such high rates of splintering have not been observed in recent laboratory experiments, the latter may have failed to detect very small splinters which could nevertheless grow in the ice-supersaturated air of a natural cloud.

A possible alternative explanation for the apparent discrepancy of 2–3 orders of magnitude between the observed concentrations of ice crystals and ice nuclei active at temperatures above about $-12°$ C is that the majority of the latter nuclei are too small to be detected by conventional cloud-chamber and Millipore-filter techniques but may nevertheless cause the freezing of cloud droplets following their Brownian capture. Sax (1970) has demonstrated that freely-falling drops of $r = 30$ to 70 μm can be made to freeze, and about 10 per cent of them to shatter, by the Brownian capture of 0·01 μm silver iodide particles active at temperatures between -12 and $-18°$ C. The question remains as to whether the atmosphere contains sufficient such small and effective ice nuclei. Calculations indicate that in a cloud containing 1 g m^{-3} of liquid water composed of drops of $\bar{r} = 15\ \mu$m, a primary ice-nucleus concentration of 1 cm^{-3} would be required to build up an ice-crystal concentration of 10/litre in 10 min. Since only about 1 in 10^4 of silver iodide particles of $r = 0·01\ \mu$m become active at $-10°$ C, it seems unlikely that natural aerosol will contain the 1 cm^{-3} of such small nuclei required to account for the glaciation of only slightly supercooled cumulus clouds, but the possibility merits further investigation.

4.6. Artificial ice nuclei

The fact that supercooled clouds are a common occurrence in the atmosphere, suggesting that efficient ice-forming nuclei are often deficient, and the possibility of inducing rain by supplying artificial nuclei, have stimulated a search for substances which may be dispersed into the atmosphere as fine smokes to produce vast numbers of effective and durable ice nuclei.

The most important discovery was made in November 1946 by Vonnegut, who found that silver iodide was highly effective in this respect. One gramme of this substance dissolved in acetone may be vaporized in a hot flame to produce about 10^{15} ice nuclei active at $-15°$ C. In recent years, much effort has been devoted to developing efficient methods of generating silver iodide smokes and to studying their ice-nucleating properties both in the laboratory and in the free atmosphere. Many other chemical compounds have also been tested, but so far no entirely satisfactory substitute for silver iodide has been found. A critical review of this work will now be given.

4.6.1. *Ice-nucleating properties of inorganic compounds*

(a) *Activity in cloud chambers.* The properties of artificial ice nuclei have been studied mainly in cloud chambers of the mixing and expansion types and also in the continuously-sensitive diffusion chamber described on p. 258. The investigation usually takes the form of introducing the chemical, either as a smoke, spray, or fine dust into the chamber and determining the highest temperature at which a detectable number of ice crystals appear. In general, this threshold temperature will be determined not only by the properties of the particles themselves, but also by the experimental conditions under which they are tested. Discrepancies between the results of different workers may arise from differences in one or more of the following factors: the criterion adopted for the onset of nucleation, and the sensitivity and reliability of the technique used for ice-crystal detection; the concentration and sizes of the nuclei; the volume of the observational space and the total number of nuclei observed during a test; the magnitude and duration of the supersaturation to which the particles are subjected and which may determine the time-lag for nucleation and whether or not the nuclei first become centres of condensation; the solubility and rate of solution of the nuclei in water; and the crystalline and surface structure of particles, the presence of favourable nucleation sites, impurities etc., all of which may be dependent upon the method of preparation. However, when due allowance is made for such effects, it is difficult to reconcile the very conflicting results that have appeared in the literature; for example,

TABLE 4.5

The ice-nucleating properties of some chemical compounds

	Ice	Insoluble					Soluble				
Substance	Ice	AgI	CuS	HgI_2	Ag_2S	Ag_2O	PbI_2	NH_4F	CdI_2	V_2O_5	I_2
Crystal symmetry	Hex.	Hex.	Hex.	Tetrag.	Monoclinic	Cubic	Hex.	Hex.	Hex.	Orthorhombic	Orthorhombic
Lattice spacings (Å)											
(a)	4·52	4·58	3·80	4·36	4·20	4·72	4·54	4·39	4·24	11·48	4·78
(b)				4·36	6·93					4·36	7·25
(c)	7·36	7·49	16·43	12·34	9·50		6·86	7·02	6·84	3·55	9·77
Solubility (g/100 cm^3)	—	8×10^{-6}	Insol.	3×10^{-3}	2×10^{-5}	Insol.	4×10^{-2}	v. sol.	79·8/0° C	0·5	3×10^{-2}
Threshold nucleation temp. (° C)											
Mixing chamber—soap film		−4	−6	−8	−8	−11	−6	−9	−12	−15†	−14
Diffusion chamber		−5	−5	−8	−8	−12	−7			−17†	
Bulk water		−4	−7	−7	−9	−10	−6				
Orientation on single crystals*		−4	−6				−6		−12	−12	−10
Threshold temperatures of other workers:											
Hosler (1951) (M)		−6		−18	−4		−4		−15		−12
Mossop (1956b) (E)		−6		−17			−7				
Pruppacher and Sanger (1955) (M)		−4	−4	−13	−4	−6	−5	×	−15	−6	×
Mason and Hallett (1956, 1957) (D)		−5	−6	×			−7		×	−15	−14
Fukuta (1958) (M)		−3	−7	−12	−7	−5	−1	−13	−12		×
Sano et al. (1960) (D)				−14		×	−18		−18		−35
Katz (1960) (M)		−4	−6								

† A much lower degree of activity is observed at about −10° C.
× Tested and found inactive.
M denotes mixing chamber, D denotes diffusion chamber, E denotes expansion chamber.
* According to Bryant, Hallett, and Mason (1959).

substances reported as highly effective by one laboratory have been found completely inactive by another.

After some very careful experiments, performed in very clean apparatus and using highly pure air and chemicals, Mason and Hallett (1956, 1957) concluded that some of the earlier results must have been spurious because of the presence, in the air or the chemicals, of small traces of silver iodide. If all such traces were removed, many of the substances hitherto claimed as highly effective nucleators were then found to be inactive. The same authors also pointed out the dangers of the conventional method of inferring the onset of ice-crystal formation solely from the appearance in the cloud of twinkling particles, because crystalline particles of the seeding agent itself may glitter and be optically indistinguishable from ice crystals. Furthermore, no reliance could be placed on the results of tests on unstable compounds, such as some of the iodides, which may decompose either at room temperature or during the dispersion process to produce free iodine.

Mason and Hallett tested a considerable number of substances that had been reported effective by other workers but found only AgI ($-5°$ C), CuS($-6°$ C), PbI$_2$($-7°$ C), Cu$_2$I$_2$($-15°$ C), V$_2$O$_5$($-15°$ C) to be definitely active, about one particle in 10^4 forming an ice crystal at the temperatures indicated when introduced into a supercooled water cloud in both a diffusion- and a mixing-cloud chamber. The inactivity of small particles of a wide range of highly soluble salts was attributed to their going into solution. In fact, it was demonstrated that, while ice sublimed directly on to iodine crystals at temperatures below $-14°$C in an atmosphere supersaturated relative to ice but sub-saturated relative to water, on entering a water cloud, the minute iodine crystals quickly dissolved and lost their ice-nucleating ability.

Further tests on a number of stable, insoluble or only very slightly soluble substances, dispersed as fine powders, were carried out by Mason and Van den Heuvel (1959) using a mixing chamber and a soap-film detector. The results for materials having threshold temperatures of $-15°$ C or above are shown in Table 4.5, together with the threshold temperatures at which a similar level of activity appeared in diffusion chamber experiments. In all cases, the two sets of results are in good agreement. To determine whether or not the powdered materials would nucleate bulk water at these same temperatures, aqueous suspensions containing 1 per cent by weight of the material were made with highly-purified water, 1-mm drops of which could be readily supercooled to $-20°$ C. When 1-mm drops of the suspensions were cooled at the rate of 1–5° C min^{-1}, Table 4.5 shows that they froze at temperatures very close to the threshold temperatures at which the same substance was active in a cloud chamber.

Table 4.5, which also shows the temperatures obtained in other laboratories in various parts of the world, indicates that there is now widespread agreement on the threshold activity of AgI($-4°$ C), CuS ($-6°$ C), PbI$_2$($-6°$ C), CdI$_2$($-12°$ C), and I$_2$($-13°$ C). The last three substances, being soluble in water, behave as sorption or sublimation nuclei in the sense that they act if the environment is sub-saturated relative to liquid water but lose their potency if subjected to high supersaturations such as probably existed in the diffusion cloud chamber of Sano, Fujitani, and Maena (1960). Fukuta's threshold temperature of $-1\cdot3°$ C for PbI$_2$ has not been confirmed and it appears that he was misled by the PbI$_2$ crystals themselves glittering in the light beam, which occurs even at positive temperatures if the humidity is below 100 per cent. In testing HgI$_2$ as a red powder, with their three different techniques, Mason and Van den Heuvel always obtained threshold temperatures of -7 or $-8°$ C, which are considerably higher than those reported by Hosler ($-18°$ C), Pruppacher and Sanger ($-13°$ C), Mossop ($-17°$ C), Fukuta ($-12°$ C), and Sano et al. ($-14°$ C). This rather large discrepancy may have arisen because four of these authors produced their aerosol by heating and caused partial decomposition and transformation to the much less active orthorhombic yellow form. Silver oxide and silver sulphide appear genuinely active at about -8 and $-11°$ C respectively; the activity at $-4°$ C reported by Fukuta and by Pruppacher and Sanger may have been caused by small traces of iodine leading to the formation of silver iodide. Vanadium pentoxide shows very slight activity at about $-10°$ C and thereafter no appreciable increase until the temperature falls below $-15°$ C. The existence of these two threshold temperatures appears to be associated with the orientation of the prism plane of ice on the (100) face of V$_2$O$_5$ at $-10°$ C and the orientation of the basal ice plane at $-15°$ C.

Mason and Van den Heuvel (1959) found that spectroscopically-pure oxides of aluminium, cadmium, copper, manganese, and tin all showed similar behaviour, in that a small fraction of the particles became effective ice nuclei between -6 and $-12°$ C but that their activity thereafter increased only very slowly with falling temperatures and only a very small fraction of the particles are activated even at $-30°$ C. This behaviour, which is illustrated in Fig. 4.13 for the copper oxides, is quite different from that of AgI and CuS, and underlines the limitations

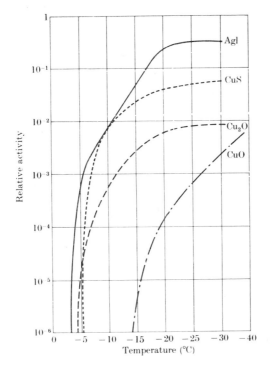

Fig. 4.13. The relative ice-nucleating ability of CuS, Cu_2O, and CuO compared with that of AgI as a function of temperature. (From Katz (1960).)

of the threshold temperature as a measure of the true efficacy of a nucleating agent. A proper comparison of two agents must be made in terms of their activity over a wide range of temperature and supersaturation. The differences between the threshold temperatures reported by different authors for the oxides shown in Table 4.6 may be due largely to the fact that they refer to different levels of activity. The nucleating ability of these oxides is considerably reduced if they become centres of condensation in a diffusion chamber, when they are immersed in bulk water, or when they are subjected to repeated cycles of freezing and melting, or evaporation above 0° C (Serpolay 1959). The reasons for this loss of potency are not clear, but it may, perhaps, be associated with the destruction of the relatively few active sites on the particle surfaces or the formation of hydration compounds.

(b) *Effect of particle size.* Using arguments very similar to those given in § 1.3 for the rate of condensation on spherical insoluble particles, Fletcher (1958) gives the same formula as (1.49) for the rate

TABLE 4.6
Nucleating properties of metallic oxides

Substance Threshold nucleation temp. (°C):		Al_2O_3	CdO	CuO	Cu_2O	Mn_3O_4	NiO	SnO_2
Mixing chamber (1 in 10^4)	Mason & Van den Heuvel (1959)	−12	<−16	−7	−6	−10	<−16	−7
Diffusion chamber (1 in 10^5)		−8	−9	−7	−6	−8	−11	−8
In aqueous suspension				<−15	−10	−14	<−17	<−16
On surface of water drops		−6	−9	−6	−7	−7	−10	−7
Pruppacher and Sanger (1955) (M)				−12	−5			
Fukuta (1958) (M)		−7	−10	−4	−5		−6	−14
Serpolay (1958) (M)		−10		−9			−16	−18
Serpolay (1959) (M)				−11	−10		−18	
Sano et al. (1960) (D)		−12	−20	−18				
Katz (1960) (M)					−7		−7	

(M) = Mixing chamber, (D) = diffusion chamber.

§ 4.6 INITIATION OF THE ICE PHASE IN CLOUDS

of formation of ice nuclei by sublimation, viz.

$$I \simeq \frac{p}{(2\pi MRT)^{\frac{1}{2}}} 4\pi r^2 r^{*2} n' \exp(-\Delta G^*/kT), \qquad (4.19)$$

but with σ_{LV} replaced by σ_{SV}, the interfacial surface energy between the ice nucleus and its vapour, in the expression for

$$\Delta G^* = \frac{16\pi M^2 \sigma_{\mathrm{SV}}^3 f(m,x)}{3\{\rho_{\mathrm{L}} RT \ln(p/p_{\infty,\mathrm{i}})\}^2} \qquad (4.20)$$

and $p/p_{\infty,\mathrm{i}}$ is now the saturation ratio of the environment relative to a plane surface of ice.

By analogy with eqn (4.6), Fletcher writes for the rate of formation of ice nuclei by the freezing of supercooled water condensed on the particle

$$I \simeq \frac{n_c kT}{h} 4\pi r^2 \exp\{-(U+\Delta G^*)/kT\}, \qquad (4.21)$$

where
$$\Delta G^* = \frac{16\pi M^2 \sigma_{\mathrm{SL}}^3 f(m,x)}{3\{\rho_{\mathrm{L}} RT \ln(p_{\infty,\mathrm{w}}/p_{\infty,\mathrm{i}})\}^2} \qquad (4.22)$$

and n_c is the number of liquid molecules in contact with unit area of the nucleating particle. $p_{\infty,\mathrm{w}}$ and $p_{\infty,\mathrm{i}}$ now refer to the equilibrium vapour pressures over plane surfaces of liquid water and ice respectively. Using eqns (4.19) and (4.21), with $\sigma_{\mathrm{SV}} = 100$ erg cm^{-2} and $\sigma_{\mathrm{SL}} = 20$ erg cm^{-2}, Fletcher calculates that the temperature at which a spherical particle will nucleate an ice crystal within one second by sublimation (sorption) in a water-saturated environment is not very sensitive to the particle radius until this falls below 0·1 μm but, below this size, the nucleation efficiency falls off rapidly. In the case of nucleation by freezing, this change occurs at a radius of about 0·03 μm—see Fig. 4.14.

This theory is probably much too simple to represent the processes that take place on the surface of a real particle, where nucleation may occur at only a few active sites, and does not seem capable of predicting the relative efficiencies of the sorption and freezing mechanisms in real cases. However, it is probably qualitatively correct in suggesting a lower limit for particle size below which the nucleating ability decreases rapidly. Experimentally, Sano et al., (1960) found that for AgI, HgI$_2$, PbI$_2$, and various metallic oxides, the nucleating efficiency did indeed decrease markedly for particles of $r < 0\cdot1$ μm, while Katz (1962) found that for Cu$_2$O and FeS particles of $d > 0\cdot5$ μm the number of particles activated at a given temperature is proportional to their

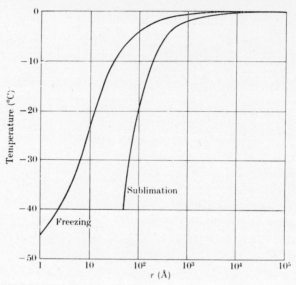

FIG. 4.14. The temperature at which a spherical particle of radius r and surface parameter $m = 1$ will nucleate an ice crystal in 1 second by the freezing of water and by sublimation from an environment at water saturation according to the theory of Fletcher (1958).

surface area. We also recall that Edwards and Evans (1960, 1968) found that, while 2–3 per cent of 10-μm silver iodide particles acted at ice saturation at temperatures between $-5°$ C and $-10°$ C, and all were active at water saturation at $-7°$ C, very few particles of $r < 0\cdot 01$ μm were active as sorption nuclei even at $-18°$ C and most remained ineffective until the humidity relative to water exceeded 110 per cent. An additional factor in determining the minimum size of effective freezing nuclei arises in that particles, which are normally regarded as insoluble in water, may not be immune when they are very small, say, of $r < 0\cdot 01$ μm.

(c) *Epitaxial growth of ice on single-crystalline substrates.* The most direct and convincing evidence for a particular crystal acting as an ice nucleus is afforded by the appearance on its surfaces of similarly-oriented ice crystals. Epitaxial deposits of ice have been observed on AgI, PbI$_2$, and mica by Montmory (1956) and Jaffray and Montmory (1956, 1957), on AgI and PbI$_2$ by Mason (1958), and on AgI, PbI$_2$, CdI$_2$, and biotite by Kleber and Weis (1958).

A detailed study of the conditions under which the crystals appeared was made by Bryant, Hallett, and Mason (1959), who observed oriented ice crystals on the basal (0001) faces of hexagonal AgI, PbI$_2$,

CuS, CdI_2, and brucite, on the (001) face of freshly-cleaved muscovite, on the (010) face of orthorhombic HgI_2, on the (001) face of orthorhombic iodine, on the (100) plane of V_2O_5, and on the surfaces of rhombohedral calcite. The crystals were grown in a modified version of the Shaw–Mason apparatus described on p. 256, in which the temperature of the nucleating substrate and the supersaturation of the surrounding water vapour were accurately controlled. The results for silver iodide are shown in Fig. 4.15. At temperatures above $-4°$ C, only water droplets appeared. As the temperature was lowered from -4 to $-12°$ C, increasing numbers of ice crystals appeared at selected sites, provided that the air surpassed saturation relative to liquid water. At temperatures below $-12°$ C, however, crystals appeared when the air was sub-saturated relative to water but supersaturated with respect to ice by at least 12 per cent. Similar results were obtained for cupric sulphide (covellite), lead iodide, and cadmium iodide, with critical supersaturations of 13, 15, and 22 per cent at -13, -15, and $-21°$ C, respectively.

The highest temperatures at which ice crystals appeared on the various host crystals are shown in the last column of Table 4.7 and agree well with the corresponding threshold temperatures obtained in

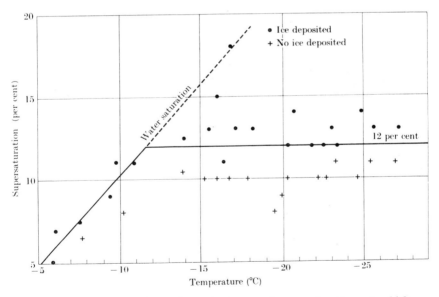

Fig. 4.15. The conditions of temperature and supersaturation at which oriented ice crystals appeared at special sites on the surface of a single crystal of silver iodide. (From Bryant, Hallett, and Mason (1959).)

TABLE 4.7

Data on substances active as ice nuclei at temperatures above $-16°C$ (Mason and Van den Heuvel 1959)

Substance	Crystal symmetry	Lattice constants (Å) a	b	c	Substrate plane	% Misfit between substrate and ice lattice in directions: (1$\bar{2}$10)	(10$\bar{1}$0)	(0001)	Threshold temp. (°C)
Ice (0°C)	Hex.	4·52		7·36					
(a) Insoluble substances									
AgI	Hex.	4·58		7·49	(0001)	$+1\cdot3_{(1:1)}$	$+1\cdot3_{(1:1)}$	$+1\cdot8_{(1:1)}$	-4
CuS	Hex.	3·80		16·43	(0001)	$-2\cdot8_{(3:2)}$	$+2\cdot8_{(1:2)}$	$+11\cdot5_{(2:1)}$	-6
HgI$_2$	Tetrag.	4·36	4·36	12·34	(001)	$-3\cdot5_{(1:1)}$	$+11\cdot5_{(1:2)}$	$-16\cdot2_{(2:1)}$	-8
Ag$_2$S	O. Rhombic	4·67	13·76	7·32	(010)	$+3\cdot3_{(1:1)}$	$-6\cdot4_{(1:1)}$	$-6\cdot5_{(2:1)}$	
	Monoclinic $\beta = 55°$	4·20	6·93	9·50	(010)	$-7\cdot1_{(1:1)}$	$-0\cdot3_{(1:1)}$	$-5\cdot9_{(1:1)}$	-8
Ag$_2$O	Cubic	4·72			(001)	$+4\cdot4_{(1:1)}$	$-9\cdot5_{(2:3)}$	$-3\cdot8_{(2:3)}$	-11 (prisms)
(b) Soluble substances									
PbI$_2$	Hex.	4·54		6·86	(0001)	$+0\cdot4_{(1:1)}$	$+0\cdot4_{(1:1)}$	$+6\cdot8_{(1:1)}$	-6
NH$_4$F	Hex.	4·39		7·02	(0001)	$-2\cdot9_{(1:1)}$	$-2\cdot9_{(1:1)}$	$-4\cdot6_{(1:1)}$	-9
V$_2$O$_5$	O. Rhombic	11·48	4·36	3·55	(100)	$-3\cdot5_{(1:1)}$	$-2\cdot1_{(3:2)}$	$-3\cdot5_{(1:2)}$	-10 (prisms)
					(001)	$-3\cdot5_{(1:1)}$	$-2\cdot1_{(3:2)}$	$-3\cdot5_{(1:2)}$	-15 (plates)
CdI$_2$	Hex.	4·24		6·84	(0001)	$-6\cdot2_{(1:1)}$	$-6\cdot2_{(1:1)}$	$-7\cdot1_{(1:1)}$	-12
I$_2$	O. Rhombic	4·78	7·25	9·77	(001)	$+5\cdot8_{(1:1)}$	$-16\cdot3_{(3:2)}$	$-1\cdot5_{(1:1)}$	-12 (prisms)
						$+5\cdot8_{(1:1)}$	$-7\cdot8_{(1:1)}$	$-11\cdot6_{(3:2)}$	-14 (plates)
(c) Metallic oxides									
Cu$_2$O	Cubic	4·25			(100)	$-6\cdot0_{(1:1)}$	$+8\cdot7_{(1:2)}$	$+15\cdot5_{(1:2)}$	-6
CuO	Monoclinic $\alpha = 99°$	4·65	3·41	5·11	(001)	$+2\cdot9_{(1:1)}$		$-7\cdot35_{(1:2)}$	-7 (prisms) (plates)
					(010)	$+2\cdot9_{(1:1)}$	$-2\cdot0_{(2:3)}$		
SnO$_2$	Tetrag.	4·72	4·72	3·16	(001)	$+4\cdot4_{(1:1)}$	$-14\cdot2_{(1:2)}$	$-3\cdot8_{(2:3)}$	-8
α-Al$_2$O$_3$	Rhombohed.	5·13	$\alpha = \beta = \gamma = 55°6'$		Any face	$+13\cdot5_{(1:1)}$	$+7\cdot7_{(1:2)}$		-8
Mn$_3$O$_4$	Tetrag.	5·75	5·75	9·42	(100)	$+4\cdot2_{(1:2)}$	$+10\cdot3_{(2:3)}$	$+17\cdot2_{(2:3)}$	-8
CdO	Cubic	4·69			(100)	$+3\cdot8_{(1:1)}$	$-10\cdot0_{(2:3)}$	$-4\cdot4_{(2:3)}$	-9
					(100)	$-7\cdot7_{(1:1)}$	$+6\cdot6_{(1:1)}$	$+13\cdot3_{(1:1)}$	-11

cloud chambers. The typical photographs of Fig. 4.16 illustrate that the ice crystals show a strong tendency to form on specific sites such as etch pits, cleavage steps, cracks, and ledges of growth spirals; they appeared on the flat, rather perfect areas of the substrate only at very high supersaturations, usually exceeding 100 per cent. With soluble substances, such as PbI_2 and CdI_2, melting and evaporation of the ice deposit appeared to damage the substrate and reduce its nucleating ability.

4.6.2. *Ice-nucleating properties of organic compounds*

Although Bashkirov and Krasikov (1957) reported that phloroglucinol was an effective ice nucleator at $-6°$ C, little interest in organic compounds was aroused until Head (1961) discovered that seven out of thirty steroid compounds, which he tested by observing the formation of ice crystals around particles exposed to moist air on the cold stage of a microscope, were effective at temperatures as high as $-5°$ C. To this list (see Table 4.8), Head (1962a) later added several more organic compounds that were active between -2.5 and $-7°$ C.

Komabayasi and Ikebe (1961) introduced a number of finely powdered aromatic compounds into a mixing chamber and used a supercooled sugar solution to determine the highest temperature at which one particle in about 10^4 produced an ice crystal. Using soap-film and sugar-solution detectors, Power and Power (1962) found a number of amino acids to be active at temperatures above $-10°$ C. All these results are summarized in Table 4.8.

Using a mixing-cloud chamber and a soap-film detector, Fukuta and Mason (1963) isolated several substances that were effective above $-5°$ C, most of them having been discovered meanwhile by Head. They then produced large single crystals of these compounds, usually by slow crystallization from a melt or an alcohol solution, and then used the technique described on p. 220 to study the epitaxial growth of ice crystals on their surfaces. Table 4.8 shows that epitaxial deposits were observed at temperatures as high as $-1.0°$ C on the steroids cholesterol, pregnenolone, diosgenin, and stigmasterol, and on testosterone at $-1.3°$ C. These substances also nucleated the soap film at about $-1.5°$ C, except stigmasterol, which was effective only below $-8°$ C. The photographs, examples of which are shown in Fig. 4.17, revealed that nucleation occurred preferentially on cracks in the substrate and, in some cases, both oriented ice prisms and plates formed on the same crystal.

TABLE 4.8
Ice-nucleating properties of some organic compounds

Substance	Threshold temp. (° C) Cold stage	Soap/sugar film	Epitaxy	Author
Metaldehyde		−1		F
Cholesterol		−2	−1	F & M
	−5			H_1
Testosterone		−2	−1	F & M
	−5			H_1
Pregnenolone		−2	−1	F & M
	−5			H_1
Diosgenin		−1·5	−1	F & M
Stigmasterol	−3			H_2
		−8	−1	F & M
Methyl androstan-3, 17-diol	−2·5			H_2
α-phenazine	−3·5			H_3
9-fluorenone	−4			H_2
17-Oestradiol	−4			H_2
Androsterone	−5			H_1
Androstenolone	−5			H_1
Androstan-3, 17 dione	−5			H_1
isoleucine (d or l)	−6	−6		B & M
	−1			P & L
dl-isoleucine	−10	−10		B & M
	−6			P & L
leucine (d or l)	−5	−5		B & M
	−3/−6			E
	−2			P & L
dl-leucine		−5		P & P
		−5	−5	F & M
	−11	−12		B & M
	−10			P & L
tryptophan (d or l)		−6		P & P
	−6			B & M
	−2			P & L
dl-tryptophan		−10		B & M
	−8			P & L
aspartic acid (d or l)	−2			P & L
	−3			E
dl-aspartic acid		−6		P & P
	−7			P & L
dl-alanine		−7		P & P
	−8			P & L
valine (d or l)	−8	−6		B & M
	−2			E; P & L
dl-valine	−16	−18		B & M
	−10			P & L
phloroglucinol	−6			B & K
fluorenol	−6			H_2
β-naphthol		−9		F & M
β-naphthoquinone		−10		K & I
terephthalic acid		−10		K & I
phthalic acid		−10		K & I

Two organic compounds that are both effective and relatively cheap are α-phenazine, discovered by Head (1962b), and metaldehyde, discovered by Fukuta (1963). Head found that 10^9 nuclei/g active at $-6°$ C could be obtained from finely-crushed phenazine crystals and that 10^{11} nuclei/g active at $-12°$ C were produced by a smoke of sublimed crystals. The threshold temperature in cloud-chamber experiments was $-3.5°$ C. This high activity is attributed to the fact that the misfit between the (001), (100), and (202) planes of α-phenazine and the ice structure is <4 per cent. Fukuta reports that finely-powdered metaldehyde can nucleate ice crystals at temperatures as high as $-0.4°$ C, and that a smoke of sublimed crystals can produce 10^{11} nuclei/g active at $-10°$ C, but it is difficult to produce a high yield of small crystals because of the high vapour pressure. Parungo and Lodge (1965) tested a series of substituted phenols and benzoic acids by placing freshly-ground materials on a microscope cold stage and noting the highest temperatures at which water droplets formed by condensation on the particles froze, and by spraying a suspension of the pulverized material into a cold chamber and noting the highest temperatures at which ice crystals appeared in a tray of supercooled sugar solution. Although ice-nucleating ability appeared to be strongly correlated with the molecular dipole moment in the case of phenols, (but not in the case of benzoic acids), a much better correlation was obtained with the potential strength of hydrogen bonding between adsorbed water molecules and exposed hydroxyl or carboxyl groups on the surface, thus confirming the original suggestion of Garten and Head (1964, 1965). The behaviour of these substances showed no apparent correlation with crystal geometry and seemed to depend only on the molecular structure.

Further support for this view was provided by Parungo and Lodge (1967) when they confirmed an earlier finding of Barthakur and

---

B & K = Bashkirov and Krasikov (1957).
H_1 = Head (1961) H_2 = Head (1962a).
F & M = Fukuta and Mason (1963).
P & P = Power and Power (1962).
E = Evans (1966).
K & I = Komabayasi and Ikebe (1961).
H_3 = Head (1962b).
F = Fukuta (1963).
B & M = Barthakur and Maybank (1965).
P & L = Parungo and Lodge (1967).

Maybank (1963) that the optically active dextro-rotatory (d) and laevo-rotatory (l) forms of certain amino acids, for example leucine, tryptophan, and valine, were more effective ice nuclei than the inactive dl forms. Whereas Barthakur and Maybank (1965), in contradiction of their earlier report, found that the nucleating efficiency of amino acids was not very sensitive to the mode of preparation and testing, Evans (1966) found that freshly-crushed or etched l-leucine was effective at $-3°$ C but soon deteriorated, probably due to rearrangement of the exposed ionic groups on the surface. On the other hand, he found that particles of d-valine and l-aspartic acid showed no dependence on etching or crushing, and were effective at $-2°$ C and $-3°$ C respectively.

In general, explanation of the high ice-nucleating ability of organic compounds is complicated by the existence of polymorphic forms, by lack of information on their crystalline structure, and by the fact that the density of nucleating sites is often strongly dependent on the method of preparation. However, the balance of evidence suggests that molecular bonding at specific sites, where small clusters of bonding groups fit the oxygen atoms of a close-packed ice plane, is of major importance, and that the crystal structure, as such, plays only a minor role. Steps, cracks, and cavities on the surface are often preferred sites, probably because of the tendency for adsorbed molecules to become trapped there.

4.6.3. *Relation between nucleating properties and crystalline structure*

In attempting to correlate the nucleating ability with some physical property of the particles, it is natural to compare their crystal structures with that of ice. Table 4.7 shows that although there is a tendency for the more effective inorganic nucleators to possess hexagonal crystalline symmetry and lattice parameters reasonably close to those of ice, there are a number of exceptions. Nevertheless, for all simple inorganic substances effective at temperatures above $-15°$ C, it is possible to find a low-index crystal face on which the atomic spacings differ from those in either the basal or prism faces of ice by not more than a few per cent. If, in calculating the minimum misfit, one permits the matching of atomic rows in the ice and substrate lattices in ratios of up to 3:1, the values shown in the heavy type in Table 4.7 are the percentage misfits in the two most favoured of the three mutually perpendicular directions, $(1\bar{2}10)$, $(10\bar{1}0)$, and (0001), of the ice lattice.

Although epitaxy is usually discussed in terms of the misfit in only the most favourable direction, it seems reasonable to regard the

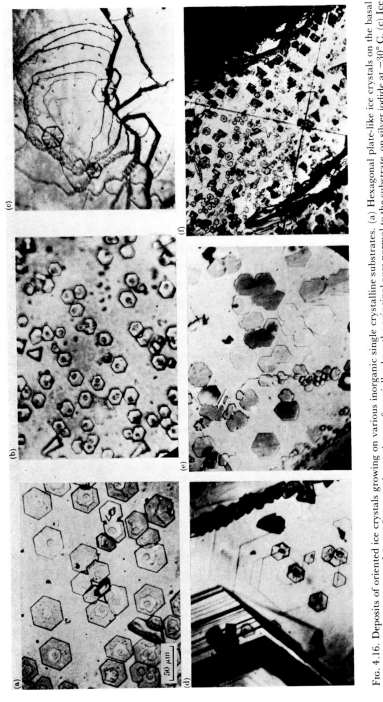

FIG. 4.16. Deposits of oriented ice crystals growing on various inorganic single crystalline substrates. (a) Hexagonal plate-like ice crystals on the basal plane of silver iodide at $-15°$ C. (b) Ice crystals growing preferentially along the principal axes normal to the substrate, on silver iodide at $-30°$ C. (c) Ice crystals growing on the edges of deep steps (>0.1 μm) on the basal plane of lead iodide. (d) Ice crystals growing preferentially on the steps of a hexagonal growth spiral on cadmium iodide. (e) Ice crystals growing on the basal plane of covellite. (f) Oriented ice plates and prisms on the (001) face of an iodine crystal. From Bryant, Hallett, and Mason (1959).

Fig. 4.17. Deposits of oriented ice crystals growing on organic single crystalline substrates. (a) Ice plates on cholesterol at −1·5°C. (b) Ice prisms on cholesterol at −2·3°C. (c) Trigonal ice crystals on pregnenolone at −2·1°C. (d) Ice plates on diosgenin at −1·8°C. (e) Plates and prisms on diosgenin at −1·1°C. (f) Ice prisms growing at cracks on stigmasterol at −2·3°C. (From Fukuta and Mason (1963).)

formation of two or three-dimensional nuclei in terms of the misfit or two or three mutually perpendicular directions. The misfit in the third direction is therefore printed in lighter type in Table 4.7.

For simple inorganic structures, a reasonable correlation exists between the threshold nucleation temperature and the sum of the misfits in the three mutually perpendicular directions. An interesting case is shown in Fig. 4.16(e), where oriented ice crystals in the form of both prisms and plates appear on the (001) face of iodine. The misfits shown in Table 4.7 suggest that the prisms should be slightly favoured and this is borne out by the fact that only prisms appear at temperatures above $-14°$ C but that plates become more frequent as the temperature is lowered.

Much poorer correlations exist for metallic oxides and for the rather complex structures of the naturally occurring minerals, while for organic crystals the lattice constants appear largely irrelevant. In general, the orientation and stability of the ice germs will be determined by their interaction with the surface force field of the substrate and its ability to make adsorbed or condensed water assume an ice-like configuration. In this respect there is probably no sharp distinction between the behaviour of inorganic and organic nuclei. In the simple inorganic structures the surface force field will usually be determined by the geometrical arrangement of the atoms, which also determines the crystal symmetry, and so a high correlation between ice-nucleating ability and lattice geometry is to be expected. In the more complex organic structures, the surface force field will be determined largely by the molecular structure, and particularly by the arrangement and nature of exposed bonding groups and this may not be directly related to the crystal structure. The ice-nucleating properties of a relatively complex inorganic crystal, such as kaolinite, may be attributed to the fact that the OH bonds on the basal surface match the oxygen atoms in ice with a misfit of only 1·2 per cent, but here the character of the binding forces is probably at least as important as their geometrical arrangement.

4.6.4. *The production and behaviour of silver iodide smokes*

In a search for efficient artificial ice-forming nuclei, Vonnegut (1947) scanned the crystallographic data for substances resembling ice as closely as possible in crystal system, space group, and dimensions of the unit cell. The two most suitable substances appeared to be silver iodide and lead iodide, but the introduction of these salts in finely powdered form into a cold chamber containing a supercooled cloud produced only small numbers of ice crystals. It was discovered subsequently, however, that a single spark passed between silver electrodes in the presence of iodine vapour produced very large numbers of ice crystals, and that the heating of silver iodide crystals to well above their melting point on a hot wire produced ice crystals at temperatures below $-4°$ C.

TABLE 4.9

Crystal form	Temperature range (° C)	Unit dimensions (Å)	Molecules per unit cell
α-AgI (cubic)	146–555	$a = 5·03$	2
β-AgI (hexagonal)	135–146	$a = 4·58, c = 7·49$	2
γ-AgI (cubic ZnS)	< -135	$a = 6·47$	4
Ice (hexagonal)	0° C	$a = 4·52, c = 7·36$	4

Vonnegut pointed out that the unit-cell dimensions of the hexagonal form of silver iodide and of ice are the same to within 1 per cent, and that the arrangement of the atoms in the cells is very similar in the two cases, each silver atom in silver iodide being bonded tetrahedrally to four iodine atoms. In fact, silver iodide occurs in two crystal forms at room temperature—a hexagonal arrangement (wurtzite structure) and a cubic arrangement (zinc blende structure), the latter form being regarded as the stable structure at temperatures below 135° C. The hexagonal form is thermodynamically stable between 135 and 146° C. Above 146° C both forms transform into a new 'high temperature' cubic form. The unit-cell dimensions are as shown in Table 4.9. It is also worth noting that the atomic array in the (111) planes of the cold cubic structure and the (0001) planes of the hexagonal form are geometrically similar, the lattice spacing being 4·58 Å in each case.

When either the crystals or a solution of silver iodide are vaporized at a temperature of several hundred degrees centigrade, X-ray and electron-diffraction patterns of the smoke show that it consists of a mixture of the hexagonal and cold cubic forms, the hexagonal structure being more dominant the higher the temperature of the source. With a source temperature of 800° C, Manson (1955) found that 95 per cent of the smoke consisted of the hexagonal form. Lisgarten (see Mason and Hallett (1956)) analysed electron-diffraction patterns of smokes produced by the vaporization of AgI–KI–acetone solution from a hot wire and found little or no trace of either AgI or KI, but suggested that mixed crystals were formed. However, when the particles became immersed in water, AgI was precipitated. AgI–NaI smokes revealed no trace of hexagonal silver iodide, the patterns being consistent with those of either the cubic form or of NaI, which have almost identical lattice spacings; alternatively, a mixture of both substances may have been formed. On the other hand, Mossop and Tuck-Lee (1968) report that the smoke from a Warren–Nesbitt generator burning a solution of AgI and NaI in acetone produced particles that contained both hexagonal AgI and NaI and were hygroscopic.

The X-ray studies of Davis (1969) indicate that an aerosol produced by the burning of AgI–NaI solution in a humid atmosphere contains one or more hydrated phases of the type AgI–NaI . nH$_2$O, depending upon the humidity. When the particles contain more than about 7 per cent by weight of water, the hydrates disappear and the aerosol consists of particles of AgI surrounded by NaI solution. According to Mossop and Jayaweera (1969) 50 per cent of such particles with mean diameter 0·09 μm acted as condensation nuclei at 1 per cent supersaturation, compared with only 3 per cent of particles produced by the vaporization of pure silver iodide. The hygroscopic aerosols were also more active, by an order of magnitude, as ice nuclei in a mixing-cloud chamber at −15° C. The authors conclude that, even so, the high supersaturations prevailing in the cloud chamber probably exaggerated the likely efficiency of 'clean', hydrophobic, silver iodide particles in natural clouds.

Silver iodide is almost insoluble in water but is readily soluble in liquid NH$_3$, aqueous solutions of KI or NaI, and acetone solutions of these iodides. Most silver iodide generators used in cloud seeding (see Chapter 7) make use of silver iodide in acetone–sodium iodide solution. A stock solution is made by placing 600 g of granulated silver in a Pyrex flask, adding 100 g of NaI, 200 g of iodine crystals, and 1 litre of acetone. After the initial chemical reaction has died down, the solution is boiled gently for several hours until it loses most of its iodine colour. It is then cooled, filtered, and diluted by a factor of about 10 with acetone for use in a generator.

There are two types of generator in common use. In one of these, a 2% solution of silver iodide is atomized through a nozzle with either compressed hydrogen, or propane, and the spray is ignited. The vapour rapidly condenses on mixing with the cool atmosphere to form a smoke of very small particles, the size of which depends upon the concentration and rate of flow of the solution, but is mostly in the range 50–1500 Å. A compact burner using only acetone as the fuel was designed for use on an aircraft by Warren and Nesbitt (1955). The acetone solution of AgI is sprayed through a swirl-type atomizer into a combustion chamber, where it is ignited by a sparking plug. The second type of generator burns small coke nuts impregnated with a 1–2 per cent silver

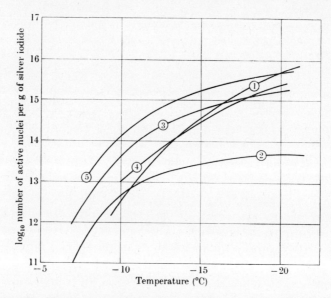

Fig. 4.18. The ice nucleating activity of smokes produced by various silver iodide burners: (1) hydrogen burner (Vonnegut 1949b); (2) kerosene burner; and (3) hydrogen burner (Smith and Heffernan 1954); (4) charcoal burner (Soulage 1955); (5) 'Skyfire' solution burner (Fuquay and Wells 1957).

iodide solution in an air-stream of a few metres per second, which assists combustion and prevents the formation of a layer of ash.

The output and efficiency of a generator may be assessed by determining the number of nuclei that become active at different temperatures for the consumption of 1 g of silver iodide. The performances of several different types of generator, characterized in this way, are represented in Fig. 4.18. All the curves show a rapid increase in activity of the smoke as the temperature falls from -5 to $-10°$ C, and a much slower increase thereafter. An efficient generator, such as the solution burner of Fuquay and Wells (1957), using propane as the fuel, yields 10^{13} nuclei/g active at $-7°$ C, 10^{14}/g at $-10°$ C, and 10^{15}/g active at $-15°$ C.

The size and therefore the total number of silver iodide particles produced by a generator is largely determined by the temperature at which the smoke is formed and the rate of quenching; as Balabanova and Zhigalovskaya (1962) have shown, the particle size decreases with increasing temperature of the source. Heating produces some decomposition of the AgI particles into silver and iodine but, rather surprisingly, Balabanova, Maleev, and Zhigalovskaya (1960) report that the decomposition products increase only from 3 to 5 per cent when the temperature is raised from 650 to 3000° C. The efficiency of the particles as ice nuclei is also determined by their internal and surface structure, the density of dislocations and other imperfections, which are also affected by the heat treatment and the presence of contaminants. The overall efficiency of a generator thus depends upon several factors and some of the differences in Fig. 4.17 may be due to differences in methods of calibration. Careful test procedures and the assessment of errors are described by Grant and Steele (1966). A realistic assessment must also take cognizance of the time for which the nuclei remain effective after dispersal in the atmosphere.

That the ice-nucleating ability of a silver-iodide smoke decreases after irradiation by natural or artificial sunlight has been established in laboratory tests by Reynolds, Hume, Vonnegut, and Schaefer (1951), Inn (1951), Vonnegut and Neubauer (1951), Birstein (1952), and Shimada, Sekihara, and Kawamura (1955). The de-activation is caused by photolysis of the silver iodide surfaces and their reduction to metallic silver. However, as Table 4.10 indicates, there is no general agreement as to the rate at which de-activation occurs for a given intensity of radiation. The decay rate appears to depend on the manner in which the silver-iodide nuclei are produced as underlined by the results of Smith, Seely, and Heffernan (1955), who released silver iodide smoke and cadmium-zinc sulphide particles simultaneously from a ground site and traced these up to distances of 25 miles in an aircraft. The ratio of the concentration of AgI nuclei active at $-17°$ C in a mixing-cloud chamber to that of the zinc sulphide particles, determined at various times (distances) after release, was used to measure the deterioration of the silver iodide. Smoke produced by the vaporization of an acetone solution of silver iodide in a hydrogen flame retained only 1 per cent of its initial activity after 16 min exposure in daylight, whereas that from a kerosene burner deteriorated to the same degree only after 100 min. The rates of decay appeared not to be greatly dependent upon the altitude, cloud cover, or the humidity. Later experiments by Smith

TABLE 4.10

Summary of results on photolytic deactivation of silver iodide

Author	Method of production	Method of testing	Decay rate	Remarks
Reynolds et al. (1951)	Propane burner	Smoke stored in tank; irradiated with strong sunlight; tested in cold chamber at $-20°$ C	$10^{-2}/24$ h	Considerable variation with cloud cover, time of day, particle size
Inn (1951) Birstein (1952)	Vaporized fused crystals from nichrome wire	Irradiated smoke with u.v. lamp of intensity comparable to sunlight on clear, summer afternoon. Tested in cold chamber at $-20°$ C	10^{-6} to 10^{-9}/h	
Vonnegut and Neubauer (1951)	Ignited charcoal impregnated with AgI	As above	0.5 to 0.2/h	
Shimada et al. (1955)	Vaporization at 600–700° C	Irradiated with strong sunlight and mercury-arc lamp. Tested in cold chamber at -14 to $-16°$ C	2×10^{-3}/h in strong sunlight	
Smith, Seely, and Heffernan (1955)	Hydrogen generator Kerosene generator (1440° C)	Smoke dispersed into atmosphere and sampled in airborne cold chamber at $-17°$ C	10^{-7}/h 10^{-1}/h	Activity decreased logarithmically with exposure time
Smith and Heffernan (1956)	Kerosene generator	As above	10^{-1} to 10^{-2}/h negligible	By day By night
Smith, Heffernan, and Thompson (1958)	Acetone solution of AgI atomized in combustion chamber of aircraft	As above	3×10^{-2}/h $10^{-3}/2$ h negligible	By day By night

TABLE 5.1. *Classification*

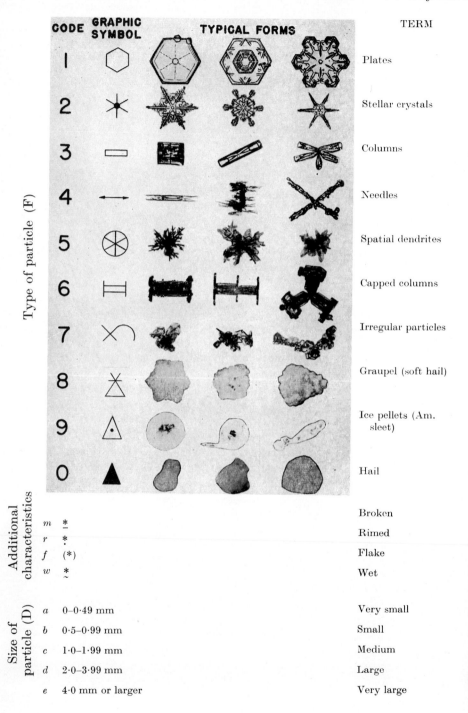

of Solid Precipitation

REMARKS

also combinations of plates with or without very short connecting columns

also parallel stars with very short connecting columns

and combinations of columns

and combinations of needles

spatial combinations of feathery crystals

columns with plates on either (or one) side

irregular aggregates of microscopic crystals

isometric shape, central crystal cannot be recognized

ice shell, inside mostly wet

The size of the particle means the greatest extension of a particle (or average when many are considered). For a cluster of crystals it refers to the average size of the crystals composing the flake

and Heffernan (1956), in which the smoke was released from a mountain top both by day and night, showed that decay proceeded much more slowly by night and, in the case of the kerosene burner, was negligible after exposure to the atmosphere for more than 2 h.

While most of the workers have found the nucleating ability to decay logarithmically with increasing time of exposure to light, the rates of decay appear to depend on the manner and the temperature of nucleus production and the presence of impurities. Because of these uncertainties and the difficulty of sampling and testing smokes, Bryant and Mason (1960a) studied large, irradiated, single crystals of silver iodide on which ice formation was directly observed under the microscope, and a direct, quantitative measure of the deactivation was obtained by counting the number of nucleating centres per unit area as a function of wavelength, intensity, and duration of the radiation.

Freshly-formed large surfaces of silver iodide were irradiated by a 250-W high-pressure mercury-arc lamp for periods varying from 5 min to 10 h. The number of oriented ice crystals appearing on unit area of the surface at $-17°$ C and water saturation was taken as a measure of its activity. After only 5 min of irradiation, the density of oriented ice crystals actually increased about ten fold but, with longer exposures of up to 3 h, the density fell quite rapidly, some parts of the surface areas became completely inactive while others showed little change. By the successive use of filters cutting off at 4500, 4300, and 4000 Å, it was established that wavelengths greater than 4400 Å caused very little photolysis and decay of nucleation. This is in accord with the absorption band of silver iodide having a sharp cut-off at 4300 Å. For radiation doses equivalent to up to 30 min exposure in strong sunlight, the nucleating activity decreased roughly logarithmically with a time constant of 20 min and thereafter decayed rather more slowly. After an equivalent exposure of 1 h, the ice crystal density fell to about one-thirteenth of its maximum value, a result that is in reasonable agreement with decay factors of 10^{-2}–10^{-1}/h obtained by the Australian workers for smokes produced by a kerosene burner. These results suggest that nuclei released for rain-making purposes will lose much of their potency if they have to spend several minutes in bright sunshine before entering cloud. But since solar radiation in the waveband 3000–4400 Å is reduced about fivefold by an overcast stratocumulus deck, the presence of an extensive deep-cloud system should afford considerable protection.

Birstein (1952) sought to prevent photolysis of silver iodide by ultraviolet light by forming a water film on the surfaces of the particles before they were exposed to the radiation. Fused crystals of silver iodide were vaporized from a heated nichrome wire in a nitrogen atmosphere of controlled humidity. The nuclei, of mean diameter 0·06 μm, were collected in a glass cell and irradiated with an ultraviolet lamp having much the same intensity as strong sunlight. The activity of silver iodide prepared in a dry atmosphere was almost completely destroyed after 10 min exposure, but nuclei prepared in atmospheres of progressively higher humidity could withstand longer exposures without suffering serious deterioration. The adsorption of water vapour by the silver iodide, plotted as a function of the relative humidity, showed a steep rise when the latter was increased beyond 50 per cent and continued to increase quite rapidly as the humidity was raised to 80 per cent. A curve of very similar shape was obtained when the maximum periods for which the nuclei could be exposed to ultraviolet radiation without apparent deterioration were plotted against the relative humidity. The implication of this was that the photolysis of the surfaces of silver iodide particles, and hence the inhibition of their ice-nucleating properties, was greatly reduced once they had adsorbed a water film some tens of molecules thick, but this was not confirmed by St Louis and Steele (1968). Birstein's (1955) result that the quantity of water adsorbed at 0° C and 84 per cent humidity, and at $-20°$ C and 76 per cent humidity, was equivalent to a layer 130 molecules thick, is surprising because the surface forces fall off sharply with distance from the substrate. Birstein suggested that the adsorbed water existed as a highly-oriented ice-like layer as a result of the strong dipole-dipole interactions between the water molecules themselves, but it seems likely that the strong adsorption was due to the presence of hygroscopic contaminants. Indeed the work of Karasz, Champion, and Halsey (1956) suggests that, after a monolayer has been completed, the effect of the substrate on the adsorption properties is very small. More recent adsorption measurements by Zettlemoyer, Tcheurekdjian, and Chessick (1961) and Hall and Tompkins (1962) show that the surfaces of AgI are largely hydrophobic, that adsorption of single molecules occurs at low coverage but, with increasing adsorption, small clusters of a few molecules form within the monolayer at specific sites on the surface as the result of strong lateral attraction forces between the water molecules themselves. These clusters may well be formed around specks of contaminant because Corrin, Edwards, Nelson, (1964) find that for very pure silver iodide prepared *in vacuo* by direct reaction between silver and iodine, the quantity of adsorbed water is less than a statistical monolayer at humidities of up to 70 per cent. The clustering of water molecules at hydrophilic sites may make the latter preferred centres of ice nucleation.

We have seen on p. 194 that the ice-nucleating properties of AgI are greatly inhibited by the chemisorption of certain vapours, notably amines, but Burley and Herrin (1962) report that photolysis may be prevented almost completely by the addition of such substances as β-naphthol or diphenyl thiourea. Clearly the mechanisms by which photolytic deactivation is produced and may be prevented are complex and require much further study.

5

The Formation Of Snow Crystals

IN the last chapter we were concerned with the nucleation of supercooled water and the initiation of the ice phase. Now, we have to trace the growth of a crystal from the formation of the ice nucleus and examine the factors that may determine its subsequent development. The collection, photography, and classification of snow crystals have long been attractive pastimes, the books by Bentley and Humphreys (1931) and Nakaya (1954) being particularly famous for their thousands of beautiful photo-micrographs. Furthermore, valuable data have been obtained on the masses, dimensions, and terminal velocities of the different crystal forms. Also, in recent years, some real progress has been made in relating the shape, structure, and growth rate of the crystals to the environmental conditions in which they are formed. These studies, which have revealed that the growth mechanisms of ice crystals are very complicated, and with many puzzling features that are apparently peculiar to ice, are now reviewed.

5.1. The classification of solid precipitation

The ice elements formed in natural clouds are of four main types: individual ice crystals (or groups of crystals having a common nucleus), snowflakes, ice pellets, and hailstones. The ice crystals grow in an ice-supersaturated atmosphere by the diffusion of water vapour to their growing surfaces, and may exist as individual units of simple geometrical shape, for example hexagonal prismatic columns (prisms) and hexagonal plates or, under suitable conditions, may grow into complex, richly-branched forms. Several of these snow crystals may coagulate to form a snowflake. Graupel (or soft hail) pellets and true hailstones may originate from ice crystals or from frozen drops, but their subsequent growth proceeds predominantly by collision with supercooled cloud droplets. Snow crystals and snowflakes may also collect supercooled droplets that freeze on impact and endow the crystal surface with a rimed appearance.

Although solid precipitation elements can be generally allocated to one of about ten main classes, detailed observations reveal many variations on these main themes. In order to correlate observations made in different parts of the world, it is necessary to work in terms of an internationally accepted classification. The classification of snow agreed by the International Commission on Snow and Ice in 1949 is shown in Table 5.1 (*between* pp. 232–3), in which each class and basic feature of snow is designated by a code figure or a symbol. Additional letters and figures indicate additional characteristics such as the physical state, size, and degree of aggregation of the particles. Thus, F1rD1·5 or ⬡ 1·5 means plate crystals with water drops attached, the average diameter being 1·5 mm, while F2fwDd or (✳)d denotes a cluster of stellar crystals partially melted, the average size of the crystals composing the flake being large, i.e. between 2·0 and 3·9 mm.

5.2. The mass, dimensions, and fall velocities of snow crystals

The most comprehensive set of measurements on the fall velocity of individual snow crystals, as a function of their mass and linear dimensions, has been published by Nakaya and Terada (1935). The observations were made on Mt. Tokati, Japan, when the air temperature was between -8 and $-15°$ C. The crystals were allowed to fall through a long vertical cylinder containing windows and their fall velocities determined by measuring with a stop-watch the time taken for them to fall a distance of 2 m. In the case of the faster-falling graupel and snow pellets, a photographic method, in which the fall velocity was determined from the displacement of the image of the crystal on the film, was used. The procedure was to select a crystal from those collected on a glass plate, photograph it for later determination of its linear dimensions, drop it into the top of the vertical cylinder, measure its fall speed, and, finally, catch it on a paraffined slide, allow it to melt, and determine its mass from the diameter of the resulting drop.

The relations which were obtained between the masses and maximum linear dimensions of the various crystal forms are shown in Fig. 5.1, where the smooth curves can be represented by the following empirical formulae:

Graupel particles	$m = 0·065d^3$
Rimed plates and stellar dendrites	$m = 0·027d^2$
Powder snow and spatial dendrites	$m = 0·010d^2$
Plane dendrites	$m = 0·0038d^2$
Needles	$m = 0·0029d$ (d = length)

m being measured in milligrammes and d, the diameter of the sphere

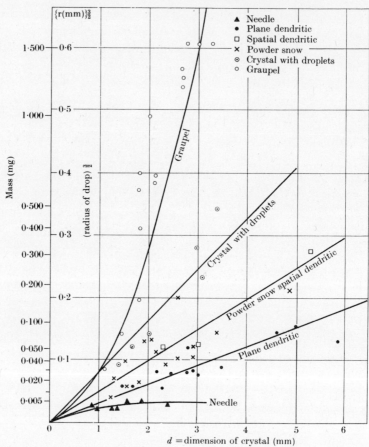

Fig. 5.1. The masses of snow crystals as a function of their linear dimensions. (From Nakaya and Terada (1935).)

which just contains the crystal, in millimetres. Of course, these relations apply only to mean values, and measurements made on individual crystals show some scatter about the mean as illustrated in Fig. 5.1. They were confirmed, in general, by Higuchi (1956) except that he obtained $m = 0.0059d^2$ for plane dendrites and, in addition, $m = 0.3\,a^2l$ for prismatic columns, l being the length and a the half-width. Nakaya and Terada, from measurements of the masses and surface areas of a number of plane dendrites, found that the crystal thickness varied very little with the diameter and, on the average, was about 11 μm. The mean density of the graupel particles represented in Fig. 5.1 was found to be 0·125 g cm⁻³, the maximum value for individual particles being 0·3 g cm⁻³.

The relations between the fall velocity and the dimensions of the crystals are shown in Fig. 5.2. The outstanding feature of this diagram is that the velocities of the plane dendritic, spatial dendritic, and powder-snow crystals are almost independent of their dimensions; that is, the velocity is nearly 30 cm s^{-1} for all sizes of plane dendrites, 50 cm s^{-1} for powder-snow aggregates, and 57 cm s^{-1} for spatial dendrites. When frozen droplets are attached to the crystals, the fall velocities increase to about 100 cm s^{-1}, and, in this case, the velocity tends to increase with the dimensions, the tendency being most marked in the case of graupel.

Mean values of diameter, mass, and fall velocity of the six different crystal types are given in Table 5.2 which is reproduced from Nakaya and Terada's paper.

Measurements of the fall velocities of crystals and small and large snowflakes have been made by Magono (1953). Using a stroboscopic method, he photographed the shape and attitude of the smaller aggregates during their fall through a cylinder and determined their fall velocity from the displacement of the image on the film between successive exposures. The results for rimed crystals and small rimed snowflakes are shown in Fig. 5.3. The larger snowflakes were timed with a stop-watch over a vertical fall path of 4 m and the results are given in Fig. 5.4. It will be seen that the relation between the fall speed and the dimensions of the flake depends upon the shape of the flake and whether or not it is rimed, but that for a particular type of snowflake, there is little variation of fall speed with diameter, especially when the latter exceeds 2 cm.

A more detailed investigation on the fall velocities of snowflakes was carried out by Langleben (1954), who photographed them falling against a dark background with a ciné-camera. The flakes were subsequently

TABLE 5.2

The diameters, masses, and fall velocities of snow crystals

	d (mm)	m (mg)	v (cm s^{-1})
Needle	1·53	0·004	50
Plane dendrite	3·26	0·043	31
Spatial dendrite	4·15	0·146	57
Powder snow	2·15	0·064	50
Rimed crystals	2·45	0·176	100
Graupel	2·13	0·80	180

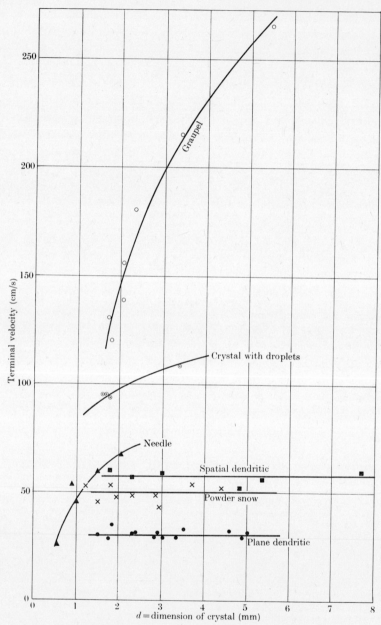

FIG. 5.2. The terminal velocities of snow crystals as a function of their linear dimensions. (From Nakaya and Terada (1935).)

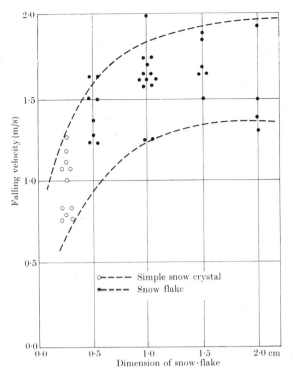

Fig. 5.3. The fall velocities of rimed snow crystals and small rimed snowflakes as a function of their linear dimensions. (From Magono (1951).)

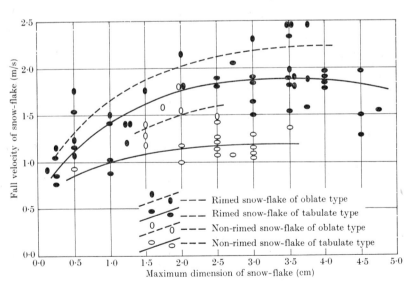

Fig. 5.4. The fall velocities of rimed and non-rimed snowflakes. (From Magono (1953).)

caught on a wool-covered disk and the diameters of the melted aggregates determined. Langleben found that the fall velocity v (cm s^{-1}) could be related to the melted diameter d (cm) by the equation $v = kd^{0\cdot 31}$, in which the numerical value of the factor k depended on the type of crystal aggregate, some typical values being as follows:

Crystal type	k (cm$^{0\cdot 69}$ s^{-1})
Prismatic columns and plates	234
Rimed dendrites	221
Dendrites ($T = 33°$ F)	203
Dendrites ($T < 32°$ F)	160

Litvinov (1956) carried out a similar set of measurements in Russia and also found a relationship $v = kd^n$, but the size dependence was much less marked than found by Langleben, the average value of n being 0·16. The values of k were also lower at about 100.

In developing an approximate theoretical expression for the terminal velocity of snowflakes, Magono (1953) considered the aerodynamic drag of the flake to consist of two parts: one represents the drag due to the air flowing round the flake and is proportional to r^2, where r is the radius of the flake; the second part, proportional to r^3, is the drag produced by air flowing through the open structure of the flake. Both terms are proportional to v^2, and this leads to an expression for the terminal velocity containing two drag coefficients. These were fitted by comparison with experiment to give

$$v = 132 \left(\frac{r}{0\cdot 40 + 0\cdot 63r} \right)^{\frac{1}{2}}$$

for non-rimed flakes, and

$$v = 194 \left(\frac{r}{0\cdot 45 + 0\cdot 60r} \right)^{\frac{1}{2}}$$

for rimed flakes, r being in cm and v in cm s^{-1}. These expressions predict $v \propto r^{\frac{1}{2}}$ for small flakes, and v independent of size for large flakes, in reasonable agreement with observation.

5.3. Fixation and photography of snow crystals

The direct photography of snow and ice crystals presents difficulties apart from the obvious one of their melting; the crystals have to be photographed in subfreezing conditions, when trouble may arise from condensation on the crystal surfaces and from frosting of lenses, etc.

Most of the difficulties were removed when Schaefer (1946) invented his technique of making permanent plastic replicas of the crystals. A solution of Formvar (polyvinyl acetal resin) in ethylene dichloride is kept at a temperature of about $-5°$ C. A clean glass microscope slide is immersed in the solution for about 30 s and then exposed horizontally to the falling crystals. A captured crystal becomes submerged in the solution, after which the slide should be kept at a sub-freezing temperature for a few minutes until the solvent evaporates, leaving the crystal encased in a thin, but tough, plastic shell. The slide may now be exposed to room temperature, when the crystals will melt, the water diffusing out of the plastic membrane as it evaporates to leave a replica that retains, in microscopic detail, all the surface configurations of the original crystal.

A little practice and care are required to get the best results. The Formvar solution should be dehydrated by shaking it up with calcium chloride or phosphorus pentoxide to remove the dissolved water, which otherwise will come out when the solution is chilled and form spurious ice crystals. It is particularly important to use solution of the right strength. If it is too viscous, small crystals will not become submerged and merely make a crater on the surface; if it is too thin, it will run off the slide and not cover a large crystal. Good replicas of natural snow crystals may be obtained with a 1–3% solution but, for small crystals less than 0·1 mm diameter, such as may be produced in a laboratory experiment, a 0·1% solution should be used.

The replicas or the crystals themselves may be photographed through a microscope by either reflected or transmitted light. The use of transmitted light produces a clear picture of the boundary and internal structure of the crystal (Fig. 5.5(a)). Reflected light increases the beauty of the photograph by producing a white image on a black background (Fig. 5.5(b)) and reveals something of the surface topography. Perhaps the best results are obtained with oblique illumination which combines the advantages of both transmitted and reflected light and reveals both the internal and surface structure (Fig. 5.5(c)). A very fine collection of several hundreds of photographs taken by this type of illumination has been published by Nakaya (1954) in his book *Snow crystals*.

5.4. The occurrence of ice crystals in natural clouds
5.4.1. *The forms of individual snow crystals*

Although extensive observations have been made of snow crystals reaching the ground, and attempts have been made to correlate the relative frequencies of the various crystal forms with the temperature at the place of observation, it proved difficult to obtain consistent results. Heim (1914), from observations made in the Antarctic, established that the occurrence of prisms was associated with lower temperatures than that of plate crystals and gave the following average temperatures for the appearance of the different crystal forms:

Prisms	$-27°$ C
Prisms predominating and plate crystals	$-23°$ C
Prisms and plates in about equal numbers	$-18°$ C
Plates predominating and prisms	$-12°$ C
Plates	$-12°$ C

Westmann (1907) reported that, on Spitsbergen, stellar crystals were frequently observed at temperatures above $-20°$ C, while plate crystals and prisms were rare, and vice versa below $-20°$ C. The large amount of data on natural snow crystals collected by Nakaya and his colleagues in Japan from 1930 to 1960, gave additional information, notably that needle-like crystals are seen only when the temperature is close to $0°$ C.

That the earlier workers did not find a clearly marked correlation between the predominant crystal habit and the temperature at the ground is, perhaps, not surprising, since only the conditions prevailing during the growth of the crystal are likely to be of major importance in determining its shape, and it is only in recent years that crystals have been collected from different types of clouds having widely different conditions of temperature, water-vapour concentration, and supersaturation relative to ice. Our knowledge on the ice-particle content of clouds is still very inadequate, particularly in the case of cumuliform clouds, where information on the predominant forms which exist at different levels and different stages of the cloud development is urgently required. The sampling of ice particles from an aircraft presents considerable difficulties, particularly at high speeds, when it becomes difficult to avoid shattering of the crystals during their collection. The problem has been partially solved by using sampling tubes in which the crystal-laden air is decelerated before it reaches the collector, but it has proved rather difficult to avoid fracture of large, delicate crystals and crystalline aggregates.

The most comprehensive collection of ice crystals from natural clouds has been obtained by Weickmann (1947) at different heights in the troposphere up to cirrus levels, the temperature of the sampling region being measured. At temperatures below about $-25°$ C the dominant crystal form was the hexagonal prism that was indigenous to cirrus and medium-level clouds. They develop as a result of preferential growth along the principal (c) axis, i.e. normal to the basal face. The crystals are typically 0·5 mm in length, the ratio of length to breadth varying from one to five. Prismatic columns appeared as single crystals, as basal twins, and in clusters (Figs. 5.6, 5.7). Clusters of linear dimensions about 2 mm were common in isolated cirrus clouds, the crystals containing very pronounced funnel-shaped cavities and having a skeleton structure in which both basal and prism faces were incomplete. The crystals constituting a cluster appear to originate from a common nucleus, probably a frozen droplet. That isolated (or convection) cirrus clouds, for example cirrocumulus, cirrus castellatus, filosus, and uncinus, originate as water clouds is indicated by their mottled appearance and iridescence in the early stages. It seems likely that when the initial droplet freezes, dendrites will shoot through the droplet and terminate in the surface to form suitable sites for the origin of the prismatic crystals. These clusters were found by Weickmann to be entirely absent from cirrostratus, where the prisms showed completed faces and no well-marked cavities, a result, no doubt, of the much slower growth that would be associated with the very slow cooling of the air and, perhaps, the formation of crystals on solid nuclei possessing a thin water film rather than from a water droplet. In cirrostratus Weickmann also found occasional prismatic crystals with a single pyramidal end (Fig. 5.7). According to Humphreys (1929), these pyramidal forms are necessary to explain the formation of certain halos that have been seen only very rarely in cirrostratus.

Weickmann reported that in proceeding from high- through medium- to low-level clouds, and therefore to higher temperatures, there was a gradual transition from prisms, through thick plates to thin hexagonal plates, the latter being the result of growth in the direction of the secondary (a) axes, i.e. preferential propagation of the prism faces. The thin, regular, hexagonal plates, represented in Fig. 5.8(a), generally occurred at temperatures above $-20°$ C. They are generally less than 500 μm in diameter and less than 50 μm in thickness. They often possess a well-marked internal and surface structure which generally shows hexagonal symmetry (Fig. 5.8(b)), but occasionally crystals with trigonal symmetry are found as shown in Fig. 5.8(c). The plate-like habit may sometimes develop in sector form and exceed 1 mm in diameter.

At temperatures between -10 and $-20°$ C, the striking star-shaped crystals are dominant. They usually possess six arms, which may grow rapidly and develop side-branches to assume the fine dendritic structure illustrated in Figs 5.9(a), (b), and attain diameters of up to 5 mm. The arms are not always coplanar (the crystal is then termed a spatial dendrite), and although they are often superficially very similar, they rarely show exact hexagonal symmetry in their detailed fine structure. These stellar crystals often develop from a hexagonal plate, which at a certain stage of its development begins to sprout at the corners (Fig. 5.9(a)), but some photographs suggest that the crystal originates from a frozen droplet that may later develop crystal faces to confer hexagonal symmetry on the crystal development.

TABLE 5.3
Weickmann's observations of predominant crystal forms in different cloud types

Level of observation	Temperature range	Cloud types	Crystal forms
Lower troposphere	0 to −15° C	Nimbostratus Stratocumulus Stratus	Thin hexagonal plates Star-shaped crystals showing dendritic structure
Middle troposphere	−15 to −30° C	Altostratus Altocumulus	Thick hexagonal plates Prismatic columns—single prisms and twins
Upper troposphere	< −30° C	Isolated cirrus	Clusters of prismatic columns containing funnel-shaped cavities. Some single hollow prisms
		Cirrostratus	Individual, complete prisms

Needle-shaped crystals, such as are shown in Fig. 5.10, are not represented in Weickmann's photographs, nor in his classification given in Table 5.3, but are often observed at the ground when the temperature is only slightly below 0° C and, according to Wall (1947) and Gold and Power (1952), originate from clouds with temperatures between −3 and −8° C. The needles, which often occur in clusters, probably represent marked preferential growth along the principal axis, since small, much more compact prisms have been produced in this temperature range in artificial clouds (see § 5.5).

So far, we have mentioned only the basic types of ice crystal, but combination forms made up of two or more of these basic structures are quite common, the prismatic column with end plates (capped column or collar stud) and the stellar-plate crystal, which are shown in Figs. 5.9 and 5.11, being typical examples. These transition forms reflect the changes in temperature and supersaturation of the environment that the crystal experiences on its journey towards the ground.

The aircraft observations of Weickmann have much in common with a similar series obtained in Russia by Borovikov (1953), who found thin plates to be predominant between 0 and $-16°$ C, thick plates between -16 and $-25°$ C, prisms from -13 to $-30°$ C with an increasing tendency for hollow cavities to appear at the lower temperatures, clusters of prisms below $-25°$ C, and pyramidal shapes between -22 and $-28°$ C. An ingenious instrument, called a 'snow-crystal sonde', has been devised by Magono and Tazawa (1966) to collect snow crystals as it ascends through a cloud at the same time as a conventional

TABLE 5.4

Temperature range	Crystal type
-3 to $-8°$ C	Needles
-8 to $-25°$ C	Plates, sector stars
-10 to $-20°$ C	Stellar dendrites
$<-20°$ C	Prisms, single crystals, twins
$<-30°$ C	Clusters of hollow prisms

radio-sonde measures the air temperature and humidity. The crystals fall into a sampling box on to a moving 35-mm tape where they are covered with Formvar replicating solution. When the balloon reaches the 500-mb level, the sampling device is released by a baroswitch and returned to the ground by parachute. These aerological observations have been supported by a great many at ground level, often from mountain observatories immersed in cloud. Among the more important are the observations of Wall (1947) in Friedrichshafen, of Gold and Power (1952, 1954) in Canada, those on the Hohenpeissenberg in Germany by Weickmann (1957) and Grunow (1960), in New England by Kuettner and Boucher (1958), and in Japan by Murai (1956), Magono (1960), Nakaya and Higuchi (1960), Higuchi (1962a, b), and Lee and Magono (1967) who summarized a long series of observations made at different altitudes on Mt. Teine. All these observations point to a strong correlation between air temperature and the predominant crystal shape, which is essentially the same as that given by Weickmann and summarized in Table 5.4. That plate crystals can grow in clouds with temperatures just below 0° C was first suggested by a photograph taken of an artificially seeded stratocumulus cloud during the operations of Project Cirrus. After a part of the cloud, whose temperature was about $-4°$ C, was converted into ice crystals, a brilliant sub-sun and sub-mock-sun appeared, these optical phenomena being possible only if plate-like crystals are present. The existence of these crystals at temperatures above $-3°$ C has been confirmed by laboratory experiments as described in § 5.5.

5.4.2. *Aggregation of ice crystals to form snowflakes*

Snowflakes are agglomerates of individual crystals in which the star-shaped dendrites are generally prominent (see Fig. 5.12), but in which needle and plate forms may also be found. The growth of these aggregates is governed by the collision and aggregation efficiencies of the crystals and by their relative motions which, because of the aerodynamic problems posed by the complexity and variability of the ice-crystal geometry, are not amenable to quantitative computation.

Clusters composed of a few individual crystals of the same shape may arise as the result of collision and aggregation of the component crystals, or, alternatively, several crystals may grow from a single nucleus, frozen droplet, or host crystal. Thus spatial dendrites, which usually grow only at temperatures between -12 and $-16°$ C in mixed clouds, probably develop as the result of the primary crystal(s) collecting supercooled droplets, some of which freeze and grow into crystals. These crystals will not usually grow in the same orientation as the host crystal because, according to Hallett (1964), freezing droplets tend not to take the same orientation as the substrate if the temperature of the latter is below $-10°$ C. On the other hand, rimed needles may grow as bundles of parallel rods at temperatures above $-5°$ C. At temperatures below $-25°$ C, spatial clusters of prisms arise in convective cirrus clouds from single frozen drops, but only single prisms appear in cirrostratus because of the absence of supercooled droplets.

Aggregates consisting of only two or three plates or needles, such as those reported by Magono (1968), may result from collisions of the kind that have been studied by Jayaweera and Mason (1965, 1966) for discs and cylinders falling in a viscous liquid. When two equal discs of Reynolds number $Re > 1$ fall horizontally and one directly behind the other, attraction is apparent at separations exceeding 40 diameters. When the centres of the two discs are displaced horizontally by less than a radius, the rear disc catches up and comes to rest at an angle to the leader, which remains horizontal. At $Re > 100$, the angle between the two discs is $30°$, but this inclination increases with decreasing Reynolds number and may approach $90°$ at $Re \approx 5$. A cluster of three equal

Fig. 5.5. Snow crystals photographed; (a) by transmitted light; (b) by reflected light; (c) by oblique illumination.

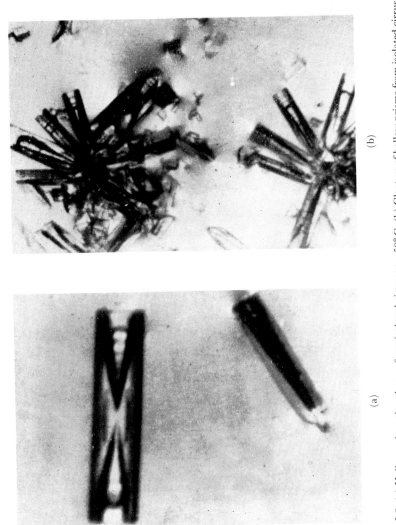

FIG. 5.6. (a) Hollow prismatic columns from isolated cirrus at −50° C. (b) Clusters of hollow prisms from isolated cirrus at −44° C. (From Weickmann (1947).)

FIG. 5.7. Complete prismatic columns from cirrostatus at −26° C. (From Weickmann (1947).)

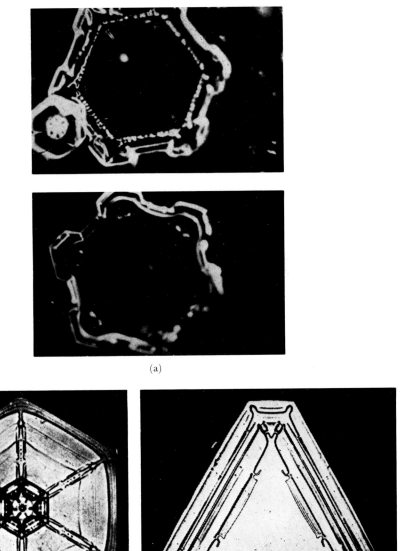

FIG. 5.8. (a) Thin hexagonal plates with no internal structure. (b) Hexagonal plates showing internal structure. (c) A plate crystal of trigonal symmetry. (From Bentley and Humphreys (1931).)

(a)

(b)

FIG. 5.9. (a), (b), Hexagonal plates that have developed into richly branched (dendritic) stellar crystals. (From Bentley and Humphreys (1931).)

Fig. 5.10. Needle-like crystals (From Nakaya and Hasikura (1934).)

Fig. 5.11. A hexagonal column capped with hexagonal plates. (From Bentley and Humphreys (1931).)

Fig. 5.12. Snowflakes composed of stellar crystals.

discs forms a 'butterfly' configuration in which one member remains horizontal and the other two form a symmetrical vee in its rear. Such a cluster is very stable when the smaller included angle between the discs is <30°.

Two equal cylinders, with $0.1 < Re < 10$, initially non-parallel and vertically separated by up to 50 diameters, catch up, slide along each other, and finally fall as a crossed-pair bisecting at right-angles. When several identical cylinders are released in random orientations, they tend to cluster and then separate into pairs, and into triplets in the form of a symmetrical #.

An attempt to measure the collection efficiencies of ice particles was made by Hosler and Hallgren (1961) by suspending ice spheres of original diameter 127 and 360 μm in a wind tunnel and drawing ice crystals of typical dimensions 8–18 μm past them. The collection efficiencies, which were mostly in the range 0·03–0·20, were found to increase with increasing temperature over the range -26 to $-11°$ C when the ice crystals were plates, and to decrease again at temperatures between -10 and $-6°$ C when the crystals were prisms. These results are difficult to interpret because they are an overall measure of both collision and adhesion efficiencies, and the target may have become strongly charged. The effect of electrical forces on the aggregation of ice crystals has been demonstrated by Latham and Saunders (1964) and Saunders (1968), who showed that the collection efficiency of an ice sphere increased rapidly when electric fields stronger than about 300 V cm^{-1} were applied parallel to the direction of flow of the crystals. In the presence of such fields, aggregation occurred at temperatures as low as $-50°$ C, whereas in the absence of a field, aggregation was negligible below $-30°$ C. It appears that the electrical forces increased the adhesion rather than the collision efficiency of the crystals and that although they may play a role in the aggregation of ice particles in thunderstorms, they are unlikely to be important in the growth of snowflakes in layer clouds.

In natural snowflakes, adhesion is effected partly by interlocking of the crystals and by sintering of the ice at the points of contact, although deposition of water vapour and the freezing of collected supercooled droplets may act as a cement.

When two ice particles come into contact, adhesion occurs as the bridge between them grows in order to minimize the surface free energy of the system. The adhesion of ice spheres, measured at temperatures down to $-20°$ C by Nakaya and Matsumoto (1954), and down to

−80° C by Hosler, Jensen, and Goldshlak (1957), has been attributed by the latter and by Jellinek (1961, 1962) to the existence of a liquid-like layer on the surface of ice which is supposed to solidify at the points of contact between the spheres.

An alternative explanation was given by Kingery (1960), who observed that, when two ice spheres were pushed together to touch at a point, the area of contact grew with time, even when the original force of contact was removed. By measuring the growth rate of the neck between the two spheres, Kingery claimed to show that, for spheres between 0·2 and 6 mm in diameter and in the temperature range −2 to −25° C, the material was transferred from the convex surfaces of the spheres into the concave neck by surface diffusion. However, from similar experiments with ice spheres of $d < 100$ μm, Kuroiwa (1961) claimed that volume diffusion was dominant between 0° C and −10° C, with surface diffusion becoming of increasing importance at lower temperatures and dominating below −15° C. But, in a comprehensive theoretical and experimental investigation, in which measurements were made on polycrystalline and single-crystal spheres of ice ranging in diameter from 50 to 700 μm and at temperatures ranging from −3 to −20° C, Hobbs and Mason (1964) demonstrated that, under very clean conditions, sintering in an ice-saturated atmosphere occurred by the evaporation of material from the convex surface of the spheres and its condensation into the concave neck, and that this evaporation–condensation mechanism was likely to be faster by four orders of magnitude than either volume or surface diffuson. Even so, with pure ice, the growth rate of the neck was quite slow; typically it took some hours for the radius of the neck to grow to one-quarter of the radius of the sphere. Under these conditions, there was no evidence for a liquid-like layer on the ice surfaces, but if the ice contained dissolved salts, or if the surfaces of the spheres became otherwise contaminated, the neck grew much more rapidly by the flow into it of a film of contaminated liquid.

In assessing the relative importance of sintering, vapour deposition, and riming in forming bridges between adjacent crystals in an aggregate, we recall that the rate of sintering increases rapidly with decreasing particle size and, as Hobbs (1965) points out, ice particles of radius 10 μm need be in contact for only 10 s in order to become firmly bonded together. We also recall that Hosler, Jensen, and Goldshlak (1957) found that, while in an ice-saturated environment, plate-like and columnar crystals did not aggregate at temperatures below −25° C, in an atmosphere supersaturated relative to ice, aggregation occurred at all temperatures down to $-36°$ C. In a layer cloud releasing persistent snow, only the first kilometre or so above the $0°$ C level may be appreciably supersaturated and contain supercooled droplets, and this may explain why aggregation of snow crystals to form large flakes occurs most readily at temperatures just below $0°$ C. Accordingly, it would appear that vapour deposition and riming are more effective than sintering in causing the adhesion of crystals, especially in mixed clouds, but further work is required to establish this firmly.

5.5. Studies of ice crystal growth in the laboratory

The existence of at least four basic forms of natural snow crystals, and of an almost infinite number of variations on each of these main themes, suggests that their growth and development are complicated matters which may best be studied in the laboratory, where the whole life history of a crystal can be observed under controlled conditions.

5.5.1. *Variation of crystal habit with temperature and supersaturation*

The growth of ice crystals in supercooled water clouds, the latter being produced in room-size cold chambers whose temperature could be controlled, has been studied by Aufm Kampe, Weickmann, and Kelly (1951) and Mason (1953*b*). The clouds were produced by the introduction of steam into the chamber. Within a few seconds, the fog cooled to the temperature of the chamber and was then seeded with dry ice or silver iodide if insufficient natural ice nuclei were present. Mason projected a small pellet of dry ice to produce a horizontal seeded track near the top of the chamber, and caught the crystals on plastic-coated slides near the floor after they had fallen through about 3 metres of supercooled cloud. The plastic replicas of the crystals were subsequently studied and photographed under the microscope.

The observed changes of crystal habit with temperature were very similar in both sets of experiments and are summarized in Table 5.5.

The classification of these laboratory-produced crystals according to the temperature of formation bears a marked similarity to that of natural snow crystals (Table 5.4), showing that it is possible to simulate quite well the early stages of growth of snow crystals in the laboratory and, at the same time, determine the transition temperatures for the different crystal forms more precisely than can be done in the atmosphere. The most striking feature of Table 5.5 is the remarkable sequence of habit, plates → prisms → plates (and stars) → prisms, which occurs as the temperature is lowered from 0° C to −25° C.

A series of experiments to study the growth of individual ice crystals in different environmental conditions has been described by Nakaya (1951, 1954). The crystals were grown on a fine rabbit hair stretched on a frame and suspended in an air-stream whose temperature and water-vapour content could be varied. The crystal development was followed over a period of 30 min or so by time-lapse photography. The apparatus, shown in Fig. 5.13, consisted of two concentric glass cylinders, the warm vapour from an electrically-heated reservoir being convected upwards inside the inner tube, cooled on its way up, and returned through the annular space. The whole apparatus was placed in a thermostat and located in a cold chamber maintained at about −30° C. The degree of supersaturation in the experimental space was varied by altering the temperature, T_w, of the water in the reservoir R. The temperature, T_a, of the air in the immediate neighbourhood of the growing crystal was a function of both T_w and the temperature, T_t, of the thermostat. For a given value of T_w, the air temperature was regulated by adjusting T_t. Crystals were produced at various combinations of T_a and T_w, with a view to studying the relationship between the crystal form and the external conditions.

Unfortunately, the conditions in Nakaya's apparatus were not steady nor well defined because the strong convection gave rise to large fluctuations in both temperature and supersaturation. The average temperature of the air in the neighbourhood of the crystal was measured with an alcohol-in-glass thermometer. Rather crude relative estimates

TABLE 5.5

Changes of crystal habit with temperature in artificially produced water clouds. (After Aufm Kampe *et al*, and Mason)

Temperature range	Crystal habit
0 to −5° C	Simple, clear hexagonal plates with no surface markings; some trigonal shapes
−4 to −9° C	Prisms, some showing marked cavities and similarity to needles
−10 to −25° C	Hexagonal plates showing ribs, surface markings, and tendency to sprout at corners
	Sector stars
	Dendritic stars, most prominent around −15° C
−25 to −40° C	Single prisms, twins, and hollow prisms
	Aggregates of prisms and irregular crystals, (Aufm Kampe *et al*.)

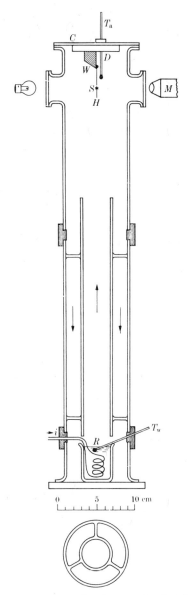

FIG. 5.13. Nakaya's apparatus for the growth of artificial snow crystals.

of the supersaturation were made by sucking the air through a filter containing P_2O_5 and measuring the increase in weight caused by absorption of both liquid and vapour. Nevertheless, it emerged that T_a and T_w were the two main parameters controlling the growth forms of the crystals; and Nakaya's classification of habit in relation to the air temperature T_a, shown in Figs. 5.14 and 5.15, is in broad agreement with the data of Table 5.5. Nakaya was of the opinion that the temperature rather than the supersaturation of the environment was the main factor controlling the crystal shape, except for dendritic growth, which occurred only at relatively high supersaturations and at air temperatures between -14 and $-17°$ C. However, Marshall and Langleben (1954) interpreted Nakaya's results as showing that the habit is principally determined by the excess of the ambient vapour density over that which would be in equilibrium with the surface of the ice crystal, that is, by a quantity closely related to the supersaturation and proportional to the flux of vapour directed towards the crystal. A

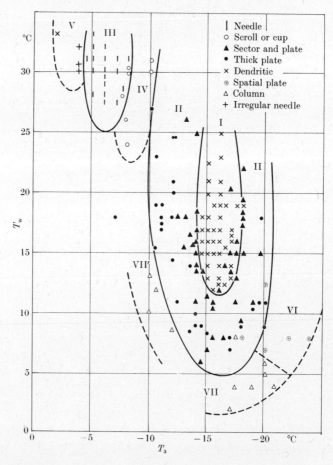

FIG. 5.14. Relation between Nakaya's crystal forms and the temperatures T_a and T_w.

§ 5.5   THE FORMATION OF SNOW CRYSTALS

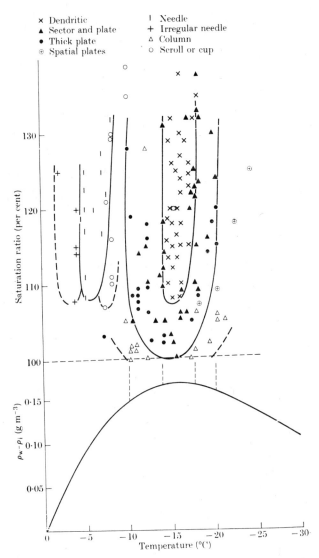

Fig. 5.15. Upper part: relation between crystal type, saturation ratio and air temperature as observed by Nakaya. Lower part: a curve of vapour-density excess against temperature.

similar suggestion was made earlier by Weickmann (1950). Marshall and Langleben hypothesize that resistance to growth will be greatest at the crystal corners, and greater on the prism faces than on the basal faces, so that growth of the corners and the prism faces will occur only when the excess vapour density, $\Delta\rho$, becomes sufficiently large to overcome these inhibitions. Thus, prismatic columns would be expected to develop at relatively low values of $\Delta\rho$, plates only when $\Delta\rho$ is relatively large, and the corners (dendritic growth) only when $\Delta\rho$ achieves very high values.

Strong experimental evidence against such an interpretation was

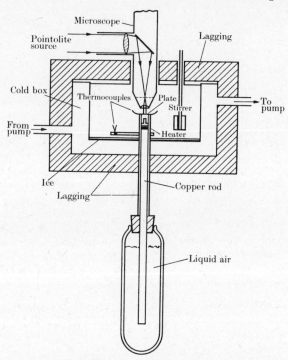

FIG. 5.16. The apparatus of Shaw and Mason for the growth of ice crystals from the vapour. (From Shaw and Mason (1955).)

obtained by Shaw and Mason (1955), who studied the growth of individual ice crystals growing on a metal surface under conditions such that the temperature and supersaturation of the surrounding air could be controlled independently. The clean, smooth metal surface was located in the centre of a cylindrical metal chamber, (Fig. 5.16), the hollow walls of which were cooled to the desired temperature by circulating chilled petrol through the annular space. The inner walls and base of the chamber were coated with ice, so that the air in the experimental space, which was stirred by a small fan, was saturated with ice at a temperature T_1, indicated by a thermocouple embedded in the surface of the ice layer in the bottom of the chamber. The metal plate on which the crystals grew was supported on a copper rod, which was insulated from the chamber and dipped into a Dewar flask of liquid air. The plate was then cooled to a temperature T_2, lower than that of the surrounding air, which could be controlled by varying the current through a heating coil. The temperature, T_2, of the crystals was recorded by a thermocouple placed immediately below the metal surface, and the saturation ratio of the air in the immediate vicinity of the crystals was given by the ratio of the equilibrium vapour pressures over ice at the temperatures T_1 and T_2 respectively. The crystals were viewed through a metallurgical microscope and photographed at 1-min intervals. The growth rates of individual crystal faces were determined from measurements made on the negatives with a micrometer eyepiece.

In order to determine whether the various crystal forms appearing in natural and laboratory-produced clouds could be produced at the same temperatures on the metal plate (the plate temperature was held constant at various temperatures between -5 and $-40°$ C), the crystals were grown under conditions of saturation relative to liquid water. The crystal habit was found to vary as follows:

-5 to $-9°$ C prisms,
-9 to $-25°$ C plates,
Below $-25°$ C prisms,

in a manner very similar to that observed in the cloud experiments. Stars were absent, but a new crystal form appeared quite often on the plate at temperatures between -4 and $-8°$ C and also below $-22°$ C. This took the form of a prism terminated at one end only by a pyramid. The appearance of such hemimorphic forms, which have been found only very occasionally in natural clouds (usually cirrostratus), but not at all in small-scale laboratory clouds, is of considerable interest in connection with the possibility that ice may possess a polar lattice.

When they were maintained at constant temperature and supersaturation, the crystals grew towards a limiting habit defined by

$$\Gamma = \left|\frac{c}{a}\right|_{\text{lim}} = \sqrt{\left(\frac{dc^2}{dt} \bigg/ \frac{da^2}{dt}\right)},$$

where c and a are respectively the principal and secondary crystal axes. Shaw and Mason established that this habit, and therefore whether a crystal developed as a prism or a plate, was determined very largely by the temperature, the supersaturation, although varied over wide limits, having no systematic effect. It was also established that a minimum critical value of the supersaturation was required to start growth on any crystal face, but this varied in an apparently random manner from face to face and from crystal to crystal; there was no systematic difference between the critical supersaturations required for growth on the basal and prism faces as required by the hypotheses of Marshall and Langleben.

It must be admitted, however, that conditions for crystal growth on the metal surface may not have simulated fairly those occurring in the free air. To meet this point and to study the whole problem in more detail, a new series of experiments were begun by Mason and Hallett in 1955 (see Hallett and Mason (1958a)).

The crystals are grown on a thin nylon or glass fibre running vertically through the centre of a water vapour diffusion chamber, 50 cm high and 30 cm diameter, constructed of Perspex and resting on a solid aluminium block maintained at about −60° C by dry ice. The apparatus is shown diagrammatically in Fig. 5.17. Cooled from below, with its top

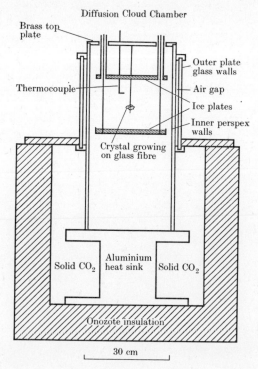

FIG. 5.17. The diffusion cloud chamber used by Mason and Hallett for growing ice crystals at various temperatures and supersaturations.

maintained either at or above room temperature, the chamber encloses a thermally-stratified, convectively stable atmosphere in which steady-state conditions are readily achieved. Water vapour, evaporated from an extended source, diffuses downwards through the chamber towards the low-temperature sink, the supersaturation regime being largely determined by the temperatures of the source and sink, whose vertical separation can be varied. In such a chamber containing room air, condensation of water vapour on the aerosol particles produces a dense cloud of tiny water droplets which freeze spontaneously on falling beneath the $-40°$ C level. In the presence of a persistent droplet cloud, the air may be regarded as saturated relative to liquid water, the supersaturation relative to ice at any level being determined solely by the temperature at that level. The variation of temperature with height is measured with a thermocouple, the separation of the $0°$ C and $-40°$ C levels being about 10 cm.

If the chamber is sealed and left for some hours, the condensation nuclei are progressively removed by sedimentation, leaving clean, highly supersaturated air in which condensation tracks produced by cosmic rays may be seen.

In order to simulate conditions in natural clouds, where ice crystals may grow at humidities below water saturation and therefore at saturations of only a few per cent relative to ice, the chamber was modified so that the crystals are grown on a fibre suspended centrally between two plane parallel sheets of ice. The vertical profiles of both temperature and supersaturation between the ice plates, the upper of which serves as the vapour source, is determined by their positions in the chamber, which are adjustable; additional control is provided by electrical heating of the top plate. A thermocouple passing through a small hole in this plate measures the temperature profile between the plates, and this, together with the solution of the diffusion equation for the profile of water-vapour concentration, allows the supersaturation at each level to be calculated.

With these facilities, Mason and Hallett were able to grow crystals over the temperature range $0°$ C to $-50°$ C and under supersaturations ranging from a few per cent to about 300 per cent. The results of many experiments are summarized in Fig. 5.18. Consistently, the crystal habit varied along the length of the fibre as shown in Table 5.6.

This scheme is very similar to that of Table 5.5, but the simultaneous growth of all the crystal forms on the same fibre brought to light the sharpness of the boundaries between one habit and another. For example,

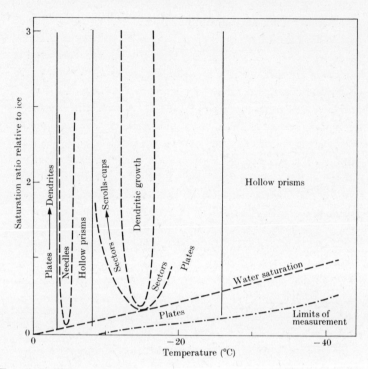

FIG. 5.18. The growth habits of ice crystals in relation to the temperature and supersaturation of the environment as observed by Hallett and Mason (1958a).

the transition between the plates and needles at −3° C, and that between hollow prisms and plates at −8° C, occurred within temperature intervals of less than one degree. A photograph showing the variation of crystal habit along the length of the fibre is reproduced in Fig. 5.19.

Fig. 5.18 is similar, in many respects, to the diagram published by Nakaya (1954), but it covers much wider ranges of temperature and

TABLE 5.6
Variation of crystal habit with temperature (Hallett and Mason 1958a)

0 to −3° C	Thin hexagonal plates
−3 to −5° C	Needles
−5 to −8° C	Hollow prisms
−8 to −12° C	Hexagonal plates
−12 to −16° C	Dendritic crystals
−16 to −25° C	Plates
−25 to −50° C	Hollow prisms

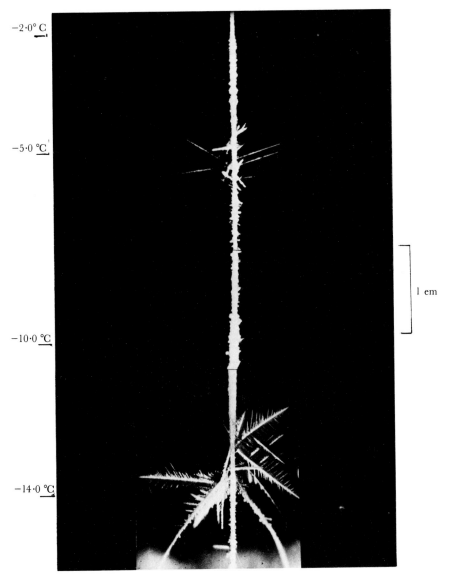

Fig. 5.19. The variation of crystal habit along part of the fibre suspended in the diffusion cloud chamber. The temperature varied from $-2°$ C to $-18°$ C and the saturation ratios everywhere exceeded 2·0. The sequences from top to bottom is plates→needles→hollow prisms→plates→dendrites. (From Hallett and Mason (1958a).)

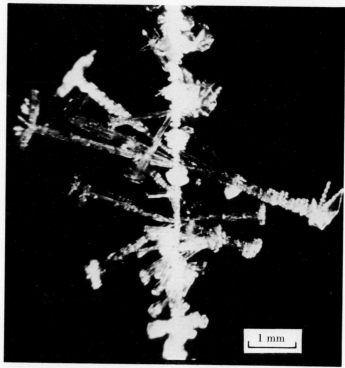

FIG. 5.20. Crystal hybrids showing how the crystal habit is dictated by temperature. Needles grown at −5° C developed plates on their ends when shifted to a temperature of −2° C. (From Hallett and Mason (1958a).)

FIG. 5.21. Needles grown at −5° C developed stars on their ends when shifted to −14° C. (From Hallett and Mason (1958a).)

supersaturation. It differs from his in that it shows the existence of plates between 0° C and −3° C, of hollow prisms rather than needles between −5 and −8° C, and does not show prisms to occur between −10 and −15° C at low supersaturations.

Crystals having an almost identical variation of habit with temperature have also been grown by Mason and Hallett from the vapour of heavy water (99·75 per cent pure) but with the transition temperatures all shifted upwards by nearly 4°, in conformity with the difference between the melting points of H_2O and D_2O.

These experiments appear conclusive in showing that very large variations of supersaturation do not change the basic crystal habit as between prism and plate-like growth although, of course, the growth rates are profoundly affected. On the other hand, the supersaturation appears to govern the development of various secondary features such as the needle-like extensions of hollow prisms, the growth of spikes and sectors at the corners of hexagonal plates, and the fern-like development of the star-shaped crystals, all of which occur only if the supersaturation exceeds values which, in these experiments, correspond roughly to saturation relative to liquid water.

The effect of suddenly changing the temperature and supersaturation on the growth form of a particular crystal could be observed simply by raising or lowering the fibre in the chamber. Whenever a crystal was thus transferred into a new environment, the continued growth assumed a new habit characteristic of the new conditions. Thus, when needles grown at temperatures between −3 and −5° C were suddenly moved up in the chamber, to about −2° C, plates developed on their ends (Fig. 5.20), and when similar needles were lowered to about −14° C, they gave way to star-shaped crystals as in Fig. 5.21. These are only some examples of metamorphoses that were observed when crystals were transferred to a new environment; in fact, combination forms at all the basic crystal types shown in Fig. 5.18 were readily produced this way. Such radical changes in crystal shape could not be produced by varying the supersaturation at constant temperature, but in some cases were produced by only a degree or two change in temperature at constant supersaturation.

Kobayashi (1957) also used a diffusion cloud chamber to grow ice crystals on a rabbit hair at temperatures ranging from 0° C to −30° C and supersaturations up to about 90 per cent relative to ice. He also concluded that the temperature was the main factor controlling the crystal habit and obtained a very similar scheme to that of Table 5.6.

Kobayashi's technique of estimating the supersaturation in his chamber, by observing the rate of growth of an ice sphere, does not appear to be satisfactory, but he also finds that sector plates and dendrites do not often occur unless the air is supersaturated relative to liquid water, that is, supersaturated by at least 12 per cent relative to ice at $-12°$ C.

Kobayashi (1965) has also succeeded in growing ice crystals from the vapour at temperatures down to $-90°$ C. In the temperature range -40 to $-90°$ C they appeared as solid prisms at low supersaturations. Long, thin, prismatic columns and whisker-like crystals were observed with air temperatures between -45 and $-50°$ C and at low supersaturations relative to ice. Pyramidal faces appeared on the ends of the prisms in the temperature range -50 to $-90°$ C.

It is hardly possible to produce and measure very small supersaturations of only a few per cent in the diffusion chamber. Kobayashi (1960) attempted to achieve this in a convective-mixing chamber of the Nakaya type, the supersaturation being measured by sucking a known volume (about 2 litres) of air from the chamber through dried methanol and determining the total water content by titration with Karl Fischer reagent. Although an accuracy of ± 5 per cent was claimed for this method, it has the serious disadvantages of disturbing conditions in the chamber and that it can give only a mean value for the whole chamber and not that prevailing in the neighbourhood of the growing crystal. The errors involved in determining the supersaturation are indicated by the fact that crystals continued to grow although the measurements indicated a subsaturation of 10 per cent. Nevertheless, at the higher supersaturations, Kobayashi obtained a very similar variation of crystal habit with temperature to that given in Table 5.6, except that he found dendrites only between -12 and $-14°$ C. At very low supersaturations, he claimed that only short, solid columns occurred at all temperatures between -8 and $-23°$ C, the implication being that the temperature-dependent variation of habit shown in Table 5.6 no longer held under these conditions.

This apparent contradiction was resolved by Kobayashi (1961) when working in the author's laboratory with an improved technique in which two air-streams saturated with respect to ice at different temperatures were mixed to produce a supersaturated mixture at an intermediate temperature. The ice crystals were grown on a fibre suspended in the mixing air currents, the temperature of which was measured by a thermistor and the supersaturation by a frost-point hygrometer. Careful measurement of time-lapse photographs of the crystals, taken over periods of about 1 hour, revealed that at very low supersaturations of a few per cent, the crystals tended to grow as nearly isometric prisms, but in the temperature ranges at which plates usually appear (Table 5.6), the crystals grew as very thick plates and approached a limiting c/a ratio of 0·8, whereas at temperatures normally associated with prisms, they approached a limiting c/a ratio of 1·4.

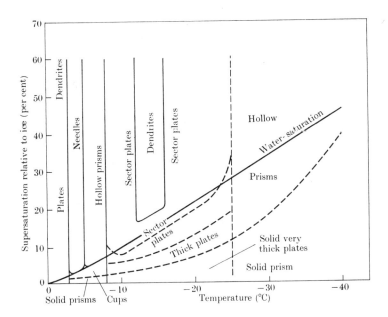

FIG. 5.22. The variation of crystal habit with temperature and supersaturation according to the experiments of Mason and his collaborators including results obtained at low supersaturations.

The results of all these laboratory experiments on the variation of ice crystal habit with temperature, and supersaturation or excess vapour density, are consolidated in Fig. 5.22. It is clear that the principal factor controlling the basic habit, as determined by the relative growth rates along the c- and a-axes, is the temperature. But as the diagrams show equally clearly, the secondary growth features are determined by the supersaturation or flux of vapour. Increasing supersaturation causes transitions from very thick plate → thick plate → sector plate → dendrite, or from solid prismatic column → hollow prism → needle.

If, as Marshall and Langleben suggested, the basic habit was determined by the flux of vapour towards the growing crystal, that is, by

$D\Delta\rho$, where D is the diffusion coefficient of water vapour and $\Delta\rho$ the excess of the ambient vapour density over that in equilibrium at the crystal surface, it might be expected to change if the crystals were grown in other gases or in air at reduced pressure. In fact, Isono, Komabayasi, and Ono (1957) and Isono (1958) reported that when crystals were grown in hydrogen and in air at low pressure, they observed nearly isometric growth ($c/a \simeq 1$) throughout the temperature range -7 to $-16°$ C, a region in which plates and dendrites are formed in air at normal pressure. These authors claimed that the environment was kept saturated relative to liquid water and attributed the observed habit changes to the more rapid diffusion of vapour in the hydrogen and low-pressure atmospheres. On the other hand, Van den Heuvel and Mason (1959) found that the variation of habit with temperature, involving plates, needles, hollow prisms, and dendrites, was unaffected by reducing the air pressure to 20 torr., or by replacing the air by carbon dioxide, hydrogen, or helium; only the growth rate of the crystals was affected as was anticipated in view of the differences in the diffusion coefficients and thermal conductivities of the various gases.

Kobayashi (1958) has also investigated the effect of lowering the air pressure on the growth habit. At pressures greater than 300 torr., the normal variation of habit with temperature was observed; at pressures between 300 and 70 torr., hollow prisms were observed in the temperature range -10 to $-22°$ C, but below 70 torr., these gave way to solid prisms. These latter crystals grew very slowly, suggesting that the increased value of D was more than offset by a reduction in $\Delta\rho$. Kobayashi has concluded that his experimental arrangement was such that a large reduction in the air pressure caused the supply of water vapour to the crystals to be much reduced, something that did not occur in the experiments of Van den Heuvel and Mason. It seems therefore that the latter were correct in stating that the basic crystal habit is not modified by the nature and pressure of the carrier gas.

5.5.2. *The influence of impurities on ice crystal habit*

It was reported by Nakaya (1955) that the growth forms of ice crystals may be modified by the presence of atmospheric aerosols. For example, he stated that when the air was filtered to remove the suspended particles, the plate and dendritic crystals, which normally occur in the temperature range -10 to $-20°$ C, were replaced by hollow prisms. Hallett and Mason (1958a) were unable to confirm this result.

Crystals growing in their diffusion chamber showed exactly the same variation of habit with respect to temperature, irrespective of whether they grew in the presence of a water cloud formed by condensation on atmospheric aerosol particles, or in very clean air from which all the particles had been removed by sedimentation.

Later, Nakaya, Hanajima, and Mugurama (1958) reported that the changes of habit had nothing to do with the removal of aerosol but were caused by a trace of silicone vapour leaking into the chamber. This recalls the earlier findings of Schaefer (1949b) that the habits of ice crystals produced by seeding a supercooled cloud were changed by the presence of vapours of ketones, fatty acids, silicones, aldehydes, and alcohols.

Hallett and Mason (1958b) also found that traces of organic vapours profoundly influenced the habit of ice crystals grown in their diffusion chamber. For example, needle-like crystals appeared at all temperatures between 0° C and −40° C when a small trace of camphor vapour was introduced into the chamber (see Fig. 5.23). Even more remarkable changes followed the introduction of isobutyl alcohol. In the temperature range −12 to −16° C, when the partial pressure of alcohol exceeded about 10^{-3} mb, the normal dendritic growth was suppressed and replaced by plate-like crystals. A further rather small increase in alcohol concentration caused these to be replaced by either hollow prisms or needles that persisted until the concentration was raised beyond 0·1 mb, above which there was a further transition to plates, and finally at higher concentrations, a reversion to a rather malformed type of dendritic growth.

An earlier observation by Vonnegut (1948b), that a cloud of tiny hexagonal plate-like crystals growing at −20° C was transformed into prismatic columns by the addition of butyl alcohol at about 10^{-2} mb partial pressure, was confirmed by Hallett and Mason, with a reversion to plates if the concentration exceeded about 0·2 mb. Similar transitions were also observed in the temperature range −8 to −12° C.

5.5.3. *The growth of ice crystals in an electric field*

The effect of electric fields on ice crystals growing from the vapour seems worthy of study for two main reasons. First, it is interesting to know whether the strong fields experienced in thunder-clouds (up to, perhaps, 10^4 V cm^{-1}) can influence the growth of ice crystals in these clouds. Second, the effect of an electric field on the habit of the crystals may help in understanding the normal mode of growth and the remarkable variation of habit with temperature that is observed in the absence of an electric field.

This subject has been investigated by Bartlett, Van den Heuvel, and Mason (1963). The crystals were grown on an insulated metal fibre in a diffusion cloud chamber in both uniform and radial electric fields of up to 1000 V cm^{-1}. When an electric field in excess of a certain minimum value is applied, long, thin, needle-like crystals are observed to grow rapidly from the tips of needles or dendrites that are growing already on the fibre. Linear growth rates as high as 5 mm/min are observed, this being 10–100 times faster than the growth rates of ordinary needles and dendrites under similar conditions of temperature and supersaturation. The onset of this new growth is determined by the *local* electric field at the tip of the needle and this is very sensitive to the shape of the growing tip and therefore varies from crystal to crystal.

Three main habits are observed for 'electric' needles. The principal forms are long thin needles with their axes either parallel to the optic axis $\langle 0001 \rangle$, or parallel to the close-packed rows in the basal plane $\langle 11\bar{2}0 \rangle$, or perpendicular to the close-packed rows $\langle 10\bar{1}0 \rangle$. The first type of 'electric' needle is usually observed growing on ordinary needles at about $-4°$ C as an extension of the normal growth. The second type is most commonly found growing on dendritic crystals at about $-12°$ C, again as an extension of normal growth but with the difference that, when the electric field is applied, the new growth develops few, if any, side branches. As soon as the field is removed, the 'electric' crystals begin to develop side branches in the usual way. Good examples of the third, and less common, type of growth are shown in Figs. 5.24(*a*), (*b*), where the crystals have been allowed to develop side branches. They show 'electric growth' along the $\langle 10\bar{1}0 \rangle$ directions which is never observed in the absence of an electric field. The 'electric' crystals do not show the marked variation of habit with temperature that is exhibited by ordinary crystals; once an 'electric' crystal has begun to grow with a certain habit, it continues to grow with the same habit even if it is moved to a different temperature.

Electric growth was never observed in a 50-Hz alternating field, even when this was five times greater than the steady fields required to initiate electric growth under the same conditions. Bartlett, Van den Heuvel, and Mason give a number of reasons for believing that the rapid electric growth occurs by the accretion of electrically-neutral (uncharged) material and that, although the applied electric fields are not strong enough to orient the molecules in the vapour phase, the field at the

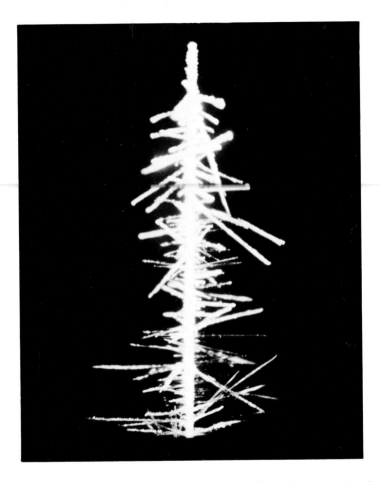

FIG. 5.23. The effect of introducing a small quantity of camphor vapour into the diffusion cloud chamber; the crystals then assumed a needle-like form at all temperatures. (From Hallett and Mason (1958a).)

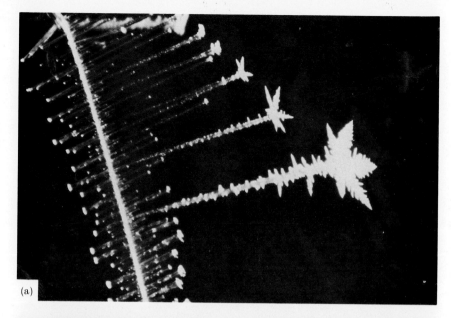

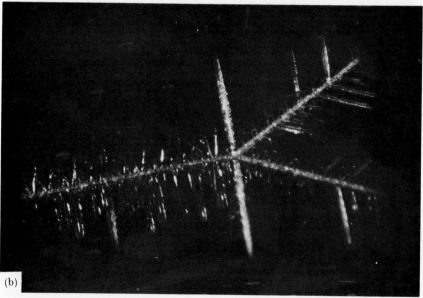

Fig. 5.24. (a) A needle-like ice crystal growing in an electric field at about $-4°$ C was allowed to develop side branches in the absence of a field. These consist of hollow prisms with their long axes parallel to the c axis, and star-shaped crystals with their arms parallel to the $\langle 11\cdot\bar{2}0\rangle$ directions. These side branches show that the original needle has its axis parallel to a $\langle 10\cdot\bar{1}0\rangle$ direction. (b) An 'electric' needle growing at about $-12°$ C with side branches. The cross-arms are at $90°$ to the axis of the electric needle (left), but at $60°$ to the dendritic arms which developed when the field was removed (right). This shows that the 'electric' needle grew parallel to a $\langle 10\cdot\bar{1}0\rangle$ direction. (From Bartlett, Van den Heuvel, and Mason (1963).)

surface of the crystal may be sufficient to modify their migration length when they land on the crystal. The exact mechanism is not, however, understood.

5.6. The mechanism of habit change

It can be seen from Table 5.6 that the habit of ice crystals changes from being essentially plate-like to being column-like four times over the narrow temperature range 0° C to −25° C; these changes, which reflect changes in the relative growth rates of the basal and prism faces of the crystals, are governed by the temperature and may be induced when this changes by as little as one degree. There has been no convincing explanation of these habit changes, which appear peculiar to ice, because present theories of crystal growth do not contain parameters that are likely to be strongly face-dependent and also sensitive to temperature changes of only a few degrees. It seems likely that the variations of habit are to be attributed to a surface property of ice (a view that receives some support from the fact that they may also be induced by the adsorption of traces of certain impurities), and that the surface diffusion of molecules on the growing faces is most likely to be the responsible parameter.

The first direct experimental evidence that water molecules arriving on the surface of a growing ice crystal may travel considerable distances before being built into the crystal structure was obtained by Bryant, Hallett, and Mason (1959) when studying the epitaxial growth of ice crystals on the basal faces of crystals of natural cupric sulphide (covellite). The ice crystals, being only a few thousand angstroms high, exhibited interference colours when viewed in reflected white light. The colours gave a measure of the crystal thickness, changes of which could be measured to within an accuracy of about 150 Å. At low supersaturations, some crystals were observed to grow considerably in diameter with no discernible change of thickness, suggesting that molecules arriving on the upper basal surface were not being assimilated, but were migrating over the surface and being built into the prism faces. This interpretation was reinforced by the observation that the lateral growth rate of two neighbouring plate-like crystals, of constant, nearly equal thickness, did not decrease as they approached each other, even when the intervening gap narrowed to about 1 μm. Had the crystals been growing mainly by direct deposition of vapour molecules on their edges, they would have slowed down as they approached and shadowed each other, but this would not be expected if growth occurred mainly by migration of molecules from the top surface. It was frequently observed that a small plate-like crystal did not thicken until it collided with a neighbouring crystal or a cleavage step on the substrate; coloured growth layers, often originating from the point of contact (see Fig. 5.25), then spread across the crystal surface. Hallett (1961) established that, at constant temperature and supersaturation, the layers travel with a velocity inversely proportional to their thickness.

We may consider the case of a straight step of height h growing parallel to the basal plane by collecting material directly from the vapour phase and by surface diffusion. If x_s is the average distance which a molecule travels on the ice surface before re-evaporating, all molecules within a distance x_s on either side of the growing step will contribute to its growth. The velocity v of the step is then

$$v = (h+2x_s)F/h\rho_i,$$

where ρ_i is the crystal density and F is the net mass flux of vapour per unit surface area. If $x_s \gg h$, the step will grow largely by surface diffusion with $v \propto h^{-1}$. Hallett's measurements, therefore, suggest that the layers advance mainly by surface diffusion and not by direct deposition of water molecules from the air. Hallett also found that v increased steadily from -40 to $-6°$ C, decreased sharply with increasing temperature from -6 to $-3°$ C, and thereafter increased again as the melting point was approached. He assumed, on the basis of the preceding argument, that x_s will vary in temperature in a similar manner.

Mason, Bryant, and Van den Heuvel (1963) made more direct measurements of x_s by measuring the velocity of approach of two neighbouring growth layers. They assumed the critical separation at which the velocity slowed down to be equal to $2x_s$. In this way, they obtained the curve of x_s against temperature illustrated in Fig. 5.26. The general form of their curve is similar to that obtained by Hallett, but whereas Hallett's maximum occurred at $-6°$ C, that obtained by Mason et al. is seen from the figure to be located at $-11°$ C. In an attempt to resolve this discrepancy, Kobayashi (1969)* employed Hallett's technique to measure the velocity of step propagation, v, over the basal surface of ice grown on a covellite substrate. The measured variation of v with temperature obtained in these experiments was very similar in form to the curve of x_s against temperature obtained by Mason et al., with a maximum occurring at a temperature of

* Private communication.

Fig. 5.25. Growth layers spreading across ice crystals growing epitaxially upon a basal surface of covellite. The layers often spread out from the point of contact with another crystal. The ice crystals develop preferentially along cleavage steps on the substrate surface.

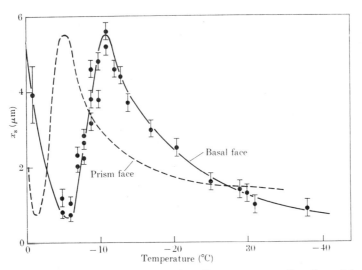

Fig. 5.26. The mean surface migration distance x_s as a function of temperature. ——— basal surface − − − − − − suggested curve for prism face. (After Mason, Bryant, and Van den Heuvel (1963).)

−11° C. It appears reasonable to assume, therefore, that Hallett's results are a less accurate description of the variation of x_s with temperature than that provided by the experiments of Mason *et al.* who, on the basis of their experimental curve illustrated in Fig. 5.26, have suggested an explanation for the variation of crystal habit with temperature. Since molecules arriving on the crystal surface have a surface migration length of several microns, they may reach an adjacent face before becoming incorporated into the lattice. The initial habit development from the embryo stage may therefore be determined by the relative values of x_s for the basal and prism faces. If x_s is greater for the basal face, there will be a net transport of material by surface diffusion to the prism faces, and the crystal will start to develop a plate-like habit. The reverse will be true if x_s is greater for the prism faces. At present it has not proved possible to measure values of x_s for prism faces, but Mason *et al.* have suggested that the curve of x_s against temperature for the prism faces may be of the same general shape as that for the basal faces, but displaced slightly along the temperature axis as indicated by the broken curve in Fig. 5.26. If this suggestion is correct, the two curves will intersect at three places, thereby providing four temperature ranges in which the ratio of x_s for the basal and prism faces alternates between values greater than, and less than, unity. When this ratio exceeds 1, the early development will be platelike, whereas prismatic columns will develop when the ratio is less than 1. If the actual curve of x_s for the prism faces against temperature provided similar intersection points to those assumed in Fig. 5.26, the sequence of habit changes predicted from this diagram would be consistent with that presented in Table 5.6. In addition, analysis of Fig. 5.26 suggests that habit transitions will occur very sharply at -3 and $-8°$ C, but more gradually at $-25°$ C. These predictions are entirely in accord with observation.

Once the crystal dimensions become large compared with x_s, surface diffusion will have little further effect on habit development, which will now be controlled primarily by the three-dimensional diffusion field. However, once a habit has been established in the early stages of growth, the diffusion field around the crystal will orient itself to conform to the crystal geometry and tend to maintain it. The observation that, at moderate supersaturations, crystals continue to develop as polyhedra suggests that the excess material arriving at the edges and corners is redistributed over the surface by surface diffusion. However, as the supersaturation is increased, surface diffusion cannot redistribute material sufficiently rapidly to compensate for its nonuniform deposition, and growth occurs preferentially at the corners, resulting in the formation of sector plates, dendrites, hopper crystals, and other skeletal forms.

It appears, therefore, that the crystal habit is determined by the interaction between the surface migration of molecules and the non-uniform flow from the diffusion field. The relative rates of surface diffusion on the basal and prism faces are responsible for determining the habit in the early stages of growth, and this is later maintained and accentuated by diffusion of vapour to, and conduction of latent heat of crystallization away from, the crystal surface.

5.7. The surface structure of ice crystals

Crystals grown under moderate and high supersaturations, both plates and prismatic columns, show hopper development with terraced faces as shown in Figs. 5.28 and 5.29. When the thin hexagonal plates are viewed in transmitted light, this development is barely discernable but, when seen in reflected light, the surfaces are covered with transverse and radial ridges as shown in Fig. 5.27. The formation of these terraces has been studied by Mason *et al.* (1963) using time-lapse photography, and their contours examined by making Formvar

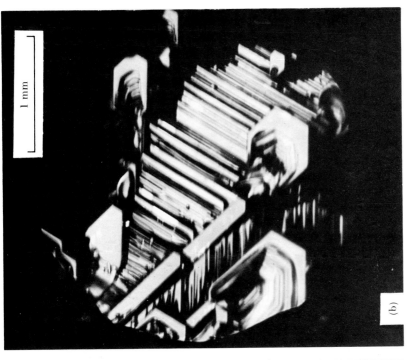

FIG. 5.27. The stepped-surface structure of a plate-like ice crystal grown from the vapour. (a) The most rapidly growing face with steps about 5 μm high and 25–50 μm across. (b) This face was partly shielded and grew more slowly, the steps now being at least 10 μm high and usually 50–150 μm across. (From Mason, Bryant, and Van den Heuvel (1963).)

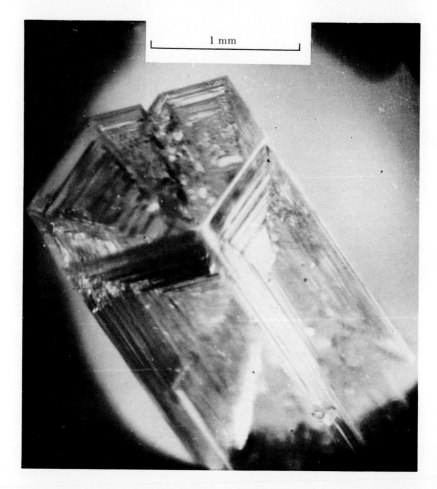

Fig. 5.28. A hopper prismatic column grown from the vapour showing incomplete and stepped basal and prism faces.

THE FORMATION OF SNOW CRYSTALS

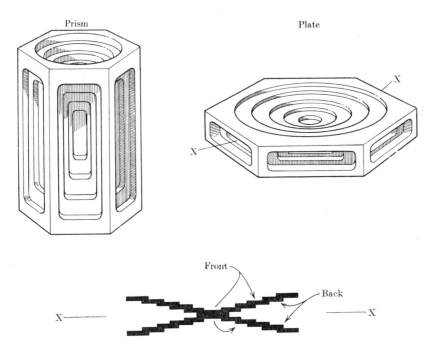

FIG. 5.29. A schematic diagram illustrating the hopper development of prismatic columns and hexagonal plates.

replicas of the surface, which were then stripped, silvered, and examined under the microscope. The ridges shown in Figs. 5.27 and 5.28 are several microns high; visible steps appear only when the crystal diameter exceeds a few hundreds of microns. The author believes that they result from the bunching of much thinner growth layers spreading out from the leading edge of the crystal where they have been nucleated under the high local supersaturation. The growth and amalgamation of layers, 150–5000 Å high, have been directly observed on ice crystals by the interference technique described in § 5.6.

Mason has shown (see, for example, Mason, Bryant, and Van den Heuvel 1963) that if, for example, a random fluctuation in supersaturation causes two successive members of a family of growing steps to be nucleated with separations less than $2x_s$, then they may close up, amalgamate, and form a double step travelling at half the speed. This, in turn, may be caught by a third layer and result in the formation of a triple step travelling at one-third the speed of a unit step. This bunching process will continue to produce progressively thicker and slower-moving steps against which thinner layers will pile up as in a traffic jam.

It may be shown that the time taken for two layers of comparable but unequal thickness to close, having reached a separation of $2x_s$, is given by $t = (4x_s \ln 2)/v$, where v is the uniform rate-of-change of separation beforehand. The distance travelled by an m step in growing to $(m+1)$ units is $2x_s/(m-1)$. The horizontal displacement of a step while growing to N units is $2x_s \sum_{1}^{N-2} \frac{1}{n}$, and the time taken is

$$\{(4x_s \ln 2)/v_0\} . \sum_{1}^{N-2} \frac{1}{1-\frac{1}{n}} \simeq N . (4x_s \ln 2)/v_0,$$

if N is large, v_0 being the velocity of a unit step. The minimum distance from the edge of a crystal at which a step many units high may appear is about $20x_s$. These calculations are found to be in good agreement with measurements made from ciné-films of growing ice crystals.

If the crystal is practically stationary relative to the surrounding vapour, layers will usually be nucleated near the edge of the crystal, where the supersaturation is highest, and will then spread more slowly as they approach the centre of face where the supersaturation is lower. This gradient of supersaturation, together with the onset of bunching, allows new growth layers to form at the edge before earlier ones have completed their travel to the face centre. This leads to a preferential thickening at the periphery, more pronounced starvation of the face centre, and thus to hopper development, either in the form of shallow hexagonal dishes (plate regime) or as hollow prisms (see Fig. 5.29). Simultaneous hopper development of basal and prism faces is shown in Fig. 5.28, where the incomplete development of basal planes leads to the formation of hollow funnel-shaped cavities, which are also in evidence in the natural ice crystals of Fig. 5.6. If, on the other hand, the crystal grows in a well-stirred medium, the supersaturation tends to be more uniform over the crystal faces, and hopper development is less marked.

We have seen that both the growth rate and growth habit of ice crystals are governed partly by the environmental conditions—the temperature, supersaturation, disposition of concentration and thermal gradients; and partly by conditions at the crystal surface—the surface concentration of material and its migration across the surface to the growth sites, where it becomes built into the crystal lattice. Growth involves the initiation and spreading of new layers. It is now a generally recognized feature of crystal growth that new layers cannot be formed at an observable rate by two-dimensional nucleation on a molecularly flat surface except at very high supersaturations. Observable growth rates can be achieved only if the free-energy requirement for successive two-dimensional nucleation can be avoided. A way out was proposed by Frank (1949), who suggested that screw dislocations might terminate in a crystal face to produce a spiral terrace of steps, each of which would be capable of acting as a two-dimensional nucleus, and which would not disappear during the deposition and evaporation of molecules. Growth fronts spreading out from the dislocation would form a growth pyramid.

Spiral growth fronts have since been identified on crystals of many different substances, while the dislocations themselves have been revealed by decoration techniques that are not applicable to ice. However, it has recently been demonstrated by several workers that etch pits mark the location of emergent dislocations in many other substances; this method has now been used successfully with ice. The appearance of etch pits on the surfaces of ice crystals was first reported by Truby (1955). Hexagonal pyramidal features of base diameter 0·5–20 μm and depth 0·25–0·5 μm were observed, and sometimes exhibited a stepped-layer concentric with the c-axis. Higuchi (1958) and Mugurama and Higuchi (1959) discovered that etch pits could be produced on the surfaces of ice and snow crystals by etching with a Formvar solution. Hexagonal pits appeared on the basal surface of a snow crystal in concentrations up to 10^5 cm^{-2}. After prolonged etching, the bottom surfaces of the pits revealed the layer-type structure noted by Truby, the height of the individual steps varying between 3 and 16 μm.

A detailed investigation of etch pits on ice crystals and an attempt to relate them to the dislocation structure was made by Bryant and Mason (1960b). The surfaces of crystals grown from both the vapour and the melt were etched with a 1% solution of Formvar in ethylene dichloride. Replicas of etch pits were studied at successive stages of development by peeling off the plastic film with adhesive tape, evaporating a thin layer of silver on to the film, and viewing it in reflected light with magnifications of × 400.

Replicas of ice crystals grown from the melt showed many small pyramidal etch pits, 5–10 μm in diameter, occurring in concentrations up to 10^5 cm^{-2}. They were mostly hexagonal in cross-section, their sides being built up of several concentric steps each a few tenths of a micron in height. On many of the larger pyramids ($d > 10$ μm) these were accompanied by one, and sometimes two, large concentric steps of height up to 1 μm. A few per cent of these pits exhibited hexagonal spirals of similar step height as shown in Fig. 5.30. A silvered replica showing an (inverted) line of pyramidal pits formed at a grain boundary is shown in Fig. 5.31.

Ice crystals were grown from the vapour in the diffusion cloud chamber as described in § 5.5.1. Replicas of previously unetched basal faces of plate-like crystals revealed, instead of etch pits, raised hexagonal pyramids, usually less than 10 μm in diameter. Under high magnification, these exhibited concentric rings and spirals similar to those observed in etch pits. These hillocks were eroded by subsequent etching; they may have been formed by growth rather than by dissolution at preferred sites on the crystal surface.

Etching with Formvar solution produced hexagonal etch pits on the basal surfaces of plate-like crystals in concentrations of order 10^4 cm^{-2}, the edges of the pits being parallel to the crystallographic $\langle 11\bar{2}0 \rangle$ directions. Pits formed on the prism faces of ice needles were rectangular in cross-section and mostly flat at the bottom as shown in Fig. 5.32.

The fact that the surface densities of the smaller etch pits on ice lie in the range 10^5 to 10^6 cm^{-2}, and are comparable with the observed densities of dislocations in other materials, suggests that they probably arose by the etching of emergent dislocations. A close association between dislocations and etch pits is also suggested by the latter tending to group preferentially along grain boundaries and slip lines. Bryant and Mason did not find it possible to say whether etch pits were generated about edge or screw dislocations, but were best able to interpret some of their etching patterns in terms of the etching of helical dislocations.

5.8. The growth rates of ice crystals

The maximum rate at which the mass of an ice crystal will increase when placed in an atmosphere of given supersaturation and temperature can be calculated in terms of the diffusion of vapour to, and heat away from, the crystal surface. It is assumed that the crystal is at rest relative to the air and that the diffusion field has achieved a steady stage, the vapour density, ρ, in space is given by the solution of Laplace's equation $\nabla^2 \rho = 0$, where ρ satisfies the boundary conditions at the surface of the crystal and at infinity. We shall now follow the

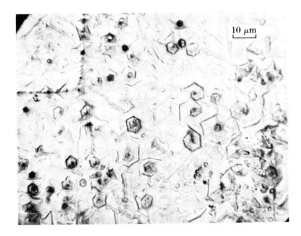

FIG. 5.30. Replica of a basal face of ice showing spiral steps on etch pits. (From Bryant and Mason (1960b).)

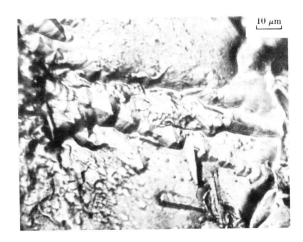

FIG. 5.31. A line of pyramidal etch pits along a low-angle grain boundary. (From Bryant and Mason (1960b).)

FIG. 5.32. Rectangular etch pits on the prism faces of an ice needle. (From Bryant and Mason (1960b).)

technique of Jeffreys (1918) and use an electrostatic analogue to calculate the rate of mass increase of a crystal, the vapour density, ρ, being analogous to the electrostatic potential V. The 'charge' on the crystal is $q = CV_c = C(\rho_c - \rho)$, where C is the electrostatic capacity, V_c the surface potential, and ρ_c, ρ are respectively the vapour densities in the immediate vicinity of the crystal surface and at infinity. The flux of water vapour towards the crystal surface is given by

$$\iint_A D\, \partial\rho/\partial n \cdot \mathrm{d}A \equiv D \iint_A \partial V/\partial n \cdot \mathrm{d}A,$$

where D is the coefficient of diffusion of water vapour in air, n a normal vector directed outwards from the crystal surface, and A the crystal surface area. But, by Gauss's theorem,

$$\iint_A \partial V/\partial n \cdot \mathrm{d}A = -4\pi q = 4\pi C(\rho - \rho_c),$$

so that the rate-of-increase of the crystal mass, m, is

$$\mathrm{d}m/\mathrm{d}t = 4\pi C D(\rho - \rho_c). \tag{5.1}$$

This equation was first used by Houghton (1950) to calculate the growth rate of crystals by a trial-and-error method based on a table of values of ρ_c against crystal-surface temperature. Mason (1953b) has derived a more convenient analytical expression for $\mathrm{d}m/\mathrm{d}t$ in terms of parameters which, for a water saturated environment, are determined only by the temperature and the geometry of the crystal.

If it is assumed that the vapour pressure p_c in the immediate vicinity of the crystal surface is the equilibrium vapour pressure with respect to ice at the surface temperature T_c of the crystal, i.e. $p_c = p_s(T_c)$, eqn (5.1) can be written

$$\frac{p - p_s(T_c)}{p_s(T)} = \frac{RT}{4\pi DCM p_s(T)} \cdot \frac{\mathrm{d}m}{\mathrm{d}t}, \tag{5.2}$$

where $p_s(T)$ is the saturation vapour pressure over a plane ice surface at the ambient temperature T, p is the ambient vapour pressure, M the molecular weight of ice, and $\mathbf{R}$ the universal gas constant.

While the crystal is growing, the surface temperature T_c will be higher than that of the surrounding air. If it is assumed that the latent heat of sublimation, L_s, is dispersed solely by conduction through the air, we have

$$L_s\, \mathrm{d}m/\mathrm{d}t = 4\pi C K(T_c - T), \tag{5.3}$$

where K is the thermal conductivity of air. These last two equations, together with the Clausius–Clapeyron equation

$$(1/p_s)\,\mathrm{d}p_s/\mathrm{d}T = L_s M/\mathbf{R}T^2,$$

may be integrated as on p. 123 to give

$$\mathrm{d}m/\mathrm{d}t = 4\pi C\sigma \bigg/ \left\{ \frac{L_s}{KT}\left(\frac{L_s M}{\mathbf{R}T}-1\right) + \frac{\mathbf{R}T}{DMp_s(T)} \right\}, \qquad (5.4)$$

where $\sigma = \{p/p_s(T)\} - 1$ is the supersaturation of the environment relative to ice, and the term in curly brackets is a function of temperature only, at constant air pressure.

Eqn (5.4) can be applied to calculate the growth rate of ice crystals whose shapes approximate to those of conductors of known capacity. Accordingly, we may follow Houghton and treat the thin hexagonal prism as a prolate spheroid of revolution of large eccentricity; a short prism or thick plate may be treated as an oblate spheroid. The following solutions are then relevant:

(a) Sphere. $C = r$, $m = \tfrac{4}{3}\pi r^3 \rho_i$, where ρ_i is the crystal density. Thus

$$r\,\mathrm{d}r/\mathrm{d}t = \sigma/\rho_i f(T). \qquad (5.5)$$

(b) Circular disk. $C = 2r/\pi$,

$$\mathrm{d}m/\mathrm{d}t = 8\sigma r/f(T). \qquad (5.6)$$

(c) Prolate spheroid. $C = 2ae/\ln\{(1+e)/(1-e)\}$, where the eccentricity $e = (1-b^2/a^2)^{\frac{1}{2}}$ and a and b are the major and minor semi-axes respectively. Writing $b/a = \zeta$ we have

$$\frac{\mathrm{d}m}{\mathrm{d}t} = \frac{8\pi(1-\zeta^2)^{\frac{1}{2}}a\sigma}{\ln[\{1+(1-\zeta^2)^{\frac{1}{2}}\}/\{1-(1-\zeta^2)^{\frac{1}{2}}\}]f(T)} = \frac{8\pi\phi(\zeta)a\sigma}{f(T)}. \qquad (5.7)$$

(d) Oblate spheroid. $C = (a^2-c^2)^{\frac{1}{2}}\{\cot^{-1}c/(a^2-c^2)^{\frac{1}{2}}\}^{-1}$,

$$\frac{\mathrm{d}m}{\mathrm{d}t} = \frac{4\pi(a^2-c^2)^{\frac{1}{2}}\sigma}{\{\cot^{-1}c/(a^2-c^2)^{\frac{1}{2}}\}f(T)} \qquad (5.8)$$

McDonald (1963) has determined experimentally the capacity of these and more complicated shapes by modelling the crystal forms in brass and measuring the change in capacity of the system when the model was suspended in the centre of a large Faraday cage. He found that hexagonal plates had very nearly the same capacity as ideal thin disks of equal area provided that their thickness was <10 per cent of the edge length. The capacity of long thin cylinders (needles) was well represented by the formula $C = a/\ln(2a/b)$ for a long, thin prolate spheroid, a being the half length and b the radius of the mid-section, if $a/b \geqslant 20$. McDonald made models of fourteen different dendritic forms and found that, even for very exaggerated shapes, the capacity varied from only 0·77 to 1·26 times that for a regular hexagonal shape of the same diameter.

When an ice crystal in the atmosphere grows to such a size that it attains an appreciable fall speed, its growth becomes complicated because it can no longer be considered stationary with respect to the air and the diffusion field. The crystal will always be moving into new surroundings and the diffusion field will be limited to a boundary layer around the crystal in which the vapour concentration and temperature gradients will be steeper than those directed towards a stationary crystal. For a spherical crystal of radius r, Mason (1953b) shows the concentration gradient at the crystal surface at a time t after being introduced into a new environment to be

$$\frac{\partial \rho}{\partial r} = (\rho - \rho_c)\left\{\frac{1}{r} + \frac{1}{(\pi D t)^{\frac{1}{2}}}\right\}, \qquad (5.9)$$

where the symbols have the same meaning as in eqn (5.1). The second term in the curly brackets is a measure of the enhancement of the concentration gradient at the crystal surface relative to the steady-state value, and the effect will be marked if $(\pi D t)^{\frac{1}{2}} \ll r$. $(\pi D t)^{\frac{1}{2}}$ has the dimensions of a length and may be defined as the thickness of the diffusion boundary layer. If we assume that $t = 2r/v$, eqn (5.9) implies that the mass growth rate of a falling ice sphere will be increased by a factor
$$1 + (vr/2\pi D)^{\frac{1}{2}} = 1 + 0\cdot28 Sc^{\frac{1}{2}} Re^{\frac{1}{2}} = 1 + 0\cdot23 Re^{\frac{1}{2}},$$

where Re is the Reynolds number and $Sc = \nu/D = 0\cdot66$ (at 0° C) is the Schmidt number, ν being the kinematic viscosity of the air. Accordingly, the growth or evaporation of a ventilated ice sphere may be represented by the equation

$$\frac{dm}{dt} = \frac{4\pi r \sigma (1 + 0\cdot23 Re^{\frac{1}{2}})}{\left\{\dfrac{L_s}{KT}\left(\dfrac{L_s M}{RT} - 1\right) + \dfrac{RT}{DMp_s(T)}\right\}}. \qquad (5.10)$$

By measuring the rates of evaporation of ice spheres, suspended from a sensitive quartz microbalance in air-streams of controlled humidity, temperature, and wind speed, Thorpe and Mason (1966) confirmed the general validity of eqn (5.10) for Reynolds numbers ranging from 10 to 200 and with air temperatures ranging from 0° C to −20° C, but their actual experimental data were best represented by a ventilation factor of the form $(0\cdot94+0\cdot29Re^{\frac{1}{2}})$.

Experiments designed to measure the rate of mass increase of crystals growing in a supercooled water cloud have been described by Reynolds (1952) and Mason (1953b). In both cases, a supercooled cloud was produced in a large cold chamber and seeded with dry ice, the crystals being collected near the floor at successive time intervals, measured under the microscope, and the masses calculated from their linear dimensions. Mason made measurements on simple, regular, hexagonal plates grown at $-2\cdot5°$ C and on prisms grown at $-5°$ C, for growth periods up to 3 min, during which time the crystals achieved maximum linear dimensions of order 100 μm. The experimentally determined values of dm/dt agreed with those predicted by eqn (5.4) to within about 10 per cent, which was as good as could have been expected in view of the experimental conditions and the difficulty of making very accurate measurements on such small crystals. Reynolds used much the same procedure, but made his measurements on plane dendritic stellar-crystals grown at $-18°$ C, the masses of which are more difficult to determine from their dimensions because of their irregular outline and non-uniform thickness. Because of this, and also because the temperature and supersaturation in his chamber varied considerably during the course of the experiment, Reynolds obtained rather poorer agreement between his measurements and the theory.

The growth rates of ice crystals produced by seeding an outdoor supercooled fog have been measured by Yamamoto et al. (1952), Okita and Kimura (1954), and Isono, Komabayasi, Yamanaka, and Fujita (1956). Crystals carried away from the seeding site by the wind were collected on slides at various distances downwind corresponding to various growth times in the cloud. In the first mentioned papers, the authors claim that the growth rates obtained from their measurements were in good agreement with those calculated from eqns (5.1) and (5.3). However, their plotted data showed a good deal of scatter, which may be attributed to spatial variations in the cloud of supersaturation and wind velocity, and the fact that the larger crystals would tend to settle out preferentially. Isono et al., who assumed that the largest crystals caught on their slides has grown under conditions of water saturation, reported that the mass growth rate of prisms growing at $-4°$ C was nearly twice that given by eqn (5.4). However, since the crystals were typically 300–400 μm in length, falling at, say 20 cm s^{-1}, the correction term $(1+0\cdot23Re^{\frac{1}{2}})$ could largely account for the discrepancy.

Although eqn (5.4) may predict quite well the rate of mass increase of stationary ice crystals, and of small crystals falling only very slowly through a supersaturated environment, it cannot predict how the mass will be distributed, and hence the detailed shapes of the crystals. Measurements on the growth rates of individual faces were made by Shaw and Mason (1955) for crystals growing on their metal plate. Under conditions of constant temperature and supersaturation, the square of the linear dimensions of a crystal increased linearly with time, this relationship being in accord with theory for a crystal growing slowly in a steady-state diffusion field. However, the growth rates of both prism and basal faces were often different in different crystals, whereas crystallographically similar faces of the same crystal sometimes grew at different rates. Furthermore, the growth rate of a particular face sometimes changed abruptly, even though the external conditions remained constant.

Measurements have also been made in the author's laboratory on needles, plates, and dendrites growing on a fine fibre in the diffusion cloud chamber. Under fairly constant conditions of temperature and supersaturation, the lengths of needles and of the arms of dendrites increased linearly and uniformly with time, at rates of a few microns per second, over periods up to half an hour. Very small plates, of diameter <100 μm, growing at constant thickness on the surface of covellite, also increased in diameter at a uniform rate, but plates larger than 0·5 mm across, growing in the diffusion chamber, did not show a linear or uniform increase in diameter. Plots of the distance of a growing edge from the crystal centre against time revealed pauses in growth at intervals of usually between 2 and 4 min, pauses coinciding with the appearance of visible ridges on the surface such as are shown in Fig. 5.27, and discussed in § 5.7. These results indicate that the growth of crystal faces is not determined solely by the flux of material to, and heat away, from the crystal, but is partly influenced by certain individual characteristics of the faces themselves which control the rate at which the material can be built into the crystal structure.

6
The Physics of Natural Precipitation Processes

PRECIPITATION, which may reach the ground in the form of rain, snow, or hail, is one of the most marked characteristics of weather and is the end product of the physical processes described in this book.

The occurrence of precipitation is strongly controlled by the motion of the cloud air. We have seen in previous chapters how the air motion during the formation of the cloud, in combination with the properties of those aerosols it contains and which act as condensation and freezing nuclei, determines the concentration, initial size distribution, and nature of the cloud particles. As soon as these have formed, the microphysical processes of condensation and aggregation begin to deform the spectrum, the production of precipitation elements now being under way. Because the air motion governs the dimensions, water content, and duration of the cloud, it controls not only the rates of these processes, but the period during which they operate and thus the maximum size that the largest of the cloud particles can attain. If precipitation should result, it is again the air motion that determines its distribution, intensity, and duration. Conversely, the growth and evaporation of particle populations, accompanied by changes of phase and water concentration, provide sources and sinks of heat that can profoundly influence the air motion. Thus the release of latent heat during the growth and freezing of cloud droplets, by providing additional buoyancy to the cloud, may promote and sustain its growth until the updraught is destroyed, either by mixing with the drier surroundings, or by accumulation of condensed water and the formation of a downdraught sustained by evaporation of the precipitation.

The factors of prime importance therefore appear to be the air motion and its aerosol content, and from specifications of them it should be possible, in principle, to calculate quantitatively the course of cloud and precipitation development. However, promising techniques for the observation, measurement, and computation of the air motions in and around clouds are only beginning to be developed, and the laws governing the formation, growth, and aggregation of cloud and precipitation particles are not yet fully established. Moreover, the great variations in the concentration and properties of atmospheric aerosol, and the great complexity of atmospheric motions (as revealed by the great variety in the scale and pattern of cloud development), add to the difficulty of constructing a general quantitative theory of precipitation development. As a result, little progress has been made in formulating realistic dynamical models of cloud systems, which are usually influenced by motions on scales both larger and smaller than the system itself. So far, we have not progressed much beyond constructing simple kinematic models to follow the history of condensed water and precipitation in a specified (rather than derived) field of vertical motion, and the construction of simple one-dimensional models of cumulus growth that allow the vertical motion to be *computed* in terms of buoyancy forces modified by the formation of condensed water, its partial conversion to precipitation, and the growth, fall-out, and evaporation of hydrometeors. These models, in which both the cloud physics and the interaction between the cloud and its surroundings are only rather crudely represented by semi-empirical relationships based on limited observational or laboratory data, have proved useful in illustrating the relative importance of various factors that can influence the maximum vertical development of the cloud and the initiation and spread of a downdraught, but they are inadequate to predict the intensity, distribution, and duration of precipitation within and beneath the cloud. This must await the development of detailed three-dimensional, time-dependent models, in which mixing and interaction with the environment arises as a direct consequence of the computed three-dimensional wind field and of turbulent mixing. Here we may note the encouraging progress made by Bushby and Timpson (1967) and Benwell and Timpson (1968) in the numerical prediction of rainfall associated with fronts and depressions, with scant attention to the cloud physics, but the formulation of realistic three-dimensional dynamical models of a cumulonimbus, involving strong interactions between the precipitation and the dynamics, is likely to prove much more difficult.

We shall not penetrate much further into the realm of cloud dynamics, but rather concern ourselves with the microphysical processes involved in the formation, growth, break-up, and evaporation of precipitation elements. A good deal of insight has been gained into the limiting conditions under which hydrometeors of various types may grow in clouds of individual particles in simple, specified updraughts. We shall see that such calculations provide at least a reasonable qualitative explanation of the development of the various forms of precipitation in different types of clouds.

6.1. Forms of precipitation

We have seen that a non-precipitating water cloud is composed of minute droplets rarely exceeding a few tens of microns in diameter and possessing terminal velocities of only a few centimetres per second. Thus, considerable growth of a cloud particle must occur before it can gain a falling speed sufficient for it to fall out of the parent cloud, survive evaporation in the unsaturated air beneath, and reach the ground as a precipitation element.

Clearly, the size of precipitation elements will be determined, to some extent, by the strength of the upward air current producing the cloud and by the humidity in the sub-cloud layer. Widespread layer clouds are associated with upcurrents of speed usually less than 0.5 m s^{-1}, so that droplets with a radius approaching 80 μm can fall out of them and approach the ground. Findeisen (1939a) estimated the distance a drop can fall through unsaturated air before completely evaporating, and has shown it to be proportional to almost the fourth power of the initial radius. In an atmosphere of 90 per cent relative humidity, he found the distance to be 3·3 cm when the radius is 10 μm, 150 m when it is 100 μm, and 42 km when the radius is 1 mm. Since (except in special circumstances) the bases of dense clouds lie at least a few hundred metres above the ground, a value for the radius of about 100 μm may be regarded as a lower limit to the size of precipitation elements.

Precipitation composed entirely of drops little larger than this commonly falls in damp weather from layer clouds not far above the ground, and is called *drizzle*. Drizzle (and, in cold weather, snow crystals or small flakes) may fall from shallow clouds whose thickness amounts to only a few hundred metres. *Rain*, consisting of drops which are larger, having radii up to about 3 mm, is produced by layer clouds some thousands of metres deep, such as are associated with fronts and depressions. The heaviest rains, composed of relatively large drops, fall from convective (cumulus) clouds whose depth may reach 10 km and which contain powerful upcurrents of several metres per second. Precipitation from these clouds is sporadic, for their horizontal extent is only a few kilometres and the active life of individuals is less than 1 hour. These rains are described as *showers*.

Shower clouds that extend well above the level of the 0° C isotherm

§6.2   PHYSICS OF NATURAL PRECIPITATION PROCESSES   283

are believed to contain ice particles, mainly in the form of *graupel* or *hail pellets*, in their upper regions, but these may often melt before reaching the ground. However, very deep and vigorous clouds may produce very large hailstones which, in hot climates, are occasionally as large as oranges. In cold weather the precipitation may reach the ground entirely as small hail.

Even the largest snowflakes have fall speeds hardly exceeding 1 m s^{-1}, and so *snow* falls only from layer clouds and from weak and decaying convective clouds which contain small vertical air currents.

Ice clouds, even at levels where the low temperature restricts the total amount of water present to less than 1 g m^{-3}, are often composed of particles large enough to have fall speeds of nearly 1 m s^{-1}. This is partly because the rare crystals appear only when the vapour pressure approximates to saturation over liquid water, so that beneath the condensation region there is a layer of air which, although not saturated with respect to liquid water, is nevertheless supersaturated with respect to ice. The crystals therefore continue to grow while they fall through this layer, which is often a kilometre deep and may be much more. In this way, a few crystals draw on the vapour in a great volume of air and become much larger than the droplets of unfrozen clouds. Ice clouds therefore often appear as fibrous streaks known as Fallstreifen or virga, which may be regarded as precipitation.

6.2. Physical processes responsible for release of precipitation

In droplet clouds the condensation process alone is usually ineffective in producing precipitation. In the lower and middle troposphere, where the droplet concentrations established during cloud formation are believed always to be of order 100 cm^{-3}, and the quantity of vapour condensed cannot exceed 7 g m^{-3}, the *average* radius of cloud droplets must be rather less than 30 μm. Droplet concentrations arising during condensation in very clean air at high levels may be one order of magnitude less, but the amount of vapour to be condensed is also greatly reduced, and so the conclusion that condensation alone is unlikely to produce droplets of *average* radius greater than 30 μm is not affected. Droplets of this size evaporate completely after falling a few metres in unsaturated air, and so are incapable of reaching the ground as precipitation. Condensation on giant hygroscopic nuclei may produce *some* larger droplets but these will attain radii of 100 μm only after several hours. In the layer-type clouds that have this duration the vertical air velocities are considerably less than 1 m s^{-1}, so that the droplets would fall out of them before reaching even this size.

The search for an adequate physical explanation of the formation of precipitation elements was apparently concluded with the publication of a classical paper by Bergeron (1935). In this paper Bergeron discussed the following factors that might possibly lead to non-uniformity in an initially nearly homogeneous cloud of water droplets: (*a*) the attractive forces between cloud droplets by virtue of their having either unlike electric charges or induced charges in the earth's electric field; (*b*) hydrodynamic attraction forces between drops of nearly the same fall velocity; (*c*) differences of capillary and hygroscopic forces at the surfaces of droplets of slightly different sizes; (*d*) temperature differences between drops of different origin leading to distillation of vapour from warmer to colder droplets; (*e*) the effect of turbulence on causing collisions between droplets. Bergeron concluded that all the above-mentioned mechanisms were inefficient or much too slow, and that the only factor that could account for the release of precipitation was the appearance of a few ice crystals (formed either by the freezing of droplets or, less likely, by the sublimation of vapour on special nuclei) among a much larger population of supercooled droplets in those parts of the cloud where the temperature was below $-10°$ C. In such mixed clouds, the state of approximate water saturation represents a supersaturation with respect to ice of 10 per cent at $-10°$ C and 21 per cent at $-20°$ C, so that rapid growth of the crystals would be expected. The physical mechanism discussed by Bergeron was just that put forward earlier by Wegener (1911), viz. 'The vapour tension will adjust itself to a value in between the saturation values over ice and over water. The effect of this must then be, that condensation will take place continually on the ice, whereas at the same time liquid water evaporates [from the droplets], and this process must go on until the liquid phase is entirely consumed.' Bergeron postulated that almost every raindrop ($d >$ 0·5 mm) originates as an ice particle that 'grows copiously' in this way, and therefore that all rain clouds must extend well above the level of the 0° C isotherm. He claimed to have conclusive proof of this in the observed behaviour of cumulus clouds in Europe which, he said, do not produce showers until their summits reach above the 0° C level and become transformed into fibrous ice clouds.

While direct evidence for the dominance of the Wegener–Bergeron mechanism in precipitating clouds was meagre at this time, it received strong support from Findeisen who, however, believed that the ice crystals arose by the sublimation of vapour on to special sublimation nuclei rather than by the freezing of droplets. On the basis of aerological observations made over Germany, Findeisen (1939b) stated categorically that all rain composed of large drops originates as snow or hail. Further data on the temperatures of cloud summits (see § 6.3) appeared to support this thesis, so that the Bergeron hypothesis came to be generally accepted during the following decade; in the absence of detailed quantitative arguments, all other possible processes were deemed to be much too slow.

However, it must have been obvious to both meteorologists and aviators that, in tropical regions, showers often fell from clouds whose tops did not reach within thousands of feet of the 0° C level. Rumours to this effect were current from time to time, but for some peculiar reason, very few reports of detailed observations were published and the views of the European meteorologists on the formation of rain were not seriously challenged. The possibility that the larger droplets in a wholly water cloud (which are never found to be as uniform as imagined by Bergeron) might continue to grow by collision with their small neighbours was not entirely dismissed. Indeed, Findeisen (1939b) calculated the radii with which drops of initial radius 30 μm would be expected to fall from a layer cloud of liquid-water content 1·25 g m^{-3} and found values of 100, 200, and 400 μm for cloud thicknesses of 150, 400, and 1000 m respectively. However, he rejected this result as contrary to experience, and believed his theory to be in error in that it assumed that a larger falling drop would collect all the droplets in its path, none being swept aside. He was firmly of the opinion that this coalescence mechanism was a slow and inefficient method of precipitation release capable of causing only falls of drizzle—an opinion endorsed by Bergeron (1949) in the case of all extra-tropical clouds. But reviewing the subject in 1941, Simpson (1941) emphasized that the belief that raindrops necessarily originate as ice particles rested on statements that rain was never observed to fall from pure droplet clouds, even when they were several thousand metres thick. Simpson would not accept these statements, and suggested that copious rain could be produced by wholly liquid clouds, providing only that the original droplets were of different sizes so that the coagulation process could start, and that the cloud was of sufficient depth.

In recent years, careful visual and radar observations have accumulated which show beyond all doubt that in tropical regions heavy rain falls from clouds that are entirely beneath the level of the 0° C isotherm.

In these regions this level may be above 5 km whereas in temperate latitudes it is often below 2 km, and here equally deep clouds commonly reach heights where the temperature is below $-20°$ C, so that their tops would almost certainly contain ice crystals. It does not follow, however, that their presence is essential to the production of rain, and there is increasing evidence that showers in temperate latitudes are often *initiated* by the coalescence process.

A more detailed evaluation of droplet growth by coalescence in wholly water clouds was made possible by the publication of a paper by Langmuir (1948) in which he calculated the efficiency with which a drop of given size would collect smaller droplets lying in its fall path (see Appendix A). He went on to calculate the growth of large water drops by this coalescence process following their introduction into the upper parts of convective clouds. He did not suggest that shower rains could arise naturally by this mechanism alone, but thought that melting hail might be an important agent in introducing large drops into the lower parts of shower clouds.

Houghton (1950) used Langmuir's values of collision efficiency to compare the growth rates of drops by coalescence with those of ice crystals by sublimation. He concluded that the growth of precipitation elements by sublimation is initially much more rapid than by coalescence, the two processes becoming equally effective when the particles are of a mass comparable to drizzle drops, and that for larger particles, the accretion process becomes dominant.

More detailed treatments by Bowen (1950) and Ludlam (1951*a*) start from the assumption of the presence, near the cloud base, of a few particles larger than the main population of cloud droplets and of the occurrence of a significant updraught throughout a considerable depth of cloud. Bowen, who assumed these larger drops to arise by chance combinations of droplets in the lower parts of the cloud, calculated their subsequent growth by condensation and coalescence in clouds of constant updraught and constant liquid-water content. He demonstrated that the process may reasonably result in the production of raindrops in clouds several thousands of feet deep, and described observations which show that, in Australia, showers frequently develop by this process.

The suggestion was made by Woodcock (1950*b*) that the larger droplets required to initiate the coalescence mechanism might be supplied by the giant salt nuclei found by him (see Chapter 3) in concentrations of order 1/litre in the sub-cloud layer over the sea, and which originate

§ 6.2   PHYSICS OF NATURAL PRECIPITATION PROCESSES

as spray droplets of radius 20–30 μm. It was shown by Ludlam (1951a), that in convective weather over the sea, the spray droplets may be carried up to the cloud without appreciable evaporation and thereafter be capable of releasing a shower.

A larger droplet, whatever its origin, will be carried up in the ascending air, but will be falling relative to the ordinary cloud droplets and so will collide and (presumably) coalesce with them. The drop thus grows at an ever-increasing rate and, ultimately, if the updraught extends to a sufficient height, its rate of fall will exceed the speed of the updraught and the drop will return to the cloud base, growing as it descends. The results of the calculations presented in § 6.5 indicate that this process could produce raindrops from convective clouds a few kilometres deep and sustained by updraughts of a few metres per second. In temperate regions, this will usually mean that the top of the cloud will extend above the 0° C level and so may contain ice crystals. It may therefore be a difficult matter to decide which of the two possible mechanisms is responsible for the initial release of precipitation, since both may be active.

Once an ice crystal growing in a dense supercooled cloud has become appreciably larger than a cloud droplet, its growth will be greatly accelerated by accretion of these droplets, which will freeze on impact to form ice pellets of irregular shape. This was pointed out by Findeisen, the implication being that, in convective clouds, the Bergeron sublimation process would be of importance only in the early development of a precipitation element. Some computations on the growth of such ice pellets in model clouds have been made by Ludlam (1952). His calculations suggest that, in clouds with bases warmer than 10° C, and containing updraughts of less than 5 m s^{-1}, droplets of radius 20–30 μm, introduced into the base of the cloud, may grow to precipitation size earlier than will ice particles originating at temperatures below $-10°$ C; in colder clouds, the ice particles will become favoured.

The growth and structure of large hailstones now find satisfactory quantitative explanations in terms of their accretion of supercooled water and dissipation of latent heat while being lifted and re-cycled in the strong, steady, tilted updraughts of severe storms, largely as the result of important contributions by Schumann (1938), Ludlam (1950, 1958), Browning and Ludlam (1962), Browning, Ludlam, and Macklin (1963), List (1960, 1963), Browning (1966), and Bailey and Macklin (1968a, b), all of which are discussed in detail in § 6.6.

The growth of drops by coalescence in layer clouds containing updraughts of only a few centimetres per second was investigated by Mason (1952a). He showed that, because these clouds generally exist for much longer than shower clouds, the larger droplets necessary to initiate coalescence may develop *in situ* by condensation, but that they cannot grow to much larger than drizzle size by coalescence, unless water contents considerably larger than those usually measured are assumed. Thus it appears that the rain that falls from the thick layer-cloud systems of extra-tropical regions is probably initiated almost entirely by the growth of ice crystals which, however, must aggregate to form snowflakes if they are to produce raindrops of the observed size. The predominance of the Bergeron process in precipitating layer clouds is borne out by radar studies which reveal, almost invariably. the presence of a strong band of echo (the melting band) just below the $0°$ C level and which is associated with the melting of snowflakes (see Chapter 8). Radar can be used to provide statistics on the frequency of rainfall associated with this process, since the presence of the melting band implies the absence of updraughts stronger than 1 m/s, which are necessary for the formation of rain by the coalescence process.

In this section we have traced, in outline, the development of modern ideas on the release of precipitation. The theoretical arguments will be presented in more detail in § 6.4 and § 6.5, but first it will be as well to review the observational evidence against which the theories must be judged.

6.3. Observational data on the characteristics of precipitating clouds

Three parameters that will obviously have an important bearing on which particular mechanism is likely to be dominant in the release of precipitation from a particular cloud or cloud system are the temperature of the cloud base, the cloud thickness, and the temperature of the cloud summit. The temperature of the base will largely control the liquid-water content of the cloud which, together with the cloud thickness, will largely determine whether or not raindrops may be produced by coalescence, while the cloud-top temperature will indicate the likely presence or absence of ice crystals. These three parameters can be readily measured from an aircraft fitted with a suitable thermometer and altimeter and, in the case of cumulus clouds in the general vicinity of a radio-sonde ascent, from theodolite readings taken on the ground. Unfortunately, reliable observational data of this kind are rather scanty, but those which are available will now be summarized.

6.3.1. *Observations on clouds in middle latitudes*

Aircraft observations of layer clouds in middle latitudes, as reported, for example, by Mann (1940), Stickley (1940), and Peppler (1940a) reveal that although rain may occasionally fall from non-freezing layer clouds, it is usually initiated by the ice-crystal process. On the other hand, drizzle may often fall from clouds that are warmer than 0° C and contain no ice crystals, and is therefore produced by the coalescence of droplets. Of nearly 1000 clouds analysed by Peppler whose tops were reached, and in which ice particles were detected, 20 per cent had summit temperatures higher than $-12°$ C, nearly 50 per cent above $-16°$ C, and 70 per cent above $-20°$ C. This, together with the fact that of the supercooled clouds in which no crystals were detected, about 30 per cent were colder than $-12°$ C, only 12 per cent colder than $-16°$ C, and only 5 per cent colder than $-20°$ C, suggests that, on average, a sharp rise in the production of ice crystals occurs in the temperature range -12 to $-16°$ C. In only 22 per cent of the clouds containing ice crystals was icing encountered at temperatures below $-8°$ C, but it was observed on 63 per cent of occasions at temperatures between 0° C and $-4°$ C.

Apart from the abundance of ice crystals, the probability of the development of precipitation will depend on the cloud thickness, the height and temperature of the cloud base, the magnitude of the vertical air currents, and the lifetime of the cloud, all of which will be governed by the large-scale dynamical and thermodynamical processes. Continuous precipitation can be maintained only if the cloud water is replenished by a persistent vertical motion fed, perhaps, over a large region, by low-level convergence in the wind field.

Peppler reported that about 70 per cent of the snow clouds had thicknesses of 1–4 km and bases below 1·2 km and were warmer than $-5°$ C. On the other hand, all the clouds producing drizzle were warmer than $-5°$ C and most of them were less than 2 km deep.

Peppler's observations were confirmed by Mason and Howorth (1952) in a detailed analysis of the data obtained from aircraft ascents made thrice daily to a level of 400 mb during 1942–4 over Northern Ireland. Of the clouds producing rain or snow whose tops were actually reached, only 28 per cent were warmer than $-12°$ C, but 55 per cent were warmer than $-20°$ C. In only 4 of the 99 well-substantiated cases of rain was the cloud-top temperature above 0° C, and in only 8 cases was it above $-5°$ C. Of these 8 cases, 3 produced moderate and 5 only light rain. The only case of continuous moderate rain was from a layer with base temperature 16° C, top temperature 5° C, and thickness 8500 ft (2·6 km). All the precipitating clouds had bases warmer than −5° C. Of the 45 reported cases of drizzle reaching the ground, in 26 (58 per cent) the cloud-top temperature was above 0° C, and in 39 cases (87 per cent) it was above −5° C. Clouds that produced snow at the ground were, in every case, colder than −12° C at their upper boundaries. Rain/snow fell only from clouds of thickness greater than 1 km, and continuous rain only from those of depth greater than 2·25 km, while no precipitation reached the ground from clouds less than 500 m in depth.

Rather similar results emerge from an analysis by Stewart (1964) of aircraft ascents made over an 8-year period over south-western England. Of 64 cases of layer clouds with their bases <1 km above the 0° C level (i.e. cloud base temperature, T_B, above −5° C), 66 per cent gave precipitation at the ground, whereas of the 135 cases with bases more than 1 km above the 0° C level ($T_B < -5°$ C), only 5 per cent gave precipitation at the ground. If only clouds of thickness greater then 2 km were considered, the corresponding figures were 88 per cent, and <4 per cent, respectively. Calculations indicated that snow crystals were unlikely to survive evaporation in falling more than a kilometre. No precipitation fell from clouds <400 m deep, and there were no cases of continuous slight or continuous moderate rain unless the thickness exceeded 1 km and 3 km respectively. Similar results were obtained for clouds whose bases lay below the 0° C level; at least 60 per cent of these clouds were entirely below this level, were unaffected by higher clouds, and therefore must have produced precipitation by the coalescence process.

We turn now to observations on *shower clouds* in middle latitudes. Peppler (1940b), from flights into 79 clouds producing showers over Germany, reported that 75 per cent had summit temperatures below −12° C and there was none lying entirely below the 0° C level. The bases of clouds that ultimately precipitated were warmer than 0° C on 90 per cent of occasions, the average temperature being 7° C. Of the non-precipitating cumuli, 99 per cent never reached the −12° C level, their average base temperature being 2° C.

Findeisen (1942b), from a rather small number of aircraft observations made over the North Atlantic, inferred that the summit temperatures of maritime cumulus clouds containing ice crystals, though varying considerably from day to day, were, on average, 6–8° C higher than those reported by Peppler for clouds occurring over the land.

However, it seems that, in those observations, glaciation was sometimes inferred from the external appearance of the cloud—a most difficult and uncertain procedure, since the presence of large drops in the upper regions of the cloud can produce the fibrous appearance that is generally associated with the presence of ice crystals. That showers falling over the sea tend to be associated with appreciably warmer clouds than is the case over land, also emerges from the work of Schwerdtfeger (1948), who analysed the observations made during 823 flights over the ocean during 1942–4 near 60° N. 13° W. Rain or snow showers were observed on 250 occasions and the temperatures of the cloud summits determined; 8 per cent of all the shower clouds were warmer than 0° C, and 45 per cent warmer than −12° C. These proportions were, however, considerably higher in the summer months, being 20 and 64 per cent, respectively. These statistics suggest very strongly that showers may be produced by the coalescence process in the maritime cumulus of middle latitudes, particularly in the summer when the cloud-base temperatures are relatively high.

Braham, Reynolds, and Harrell (1951), observing summer cumulus in New Mexico by double theodolites and radar, found that radar echoes never developed when the minimum summit-level temperature, was higher than −12° C. At temperatures between −12 and −24° C, only about one-fifth of the clouds produced echoes, and though the fraction was larger at lower temperatures, a few clouds gave no echoes even at temperatures below −30° C. The cumulus in this region have high bases, often above 10 000 ft (3 km), so that even in summer the base temperature is generally not greater than about 10° C. Of the non-precipitating cumulus studied in New Mexico, 75 per cent were colder than −12° C and 25 per cent colder than −20° C, in striking contrast to those forming over Germany of which only 1 per cent were colder than −12° C.

These geographical differences in cloud behaviour are further underlined by the work of Battan and Braham (1956), who used aircraft equipped with radar to make a census of both precipitating and non-precipitating cumulus, selected at random, over the Central United States. During the summer, when the 0° C level was located on average at about 15 000 ft, and cloud base at about 5000 ft, no precipitation was detected in those clouds with tops below 12 000 ft and warmer than +6° C, and only two of the twenty-nine clouds lying entirely below the 0° C level produced radar echoes. Of the clouds reaching the −12° C level (23 000 ft), about 30 per cent produced echoes, the probability of shower production increasing rather sharply to over 80 per cent when the cloud surpassed the $-20°$ C level at about 27 000 ft. These statistics contrast rather strongly with those for the New Mexico clouds, and suggest that shower formation proceeds more easily in the humid Central United States than in the arid south-west.

Because the majority of the precipitating clouds were appreciably supercooled, one might infer that the ice-crystal process plays an important role in the release of showers in the Central United States. However, Battan and Braham found in these clouds appreciable numbers of water drops exceeding 100 μm in diameter, which must have been produced by condensation and coalescence. Further evidence in favour of the coalescence process was published earlier by Battan (1953) from an analysis of the ground-radar records of the Thunderstorm Project obtained in Ohio during the summer of 1947. Although the tops of the visual clouds invariably reached the $-20°$ C level, and often a great deal higher, in about 60 per cent of the 123 cases the whole of the initial radar echo was located at temperatures above $0°$ C. The warmest initial echo had its base at $20°$ C and its top at $14°$ C, the corresponding temperatures for the coldest echo being $10°$ C and $-6°$ respectively. There was a definite tendency for the colder echoes to be associated with lower freezing levels, so that the cloud thickness appeared to be a more important parameter than the temperature difference in determining the onset of precipitation. Battan is of the opinion that, in most cases, the initial radar echo was produced as the result of droplet growth by coalescence, starting, perhaps, on large hygroscopic nuclei. From the appearance and subsequent development of the echoes, and the fact that updraughts of several metres per second existed in these clouds, he argues plausibly against the possibility of snowflakes formed higher up falling, melting, and growing by accretion to produce the first radar echo below the $0°$ C level.

One obvious difference between the clouds studied by Battan and Braham and those observed in New Mexico is that with the $0°$ C level at about 15 000 ft in both regions, the cloud base was located at about 5000 ft in the former case and at about 10 000 ft in the latter; thus the clouds in the Central United States had an extra 5000 ft of non-supercooled cloud in which to accommodate drops growing by coalescence. On the other hand, Battan (1963) found that while many of the initial radar echoes from convective clouds in Arizona appeared at temperatures above $0°$ C, the altitudes of the centres of the echoes were positively correlated with the altitudes of the cloud bases in such a way

§ 6.3 PHYSICS OF NATURAL PRECIPITATION PROCESSES

that, the lower the cloud base, the smaller the vertical separation of the cloud base and the initial echo. He took this as evidence for the initiation of rain by the coalescence process on the argument that the higher the liquid-water concentration associated with lower and warmer cloud bases, the lower would be the region of early raindrop growth. However, it is most unlikely that the observed differences in behaviour between cumulus clouds occurring in different regions and under different meteorological and topographical conditions are to be interpreted in terms of microphysical processes alone. The dynamical and thermodynamical processes that will govern such factors as the lifetimes, dimensions, and vertical motions of the clouds; their degree of mixing with the environment and hence their liquid-water content and buoyancy; the temperature, humidity, and wind structure of the atmosphere as manifested by such factors as stability and wind shear, must surely be of fundamental importance.

Indeed there is evidence for this in Battan's (1965) later radar and photogrammetric studies of clouds in Arizona. Finding that the correlation between the percentage of clouds producing radar echoes and the cloud-summit temperature was virtually the same on days of both heavy and light rain, he concluded that the duration and intensity of the rain reaching the ground were determined mainly by the size, strength, and duration of the updraughts.

Turning now to observations made in Australia, Bowen and his colleagues, working in Sydney, have clearly shown that rain may be released by either the ice-crystal process, or by coalescence from non-supercooled clouds, in this warm temperate region. Day (1953) classified the radar echoes observed throughout the year 1950 at Sydney into three categories: those showing melting bands indicative of the Bergeron process (30 cases), non-freezing showers (17 cases), and complex echoes such as are produced by cold-frontal showers, thunderstorms, and ill-defined cloud systems (24 cases). Day found that the Bergeron-type rain was associated predominantly with low-pressure systems and unstable air, and the non-freezing rain with high pressures and stable air when the clouds were unable to surpass the 0° C level.

A well-instrumented aircraft fitted with a radar set has been used with good effect by Bowen (1951), Smith (1950, 1951), and Styles and Campbell (1953), who give case histories of the precipitating clouds compiled from visual and radar observations. There were twelve occasions on which showers were observed to fall from clouds entirely below the 0° C level. All these clouds had base temperatures above

10° C, were (with one exception) more than 5000 ft thick, and were located either over the sea or near the coast where large salt nuclei were probably present in sufficient quantities to start the coalescence process.

Radar evidence for the production of showers by the coalescence process in England was obtained by Feteris and Mason (1956). On 13 days during the summer of 1955, 48 separate echoes were observed to originate below the $-2°$ C level and 26 of these remained there during the whole life of the shower. These echoes, which were often associated with mixed formations of cumulus and stratocumulus cloud, sometimes extended above the 0° C level at their maximum state of development, but surpassed the $-8°$ C level on only one occasion. On 9 of the 13 days, vertical development of the clouds was checked by the existence of stable layers usually below the $-5°$ C level. The authors point out the limitations of using radar alone to study the initial stages of precipitation development; in particular, the stage at which a radar echo will first appear will obviously depend upon the range and reflectivity of the cloud and upon the performance characteristics of the particular radar. In order to obtain really conclusive information, the radar observations should be closely coordinated with measurements on the visual cloud.

6.3.2. *Observations on clouds in tropical and subtropical latitudes*

The paucity of well-documented information on the characteristics of the clouds that produce precipitation in tropical and subtropical regions has already been mentioned. However, during recent years, observations have slowly accumulated which show that heavy rain may fall from clouds which are entirely below the 0° C level, while the theoretical developments described in a later section suggest that conditions in the tropics may be more favourable for the coalescence process than elsewhere. There are therefore good reasons for believing that drop growth by coalescence may be predominant mechanism for releasing showers from tropical clouds.

Isolated reports of rain falling from non-supercooled clouds have been published by Heywood (1940), Wexler (1945), Kotsch (1947), Hunt (1949), Craddock (1949), and Davies (1950), and are summarized in Table 6.1, while Leopold and Halstead (1948) stated categorically that 'the majority of tropical convective showers come from clouds which do not reach into sub-freezing temperatures'.

Virgo (1950) reports that, in the Bahamas, appreciable amounts of

TABLE 6.1
Reports of rain from non-supercooled tropical clouds

Author	Location	Cloud base (ft)	Cloud thickness (ft)	Base temperature (° C)	Summit temperature (° C)	Remarks
Heywood (1940)	Hong Kong	1000	7000	23	13·5	Nimbostratus giving 3·2 mm rain at ground
Wexler (1945)	Cape Henry, Virginia	1500	500	. .	21	Very light rain
Kotsch (1947)	Guam	1000	4500	. .	17·5	Nimbostratus giving moderate rain. 0° C isotherm at 17 300 ft
Hunt (1949)	Seychelles	1200–1500	7000–10 000	. .	11–12	Cumulonimbus giving moderate to heavy showers
Davies (1950)	East Africa	Few hundred feet	10 000	24	5	

Davies also quotes three other cases of showers falling from cumulonimbus with tops at 12 000 ft, 8000 ft, and 6000 ft, respectively, where temperatures were well above 0° C.

rain fall from cloud layers that are a mixture of cumulus and stratocumulus and have very little vertical development. He cites a case of a layer 3800 ft thick, with its top at 5300 ft, producing 3·3 mm of rain; the 0° C level was located at 15 000 ft. He also reports heavy rainfalls in summer from cumulonimbus that do not reach the freezing level. An analysis of the reports on the heights of cloud tops made by pilots flying the triangle Trinidad–Barbados–Grenada revealed that, on 75 per cent of the days in 1946, the clouds did not reach the 0° C level, and yet Barbados had 184 days with rain, Trinidad 236 days, and Grenada 210 days.

The occurrence of light rain or virga from very thin stratiform clouds, only a few hundreds of feet thick, in Puerto Rico, has been reported by Schaefer (1951), while showers were observed to fall from clouds lying entirely below the 0° C level at about 14 000 ft by Schaefer (1953) and Alpert (1955). Valuable information on the behaviour of cumulus clouds in this area was obtained by Byers and Hall (1955) during 15 aircraft flights in the winter and spring of 1953–4, when 495 clouds growing over the ocean and 192 growing over the island, all randomly selected, were examined by two aircraft, one skimming the cloud tops, and one flying into their centres. Information was obtained on the vertical and horizontal dimensions of the clouds, on the temperatures of their summits, and whether or not they produced a precipitation echo on a radar set of constant performance. All the clouds examined lay wholly below the 0° C level, which varied between 13 000 and 15 000 ft. Over the water, the fraction of clouds giving a radar echo varied from zero if the tops were below 6000 ft, rising to nearly 50 per cent if they reached 10 000 ft (8° C) and to 100 per cent if they surpassed 12 000 ft (5° C). Clouds forming over the island precipitated less readily, the probability of their doing so being zero if the cloud tops were below 7000 ft, increasing to only 20 per cent for those reaching 10 000 ft, and was still less than 50 per cent even for clouds surpassing 12 000 ft. This difference may have been partly due to the fact that the clouds grew more rapidly over the heated island than over the sea and so attained greater heights during the time taken for drops to grow by coalescence to precipitation size. It is interesting to note that all the clouds contained giant salt nuclei in concentrations of several hundreds per cubic metre and that these could apparently grow to raindrop size in clouds of thickness only 4000 ft. But Byers and Hall report that while on some days clouds of a given vertical thickness regularly precipitated, on other days none would give rain although some of them rose to 11 000 ft. Analysing the same observations, Batten (1958) draws attention to the importance of environmental and dynamical factors, and demonstrates that the formation of rain in small cumulus is inhibited in only shallow layers of moist air and in the presence of strong instability (updraughts) and strong vertical wind shears.

An interesting study of rain-producing clouds in Hawaii was made by Mordy and Eber (1954). During the trade-wind season, clouds are ever present over the Koolau Mountain Range which rises to a mean height of 2500 ft above sea level. The trade-wind cumuli are often surmounted by an orographic canopy of stratiform cloud. In general, rain occurred when the clouds were 5000 ft or more in depth. The lowest cloud-summit temperature observed was 7° C, the 0° C level being always above 14 000 ft. The tops of the clouds coincided roughly with the base of the trade-wind inversion; occasionally cloud turrets penetrated it to a depth of, perhaps, 500 ft, but these soon evaporated in the dry air. Rainfall intensities as high as 10 mm h^{-1} were often observed, the drop diameter was commonly 1 mm and sometimes exceeded 2 mm. The orographic stratus layer was never more than a few hundreds of feet thick, but its presence appeared essential for the production of rain in the cumuli beneath. It may be that large droplets were produced in the orographic canopy by the growth of large salt nuclei carried up from the sea, and that these seeded the underlying cumulus. Also, the layer cloud may have prevented evaporation of the cumulus tops and any incipient raindrops which may have been carried up in them.

6.4. The release of precipitation from layer clouds
6.4.1. *Characteristics and structure of precipitating layer clouds*

The extensive layer clouds from which falls precipitation of a steady, persistent character are generally formed in cyclonic depressions and near fronts. They are formed by slow but prolonged ascent throughout a deep layer of air, a rise of a few centimetres per second being maintained for at least several hours. Precipitation from these clouds is generally more widespread, of longer duration, lower intensity, and consists of smaller elements, than that from shower clouds. This suggests that the precipitation-release mechanisms act much more slowly than in cumuliform clouds, a consequence, no doubt, of the much weaker vertical currents and the lower concentrations of liquid water in stratiform clouds. The structure of these great rain-cloud systems can be examined satisfactorily only by aircraft and by radar and their systematic exploration has hardly begun.

The very valuable information that may be obtained by the simultaneous use of an instrumented aircraft and radar is illustrated in a report of four very interesting flights made in Massachusetts by Cunningham (1951, 1952) in which the airborne measurements were accurately coordinated in space and time with ground-radar observations. The aircraft was equipped with instruments for the measurement of temperature, pressure, humidity, and liquid-water content, and with an accelerometer, rate-of-icing meter, and a device for collecting ice crystals. The observational data are presented on a diagram on which the basic coordinates are altitude and the distance from the ground-radar station, the exact position of the aircraft being recorded by the radar every 15 s. Thus the observations of cloud structure, the forms of precipitation, and the readings of the aircraft instruments are plotted to give a composite vertical cross-section of the lower atmosphere for each flight, with the radar echo, as determined from the RHI-scope, superimposed.

In a typical intense cyclonic cloud system, the precipitation was heaviest near the centre of the storm, where there were usually localized regions of strong vertical motions producing clouds of high liquid-water content in which soft hail and raindrops grew rapidly by accretion. The intensities of the vertical motion and of the precipitation fell off

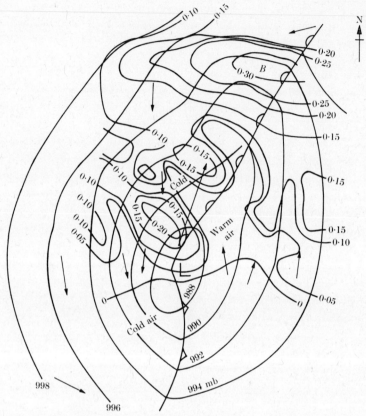

Fig. 6.1. The distribution of rainfall in a depression. The thick lines are isobars drawn at intervals of 2 mb showing the distribution of surface pressure. The thinner lines show the surface distribution of rainfall during the preceding hour in intervals of 0·05 in. The arrows show the directions of the surface winds.

inversely as the distance from the storm centre and, in the outer regions, weak, widespread, uniform ascending motions gave rise to uniform layer-type radar echoes and precipitation of low intensity. But these cyclonic cloud systems are reluctant to conform to a 'textbook' pattern; one such, which was explored with radar and aircraft over a period of nearly 4 h, is depicted in Figs. 6.1–6.4.

Fig. 6.1 shows the pressure and wind distribution and the position of the surface fronts at the middle of the period, together with the distribution of rain that fell within the previous hour. Low clouds covered most of the storm area, but instead of the rain being more intense in the centre of the depression, it actually ceased along a line through the centre and perpendicular to its direction of travel, and no precipitation

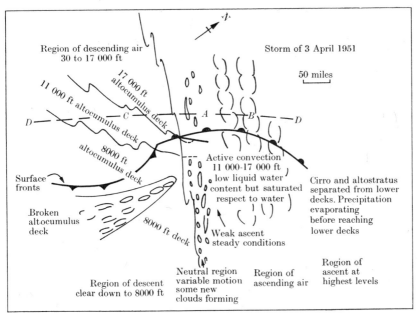

FIG. 6.2. Sketch map of the storm looking down from above showing the distribution of cloud relative to the surface fronts and the flight paths of the aircraft. (After Cunningham (1952).)

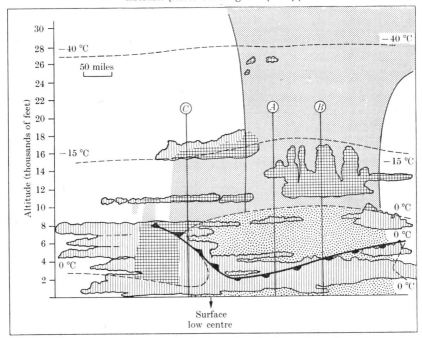

FIG. 6.3. Vertical cross-section, based on aircraft observations, through an extensive cyclonic layer-cloud system. The flight path is that marked DD in Fig. 6.2. The close dotting represents ice crystals or snow, the vertical hatching water cloud, the cross hatching areas of both supercooled water and snow, and the open dotting rain. (After Cunningham (1952).)

occurred along the cold front. The heaviest rain occurred well ahead of the centre, about 200 miles to the north.

In the south-western quadrant of the depression, which was moving NNE, there were shallow altocumulus layers at 17 000 ft and 11 000 ft and another at 8000 ft which merged with the lower-level cloud layers. Between 30 000 and 17 000 ft, the air was descending and was relatively dry. In the south-eastern quadrant there was clear, dry, descending air above 8000 ft and a broken altocumulus deck just below. The northern quadrants were regions of ascending air; the vertical motion was rather weak near the centre of the depression but, farther north, there was active convection at middle levels between 11 000 and 17 000 ft surmounted by extensive canopies of cirrus and cirrostratus. It was here, the region B in Fig. 6.1, that the heaviest rains, amounting to 0·3 in. in an hour, occurred. Further ahead, some 400 miles from the centre, the ascending motions occurred at high levels to form cirrostratus and altostratus decks whose precipitation evaporated before reaching the low cloud layers.

Fig. 6.3 shows a vertical cross-section of the cloud system along the flight path marked DD in Fig. 6.2. An aircraft ascent along the vertical marked A revealed the presence of a uniform sheet of cirrostratus at 25 000 ft through which the sun was dimly visible and surrounded by a

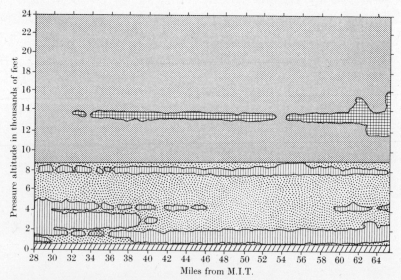

FIG. 6.4. Vertical cross-section through the cloud system along the flight path EE in the warm sector of the depression. The key to the shading is the same as for Fig. 6.3. (After Cunningham (1952).)

faint halo. Hexagonal ice plates and prismatic crystals, of diameter about 200 μm, were present in roughly equal concentrations. Between 13 500 and 14 000 ft, where the temperature was about $-10°$ C, there was a narrow band of altocumulus containing liquid water in quantities of about 0·05 g m^{-3}. Below 14 000 ft there were dendritic ice crystals and small snowflakes, but in the layer between 13 000 ft and the melting layer at 9000 ft, no liquid water was detected. The snow melted rapidly below 9000 ft to give a very shallow radar melting band; the absence of large wet flakes indicated that there was little aggregation. Just below the base of the melting layer, a thin stratocumulus (or nimbostratus) deck, broken in places, contained liquid water in concentrations of 0·3–0·4 g m^{-3}. Clouds were absent between 7700 and 5000 ft but, below this, there was a dense layer formed by mixing, convergence, and lifting of the air along the frontal zone. The average water content of this sheet was 0·8 g m^{-3}, so that the melted snowflakes grew appreciably by sweeping up the cloud droplets in this layer on the last stage of their journey to the ground, where they arrived with an average diameter of 0·8 mm.

The relatively heavy rain in the region B was due largely to convective clouds, nearly 6000 ft deep, embedded in the altostratus layer. Ice crystals falling into this cloud from above grew rather rapidly both by condensation and accretion of cloud droplets, and there was evidence for considerable aggregation of snowflakes in the melting layer. Further growth again took place in the lower cloud decks. The 'seeding' of lower clouds by ice crystals falling from altocumulus is shown in Fig. 6.3 to be occurring in the cold air behind the cold front.

Fig. 6.3 is hardly consistent with the idea of slow, uniform ascent of air that is usually associated with the upglide along the surface of a warm front. But more uniform conditions, depicted in Fig. 6.4, were found during the flight EE in the north-eastern part of the warm sector. Here the upcurrents were weak and fairly steady to produce extensive sheets of altostratus and cirrostratus, and shallow altocumulus layers at 8000 and 14 000 ft from which continuous, moderate precipitation reached the ground.

These and other radar investigations have revealed that cyclonic cloud systems, even warm-frontal systems well removed from the active centres of depressions and sustained largely by widespread and prolonged ascent, usually contain much small-scale structure in the form of convective elements, precipitation streamers, topographic disturbances, etc., which may contribute significantly to the water balance of the cloud and certainly influence the intensity and local distribution of rainfall.

An important step in relating some of the small-scale factors influencing the growth and release of precipitation to larger scale patterns of temperature and air motion in a depression has been made by Browning and Harrold (1969). The large-scale three-dimensional flow is revealed by isentropic analysis of radio-sonde ascents, using the wet-bulb potential temperature, which remains invariant whether or not evaporation/condensation takes place, as the principal parameter. This technique has the great advantage over isobaric analysis of providing a direct indication of vertical motions (since adiabatic ascent/descent occurs on isentropic surfaces), and also facilitates

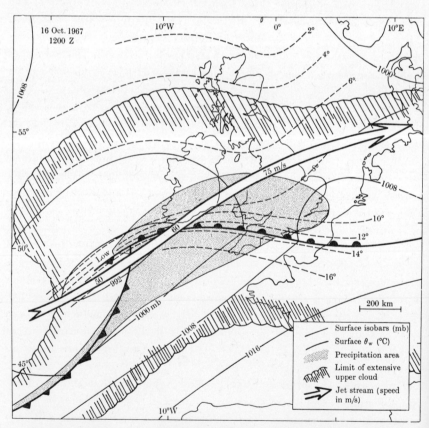

FIG. 6.5. The synoptic situation at 1200Z on 16 October 1967, showing an open-wave depression on the surface chart, which carries isobars and lines of constant θ_w, and shows the position of the jet-stream axis and the areas of precipitation and upper cloud. (From Browning and Harrold (1969).)

representation of the moisture field and the visualization of non-adiabatic motions. The winds relative to the synoptic system are plotted on surfaces of constant wet-bulb potential temperature, θ_w. The resulting charts enable the three-dimensional motion to be evaluated provided that the motion is nearly adiabatic and the system is in practically a steady state while the air travels through it.

Such an analysis for the synoptic situation shown in Fig. 6.5 revealed a region of widespread ascent and vertical velocities >5 cm s^{-1} which coincided with an area of fairly uniform rain falling at about 2 mm h^{-1} ahead of the surface warm front, and also a region of descending motion near the base of the front which was associated with a dry but potentially unstable layer of air near the ground.

As the coarseness of the radio-sonde network does not permit isentropic analysis on scales less than 100 km, motions on the scale of 10 km were investigated by using Doppler radar in the conical VAD mode (see p. 454), scanning at elevations of 5, 10, and 20 degrees during the passage of the warm front and into the warm sector. The radar was also used in the vertically-pointing mode to obtain particle fall-speeds, and vertical motions if these were strong enough. Measurement of the horizontal wind components, u and v, parallel to, and transverse to the warm front, and computation of the divergence, div $\mathbf{V} = \partial u/\partial x + \partial v/\partial y$, resulted in the vertical motion being obtained from the integrated convergence. Fig. 6.6 shows a height-time section of the vertical velocity (dashed lines) the shaded areas denoting downward motion, and the transverse circulation by the streamlines. The descending motions of a few cm s^{-1} in the cold air beneath the frontal zone and up-gliding currents of some tens of cm s^{-1} over the frontal surface were consistent with the results of the larger-scale isentropic analyses.

Detailed analysis of a close network of raingauges revealed, besides the area of uniform moderate rain well ahead of the surface warm front, narrow bands of heavier rain parallel to, and just ahead of, the warm front and, in the warm sector, similar bands aligned along the wind direction prevailing at the 3–6 km level. The bands contained areas of heavy rain, some 50 km in horizontal extent, in which the rainfall rate reached 12 mm h^{-1}. They probably originated in convective generating cells located between the 4- and 6-km levels as indicated in the time-height section of Fig. 6.7, which shows the vertical distribution of the precipitation relative to the warm front, as seen by radar at Pershore.

The heavy solid contours represent the extent of the echo detected

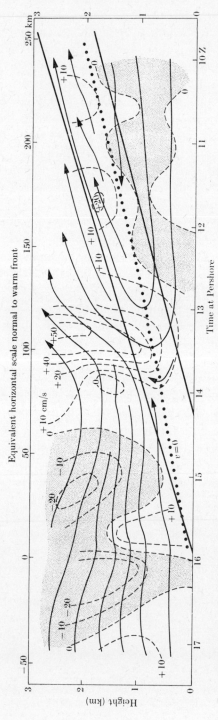

FIG. 6.6. Time-height section showing vertical air velocity U (dashed lines) and the transverse circulation (solid streamlines) in the vicinity of the warm front. Isopleths of U are shown at intervals of 10 cm s^{-1}, regions of downward motion being stippled. The warm frontal zone is indicated by 2 straight lines. (From Browning and Harrold (1969).)

§6.4 PHYSICS OF NATURAL PRECIPITATION PROCESSES

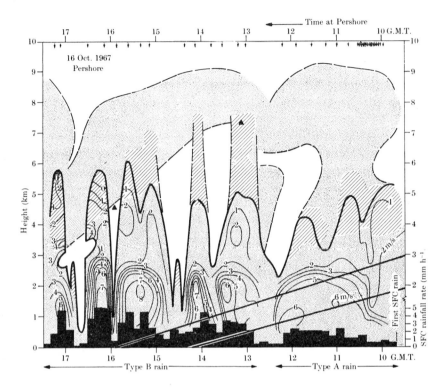

FIG. 6.7. Time–height section showing the vertical distribution of precipitation echo in the vicinity of the warm front at Pershore. The warm frontal zone is indicated by the two straight lines. The heavy solid contours represent the extent of echo detected by the Doppler radar at Pershore when its aerial was pointing vertically. The thin solid isopleths at intervals of 1 m s^{-1} represent the reflectivity-weighted mean particle fall-speed, V_f, measured by the Doppler radar. The melting level is situated where V_f changes sharply in the vertical from 2 to 5 m s^{-1}. The heavy dashed contours represent the extent of relatively weak echo detected at higher levels by a conventional radar. Areas where clusters of precipitation generators were observed are shown hatched. Surface rainfall rate at Pershore is plotted at the bottom of the diagram. (From Browning and Harrold (1969).)

by the Doppler radar when its aerial was pointing vertically, and the heavy dashed lines represent the extent of relatively weak echo detected at higher levels by a conventional radar. The thin solid isopleths at intervals of 1 m s^{-1} represents the reflectivity-weighted mean particle fall-speed V_f measured by the Doppler radar. The melting level is situated where V_f changes sharply in the vertical from 2 to 5 m s^{-1}. Areas in which clusters of slanting precipitation streamers, generated at middle and high levels and composed of snow particles, seeded the underlying stratiform cloud deck, are shown hatched. The areas occupied by streamers, which were typically 50 km across, and the vertical extent of the radar echo detected by the Doppler radar, appear to be correlated positively with the rainfall rate measured at the surface and plotted at the bottom of the diagram, except in the uniform steady rain ahead of the surface front where considerable evaporation may have occurred in the dry subsiding air beneath the frontal zone. The last-mentioned rain was certainly less intense, and composed of slower-falling (smaller) drops, than the more convective rain falling near, or just behind, the surface warm front.

6.4.2. *The growth of precipitation elements in layer clouds*

We have seen that the observational evidence, particularly that provided by radar, indicates very strongly that the Wegener–Bergeron process is mainly responsible for the release of rain from stratiform clouds, the coalescence mechanism giving rise to only drizzle or light rain.

In his original paper, Bergeron (1935) estimated that in a mixed cloud at a temperature of about $-12°$ C, where the difference of the equilibrium vapour pressures over water and ice approaches a maximum, all the supercooled water could be transferred by evaporation-sublimation to the ice crystals in a period of 10–20 min. His calculation was, however, unrealistic in that no account was taken of the fact that vapour is continually being released for condensation by lifting of the air and that the growth rate of the crystals is limited by the rate at which the latent heat of crystallization is liberated at their surfaces.

Detailed calculations on the growth of ice crystals in a layer cloud were made by Wexler (1952), with the object of computing the variation of radar-echo intensity with height above the melting level; they are therefore discussed in more detail in § 8.7.2. Wexler calculated the growth of crystals originating at 3 km above the $0°$ C isotherm in a cloud with base at $0°$ C and 805 mb and sustained by a constant updraught of 10 cm s^{-1}. They were assumed to grow as hexagonal plates having a constant thickness of either 15 or 40 μm, according to equations that are analogous to eqns (5.1)–(5.4). If the crystals were assumed to grow solely by diffusion in a water-saturated atmosphere, their computed masses on arriving at the $0°$ C level were 2·96 and 0·39 mg respectively. Because these masses are much larger than those normally observed for single crystals, and because the corresponding radar-echo intensity increased with decreasing height much more rapidly than is observed, Wexler concluded that, in layer clouds with small updraughts, ice crystals do not grow at water saturation. He therefore examined the case in which all the water created by the updraught within each height interval is deposited directly on the growing crystals within that region, so that there is no storage of liquid water in the form of cloud droplets. On this assumption, the computed variation of radar-echo intensity with height was in much better agreement with observation, suggesting, perhaps, that crystals usually grow in conditions intermediate between ice and water saturation.

However, the results that emerge from such calculations depend a good deal on the assumptions made concerning the geometry of the growing crystals. The thinner the crystals, the larger their radius for a given mass, and the lower the supersaturation needed for a given rate of growth. Thus Browne (1952a), who assumed that, instead of remaining nearly constant, the thickness of a crystal was always one-fifth of its radius, concluded that the crystals would not grow fast enough, even at water saturation, to account for observed echo intensities unless additional growth by accretion of supercooled droplets were postulated—a deduction directly opposed to that of Wexler. Whether or not it is legitimate to assume a steady state, in which all the water released by the updraught at any level is removed at the same rate by the growth of the precipitation particles falling through that level, has been questioned by Browne, Palmer, and Wormell (1954). They point out that, in the case of a pure ice cloud, the supersaturation can change to maintain the equilibrium rate of growth of the crystals. For example, should the crystal concentration fall, equilibrium could be maintained by a rise in the relative humidity which, in turn, would lead to an increase in the growth rate of the crystals and, perhaps, an increase in their numbers through an enhanced rate of splintering (see p. 207). But, if liquid droplets are abundant in the cloud, the relative humidity is held close to water saturation, equilibrium can no longer be maintained by variations in humidity, and a steady state may no longer prevail. Calculations by Wexler and Atlas (1958) of a steady-state water budget for a stratiform cloud, which follows a saturated adiabatic, has a parabolic distribution of vertical velocity, and contains particles with an exponential size distribution, predicted that while production of liquid water is likely to exceed consumption just above the 0° C level, at temperatures below $-2°$ C the rate of growth of ice crystals by deposition is such that the cloud liquid water vanishes above the $-4°$ C level for a surface precipitation rate of 2 mm h^{-1}, and above the $-10°$ C level for 10 mm h^{-1}. It therefore seems highly desirable to establish whether or not warm-frontal nimbostratus giving continuous precipitation contains appreciable concentrations of supercooled droplets above the 0° C level. Observations made from aircraft appear to indicate that, on the whole, deep, precipitating, layer-cloud systems contain rather little supercooled water except, perhaps, close to the 0° C level. The observation of Peppler described on p. 289, in which icing was rarely encountered at temperatures below $-8°$ C, is particularly relevant in this context. There may, however, be local concentrations of supercooled water in regions where stronger vertical motions are produced by orographic or convective disturbances.

The fact that snow reaching the ground with temperatures a few degrees below freezing often contains aggregates of several individual crystals suggests that aggregation of crystals generally occurs above the 0° C level in layer clouds. There is also some radar evidence for this (cf. § 8.7.2), though the distribution of radar-echo intensity suggests that the process is greatly accelerated in the melting zone, which may be only 150 m deep. That the variation of radar-echo intensity with height in steady precipitation can be interpreted in terms of the growth, aggregation, and melting of snow crystals is shown in § 8.7.7.

It is now appropriate to consider whether the growth of water drops by coalescence may lead to the release of rain from layer clouds. We have already mentioned the calculations of Findeisen (1939b) which indicated that a drop of initial radius 30 μm would grow to 400 μm after falling 1 km through a cloud of liquid-water content 1·25 g m^{-3}. This result, which Findeisen states was incompatible with observation, may be rather unrealistic in that he assumed appreciably higher values of water content and median-volume droplet radius (15 μm) than are normally found in non-precipitating layer clouds. A more detailed treatment of the problem has been given by Mason (1952a) and will be summarized in the following paragraphs.

Initiation of the coalescence process will be dependent on the appearance of some droplets having radii and fall speeds larger than average, and an appreciable collection efficiency for the smaller cloud droplets. Mason assumed that the smallest drop that would have a finite collection efficiency for cloud droplets of mean-volume radius about 6 μm was one of radius 20 μm, a guess that has been largely confirmed by the recent calculations and experiments described in Appendix A. Hence it appears necessary to produce some droplets of radius about 20 μm by condensation before the coalescence mechanism can be brought into play. Frith (1951) found droplets of this size in concentrations of at least 100/litre, i.e. one for about every 5000 cloud droplets, in thin layers of stratocumulus: if each of these were to grow to a radius of 100 μm in an updraught of 10 cm s⁻¹, drizzle would fall at a rate exceeding 1 mm h⁻¹.

Mason assumed that the life histories and hence the sizes of the cloud droplets are governed by their turbulent transfer between the interior of the cloud and its boundaries, and that it is the few droplets that remain in the cloud for times much longer than average that are responsible for initiating the coalescence process. A droplet originating on a nucleus containing 10^{-14} g of NaCl would grow in a supersaturation of 0·05 per cent and a temperature of 273 K, to a radius of 5 μm in 1000 s and to a radius of 20 μm in 8500 s, while a nucleus of 10^{-12} g would, in the same conditions, grow to a radius of 20 μm in about 6000 s. Thus, about 1 droplet in 5000 would have to remain inside the cloud 6 to 9 times longer than the mean period for all droplets. Assuming the eddies to be small compared with the thickness of the cloud, the motion of the droplets in the eddies is treated as a random-walk in which the upper and lower boundaries of the cloud act as absorbing barriers. It is shown that if 50 per cent of the droplets remain in the cloud for 1000 s, 0·6 per cent remain for 6000 s and 0·01 per cent for 10 000 s—ratios that appear sufficient for 1 droplet in 5000 to attain a radius of 20 μm if the mean radius is 6 μm.

In following the career of a droplet growing both by diffusion, and by coalescence with smaller droplets, Mason used the growth equation

$$\frac{dm}{dt} = 4\pi R^2 \rho_L \frac{dR}{dt} = \frac{4\pi R \sigma}{\left\{\frac{L^2 M}{KRT^2} + \frac{RT}{DMp_s(T)}\right\}} + E'\pi R^2 V'w, \quad (6.1)$$

where dm/dt is the rate of mass increase, R the radius and ρ_L the density of the droplet, σ the supersaturation and $p_s(T)$ the saturation vapour pressure of the vapour, D the diffusion coefficient, K the thermal conductivity of air, L the latent heat of condensation, w the concentration of liquid water in the form of smaller droplets, V' the fall velocity of the collecting drop relative to the smaller droplets, and E' the collection cross-section† of the drop. It was assumed that for $R \leqslant 20$ μm, the coalescence term of eqn (6.1) was zero, the growth rate being determined by the mass of salt in the nucleus, the temperature, and the supersaturation of the environment. On the other hand,

† E' is related to the collection efficiency E by $E' = E(R+r)^2/R^2$.

for $R > 60$ μm, the condensation term may be neglected and growth is dominated by coalescence.

Assuming the nucleus to enter the cloud base and to be carried up in a constant updraught U, its vertical distance above the cloud base after growing from radius R_0 to radius R (ignoring motion in eddies) is given by

$$z = \int_{R_0}^{R} (U-V) \, dR/(dR/dt), \qquad (6.2)$$

where V is the fall velocity relative to the air.

Graphs of the radius of a drop as functions of its height above cloud base, and of time, are shown in Figs. 6.8 and 6.9. These calculations, based on the drop falling through a uniform updraught with no diversions caused by eddies, predict the minimum radius to which it could grow during the time elapsing between its initial entry into the cloud base and its falling out through the updraught. The sizes attained

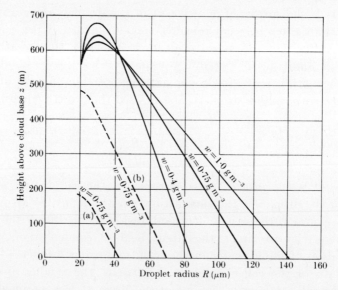

FIG. 6.8. Growth of a droplet by coalescence in a layer cloud. Radius of droplet as a function of its height above cloud base for different liquid-water contents w. The effect of varying the nuclear mass, the updraught speed and the supersaturation on the droplet trajectories. Full curves: nuclear mass = 10^{-13} g; $T = 273$ K; supersaturation = 0·05 per cent; $U = 10$ cm s^{-1}. Pecked curves: nuclear mass = 10^{-13} g; $T = 273$ K; supersaturation (a) = 0·05 per cent, (b) = 0·025 per cent; $U = 5$ cm s^{-1}. Water contents as indicated. (From Mason (1952a).)

§ 6.4   PHYSICS OF NATURAL PRECIPITATION PROCESSES

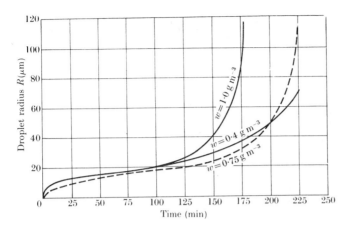

FIG. 6.9. Radius of droplets growing by condensation and coalescence in a layer cloud as a function of time for different liquid-water contents w. Full lines: nuclear mass 10^{-12} g. Pecked line: nuclear mass 10^{-13} g. $T = 273$ K; supersaturation = 0·05 per cent; $U = 10$ cm s^{-1}. (From Mason, loc. cit.)

by the drops in Fig. 6.8 are about one-third smaller than those calculated by Findeisen for the same values of initial radius (30 μm), water content, and distance of fall. But, for liquid-water contents of about 0·4 g m^{-3}, which are rarely exceeded in layer clouds, Mason's calculations show that only drops of drizzle size are likely to be produced by coalescence.

These calculations also suggest that larger drops will fall from a deeper cloud sustained by a stronger updraught, in which a drop will have a longer fall-path relative to the small cloud droplets. The speed of the updraught will largely determine the ultimate size of a drop falling from a cloud in the absence of turbulent motions. For given values of the mean liquid-water content and the updraught velocity, the indications are that larger drops result if the early stages of the growth are slow, as this also results in a longer fall path. One cannot conclude from this that in natural clouds the larger drops do, in fact, arise from small nuclei, or that low initial supersaturations favour precipitation release, because the slow growing particles may be carried out of the cloud before reaching the critical size at which they can grow by coalescence.

The extent to which a simple cloud model with constant updraught is capable of accounting for the release of drizzle and rain by coalescence from layer clouds can best be assessed by comparing the results predicted by the model with observation of precipitation from natural clouds. But first it is necessary to calculate how the drops will shrink by evaporation in the unsaturated air between the cloud base and the ground. Mason, using a slightly less accurate form of eqn (3.11) for the rate of evaporation of a drop of radius R falling in an atmosphere of temperature T, equilibrium vapour pressure p_s, and sub-saturation σ, calculated that the rate-of-change of radius, in terms of the distance of fall h through a constant updraught U, from

$$\left(\frac{L^2 M \rho_L}{KRT^2} + \frac{RT\rho_L}{DMp_s(T)}\right) R \frac{dR}{dh} = -\sigma \frac{(1+FRe^{\frac{1}{2}})}{(V-U)}. \quad (6.3)$$

Hence the radius with which a drop will have to leave the cloud base at height h to reach the ground with radius 100 μm (taken as the lower limit for a drizzle drop) was computed by integration of eqn (6.3), viz.

$$-\int_R^{100\mu m} \frac{(V-U)R\, dR}{1+FRe^{\frac{1}{2}}} = \int_0^h \left(\frac{L^2 M \rho_L}{KRT^2} + \frac{RT\rho_L}{DMp_s(T)}\right)^{-1} \sigma\, dh. \quad (6.4)$$

The integral on the left-hand side of eqn (6.4) in which V, F, and Re are all functions of R, was evaluated using the experimental values of Kinzer and Gunn (1951) for F, and with $U = 10$ cm s^{-1}. The right-hand side was computed for various mean values of σ, assuming in all cases a ground temperature of 15° C and a lapse rate of 6·5° C/km between ground and cloud base. Values of R as a function of h, for mean relative humidities of 98, 95, 90, and 80 per cent respectively, are shown in Fig. 6.10.

From the data analysed by Mason and Howorth (1952) and described in § 6.3, details were extracted of the cases in which continuous drizzle reached the ground from clouds warmer than $-5°$ C. Knowing the height of the cloud base and the average relative humidity below it, the radius with which a drop would have to leave the base in order to arrive at the ground with radius 100 μm was calculated. These values ranged from 110 to 170 μm. According to the foregoing calculations, in clouds with updraught velocities of 10 cm s^{-1} and supersaturations of 0·05 per cent, mean water contents ranging from 0·4 to 1·25 g m^{-3} would have been required to produce even the smallest drizzle drops. These values are considerably higher than most published measurements for rather thin stratiform clouds.

Mason suggested that the turbulent motions in the cloud, which appear necessary to account for the production of the larger droplets

§ 6.4   PHYSICS OF NATURAL PRECIPITATION PROCESSES   313

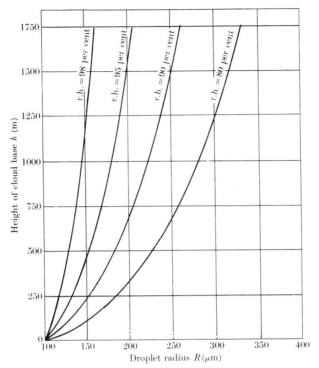

FIG. 6.10. Evaporation of drops falling from cloud base to ground through atmospheres of given mean relative humidity and temperature distribution. The drop radius is shown as a function of the fall path for various values of the mean relative humidity. (From Mason (1952a).)

which initiate the coalescence process, will also increase the effective fall path of some of the larger drops. Calculations show that in order to produce a drop of radius 120 μm, the required water content could be reduced from just over 1·0 to 0·6 g m^{-3} if the fall path were increased from 500 to 800 m, and from 1·0 to 0·4 g m^{-3} if the fall path were increased to 1300 m but, of course, extra time is required for this additional growth.

Fig. 6.9, showing the drop radius as a function of elapsed time, would suggest that 3 to 4 hours must elapse before a drizzle drop can fall out of the cloud base through a uniform updraught of 10 cm s^{-1}, about 2 hours of this being required for the droplets to grow to radius 20 μm by condensation. This indicates that the cloud as a whole must exist for such a period if drizzle is to be produced; but the individual droplets, except the very small proportion from which the drizzle drops originate, are continually being renewed and have much shorter lives.

However, these calculations were based on the simplifying assumption that all the drops grow by coalescence at the average 'continuous' rate given by eqn (6.1). But, as discussed in Chapter 3, it is likely that a small but sufficient proportion of the drops grow at rates considerably faster than the average rate because of the stochastic nature of the coalescence process, and it is these that become precipitation elements. Indeed, Berry (1968), after following the growth of several simulated populations of droplets on a computer, found that the evolution of the drop spectrum becomes largely independent of the initial size distribution when the mode-mass radius exceeds 40 μm, and that over the range 40–400 μm, the rate-of-change of the mode-mass radius R_p is given by

$$\frac{dR_p}{dt} = 4 \times 10^{-3} w R_p, \tag{6.5}$$

where the liquid-water concentration, w, is measured in g m^{-3}. According to (6.5), R_p would grow from 40 μm to 120 μm in a cloud of $w = 1$ g m^{-3} in about 5 min, compared with the 25 min indicated in Fig. 6.9, and therefore it is not now difficult to explain the release of drizzle from shallow layer clouds containing liquid-water concentrations of only a few tenths of a gramme per m^3.

We may summarize the conclusions of this section by saying that well-stirred, low-level, layer clouds of thickness a kilometre or a little less, and sustained by feeble updraughts of a few centimetres per second, may produce drizzle if they have existed for at least 2 hours. Apparently similar clouds occur frequently without causing precipitation; it may be supposed that these are clouds that have formed only very recently, or contain unusually little condensed water, or in which the large droplets required to initiate coalescence did not form. Near hills and coasts, where a more pronounced vertical motion may be produced by orography, drizzle and light rain may fall more readily from thin clouds because the growing droplets are held for a longer time in the cloud. Layer clouds of thickness greater than a kilometre rarely occur well above the ground except in storm regions where the great thickness and extent of the layers, and persistent vertical motions of 10 cm s^{-1} or more, make the production of widespread rains by the ice-crystal process almost inevitable.

6.5. The release of precipitation from shower clouds
6.5.1. *Characteristics and structure of shower clouds*

Precipitation from shower clouds is generally of greater intensity and shorter duration than from layer clouds, the elements generally being

§ 6.5   PHYSICS OF NATURAL PRECIPITATION PROCESSES   315

of larger mass. The greater vertical depth of the clouds, their stronger updraughts and higher concentrations of liquid water, all favour rapid growth of the precipitation elements by accretion. Precipitation from vigorous shower clouds and thunderstorms may reach the ground in the form of raindrops, pellets of soft hail, or as hailstones; snow showers are generally associated with weak or decaying clouds. Before discussing the growth of precipitation particles it is necessary first to consider the large-scale features of cumulus, and how they determine conditions that must be fulfilled before shower production can begin.

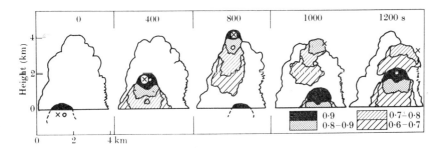

Fig. 6.11. Schematic representation of the structure of cumulus. The approximation of the liquid-water content to that in air rising from the bases *unmixed* is indicated for two successive rising 'bubbles' of cloud air. The successive positions of two particles, one of which grows into a precipitation element, are also shown. (From Ludlam (1952).)

The careful observation of cumulus and the experience of glider pilots in using the updraughts both in and beneath the clouds suggest that they are composed of rising bubbles of buoyant air, called thermals, which have a diameter of about 1 km. These thermals seem to have well-defined upper surfaces, but in constructing a model of the cloud structure, Scorer and Ludlam (1953) suppose them to have extensive wakes in which there is mixing with the outside air. At any moment a large cloud may contain a number of thermals, and for the rest, be composed of their wakes or the residues of earlier thermals, as suggested schematically in Fig. 6.11. Once a thermal reaches the periphery—the summit or the flanks—of the cloud, the mixing and wastage into a wake is accelerated; the rapid evaporation at the cloud borders chills the air there and destroys its buoyancy or even produces sinking, whereas in the interior of the cloud, the thermal is to some extent protected and is able to rise much farther than if it ascended through clear air. A cumulus therefore has a characteristic pyramidal shape and viewed from a distance appears to have an unfolding motion, fresh cloud masses continually emerging from the interior to form the summit, and then sinking aside and evaporating. The size of the cloud depends not only on the properties of the environment, for example temperature, humidity, and stability, but upon the rate of supply of warm thermals into the cloud base, which is determined by the rate of heating of the surface air and other factors which are, as yet, hardly understood.

Fig. 6.11 is taken from a paper by Ludlam (1952) and indicates hypothetically the approximation of the liquid-water content to the value corresponding to ascent from the cloud base without mixing. The strength of the updraught is similarly distributed, and it is clear that an aircraft making a high-speed traverse of the cloud would experience turbulence in passing through several regions of varying updraught, whereas a glider pilot ascending in a region of maximum updraught would regard the air as smooth, for during the time the glider takes to sink at about 1 m/s completely through a thermal it may be lifted into the cloud summits. In considering the formation of showers in large cumulus, Ludlam has assumed that the precipitation elements begin their growth in the thermal cores that rise without mixing from the cloud base to near its top. (Recent measurements indicate that liquid-water contents very close to the theoretical maximum values are sometimes encountered inside large cumulus clouds.) In other words, we assume the most favourable conditions for shower formation, so that the effect of any dilution of the cloudy air must be to increase our subsequent estimate of the size of a cloud which is capable of producing a shower.

In Fig. 6.11 are marked the positions within a thermal of two particles inside the cloud that begin to grow into precipitation elements; these may be regarded as large condensation nuclei that grow within the cloud by coalescence, or as ice nuclei that grow by accretion into ice pellets in the supercooled part of the cloud. One of these particles grows faster, being either larger initially or active as an ice nucleus at a higher temperature, and is shown as settling out of the air that carried it up into the cloud; it remains in the interior of the cloud, enters a newly rising thermal, and completes its growth as a raindrop or hailstone. Meanwhile the other fails to acquire a substantial fall speed before being carried to the outside of the cloud, where it is quickly evaporated together with the neighbouring cloud droplets.

In order to become a precipitation element, a particle must reach some rather critical size during the time taken for cloud air to rise from the base to the summit, such that it shall have settled some hundreds of metres or more relative to the air, gained a fall speed about equal to that of the updraughts near the cloud top (c. 1 m s^{-1}), and have the capacity for falling with insignificant shrinking through several hundreds of metres of unsaturated air—and hence the opportunity to survive even if caught in the general evaporation of a cloud turret— and so resume its growth by settling into a newly rising mass of air within the cloud bulk. Ludlam (1951a, 1952) shows that these conditions are simultaneously satisfied when the particle has a radius of about 150 μm in the case of a water droplet, and a radius of either 550 or 300 μm in the case of an ice particle, depending on whether a mean density of 0·1 or 0·3 g cm^{-3} is assumed for the ice. Whether or not particles can attain this critical size, and hence the probability of shower formation, will depend on the depth of cloud traversed, in turn determined by the speed of the updraught, and the liquid-water concentration which is largely governed by the temperature of the cloud base and the degree of mixing of the cloudy air with the drier environment. If the entire cloud is small, its life may be only a little longer than the time taken for particles to reach the critical size, so that the particles are prevented from growing into precipitation elements; but if the cloud is large, a shower is almost certain to develop, although a strong wind shear may intervene both by reducing the life of the cloud and by causing the large particles settling out of the cloud tops to fall into clear air rather than into succeeding thermals where they may continue their growth.

We shall now discuss various mechanisms by which precipitation may be released from shower clouds and present the results of some calculations on the growth of precipitation elements.

6.5.2. *Production of showers by coalescence of cloud droplets*

The fundamental equations for the growth rate of a drop by coalescence with smaller cloud droplets may be written as follows. The rate of mass increase of a drop of radius R, falling with a velocity V through a cloud of liquid-water content w, composed of smaller droplets having terminal velocity v, is given by

$$\mathrm{d}m/\mathrm{d}t = 4\pi R^2 \rho_\mathrm{L}\, \mathrm{d}R/\mathrm{d}t = E'\pi R^2 w(V-v), \tag{6.6}$$

i.e.
$$\mathrm{d}R/\mathrm{d}t = E'w(V-v)/4\rho_\mathrm{L}, \tag{6.7}$$

where E' is the collection cross-section of the drop and ρ_L its density. Its rate of vertical displacement above a fixed level is

$$\mathrm{d}z/\mathrm{d}t = U - V, \tag{6.8}$$

where U is the updraught velocity. Thus the increase of radius with height is given by the equation

$$\mathrm{d}R/\mathrm{d}z = E'w(V-v)/4\rho_\mathrm{L}(U-V) \simeq E'wV/4\rho_\mathrm{L}(U-V). \qquad (6.9)$$

If U is considered to be constant with height, and E' and V to be functions only of R, eqn (6.9) may be integrated to give the radius of the drop at level z, viz.

$$4\rho_\mathrm{L} \int_{R_0}^{R} \frac{(U-V)\,\mathrm{d}R}{E'V} = \int_{z_0}^{z} w\,\mathrm{d}z, \qquad (6.10)$$

where R_0 is the initial radius of the drop at the starting level z_0. Evaluation of the integrals in eqn (6.10) requires a knowledge of the distribution of liquid-water content with height, and of the collection cross-section E' and the terminal velocity V as functions of the drop radius R (see Appendices A and B).

Langmuir (1948) tabulated values of

$$4\rho_\mathrm{L} \int_{R_0}^{R} \mathrm{d}R/E' \quad \text{and} \quad 4\rho_\mathrm{L} \int_{R_0}^{R} \mathrm{d}R/E'V$$

for various values of R_0 and R, and used them to compute the growth of water drops by coalescence in clouds in which the water content was constant with height and the updraught was zero. He calculated that a drop of initial radius 40 μm would have to fall 13 km through a cloud of water content 1 g m^{-3} and composed of droplets of 8 μm radius in order to attain a radius of 3 mm, the time taken being 64 min; for other values of water content, the distances and times would be inversely proportional to w. The distances calculated by Langmuir are considerably reduced if one assumes the presence of an appreciable updraught in the cloud. In this case, the incipient raindrop may be carried up in the rising air, growing by coalescence with the smaller cloud droplets until it reaches a level at which its rate of fall exceeds the speed of the updraught, when it will return towards the cloud base, growing as it descends.

Calculations on the growth of drops by coalescence in clouds with constant updraught velocity and liquid-water content were made by Bowen (1950), using values of E' computed by Langmuir. A drop, originating by the chance coalescence of two cloud droplets near the cloud base, was assumed to grow by condensation at a supersaturation which was a linear function of the updraught velocity until it reached a radius of 15 μm at which, according to Langmuir, it attained a finite collection efficiency for the cloud droplets of radius 10 μm. Assuming a uniform concentration of liquid water of 1 g m^{-3}, Bowen computed a number of height–time curves for various values of the updraught velocity (Fig. 6.12) and a corresponding series of curves of height against drop diameter (Fig. 6.13). The effect of a stronger updraught is to carry the drop higher into the cloud, and because of the increased

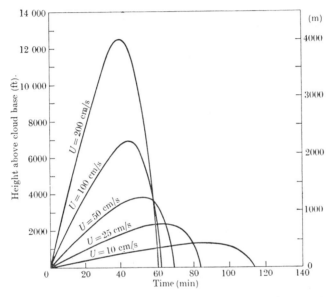

FIG. 6.12. The trajectories of drops which grow by coalescence in clouds having updraughts of 10 to 200 cm s^{-1}. Water content = 1 g m^{-3}; cloud droplet radius 10 μm; percentage supersaturations are one-thousandth of the updraught velocities. (From Bowen (1950).)

double fall path, to produce a larger drop at the cloud base. That Fig. 6.12 appears to suggest that a stronger updraught would produce a larger drop in a shorter time, is a consequence of Bowen's assumption that a stronger updraught is associated with a higher supersaturation and more rapid growth of the drop in the early stages. Bowen's calculations also indicate that a threefold increase in the water content would cause an increase of only 33 per cent in the final drop size, because although the growth rate increases with larger values of w, the length of the path over which growth takes place is reduced. It can be seen, therefore, that the final size of the drops is relatively insensitive to changes in liquid-water content. However, mainly because Langmuir's values of collision efficiency are in error for small droplet sizes,

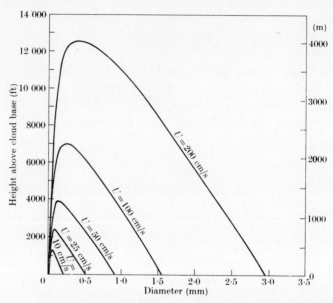

Fig. 6.13. The variation of drop diameter with height for a range of updraught velocities. Other conditions as in Fig. 6.12. (From Bowen, loc. cit.)

Bowen seriously underestimated the time taken for a cloud droplet to grow by condensation to a radius, (nearer 25 μm than 15 μm), beyond which growth by coalescence can proceed at a significant rate.

One possible source of the larger droplets required to initiate coalescence in shower clouds is the giant hygroscopic salt nuclei found over the sea, with masses exceeding 10^{-9} g and in concentrations of order 1/litre. They will leave the sea surface as spray droplets of radius greater than 20 μm, and in convective weather may be carried up into the cloud without appreciable evaporation. These particles may, therefore, be important for shower production in unstable maritime air in warm weather. As the air moves inland, they may disappear by sedimentation and shrink by evaporation towards their equilibrium radii appropriate to the prevailing relative humidity. Thus a salt nucleus of mass 10^{-9} g might shrink to a radius of about 8 μm in an atmosphere of 60 per cent relative humidity, but on re-entering a cloud, would take only 3–4 min to grow to a radius of 25 μm; a nucleus of mass 10^{-10} g would, in similar circumstances, take about 30 min to reach this size.

If the drop enters the cloud with a radius R_0 and leaves it with a radius R, U is constant with height, and w is considered independent of time (i.e. we ignore depletion of the liquid water by the growing drops), eqn (6.10) becomes

$$4\rho_L \int_{R_0}^{R} \frac{(U-V)}{E'V} \, dR = \int^{z_0} w \, dz = 0. \tag{6.11}$$

Ludlam (1951a) argued that, except in the lowest part of the cloud where, in any case, growth by coalescence is very slow, E' may be considered as a function of z only, or even to be constant with height. Eqn (6.11) then becomes

$$R - R_0 = U \int_{R_0}^{R} dR/V, \tag{6.12}$$

showing that the radius of a raindrop falling out of a cloud is a function only of the size at which it originally entered the cloud and the speed of the updraught. Values of R computed by Ludlam for various values of U and R_0 are shown in Fig. 6.14. The size of the raindrops increases with the speed of the updraught, until at a speed of between 3 and 4 m s^{-1}, they reach maximum stable size ($R \simeq 2 \cdot 5$ mm). If the critical speed is exceeded, drop-breaking occurs within the cloud. It is interesting to note that the large raindrops may develop in updraughts having speeds only a third or a quarter of their fall speed. Fig. 6.14 shows that for a given value of U, the largest raindrops result from

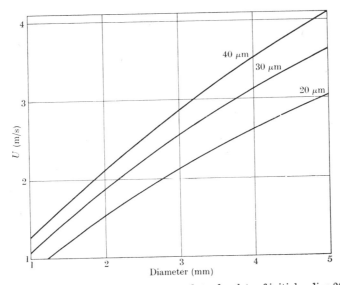

FIG. 6.14. Diameters of raindrops grown from droplets of initial radius 20, 30, and 40 μm as a function of the steady updraught U. (From Ludlam (1951a).)

smaller initial drops, as these are carried up farther into the cloud and so have a longer fall path. This does not necessarily imply that smaller droplets are more efficient in releasing rain, for although their growth is potentially greater, it is accomplished over only a longer period that may extend beyond the life of the cloud.

So far the water content w has been eliminated from consideration, but we must introduce this if we wish to know whether raindrops can actually be produced. A low value of w might mean that insufficient time is available, or that the growing droplet would be lifted to a great height at which it might freeze, and therefore merely contribute to the ice phase; similarly the cloud might not be deep enough to allow unrestricted growth and the particle might be carried out in a thermal at the top of the cloud and not be large enough to survive evaporation before falling back into a newly rising cloud mass. As explained in § 6.5.1, Ludlam takes as a criterion for shower production that air carried to the cloud summits shall contain drops of radius $R > 150$ μm. Drops of this size, if they settle into a freshly rising cloud mass at the top of the cloud (where updraughts will generally not exceed 1 m s^{-1}), will still recede from the summit level and continue to grow. Also, drops of $R \geqslant 150$ μm, if carried 300 m above the cloud top into air of relative humidity 95 per cent, will evaporate very little while falling back into the cloud, whereas appreciably smaller ones ($R < 120$ μm) will evaporate almost completely.

Thus, the necessary condition for shower formation is the growth of some drops to 150 μm radius during the time taken for air to rise from cloud base to summit. From eqn (6.10) we have

$$U \int_{R_0}^{150\mu m} \frac{dR}{E'V} - \int_{R_0}^{150\mu m} \frac{dR}{E'} = \frac{1}{4\rho_L} \int_{z_0}^{z_b} w \, dz, \qquad (6.12a)$$

where z_b is now about 300 m less than the critical summit level that must be reached if a shower is to develop. Taking a mean value of $E' = 0.85$, Ludlam calculated the critical minimum depth z_b of a cloud as a function of U for initial droplet radii $R_0 = 20, 30, 40$ μm in clouds of base temperatures $-5°$ C and $20°$ C, as shown in Fig. 6.15. It is seen that the critical depth of cloud necessary for shower formation increases considerably with the speed of the updraught and is also somewhat larger for colder clouds and smaller values of R_0. According to these computations, if droplets of initial radii 20–30 μm are present, clouds of only 1·5 to 4 km depth are required to produce showers by the

§ 6.5   PHYSICS OF NATURAL PRECIPITATION PROCESSES   323

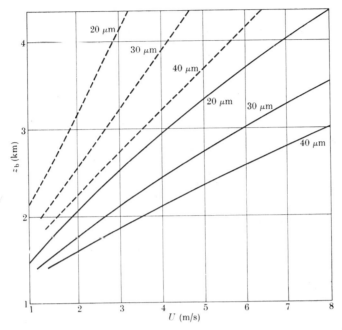

FIG. 6.15. The variation of the minimum cloud depth z_b for shower development as a function of the updraught U, for initial droplet radii of 20–40 μm in clouds with base temperatures −5° C (pecked lines) and 20° C. (From Ludlam, loc. cit.)

coalescence mechanism. The time required for the growth of a raindrop is found by adding a few minutes to the value obtained by dividing the critical depth z_b by the velocity of the updraught, and varies from about half an hour with $U = 1$ m s^{-1} to about 20 min in the stronger updraughts. In order for a considerable shower to occur, the life of the entire cloud must exceed these periods by at least several minutes. Fig. 6.16 shows the variation with base temperature (which largely controls the liquid-water content—see Fig. 6.17) of the minimum cloud depth likely to be associated with showers when the updraughts are weak ($U = 1$ m s^{-1}), and the critical depths for some other conditions shown on the right-hand side. If the cloud-base temperature exceeds 8° C, showers may fall from clouds whose summits do not reach the 0° C isotherm. If the base temperature approaches 20° C, the shower-cloud tops are not supercooled even with updraughts of 8 m s^{-1}. On the other hand, in the colder clouds, moderate updraughts may lift the growing drops to temperatures below 0° C where they may freeze, particularly if the temperature is below −20° C.

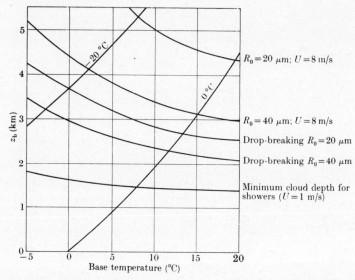

FIG. 6.16. Occurrence of showers produced by the coalescence mechanism in relation to the temperature of the cloud base. The curves show the minimum cloud depths for the various conditions described on the right of the diagram. (From Ludlam (1951a).)

However, these calculations by Ludlam were also based on the assumption that the drops grow by coalescence at the average 'continuous' rate, and further assumed that the cloud contained the undiluted adiabatic concentration of liquid water in the form of small cloud droplets. If, instead, we use eqn (6.5) for the rate-of-change of the mode-mass radius, eqn (6.12a) takes the form

$$U \int_{R_0}^{150\mu m} \frac{dR_p}{R_p} - \int_{R_0}^{150\mu m} \frac{V \, dR_p}{R_p} = 4 \times 10^{-3} \int_{z_0}^{z_b} w \, dz, \qquad (6.12b)$$

and predicts that if the droplets were carried in a steady updraught of 2 m s^{-1} from the base of a cloud at 900 mb, $-5°$ C, R would increase from 40 to 150 μm within 500 s, and the minimum cloud depth, z_b, would be only 700 m compared with about 2·2 km as in Fig. 6.16. In an updraught of 6 m s^{-1}, R would reach 150 μm after 270 s, with $z_b = 1\cdot4$ km instead of 4 km as in Fig. 6.16. This implies that Ludlam's calculations underestimate the growth rates in this size range by a factor of 3. However this may be largely offset by the fact that actual liquid-water concentrations in small cumulus are usually considerably smaller than the adiabatic values assumed by Ludlam.

§ 6.5   PHYSICS OF NATURAL PRECIPITATION PROCESSES   325

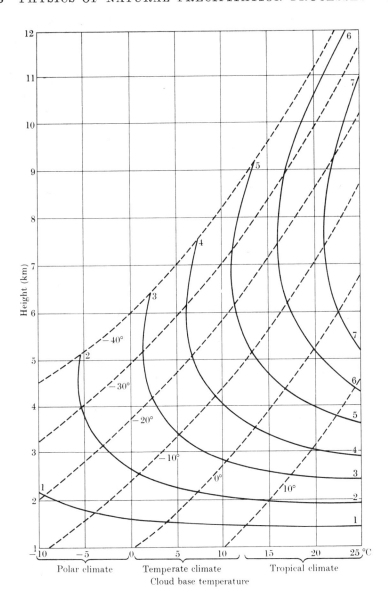

Fig. 6.17. Concentration (in g/m³) of liquid water condensed during adiabatic ascent from a condensation level at 900 mb (about 1 km above the ground), as a function of height and temperature at the condensation level. The indicated values represent the maxima which can occur in convective clouds in the absence of precipitation processes.

It is instructive to compare the theoretical criteria for the production of coalescence showers with observational data.

From careful visual observations of *large*, growing, summer cumulus clouds in central Sweden, Ludlam and Saunders (1956) noticed that the onset of showers was marked by the appearance of evanescent fibrous streaks in the evaporating cloud towers. They appeared when the cloud tops reached a rather well defined level or zone, which the authors called the *fibrillation zone*. Its height above cloud base varied from day to day, ranging from 4 to 6·9 km, at temperatures ranging from -9 to $-23°$ C with a median value of $-15°$ C. The height of the fibrillation level was also strongly correlated with the vigour of the convection, being great when the cloud towers ascended rapidly and small in the feeble clouds containing only weak updraughts. This circumstance, and the high temperatures of the lower fibrillation levels, led Ludlam and Saunders to conclude that it was the coalescence process, rather than the growth of ice particles, which produced the precipitation elements in the fibrillating cloud tops. If the clouds continued to grow and rise appreciably above the fibrillation level, they glaciated if they were supercooled, and acquired the lustre and persistent fibrous texture that is characteristic of ice clouds.

Ludlam and Saunders used eqn (6.12a), replacing $(U-V)$ by U', the speed of ascent of the thermal containing the droplet, to calculate the growth of the drops in models of their large Swedish clouds in which U' and its variation with height were derived from measurements of the rates of ascent of individual cloud towers, and the water content was given the adiabatic value at all levels. For the vigorous clouds containing maximum updraughts of 5–7 m s^{-1}, the calculated values of z_b, the minimum depth of cloud necessary to grow drops from radius 20 μm to radius 150 μm, agreed quite well with the observed values of about 4 km for the height of the fibrillation levels above cloud base. The calculations also lead one to expect most rapid growth of the drops to occur as the thermal approaches the cloud summit, and that after a further ascent of 200 m above the critical level, z_b, the towers to become charged with raindrops. This accords with the fibrillation level being well defined, and the fact that showers developed rather suddenly when the clouds grew above this critical limit but, as we have seen, this agreement may have been fortuitous in view of the highly simplified nature of the calculations.

In the case of feeble clouds containing updraughts of <2 m s^{-1}, the calculations pointed to the difficulty of growing drops from 30 to 150 μm

§ 6.5 PHYSICS OF NATURAL PRECIPITATION PROCESSES 327

radius in clouds of the observed depth and, indeed, fibrillation was not often observed. Nevertheless, small cumulus sometimes produced virga and light showers even when the cloud depth was as little as 900 m and the summit temperature as high as $+3°$ C. On such occasions, the cloud summits entered a very stable layer or an inversion and spread out to form expanding shelves of stratocumulus cumulogenitus; the evaporation of the cloud summits was then very slow, and the growth of the precipitation elements probably continued over a long period within a succession of thermals.

6.5.3. *Raindrop multiplication by break-up of large drops*

Fig. 6.14 shows that raindrops of radius 2·5 mm may be produced if the updraught exceeds 3 or 4 m s^{-1}. Falling drops with radii greater than 1 mm suffer a distortion in shape, which becomes ever more pronounced until a stage is reached at which any further increase in mass results in disruption of the drops. Experiments, described in Appendix B, in which water drops are suspended in a vertical airstream, indicate that break-up generally occurs when the drop diameter exceeds 6 mm, thus accounting for the fact that drops larger than this rarely occur in rain. Although drops of up to 8 mm diameter may remain stable for periods of many minutes in a non-turbulent, uniform air-stream, drops of diameter greater than 5 mm are extremely sensitive to the degree of micro-turbulence and to any sudden changes in the air velocity, and in a non-laminar current, will break up within a very short time. Break-up may also follow the collision of two raindrops, the stability of the resultant mass depending on the size, shape, and phase of oscillation of the two drops at the moment of impact. The number and size of the fragments produced during disruption also depend very much on these last-mentioned factors and on the velocity fluctuations of the air stream, but, under the conditions which might be expected to exist in a shower cloud, usually several drops of about 1 mm radius and much larger numbers of much smaller ones are produced.

Now each of the large fragments may continue to grow by coalescence and may reach a size at which it, in turn, breaks up. In the case of a very deep cloud with high water content, it is conceivable that the process might repeat itself a number of times, so that the number of raindrops is rapidly increased by a kind of chain reaction. Eventually, however, their accumulation will overcome the buoyancy of the air which sustains the upcurrent, so that they will be released as a heavy shower which, by its drag, and by chilling due to partial evaporation, may generate an accompanying downcurrent as strong as the original updraught.

Langmuir (1948) examined the critical conditions under which the fragments of a breaking drop may themselves reach an unstable size before they have fallen back again to their level of origin, the argument being that the second disruption must occur at least as high in the cloud as the first, if there are to be a number of successive drop breakings enabling the chain reaction to build up to major proportions. The critical condition is then given by the equation

$$4\rho_\mathrm{L} \int_{R_1}^{R_\mathrm{m}} \frac{(U_\mathrm{c} \sim V)}{EV}\,\mathrm{d}R = \int_{z_1}^{z_1} w\,\mathrm{d}z = 0, \qquad (6.13)$$

where R_1 is the radius of the fragments, R_m the maximum stable drop radius, and z_1 the level at which the first disruption occurs. Langmuir assumed $R_\mathrm{m} = 3$ mm and computed the critical updraught velocity U_c for various values of R_1 as follows:

R_1 (mm)	U_c (cm s^{-1})
0·2	632
0·6	815
1·0	890
1·4	900

the cloud-droplet radius being 8 μm.

But, as pointed out earlier, eqn (6.13) cannot tell us whether drops of unstable size can actually be produced in a cloud containing these critical updraughts, because it does not contain the depth of the cloud and the liquid-water content. Neither do Langmuir's calculations tell us how long it takes for a drop to grow to break-up size and therefore how many successive drop breakings are possible within the lifetime of the cloud.

In Fig. 6.16, Ludlam has calculated the *minimum* depths of cloud necessary for droplets of initial radii 20 and 40 μm entering the cloud base to attain break-up size (radius 2·5 mm) before falling out of the cloud; the critical updraughts are 3·0 and 4·1 m s^{-1} respectively. It would appear from this diagram that, if the base temperature of the clouds is below 0° C, the growing raindrops are lifted to levels where they are likely to freeze before reaching the unstable size. Thus raindrop breaking is unlikely to occur in such cold clouds but will be favoured in warm clouds of high water content.

Some unpublished calculations by the author show that in a cloud with base at 10° C and 900 mb and sustained by a uniform updraught of 7 m s^{-1}, a droplet of 40 μm entering the cloud base would be carried up to a height of 3·5 km above cloud base (temperature $-10°$ C) before commencing to fall through the updraught, and would attain break-up size (radius 2·5 mm) at the 3·2 km level after a total period of 12 min. If the disrupting drop is assumed to produce fragments of 1 mm radius, these, in turn, will grow to break-up size about 3 min later, after a fall of about 400 m. Successive disruptions will take progressively longer because of the lower water contents at lower levels, so that in the above cloud model, the maximum possible number would be six. However, these calculations are made on the assumption that the cloud contains the adiabatic liquid-water content and that this remains constant during growth of the drops. In practice, it is likely to be progressively depleted, and in consequence the chain reaction will gradually run down. Thus even in the centre of a large, warm cumulonimbus, it is doubtful whether a drop may undergo more than three successive breakings; these would require more than 10 min, by which time the original number of raindrops may be increased a thousandfold, building up in a localized region a concentration of rain-water sufficient to destroy the updraught and release a heavy shower.

6.5.4. *The release of showers by the growth of ice particles*

In a further paper, Ludlam (1952) made analogous calculations to those described in § 6.5.2 for the case when showers are initiated by the growth of ice particles.

The newly-formed ice particles grow at first by deposition of water vapour but, at temperatures between -10 and $-20°$ C, soon attain a size and fall speed at which their growth continues predominantly by the capture and freezing of cloud droplets. Thus at $-14°$ C, when the difference between the saturation vapour densities of water and ice is a maximum, a plate-like crystal of thickness 10 μm will grow in a water-saturated environment to a radius of 50 μm in 100 s, while a droplet growing on a giant salt nucleus of 10^{-9} g will achieve a radius of only 20 μm in this time. Subsequent growth of the ice particle is partly by sublimation, and partly by the accretion of supercooled droplets that freeze on contact to form a loosely-packed aggregate with air spaces in between. The density of this rimed particle, ρ_i, may therefore be quite low; Ludlam assumed that for $R < 65$ μm, $\rho_i = 0.9$ g cm^{-3}; for $65 < R < 200$ μm material is added at a density of 0.6 g cm^{-3}; and for $R > 200$ μm, at a density of 0.1 g cm^{-3}. Again he regards a particle as a potential precipitation element if, on reaching the cloud summit, it attains a size such that (a) during its growth it has settled a 100 m or more relative to the air; (b) it has acquired a fall speed exceeding 1 m s^{-1}; (c) it can fall through several hundred metres of unsaturated air with insignificant shrinking. Computations on the evaporation of ice pellets of mean density 0.1 and 0.3 g cm^{-3} in falling 300 m through air of 90 per cent humidity show that conditions (b) and (c) are fulfilled for radii of about 500 μm and 300 μm respectively. Ludlam calculated the growth of the ice particles from an equation equivalent to

$$\frac{dm}{dt} = 4\pi R^2 \rho_i \frac{dR}{dt} = \frac{4\pi R \sigma (1 + F Re^{\frac{1}{2}})}{\left\{\frac{L_s}{KT}\left(\frac{L_s M}{RT} - 1\right) + \frac{RT}{DMp_s(T)}\right\}} + \pi R^2 E' V w, \quad (6.14)$$

where the first term on the right-hand side represents the mass added by sublimation, and the other, the mass added by accretion. The symbols have the same definitions as in eqn (6.1) except that L_s and ρ_i now refer to the ice phase. F is a ventilation coefficient to allow for the increased rate of sublimation due to the motion of the particle through the air, and Re is the Reynolds number.

The curves B, B' of Fig. 6.18 show the minimum depths that shower clouds of various base temperatures would have to reach if they were to produce ice particles of the minimum critical size, starting at -10 and $-20°$ C in a steady updraught of 5 m s^{-1}. For comparison, the curve A shows the minimum depths of cloud required to produce showers by coalescence initiated by salt droplets of radius 25 μm entering the cloud base.

The diagram suggests that for clouds with base temperatures above 5° C, the coalescence process may forestall the ice particles in initiating precipitation, and even though the cloud summits may eventually rise well above the 0° C isotherm, this is no certain indication that the presence of the ice phase is essential to the formation of a shower. On the other hand, in clouds with cold bases, the initial rapid growth of the ice particles by sublimation may allow them to release a shower before the coalescence process can get underway. In clouds of intermediate base-temperature, say -5 to $+5°$ C, either the one or the other process may be the more important depending upon the updraught strength and aerosol content, and indeed both may contribute to the

§ 6.5 PHYSICS OF NATURAL PRECIPITATION PROCESSES

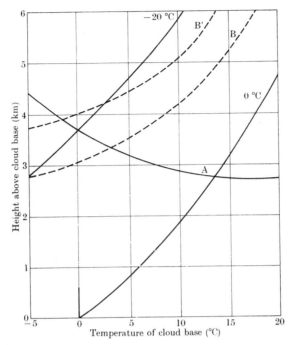

FIG. 6.18. The variation of the minimum cloud depth for shower development by the growth of ice particles as a function of cloud-base temperature. Curves B, B' refer to the growth of ice particles starting at $-10°$ C and $-20°$ C in a steady updraught of 5 m s^{-1}. For comparison, curve A refers to the growth of droplets by coalescence. (After Ludlam.)

formation of precipitation elements inside a cloud of more than the minimum size. In such clouds, the complete description of shower formation and development must be very complicated, and it is scarcely surprising that a *quantitative* theory of shower rains has not yet been attempted.

In summary, the elementary calculations made by Ludlam and represented in Figs. 6.15–6.18, probably describe quite well the relations between the updraught, cloud temperature, and minimum critical depth of the cloud necessary for shower production, but only because, on the one hand, they underestimate the growth rates of incipient precipitation elements by treating coalescence and accretion as continuous rather than as stochastic processes and, on the other hand, they overestimate the liquid-water concentrations of the cloud by neglecting mixing with the environment. There is a need to revise these calculations using the stochastic approach to the growth process, using more recent and reliable computations of collection efficiency, and cloud models in which the updraught and liquid-water content vary with height in a realistic manner.

6.6. Hail

6.6.1. *Occurrence of hail*

Hailstorms occur most frequently in the continental interiors of middle latitudes and diminish towards the poles and the equator and over the sea. In cold climates, the clouds are not sufficiently vigorous, neither do they contain large enough concentrations of cloud water to produce large hailstones. Over the oceans, in the absence of intense surface heating, the cumulonimbus do not usually become organized into the large cloud systems that are usually associated with large hail. In the tropics, the strong horizontal temperature gradients and strong vertical wind shears, which favour the development of damaging hailstorms (see p. 359), do not usually extend into the high troposphere.

Although the incidence of hail, particularly of large stones, is usually associated with thunderstorms, small hail may fall from clouds that do not reach thunderstorm proportions. Beckwith (1960), from an analysis of about 300 hailstorms occurring on 225 hail-days over the 10-year period 1949–58 in the Denver area of Colorado, concluded that hail fell on only about 1 in 8 of the days on which thunderstorms were reported by a given station, but that the numbers of hail incidents reported by the network of forty stations was four times that reported by any one single station. But none of these figures may be very reliable in indicating the proportion of thunderstorms that contain hail at some stage of their development, and Beckwith expresses the opinion of his forecasters and pilots that 'nearly every thunderstorm developing off the Rocky Mountains contains hail in some stage of its development.'

Large hail usually falls from only rather localized regions of the storm and for a brief period. The areas of severe hail damage on the ground may vary in width from a few yards to several miles, commonly 1 mile. The duration of a hail fall may vary from as little as 10 s to as much as 30–40 min, a median value being, perhaps, 3–5 min.

6.6.2. *Classification of hail*

The term 'hail' is applied to a variety of solid hydrometeors that fall from supercooled clouds. The International Commission for Ice and Snow recommended, in 1956, the following definitions which broadly categorize the various forms of such precipitation.

Snow pellets (soft hail, graupel). These are white, opaque, rounded or conical pellets of diameter up to about 6 mm. They are composed largely of small cloud droplets individually frozen together, have a low density, and are readily crushed. They may break up on striking a hard surface.

Ice pellets (small hail, grains of ice). These are transparent or translucent pellets of ice, spherical, spheroidal, conical, or irregular in shape, and having diameters of a few millimetres. There are two main types: (a), frozen raindrops or largely melted and refrozen snowflakes, the freezing process usually taking place near the ground and (b), pellets of snow encased in a thin layer of ice which has formed by the freezing either of droplets intercepted by the pellets or of water resulting from the partial melting of the pellets.

Hail. Small balls or pieces of ice (hailstones) upwards of 5 mm in diameter. A moderately severe storm may produce stones a few centimetres in diameter, while a very severe storm may give rise to stones with maximum diameters of 10 cm or more. Hailstones are composed almost entirely of transparent ice, or of alternate layers of clear and opaque ice.

This classification could usefully be revised and extended in the light of the recent researches described in the following pages.

6.6.3. *The size and shape of hailstones*

Hailstones exhibit a wide diversity of shape—spherical, spheroidal, conical, discoidal, and irregular forms being found, sometimes several of these together in one storm. The roughly spherical form is perhaps the most common when the hail is small, but most large hailstones are not spherical. According to Macklin (1963), 60–70 per cent of the stones from a severe storm in SE. England were oblate spheroids and only a few per cent were spherical. Browning and Beimers (1967) report that the oblateness of large hailstones tends to increase during growth. Moreover, a further increase may be produced by preferential melting at the two ends of the minor axis during descent to the ground and, in extreme cases, this leads to the development of apple-shaped stones. The surfaces of the stones may be smoothly curved, or irregular with protuberances and spikes.

There have been few systematic observations of hailstone sizes; observers tend to report the maximum and mean sizes, but not the smaller sizes which, in any case, may melt on their journey to the ground, so that we have very little reliable information on the size distribution of hailstones from individual storms. The size distributions of the pellets in three falls of soft hail on Weissfluhjoch have been reported by List (1958). They were mostly conical pellets with apex angles ranging from 70 to 90°. Their diameters ranged from 0·5 to 5·5 mm, the most frequent diameter being 1·5 mm in two of the falls, and 2·5 mm in the third. Many seemed to originate from an ice crystal that could be identified in the apex of the particle and then grew by the collection and rapid freezing of supercooled cloud droplets. List also gives size distribution curves for two falls of *small hail*. Here the particle diameters range from 2·5 to 7·5 mm, the most frequent diameter being 4 mm for conical and irregular pellets and 6·5 mm for spherical pellets. The frequency distribution of the sizes of the *largest* hailstones observed in storms in the Denver area during 1949–55 is given by Beckwith (1956) as follows:

Size of largest hailstones	Number of cases	Percentage
Grain ($<\frac{1}{4}$ in)	10	1·4
Currant ($\frac{1}{4}$ in)	122	19·4
Pea ($\frac{1}{2}$ in)	282	45·0
Grape ($\frac{3}{4}$ in)	149	23·5
Walnut (1–1$\frac{1}{4}$ in)	38	6·0
Golf ball (1$\frac{3}{4}$–2 in)	26	4·1
Tennis ball (2$\frac{1}{2}$–3 in)	4	0·6

Hailstones of at least walnut size fell on at least 2 days per year—once for about every sixty thunderstorms.

The largest hailstones ever reported from the United States fell in Nebraska, the largest being 5·4 in (13·8 cm) in diameter and weighing 1$\frac{1}{2}$ lb.

A complete spectrum of the sizes of hailstones, ranging from 0·7 to 8·5 cm in diameter, that fell from a particular severe storm has been published by Ludlam and Macklin (1959), to which Atlas and Ludlam (1961) fitted the formulae $N(R)\,\mathrm{d}R = N_0 \exp(-\Lambda R)\,\mathrm{d}R$ for the spatial concentration, N, of the stones in the cloud, where $N_0 = 8 \times 10^{-5}$ cm^{-4} and $\Lambda = 4\cdot54$ cm^{-1} are constants.

An analysis by Carte (1963) of over 1100 reports of hail, made during 56 days of 1962–3 in Pretoria, South Africa, revealed that the largest stones reported were pea or grape size ($d < 1$ cm) on 56 per cent of occasions, and golf-ball size or larger ($d > 3$ cm) on 3 per cent of occasions. More recent analyses by Carte (1966), and Carte and Kidder (1966) of the reports of about 800 voluntary observers from an area of 1000 miles^2, where hail fell on 60–90 days per year, revealed that stones of $d > 3$ cm fell on <5 per cent of occasions, and that these were often composed of clear ice having a knobbly structure although smooth and opaque stones sometimes fell at the same time. Although some large stones consisted of soft slushy ice, they were more frequently hard, or had hard shells that may have formed as the result of freezing during storage. On the other hand, hail of $d < 1$ cm were frequently soft and composed of rime ice into which melted or accreted water had permeated. Classification of 3514 hailstones into five categories revealed that spheroidal stones outnumbered all other shapes in all size groups below 5-cm diameter, but both spheroidal and conical shapes were less dominant among the bigger stones, where ellipsoids, prolate spheroids, and irregular shapes were well represented. Disc- and lens-shaped stones were always rare, but about one-quarter of all the stones had a flattened profile.

6.6.4. *The structure of hail*
6.6.4.1. Types of ice structure
The internal structure of a hailstone may best be studied by examining thin sections of the stone under the microscope with transmitted and polarized light. Thin slices, preferably 0·3–0·4 mm in thickness, may be prepared either by cutting with a special horizontal saw described by List (1961) or by melting down a stone to the required thickness between two warm metal plates. In this way, one can study the size and arrangement of the individual ice crystals that compose the stone, the size and distribution of the air bubbles in the ice, and the location of the growth centres. The orientation of the larger crystals can also be determined from the orientation of either etch pits (Painter and Schaefer (1960), Koenig (1962), Aufdermaur, List, Mayes, and de Quervain (1963)), or expitaxial deposits of ice crystals (Hallett 1964), formed on their surfaces. There is a danger, however, that the storage, cutting and polishing of the stones may cause some re-crystallization and structural changes.

Within the complex structure of hail particles it is possible to distinguish three kinds of ice deposit:

(a) *Porous (rime) ice* is formed when the deposited water freezes rapidly, mainly as individual droplets, and produces a rimed structure. The air spaces between the frozen droplets give the ice a low density, perhaps as low as 0·1 g cm^{-3} in some cases. The frozen droplets contain large numbers of tiny air bubbles which, because of their high scattering power for light, give the ice an opaque, white appearance. Soft-hail pellets are largely composed of low-density rime ice and this is also found in the cores of some hailstones.

(b) *Compact ice* having density close to 0·9 g cm⁻³ is formed when the droplets have time to spread over the surface and form a continuous film before freezing. This happens when the transfer of heat between the deposit and the environment is just sufficiently rapid to allow all of the accreted water to freeze and maintain the surface in a just-wet condition at 0° C. Since only rather small concentrations of air bubbles are released during freezing under these conditions, the ice is usually clear. Compact ice may also be formed on colder dry surfaces if the droplets impact at high speeds and spread before freezing. In these circumstances the ice may be opaque.

(c) *Spongy ice* is produced when the heat exchange between the hailstone and the environment is not sufficiently rapid to allow all of the deposited water to freeze. For a stone of given size and shape falling through a specified environment, there exists a critical value of the concentration of liquid water above which the surface becomes wet. Only a fraction of the collected water freezes immediately to produce a skeletal framework of ice that retains the unfrozen water, the whole mixture being maintained at 0° C. This wet coating of 'spongy' ice, of density 0·9–1·0 g cm⁻³, is often quite transparent, but sometimes contains sufficient concentrations of small air bubbles to give it a milky appearance. In this wet condition, the hailstone may also collect ice crystals and snowflakes and grow rapidly in size.

6.6.4.2. *The density, crystal structure, and air content of accreted ice*

The conditions under which these different structural forms of ice are produced have been elucidated in the laboratory experiments of List (1959a, 1960) and Macklin (1961, 1962). Macklin examined the ice deposited on cylinders under a wide range of conditions: air temperatures from -5 to $-30°$ C, air speeds from $2\frac{1}{2}$ to 12 m s⁻¹, cloud-water concentrations from 1 to 7 g m⁻³, cloud droplets of mean-volume radii 11–32 μm, and cylinder diameters from 1 to 14 mm. In the regime corresponding to the linear part of the graph in Fig. 6.19, Macklin found that the density of the rime deposit fitted the empirical formula

$$\rho_i = 0\cdot11(-rv_0/T_s)^{0\cdot76},$$

where r is the mean-volume droplet radius in microns, v_0 the impact velocity of the droplets in m s⁻¹, and $T_s°$ C the surface temperature of

§ 6.6 PHYSICS OF NATURAL PRECIPITATION PROCESSES 337

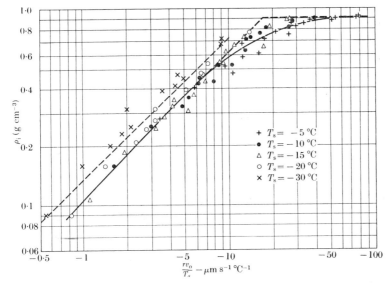

FIG. 6.19. The density of a rime deposit as a function of the parameter rv_0/T_s. (From Macklin (1961).)

the deposit. At temperatures between $-5°$ and $0°$ C, the relationship was more complicated because the density of the ice was also influenced by the air temperature. At temperatures below $-20°$ C, the density was only a function of rv_0.

Microphotographs showed that at low temperatures and with small impact velocities, the droplets tend to freeze rapidly as individual spheres in loosely woven chains and feathery structures of low density—perhaps as low as 0.1 g cm^{-3}. Similar results are reported by Macklin and Payne (1968). At higher velocities and higher temperatures (longer freezing times), the drops become distorted and tend to pack more closely. When the impact velocity was increased from 2 to 11.4 m s^{-1} at temperatures below $-16°$ C, ρ_i increased from 0.28 to 0.80 g cm^{-3}; at $-4°$ C, the corresponding change was from 0.62 to 0.89 g cm^{-3}. The density of a structure composed of regularly packed spheres of density 0.9 g cm^{-3} is 0.67 g cm^{-3}, while the corresponding figure for randomly close-packed spheres is 0.57 g cm^{-3}. The latter is the highest density that one would expect for rime unless, at high speeds, there is spreading and distortion of the droplets. When the surface temperature of the ice is near $0°$ C, the droplets spread over the surface and tend to fuse before freezing and produce a glaze, rather than a rime, of density close to 0.9 g cm^{-3}. Macklin found that demarcation between the regimes of wet and dry growth was reasonably well marked by the critical value, w_c, of the concentration of liquid-water in the cloud at which the ice surface just became wet with $T_s = 0°$ C. However, if the ice surface became irregular, this increased its surface area, promoted turbulence and so increased the rate of heat transfer. The actual value of w at which the surface became wet was then probably higher than w_c calculated for a smooth surface. When the water content of the cloud exceeded the critical value w_c, the deposit acquired an excess of liquid water, only T eightieths of which could freeze immediately as the temperature of the mixture rose from $-T°$ C to $0°$ C. Laboratory studies on the freezing of bulk water by Macklin and Ryan (1962, 1965) indicate that, except at temperatures within a few degrees of $0°$ C when the ice tends to grow in the form of flat dendritic sheets, this fraction of the water will freeze in a three-dimensional open mesh of dendrites that may retain three or four times its own mass of unfrozen water.

In a mixed cloud, composed of supercooled droplets and ice crystals, the rate of freezing of the liquid is increased because an ice crystal landing on the liquid skin can freeze $-T/160$ of its own mass. Macklin introduced into his cloud ice crystals of $d = 30$–300 μm in concentrations of up to 5 g m^{-3}. He found that the maximum quantity of ice accreted by the cylinders was a function of the excess liquid-water concentration, $(w-w_c)$, and established an approximate relationship between the concentration of the accreted ice, w_i (g m^{-3}), and $(w-w_c)/w_c$, the ratio of unfrozen to frozen liquid, of the form $w_i \simeq 1\cdot4$ $(w-w_c)/w_c$.

The crystalline structure of the ice deposit, as revealed by the size, shape, spacing, and orientation of the component crystals, is determined by several factors such as the concentration, size, and impact velocity of the supercooled droplets, the temperatures of the air and hailstone surface, and the detailed mechanisms of freezing. By examining in polarized light the freezing of 1 mm water drops on single crystals of ice, Hallett (1964) found that if both the drop and substrate were warmer than $-5°$ C, the drop froze as a single crystal having the same orientation as the substrate but, at lower temperatures, new orientations appeared, with more crystals at lower temperatures, and as many as 50 at $-20°$ C. Brownscombe and Hallett (1967) report that new crystal orientations form when the *air* temperatures fall below a critical value which is lower for smaller droplets, being about $-5°$ C for 1-mm drops and about $-15°$ C for droplets of 20-μm radius.

The implication is that, in the dry-growth regime, the supercooled droplets freezing on impact at temperatures above the critical value will tend to take the orientation of the underlying surface and produce a fabric of large crystals. This tendency will be even more pronounced if the droplets impact at high speed and spread out before freezing to form a more compact glaze ice. On the other hand, at temperatures below $-15°$ C, most droplets will freeze individually in random orientations and produce a low-density matrix of small crystals.

The appearance of an ice deposit, its opacity and whiteness, is due largely to the scattering of light by minute air bubbles trapped between, and especially inside, the frozen droplets. The opacity is determined by both the ambient temperature and the surface temperature of the deposit. The ambient temperature governs the concentration of air dissolved in the impinging droplets, the concentration increasing with decreasing temperature, and hence the quantity of air released during freezing. The temperature of the deposit determines the rate of freezing and hence whether the air bubbles become trapped in the ice rather than escape by migrating to the surface. Low temperatures favour the formation and retention of high concentrations of small air bubbles and hence the growth of opaque ice. The deposition and slow freezing of water at temperatures close to $0°$ C tend to produce clear ice, but the freezing of ice–water mixtures at low ambient temperatures often produces milky ice containing modest concentrations of air bubbles that are considerably larger than those appearing in opaque ice.

Carte (1961) has studied the formation of air bubbles released during the freezing of bulk water in some detail. Circular disks of ice, 0·5 mm in thickness, were formed by freezing a film of water held between two separated glass disks. A cold copper needle made contact near to the centre of the disk and caused radial freezing to start from that point. Clear or opaque rings of ice could be formed by changing the rate of cooling during freezing, the opacity being due to tiny bubbles and threads of air.

There was a tendency for bubbles to form in waves even when the ice grew at constant rate. Evidently bubbles may reduce the local supply of dissolved air to such an extent that further bubble formation is inhibited until the concentration is built up again. The rapid growth of bubbles at the ice–water boundary indicated that air supersaturations of up to thirtyfold were achieved for rates of freezing greater than 2 mm min^{-1}. Clear ice followed immediately after a wave of bubbles. Thereafter, the density of small bubbles increased and the next wave followed. Agitation of the water was able to prevent the critical supersaturation of air from being attained and clear ice then formed. Average bubble sizes and concentrations were determined not only by the rate of freezing, but also by the quantity of dissolved air, changes of pressure during freezing, movement of the water, and the escape of bubbles by buoyancy. These laboratory experiments, though valuable in helping to understand the physics of air-bubble formation in freezing water, do not provide complete explanations for the complex distributions of air bubbles in hailstones and their relation to the crystalline structure and growth conditions, matters which require further study.

6.6.4.3. Structure of soft hail

Snow pellets or pellets of soft hail may originate either on an ice crystal or on a small frozen raindrop, the embryo continuing to grow by the accretion of supercooled droplets that freeze to form a porous, rime-like structure. The ice crystal tends to collect droplets preferentially on its undersurface and produce a conical structure (see Fig. 6.20), which falls with its apex pointing upwards. At low temperatures and impact velocities, the droplets freeze individually with air spaces in between to form a low-density rime structure composed of small crystals. The density of the rime and the size of the individual crystals increase with increasing temperature and impact velocity, and variations in these parameters may produce a number of zones of different structure and density.

The density of such a particle cannot be determined by weighing it in air and then immersed in a liquid because the liquid may penetrate the air spaces between the frozen droplets. By weighing and simple mensuration, List (1958) obtained values of 0·5–0·7 g cm^{-3} for the density of soft hail pellets falling in warm weather, but values of between 0·04 and 0·24 g cm^{-3} have been reported by Nakaya and Terada (1935) and Magono (1954a) for particles falling from cold-weather showers. Another technique for determining the density of these porous particles is described by List (1961). Phthalic acid, a liquid which freezes at −7° C, is injected into the particles with a hypodermic syringe. The density of the original particle can then be deduced from its weight, and from the weights of the filled particle in air and when immersed in a liquid, such as tetralin, that does not dissolve phthalic acid.

The white, opaque appearance of soft hail is produced by the high concentrations of tiny air bubbles released during the rapid freezing of the droplets.

Fig. 6.20. Section through a conical pellet of soft hail with base diameter about 5.5 mm photographed under translucent light (Photograph by courtesy of Dr. R. List).

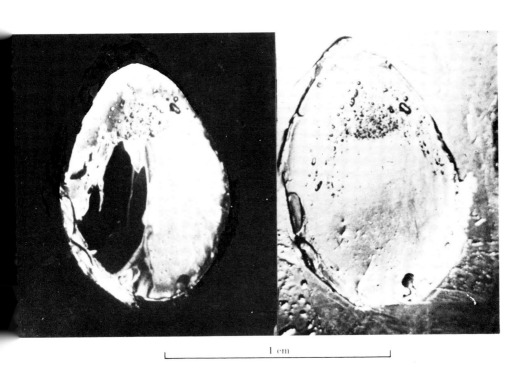

Fig. 6.21. Sections of conical hail seen in polarized and transmitted light. (Photograph by courtesy of Dr. J. Hallett.)

6.6.4.4. Structure of small hail pellets

Small hail pellets are formed by the accretion of water droplets, at temperatures only slightly below 0° C, on pellets of soft hail. The freezing of the water takes place slowly so that it has time to spread over the surface and may percolate into the air channels of the original soft hail pellet. The penetration of water during the slow-freezing stage causes most of the air to escape and produces only a few large bubbles. This can be seen in the thin sections of Fig. 6.21. In the later stages of growth, the drops may continue to freeze in the same orientation as the underlying ice structure and form fairly large single crystals with their optic (c) axes along the direction of growth.

A glazed surface, occurring in zones round a partly opaque core, is therefore the chief characteristic of small hail pellets. Their density is usually $0 \cdot 8$–$0 \cdot 9$ g cm^{-3}—close to that of solid ice. A rapid and fairly accurate estimate of the density may be obtained by placing a particle in each of a number of glass cylinders filled with cold liquids of different density. By noting the liquid in which the hailstone under observation just begins to sink, one may obtain a measure of its density to within an accuracy of about 1 per cent.

6.6.4.5. Structure of hailstones

Hailstones may originate either as snow pellets, particles of small hail, or as particles of clear ice (perhaps frozen raindrops), and grown by the deposition and freezing of water droplets. Schumann (1938) and Ludlam (1950) have pointed out that there is a limit beyond which water deposited on an ice particle can no longer freeze completely because the latent heat of fusion cannot be dissipated sufficiently rapidly to the environment by forced convection and evaporation. Ludlam calculated these limits *using data for smooth spheres* and deduced that large hailstones cannot grow in the dry state, but that their surfaces must acquire a wet coat. He suggested, moreover, that only that fraction of the water that actually freezes can contribute to the growth of the stone and that the rest is shed in the wake of the hailstone. Shedding of the unfrozen water would certainly impose a severe limitation on the growth of large stones and make it very difficult to account for the formation of the very large ones. The situation changed when List (1959a) and Macklin (1961) found, in growing large, stationary wet aggregates of accreted ice in a wind tunnel, that the excess surface water was not carried away but was retained by a skeletal framework of ice, the proportion of liquid water in this 'spongy' ice being determined by the heat exchange between the target and the air-stream. By weighing such artificial hailstones before and after spin-drying them in a centrifuge, List (1960) found that, in some cases, they contained up to 70 per cent by weight of liquid water. However, these stones were grown in air streams of only 6–12 m s^{-1}; working with air velocities of 40 m s^{-1}, Bailey and Macklin (1968a) observed that fixed spheres of spongy ice shed some of their water even with air temperatures as low as $-20°$ C. This suggests that very large hailstones may not be able to retain all of the accreted water even if the fabric consists largely of spongy ice.

The crystalline structures that result from the complete freezing of spongy ice have been studied and discussed by Aufdermaur *et al.* (1963), Hallett (1964), Macklin and Ryan (1965), Aufdermaur and Mayes (1965), Knight (1968), and Knight and Knight (1968), but still await detailed and consistent interpretation. Here it is important to realize that the crystallization of spongy hail could well differ in some important respects from the free growth of ice in bulk supercooled water. Nevertheless, present evidence suggests that if the water is deposited at temperatures between -5 and $-20°$ C, the initial freezing of the fraction required to raise the mixture to $0°$ C is accomplished by the growth of a three-dimensional array of dendrites. Later freezing of the water retained in this skeletal framework produces rather large single crystals, many with optic axes roughly tangential to the hailstone surface. When the deposited water is deeply supercooled, the dendritic growth is less orderly and freezing may start on the outside and spread inwards. The situation may be further complicated by fracture of the

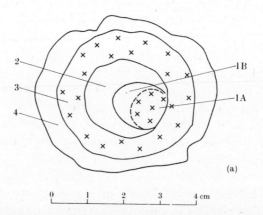

FIG. 6.22. (a) Diagrammatic sketch showing the structure of a thin section of a large hailstone. (b) The hailstone section in transmitted light. (c) The same section in transmitted polarized light. (From Mossop and Kidder (1961).)

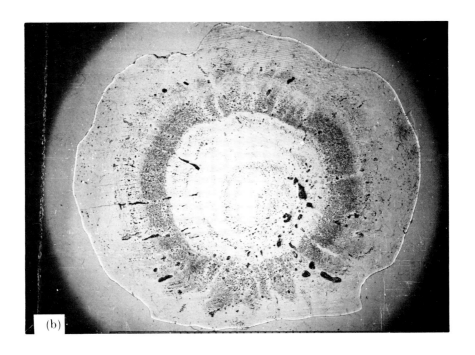

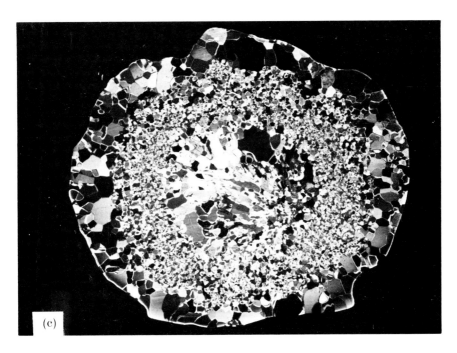

Fig. 6.22.

fragile dendrites and the capture of ice crystals, both of which give rise to new nucleation centres and the formation of a matrix of medium or small crystals with random orientations.

The growth of large crystals of clear ice, with their c axes oriented along the growth direction of the hailstone, is more likely to result from the orderly growth of compact ice in the just-wet condition than from the freezing of wet spongy ice. In this case, laboratory evidence suggests that cloud droplets warmer than $-15°$ C, on impinging, spreading, and freezing on the surface of the hailstone, will form single crystals with their c axes perpendicular to the surface. Later droplets will assume the same orientation and so form large single crystals with their c axes parallel to the growth direction, i.e. along the radius of the hailstone.

The unit crystals, growing out symmetrically from the centre of the hailstone, exhibit a pyramidal or truncated pyramidal form. In longitudinal section they appear triangular or trapezoidal (see Fig. 6.22(c)), while sections at right angles to the growth direction are polygonal. It is usually possible to determine the centre of growth from the form and arrangement of the single crystals because these continue to radiate from the same centre so long as the growth conditions, shape, and orientation of the stone remain much the same. But, should the falling stone change its attitude or cease to rotate, a new centre of symmetry may be formed and a second generation of crystals radiate from this.

We shall now summarize some rather general features of the structure of large hailstones as revealed by thin sections without attempting to describe and explain the detailed microstructure. They have a growth centre or 'embryo', usually a few millimetres to 1 cm in diameter, followed by a number of alternate zones of transparent and relatively opaque ice. Carte and Kidder (1966), reporting on the structure of Transvaal hailstones, found that the centres were most commonly opaque spheroids composed of rime ice, clear spheroids being the next most common centre, except in the largest stones of $d > 5$ cm, which often contained conical embryos.

Fig. 6.22(a) shows the structure of a spheroidal stone of mass 45 g and maximum diameter about 5 cm as deduced from the thin section shown in Figs. 6.22(b), (c). Zone 1, the embryo, began as a pellet of soft hail, 1A, and developed into a small hail pellet by the addition of growth in the wet stage 1B. The boundaries of this zone are indicated in Fig. 6.22(b) by rows of bubbles and, in Fig. 6.22(c), by discontinuities in the crystal structure. There follow three further zones which can be delineated by study and comparison of Figs. 6.22(b), (c); these show that the comparatively clear ice of Zones 2 and 4 is composed of large crystals, up to 1 mm in width and 5 mm in length, while the opaque ice of Zone 3 is rich in air bubbles and is composed of much smaller crystals, 0·1 to 0·5 mm in diameter. The outer boundary of the opaque layer is often less sharp than the inner because of subsequent filling up of the interstices during the subsequent wet stage. Fissures of clear ice extending right through this layer are seen in Fig. 6.22(b).

Mossop and Kidder (1961) reported that six out of fifty-five large stones had more than four well-defined zones, while Carte and Kidder (1966) found that their largest stones contained as many as twenty-eight layers of differing opacity, nine being the most frequent number, but that the number of layers of distinctly different crystal structure never exceeded eight and was most frequently only four. Fig. 6.23(a) shows a sketch of a stone having eight zones, the seventh being incomplete. Figs. 6.22 and 6.23 are examples of stones with fairly simple symmetrical structures developed from only one centre of growth but, even so, it is often difficult to interpret the growth history of a particular stone from its structure, especially when two different phases of growth become superposed.

Although the appearance of rings of air bubbles indicate the direction of growth, an inner ring indicating an early stage of growth, Mossop

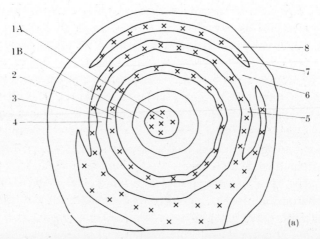

Fig. 6.23. (a) Sketch showing the structure of a large hailstone with 7 zones of growth. (b) Thin section of the hailstone in transmitted light. (c) The same section in transmitted polarized light. (From Mossop and Kidder (1961).)

(b)

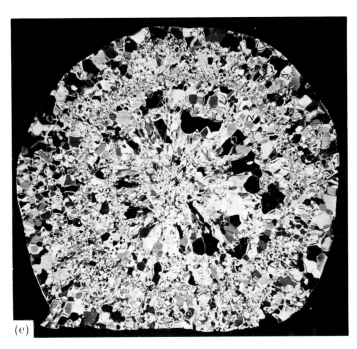

(c)

Fig. 6.23.

Fig. 6.24. A large artifical lobed hailstone grown in a wind tunnel. (Photograph by courtesy of Dr. W. C. Macklin.)

§ 6.6   PHYSICS OF NATURAL PRECIPITATION PROCESSES   345

and Kidder could not interpret the arrangement, size, and concentration of the air bubbles in their large hailstones in terms of Carte's (1961) laboratory observations. They found that the transparent zones were either free of bubbles or that the bubbles were quite erratic as to size and concentration. On the other hand, Browning (1966) has drawn attention to the existence of 'hyperfine' growth layers in the transparent zones of giant hailstones which are delineated by fronts of air bubbles little more than 100 μm thick. Since the zones were formed near the wet-growth limit, it is possible that these rings of bubbles were formed by the repetitive build up of the concentration of dissolved air as described by Carte for water films.

Detailed studies of the internal bubble and crystalline structures of large hailstones of up to 9 cm maximum dimension have been reported by Sarrica (1965) and by Browning (1966) and reveal that large hailstones are often composed of a three-dimensional array of lobes that give the surface a convoluted appearance. Browning suggests that the lobes arise from surface protuberances which, as pointed out by Macklin and Bailey (1966), will capture cloud droplets more efficiently than the adjacent depressed areas. Thus the lobes tend to grow into the airstream and, if the hailstone tumbles constantly during growth, the lobes develop into a three-dimensional array with radial symmetry (see Figs. 6.24 and 6.25. The spaces between neighbouring lobes are usually filled with transparent ice containing radial lines of air bubbles, probably as the result of slow freezing of water that collects there during subsequent stages of wet growth. The strongly convex growing surfaces of the lobes cause successive growth layers to appear convoluted or scalloped, and give rise to surface knobs that become more pronounced with increasing hailstone size. Browning believes that this structure is of great importance in determining the growth of large hail because the surface irregularities are likely to produce a significant increase in both the drag and heat transfer coefficients and allow the hailstones to grow to a large size without becoming excessively wet and spongy.

These views have been largely confirmed by Bailey and Macklin (1968a), who have used a rather simple but very effective icing tunnel to grow artificial hailstones of diameter up to 12 cm, freely supported in a vertical air-stream whose temperature can be varied from 0° C to 30° C and its speed raised to 40 m s^{-1}. They have been able to grow non-spongy lobed hailstones of average diameter 8 cm at air temperatures of $-10°$ C when the median-volume diameter of the drops was about 30 μm. The lobe structure, which sometimes appeared quite early on in the development of the stones, was most pronounced when growth was taking place near the wet limit (i.e. in effective liquid-water concentrations of a few g m^{-3}) and when the accreted droplets were small. Lobes were produced also in dry growth with deep fissures between them. With larger droplets, and when the growth was very spongy, the surface irregularities were far less marked, these less convoluted shapes being due to water movement over the surface and seepage during freezing. Three-dimensional arrays of lobes were obtained only when the hailstone was freely rotating. The fact that the lobes were more pronounced with small droplets suggest that their formation is due to a collision efficiency effect, and implies that natural lobed hailstones grow in strong updraughts containing small cloud droplets and not in regions containing high concentrations of supercooled raindrops that would almost certainly produce spongy growth. The fact that the effective liquid-water concentrations required to produce wet growth in Macklin's tunnel were 1·5 to 3 times greater for lobed stones than for smooth spheres of the same mass, demonstrates that heat transfer was much more effective from the irregular surfaces. Confirmation is provided by the measurements of Bailey and Macklin (1968b) which show that the heat-transfer coefficient for lobed stones of $d > 6$ cm are 1·2 to 3 times greater than those for smooth spheres of comparable size.

6.6.5. *The density of hailstones*

There is surprisingly little accurate information on the density of hailstones. Steyn (1950) gives the specific gravities of nine stones, 5 to 9 cm in diameter, as lying between 0·89 and 0·91.

Vittori and Di Caporiacco (1959) used a very accurate weighing method to obtain the specific gravities of forty stones ranging in mass from 0·5 to 13·6 g. A stone was first weighed in air, then its buoyancy found by means of a wire device which, when immersed to a fixed point, submerged the hailstone beneath a liquid such as tetralin. Three stones had specific gravities of 0·80; the others had values between 0·873 and 0·915.

Macklin, Strauch, and Ludlam (1960) immersed each hailstone in turn in a series of chilled liquids of varying densities. The liquids were mixtures of kerosene and carbon tetrachloride, their specific gravities ranging from 0·850 to 0·910 in steps of 0·010. By observing the velocity of rise or fall of the stones in the liquid, it was possible to estimate their specific gravity to within ±0·003. The values of 169 stones (mass 0·1

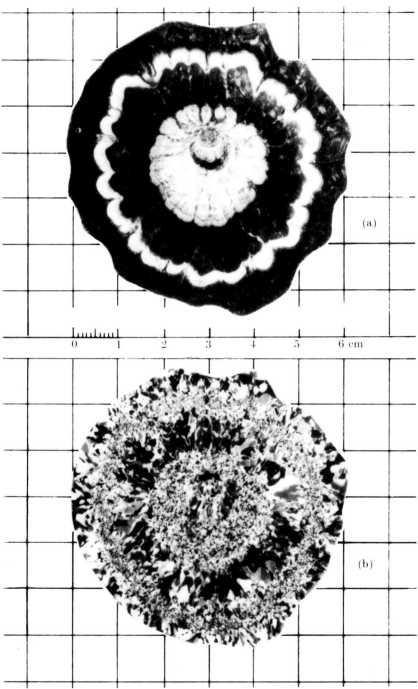

FIG. 6.25. Thin section through the centre of a large lobed hailstone: (a) shows the bubble structure in reflected light; regions of clear ice appear black, and milky or opaque ice appears white; (b) shows the crystal fabric photographed in transmitted polarized light. (From Browning (1966).)

to 17·3 g) from a single storm ranged from 0·875 to 0·912 and supported the high values found by other workers. Over 90 per cent of these stones had transparent centres, the diameters of which ranged from 1·5 mm to 1 cm. They were probably frozen raindrops that had continued to grow in the wet stage.

The specific gravities of 16 representative stones, with masses ranging from 7 to 50 g, from a violent South African storm, are quoted by Mossop and Kidder (1961) as lying between 0·87 and 0·90, with an average of 0·89.

It is apparent that the densities of hailstones are within about 5 per cent of that of clear ice. This is to be expected in view of the high impact speeds and surface temperatures associated with their growth.

6.6.6. *The aerodynamics of hailstones*

The aerodynamic resistance, or drag force, F, exerted by a bluff body in an air-stream of velocity V is given by

$$F = \tfrac{1}{2} A C_D \rho_a V^2, \qquad (6.15)$$

where A is the projected cross-sectional area of the body normal to the flow, ρ_a the air density, and C_D is the drag coefficient. A body in free fall attains its terminal velocity when the drag force is balanced by the gravitational minus the buoyancy forces. For *smooth spheres*, C_D remains constant at about 0·45 over the Reynolds number range 10^3 to 4×10^5 but, at this latter critical value, C_D decreases abruptly to about 0·1 because the whole of the boundary layer of the sphere becomes turbulent.

Measurements of the drag coefficients of natural hailstones, hailstone replicas, and objects resembling hailstones up to 6 cm in diameter, have been made by a number of workers. List's (1959*b*) measurements on wooden models in a wind tunnel suggested that the drag coefficient of a small conical hail pellet, falling with its apex pointing upwards, would be about 1·0. He also made measurements on four large hailstones, three ellipsoidal and one spherical, varying in mass from 18·6 to 32·4 g. When they were oriented with the shortest axis in the direction of the wind, (this is the stable position in free fall), the drag coefficients were about 0·65 compared with 0·45 for a smooth sphere. These results imply that the fall speeds of ellipsoidal stones and of spherical stones with a rough surface must be considerably lower than those of smooth spheres having the same mass.

List's measurements were confirmed in a general way by Macklin and Ludlam (1961), who measured the fall speeds of seventeen large model hailstones made by freezing water in moulds. The stones were placed in a box, carried aloft on a balloon, tipped out at a suitable height, and their falling speeds measured with radar. The corresponding drag coefficients for smooth, spherical stones, of Re up to $1\cdot 4 \times 10^5$, were close to 0·45, but oblate spheroids had values between 0·7 and 0·8 and much smaller falling speeds than spheres of the same mass. The drag coefficients of three real hailstones, of masses between 14·7 and 21·8 g, were measured in a wind tunnel, the values being 0·5–0·6 for spherical stones with $Re = 2\cdot 5 - 6 \times 10^4$, and about 0·7 for oblate spheroids oriented with their shortest axis along the wind direction.

Young and Browning (1967) investigated the effect of surface roughness on the drag coefficients of spheres in a wind tunnel, having a very low level of free-stream turbulence. While the drag coefficient of smooth spheres rose from 0·43 to 0·53 as the Reynolds number, Re, was increased from 4 to 20×10^4, spheres with rough surfaces experienced a sharp drop in drag at critical values of Re ranging from 7 to 17×10^4 as the roughness parameter (ratio of height of roughness elements to sphere diameter) was increased from 1·3 to 13×10^{-3}. These results are consistent with those of Willis, Browning, and Atlas (1964), who made simultaneous measurements of the fall speeds and radar cross-sections of ice spheres of $d \simeq 5$ cm falling in free air from about 20 000 ft (6 km). While they were dry, the spheres fell with slightly supercritical Reynolds numbers ($Re > 1\cdot 3 \times 10^5$) and with drag coefficients of only 0·24 to 0·30. The surface of one sphere, 5·1 cm in diameter, became wet during its fall, whereupon the Reynolds number fell to $0\cdot 98 \times 10^5$ and the drag coefficient increased to 0·56, probably as a result of the boundary layer undergoing a transition from turbulent to laminar flow when the first layer on the sphere melted.

Bailey and Macklin (1968a) have confirmed that the drag coefficients of roughly spherical artificial hailstones, 4–6 cm in diameter and with moderately rough surfaces, may drop quite sharply when Re exceeds 10^5, but finds that this is not the case for larger stones with pronounced lobes, which have drag coefficients of 0·4 to 0·5 even when the Reynolds numbers exceed 3×10^5. It seems, then, that large hailstones, unlike smooth spheres of similar size, do not enter the supercritical flow regime.

6.6.7. *Theories of hailstone growth*

The early theories of hail formation held that hailstones grew either by the aggregation of cloud elements at temperatures below 0° C, or by deposition of water vapour directly into ice. When the multi-layered structure of relatively large stones was discovered, the formation of the opaque low-density layers was usually explained in terms of the collection of snow or the accretion and rapid freezing of supercooled droplets in cold regions of the cloud, while that of the transparent layers was attributed to the acquisition of a coating of liquid water in the non-freezing regions which could freeze only when the hailstone was again carried above the 0° C level. The occurrence of alternate transparent and opaque layers was then explained by its oscillation above and below the 0° C level under the influence of a pulsating updraught. A necessary consequence of this theory was that updraughts of the same magnitude as the terminal velocity of the largest formed hailstone should exist in the cloud in order to carry it up during the penultimate stage of its growth. The appearance of hailstones 5 cm in diameter would therefore imply updraughts of about 30 m s^{-1}.

An important advance in the theory was made by Schumann (1938), who produced the first really quantitative treatment of hailstone growth. He assumed a hailstone to grow as a spherical particle by sweeping up all the supercooled water droplets lying in its fall path. Ignoring the early stages of growth, he followed its history only after it had attained a diameter of 0·5 cm and calculated the ultimate radius in terms of the distance travelled, the liquid-water content w, and the updraught velocity U in the cloud. He used equations equivalent to eqns (6.6)–(6.9), integrated and written in the form

$$4\rho_i \int_{R=0\cdot 25\,\text{cm}}^{R} dR = \int_0^h w\, dh$$

or
$$\rho_i(4R-1) = wh, \tag{6.16}$$

where ρ_i is the particle density and h the distance travelled by the particle relative to the air. Also,

$$4\rho_i\, dR/dt = w\, dh/dt = wV, \tag{6.17}$$

where V is the terminal velocity of the hailstone.

Using the relation $V = 2000\,(\rho_i R/\delta)^{\frac{1}{2}}$, which is based on the calculations of Bilham and Relf (1937), and where δ is the specific gravity of air with respect to air at 0° C and 1000 mb pressure, Schumann used eqn (6.17) to calculate the distance $z = h - Ut$ travelled by the hailstone relative to a fixed level, viz.

$$z = \frac{\rho_i}{w}(4R-1) - \frac{U(\rho_i\delta)^{\frac{1}{2}}}{1000w}(4R^{\frac{1}{2}}-2). \tag{6.18}$$

He considered, as an example, a particle starting at the $-20°$ C level (pressure 420 mb) with radius 0·25 cm and followed its subsequent growth over a fall path of $z = 3$ km, the average density ρ_i being 0·6 g cm^{-3}. The final radius of the hailstone as a function of the liquid-water content and the updraught velocity is shown in Fig. 6.26, from which we may infer that a hailstone of 2·5-cm radius could fall from a cloud of liquid-water content 6 g m^{-3} and an average updraught of only 15 m s^{-1}.

From these and other similar calculations, Schumann concluded that, for the growth of large hailstones, the clouds must extend to heights of

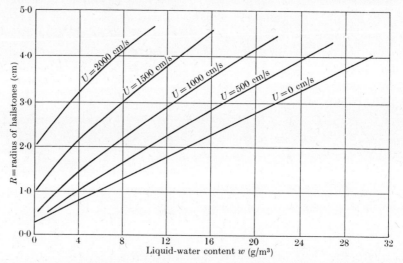

Fig. 6.26. The final radius attained by a hailstone of mean density 0·6 g cm^{-3} falling from a height of 7 km for different values of liquid-water content and updraught velocity. (From Schumann (1938).)

several kilometres, and that they must contain either very high concentrations of liquid water or very strong updraughts. However, it is noteworthy that his calculations indicate that clouds of moderate water content may produce hailstones exceeding 1 inch in diameter and that the updraught velocities need be only about one-half the terminal velocities of the stones. But one must be careful in drawing detailed conclusions from these calculations in which Schumann did not investigate the early, slow stages of growth, and assumed that all the water collected by the hailstone would freeze on its surface.

In the second part of his paper, Schumann realized that the growth rate of a hailstone will be limited by the rate at which it can dissipate the latent heat of fusion. He therefore considered the heat balance of the hailstone and formulated relationships to determine, for given values of the ambient temperature, limiting values of the liquid-water content above which the water captured by the hailstone cannot all freeze and some may be shed in the form of droplets. His general treatment of the heat balance is sound although his theoretical estimates of the transfer coefficients for heat and water vapour are unreliable. Nevertheless, Schumann was able to show that a hailstone might not be able to freeze all the water collected, even with air temperatures well below 0° C, and that at higher ambient temperatures, the growth of a hailstone might be even more restricted.

The validity of these conclusions was demonstrated by Ludlam (1950, 1958), who calculated the rate of heat transfer from a hailstone, using the empirical relations obtained by Kramers (1946) for heat conduction and by Frössling (1938) for mass transfer from spheres under forced convection. Subsequently, slightly different formulations of the problem have been given by List (1960, 1963), Macklin (1963), List, Schuepp, and Methot (1965), and List and Dussault (1967), but Ludlam's original analysis is essentially correct for *smooth ice spheres of $d < 3$ cm*, and brings out clearly the physics of the problem. We shall now follow a slightly more general version of Ludlam's treatment and point out how this has to be modified to deal with large natural hailstones having rough, non-spherical surfaces.

Assuming that the hailstone grows by the accretion and freezing of water that spreads uniformly over its surface, the mass of water collected per unit time by a spherical stone of radius R, falling at velocity V relative to the cloud droplets, will be $E\pi R^2 V w$, where E is the average collection efficiency and w is the concentration of cloud water. If this is to be frozen completely, then the rate of release of heat is

$$\dot{Q}_1 = E\pi R^2 V w \{L_f + c_w(T_a - T_0) + c_i(T_0 - T_s)\}, \qquad (6.19)$$

where T_0 is the melting temperature of ice, T_a the ambient air temperature, T_s the mean surface temperature of the hailstone, L_f the latent heat of fusion, and c_w, c_i are respectively the specific heats of water and ice.

The release of latent heat by the freezing of the accreted water causes the surface temperature of the hailstone to rise above that of its surroundings. The rate at which heat is transferred by conduction and forced convection from such a ventilated sphere is given by

$$\dot{Q}_2 = 2\pi K R(T_s - T_a)(Nu), \qquad (6.20)$$

K being the thermal conductivity of air, and Nu the Nusselt number which, according to the measurements of Ranz and Marshall (1952) on water drops of $0 < Re < 200$, may be written

$$Nu = 2 \cdot 0 + 0 \cdot 60 P_r^{\frac{1}{3}} Re^{\frac{1}{2}}, \qquad (6.21)$$

where the Prandtl number $P_r = \mu c_p / K$, μ being the dynamic viscosity and c_p the specific heat of air.

The rate of heat transfer due to the evaporation of water from the surface of the hailstone is given by

$$\dot{Q}_3 = L_v 2\pi R D \Delta \rho_v (Sh), \qquad (6.22)$$

L_v being the latent heat of vaporization, D the coefficient of diffusion of water vapour in air, $\Delta \rho_v$ the difference between the vapour concentration at the surface of the sphere and that in the remote environment, and Sh the Sherwood number which, according to Ranz and Marshall (1952), is given by

$$Sh = 2 \cdot 0 + 0 \cdot 60 Sc^{\frac{1}{3}} Re^{\frac{1}{2}}, \qquad (6.23)$$

where the Schmidt number $Sc = \mu / \rho_a D$, ρ_a being the air density.

Under the conditions of hailstone growth, the constants 2·0 may be neglected compared with the other terms in the expressions for Nu and Sh. The recent measurements of Thorpe and Mason (1966) on evaporating ice spheres give $P_r^{\frac{1}{3}} = Sc^{\frac{1}{3}} = 0 \cdot 90$ and $Nu = Sh = xRe^{\frac{1}{2}} = 0 \cdot 58 Re^{\frac{1}{2}}$, the corresponding result from Ranz and Marshall's measurements being $Nu = Sh = 0 \cdot 54 Re^{\frac{1}{2}}$. Ludlam assumed that both coefficients were equal to $0 \cdot 60 Re^{\frac{1}{2}}$.

By equating the expressions $\dot{Q}_1$ and $(\dot{Q}_2 + \dot{Q}_3)$ for the heat balance of the hailstone, we may follow Ludlam and obtain a critical value, w_c, for the concentration of cloud water for which the maximum rate of heat loss is just that required to allow all of the collected water to freeze and maintain the surface temperature of the hailstone at 0° C. Thus

$$E\pi R^2 V w_c \{L_f + c_w(T_a - T_0)\} = 2\pi R x Re^{\frac{1}{2}} \{L_v D \Delta \rho_v + K(T_s - T_a)\} \qquad (6.24)$$

or

$$w_c = \frac{2xRe^{\frac{1}{2}}}{RVE} \frac{\{L_v D \Delta \rho_v + K(T_s - T_a)\}}{\{L_f + c_w(T_a - T_0)\}}. \qquad (6.25)$$

Since $Re = 2VR\rho_a / \mu$, the condition for the stone to become wet is

$$RV > \frac{8x^2 \rho_a}{\mu E^2 w^2} \left\{ \frac{L_v D \Delta \rho_v + K(T_s - T_a)}{L_f + c_w(T_a - T_0)} \right\}^2 = f_1(z), \qquad (6.26)$$

where μ and ρ_a are the dynamic viscosity and density of the air, and $T_s = T_0 = 273°$ K. Values of $f_1(z)$, which are a function of height, z,

§6.6 PHYSICS OF NATURAL PRECIPITATION PROCESSES

in the cloud, have been calculated by Ludlam (1958) with $x = 0.6$ and are therefore applicable to smooth spheres. They are given in Table 6.2. Eqn (6.26) does not allow for conduction of heat into the interior of the stone but, as Ludlam has shown, this may be neglected.

When, for a hailstone of given radius, and for given ambient conditions, the concentration of cloud water is $<w_c$, the surface temperature of the stone is below 0° C, all the accreted water is frozen, and Ludlam speaks of growth in the 'dry' regime. The growth rate is then given by

$$dR_i/dt = EVw/4\rho_i, \qquad (6.27)$$

where ρ_i is the density of the deposited ice. On the other hand, when $w > w_c$, more water is collected than can be frozen; under these conditions, Ludlam assumes that the excess accumulates as a liquid film

TABLE 6.2

Parameters $f_1(z)$ and $f_2(z)$ as functions of height, z, for a smooth spherical hailstone following an adiabatic ascent above a cloud base at 900 mb and 20°C

z (km)	5	6	7	8	9	10
$10^{-2}f_1(z)$	0.1	0.6	1.8	4.2	9.5	20.0 c.g.s.
$f_2(z)$	0.1	1.0	3.0	6.1	10.6	16.4 c.g.s.

or is shed, and refers to growth in the 'wet' regime in which transparent ice of density about 0.9 g cm^{-3} is assumed to be deposited beneath the liquid film at a rate governed by the rate of heat transfer, viz.

$$4\pi R^2 \rho_i \{L_f + c_w(T_a - T_0)\} \frac{dR}{dt} = 2\pi RxRe^{\frac{1}{2}}\{L_v D\Delta\rho_v + K(T_s - T_a)\}. \qquad (6.28)$$

Since the two terms on the right-hand side of eqn (6.28) are comparable in magnitude, the rate of change of mass due to evaporation is approximately the fraction $\{L_f + c_w(T_a - T_0)\}/2L_v$ of the growth rate. This is only a few per cent and may be ignored. Thus eqn (6.28) becomes

$$\left(\frac{R}{V}\right)^{\frac{1}{2}} \frac{dR}{dt} = \left(\frac{2\rho_a}{\mu}\right)^{\frac{1}{2}} \frac{x}{2\rho_i} \left\{\frac{L_v D\Delta\rho_v + K(T_s - T_a)}{L_f + c_w(T_a - T_0)}\right\} = f_2(z), \qquad (6.29)$$

values of $f_2(z)$, with $x = 0.6$ and $\rho_i = 0.9$ g cm^{-3}, being given in Table 6.2. Eqn (6.28) may be combined with that for the vertical motion of the hailstone in an updraught, U, i.e.

$$(U - V)\left(\frac{R}{V}\right)^{\frac{1}{2}} dR = f_2(z) \, dz \qquad (6.30)$$

for the variation of R with height when the hailstone is growing in the wet condition and shedding its excess liquid water.

Using the adiabatic values of temperature, air density, and liquid-water concentration, shown as functions of height in Fig. 6.27 for a model cloud with base at 900 mb, 20° C, and the two growth equations (6.27) and (6.28), Ludlam obtained Fig. 6.28, which depicts the subsequent growth of two small cloud drops, frozen at temperatures of $-5°$ and $-10°$ C respectively, in a persistent and constant updraught of 20 m s^{-1}. They develop into pellets of soft hail about 1 mm in diameter within about 2 min. The density of the stones during dry growth, which is represented by the dashed portions of the curves, is assumed to be 0·3 g cm^{-3}, the stones become wet at about 9 and 10 km respectively, with radii of 0·6 and 0·9 cm, and they finally attain radii at the 0° C level of about 1·2 and 1·5 cm. This is about the maximum size attainable on this basis for, if the updraught is increased to increase the growth time, the stones are carried above the $-40°$ C level where growth

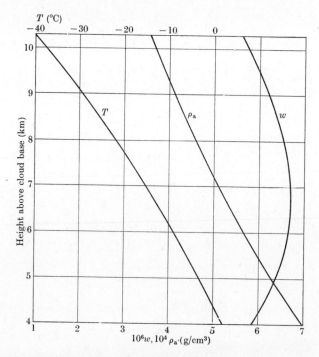

FIG. 6.27. Adiabatic values of temperature (T), air density (ρ_a) and liquid-water content (w) as functions of height in a model cloud with base at 900 mb, 20° C. (From Ludlam (1958).)

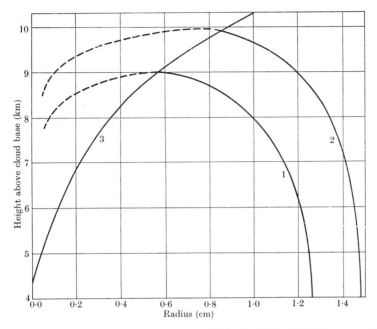

Fig. 6.28. Growth of hailstones in the model cloud of Fig. 6.27. The curves (1) and (2) show the growth in a constant updraught of 20 m s^{-1} following the freezing of a small cloud droplet at $-5°$ C and $-10°$ C respectively. Curve (3) shows the conditions under which a hailstone of density 0·9 g cm^{-3} becomes wet. (From Ludlam) 1958).)

ceases. This maximum value is small by comparison with the diameters of 6 to 10 cm with which hailstones occasionally reach the ground.

Further computations of the growth of hail pellets in both cold and warm clouds have been published by Browning, Ludlam, and Macklin (1963). Starting with frozen drops of radii 0·1, 0·5, and 1·0 mm and density $\rho_i = 0·9$ g cm^{-3}, they use the heat balance equation to calculate the surface temperature, T_s, of the stones as they grow at -10, -20, and $-30°$ C in a cloud containing the adiabatic concentration of liquid water. They then use this value of T_s, together with an assumed value for the volume-mean radius, r, of the cloud droplets, and appropriate values for the fall speeds, V, of the particles, to calculate the density of the growing particle according to Macklin's formula $\rho_i = f(-rv_0/T_s)$.

In cold clouds with base at 0° C and 900 mb, and $r = 10$ μm, they find that the surface temperatures of hail only a few millimetres in diameter are within a few degrees of the ambient temperature, and all the droplets freeze rapidly to produce pellets of soft hail. Most of the growth occurs at temperatures below $-10°$ C and produces particles 2–3 mm in diameter of $\rho_i = 0.1-0.3$ g cm^{-3}. Such particles have fall speeds of several metres per second and are the largest that may be expected from showers in polar air masses.

In a warm cloud, with base at 20° C, 900 mb, containing a mean liquid-water content of 6 g m^{-3} and composed of droplets of $r = 15$ μm, the mean densities of the hail particles are a good deal higher. An embryo consisting of a frozen drop of radius 0·1 mm may grow into a low-density core, a few mm in diameter, but further growth occurs in

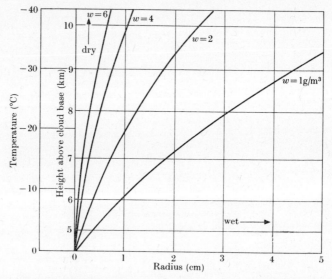

FIG. 6.29. The conditions for the entry of smooth spherical hailstones into the wet growth regime in the model cloud of Fig. 6.27. (From Ludlam (1958).)

the wet stage so that stones of diameter >1 cm inevitably have densities close to 0·9 g cm^{-3}.

The conditions for the entry of hailstones into the wet growth phase in the model cloud of Fig. 6.27 are shown in Fig. 6.29. If the position determined on the diagram by the ambient temperature and the radius of the stone is to the left of the appropriate w-isopleth, the stone is in the dry growth regime; if the position is on the right-hand side, the surface of the stone is wet. The isopleth for $w = 6$ g m^{-3} corresponds approximately to the adiabatic liquid-water content for our model cloud and, under these conditions, we note that stones of $R > 6$ mm would become wet. On the basis of these calculations, which treat hailstones as smooth spheres, it is difficult to account for the growth of stones larger than about 1 cm radius if, during wet growth, the excess water is shed. But, except in the case of very large stones, the excess water is likely to be retained in a framework of spongy ice and so, in practice, the limitations of Fig. 6.29 are removed, and hailstones may grow to diameters of several centimetres in the wet spongy state.

Moreover, real hailstones of $R > 1$ cm are often markedly aspherical and have rough surfaces, so that both the drag and heat transfer coefficients are considerably greater than those for smooth spheres of the same mass. For a spheroidal hailstone, we may follow Macklin (1963) and write the heat balance equation in the form

$$E \cdot \pi a^2 V w_c \{L_f + c_w(T_a - T_0)\} = \frac{x Re^{\frac{1}{2}} A}{2a} \{L_v D \Delta \rho_v + K(T_s - T_a)\}, \quad (6.31)$$

where a is the semi-major axis of the hailstone and A its total surface area. The value of the ventilation coefficient, x,† will depend on the shape and surface roughness of the stone. Macklin (1963), from his measurements of the rate of melting of *smooth* ice spheroids, found x to vary roughly linearly from 0·68 for spheres, to 0·80 for oblate spheroids of eccentricity $\xi = b/a = 0·5$. Writing the volume of the spheroid $\frac{4}{3}\pi a^3 \xi$, $A = \gamma \cdot 4\pi a^2$ and the fall speed $V = (8a\xi \rho_i g / 3 C_D \rho_a)^{\frac{1}{2}}$, the condition for the hailstone just to freeze all the water and maintain its surface temperature at 0° C is

$$w_c = \frac{2\sqrt{(2)}}{E\mu^{\frac{1}{2}}} \left(\frac{\rho_a}{a}\right)^{\frac{3}{4}} x\gamma \left(\frac{3C_D}{8\rho_i \xi g}\right)^{\frac{1}{4}} \left\{\frac{L_v D \Delta \rho_v + K(T_s - T_a)}{L_f + c_w(T_a - T_0)}\right\}. \quad (6.32)$$

Macklin shows that the combination of shape-dependent terms $(x\gamma C_D^{\frac{1}{4}} \xi^{-\frac{1}{4}})$ is practically independent of the shape, so that the conditions for the onset of wet growth on *smooth* spheroids are virtually the same whatever their eccentricity. The curves of Fig. 6.29 therefore apply equally well to smooth non-spherical stones and show that, even in modest concentrations of cloud water, large stones of this shape would become wet if they had smooth surfaces. However, some recent heat-transfer measurements by Bailey and Macklin (1968b) show that the situation may be quite different for stones having the characteristic lobe structure and the knobbly surfaces shown in Fig. 6.24. Since the heat transfer coefficients of such stones with diameters 6–10 cm were found to be 1·2 to 3 times greater than those for smooth ice spheres of the same size, the critical liquid-water concentrations required to cause wet

† Macklin's parameter χ is related to our x by $x = \chi P_r^{\frac{1}{3}} \simeq 0·9 \chi$.

growth are also likely to be 1·2 to 3 times those computed previously. This implies that quite large stones may grow in liquid-water concentrations of a few g m^{-3} without becoming excessively wet or spongy. This is illustrated in Fig. 6.30 where Bailey and Macklin have plotted the effective critical liquid-water content, Ew_c, (because the collision efficiency E has not been accurately determined), as a function of the radius for both smooth ice spheres and for their artificial hailstones,

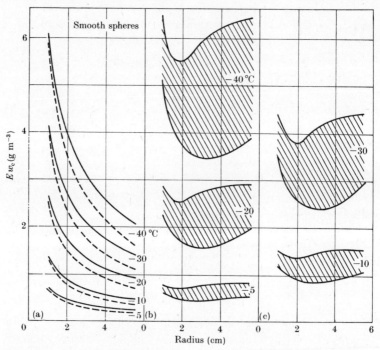

FIG. 6.30. Effective critical liquid water concentrations, Ew_c, as a function of hailstone radius: (a) shows the curves for smooth spheres, the full lines referring to the experimental data of Bailey and Macklin (1968b), and the dashed lines for the previously assumed value of $x = 0·68$; (b) and (c) show the curves for hailstones, the two lines drawn for each value of ambient temperature indicating the spread in measured values of the transfer coefficients. (From Bailey and Macklin (1968b).)

using both their experimentally-determined heat transfer coefficients and constant values of $x = 0·68$.

The fraction, ϕ, of the deposited water that becomes frozen during wet growth may be calculated from eqn (6.24) with L_f replaced by ϕL_f, viz.

$$\phi = \frac{2\sqrt{2}}{Ew\mu^{\frac{1}{2}}}\left(\frac{\rho_a}{a}\right)^{\frac{3}{4}}\frac{x\gamma}{L_t}\left(\frac{3C_D}{8\rho_i\xi g}\right)^{\frac{1}{4}}\{L_v D\Delta\rho_v + K(T_s - T_a)\} - \frac{c_w(T_a - T_0)}{L_t}.$$

(6.33)

According to this equation, a smooth spherical stone of radius 1 cm falling through a cloud containing 10 g m^{-3} of liquid water, at temperature $T_a = -20°$ C and pressure 833 mb, would freeze only half of the accreted water. This result is also largely independent of the shape of the stone, but marked surface roughness, leading to an increase in the heat transfer coefficient x, will enable a higher fraction of the water to freeze. Indeed this may be an important factor in the freezing and consolidation of the outer layers of large stones, and may account for Browning, Hallett, Harrold, and Johnson (1968) finding that, while the outer scalloped layers of large, freshly-fallen stones were completely frozen, shells only 1 cm in diameter in the centre of the stone were composed of spongy ice.

If a hailstone sheds no liquid water, its rate of growth is given by the equation

$$\frac{dm}{dt} = 4\pi R^2 \rho_i \frac{dR}{dt} = E \cdot \pi R^2 V w$$

or

$$\frac{dR}{dt} = \frac{Ew}{4\rho_i}\left(\frac{8R\rho_i g}{3C_D \rho_a}\right)^{\frac{1}{2}}. \tag{6.34}$$

According to some measurements by Macklin and Bailey (1968), the collection efficiencies of large artificial hailstones for cloud droplets roughly obey the relation $ER^{\frac{1}{2}} = 0.6$. Substitution of this in eqn (6.34), together with $C_D = 0.5$, $\rho_i = 0.9$ g cm^{-3}, $\rho_a = 4 \times 10^{-4}$ g cm^{-3}, yields after integration

$$t = \frac{575}{w}(R - R_0) \tag{6.35}$$

for the time taken for the stone to grow from radius R_0 to redius R in a liquid-water concentration of w g m^{-3}. Thus a stone could grow to radius of 3 cm in a water content of 3 g m^{-3} in about 10 min.

Large hailstones have falling speeds of 30 m s^{-1} or more, so any theory of their production requires that the cloud shall contain updraughts of closely comparable speed in order to keep the stones suspended for periods of about 10 min. In its early stages of growth, the fall-speed of the hailstone increases rather slowly so that a strong steady updraught will carry it up through the supercooled region before it can attain a large size. It therefore appears necessary to allow it to re-enter this zone and continue its growth on a second journey. If the updraught were intermittent, the hailstone might repeatedly fall from a high level and then be carried up again, but there is no evidence that strong organized updraughts are intermittent. Instead, Browning and Ludlam (1962) suggested that a severe hailstorm may contain an almost *steady* strong updraught which, because of wind shear, is inclined to the vertical near the ground, so that particles falling from the high-level outflow may re-enter the updraught at low level and make a further ascent.

Particles arising on small embryos will grow into only small hail pellets on their first passage through the updraught and will be swept upward and forward along a trajectory such as AA in Fig. 6.31†, and

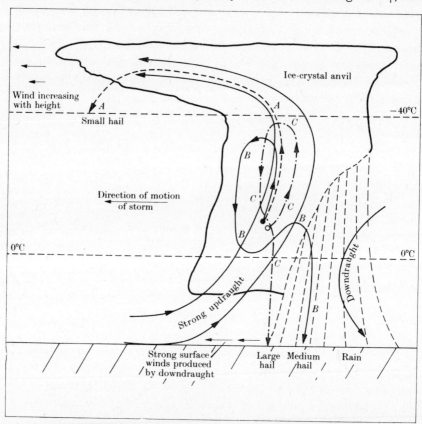

FIG. 6.31. Schematic representation of the structure of a hailstorm.

may fall, melt, and reach the ground as rain ahead, and to one side, of the core of the updraught. Rather larger particles may be thrown out of the core of the updraught but follow such paths as BB, grow on their downward journey, and re-enter the updraught at a lower level. If we now suppose that the updraught speed increases with height, then a small proportion of the re-entering particles that have a favourable size, may be lifted slowly by the updraught, growing at such a rate

† See also Fig. 8.34.

§ 6.6   PHYSICS OF NATURAL PRECIPITATION PROCESSES

that the increase in their fall speed closely matches the increase in the speed of the updraught. Finally, they acquire a fall speed very nearly equal to the speed of the updraught where it is strongest, move forward near the cloud tops, and then fall downwards, briefly passing through the updraught along some such path as CC, before reaching the ground as very large hailstones.

Re-entering particles which are too small are carried up too quickly to reach the maximum possible size, and may be re-cycled again, while those that are too large grow more quickly and cannot be carried up very far by the updraught before falling out again as medium-size hailstones, perhaps 1–2 cm diameter, towards the rear of the storm. In other words, the storm acts as a sorting machine, winnowing out most of the stones and selecting only a few for growth to the largest sizes.

Browning, Ludlam, and Macklin (1963) have traced the growth of a re-entrant hailstone in a quantitative calculation in which they use eqn (6.34), together with

$$U = \alpha z, \qquad dz/dt = U - V,$$

for the variation of updraught velocity with height, to yield

$$\frac{d}{dz}(U-V) = \alpha - \frac{dV}{dR} \cdot \frac{dR}{dt} \cdot \frac{dt}{dz} = \alpha - \frac{g(Ew)}{3\rho_a C_D (U-V)}. \qquad (6.36)$$

Thus if $(U-V)$ at the place where the hailstone re-enters the updraught has the critical value $gEw/3\rho_a C_D \alpha$, then the hailstone may ascend slowly and steadily at constant speed {as $(d/dz)(U-V) = 0$}, the increase in V keeping pace with the increase in U. In these circumstances, $(U-V)$ in m s^{-1} $\simeq 2\,Ew$ (in g m^{-3}) and so amounts to a few metres/s. Consequently, near the $-40°$ C level, where $w = 0$, the hailstone will have $V \simeq U$ and a size that is practically the maximum that can be achieved in this type of updraught. But this mode of growth is not stable. If $(U-V)$ falls a little below the critical value $gEw/3\rho_a C_D \alpha$, the difference increases with increasing z, the stone fails to reach the top of the supercooled zone and therefore falls out of the cloud with a smaller final size. On the other hand, if $(U-V)$ on re-entry exceeds the critical value, the difference again continues to increase, the stone is accelerated upwards and passes out through the top of the supercooled zone, again with a smaller final size. The stringency of the condition to be met by U and V at re-entry into the updraught suggests a reason for the concentration of large stones being always small.

Browning, Ludlam, and Macklin go on to show that when Ew is a continuous function of height, such that the value of V for which the surface temperature of the stone is just 0° C increases with increasing height, (as will always be the case in adiabatic updraughts), there is a particular profile of U, the updraught speed, for which a hailstone growing during its second ascent has a surface temperature that is always just 0° C. In such circumstances, it is reasonable to suppose that rather small fluctuations in the water content, updraught velocity, or in the size, shape, surface roughness, or fall speed of the stones could cause even quite large stones to oscillate between the wet and dry growth regimes and so form layers of differing structure and opacity. That changes of structure may result from variations in the concentration of liquid water is also inferred by List, Charlton, and Buttuls (1968), who calculate the growth of a constant flux of hailstones injected at the 0° C level and carried up in a steady adiabatic updraught that increases with height in inverse proportion to the air density. The number concentration and diameter of the hailstones, and the liquid-water concentration are computed, as functions of height, for specified updraught profiles and for given concentrations and sizes of particles injected at the 0° C level. The authors show that the cloud water is seriously depleted by the growing stones if their initial concentration exceeds about 1 m^{-3}, in which circumstances it is difficult to grow stones of diameter >2 cm. List *et al.* suggest that variations in updraught velocity may cause variations in the number concentration of hailstones and hence variations in the concentration of liquid water which, in turn, may account for the layered structure of hailstones without invoking their re-entry into the updraught. But, as we shall see presently, the growth of well-defined layers marking sharp transitions from dry to wet growth are more convincingly explained in terms of the stone making two (or possibly more) ascents in the core of the updraught. It is quite possible, however, that fluctuations in the heat and water balance of the stone, engendered by changes in the environmental factors or the aerodynamics of the stone itself, may cause less-marked but discernible changes in structure and opacity.

In order to summarize and conclude this account of the growth of largest hailstones, we shall describe the important results of a recent detailed study, by Macklin, Merlivat, and Stevenson (1970), of the structure and life history of a single large stone growing in the strong updraught of a severe storm. This represents the most successful attempt, so far, to deduce the thermal history of a hailstone from the

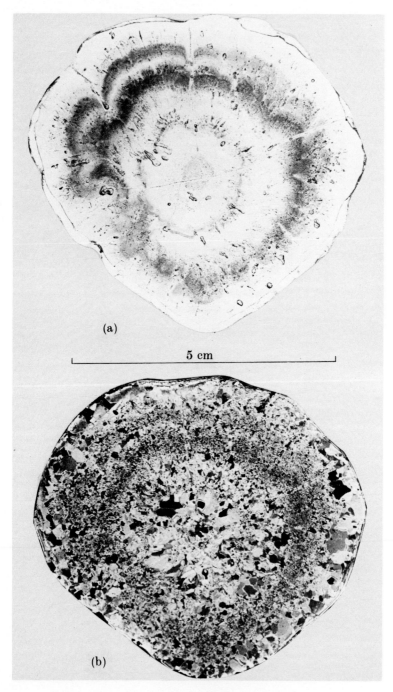

FIG. 6.32. Photographs of thin sections of a large hailstone that fell in Cardiff on 1 July 1968 showing (a) the bubble structure in transmitted light; (b) the crystal fabric in polarized light. (From Macklin, Merlivat, and Stevenson (1970).)

§ 6.6   PHYSICS OF NATURAL PRECIPITATION PROCESSES   363

isotopic composition of the ice in the various growth layers, to compute in consequence the trajectory of the stone relative to the updraught, and to relate these in a consistent manner to the crystal and air-bubble structure. The hailstone in question fell on 1 July 1968 near Cardiff, Wales, from a severe thunderstorm that produced some stones more than 7 cm in maximum dimension. A thin section, taken from near the centre of the stone, was photographed in transmitted and polarized light to provide Fig. 6.32(a) and (b). It is seen that, apart from the central core, the stone is divided into five main layers of alternatively clear and opaque ice. The first opaque layer is quite thin, while the second densely opaque layer has a clear layer running through it on one side of the stone. The lobe structure of the stone is also apparent. The crystal structure, shown in Fig. 6.32(b), coincides exactly with the bubble layers, the crystal sizes changing from a few millimetres to only a few tenths of millimetre as the ice goes from clear to opaque. Even in the thin clear layer embedded in the second opaque layer, there is a temporary reversion to relatively large crystals.

Small samples, about 20 mm³ in volume, were cut out of each layer along a number of radial directions, and their deuterium and ^{18}O composition was determined by mass spectrometry using the technique described by Merlivat, Nief, and Roth (1965). Results of the deuterium analysis are shown in Fig. 6.33; the ^{18}O analysis was very similar and showed the same variation of concentration with hailstone radius. Assuming that the hailstone grows in air making a saturated adiabatic ascent, the molecular species $H_2{}^{16}O$, $HD^{16}O$, and $H_2{}^{18}O$ are conserved in the air but, during condensation, the cloud droplets become enriched, relative to the vapour, in the heavier isotopes. Because the droplets achieve very rapid equilibrium with the vapour phase, the molar fraction of deuterium (or ^{18}O) in the drops is a function only of the vapour pressure and hence the temperature of the environment. (In fact, the quantity δD in Fig. 6.33, defined by

$$\delta D \simeq \left(\frac{N_L}{N_L^0} - 1\right) \times 1000,$$

where N_L is the molar fraction of HDO molecules in the cloud droplets and N_L^0 that in water initially condensed at the cloud base, decreases with decreasing temperature.) Thus a measurement of the deuterium (or ^{18}O) content in a melted sample of ice taken from a particular layer of a hailstone determines the temperature of the cloud droplets that were accreted to form this layer, this being also the temperature of the surrounding air at that time.

The data in Fig. 6.23 indicate the existence of two main growth zones where the deuterium concentration steadily changes with radius. The first begins at the onset of the first clear layer and the δD value decreases sharply with increasing radius until the first opaque layer is reached. Then there is an abrupt rise at the commencement of the second clear layer, with a subsequent fall to the thick outer opaque layer. There is also an indication of a final rise in the curve corresponding to the formation of the outermost clear layer. Using the curves of Fig. 6.33, curves showing how δD varies with temperature, and the temperature–height relationship for the adiabatic ascent, it is possible to determine how the height of the hailstone varied with its radius during growth, and also the air temperatures at these heights (see Fig. 6.34). This shows clearly the two main growth zones in the cloud, each of which gives rise to two growth layers in the stone. The layers are

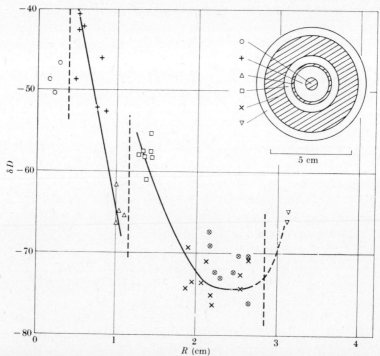

FIG. 6.33. The deuterium values as a function of hailstone radius. The equivalent spherical model is depicted in the upper right-hand corner. (From Macklin *et al.* (1970).)

§ 6.6 PHYSICS OF NATURAL PRECIPITATION PROCESSES   365

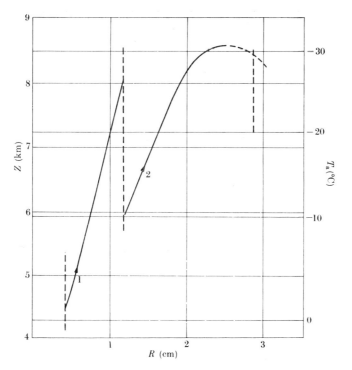

FIG. 6.34. The height of the hailstone as a function of its radius showing the two ascents. (From Macklin *et al.* (1970).)

alternately clear and opaque, with concomitant changes in the crystal size from large to small. These events are readily interpreted in terms of the stone making a first ascent from near the 0° C level to about the 8 km ($-25°$ C) level, where it was carried out of the updraught, fell down, and re-entered the updraught at a higher level (6 km, $-10°$ C) than previously, and made a second ascent to the $-30°$ C level. The flattening of this second part of the curve results from the fall-speed of the stone becoming comparable with the updraught. A spherical stone of this equivalent radius (2·3 cm), at this height, and having a drag coefficient of 0·55, would have a fall-speed of 45 m s^{-1}.

The updraught profiles were calculated from eqn (6·9), viz.

$$(U-V)\frac{dR}{dz} = \frac{EVw}{4\rho_i}.$$

With values of dR/dz taken from Fig. 6.33, the terminal velocity of the stone calculated from $V = (8R\rho_i g/3C_D\rho_a)^{\frac{1}{2}}$, and w taken as the adiabatic value of the liquid-water concentration, the updraught speed U

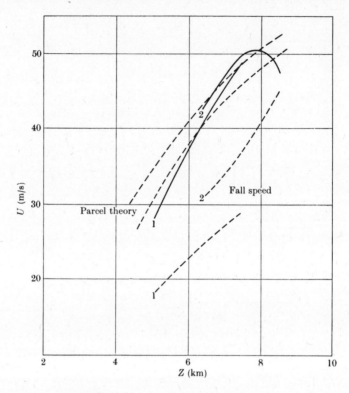

Fig. 6.35. Updraught profiles deduced for the two hailstone ascents. Included also are the updraughts deduced from the parcel method and assumed fall-speeds of the stone at various heights. (From Macklin et al. (1970).)

was calculated as a function of height, and is shown in Fig. 6.35. Where they overlap, the profiles deduced from the first and second ascents are in good agreement, and both agree quite well with two theoretical computations based on variants of the parcel theory of adiabatic ascent.

6.6.8. *The melting of hailstones*

In their treatment of the melting of large hailstones, Ludlam (1958) and Macklin (1963) consider that most of the melt-water is shed so that, at any stage, the stone is covered with a film of water so thin that the surface temperature may be taken to be 0° C. The rate of melting of a smooth, spheroidal stone, falling in still, clear air, with its shortest axis vertical, may then be expressed by

$$L_\text{f} \frac{dm}{dt} = \frac{xRe^{\frac{1}{2}}A}{2a}\{L_\text{v}D\Delta\rho_\text{v}+K(T_s-T_\text{a})\} = \frac{xRe^{\frac{1}{2}}A\beta}{2a}, \qquad (6.37)$$

where $\Delta\rho_v$ is now the difference between the vapour density in the environment and the saturation vapour density at 0° C. Integration of (6.37), with $m = \tfrac{4}{3}\pi\rho_i a^3 \xi$, $\gamma = A/4\pi a^2$, $V^2 = 8\rho_i g a \xi/3\rho_a C_D$,

$$da/dt = V\, da/dz, \qquad \rho_i = 0{\cdot}9 \text{ g cm}^{-3},$$

yields

$$a_0^{\tfrac{7}{4}} - a_g^{\tfrac{7}{4}} = 2{\cdot}23 \times 10^{-3}\, x\gamma C_D^{\tfrac{1}{4}} \xi^{-\tfrac{5}{4}} \int_0^{z_0} \beta \rho_a^{\tfrac{3}{4}} \mu^{-\tfrac{1}{2}}\, dz, \tag{6.38}$$

a_0, a_g being the major semi-axes of the stone at the 0° C level and at the ground respectively. The effect of the shape of the stone on its melting is therefore dependent on the term $x\gamma C_D^{\tfrac{1}{4}} \xi^{-\tfrac{5}{4}}$, which is evaluated in Table 6.3. This shows that the change in the maximum dimension during melting is more pronounced the smaller the value of ξ. The fifth line of Table 6.3 shows the minimum size that the hailstone must possess at the 0° C level in order to just reach the ground after falling through still, clear air. In order to survive, the major axis of a spheroidal stone of $\xi = 0{\cdot}4$ must be about twice the diameter of a spherical stone, but its mass would be only some 17 per cent larger.

For spherical stones falling through still, clear air in conditions appropriate to his Wokingham storm, Ludlam (1958) showed that eqn (6.38) takes the form

$$R_0^{\tfrac{7}{4}} - R_g^{\tfrac{7}{4}} = 0.51 \tag{6.39}$$

with solutions

R_g(cm)	0	1	2	3	5
R_0(cm)	0·73	1·3	2·2	3·2	5·1

These figures show that a smooth, spherical hailstone must have a diameter rather larger than 1 cm at the 0° C level if it is to reach the ground in hot weather. On the other hand, if the size at this level is

TABLE 6.3

The influence of shape on the melting of smooth hailstones (after Macklin 1963)

ξ	0·4	0·6	0·8	1·0
C_D	0·73	0·67	0·61	0·55
x	0·80	0·76	0·72	0·68
$x\gamma C_D^{\tfrac{1}{4}} \xi^{-\tfrac{5}{4}}$	1·52	0·99	0·75	0·60
a_0 (min)	1·20	0·94	0·80	0·70
mass spheroid / mass sphere	1·17	1·08	1·04	1·00

appreciably greater than this minimum value, the shrinkage during the fall to the ground is not very great.

However, these conclusions require modification for stones having a lobed structure and correspondingly higher values of the heat transfer coefficient. If, as the experiments of Bailey and Macklin (1968b) suggest, heat transfer from such stones of radius >3 cm is three times that assumed by Ludlam, then eqn (6.39) should read

$$R_0^{\frac{7}{4}} - R_g^{\frac{7}{4}} = 1{\cdot}53 \tag{6.40}$$

to give the following values:

R_g (cm)	3	4	6
R_0 (cm)	3·37	4·30	6·23
% change in R	12	7·5	3·8
% change in mass	41	24	12

which imply that even very large stones can undergo appreciable melting during their fall to the ground if they are lobed.

The rate of melting of small hailstones, spheres of solid or rime ice, in falling from the 0° C level to the ground, was computed by Mason (1956b). The calculations showed that, in a saturated atmosphere with a temperature gradient of 6·5° C/km, solid ice spheres of $R = 3$ mm would melt completely after falling 2·5 km, and soft hail pellets of the same size, but density 0·3 g cm^{-3}, after falling only 1 km from the 0° C level. Laboratory measurements by Drake and Mason (1966) on ice particles suspended in airstreams of controlled temperature, humidity, and velocity, gave melting times that agreed well with the calculated values especially if allowance was made in the latter for the heat content of the water retained by the particles during melting. The melting times of small ice cones of various shapes did not differ from those of spheres of equal volume by more than 10 per cent.

7
Artificial Modification of Clouds and Precipitation

7.1. Historical introduction

AN examination of the literature reveals a number of early attempts by man to increase the rainfall by a variety of interesting methods among which may be cited the lighting of fires, the firing of cannon, the production of electric discharges by kites, and the spraying of liquid air and dust from aircraft. It is only since the Second World War, however, that methods based on a knowledge of the physical processes of rain formation have been used.

The demonstration in the early 1950s that suitable clouds may be modified and sometimes made to release their precipitation by introducing artificial nuclei, and the possibility of extending this technique to produce economically important increases of rainfall, created the greatest interest and, naturally, a good deal of controversy. The hopes raised by these early developments have not been realized but they have given much impetus to researches on the fundamental physics of clouds and precipitation.

Modern attempts at cloud modification are based on four main assumptions:

(1) that either the presence of ice crystals in a supercooled cloud is necessary to release rain by the Wegener–Bergeron process, or the presence of comparatively large water droplets is essential to initiate the coalescence mechanism;

(2) that some clouds precipitate inefficiently, or not at all, because these agents are naturally deficient;

(3) that this deficiency can be remedied by seeding the clouds artificially with either solid carbon dioxide (dry ice) or silver iodide to produce ice crystals, or by introducing water droplets or large hygroscopic nuclei;

(4) 'overseeding' with massive doses of nuclei will result in large concentrations of crystals that will be unable to grow sufficiently large to fall out and reach the ground. Such an operation may therefore retard or prevent the development of precipitation and, in particular, suppress the growth of large, damaging hailstones.

The possibility of producing rain and suppressing hail from supercooled clouds by the introduction of artificial sublimation nuclei was foreseen by Findeisen (1938), but it was not until 1946 that a satisfactory method of supplying nuclei in the required large quantities was discovered. Earlier experiments by Veraart (1931) in Holland, in which he dropped dry ice, among other things, into supercooled clouds, must have produced such nuclei, but Veraart was not aware of this. It appears that he may have produced slight amounts of rain on several occasions, but because of his sweeping claims, all his attempts were discredited.

Not until Schaefer (1946) discovered that a tiny fragment of dry ice, when dropped into a cold chamber filled with supercooled cloud, resulted in the formation of millions of ice crystals, was it feasible to put Findeisen's ideas to the test. Schaefer made the first field trial on 13 November 1946 when 3 lb of crushed dry ice were dropped along a line about 3 miles long into an altocumulus deck whose temperature was about $-20°$ C. Langmuir, observing from the ground, saw snow fall from the seeded cloud for a distance of about 2000 ft before evaporating in the dry air. A number of such tests in the following months gave very similar results, i.e. large areas of supercooled stratiform cloud were converted into ice cloud. On some occasions, snow fell from the seeded part of the cloud to leave a clear lane. These experiments were described by Langmuir (1947).

The seeding of large, supercooled cumulus clouds with dry ice was first carried out by Kraus and Squires (1947) in Australia; they reported that 6 out of 8 clouds gave radar echoes after seeding, and in 4 of these, heavy rain reached the ground. The most spectacular case is illustrated in Fig. 7.1. A cloud with base at 11 000 ft, the $0°$ C level at 18 000 ft, and the summit at 23 000 ft was seeded with 150 lb of dry ice, whereupon it grew to 40 000 ft in 13 min. A radar echo appeared within 5 min of seeding, and after 21 min heavy rain was observed to be falling from the base of the cloud. No other radar echoes were detected within a range of 100 miles.

The next important step in the history of cloud seeding was the discovery by Vonnegut in November 1946 that minute crystals of silver iodide, produced in the form of a smoke, acted as ice-forming nuclei at temperatures below $-5°$ C. The fact that enormous numbers of nuclei (about $10^{15}/g$ of silver iodide) could be produced by vaporizing an acetone solution of silver iodide in a hot flame, immediately suggested the possibility of dispersing them from the ground, since the

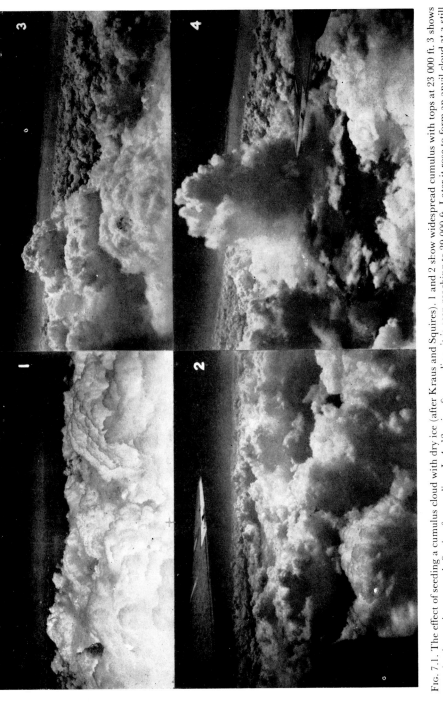

Fig. 7.1. The effect of seeding a cumulus cloud with dry ice (after Kraus and Squires). 1 and 2 show widespread cumulus with tops at 23 000 ft. 3 shows one cloud towering upwards 9 min after seeding. In 4, 13 min after seeding, it is seen reaching to 29 000 ft. Later it rose to form an anvil cloud at a still higher level and rain fell for 2½ hours. (Photo by courtesy of Commonwealth Scientific and Industrial Research Organization, Sydney, Australia.)

minute particles could remain in the atmosphere for long periods until carried up by convection and diffusion into the supercooled regions of the cloud.

The first clear-cut evidence that silver iodide smokes could modify natural, supercooled clouds was obtained by Project Cirrus on 21 December 1948. By dropping lumps of burning charcoal impregnated with silver iodide from an aircraft, approximately 6 square miles of supercooled stratus, having a thickness of 1000 ft and a temperature of $-10°$ C, was converted into ice crystals by less than 1 oz of silver iodide.

The first large-scale ground seeding trials to be undertaken as a commercial enterprise are believed to have been carried out in Arizona and in the eastern part of Washington State in June 1950.

7.2. The experimental seeding of cumuliform clouds

The most satisfactory method of seeding individual cumulus clouds and of observing subsequent changes is to disperse the seeding agent from an aircraft equipped with radar and under the control of a ground radar station. In this way one can be quite certain of the artificial nuclei entering the cloud, and can observe changes in cloud structure, the onset of precipitation and when this falls from the cloud base, all in relation to the time of seeding. Close comparisons can also be made with neighbouring unseeded clouds.

7.2.1. Experiments with dry ice

The dry ice is usually crushed into pellets of about 1 cm diameter and dispensed from a chute at the rate of a few pounds per mile of flight.

We have already described the results of the successful experiment by Kraus and Squires (1947) in which a large cumulus congestus was seeded with 150 lb of dry ice. The spectacular growth of the cloud that followed appears to be a rather rare phenomenon and merits some discussion. Kraus and Squires suggested that it may have been due to the latent heat liberated by the transformation of supercooled water into ice, the atmosphere above the cloud summit being presumably unstable for changes involving the transformation from vapour to ice, but stable with regard to condensation processes. At a temperature of $-10°$ C and pressure 650 mb, the lapse rate of temperature of saturated air in the first case is $6·4°$ C km^{-1}, and in the latter case, $6·9°$ C km^{-1}; thus, glaciation of the upper part of the cloud can produce further growth only if the lapse rate of the environment lies between rather narrow limits. The warming effect of the liberated latent heat is probably less than 1° C, but under favourable conditions the resulting increase in buoyancy might be sufficient to cause considerable further growth of the cloud.

Similar occurrences of rapid growth after seeding have been reported for example, by Vonnegut and Maynard (1952), from New Mexico. Prior to seeding, the only clouds to be seen were small cumulus with bases at 15 000 ft and temperatures −2° C, no more than a few thousand feet thick and less than a mile across (Fig. 7.2(a)). One of these small clouds was seeded with dry ice (and not silver iodide as stated in the paper). A few minutes later the cloud top was observed to glaciate, after 20 min the cloud had grown greatly in size and produced a radar echo. A photograph taken at this stage is shown in Fig. 7.2(b). The cloud continued to grow and, 75 min after seeding, became a large thunderstorm.

Experiments in which a few pounds of crushed dry ice are dropped into the tops of supercooled cumulus have been carried out in many countries, for example, Australia, Canada, England, South Africa, and the United States. In the South African experiments, carried out in the summer of 1947–8 (Anon. 1948), dry ice was dropped into the tops of 36 supercooled cumulus clouds, and the results observed on a 3-cm ground radar. Ten of the seeded clouds produced radar echoes that persisted for <15 min, and 14 gave echoes of longer duration. Four clouds gave no echoes, and there were 8 doubtful cases. On only 11 of the 24 occasions did the echoes appear within 30 min of seeding, while in 6 cases the delay was 30–60 min. The remaining 7 echoes, which appeared only after delays of 90 min or more, probably developed independently of the seeding. All the clouds that gave radar echoes after seeding extended more than 5000 ft (1·5 km) above the 0° C level so that their summit temperatures were probably below −7° C.

Similar investigations were carried out in Ohio in 1948 and in the Gulf States in 1949 (Coons, Jones, and Gunn, 1948, 1949). The aircraft, which were fully equipped to make meteorological measurements, carried radar and were also controlled by radar from the ground. It was possible to detect whether the clouds were in the precipitation stage before seeding, to follow their development after seeding, and to compare them with unseeded clouds occurring within a radius of 30 miles. Although during the Ohio operations precipitation fell from a total of 16 seeded clouds which gave no radar echo before seeding, in all but one case the amount was trivial, and on only 5 of these occasions were

(a)

(b)

Fig. 7.2. (a) Area of New Mexico at 09.45 a.m. before cloud seeding operation. The cumulus clouds are small with bases at about 15 000 ft. their tops being supercooled. (b) Towering cumulonimbus rising from the region where a cloud was seeded 20 min earlier. (Photos copyright—Project Cirrus, Signal Corps Engineering Laboratory.)

there no natural showers within 30 miles. Although most of the clouds were growing at the time of seeding, only 6 continued to grow afterwards, the remainder showing a marked tendency to dissipate. In 6 cases the dissipation was complete.

The results of the tests in the Gulf States were of much the same general pattern. Of the 24 clouds that showed no radar echo before seeding, 2 produced rain and 10 produced virga after seeding, but in all cases there were natural showers within 30 miles. All the treated clouds started to dissipate after seeding.

The outcome of 20 trials in Australia, which gave clear-cut results on days when no natural showers were observed within 25 miles, has been described by Smith (1949) and Squires and Smith (1949). Dry ice was dropped at the rate of 10 to 30 lb per mile and subsequent developments were followed by radars installed in the aircraft and on the ground. The general procedure was to make aircraft soundings of height, temperature, etc., to drop the dry ice, and then to wait near the cloud base in order to observe the precipitation. The intensity and character of the precipitation was investigated by flying through it and observing the rain or snow striking the windscreen. Showers were believed to have been induced by seeding in 15 of the 20 clouds because similar unseeded clouds within 25 miles remained unaffected. In most of the 15 positive tests the cloud summit extended above the $-7°$ C isotherm. These clouds were stable, solid and compact, and persisted for considerable periods. In 10 of the 15 cases, precipitation was seen to reach the ground, and in all these, the cloud depth was greater than 4000 ft and the cloud base below 8000 ft. On four occasions radar echoes were observed between 16 and 34 min after seeding; the times that elapsed before precipitation was observed below cloud base were between 9 and 60 min, the delay being greater for clouds of greater depth. The 5 clouds in which precipitation was not induced remained quite dense and compact, cases where the cloud quickly evaporated being excluded by the authors. In 4 of these 5, the cloud top was warmer than $-8°$ C.

Canadian operations, in which 41 cumulus clouds were seeded with dry ice, are described by Orr, Fraser, and Pettit (1949). The seeding rate varied from 2 to 10 lb per mile and the results were observed visually. Of the 35 supercooled clouds that were treated, 12 produced precipitation that reached the ground and 11 showed virga, but on 15 of these 23 occasions, precipitation was occurring naturally within 25 miles. None of the 6 non-supercooled clouds precipitated after seeding—on the contrary, 4 of them started to evaporate.

The behaviour of the South African, Australian, and Canadian clouds was very similar in that all three sets of trials indicated that the chance of inducing precipitation by the introduction of dry ice is high if the cloud top is colder than $-7°$ C and the depth of the supercooled region exceeds 4000 ft. These results were confirmed and amplified by Bowen (1952a) in a review of the Australian work in which, on the basis of about 100 dry-ice experiments performed in the period 1947–51, he drew the following conclusions.

(1) With cloud summit temperatures of $-7°$ C and colder, there is a 100 per cent chance of producing precipitation. At temperatures between -7 and $0°$ C, the chances of success fall off progressively, tending to zero at $0°$ C. At temperatures of approximately $-15°$ C and below, the results lose their significance because of the high probability of clouds producing rain naturally.

(2) The time at which precipitation will appear at the base of the cloud depends mainly on the thickness of the cloud; there is a gestation time of about 10 min, followed by an additional time of about 1 min for every 800 ft of cloud.

(3) The intensity of the precipitation from the base of the cloud increases with increasing cloud thickness.

(4) A considerable fraction of the precipitation will reach the ground if the height of the cloud base is equal to, or less than, the cloud thickness.

However, the summit temperature and cloud depth cannot be the only factors that will determine the chances of successful seeding; for example, the strength and duration of the updraughts must be sufficient for the particles to grow to precipitation size. Herein may lie an explanation for the fact that a similar programme of dry-ice seeding of supercooled cumulus clouds in the central United States, described by Braham, Battan, and Byers (1957), produced no evidence that the dry ice initiated precipitation.

Although most of the above experiments may be criticized for not being randomized with a view to eliminating observer bias (see p. 384), they do indicate that, under some circumstances, seeding cumulus clouds with dry ice may initiate or enhance precipitation, and in others it may inhibit cloud growth and decrease precipitation. In only rather rare cases was seeding followed by more than a light shower at the ground.

7.2.2. *Seeding cumulus with silver iodide*

There are rather few data on the effect of seeding *individual* cumulus clouds with silver iodide. In order to infect the region bounded by the

−5 and −15° C levels in a cloud of cross-sectional area 5 km² with nuclei in an average concentration of 1/litre, would require about 10^{13} nuclei, so that the required quantity of silver iodide, if this were introduced from an aircraft, should be only a few grammes.

Warner and Twomey (1956) describe an experiment conducted in Australia in which 35 supercooled cumulus clouds, with summit temperatures ranging from −2·5 to −16° C and depths ranging from 4000 to 17 000 ft, were seeded with about 10 g (or 10^{14} nuclei effective at −10° C) of silver iodide from an aircraft flying through the upper levels of the cloud or just below the cloud base. Of those clouds that were colder than −5° C at their summits, 72 per cent precipitated usually within 20–25 min of seeding, 21 per cent evaporated, and 7 per cent showed no discernible effects. At first sight, these results seem very similar to those described above for seeding with dry ice, but further trials, in which clouds were selected for seeding on a randomized basis, have failed to produce such clear-cut results.

Bethwaite, Smith, Warburton, and Heffernan (1966) describe trials involving 69 isolated supercooled cumulus clouds in which either a large (20 g), small (0·2 g), or zero quantity of silver iodide was introduced into the cloud base, with random choice of treatment. A cloud was regarded as suitable for treatment if its summit was either colder than −5° C or judged likely to become colder than −5° C within 30 min, if its depth was >1000 m and exceeded half the terrain clearance, if the height of the base did not exceed 3500 m, if the cloud was fairly isolated, dense, and compact with a well-defined base and top, and was not inclined in a strong wind shear. A further condition was that neither the selected cloud not any other within a radius of 30 km should show visible signs of either precipitation or glaciation. The quantity of rain that fell from the cloud was estimated from the numbers and sizes of raindrops recorded on a foil impactor sampling device flown beneath cloud base. The authors claim that the 11 clouds with tops colder than −10° C which were treated with the larger quantity of silver iodide yielded significantly more rain than similar, untreated clouds. However, in view of the very small sampling volume of the impactor device, the large sampling errors, and the fact that the total rain falling from a cloud was based on a rain sample of only a few cm³, this claim is not at all convincing.

A particularly well-conceived experiment designed to evaluate the effect of silver-iodide seeding upon individual cumulus clouds has been described by Simpson, Brier, and Simpson (1967), who regarded it primarily as an experiment in cumulus dynamics, formulated and analyzed in terms of a numerical model.

A randomized seeding operation was carried out on 23 tropical cumulus clouds over the Caribbean Sea on 9 days in the summer of 1965. Once a cloud has been judged suitable for seeding (one of the important criteria being that its summit temperature should lie between -5 and $-20°$ C), the decision whether to seed or not was made by opening an envelope that contained either a YES or NO instruction determined by random choice made in advance of the whole operation. An aircraft seeded 14 of the clouds with 8 to 16 pyrotechnic generators each releasing about 1·2 kg of silver iodide smoke. The nine remaining clouds were studied in an identical manner as controls, using the same four instrumented aircraft to penetrate the cloud before and after the

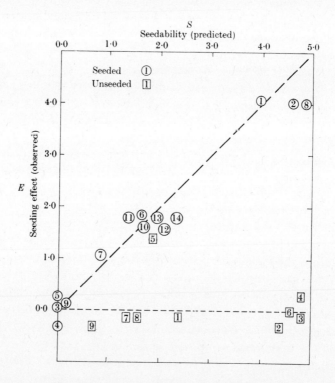

FIG. 7.3. Seeding effect versus seedability, each in kilometres, for all 23 suitable clouds in the 1965 programme. The heavy dashed line represents perfect predictability for seeded clouds, i.e. seeding effect equals seedability, while the light dashed line represents perfect predictability for control clouds, i.e. zero seeding effect independent of seedability. (From Simpson, Brier, and Simpson (1967).)

seeding run. Cloud growth was documented by using aircraft, radar, and photogrammetry.

A simplified numerical model of cumulus dynamics was specified in advance of the field programme, and was used to predict the rate-of-rise and actual heights of the cloud tops, and in-cloud properties of both seeded and unseeded cumuli, in terms of the measured vertical profiles of temperature, pressure, and humidity in the environment, the horizontal dimensions of the cloud tower, and the conditions at cloud base. The results are summarized in Fig. 7.3 where seedability, defined as the predicted difference between the seeded and unseeded top of the same cloud, is plotted against the seeding effect, defined as the difference between the observed top of the seeded cloud and the predicted top of the same cloud if unseeded. Both parameters were computed for all 23 clouds and are plotted in Fig. 7.3 to reveal that the seeded and unseeded clouds separate into two distinct classes. This diagram indicates that the model was quite successful in predicting the growth of both seeded and unseeded clouds under specified conditions, and that seeding had a clear-cut effect on cumulus growth.

In fact, 8 of the 12 properly seeded clouds (tops colder than $-5°$ C) grew vertically following seeding by amounts ranging from 2·2 to 5·3 km. Two others grew a little over 1 km, while two failed to grow at all. One of the nine control clouds grew 2·7 km, one grew 1·4 km, and the remainder grew <1 km following the 'seeding' run. The average growth of the seeded clouds was 1·6 km greater than that of the control clouds, this difference having a significance probability of 1 per cent.

7.2.3. *Seeding of cumulus with water drops and hygroscopic nuclei*

Attempts to release showers by spraying cumulus clouds with water and aqueous solutions were made before the theory of the coalescence mechanism was clearly formulated.

From South Africa there is a report (Anon. 1948) of 5 gal of calcium chloride solution being released over cumulus with tops at 18 000 ft on 20 March 1948. Three hours later, two isolated showers lasting approximately 30 min occurred in the seeded area, but in view of the long time delay, it is most unlikely that they were caused by the seeding. Similar quantities of water were released on two other occasions; on one, the seeded clouds produced radar echoes shortly after seeding, on the other, there were no observable effects.

In the Ohio operations reported by Coons, Jones, and Gunn (1948), 21 cumulus clouds, most of them growing at the time, were seeded with water, 15 of them at a rate of 1 gal/mile of flight and the remainder at 50 times that rate. It was estimated that the water broke up in the aircraft slip-stream to form drops of diameter exceeding 0·5 mm. Only two of the clouds that were not in the precipitation stage at the time of seeding produced rain after seeding. In only one case did the rain reach the ground, when the parent cloud was based at 5500 ft (temperature 17° C), with tops initially at 16 700 ft but continuing to grow after seeding. On the other hand, 17 of the clouds showed a tendency to dissipate after seeding.

Rather similar experiments, but using much larger quantities of water, were conducted in the Caribbean area by Braham, Battan, and Byers (1957). The water was released into the tops of the clouds at rates of up to 450 gal/mile through a large valve into the aircraft slip-stream. The drop size was estimated to range between 80 and 2000 μm with a predominant diameter of about 100 μm. The clouds were studied in pairs, one of each pair being treated and the other not. Elaborate precautions were taken to ensure independence between the choice of clouds for observation and the decision to seed or not to seed, the latter decision being made at random according to a pre-determined scheme. The treatment of 32 pairs of clouds at a rate of 130 gal/mile produced no marked effects; radar echoes were detected in only 6 of the seeded clouds and in 7 of the untreated ones. When 46 pairs of clouds were seeded at the rate of 450 gal/mile, 22 of the seeded, and only 11 of the unseeded clouds produced radar echoes, which suggests, perhaps, that seeding may have had some effect. The seeding of 17 clouds in the Central United States produced no apparent effect since only 4 of these gave radar echoes and precipitation was detected in 3 of the control clouds.

That the spraying of even 100 gallons of water into the *top* of a cloud cannot produce a shower unless a process of raindrop multiplication occurs, can be shown by the following simple calculation. If a shower cloud of base area 5 km^2 produces only 1 mm of rain, the equivalent volume of water is about 10^6 gal. If this were released by the introduction of 100 gal into the top of the cloud, the drops of this 100 gal would have to grow by a factor of 10^4 in mass. But even if the drops were to grow from an initial diameter of 0·5 mm to the maximum stable size of 5 mm, the mass growth factor would be only 10^3. For this to happen in a cloud with base temperature 20° C and containing the maximum theoretical liquid-water content, the minimum cloud depth would be 3 km; for the chain reaction to get under way and produce a shower, the required cloud thickness would be nearer 5 km; such clouds usually have a high probability of precipitating naturally.

A much more efficient method of inducing showers by water seeding is suggested by the theoretical treatments of the coalescence process by Bowen (1950), Ludlam (1951a), and described in § 6.5.2. They assumed the presence near the cloud base of some drops rather larger than average and traced their growth during their upward journey in the cloud and also during their return journey to the cloud base. A droplet of initial radius 30 μm, carried up from the base where the temperature is 20° C in a uniform updraught of 3 m s^{-1}, will return with a radius of 1·9 mm, the corresponding increase in mass being $2·5 \times 10^5$. A 20-μm droplet, growing by coalescence in the same conditions, will emerge with the maximum stable radius of 2·5 mm, the mass growth factor now being nearly 2×10^6. The minimum depth of cloud required to accommodate the trajectories of the drops will be about 2·5 km in each case. These calculations show that it is obviously much more economical to spray small droplets of radii 20–30 μm into the base of a growing cumulus and to capitalize on their subsequent growth during both their upward and downward journeys through the cloud than to introduce larger drops into the cloud summits. From Fig. 6.14 it is apparent that, from the point of view of obtaining maximum growth, it is advantageous to inject the smallest droplets that will have a finite collection efficiency for the cloud droplets, provided that the cloud has sufficient thickness and duration to accommodate their trajectories. The optimum droplet size for seeding will therefore depend on the speed of the updraught and the depth of the cloud but, in warm-base clouds containing updraughts of only 1 m s^{-1}, droplets of initial radius 30 μm may release a light shower within about 40 min, if the cloud depth exceeds 1·5 km (5000 ft).

Experiments along these lines have been carried out in Australia and reported by Bowen (1952b). Water droplets of median radius 25 μm were dispersed at the rate of about 30 gal/min from two horizontal spray bars fitted to an aircraft, during flights made about 1000 ft above cloud base. On six of the seven occasions when the cloud thickness was less than 5000 ft (1·5 km), light rain fell out of the cloud but evaporated before reaching the ground. Neighbouring untreated clouds did not precipitate. In the four cases where the cloud depth exceeded 5000 ft there was a considerable fall of rain or hail within a short time of seeding and, on three of these occasions, the neighbouring unseeded clouds did not rain. Thus it appears probable that the observed effects were due to seeding, but the number of experiments was small and, in view of the technical problems of dispersing sufficient quantities of water in small droplets, this technique has been abandoned.

In view of the important role which giant hygroscopic nuclei are believed to play in initiating the coalescence process, it would appear even more economical to use such nuclei rather than water droplets as a seeding agent. Since dry salt crystals of $d \sim 10$ μm will more than double their size while being carried up through the first few hundred metres of cloud, the dispersal of about 100 g of salt would be equivalent to that of a gallon of water in 50-μm diameter droplets. This method has been tried in East Africa by Davies, Hepburn, and Sansom (1952). Bombs containing a mixture of gunpowder and sodium chloride, carried aloft by balloons, were fused to explode near cloud base and disperse about 15 g of salt in particles of diameter 5–100 μm. For the 38 days when salt was released, the total rainfall in an area 6–12 miles downwind of the release point was estimated to be 6 inches in excess of that for the intermediate unseeded days, but the rainfall over an area extending 5 miles *upwind* was also greater by 2 or 3 in. In a later experiment, described by Sansom, Bargman, and England (1955), in which seeding was carried out on alternate days over a 3-month period, 24 of the 33 seeded clouds produced rain and the total rainfall directly downwind of the release point was 2–3 in. higher on the seeded days than on the unseeded days.

It appears, however, that in all these experiments, as in those carried out in Pakistan by Fournier d'Albe, Lateef, Rasool, and Zaidi (1955), and in India by Chaterjee, Murty, and Biswas (1968), with salt dispersed from ground generators, the number of salt particles injected into the clouds cannot have been sufficient to produce a detectable amount of rain even if each grew into a large raindrop. The minimum average concentration of giant nuclei likely to produce a significant effect would be about 1/litre. Thus a modest cumulus of volume 10 km³ would require some 10^{13} particles each of mass about 10^{-9} g, i.e. about 10^4 g of salt. The total output of the ground generator used by Fournier d'Albe *et al.* was only 5×10^9 particles/s so that, according to the measurements of Smith, Seely, and Heffernan (1955), the concentrations at cloud level 10 km downwind must have been two orders of magnitude below the requirement.

7.3. The experimental seeding of layer clouds

In seeding supercooled stratiform clouds from aircraft, the usual procedure has been to drop granulated dry ice at the rate of a few

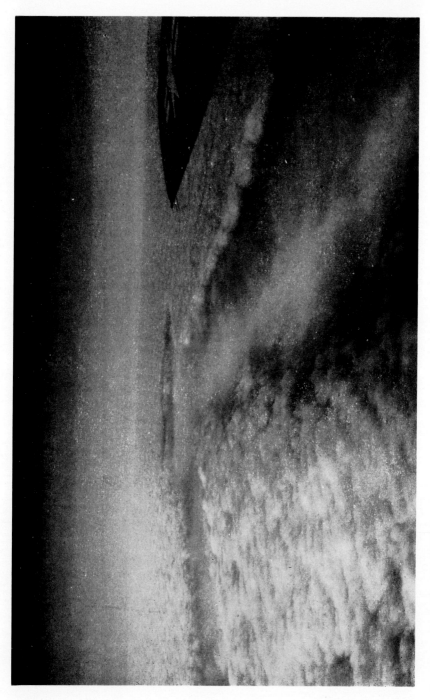

Fig. 7.4. A γ-shaped lane partly filled with ice crystals produced in a th n supercooled layer cloud by seeding with dry ice. The photograph was taken 36 min after the loop in the upper part of the picture was seeded. (Photo by courtesy of General Electric Company, Schenectady, U.S.A.)

pounds per mile of flight along a well defined track into the top of the cloud deck, and to photograph subsequent developments from above. The clouds are usually too thin to produce precipitation at the ground, but seeding often produces spectacular changes in the appearance of the cloud. Along the seeded track the cloud becomes completely transformed into ice crystals. Gradually the turbulent motions in the cloud layer diffuse these crystals laterally so that after a few minutes a long lane a few hundred yards broad becomes affected. After a while, the crystals at the edge of the lane become so diffused that each has a large share of the available water and grows large enough to fall out of the cloud layer into the drier air beneath. The crystals then evaporate, and the cloud disappears in a broad strip around the seeded track to leave a long hole in the cloud.

Fig. 7.4 shows the result of such an experiment carried out by the American Project Cirrus in November 1948. An extensive layer of stratus with base at 6000 ft, temperature $-3 \cdot 5°$ C, and top at 7300 ft, temperature $-5°$ C, was seeded with dry ice at the rate of about 0·7 lb/mile. The seeded track took the form of a letter γ with a total length of about 57 miles. The photograph shows the appearance of the pattern 36 min after the loop in the upper part of the picture was seeded. At this time the width of the track to the right was 1·7 miles. An observer on the ground saw curtains of snow falling from the edge of the open lanes. Ultimately new clouds developed in the openings, so that within 2 h they were more than half filled, but even after 4 h some openings were still visible.

Many similar experiments have been carried out in Canada, England, Russia, and elsewhere, with similar results. Whether or not a hole develops depends upon the rate of seeding, the temperature, depth, and water content of the cloud, the extent to which air is converging or diverging into the region, and the intensity of lateral mixing, the area affected depending largely on the last two factors. In a favourable situation the seeded lanes may spread to widths of 2 miles in half an hour.

The most systematic investigation was carried out by Aufm Kampe, Kelly, and Weickmann (1957) in the Maine region of the United States, where there is a high incidence of supercooled, stratiform cloud decks. Their primary object was to investigate the possibility of clearing appreciable areas of cloud by seeding. Two aircraft were used; the seeding aircraft carried equipment for measuring temperature, visibility, cloud-droplet size, liquid-water content, etc., and the second was used for photographing the clouds from above. Preliminary trials made in the Spring of 1952 on stratocumulus decks colder than $-5°$ C, of thickness

600 to 7500 ft, indicated that such layers, up to 3000 ft in depth, could be modified by dropping dry ice at rates varying between $1\frac{1}{2}$ and 4 lb/mile, so that the ground could be seen through at least part of the seeded region. Clouds forming in areas under the influence of an anticyclonic regime (regions of horizontal divergence and stable lapse rate) could be cleared fairly readily with small amounts of dry ice, the seeded lanes generally spreading to widths of 2 miles within half an hour. In cyclonic or convective conditions, however, only small regions of the cloud could be cleared. In these circumstances, clouds generally formed again in the seeded area within the hour, unless the whole cloud layer dispersed. In later experiments, particular attention was paid to the problem of overseeding and to determining ways and means of clearing an area of approximately 100 square miles. The conclusions drawn from the preliminary experiments were very largely confirmed. The authors state that thin cloud decks colder than $-5°$ C may be cleared equally well with seeding rates between 1 and 20 lb/mile, and that a thick, convective layer may be underseeded with rates as high as 10 lb/mile. In thin clouds ($<$3000 ft thick), a large area may be cleared within about 30 min by seeding along parallel lines spaced about 2 miles apart. In thicker cloud layers, holes may be made by seeding, but their size and location are much more difficult to control. The thicker the cloud deck, the more time will elapse before the ground can be seen, the minimum time being about 10 min for a cloud depth of 500 ft and about 25 min for a depth of 2000 ft.

In very cold weather it may be possible to disperse entire layers of fog and low cloud by releasing suitable seeding agents from the ground. For example, Dessens (1952) treated a fog 200 m in depth, when the ground temperature was $-7°$ C, by dispersing 100 g of silver iodide as a smoke over 1 hour. Soon after the generator was lit, ice crystals fell in great numbers and, for an hour, the fog became thin and patchy over an area of about half a mile across and a mile downwind, but re-formed when the generator was extinguished. Similar successes are reported from Russia by Borovikov et al. (1961), who also describes how supercooled fogs have been cleared over areas of tens of km^2 by a mobile dispenser that disperses fragments of CO_2 snow, produced by expansion of the gas through a nozzle, to a height of 10–12 m.

7.4. Large-scale cloud-seeding operations

The experiments just described have clearly demonstrated that the introduction of seeding agents into suitable clouds may be followed by observable, and sometimes spectacular effects. They do not, however, tell us very much about the possibilities of modifying the rainfall from widespread cloud systems extending over many thousands of square miles. For operations on this scale, it has been common practice to release silver iodide in the form of a smoke from ground generators and to rely upon the air currents to carry it up into the supercooled regions of the cloud. Trials of this type have been carried out on a large scale during the past 20 years, mainly in the United States, Canada, Japan, and South America by commercial operators employed by ranchers, farmers, power companies, and public utilities. Because of the lack of control over the concentration of nuclei reaching the clouds from ground generators, and because of the limited life of silver iodide in the atmosphere, there has been an increasing tendency in recent years to use aircraft equipped with suitable burners to release the silver iodide at cloud level. In evaluating the results of all such large-scale trials, the vital question is: is it probable that the precipitation pattern that appeared immediately following seeding would have appeared even if no seeding had taken place? The difficulty lies in the inherent variability of natural precipitation and the meteorologists' present inability to predict with sufficient accuracy what would have occurred in the absence of treatment. In places having a fairly plentiful rainfall, say 80 cm/year, the rainfall in any one year may easily deviate from the average over a 50-year period by 15 or 20 per cent, while in dry regions like Arizona the variation may be as much as 50 per cent. It is in such dry regions, where rain-making is most in demand, that it is most difficult to assess the efficiency of cloud seeding because of the large year-to-year variation in the natural rainfall.

7.4.1. *Evaluation procedures*

Since we are still far from being able to form an accurate estimate, on a purely physical basis, of how much rain will fall naturally in a given region, the effects of seeding must necessarily be assessed statistically. The target area having been decided, the usual procedure is to choose an adjacent control area, preferably of much the same size, shape, and topography and unlikely to be affected by the seeding agent. One method of evaluation is then to plot, for past years, the annual target-area rainfall against that for the control area and, knowing the rainfall in the control area during the period of seeding operations, use this graph to predict the natural rainfall in the target area. The difference between the latter and the actual amount is then attributed to the

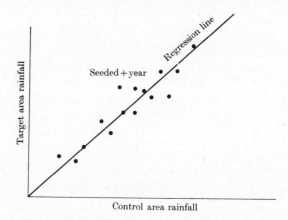

Fig. 7.5. A typical correlation between the rainfall in the target and control areas of a cloud-seeding operation.

effects of seeding with a degree of confidence determined by the scatter of the historical data on the graph (see Fig. 7.5). In other words, the assessment takes the form of determining whether there are significant departures from the normal historical relationship between the rainfall in target and control areas and, clearly, an apparent departure can be statistically significant only if there is a high correlation between past rainfall in the two areas.

Although this method of evaluation may seem very reasonable, it is open to the serious criticism that if the pattern of storms during the seeding period differs appreciably from the long-term average pattern on which the historical relationships are based, then the latter cannot be used to predict accurately the rainfall in the target area, and consequently the assessment of seeding effects may be most misleading. For these reasons, statisticians have insisted that an acceptable form of randomization should be incorporated into the operation from the beginning; for example, seeding should be carried out on only half the suitable occasions selected at random, the other half being used as a control. After it has been decided on meteorological grounds that a storm is suitable for seeding, the result of a previously performed randomized experiment (such as tossing a coin) decides whether or not seeding shall take place. Evaluation then takes the form of a comparison between those storms which were actually seeded and those which were not. The rainfall per day or per storm falling in the target area is plotted against the corresponding values for the control area, using the historical records for a large number of past storms of the same general type.

This graph is then used to predict the natural rainfall to be expected in the target area on the occasions of seeding knowing that in the control region. The difference between the actual and predicted values of the target-area rainfall may be attributed to seeding with a degree of confidence evaluated by carrying out the same procedure for the non-seeded occasions, when the differences between predicted and observed rainfall may be attributed to natural variations. If it should prove impracticable to find a suitable control area, the unseeded occasions may be used as the only control, but then the experiment will have to be carried out for a longer period in order to achieve the same degree of confidence in the results.

In another variant, the so-called cross-over design, there are two target areas, say A and B, sufficiently close to ensure a high correlation between normal precipitation amounts, and yet sufficiently distant so that seeding in A is unlikely to influence the precipitation in B and vice versa. Seeding is done on every suitable occasion, but the decision as to which of the two targets A and B is to be seeded is determined by a randomized procedure. This cross-over design is the most efficient of the procedures just described, provided that seeding in one target area does not influence the precipitation in the other. On the other hand, if there were strong interaction between the two areas, the conclusions drawn from the analysis of the cross-over design could be misleading or erroneous.

Unfortunately randomized procedures have not been adopted in the great majority of commercial rain-making operations and so it is not possible to assess their results on an objective scientific basis. Accordingly we shall not attempt to summarize the vast literature in this field, but rather select for description and discussion a representative group of carefully-conducted randomized large-scale operations that will not only illustrate the operational and evaluation procedures, but also give a fairly balanced picture of the present state of the art.

7.4.2. *Seeding with silver iodide from aircraft*

(a) *A group of operations conducted in Australia.* A 5-year randomized operation, carried out from June 1955 to the end of 1959 in the Snowy Mountains of Australia, is described by Smith, Adderley, and Walsh (1963). The objective was to determine whether seeding with silver iodide can produce an economically significant increase in precipitation over the Snowy Mountains at about 6000-ft altitude where it can be utilized for generating hydroelectric power. Aircraft seeding was employed to make sure that the silver iodide actually entered the cloud systems involved, the output of the airborne generator being about 10^{17} nuclei/h, active at $-17°$ C. Cumulus clouds were seeded at near the $-6°$ C level. The central target area was only 35 miles^2 but a surrounding area of 1100 miles^2 was thought to be affected by the seeding. The control area was 750 miles^2. Seeding was confined to the non-summer months when the dominant cloud systems were associated with cyclonic and frontal disturbances, the vertical motion of the air being intensified by orographic lifting. Operations were divided into 'seeded' and 'unseeded' periods each of not less then 8 days' duration, the division being made on the basis of a set of random numbers. This series was unknown to the individual who decided the end of each period and the beginning of the next on the basis of the passage of successive anticyclones across the region. Analysis of the results of the 5 years' operations was based on the total precipitation of the three areas as estimated by integration of isohyetal maps prepared from rain-gauge data. Over the 5 years, the ratio of the total precipitation in the target area to that in the control area during the seeded periods exceeded by 19 per cent that in the corresponding unseeded periods. The excess was judged significant at the 3 per cent level according to one statistical test, and at the 9 per cent level according to another. But, for individual years, the excess of target over control-area rainfall varied from 3 to 27 per cent. The corresponding figures for the 'seeded' area surrounding the target area were -4 to $+19$ per cent, with a 5-year average of 11 per cent judged to be significant at either 4 or 24 per cent, according to the nature of the test. In view of these figures, the strong differences between the rainfall regimes in the different years, the fact that the target area is at a higher elevation and has a considerably higher natural spring rainfall than the control area, and because of the difficulty involved in measuring the precipitation, much of which was snow driven by strong winds, the overall result of the operation must be judged inconclusive.

In three other locations, South Australia, New England, and the Warragamba Catchment, the target-control cross-over design was used, the random process determining which of two areas should be seeded, and which should not, during each period. The results of all three operations have been summarized by Smith (1967) and are condensed in Table 7.1, where the effect of seeding was calculated from the ratio

$$\frac{p(s)}{p(ns)} = \left\{\frac{p(A, s)}{p(A, ns)} \times \frac{p(B, s)}{p(B, ns)}\right\}^{\frac{1}{2}}, \qquad (7.1)$$

where $p(A, s)$, $p(B, s)$ are respectively the average totals of precipitation falling in areas A and B on days of seeding, and $p(A, ns)$, $p(B, ns)$ the corresponding totals on unseeded days.

The indicated increase of only 4 per cent in New England and the small decreases in the other two experiments are hardly significant and may well have occurred by chance.

Pointing out that the largest apparent increases in precipitation due to seeding tend to occur in the early years of an operation, and are often offset by apparent decreases in later years, Bowen (1966) has suggested that seeding may produce persistent effects that may carry over from one year to another.

TABLE 7.1
Results of Australian seeding operations

Area	Duration (years)	$p(s)/p(ns)$	Significance probability (one-sided)
Snowy Mountains	5	1·19	0·05
South Australia	3	0·95	0·7
New England	6	1·04	0·1
Warragamba	4	0·97	0·6

(b) *Seeding of orographic cumulus in Arizona.* Two experiments, conducted to investigate the possibility of modifying the precipitation from cumulus clouds that form almost daily in summer over the mountains near Tuscon, Arizona, have been described by Battan (1966) and Battan and Kassander (1967). The seeding was randomized by pairs of days, i.e. when a forecast based on objective criteria indicated that the following 2 days would have clouds suitable for seeding, the decision whether the first or second day should be seeded was made by a random process. In the first experiment, in 1957–60, silver iodide smoke was released from an aeroplane flown at the $-6°$ C level along a line perpendicular to the wind, upwind of the target area. Rainfall was measured by a network of 29 recording rain-gauges during a 5-hour period that included the seeding period lasting 2–4 hours. The results are shown in Table 7.2. Increases in rainfall were indicated in the first two years, and even larger decreases during the last two years. All the data for the whole 4-year period indicated an overall decrease of 30 per cent, but at a level of significance that suggested that this result could have occurred by chance. Indeed, much of the large decrease for 1959 was due to extremely large falls of rain on one non-seeded day; if this pair of days is excluded from the analysis, the indicated overall decrease over 4 years falls to only 7 per cent.

TABLE 7.2
Results of silver iodide seeding of convective clouds in Arizona

Year	Number of pairs	Mean rainfall per station		S/NS
		Seeded days	Non-seeded days	
1957	16	0·067	0·059	1·14
1958	16	0·059	0·041	1·44
1959	20	0·026	0·094	0·28
1960	17	0·018	0·034	0·51
All data	69	0·041	0·059	0·70
with August 17,18, 1959 excluded	68	0·042	0·045	0·93
1961	17	0·035	0·106	0·33
1962	7	0·029	0·039	0·74
1964	13	0·101	0·072	1·40
All data	37	0·057	0·082	0·70

Following a preliminary analysis of the first experiment, a new experiment was started, using more restricted criteria for selecting the pairs of days in order to reduce the frequency of experimental days with zero rainfall, seeding now being carried out at a few hundred metres below cloud base, instead of at the $-6°$ C level. The target area was reduced and the number of rain-gauges in it was increased. The results of this experiment, which was carried out in 1961, 1962, and 1964, were much the same as those of the first, and indicate a net decrease of 30 per cent in rainfall over the 3-year period but, with significance probabilities estimated to lie between 0·16 and 0·30, this could well have occurred by chance.

(c) *Project 'Whitetop' (1960–4), Missouri, U.S.A.* This was a carefully designed long-term experiment to test the effect of seeding with silver iodide from aircraft on the release of precipitation from non-orographic summer cumulus clouds in Missouri. The days were identified as seedable on the basis of objective criteria indicating the likelihood of instability showers. If the quantity of precipitable water in the atmosphere shown by the morning ascent of three nearby radio-sonde stations exceeded 3 cm, and the wind direction at 1300 m was between 170° and 340°, the day was designated operational, and a random process determined in advance decided whether seeding should take place or not. On seeding days, aircraft equipped with silver iodide burners flew to and fro along a seeding line 30 miles long and 45 miles upwind of a central radar site, the smoke being released at the level of cloud base from noon until 1800 Local Time. On both seeded and unseeded days, this line was used to define 'Chicago-plume' positions on the basis of the most divergent winds between the seeding level and 4500 m. Narrower plumes, based on the winds at the seeding level, were called 'Missouri plumes'. The target area was considered to be the area covered by these plumes, and comparisons were made between the precipitation in the plumes on seeded and on unseeded days, and between precipitation in the plumes and in the remainder of the circular experimental area, 60 miles in radius, which was labelled 'out of plume' area and considered to be uncontaminated by the silver iodide.

The rainfall data were obtained from a rather sparse network of rain-gauges (one gauge per 700 km^2) and subjected to three independent statistical analyses that are summarized in Table 7.3. All the analyses, though treating the data rather differently and applying different tests of significance, nevertheless agree that there was a significant *decrease* in the precipitation on seeded, as opposed to unseeded days, both in the Missouri plume and the Chicago plume, amounting to 60 per cent and 40 per cent respectively. Precipitation was also less in the 'out of plume' area of $>10^5$ miles^2 on the seeded days, but this result carries a lower level of statistical significance. In a later study, Neymann, Scott, and Wells (1969) extended the rainfall analysis up to a radius of 180 miles and found an overall decrease of about 20 per cent on seeded days averaged over the entire area, but the significance probability was only 0·13.

TABLE 7.3
Statistical analysis of the Whitetop experiment

Author	Specification	Missouri plume	Chicago plume	Out of plume
Decker and Schickedanz (1967)	Mean rain per expt. day	−52%	−39%	−24%
	Significance probability (P)	<0·01	0·14	0·06
Braham and Flueck (1967)	Frequency of wet days	+24%	+13%	+19%
	Mean rain per wet day	−58%	−43%	−31%
	Mean rain per expt. day	−52%	−39%	−24%
Neyman and Scott (1967)	Frequency of wet days	+17%	+9%	+9%
	P (two-sided)	0·49	0·60	0·45
	Mean rain per wet day	−57%	−44%	−29%
	P	0·014	0·040	0·12
	Mean rain per expt. day	−50%	−39%	−22%
	P	0·076	0·11	0·29

(d) *The Israeli experiment.* A randomized cross-over experiment, involving the seeding of clouds, mostly cumulus, with silver iodide from aircraft, was carried out in Israel during 1961–6, and has been described by Gabriel (1967) and Gabriel, Avichai, and Steinberg (1967). The trial involved two experimental areas, each about 50 km in maximum dimension, for which the correlation between daily amounts of precipitation was 0·81. They were separated by a buffer zone about 20 km wide. Silver iodide was released at the rate of 800 g/h from a single aircraft flying just below cloud base in an area displaced upwind of the target by a distance equivalent to 0·5 h run of the wind. Each day, one area was designated for seeding on the basis of a randomized decision taken in advance, but only if suitable clouds were present was seeding actually carried out.

The effect of seeding was expressed by the ratio of the average daily precipitation falling on seeded and non-seeded days and was calculated from eqn (7.1). The seasonal values of this ratio, calculated from data on all the experimental days, varied from 1·01 to 1·65. If only rainy days, defined as days on which rain fell in the buffer zone, were included in the analysis, the seasonal ratios ranged from 1·02 to 1·74. Over the $5\frac{1}{2}$-year period as a whole, the statistical analysis indicated a rainfall increase of 18 per cent. Most of this increase could be attributed to large effects occurring on only a few days. This is one of the very few long-term randomized experiments that appear to have produced significant increases in rainfall.

Gabriel, Avichai, and Steinberg (1967) carried out a number of statistical tests on their data but found no evidence for the persistence of seeding effects, either from day to day, or within each season, or from season to season.

7.4.3. *Seeding with silver iodide from ground generators—Project 'Grossversuch III'*

One of the very few well-designed large-scale operations using ground-based silver iodide generators was the Swiss Project 'Grossversuch III', whose primary objective was to reduce the incidence of large damaging hail from cumulonimbus clouds growing over the southern slopes of the Alps.

The experiment was fully randomized as described by Thams *et al.* (1966). Experimental days were those during the summer months of the seven years 1957–63 for which thunderstorms were forecast on the previous afternoon. The decision whether or not to seed on a particular experimental day was taken by selecting at random an envelope containing a YES or NO instruction. In this way, 292 experimental days were selected during the seven years, seeding being conducted on 145. The seeding was carried out by 20 silver iodide generators operating from 0730 to 2130 hours on a pulsating schedule of 5 min on and 10 min off. The generators were located in, and to the south of, the 3500-km^2 target area ranging in altitude from 200 to 3400 m.

According to Schmid (1967), the experiment was a failure in the sense that the results showed no significant difference in the duration, areal extent, or intensity of hail on seeded and unseeded days, but the frequency of hail on seeded days was considerably *greater* than on unseeded days, with a probability of only 4 per cent that this difference could have occurred by chance. As far as precipitation was concerned, however, the average rainfall on seeded days was 21 per cent higher than on unseeded days, with a significance level bordering on the acceptable. The data were re-analysed in greater detail by Neymann and Scott (1967), who found an average increase of 33 per cent in rainfall, significant at the 2 per cent level if all the experimental days were included in the analysis; if only rainy days were included, the average increase was 41 per cent significant at the 1 per cent level. Schmid also presents evidence that seems to suggest that seeding may have produced quite large increases in rainfall at distances of up to 100 miles from the target area. This is by no means proved, but should seeding be capable of producing such long-range effects as suggested both here, and by the analysis of Project Whitetop, the results of experiments employing the cross-over design would be open to serious doubt.

7.5. Suppression of large hail

In recent years, projects aimed at suppressing large, damaging hailstones have been carried out in several countries, for example, Austria, Argentina, Bulgaria, Canada, France, Italy, Yugoslavia, Kenya, Switzerland, Russia, and the United States. The most extensive operations are conducted in the Aquitain Basin of south-west France (Dessens 1968); started on a small scale in 1951, this operation employed 240 ground-based silver iodide generators in 1967 to treat an area of 7 million hectares. The best designed experiment, in that it was properly randomized, was the 'Grossversuch III' project in Switzerland described above. The most intensive operation was mounted in Kenya (Sansom, 1968), where 10 000 rockets, each containing about 800 g of TNT, were exploded to treat a total of about 150 storms in an area of only about

1500 hectares. The rockets contained no nucleating agent, but it has been suggested that the shock waves from the explosion might shatter the large hailstones into smaller, less damaging fragments, or produce large numbers of ice crystals by adiabatic expansion and cooling of the air. None of these three projects nor some others for which detailed reports are available provide convincing evidence of success in reducing either the frequency or intensity of hail. Indeed the scientists in charge of the operations in Switzerland and Argentina considered them failures, while the National Science Foundation annual report on weather modification for 1968 makes no claim for success in the United States.

The working principle behind most of the trials is that, if a growing supercooled cloud is seeded with a massive dose of ice nuclei, the competition for the available water between the high concentration of ice particles will prevent any of them growing into large hailstones. One of the major doubts about most of the practical trials, especially those employing ground generators, is whether artificial nuclei were able to reach the supercooled parts of the cloud in anything like sufficient concentrations. The most impressive claims for success come from the Soviet Union, where large-scale operations have been conducted in Georgia, Armenia, and the Caucasus since 1962, and in Moldavia since 1964. Recent accounts of this work are given by Gaivoronskii, Seregin, and Voronov (1968), Kartsivadze (1968), and Sulakvelidze (1968).

Soviet scientists believe that an important feature of a hail-producing cloud is the formation of an 'accumulation zone' where the concentration of liquid water becomes very high as the result of the updraught reaching its maximum speed at this level. If the accumulation zone reaches above the $0°$ C isotherm, hailstones are assumed to originate by the freezing of supercooled raindrops and thereafter are able to grow to diameters of 3 cm within 4–10 min, depending on the concentration of supercooled water and the concentration of particles competing for it. There is, in fact, no real evidence for the existence of persistent accumulation zones and it is not clear how they could be maintained for several minutes once precipitation has developed.

Nevertheless, the technique of delivering the seeding agent into the heart of an incipient storm is sound, and the organization and logistics of the Soviet experiments are impressive. They aim to introduce silver iodide in the clouds between the -6 and $-12°$ C levels, either by rocket, as in the Georgian operations, or by artillery shell, as in the Caucasus. The shells, 12·5 cm in diameter and weighing 33 kg, have a maximum range of 15 km and contain 100 g of silver iodide that yields about 10^{15} ice nuclei active at $-10°$ C. Field tests show the nuclei to be dispersed through a volume of order 10 km^3 to give an average concentration of order 100/litre. One shell is fired for every 5 km^3 of the estimated volume of the supercooled part of the cloud, so that a large cloud may receive 1 kg of AgI within about 1 h. Operations in the Caucasus employed 16 artillery batteries in 1969 to protect an area of 5 million hectares. The relative advantages of shells and rockets is a matter of some dispute, but the rockets have only half the range of the shells and cannot be controlled with the same degree of accuracy.

The probability that large hail will develop is assessed on a number of criteria based on radar measurements. These include the height and temperature of the top of the radar echo, its vertical extent, and maximum reflectivity, and the relative vertical depths of the echo above and below the 0° C level, but it is not known how these various factors are weighted. Sulakvelidze (1968) claims that by measuring the reflectivity of the radar echo on two wavelengths, 3·2 cm and 10 cm, simultaneously, it is possible to detect the presence of large concentrations of hail greater than 0·6 cm in diameter and to estimate their average diameter on the assumption that they are monodisperse, spherical, and wet. In practice, none of these three assumptions is likely to be generally true and consequently, as pointed out in § 8.2.4, it is difficult to obtain reliable information on hailstone sizes from radar data alone.

None of the Soviet experiments is randomized, so the results are assessed on subjective judgements of crop losses occurring in the protected regions compared with those in unprotected control regions. It is claimed that losses have been reduced by as much as 80–90 per cent in the Caucasus and Georgia, with economic benefits estimated at between 10 and 40 times the cost of the operations. These claims are difficult to assess in the absence of proper statistical tests, and in view of the fact that suitable comparison areas are very difficult to obtain. However, the author, in a recent visit to one of the sites in the Caucasus, was impressed by the enthusiasm of the agricultural industry which is convinced of the efficiency of these operations.

7.6. Discussion

Looking then at the world-wide attempts to achieve some measure of weather modification and control, one can find little convincing evidence that large increases in rainfall can be produced consistently over large areas. Although it is possible that marginal, but still economically important effects of, say, 10–20 per cent, may be produced, in general it has not been possible to distinguish induced changes of this magnitude from the natural variations in rainfall.

Apart from doubts about the validity of the evaluation procedures that have been used to assess massive cloud-seeding operations, these are conducted on certain basic assumptions that are insecurely based, as may be illustrated by our present inability to answer the following important questions.

(a) How often in a given locality, and under what conditions, are natural ice nuclei deficient, and when and in what quantities should they be supplied artificially?

(b) In which part of the cloud or cloud system, and at what stage of its development, should the seeding agent be introduced in order to stimulate the maximum effect? May not seeding sometimes stimulate the premature release of precipitation, suppress the natural growth of the cloud, and thereby reduce rather than increase the precipitation from it?

(c) When silver iodide smoke is released from the ground, in what concentration does it reach the supercooled regions of the cloud and how will this vary with the meteorological situation and the distance from the source?

(d) For how long does the nucleating agent remain effective under specified atmospheric conditions?

Even more important, it is difficult to escape the conclusion that, once initiated, the distribution, intensity, and duration of precipitation are largely controlled by the cloud dynamics, and influenced by air motions on scales both larger and smaller than that of the cloud itself. Perhaps the greatest weakness in our present knowledge and approach to cloud modification is our poor appreciation of the relative importance of microphysical and dynamical processes in controlling the efficiency of natural precipitation mechanisms.

At the present time, when the efficacy of large-scale seeding has yet to be firmly established despite 20 years of effort, one can only indicate some of the possibilities that may arise from improved knowledge and techniques.

In middle latitudes, much of the precipitation falls from deep layer-cloud systems whose tops usually reach to levels at which there are abundant natural ice nuclei and in which the natural precipitation processes have plenty of time to operate.

Here it is relevant to mention the outcome of Project Scud, described by Spar (1957), in which massive seeding with both dry ice and silver iodide was conducted over the east coast of the United States during the winters of 1953 and 1954 to determine the effects of seeding on the development of cyclones. On 19 out of 37 occasions of potential cyclogenesis, seeding was carried out on a randomized schedule, the other 18 occasions being used as controls. Altogether, 250 lb of silver iodide and 30 tons of dry ice were used in the operation. A careful statistical analysis revealed that if seeding produced any effects on either the total precipitation over the area or on the sea-level pressure field, they were too small to be detected against the background of the natural variance. A similar result also emerged from an operation in the State of Washington where the seeding, with dry ice, of migratory cloud systems associated with cyclonic activity produced no detectable effects. Indeed Hall (1957) states that favourable seeding opportunities were rather infrequent because of the abundance of natural ice crystals in the tops of the deep cloud layers.

The prolonged, steady release of precipitation from the deep clouds of a well-established warm front, in which there is little or no storage of supercooled water above the $-5°$ C level, suggests a quasi-steady state, dominated by the dynamics, in which the processes of nucleation and particle growth adjust themselves to match the rate at which moisture is released by the vertical motion. If this be so, seeding is unlikely to have a major effect on the intensity or duration of the precipitation from these mature storms. It is possible, however, that judicious seeding of the cloud during the early stages of its development, when thick layers of supercooled or mixed cloud may have already formed, could forestall the natural release of precipitation and thereby effect some redistribution on the ground. But the scope seems rather limited and an elaborate surveillance system would be required to achieve the best timing of such an operation.

Perhaps more promising as sources of additional rain or snow are the persistent, supercooled orographic clouds produced by the ascent of damp air over large mountain barriers. The continuous generation of an appropriate concentration of ice crystals near the windward edge might well produce persistent light snowfall to the leeward, since water vapour is continually being made available for crystal growth by the lifting process. The condensed water, once converted into snow crystals, has a much greater opportunity of reaching the mountain surface without evaporating, and might accumulate in appreciable amount if seeding were maintained for many hours.

A semi-quantitative assessment of the potentialities of inducing snowfall from orographic clouds forming over the central Swedish mountains was made by Ludlam (1955). He first of all calculated the rate of growth of ice crystals forming on silver iodide nuclei introduced into the windward edge of a simple orographic cloud at a height of several hundred metres above the ground where the temperature is about $-7°$ C. The essential requirement is that while being carried through the cloud by the wind, the crystal shall acquire a sufficient size and fall speed to fall on the mountain before the air has begun to descend in the lee when the contained crystals would tend to evaporate. Ludlam shows that for this to occur, the crystal must remain in the cloud for about 1500 s, during which time it will have acquired a size equivalent to a water drop of 150-μm radius. In winds of 5 to 15 m s^{-1}, this implies that the silver iodide nuclei should be introduced into the cloud some 8 to 23 km to the windward of the mountain. It is estimated that if all the water condensed in the orographic updraughts could be precipitated artificially in this way, then on the basis of the available data on the frequency and duration of supercooled clouds over the central Swedish mountains, an additional $2 \cdot 5 \times 10^8$ m^3 of water might be released on a 50-km front during a single winter. If this were deposited on a strip 10 km wide in the direction of the wind, the normal snow cover would thereby be doubled, the increase in precipitated water being about 60 cm. Increases of this magnitude should be easily detected, but in practice, such an operation would probably be far from 100 per cent efficient, and consequently increases in snow pack or in the spring discharge of the rivers would be much less.

However, this type of large-scale operation appears to be well suited for scientific investigation since there are good physical reasons for believing that seeding, skilfully conducted, might produce *observable* effects.

Rather different considerations apply in the case of incipient shower clouds where seeding could produce quite opposite effects depending upon the dynamical structure of the cloud, its state of organization and development, and such important environmental factors as the wind shear and humidity. In a dry environment, potentially unstable over only a limited depth, the cloud may have a high probability of dissipating without precipitating naturally; in such a case, acceleration of particle growth by seeding may increase the probability of releasing a shower, albeit only a light one. On the other hand, if a cloud growing in a moist, highly unstable environment is seeded early in its career, the premature release of precipitation may well destroy the updraught and cause the cloud to dissipate before reaching its maximum potential. This latter possibility has not received sufficient attention, but it may account, at least in part, for the fact that several large projects, for example, in Australia, Arizona, and Missouri, involving the seeding of cumuliform clouds, have led to an apparent *decrease* in the rainfall over the target area.

Another possibility is that glaciation following seeding may lead to enhanced growth of the cloud stimulated by the release of latent heat. Although explosive growth of the cloud, in the manner observed in the famous early experiment of Kraus and Squires (1947), is a rather rare event, careful observations such as those of Simpson, Brier, and Simpson (1967) suggest that modest growth may quite often follow seeding.

The feasibility of decreasing rainfall by 'overseeding' clouds has also to be considered. If the concentration of ice crystals in a cloud were to exceed 1 cm^{-3}, none would be able to grow sufficiently large to fall out of the cloud and reach the ground as precipitation. The introduction of an excessive concentration of ice nuclei into supercooled clouds might therefore retard or prevent the development of precipitation and suppress the formation of large hail and lightning. It is unlikely that 'overseeding' has been regularly achieved in any large-scale operation using ground-based generators; indeed it seems impracticable to ensure persistent over-seeding of active, moving storms by this means. Delivery of the seeding agent directly into the core of the cloud by rocket, shell, or aeroplane seems the only feasible method; accordingly, the technique pioneering by the Russian scientists is likely to be adopted elsewhere.

The greatly exaggerated claims made by the early cloud-seeding operators are now largely discredited; current claims are much more modest and the difficulties of executing and evaluating cloud-seeding operations are now much more clearly and widely recognized. Statistical assessments of these operations are continually revealing features in the rainfall patterns that are difficult to interpret and explain, even qualitatively, in meteorologically convincing terms. The formulation of reliable numerical models of cloud growth and development, capable of predicting important and readily observed characteristics of seeded and unseeded (control) clouds, offers the best long-term prospect of assessing the potentialities and results of cloud-modification experiments, and should provide deeper physical insight into the problem than the purely statistical approach. However, it would be optimistic to expect the early development of models capable of predicting precipitation amounts with sufficient accuracy to detect the changes of 10–20 per cent that are currently being attributed to cloud-seeding operations. Assessment of such experiments will therefore continue to depend on a combination of statistical and physical criteria, but it is on the latter that the greatest effort is now required. Little further progress is likely until we have acquired a much deeper understanding of the physical, and particularly the dynamical processes that control the release, duration, and intensity of precipitation from the major types of cloud systems. The nature and magnitude of the problems have recently been discussed by Mason (1969).† Their solution is essential, not only for the intelligent and efficient planning, execution, and evaluation of cloud modification experiments but, perhaps even more important, for accurate quantitative forecasting of precipitation.

† Some outstanding problems in cloud physics—the interaction of microphysical and dynamical processes. *Q. Jl. R. met. Soc.* **95**, 449.

8

Radar Studies of Clouds and Precipitation

T H E advent of radar provided the cloud physicist with a powerful new tool. The fact that hydrometeors are able to back-scatter sufficient of the incident electromagnetic energy to be detectable by a sensitive radar receiver many miles distant, enables him to study the distribution, movement, structure, and development of precipitating clouds and to obtain useful assessments of rainfall over considerable areas. The development of radar meteorology, during its early descriptive phase in the 1940s and early 1950s, and during the following decade, which was devoted largely to semi-quantitative interpretation of the character and intensity of the echoes from precipitation, has been described in comprehensive reviews by Marshall, Hitschfeld, and Gunn (1955), Marshall and Gordon (1957), and Battan (1959). The early 1960s saw a very important new development in the use of Doppler radar to determine the falling speeds and hence the size and growth of hydrometeors, and also the vertical and horizontal air motions inside precipitating clouds. The early work in this area, together with progress in other aspects of radar meteorology during the period 1958–63, were reviewed by Atlas (1964). By using Doppler and conventional radars together with measurements from aircraft, radio-sondes, and satellites, we can look forward to greatly improved knowledge of both precipitation physics and cloud dynamics and, particularly, of the interrelations between them. This is likely to be one of the most important areas of development in cloud physics during the next decade.

There has also been a great increase in the application of radar to practical meteorological problems such as storm detection and short-period forecasting. These matters are too extensive and technical to receive detailed treatment in this chapter, which will be devoted largely to the basic theory of back-scattering and attenuation of microwaves by clouds and precipitation, and to the principles of some of the more important techniques that have been used to elucidate the nature, size, and growth of hydrometeors.

8.1. Basic radar theory

8.1.1. *Calculation of back-scattered power from a target*

The basic principles of radar operation are simple. Electromagnetic energy is transmitted in the form of a regular succession of pulses of very short duration and high power from a highly directive aerial that concentrates the radiation into a narrow beam. When a pulse strikes a suitable target, part of the energy is scattered backwards towards the receiver which is generally located very close to the transmitter, both generally sharing the same aerial. The received signal, or 'echo', which is very weak compared with the transmitted pulse, is greatly amplified, detected, and displayed on a cathode-ray tube. The time base of the latter is triggered by the transmitter pulse and so locked to the pulse-recurrence-frequency of the transmitter. The echo therefore appears stationary on the time base at a distance from the origin that represents the time taken for a pulse to travel the double journey between the transmitter–receiver and the target. This distance therefore gives the range of the target, and, if the elevation and azimuth of the aerial beam are accurately known, the position of the target in space can be located. We shall now calculate the intensity of the received signal in terms of the characteristics of the radar and the properties of the hydrometeors in a cloud that form the target.

When an electromagnetic wave passes over a stationary object whose dielectric properties differ from those of the surrounding medium, some of its energy is absorbed by the object and appears as heat, while some is scattered in all directions without change of wavelength. The scattering is the result of induced oscillating electric and magnetic multipoles, or a combination of these with surface waves and reflections. In radar meteorology one is mainly interested in the intensity of the radiation back-scattered from the target in the direction of the receiver as this decides the strength of the received signal or 'echo' from the target, which usually consists of an assemblage of many individual raindrops, hailstones, or snowflakes.

In typical radar operation, a regular succession of high-frequency pulses, each of duration τ (of order 1 μs) and power level P_0 (of order 100 kW), are generated in the transmitter and radiated by the antenna. If the antenna were to radiate isotropically, the power traversing unit area (i.e. the intensity of radiation) at range R from the transmitter would be $P_0/4\pi R^2$. In practice, the radiation is concentrated in a

§8.1 RADAR STUDIES OF CLOUDS AND PRECIPITATION

narrow beam; the power incident on unit area of the target is then $P_0 G/4\pi R^2$, where G is the gain of the aerial over an isotropic radiator.

If the effective target area is T, the target will intercept power given by $TP_0 G/4\pi R^2$ and the intensity of the back-scattered radiation at the transmitter–receiver will be $TP_0 G/(4\pi R^2)^2$. If the effective area of the receiving antenna is A_e, the received power becomes

$$P_r = A_e T P_0 G / 16\pi^2 R^4. \tag{8.1}$$

The effective reflectivity of a spherical particle may be calculated by replacing the sphere by an equivalent dipole whose length is the sphere diameter. (The approximation is valid only when the diameter is much less than a wavelength; for larger dimensions, the multipole moments of the particle must also be taken into account.) The time average of the intensity of the radiation scattered by an oscillating dipole of moment μ in the direction of the source is $I_\mu = 2\pi^3 \mu^2 c / \lambda^4 R^2$, where c is the velocity of light, λ the wavelength of the radiation, and R the distance of the target from the transmitter–receiver. The dipole moment of a spherical particle of radius a is

$$\mu = \frac{\epsilon-1}{\epsilon+2} a^3 E,$$

where ϵ is the dielectric constant of the scatterer and E the electric field acting on the dipole. Thus,

$$I_\mu = 2\pi^3 \left(\frac{\epsilon-1}{\epsilon+2}\right)^2 \frac{E^2 a^6 c}{\lambda^4 R^2} = 16\pi^4 \left(\frac{\epsilon-1}{\epsilon+2}\right)^2 \frac{I_0 a^6}{\lambda^4 R^2}, \tag{8.2}$$

where the intensity of the incident radiation $I_0 = cE^2/8\pi$. The back-scattering cross-section σ is defined as the equivalent area of an isotropic scatterer that scatters energy in all directions of intensity equal to that scattered directly back by the target concerned. Hence,

$$\sigma = 4\pi R^2 I_\mu / I_0 = 64\pi^5 \left(\frac{\epsilon-1}{\epsilon+2}\right)^2 \frac{a^6}{\lambda^4}. \tag{8.3}$$

If there are n scattering particles of radius a per unit volume within the beam, and echoes arrive at the receiver from a volume V at the same time but with random phases, the effective target area is

$$T = V \sum n\sigma = \frac{64\pi^5}{\lambda^4}\left(\frac{\epsilon-1}{\epsilon+2}\right)^2 V \sum na^6, \tag{8.4}$$

and the received power is

$$P_r = \frac{4\pi^3 P_0 A_e G V}{R^4 \lambda^2} \left(\frac{\epsilon-1}{\epsilon+2}\right)^2 \sum na^6. \tag{8.5}$$

If the beam is completely intercepted by the cloud of scattering particles, the volume of space illuminated is the product of half the pulse length in space and the area illuminated by the transmitted beam, i.e.

$$V = \pi \frac{R\theta}{2} \frac{R\phi}{2} \frac{h}{2} = \frac{\pi R^2}{8} \theta\phi h,$$

where θ, ϕ are the angular beam widths measured to half power for one-way transmission, and h is the pulse length in space. For a uniformly illuminated circular aerial, Probert-Jones (1962) shows that

$$G = 4\pi A_e/\lambda^2 = \pi^2/\theta\phi,$$

so that

$$P_r = \frac{\pi^7}{8} \frac{P_0}{R^2 \lambda^2} \frac{h}{\theta\phi} \left(\frac{\epsilon-1}{\epsilon+2}\right)^2 \sum na^6. \tag{8.5a}$$

In deriving this equation we have assumed that the solid angle of the beam for the two-way transmission is the same as that for one-way transmission. In a more detailed analysis, Probert-Jones shows that this is not strictly the case and that (8.5a) thereby overestimates P_r by a factor of $2 \ln 2 = 1\cdot 4$ or $1\tfrac{1}{2}$ dB. He also shows that previous derivations of (8.5a), in taking $G = 16/\theta\phi$, overestimated P_r by a further 2 dB, or by $3\tfrac{1}{2}$ dB altogether. Eqn (8.5a) also requires modification to allow for attenuation of the signal over the two-way path between transmitter and receiver. This can be allowed for by a multiplying factor κ defined by

$$10 \log \kappa = 2 \int_0^R (g+c+p) \, dR,$$

where g, c, and p represent the attenuation (in dB/km, one way) by gases, cloud, and precipitation, respectively. The final version of (8.5) is therefore

$$P_r = \frac{\pi^7}{16 \ln 2} \frac{P_0}{R^2 \lambda^2} \frac{h}{\theta\phi} \left(\frac{\epsilon-1}{\epsilon+2}\right)^2 \kappa \sum na^6. \tag{8.5b}$$

If the entire beam is not intercepted by the target, the volume relevant to the computation is the fraction of the maximum volume that the beam illuminates effectively. Determination of this factor in a particular case is difficult and may be impossible. The echo from a weak storm that fills the beam may equal that from an intense storm that

§ 8.1   RADAR STUDIES OF CLOUDS AND PRECIPITATION   403

only partially fills it, but one cannot distinguish between these two conditions from the signal characteristics alone.

8.1.2. *Choice of radar parameters*

Neglecting the attenuation, eqn (8.5b) shows that, in order to detect weak precipitation at long ranges, the radar should have a short wavelength, a long pulse duration, high peak power, and a narrow beam.

The sensitivity of the radar set is limited by the noise generated in the receiver, and therefore the received power must be sufficient to be detectable above the inherent noise that is somewhat greater at shorter wavelengths. The minimum detectable power in most 10-cm receivers is about 2×10^{-13} W, and perhaps twice as large for 3-cm equipments.

The range resolution is determined by the pulse duration τ (or the pulse length τc); two targets in the beam will be resolved only if their range separation exceeds $\frac{1}{2}\tau c$. Thus, the scattered energy from the particles in only one-half the volume illuminated by the pulse can reach the receiver at the same time.

In order that the radar should have highly directive properties, i.e. high angular resolution in azimuth and elevation, the beam should be narrow, the radiant energy then being highly concentrated to facilitate detection of distant targets. However, very narrow beams require reflectors of large aperture and allow only a small volume of space to be illuminated by the pulse. It is therefore necessary to effect a compromise between these factors when choosing the beam width and pulse duration in order to obtain the best presentation of the storm.

The choice of the most suitable wavelength for meteorological work merits careful consideration. The back-scattered radiation by particles of linear dimension small compared with the wavelength is proportional to $1/\lambda^4$. However, the increased absorption and attenuation of the incident and scattered energy at very short wavelengths may more than compensate for the higher reflectivity and seriously restrict the maximum range, and again a compromise has to be effected. The radar sets in general use work on either 10-cm or 3-cm wavelengths, 10 cm being essential if it is desired to penetrate heavy rain. Several people have suggested that a wavelength of 5·6 cm would be the most suitable for the study of precipitation, especially snow. A radar working on 5·6 cm would be roughly 10 dB more sensitive than an otherwise similar equipment working on 10 cm; but for a path of 100 miles through rain of intensity 10 mm/h, the greater attenuation on 5·6 cm would just offset this gain.

The received power varies inversely as the square of the range of the target if the latter entirely intercepts the beam. At long ranges, therefore, the shapes and areas of the rain storms may not be truly depicted since only the heavier rain will be detectable. Also, siting of the radar is of great importance in determining the maximum range at which storms may be detected, this being restricted by the elevation of the effective horizon.

8.2. The scattering and attenuation of radar waves by meteorological particles

8.2.1. *Scattering of spherical particles of $D \ll \lambda$*

So far we have discussed only those parameters in eqn (8.5b) that are characteristic of the radar; we now have to examine the properties of the target and the intervening atmosphere that influence the intensity of the back-scattered radiation reaching the receiver. The following discussion is based largely on the original work of Ryde (1946). Postponing for the moment the question of attenuation, we shall discuss the reflectivity of the target defined as the sum of the scattering cross-sections of the particles in unit volume, i.e.

$$\eta = \frac{\pi^5}{\lambda^4}\left(\frac{\epsilon-1}{\epsilon+2}\right)^2 \sum nD^6, \qquad (D = 2a), \qquad (8.6)\dagger$$

which, for a given wavelength, is determined only by the concentration, size, and dielectric constant of the particles composing the cloud. The value of the dielectric constant depends on the material of the scatterer and on the wavelength of the incident radiation. For water,

$$K^2 = \{(\epsilon-1)/(\epsilon+2)\}^2 = 0.93 \pm 0.004,$$

for wavelengths between 3 and 10 cm and temperatures between 0 and 20° C. The corresponding value for 1·24 cm, is 0.91 ± 0.009. For ice, $\epsilon = 3.17$ in the microwave region at all temperatures, so $K^2 = 0.176$ for ice of density 0·917 g cm^{-3}. It is also useful to have a dielectric factor K^2 for ice of *unit* density; applying Debye's theory to a mixture of ice and air, and neglecting the contribution of air, K/ρ is constant, and therefore K^2 for ice of unit density is $K'^2 = 0.176/(0.917)^2 = 0.209$. By using this value of K^2 for ice, we may employ the melted diameter of an

† Eqn (8.6) holds only when $D \ll \lambda$, i.e. in the so-called Rayleigh-scattering regime. The more complex case of $D \geqslant \lambda$ is discussed later. At 3-cm wavelengths, particles of $D/\lambda < 0.05$ obey the Rayleigh law within 10 per cent.

ice particle, whatever its density, in computing η as long as the particles obey the Rayleigh law of scattering.

The function $Z = \sum nD^6$ is obviously closely related to the intensity of precipitation p, and several relations between the two have been derived empirically from the drop-size distributions of different rains. Since the drop-size spectra show considerable variations with the type and intensity of the precipitation, Z and p are not uniquely related but, for the purpose of calculating the radar reflectivities corresponding to various rainfall intensities, we shall use the relation $Z = 200\ p^{1.60}$ proposed by Marshall and Palmer (1948) as being fairly representative of 'steady' rain in temperate latitudes. The computed values are given in Table 8.1.

TABLE 8.1
Radar reflectivities $\eta = (\pi^5/\lambda^4)\ K^2 \sum nD^6$ of rain of various intensities

p (mm/h)	$\sum nD^6$ (mm^6/m^3)	$\eta_{\lambda=3.2\text{cm}}$ (cm^{-1})	$\eta_{\lambda=10\text{cm}}$ (cm^{-1})
1	220	6.00×10^{-10}	6.30×10^{-12}
2	667	1.84×10^{-9}	1.92×10^{-11}
3	1280	3.50×10^{-9}	3.66×10^{-11}
5	2900	7.90×10^{-9}	8.32×10^{-11}
7	4920	1.34×10^{-8}	1.41×10^{-10}
10	8750	2.40×10^{-8}	2.45×10^{-10}

Given the minimum detectable signal of the receiver and the other characteristics of the radar, we are now in a position to use eqn (8.5a) to calculate the maximum range at which precipitation of a given intensity can be detected. Thus a 10-cm radar having a peak-power output of 250 kW, a pulse duration of 1 μs, a conical beam of width 2°, and a minimum detectable power of 4×10^{-13} W, would, in the absence of attenuation, detect rain of 1 mm h^{-1} at a maximum range of 120 km.

8.2.2. *Scattering by non-spherical particles of $D \ll \lambda$*

For non-spherical particles, for example ice pellets, hailstones, and snowflakes, the radar reflectivity can be defined as

$$\eta = (c_1/\lambda^4)\{(\epsilon-1)/(\epsilon+2)\}^2 \sum nm^2,$$

where m is the particle mass and c_1 a numerical constant; in the unlikely event of all particles having the same mass and fall-speed, v,

$$\eta = (c_1/\lambda^4)K'^2 pm/v,$$

p being the precipitation rate. Thus, if ice pellets were to melt and form raindrops of much the same fall speed, the radar reflectivity would increase because the dielectric constant of water is greater than that of ice at centimetre wavelengths. In fact, it is not necessary for the ice to melt completely to produce an enhanced signal because, as Kerker, Langleben, and Gunn (1951) have shown, an ice sphere of $D \ll \lambda$, surrounded by a water film of thickness greater than one-fifth of its original radius, will scatter as effectively as a water sphere of the same mass. For low-density pellets and snow, the problem is more complicated because the intensity of the back-scattered radiation is influenced by the shape and orientation of the non-spherical particles—factors that we shall now consider in some detail.

The scattering by non-spherical hydrometeors was investigated theoretically by Atlas, Kerker, and Hitschfeld (1953) using the approximate treatment of Gans (1912) for ellipsoids. The scatterers are assumed to approximate to either oblate or prolate spheroids. The incident electric field intensity is resolved into components along the figure axis and two perpendicular diameters of the spheroid. These orthogonal components then induce proportional dipole moments in the same directions. The dipole moments in turn re-emit streams of radiation polarized in planes, each containing one of the chosen axes. Finally, each of the re-radiated components is resolved along the x, y, and z coordinates of the receiving system. As in Rayleigh's treatment of small spheres, the higher-order multipoles are neglected, these probably being unimportant provided the particles are small compared with the wavelength of the incident radiation.

Let the radiating antenna describe an orthogonal system of coordinates x, y, z as shown in Fig. 8.1, z being the direction of travel of a plane polarized wave, with the electric vector E vibrating at an angle β to the x-axis. Let this incident radiation have unit amplitude at the scatterer which defines another coordinate system ξ, η, ζ, where ξ lies in the direction of the figure axis of the spheroid, and η and ζ are along any pair of perpendicular diameters. The electric field at the spheroid has components E_ξ, E_η, E_ζ. The electric dipole set up in the spheroid has components $\mu_\xi, \mu_\eta, \mu_\zeta$, which are related to E by

$$\mu_\xi = \frac{(\epsilon-1)VE_\xi}{4\pi+(\epsilon-1)B}, \quad \mu_\eta = \frac{(\epsilon-1)VE_\eta}{4\pi+(\epsilon-1)B'}, \quad \mu_\zeta = \frac{(\epsilon-1)VE_\zeta}{4\pi+(\epsilon-1)B'},$$

where ϵ is the dielectric constant of the particle, V the particle volume and B, B' are geometrical factors determined by the eccentricity of the

§ 8.2   RADAR STUDIES OF CLOUDS AND PRECIPITATION   407

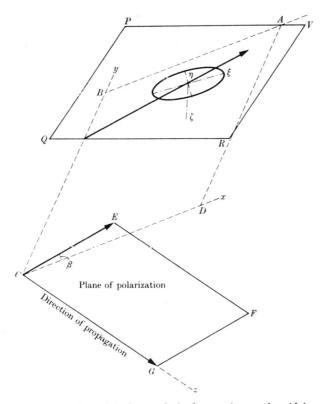

FIG. 8.1. The scattering of a plane polarized wave by a spheroidal particle.

spheroid. Thus the amplitude of the scattered radiation is proportional to the volume of the particles and increases with their dielectric constant.

The components of the scattered field are proportional to the induced dipole moments. The time average of the intensity of the total back-scattered radiation is $I = (2\pi/\lambda)^4(\mu_y^2+\mu_x^2)$, where μ_x, μ_y are the vector sums of the components of the dipole moments along the x and y directions respectively. But as the aerial can receive only the radiation polarized in the plane $CGFE$, the intensity of the received radiation will be given by the component of I in the direction of the electric vector E. We shall now consider two special cases: (1) when $\beta = 0$, i.e. the electric vector is parallel to the x-axis giving a horizontally polarized wave, only the component of the back-scattered radiation along the x-axis, I_x, will be received by the antenna and the component along the y-axis, I_y, will represent the cross-polarized component; (2) when $\beta = 90°$, i.e. the electric vector is parallel to the y-axis giving a vertically polarized wave, I_y will be the component detected by the receiver and I_x the cross-polarized component.

Now when the electric-field vector of the incident radiation is parallel to the axis or one of the diameters of a scattering spheroid, only the corresponding dipole moment is excited. Similarly, when the electric vector is perpendicular to the axis or one of the diameters, the dipole moment excited has no component in the direction of such axis or diameter.

In considering the orientation of hydrometeors relative to the radar antenna, two cases are of particular relevance.

(a) The spheroids are randomly oriented, every possible orientation of the figure axis being equally probable. In this case, the scattering is independent of the direction of polarization and of the angle of incidence of the radiation. Calculated values of both the parallel-polarized and cross-polarized components of the radiation scattered back by oblate and prolate spheroids, relative to the return from spheres of equal volume, are shown in Figs. 8.2 and 8.3. The graphs indicate that a considerable gain of back-scattered intensity, as well as of attenuation,

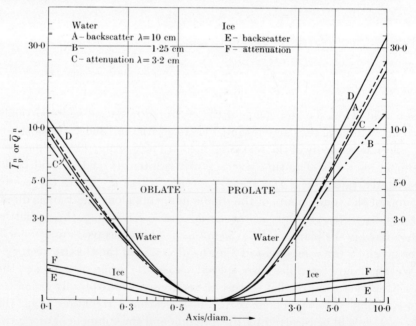

FIG. 8.2. Parallel-polarized back-scatter and attenuation by randomly oriented water and ice spheroids, relative to return and attenuation by spheres of equal volume. (From Atlas, Kerker, and Hitschfeld (1953).)

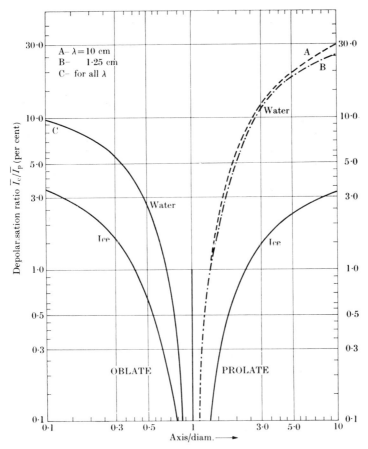

Fig. 8.3. Depolarization ratio (ratio of cross-polarized component of back-scattered energy to parallel-polarized component) for randomly oriented water and ice particles. (From Atlas, Kerker, and Hitschfeld (1953).)

is to be expected if water particles are distorted from the spherical shape. Raindrops generally suffer only rather small distortion (maximum axis/diameter ratio $\simeq 1{\cdot}5$), but the calculations may be very relevant to melting snowflakes and wet ice particles that probably scatter very much like water particles of the same shape. For dry ice particles, on the other hand, both back-scatter and attenuation vary very little with the particle shape. This is because the shape factors in the expressions for the induced dipole moments are multiplied by $(\epsilon-1)$, this factor being much greater for water than for ice. The effect of the shape on the scattering cross-section for low-density snowflakes will be even less than that for ice, because the dielectric constant of snow will

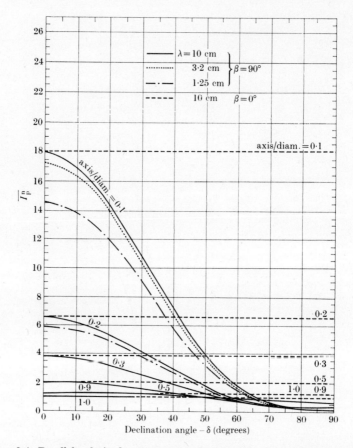

FIG. 8.4. Parallel-polarized component of intensity back-scattered from oriented water oblates (relative to back-scatter from spheres of equal volume) as a function of the angle of declination of the beam from the zenith. The horizontal lines refer to a horizontally polarized ($\beta = 0°$) antenna, the other curves ($\beta = 90°$) to a vertically polarized antenna. (From Atlas, Kerker, and Hitschfeld (1953).)

be considerably less; indeed, for purposes of their electromagnetic behaviour, dry snowflakes may be regarded as spheres. Fig. 8.3 shows that, for comparable deformations, water prolates depolarize the radiation more strongly than oblates. Ice, in both cases, returns only a very small and probably indetectable cross-polarized component.

(b) The spheroids have their figure axes all oriented either vertically or horizontally. Plate-like ice crystals, which approximate to oblate spheroids, are believed to fall with their figure axes vertical and so will always have a long diameter parallel to the incident electric vector.

§ 8.2   RADAR STUDIES OF CLOUDS AND PRECIPITATION   411

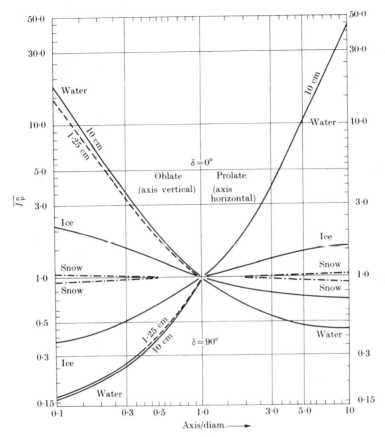

Fig. 8.5. Parallel-polarized component of back-scattered radiation from oriented particles (relative to back-scatter from spheres of equal volume) for the two extreme cases of vertically ($\beta = 0°$) and horizontally ($\beta = 90°$) pointing vertically polarized antennae. (From Atlas, Kerker, and Hitschfeld (1953).)

Since the orthogonal component of the dipole moment is not excited, there can be no cross-polarization. The parallel-polarized component, in this case also the total back-scattered intensity, from horizontally oriented water oblates is shown in Fig. 8.4. If the incident wave is horizontally polarized, the back-scattered radiation will be independent of the angle of elevation of the aerial, since the disposition of the electric vector on a horizontal plane will be the same whatever the elevation of the beam. If, however, the incident wave is vertically polarized, the intensity of the received radiation will increase as the beam moves towards the zenith. The corresponding curves for ice oblates are very similar to those of Fig. 8.4, except that the ordinate varies over a much

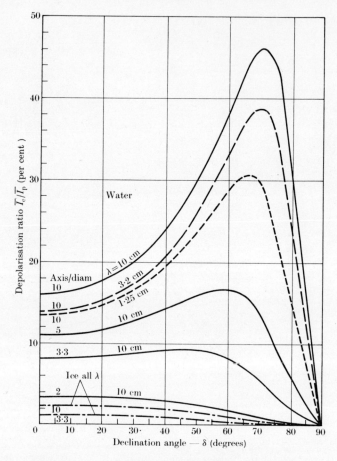

Fig. 8.6. Depolarization ratios for oriented prolates. Antenna is vertically polarized. (From Atlas, Kerker, and Hitschfeld 1953).)

narrower range (2·1–0·38) as the declination increases from 0° to 90°. This is shown in Fig. 8.5, where the parallel-polarized component of the back-scattered radiation from both oriented oblates (figure axis vertical) and prolates (axis horizontal) is plotted against the axis/diameter ratio. The results, which are plotted for a vertically polarized wave and for a vertically and a horizontally pointing radar, show that the echo intensity from ice particles is appreciably affected by their shape only when they have a preferred orientation, while very marked effects are possible when the particles are coated with a water film and scatter almost like water particles. Ice needles and prismatic columns fall with their figure axes orientated at random in the horizontal plane, so that only a small proportion will have their axes parallel to the incident electric vector. As a result, they will return a cross-polarized component, which will vary with the elevation of a vertically polarized antenna as shown in Fig. 8.6. The cross-polarization component from ice crystals is hardly detectable by present radar sets.

We shall make use of the results of these calculations on scattering by non-spherical particles in § 8.7 when interpreting the echoes received from clouds of meteorological particles.

8.2.3. *Experimental verification of scattering theory*

It was shown theoretically on p. 402 that the intensity of radiation back-scattered from a sphere should be proportional to the sixth power of its diameter, provided this is small compared with the wavelength. This relationship was tested experimentally by Labrum (1952a, b), who inserted steel spheres, and hemispherical water drops resting on a waxed mica plate, into a 10-cm wave-guide, and measured the power scattered back from them. Measurements of the scattered power were made for sets of drops having different diameters between about 4·5 and 7·5 mm, and in each case it was proportional to the square of the volume of the scatterer. Labrum also investigated the variation of the back-scattered radiation during the melting of an ice disk or rod. He found that the back-scattered power increased rapidly after the particle began to melt and became wet, passed through a maximum when it began to lose its original shape, and thereafter decreased as the particle collapsed into a nearly hemispherical drop. This experiment provided qualitative confirmation of the theoretical predictions given above in § 8.2.2.

A rather less accurate expression than (8.5b) for the power received from a precipitating cloud, viz.

$$P_\mathrm{r} = \frac{\pi^4 P_0 A_\mathrm{e} \tau c}{8R^2} \frac{1}{\lambda^4} \left(\frac{\epsilon-1}{\epsilon+2}\right)^2 \sum nD^6, \qquad (8.7)$$

was investigated experimentally by Hooper and Kippax (1950a). Taking Ryde's calculated values for the reflectivity

$$\eta = (\pi^5/\lambda^4)\{(\epsilon-1)/(\epsilon+2)\}^2 \sum nD^6$$

in terms of the precipitation rate p, they plotted calculated values of P_r against p appropriate to a range of 1000 yards and to the A_e, τ, and P_0 values of their radar set. On the same diagram they plotted measured values of the power received from actual precipitation, these being corrected to a standard range of 1000 yards. The received power was measured by feeding the output from a pulsed signal generator into the aerial wave-guide and by adjusting the calibrated attenuator of the signal generator until both the echo and the injected signal were of equal amplitude on a cathode-ray tube; the output of the generator was calibrated in terms of absolute power. The rate of precipitation was measured by an integrating rainfall recorder. After allowing for a 3-dB loss through the wet radome, the measured values of P_r were on average 1·5 dB above those expected from theory. But Austin and Williams (1951) pointed out that Hooper and Kippax calculated the 'average peak signal' rather than the average signal as required by theory and thereby overestimated the average received power by at least 5 dB. The corrected result of their experiment is therefore a loss of about 3·5 dB in received power below the theoretical value but, by applying the correction factor of $2 \ln 2 \simeq 1\cdot5$ dB mentioned on p. 402, the discrepancy is reduced to 2 dB.

To investigate the dependence of the received power on the duration of the pulse, Hooper and Kippax used a 3·2-cm radar giving pulses of either about 0·5 or 2 μs duration. The ratio of the mean transmitted powers when the two pulses were emitted at the same recurrence frequency was found to be $4\cdot83 \pm 0\cdot02$, i.e. $6\cdot84 \pm 0\cdot02$ dB. Observations were then made at half-minute intervals on precipitation at 9000 ft, with the beam directed vertically upwards, using long and short pulses on alternate occasions. A check that the precipitation rate was remaining constant was made by a separate radar with a fixed pulse duration. The mean of seventy-one readings gave a value of $6\cdot9 \pm 0\cdot2$ dB for the ratio of the echo intensities with the two pulse lengths, confirming that the echo power from the steady precipitation was proportional to the energy within the transmitted pulse.

Hooper and Kippax also set out to check that the echo intensity is proportional to $(1/\lambda^4)\{(\epsilon-1)/(\epsilon+2)\}^2$ by using vertically directed radar sets working on 1·25-, 3·2-, and 9·1-cm wavelengths, measurements being made at 2-min intervals on precipitation at heights of 4500, 6000, and 7500 ft in turn. When the echo intensities were corrected for attenuation by rain between the target and the radar set and for losses in the perspex radome, good agreement was obtained between the experimental values for the ratios of the intensities measured on two wavelengths and the corresponding values calculated from eqn (8.7) as is shown in Table 8.2.

The same authors also tested Ryde's expression for the reflectivity

TABLE 8.2

Ratios of the echo intensities received from precipitation on two wavelengths according to Hooper and Kippax (1950a)

Ratios of intensities	$\lambda_{9\cdot 1}/\lambda_{3\cdot 2}$	$\lambda_{9\cdot 1}/\lambda_{1\cdot 25}$
Theoretical values (eqn (8·7))	$-6\cdot 4 \pm 0.7$	$-3\cdot 4 \pm 1.6$
Observed values at 4500 ft	$-6\cdot 3 \pm 0.4$	$-2\cdot 5 \pm 0.4$
Observed values at 6000 ft	$-6\cdot 0 \pm 0.3$	$-2\cdot 4 \pm 0.4$
Observed values at 7500 ft	$-6\cdot 0 \pm 0.3$	$-2\cdot 3 \pm 0.5$

of falling snowflakes, viz.

$$\eta = \frac{C_1}{\lambda^4} K'^2 \sum nm^2 = \frac{C_1}{\lambda^4} K'^2 \frac{p\bar{m}}{\bar{v}}, \qquad (8.8)$$

where p is the precipitation rate, $\bar{m}$ and $\bar{v}$ are respectively the mean mass and mean fall-velocity of the flakes. The depth of the snow was measured at half-hourly intervals and, from its density, p was estimated. The echo intensity from the falling snow, as measured with a vertically directed radar, increased with decreasing height to 6000 ft and below that remained constant. Taking the mean mass of the snowflakes to be 1 mg and their mean velocity to be 1 m s⁻¹, the experimental value of the reflectivity at 3000 ft agreed with that calculated from eqn (8.8) to within about 1·5 dB, but again this must be judged in the light of the errors mentioned above.

The reflectivity of snowflakes received further attention from Langille and Thain (1951), who plotted the power received by a vertically pointing 3·2-cm radar against the precipitation rate measured by weighing the snow collected on a ground sheet at 5-min intervals. Straight lines of best fit were drawn through the experimental points and were used to determine the equivalent homogeneous particle mass. On one day they were able to determine the size distribution of the flakes by allowing them to fall and melt on dyed filter papers, their masses being found from the diameters of the stains produced. When the received power was plotted against $\sum nm^2$, the slope of the line of best fit agreed reasonably well with the slope of the locus calculated from eqn (8.8), but the observed signal strength was about 4 dB below the calculated value. Had they used the corrected form of the radar equation (8.5b), the discrepancy would have been only 0·5 dB.

A more recent experiment, carried out in England by Roberts (1959) with 3-cm and 8-mm radars, gave mean differences between calculated and observed powers of 7 dB and 5 dB respectively. Had he used the

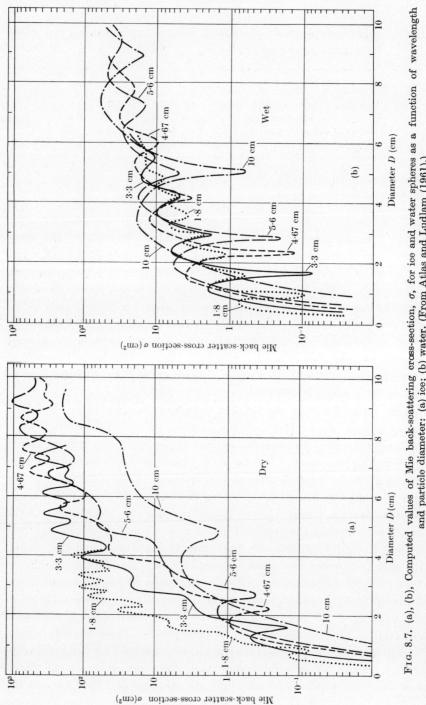

FIG. 8.7. (a), (b), Computed values of Mie back-scattering cross-section, σ, for ice and water spheres as a function of wavelength and particle diameter: (a) ice; (b) water. (From Atlas and Ludlam (1961).)

corrected radar equation and the relation $Z = 200p^{1\cdot 6}$, these discrepancies would have been only 1·5 dB and $-0\cdot 5$ dB.

8.2.4. *Scattering by large hydrometeors of $D \geqslant \lambda/20$*

In general terms the *normalized* back-scattering cross-section of a spherical particle of radius a is defined as

$$\sigma_b = \frac{\sigma}{\pi a^2} = \frac{4\pi R^2}{\pi a^2} \frac{P_s}{P_i}, \tag{8.9}$$

where P_i is the power density in the incident wave and P_s that in the scattered wave at distance R from the scattering particle. According to the Mie theory of scattering, σ_b is a function of the circumference of the particle relative to the wavelength, i.e. $\alpha = 2\pi a/\lambda$, and of the complex refractive index of the scattering material. We have already seen that in the region of Rayleigh scattering, where $a \ll \lambda$,

$$\sigma_b = \frac{16\pi^4}{\lambda^4}\left(\frac{\epsilon-1}{\epsilon+2}\right)^2 a^6, \tag{8.9a}$$

but, for larger values of α, computation of σ_b becomes much more difficult because of surface, reflection, and resonance effects. The first computations for water and ice spheres were made by Ryde (1946) for values of $\alpha \leqslant 4$ and these have been extended by Haddock (see Kerr 1951), Aden (1952) and, more recently, by Herman and Battan (1961a, b) and Stephens (1961) using the full Mie theory and electronic computers.

Stephens (1961) calculated and tabulated values of σ and σ_b for water spheres at fourteen different wavelengths between 0·43 and 16·43 cm and for $\alpha \leqslant 4\cdot 0$. Computed values of the Mie back-scattering cross-section, σ, for ice and water spheres are plotted in Fig. 8.7(a), (b). The normalized coefficients, σ_b, for water and ice spheres at $\lambda = 3\cdot 2$ cm and 0° C appear in Fig. 8.8, which shows that the reflectivities of dry ice spheres exceed those of water spheres of the same diameter when $\alpha > 2\cdot 5$, in direct contrast to the situation in the Rayleigh region, where water spheres have scattering cross-sections five times those of equivalent ice spheres. Whereas the scattering cross-sections of ice spheres are practically independent of wavelength, those of water spheres vary with both wavelength and temperature through changes in the complex refractive index of water. The temperature variations are largest in the vicinity of $\alpha = 0\cdot 4$ where, for $\lambda = 3\cdot 2$ cm, σ may change by 50 per cent over the temperature range 0° C to 18° C, but temperature changes

418 RADAR STUDIES OF CLOUDS AND PRECIPITATION 8

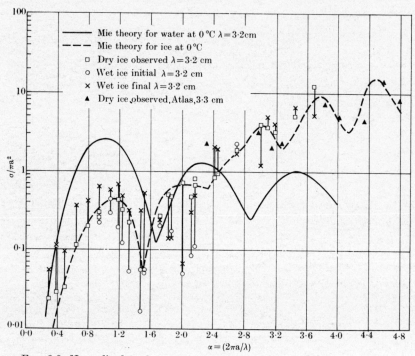

FIG. 8.8. Normalized back-scattering cross-sections, σ_b, of dry and wet ice spheres at $\lambda = 3\cdot2$ cm and $0°$ C as a function of $\alpha = 2\pi a/\lambda$. (From Gerhardt et al. (1961a).)

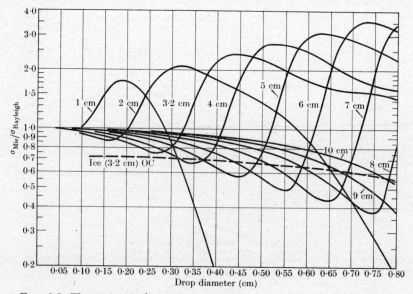

FIG. 8.9. The ratio $\sigma_{\text{Mie}}/\sigma_{\text{Rayleigh}}$ as a function of drop diameter. (From Stephens (1961).)

§ 8.2 RADAR STUDIES OF CLOUDS AND PRECIPITATION

of a few degrees will not usually be important in practice. The ranges of validity of the Rayleigh approximation, for both water and ice spheres and for wavelengths between 1 and 10 cm, are shown in Fig. 8.9. The most pronounced deviations tend to occur at the longer wavelengths but, it is at these wavelengths, that more and more hydrometeors fall within the Rayleigh regime. Thus at 10-cm wavelengths, the true scattering cross-section of even the largest raindrops differ by only 20 per cent from the Rayleigh value.

Herman and Battan (1961a) computed σ_b for ice spheres for $\lambda = 3\cdot 2$ cm and values of $\alpha \leqslant 30$, obtained close agreement with the calculations of Ryde and of Stephens over the common range of $\alpha \leqslant 4$, found that large ice spheres scatter more than an order of magnitude better than water spheres of the same diameter, and that this ratio increases with increasing α up to $\alpha \simeq 60$, when σ_b reaches a maximum value of $38\cdot 5$ for ice, and thereafter decreases. In contrast, the oscillations of the back-scattering curve for water spheres rapidly diminish with increasing α, so that for $\alpha > 10$, the drops scatter very like a plane surface of water with reflectivity $0\cdot 63$. Whereas the back-scattering from a water sphere is essentially a surface phenomenon as with a metal sphere, an ice sphere, because of its small refractive index ($1\cdot 78$) and low absorption, acts like a dielectric lens and focuses the incident radiation on to the back surface from which it is reflected backwards as a concentrated beam.

The computed radar scattering cross-sections may be compared with measurements made by Atlas, Harper, Ludlam, and Macklin (1960) at wavelengths of $3\cdot 3$ and $4\cdot 76$ cm on ice spheres covering a range of $\alpha = 0\cdot 4$–10. The ice spheres, made in plastic or rubber moulds, were placed in fine nylon nets and suspended 50 m below a balloon in the beam of the radar. The received power was measured and compared with that from a metal sphere used as a reference target. The measurements at $\lambda = 3\cdot 3$ cm, whose reliability was checked by measurements made on a series of metal and Perspex spheres, and in which individual values had a scatter of typically 1 dB about the mean, were in good agreement with theoretical values, as shown in Fig. 8.8. The scattering cross-sections of water drops of diameter 2–7 mm, and of ice spheres of diameter 3–35 mm were measured in the laboratory by Gerhardt, Tolbert, and Brunstein (1961a) as they fell through the beam of a $3\cdot 2$-cm c–w radar. Aluminium spheres were used for calibration. The ice measurements were in excellent agreement with the theoretical Mie values, except for considerable scatter at the large diameters as shown in Fig. 8.8. Measured reflectivities of the water drops were in excess of both the Mie and Rayleigh values for diameters <3·5 mm, but agreed closely with the Mie theory for larger diameters. In a later paper, Gerhardt et al. (1961b) report similar measurements on water and ice spheres at wavelengths of 1·50 and 5·72 cm and again obtain good agreement with theory, this time for the smaller drops as well.

The back-scattering cross-sections of melting ice spheres have been measured at 3·3- and 4·67-cm wavelengths by Atlas et al. (1960), who found that, whereas for ice spheres of $D < 0·8\lambda$, the formation of a liquid coat led to an increase in σ_b, for spheres of $D \gg 0·8\lambda$, melting was accompanied by a rapid drop from the ice towards, or even below, the all-water values. Thus for 3·3-cm wavelengths, as soon as ice spheres of $D \gg 3$ cm acquired a water film of thickness about 0·1 mm, σ_b fell by as little as 5 dB or by as much as 20 dB, depending on the exact D/λ ratio. At the longer, 4·67-cm, wavelength, the early stages of melting of such large stones had little effect, suggesting that the particle would require a thicker water coat before it would scatter as a water sphere.

In rather more careful measurements at $\lambda = 3·2$ cm, Harper (1962) found that as ice spheres of $D > 3$ cm melted, their cross-sections dropped rapidly as the water coat built up, fell rapidly into deep minima as the first drops of water were shed, and then rose towards the appropriate all-water values as the spheres grew smaller. Willis, Browning, and Atlas (1964), in tracking a freely-falling ice sphere of $D = 5$ cm with a 5·47-cm radar, found the initial value of the back-scattering cross-section to be within 2 dB of the theoretical value for an all-ice sphere. But, on descending below the melting level, the cross-section decreased by about 5 dB, or by about half that to be expected during transition to an all-water sphere. The fact that the reduction was also only half of that observed during the initial melting of stationary ice spheres on 3·2-cm radar by Atlas et al. (1960) may indicate that freely-falling stones cannot maintain the 0·1-mm wet coats necessary to make the particles scatter as water spheres. On the other hand, no theoretical calculations have been made for 5·47-cm wavelengths.

These measurements can be interpreted in terms of the calculations of σ_b for ice spheres surrounded by shells of liquid water made by Herman and Battan (1961b). Their curves of σ_b as a function of the thickness of the liquid coat, examples for $\lambda = 3·21$ cm being reproduced in Fig. 8.10, show that the back-scattering may either increase or

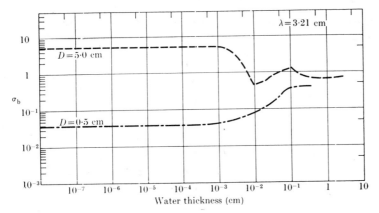

FIG. 8.10. Normalized back-scattering cross-sections of ice spheres covered by shells of liquid water of various thickness. (From Herman and Battan (1961b).)

decrease as melting proceeds depending upon the wavelength of the incident radiation and the original size of the sphere. For 3·21-cm radiation, as melting proceeds, the value of σ_b decreases for spheres of $D \geqslant 2$ cm, while it increases for spheres of $D \leqslant 1$ cm. For 4·67-cm radiation, the corresponding diameters are 3 and 4 cm respectively. For 10-cm radiation, the back-scattering always increases as melting proceeds for all particle sizes normally encountered in clouds. The thickness of the water shell necessary to make the sphere act as though it were all-water is likewise a function of wavelength. For 3·21-cm radiation, this thickness is about 5 mm, for 4·67-cm radiation about 7 mm, and for 10-cm radiation, is 1–2 cm. For spheres of initial radius less than these values, the all-water value of σ_b is not reached until melting is complete. When the water shells are thinner than the above values, the σ_b values differ from those of all-ice or all-water spheres and show maxima and minima (see Fig. 8.8), which are due to interference between the externally reflected wave at the front surface of the sphere and the internally reflected waves from the two water-ice interfaces and from the back surface of the sphere. These computations are applicable to hailstones that acquire a liquid coat either by melting, or by collecting supercooled water more rapidly than it can be frozen as described in § 6.6.7.

Battan and Herman (1962) have also calculated back-scattering cross-sections for 'spongy' ice spheres consisting of homogeneous mixtures of water and ice. At a wavelength of 3·21 cm and 0° C, they calculate that σ_b for spheres of $D > 3$ cm decreases as the fraction of liquid water increases, and falls by an order of magnitude to near the all-water value when the liquid-water content reaches 50 per cent. Spheres of $D < 3$ cm are predicted to show much smaller changes, but a tendency for σ_b to increase when the water content exceeds 50 per cent. Measurements were made at wavelengths of 3·21, 4·67, and 10 cm on ice spheres of $1·2 < D < 4·2$ cm with a spongy ice shell, and on snow spheres dipped in water to give spongy ice throughout, by Joss and Aufdermauer (1965). The results did not show systematic differences either for the three wavelengths or for the two kinds of particle and so were apparently not sensitive to how the water was distributed in the sphere. For values of $\alpha = \pi D/\lambda < 0·7$, the measured cross-sections for spheres containing 10, 20, and 30 per cent of water were respectively about 2, 3, and 3·5 times those of completely frozen particles but always less than those of equivalent all-water spheres. For $\alpha > 1·0$, a liquid-water content of only 10 per cent was sufficient to produce a mean scattering cross-section equal to, or even larger than, that of an all-water sphere. The discrepancy between these measurements and the computations of Battan and Herman were attributed to the latter's incorrect treatment of the dielectric constant of ice-water mixtures.

Recognizing that hailstones are not always spherical, Atlas and Wexler (1963) made measurements on dry, oblate, ice spheroids at 3·2- and 9·67-cm wavelengths. Their data are highly complex and not easily summarized. The only regular behaviour was shown by particles with major axes smaller than $\lambda/4$, which roughly fulfilled the predictions of the Gans theory as discussed above and summarized in Fig. 8.5. For slightly larger particles $(D/\lambda \simeq 0·5)$, it was found that the large (circular) face cross-sections are generally enhanced with respect to equivolume spheres, while small-face cross-sections are usually degraded, the more so the smaller the b/a ratio, and more with polarization perpendicular to the major axis than with parallel polarization. Thus, as in the case of small particles, parallel polarization is superior to perpendicular polarization but, unlike the smaller particles, both polarizations show smaller cross-sections than do equivolume spheres. The situation is reversed only with nearly spherical particles ($b/a > 0·8$), which show a large-face cross-section slightly smaller than its equivolume sphere, but small-face cross-sections are slightly enhanced, again more so with parallel polarization, but even then by only 2–4 dB. For still larger particles, $(D/\lambda > 0·5)$, Atlas and Wexler found that the large-face cross-sections tend to follow a behaviour indicated by geometric optics, with enhancement usually occurring for $b/a > 0·7$, but oscillating between enhancement and diminution for smaller values of b/a. This oscillation is presumed to be due to interference between the front- and rear-surface reflections. On the other hand, the small-face cross-sections are usually equal to, or less than, those of equi-volume spheres in this size regime, and in contrast to the smaller sizes, perpendicular polarization is superior to parallel polarization for the more commonly expected values of $b/a > 0.5$.

The large back-scattering cross-sections of dry hailstones with $D > \lambda$, the marked changes that may occur when hailstones acquire a liquid coat or a layer of spongy ice, and the effects of particle shape and orientation, have important implications for the interpretation of the intensities of radar echoes from cumulonimbus clouds. When the precipitation contains hydrometeors too large to obey Rayleigh scattering, the reflectivity can be defined in terms of the $Z = \sum nD^6$ value for small water drops that would have the same total back-scattering cross-section as the measured echo. This equivalent value is then given by

$$Z_e = (10^6 \lambda^4/\pi^5 K^2) \sum n\sigma \ mm^6 m^{-3}, \qquad (8.10)$$

where σ is now the Mie cross-section measured in cm^2, n the number of particles per m^3, and λ is measured in cm.

Thus $\quad Z_e = 3.68 \times 10^5 \sum n\sigma \quad$ for $\quad \lambda = 3.2$ cm, $\qquad$ (8.10a)

and $\quad Z_e = 3.52 \times 10^7 \sum n\sigma \quad$ for $\quad \lambda = 10$ cm. $\qquad$ (8.10b)

Using a 3·2-cm radar, Donaldson (1958, 1959, 1961) measured Z_e values approaching, and occasionally exceeding, 10^7 mm^{6}m^{-3} at about the 6-km level in intense storms. These large values of Z_e, which may have been underestimated because of attenuation, were usually associated with dry hail of $D > 1.3$ cm. Geotis (1963) measured values of Z_e approaching 10^7 mm^{6}m^{-3} with 10-cm radar and reported a close correlation of large hail with values of $Z_e \geqslant 10^{5.5}$ mm^{6}m^{-3}. Such high values of reflectivity imply improbably high rates of rainfall; even if the rain were to consist entirely of the largest possible drops, 6 mm in diameter, a Z_e value of 10^7 mm^{6}m^{-3} with 3·2-cm wavelengths would require a rain-water concentration of 15 g m^{-3}, while for 10 cm wavelengths, it would be 30 g m^{-3}.

Because of lack of data on the size distribution of hailstones, we shall discuss the reflectivity of hailstones in terms of spherical stones of uniform size. The concentrations of such stones required to produce Z_e values of 10^7 mm^{6}m^{-3} are shown, as a function of hailstone diameter and wavelength, in Table 8.3. This also shows the Z_e values that would be produced by 1 g m^{-3} of completely wet or dry hailstones of various sizes, and similar computations for a number of wavelengths are plotted in Fig. 8.11. The 'wet' values may be regarded as applicable to spongy hailstones containing more than about 25 per cent of liquid water. It

TABLE 8.3

The back-scattering properties of ice and water spheres simulating dry and wet hailstones

	$\lambda = 3 \cdot 2$ cm					
	Dry			Wet		
D (cm)	σ_b	log Z_e (mm^{6}m^{-3}) for 1 g m^{-3}	Water content (g m^{-3}) for $Z_e = 10^7$	σ_b	log Z_e for 1 g m^{-3}	Water content for $Z_e = 10^7$
1	0·38	5·33	47·5	2·58	6·16	7
2	0·65	5·26	55·0	0·74	5·31	49
3	3·4	5·80	15·9	0·33	4·78	164
4	6·6	5·96	10·9	0·45	4·80	160
5	5·2	5·76	17·3	0·70	4·89	128
6	11·8	6·04	9·1	0·80	4·87	160

	$\lambda = 10$ cm					
	Dry			Wet		
D (cm)	σ_b	log Z_e for 1 g m^{-3}	Water content for $Z_e = 10^7$	σ_b	log Z_e for 1 g m^{-3}	Water content for $Z_e = 10^7$
1	0·007	5·66	27·0	0·04	6·32	4·8
2	0·09	6·45	4·2	0·90	7·36	0·4
3	0·34	6·88	1·7	2·48	7·60	0·2
4	0·36	6·68	2·1	1·62	7·32	0·5
5	0·12	6·10	7·9	0·32	6·52	3·0
6	0·62	6·74	1·8	0·52	6·56	2·8

appears that Z_e values of 10^7 mm^{6}m^{-3} at 10-cm wavelengths could be produced by modest concentrations of either dry or electromagnetically 'wet' hailstones of diameter 2–4 cm, and that the requirement could be reduced even further if a large proportion of the stones were oblate spheroids rather than spheres. Such large reflectivities at 3-cm wavelengths could be produced by wet hailstones of $D \sim 1$ cm in concentrations of 7 g m^{-3}. Joss and Aufdermauer (1965) find that spongy spheres

§ 8.2   RADAR STUDIES OF CLOUDS AND PRECIPITATION   425

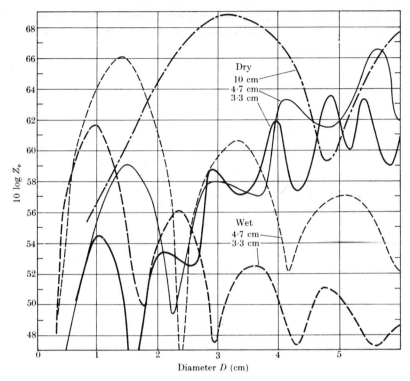

FIG. 8.11. log Z_e versus particle diameter for 1 g m^{-3} of dry and wet ice spheres at different wavelengths. (From Atlas and Ludlam (1962).)

of $D \simeq 0.8$ cm, containing 30 per cent of liquid water, scatter about 30 per cent more strongly than equivolume water spheres, but this would only reduce the required concentration to about 5 g m^{-3}.

Z_e values of 10^7 mm^{6}m^{-3} could also be produced by large dry hailstones of $D > 3$ cm in concentrations of 10–15 g m^{-3}. In this connection, we recall that Geotis (1963) found Z_e values $\geqslant 3 \times 10^5$ mm^{6}m^{-3} to be a fairly reliable indicator of hail in New England, and measured values of $Z_e > 4 \times 10^6$ mm^{6}m^{-3} only when hailstones of $D > 3$ cm were falling and only in the presence of some stones of $D = 2.5$–5.0 cm. But he found no clear-cut relationships with 3·2-radar as were reported by Donaldson, or as would be expected from the data of Douglas and Hitschfeld (1961), who calculated Z_e for each of thirty-three samples of hailstones collected in Alberta. In all but four cases, the Z_e value for $\lambda = 3.2$ cm was greater if the hail was assumed to be wet rather than dry, and this was so in all cases for 10-cm radiation. The exceptional cases for 3·2-cm radiation contained some stones of $D > 3$ cm; these made a major contribution to Z_e, which reached a highest value of 3×10^6 mm^{6}m^{-3}, the corresponding frozen-water concentration being only 1 g m^{-3}.

We may now enquire whether, in view of the wavelength dependence of the back-scattering cross-sections, information on the size, shape, and physical state of hailstones might be obtained by making simultaneous measurements on two or more wavelengths, these being first corrected for attenuation by intervening precipitation (see § 8.2.6). Since by definition, Z_e values of rainfall are essentially independent of wavelength, while those of hail vary with particle size and wavelength as shown in Table 8.3 and Fig. 8.11, it should be possible to differentiate between rain and hail by making measurements on two different wavelengths. Furthermore, if Z_e were to increase tenfold on switching from 3-cm to 10-cm wavelengths, this would indicate the presence of dry hailstones of 2–4 cm diameter. However, a hundredfold increase would suggest the presence of wet, probably spongy, hail of $D \geqslant 2$ cm. If size and wetness can be determined, then an estimate of hail concentration can be made from a measurement of the absolute value of Z_e at any one wavelength. In practice, the situation may be complicated by a spectrum of hail sizes and by shape and orientation effects, making it difficult to obtain precise and unambiguous answers.

8.2.5. *Scattering by non-precipitating clouds*

The radar reflectivity of a homogeneous non-precipitating cloud is given by $\eta = (36\pi^3/\lambda^4)K'^2 wm$, w being the liquid- or solid-water content and m the particle mass. For a water-droplet cloud with $w = 1$ g m^{-3} and $m = 10^{-9}$ g, the reflectivities at wavelengths of 10 cm and 3·2 cm are about 10^{-16} cm^{-1} and 10^{-14} cm^{-1} respectively. For an ice-crystal cloud with $w = 0·1$ g m^{-3} and $m = 10^{-6}$ g, the corresponding reflectivities are of order 10^{-15} cm^{-1} and 10^{-13} cm^{-1}. These values are several orders of magnitude smaller than those shown in Table 8.2 for precipitating clouds, and the power returned by non-precipitating clouds is generally too small to be detected by many 10-cm and 3-cm radars, although they can be detected at short ranges on millimetre wavelengths.

8.2.6. *Attenuation of radar waves by clouds and precipitation*

We now have to discuss the reduction in echo intensity brought about by attenuation in the storm itself and in the atmosphere that intervenes between the target and the transmitter–receiver. Attenuation

§ 8.2 RADAR STUDIES OF CLOUDS AND PRECIPITATION

by atmospheric gases (particularly oxygen) and by water vapour is practically negligible at wavelengths of 3 and 10 cm. Even in water-saturated air at sea level and temperature 20° C, the attenuation for 3-cm wavelengths amounts to only about 0·2 dB/km; the attenuation due to water vapour reaches a maximum at $\lambda = 1·35$ cm, but is then only 0·3 dB/km.

In fog and non-precipitating clouds, where the droplets are usually less than 100 μm in diameter, the attenuation is given by

$$\alpha_c = c_2(\lambda)w/\lambda \text{ dB/km}, \quad (8.11)$$

where $c_2(\lambda)$ is a factor depending on the wavelength λ, and w is the liquid-water content. At 0° C, and for $w = 1$ g m^{-3}, we have

$\lambda = 1·24 \quad 3·2 \quad 10·00$ cm
$\alpha_c = 0·53 \quad 0·086 \quad 0·009$ dB/km.

Thus, at $\lambda = 3·2$ cm, the attenuation amounts to 2 per cent/km g m^{-3} and is about six times as serious at $\lambda = 1·24$ cm.

Randomly oriented ice crystals attenuate less strongly than water droplets of the same volume as indicated in Fig. 8.2, their effect being negligible at 3-cm and 10-cm wavelengths.

In rain, which may contain drops up to 6 mm in diameter, the Rayleigh approximation is strictly valid only for $\lambda \geqslant 10$ cm. However, Gunn and East (1954) have derived the following empirical relation based on the Laws and Parsons (1943) drop–size data:

$$\alpha_r = c_3(\lambda, T)p^n \text{ dB/km}, \quad (8.12)$$

where p is the precipitation rate in mm h^{-1}, and c_3 and n take the following values at 18° C:

$\lambda = 1·24 \quad 3·2 \quad 10·00$ cm,
$c_3 = 0·12 \quad 0·0074 \quad 0·0003$ cm,
$n = 1·05 \quad 1·31 \quad 1·00$ cm.

If $p = 10$ mm h^{-1}, the attenuation amounts to $3\frac{1}{2}$ per cent/km for $\lambda = 3·2$ cm, but is less than 0·1 per cent/km for $\lambda = 10$ cm. Thus in viewing a precipitating cloud through intervening heavy rain, the greater reflectivity obtained with 3-cm waves may be more than offset by the increased attenuation, and it is usually advantageous to use a 10-cm radar.

The attenuation produced by hail may be calculated from the general equation

$$\alpha = 0·4343 \sum nQ_t \text{ dB/km},$$

where $\sum nQ_t$ is the total attenuation cross-section per m³ and Q_t, given by the Mie theory, is in cm². A cloud containing uniform hailstones of diameter $D = 2a$ in concentration w g m⁻³ would then produce attenuation

$$\alpha_h = 0{\cdot}65\, Q'w/D \text{ dB/km} \tag{8.13}$$

where $Q' = Q_t/\pi a^2$ is the normalized attenuation cross-section as plotted in Fig. 8.12, where attenuation is shown to be greatest for

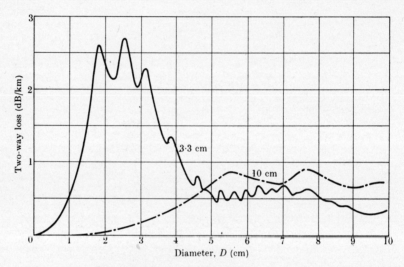

FIG. 8.12. Normalized attenuation cross-sections of ice spheres at 3·3- and 10-cm wavelengths, as a function of their diameter.

stones of $D \simeq \lambda$. At 3-cm wavelengths, greatest attenuation would be produced by stones of 3-cm diameter which, in concentrations of 1 g m⁻³, would attenuate the signal each way by 1 dB/km. For all-water spheres, Q' takes a nearly constant value of 2·0 for $\pi D/\lambda > 1$. Thus, unlike the case of small particles, where the attenuation by ice is negligible compared to that of water, attenuation by large ice spheres is comparable to that of water and sometimes exceeds it.

Experimental determinations of the attenuation of radar signals by rain have been made by Robertson and King (1946), and Anderson, Day, Freres, and Stokes (1947). The former used continuous-wave generators working on wavelengths of 1·09 and 3·2 cm, and single path lengths of about 1000 ft. The rate of rainfall was measured mid-way between transmitter and receiver. Their results were in acceptable agreement with the values calculated by Ryde (1946) and by Gunn and East (1954), except for precipitation rates of less than 10 mm/h when

FIG. 8.13. A PPI display of a cold front showing cores of high echo intensity.

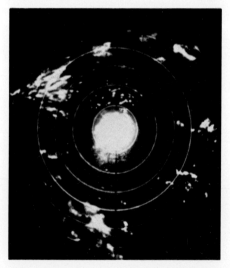

FIG. 8.14. Showers and thunderstorms on a PPI display. Range circles are 25 miles apart.

measurements became inaccurate. Anderson *et al.*, using 1·25-cm radiation, and nine rain-gauges over a path length of 6400 ft, found an average attenuation of 0·23 dB/km mm h^{-1}, which is about 50 per cent greater than the calculated value.

Tolbert and Gerhardt (1956) found that the measured attenuation in rain of 4·3-mm radiation agreed with that calculated from Mie theory if they used measured drop-size distributions but, if they used the Laws and Parsons (1943) mean data, the measured attenuation was approximately 50 per cent smaller than the theoretical values for rates of 30 to 50 mm h^{-1}. Using 2·15-mm radiation, Tolbert, Gerhardt, and Bahn (1959) also found good agreement on the average, although individual measurements sometimes differed from theory by as much as 100 per cent due to the deviation of the actual drop-size spectra from the Laws and Parsons data.

8.3. The presentation of radar information

The kind of information that the meteorologist can obtain from radar exploration of clouds includes the location of the storm in space, the general shape, extent, and movement of the precipitation areas, and the distribution of precipitation elements within the storm as revealed by variation of the echo intensity. Presentation of these data requires the use of more than one type of display.

That most suited to featuring the distribution of storms is the Plan Position Indicator (PPI) which gives a projection of the radar echoes received from all points of the compass on a horizontal plane, that is, a circular map centred on the radar set (Figs. 8.13 and 8.14). The time base is radial and starts from the centre of the cathode-ray tube. The echo causes the time base to brighten at the appropriate range, which is measured radially from the centre with reference to range-marker signals that brighten the trace at intervals of, say, 10 miles. The aerial rotates continuously in a horizontal plane in a period of 3–20 s, and the time base rotates in synchronism with it. The use of a long after-glow tube ensures that an echo persists for at least as long as the period of revolution of the aerial, so that a complete picture of the echo distribution is always presented. The PPI display thus provides the range and bearing of a storm relative to the radar set but no information concerning its height.

To study the vertical development of a cloud on a given bearing, a Range-Height Indicator (RHI) is used. The aerial is generally fixed in the azimuthal plane and scans in the vertical plane; the time base moves in synchronism with the tilt of the aerial and is brightened by the echo at a distance from the origin corresponding to the range. This produces a plot of the slant range and the angle of elevation of the target that is, in effect, a plot of slant range and height—see Figs. 8.15 and 8.16. Most RHI indicators suffer from the disadvantage that the tilt of the aerial is limited to, say, 30°, in which case, storms at horizontal distances nearer than $\sqrt{3}$ times their vertical extent cannot be completely examined. This particular limitation can be overcome by using a radar that scans from horizon to horizon through the zenith to present a vertical cross-section of the atmosphere on a fixed bearing—see Figs. 8.17 and 8.18. Difficulties arise, however, in interpreting the structure of, say, a thunderstorm echo passing nearly overhead, because changes in pattern in a particular vertical plane may be caused by the motion of the pattern in space across the plane of scanning, because the resolution in height on a RHI display is a function of both range and elevation, and because attenuation in heavy rain causes distortion of the pattern. The whole of the precipitation pattern in three dimensions can be recorded by taking a series of PPI photographs at appropriate elevation angles or a series of RHI photographs at appropriate azimuth angles, but this is a tedious process and synthesis of the many sections becomes difficult.

Many of the difficulties have been overcome by the constant plan position indicator (CAPPI) system first developed at McGill University (see for example, Marshall 1957), which presents a plan display of the echoes at one or more constant altitudes with respect to the earth's surface. Adequate vertical coverage may be achieved by obtaining PPI displays at six altitudes spaced, say, 1·5 km or 3 km apart.

After the operator has selected the particular altitude to be examined, an analogue computer 'gates' the incoming signals from a given range interval corresponding to that between which the top and bottom of the beam intercept the chosen altitude, and passes them to a storage tube. During a single scan only a torus of radar data is stored by the electrostatic image tube. During the next scan, the beam is elevated by an amount equal to the beam width and an adjacent torus is stored. In this way, the echo pattern at any one altitude is built up sequentially. The geometry of the CAPPI scan and the range gates are indicated in Fig. 8.19. Several altitudes may be recorded simultaneously by using several range gates, each operating a separate cathode-ray tube or storage device.

Using a radar with a 1° beam and scanning at 5 rev/min, a CAPPI

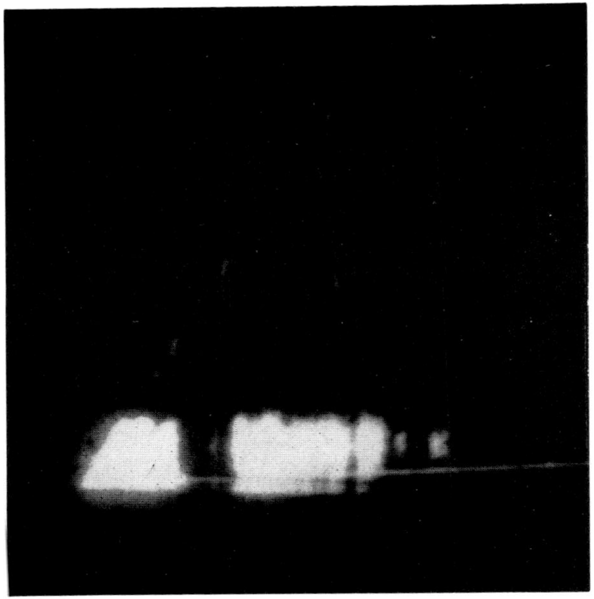

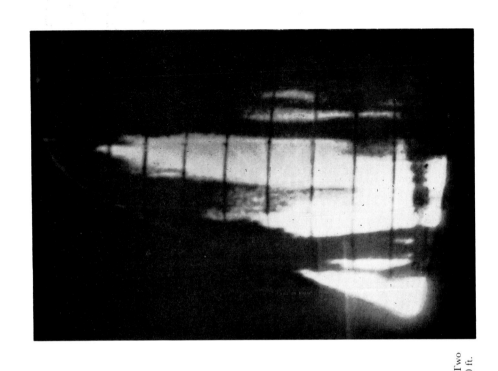

FIG. 8.15. A cold front presented on a RHI scope. The front is not very active as the echo tops are below 10 000 ft, but the individual cells are particularly well defined. The height markers (vertical scale) are at intervals of 5000 ft and the range markers are 5 miles apart.

FIG. 8.16. An echo from a very large thunderstorm shown on a RHI display. Two large cells compose the main echo, the top of which extended to a height of 42 000 ft.

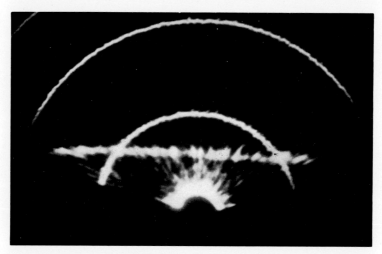

FIG. 8.17. A picture of warm front precipitation showing the melting band taken on a RHI display with the aerial sweeping in a vertical plane through the zenith. Range circles are at intervals of 2 miles. Wavelength 3 cm.

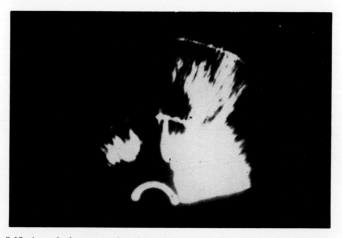

FIG. 8.18. A vertical cross-section through a thunderstorm as seen on a RHI display.

§ 8.3   RADAR STUDIES OF CLOUDS AND PRECIPITATION   431

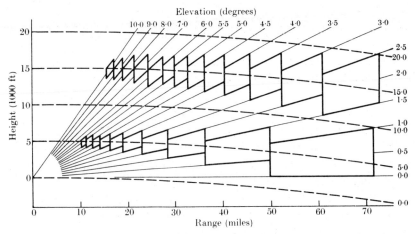

FIG. 8.19. The geometry of CAPPI scan. (From Battan, *Radar meteorology* (1959).)

display can be built up in about $3\frac{1}{2}$ min. This system has been used with considerable success to study such features as the height variations of echoes associated with hailstorms and the patterns of generating cells in stratiform precipitation.

On PPI and RHI displays, some indication of the intensity of the echo is provided by the brightness of the image on the tube, but it is extremely difficult to make accurate quantitative measurements of echo intensity from such intensity-modulated displays. However, a qualitative assessment of relative reflectivities in different parts of the storm may be obtained by reducing the gain of the receiver in a succession of calibrated steps, so that successively higher isopleths of signal intensity appear on the cathode-ray tube. Atlas (1953) developed an isoecho contour mapping technique by which two or more isoecho contours could be displayed simultaneously, by subjecting signals exceeding a predetermined threshold to an inversion process so that the region characterized by signals exceeding that threshold was blanked out on the scope. Additional isopleths can be displayed by using multiple thresholds and inverting the signals between alternate ones, thereby producing alternate white and black bands within the echo. This system operates satisfactorily only if the gradient of signal intensity with range is rather large, otherwise echo fluctuations from one pulse to the next cause the position of the visible echo boundary to fluctuate and appear fuzzy. With two or more contours, the true boundaries may then be indistinguishable. On the other hand, if the gradient of echo intensity is too great, the contours will blend into one another. Kodaira (1957) succeeded in sharpening the contours by integrating twenty-five consecutive pulses before feeding them into the contour-mapping device, thus reducing the signal variance by a factor of 5. Using a multiple contour mapper fed with the output from Kodaira's integrator, Niessen and Geotis (1963) produced maps showing as many as six isoecho lines. A similar device, in which the gain of the radar receiver is stepped over seven adjustable intervals to present the echo in seven distinct shades of gray, rather than as a continuous gradation from near white to near black, is described by Marshall and Gunn (1961).

Quantitative information on the signal amplitude is more easily obtained from a deflexion-modulated display in which the amplitude of the signal is plotted against the slant range of the target, or its height, if the aerial is pointed vertically upwards. The time base is deflected at a distance from the origin corresponding to the range, by an amount proportional to the amplitude of the received signal. This type of presentation, called an A-scope Indicator, is illustrated in Fig. 8.20. Something about the nature of the target can also be deduced from the appearance of the signal on the A-scope. Precipitation gives a very distinctive echo of rapidly fluctuating character which is caused by the interference pattern established by the precipitation elements changing from pulse to pulse, the individual scatterers being in continual motion relative to each other and to the radar.

8.4. Radar echoes from different cloud systems

From a study of radar displays in various kinds of weather, it has become clear that the cloud systems associated with different weather systems are characterized by markedly different radar echoes. Thus on a PPI display, the echo received from a cold front is very different from that associated with a warm front, while all echoes from frontal cloud systems are very different from those received from isolated showers and thunderstorms. These differences are brought out even more clearly on a RHI display giving a vertical cross-section through the cloud system. The general characteristics of radar echoes from different cloud systems have been described by Jones (1950).

A typical PPI picture of a cold front is shown in Fig. 8.13 as a long and generally very narrow band of echo, which is rarely continuous, but composed of a large number of small cores of high intensity. When seen in vertical section on the RHI scope (Fig. 8.14), the cellular structure is again in evidence, the cores of high intensity now being

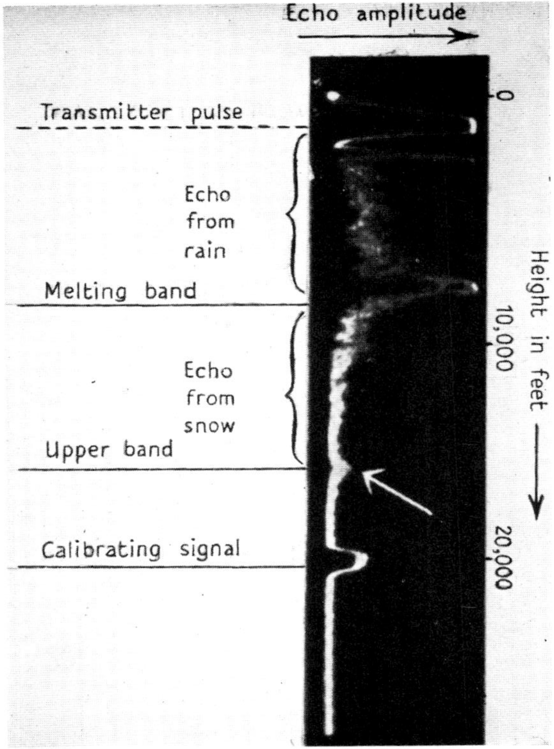

Fig. 8.21. A PPI display of a warm front on a 10-cm radar. The radial spokes in the echo are due to the screening effect of trees near the radar. The range circles are 5 miles apart.

Fig. 8.20. An A-scope presentation of a warm front, the aerial points vertically upwards to give a plot of echo amplitude against height. The picture shows a well-defined melting band and a weak upper band (shown by arrow). Wavelength 3 cm. (By courtesy of Dr. J. C. Browne.)

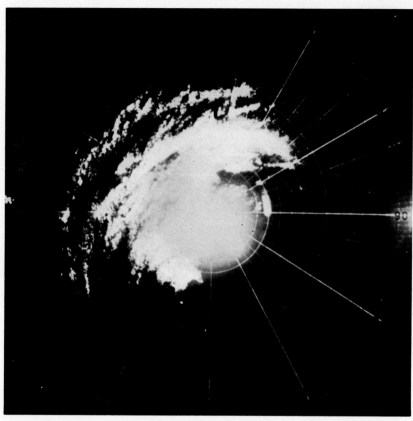

Fig. 8.22. A radar echo from Hurricane Edna 1954 appearing on the PPI display of a 23-cm radar at Cape Cod. The precipitation is asymmetrically distributed about the eye of the hurricane which is at 210° range 75 miles. Range circles are 10 miles apart.
(By courtesy of Dr. D. Atlas, Air Force Research Centre, Cambridge, Mass.)

§ 8.4 RADAR STUDIES OF CLOUDS AND PRECIPITATION

revealed as vertical columns of echo. These are no doubt produced by the cumulonimbi embedded in the frontal cloud system and are regions of strong vertical currents and heavy precipitation. Jones reports that the cumulonimbus activity is often confined to a narrow belt at the leading edge of the front perhaps only 5–10 miles wide, but a weaker, more extensive echo may be obtained from the altostratus sheet behind the surface front.

The typical PPI picture of the precipitation associated with an active warm front (Fig. 8.21) is one of widespread and diffuse echoes with no clearly marked boundaries. The cores of high intensity and rapid variations in echo intensity, which are characteristic of cold-front echoes, are generally absent from the warm-front display and the whole picture suggests a much more uniform precipitation rate over a much larger area. The first indication of the approach of a warm front on the PPI is the appearance of weak and diffuse patches of echo at fairly short ranges, which are associated with the first detectable precipitation, not necessarily reaching the ground, from the altostratus sheet ahead of the surface front. As the front continues its approach, these weak, diffuse echoes move closer to and past the station, while the heavier precipitation closer to the front produces an echo that gradually increases in size and intensity, until a PPI picture shows one large area of echo more or less symmetrically distributed around the station and fading gradually with range. In vertical cross-section, the warm-front echoes have a very characteristic structure. In almost every case, a narrow band of intense echo appears in the vicinity of the melting level, as shown in Fig. 8.17, while the remainder of the echo has a diffuse appearance and shows no marked columns or clear-cut edges. The melting band is sometimes observed in the later stages of development of a cold front when it is generally wider and more diffuse.

Mention has already been made of the tall, columnar echoes received from cumulonimbus associated with cold fronts. Isolated showers and thunderstorms of the convective type produce columnar echoes on the RHI display that on occasion may extend up to the tropopause, but show no essential differences from those of air-mass storms. Radar pictures reveal that a thunderstorm often exhibits more than one region of high echo intensity, these individual cells having horizontal dimensions of a few kilometres. Large and rapid fluctuations in intensity are characteristic of the echoes received from cumulonimbus. Fig. 8.18 is a typical example of a RHI picture.

The radar echo of a hurricane on a PPI display has characteristics which allow it to be easily identified (Fig. 8.22). The spiral bands of rain echo are observed to move slowly around the centre of the hurricane which is cloud-free, and the cells in the bands move along the bands and into the centre (in a clockwise direction in the Northern Hemisphere).

Having described the general characteristics of the precipitation echoes from different cloud systems, we shall now analyse them in greater detail and try to relate the radar information to the precipitation mechanisms described in Chapter 6.

8.5. Analysis of the radar signal

8.5.1. *Evaluation of the echo intensity*

Precipitation echoes on an A-scope show rapid fluctuations because the scattering particles move relative to one another, producing signals that sometimes reinforce and sometimes cancel each other. Such a target, in which the velocities of the component particles are not the same, is said to be incoherent; it would give a constant signal and appear coherent only if the time intervals between the radar pulses were short compared with the 're-shuffling' time of the particles. The degree of coherence or incoherence, which can be expressed in terms of the fluctuation rate or auto-correlation coefficient of the signal, can provide useful information on the relative motions of the particles in the pulse volume and this is discussed in § 8.5.3. Here we shall be concerned with the limits on the accuracy of measurement of the *average* power of a signal imposed by the random fluctuations. It is only this average power, integrated over a number of independent pulses, that is related to the reflectivity of a random array of scattering particles. In this context, the fluctuation rate or time to independence is important only in the determination of the actual number of independent samples taken from a series of consecutive pulses.

The theory of fluctuating echoes from a random array of independent scatters, well summarized by Atlas (1964), was developed by Goldstein (see Kerr 1951), Lawson and Uhlenbeck (1950), and Marshall and Hitschfeld (1953). The latter showed that the total average intensity of echoes from such a random array is equal to the sum of the intensities from the individual particles—an important result, for if it were otherwise, all deductions concerning the reflectivity of precipitation would depend not only on the particle size distribution but also on a knowledge of the particle positions and their variations with time. The probability distributions for the signal amplitude A, intensity A^2, and log (A^2) are given as follows:

amplitude: $\quad P(A)\, dA = (2A/\bar{A}^2)\exp(-A^2/\bar{A}^2)\, dA,\quad$ (8.14)

intensity: $\quad P(A^2)\, dA^2 = (1/\bar{A}^2)\exp(-A^2/\bar{A}^2)\, dA^2,\quad$ (8.15)

log (intensity):

$$P(\log A^2)\, d(\log A^2) = \frac{1}{M\bar{A}^2} \exp\left\{\frac{\log A^2}{M} - \frac{1}{\bar{A}^2} e^{(\log A^2)M}\right\} d(\log A^2),\quad (8.16)$$

where $1/M = \ln 10$, $\bar{A}^2 = \bar{P}_r$ the average received power, and $P(A)\, dA$ represents the probability that the signal amplitude will lie in the interval A to $A + dA$. In all these equations, of which the first two have been well confirmed experimentally by Austin, Bartnoff, Atlas, and Paulsen (1952), the entire population is assumed to consist of independent samples of signals, and so they apply to successive signals in time from the same pulse volume only after that volume has had time to reshuffle into another random array. This reshuffling time, or time to effectively zero correlation, depends upon the variance of the Doppler velocity spectrum, σ_v, and may be calculated from the following expression for the auto-correlation function for a Gaussian Doppler spectrum

$$\rho(\tau) = \exp -(4\pi\sigma_v\tau/\lambda)^2,$$

where τ is the time measured from when $\rho(\tau) = 1$ and λ the radar wavelength. Independence can be assumed when $\rho(\tau)$ falls to 10^{-2}, so that if $\sigma_v = 100$ cm s^{-1}, $\lambda = 3\cdot 2$ cm, the time to independence $\tau \simeq 5$ m s. In such a case, the theory of Lhermitte (1963) shows that with a radar transmitting pulses at 2 m s intervals (500/s), only 46 per cent of the pulses will be effectively independent. How many independent pulses are necessary in order to estimate the true average intensity of the signal within acceptable confidence limits? Marshall and Hitschfeld (1953) show that the measured average over twenty-five pulses will fall within ± 40 per cent of $\bar{A}^2$ for 95 per cent of the time, and since these limits ($-2\cdot 2$ to $+1\cdot 5$ dB) are comparable to the sources of random error in the experimental measurement of echo power, it is probably not necessary to average over more than about twenty-five independent pulses. In order to meet this requirement in the example just quoted, the radar signal would have to be averaged over a period of rather more than $0\cdot 1$ s. Marshall and Hitschfeld also considered the accuracy of averaging A and $\log(A^2)$, and found that averages of a given number of samples of either quantity can be made with comparable precision to those of A^2; for large numbers of pulses, the precision is identical.

On an A-scope having a persistent phosphor coating, or on photographs of the cathode-ray tube, successive echoes from the same range (abscissa position) will be superimposed, and the resulting luminance at any ordinate position will be proportional to the logarithms of the probability of occurrence of that ordinate. Whether the deflexion is proportional to A or to $\log(A^2)$, the peak luminance is close to $\bar{A}^2$, so the ordinate of maximum brightness on the scope (or maximum blackening on the film) provides a measure of the mean intensity of the signal. While the breadth of the A-luminance distribution appears much wider than that of $\log A^2$, the range of signal levels covered by the logarithmic receiver is so great that the precision with which one can estimate $\bar{A}^2$ is no better than that for the A display. However, because of its ability to handle the broad dynamic range of weather echoes, the logarithmic receiver is widely employed in modern radar.

The average signal intensity as determined from the radar scope has to be corrected for non-linearity or non-uniformity in the receiver circuits or in the sensitivity of the cathode-ray tube. This can be done by injecting a calibrating signal into the receiver from a signal generator fitted with a calibrated attenuator. At a given point on the A-scope time-base corresponding to a fixed range, the amplitude of the calibration signal is varied by the attenuator and the deflexion noted, the procedure being repeated for different ranges. From these readings, a table can be drawn up of the corrections to be applied to the measured deflexions at any particular range.

The signal/noise ratio can also be determined with the calibration signal, which is made to appear on the screen simultaneously with the received signal from precipitation. The amplitude of the calibration signal is adjusted by the attenuator, reading K_1 dB from a reference level, to give a suitable deflexion A_1 and is then re-adjusted (attenuator setting K_2) to give a deflexion equal to that of the noise A_n. If the root-mean-square amplitude of the received echo at a given range is A_0 and the received power P_r, we now have

$$P_1/P_n = (A_1/A_n)^2 = 10^{(K_2-K_1)/10};$$
$$P_r/P_n = (A_0/A_n)^2 = (A_0/A_1)^2 \text{ antilog } (K_2-K_1)/10,$$

from which P_r/P_n, the signal/noise ratio, can be found. By measuring A_0 at different ranges, we can determine P_r/P_n at different parts of the storm by correcting for range according to the inverse square law.

If absolute values of the reflectivity and hence of $\sum nD^6$ are required, the intensity of the received signal can be compared with that from a metal sphere, the scattering cross-section of the latter being known. If the radar is calibrated very carefully in this manner, the quantity $\sum nD^6$ may be determined to within about 30 per cent.

The pulse integrator. The echo-signal voltage may be measured at the output of the video amplifier, before it is presented on the cathode-ray tube, with an instrument called a pulse integrator, described by Williams (1949), which measures electronically the average amplitude of the pulsating, fluctuating signals over a short interval of time. This interval, which determines the number of pulses averaged, may be varied to suit the particular conditions, but a period of 2–4 s is generally satisfactory. The output of the pulse integrator may be fed into any of several types of visual meter or recorder, and the records are calibrated by feeding calibration signals of known power into the antenna. This method is free from observer bias and variability. Austin and Williams (1951) have compared the average signal strength as measured along the lines described in the last section and by the pulse integrator, and obtained agreement within 1 dB. The main disadvantage of using the pulse integrator is that the signal strength can be measured only for one range at a time, and unless the precipitation is very steady, errors will result in comparing the echo intensities at different levels if these are measured at different times.

8.5.2. *Signal fluctuations and Doppler radar*

So far, in dealing with the echo intensity and its relations to the back-scattering and attenuating properties of the medium, we have been concerned only with obtaining the *average* intensity of a rapidly fluctuating signal from randomly distributed scattering particles. We shall now consider what meteorological information may be extracted from the signal fluctuations themselves. The instantaneous strength of the echo depends upon whether the particles are interfering constructively or destructively, and so the *rate of fluctuation* depends on the velocities with which the particles move *relative to one another* across the constant phase surfaces that are perpendicular to the radial direction of propagation. Thus the echo fluctuations on a conventional or 'incoherent' pulsed radar are related to the *relative* motions of the particles in the line-of-sight or radial direction. In practice, the scattering particles in the finite beam of a vertically-directed radar will possess *mean* radial velocities by virtue of their terminal velocities and any vertical currents acting upon them; they will also move relative to one another because of differences in terminal velocity, because of turbulence, and because the horizontal wind will produce a small radial component that will vary across the finite width of the beam. All these effects will contribute to the fluctuations in echo intensity which may be observed on conventional incoherent radar but, as we shall see, having once obtained the absolute velocities (with a coherent Doppler radar), the relative motions and the fluctuation rate are readily computed.

The Doppler shift. The number of wavelengths contained in the two-way path, $2R$, between a point target and the radar is $2R/\lambda$, and the corresponding phase shift is $4\pi R/\lambda$. If the target moves with respect to the radar with a radial velocity $v = dR/dt$, the phase position changes proportionately at an angular frequency

$$\omega = 2\pi f = 4\pi v/\lambda$$

or
$$f = 2v/\lambda, \tag{8.17}$$

f being the Doppler shift-frequency resulting from the radial velocity v. If v is expressed in m/s and λ in cm,

$$f = 200\, v/\lambda. \tag{8.17a}$$

If $v = 10$ m s^{-1} and $\lambda = 10$ cm, the Doppler shift is only 200 Hz—a very small difference from the transmitted frequency of 3×10^9 Hz, so Doppler radars are designed to measure very small frequency differences.

Signal intensity from two scatterers. In general, the field back-scattered from a collection of point targets can be represented by

$$A(t) = \mathscr{R}\{a(t)e^{-i\omega_0 t}\}, \tag{8.18}$$

where $\mathscr{R}$ denotes the real part of the bracketed quantity, $a(t)$ is the amplitude associated with the real signal $A(t)$, and ω_0 is the transmitted angular frequency. For a single stationary target, $a(t)$ is a constant. If the target moves with velocity v_1, then $a_1(t) = a_1 e^{-i\omega_1 t}$ (with a_1 a constant), and

$$A_1(t) = \mathscr{R}\{a_1 e^{-i\omega_1 t} e^{-i\omega_0 t}\} = a_1 \cos(\omega_0 + \omega_1)t. \tag{8.19}$$

A Doppler radar measures the frequency ω_1, and hence the radial velocity can be calculated from eqn (8.17), $v_1 = \omega_1 \lambda/4\pi$.

With an incoherent radar, the quantity available for measurement is the *signal* intensity

$$I(t) = \overline{A^2(t)},$$

where the bar denotes the time average over one cycle of the carrier. In the case of a single target moving with velocity v_1,

$$I(t) = a_1^2 = \text{const.},$$

so the intensity is independent of the velocity of the target.

For two unresolved targets giving rise to two Doppler frequencies ω_1 and ω_2, the resultant field is given by

$$a(t) = a_1(t) + a_2(t) = a_1 e^{-i\omega_1 t} + a_2 e^{-i\omega_2 t}, \qquad (8.20)$$

$$A(t) = a_1 \cos(\omega_0 + \omega_1)t + a_2 \cos(\omega_0 + \omega_2)t, \qquad (8.21)$$

and

$$I(t) = \overline{A^2(t)} = a_1^2 + a_2^2 + 2a_1 a_2 \cos(\omega_2 - \omega_1)t. \qquad (8.22)$$

Thus $I(t)$ on an incoherent radar is composed of a steady term, $(a_1^2 + a_2^2)$, representing the sum of the intensities contributed by the individual particles, and a fluctuating term of angular frequency $(\omega_2 - \omega_1)$. Since the time average of this fluctuating component is zero, it does not contribute to the *average intensity* of the signal.

The frequency of the signal fluctuation

$$F = \psi/2\pi = (\omega_2 - \omega_1)/2\pi = 2(v_2 - v_1)/\lambda, \qquad (8.23)$$

results from the beating of the two Doppler frequencies with one another. Lhermitte (1960a) has derived a more general form of eqn (8.22) for any number of scatterers, viz.

$$I(t) = \sum_i a_i^2 + 2 \sum a_i a_j (\cos \psi_{ij} t). \qquad (8.24)$$

The steady part, $\sum a_i^2$, is proportional to the total reflectivity of all the scatterers, and the fluctuating portion contains all the fluctuating frequencies $\psi_{ij} = \omega_i - \omega_j$.

Eqns (8.23) and (8.24) define the basic relationship between the frequency shifts $\omega_1, \omega_2, \ldots$ as measured by Doppler radar and the signal fluctuation rates of an incoherent radar which measure the *relative* frequencies $(\omega_2 - \omega_1) \ldots$, etc.

Returning to eqn (8.21), let us consider the case where one of the two targets is fixed. Then $\omega_1 = 0$, and the frequency difference of the two components contributing to $A(t)$ is $\psi = (\omega_0 + \omega_2) - \omega_0 = \omega_2$. This is the basis of Doppler radar systems used by Barratt and Browne (1953) and Lhermitte (1960b) in which the frequency of the signal returned from a cloud was compared with that returned simultaneously from a stationary target. The returning signals were mixed in a single receiver, their beat frequency being a measure of the Doppler shift of the precipitation. This method has the disadvantage that one can make measurements only at fixed levels in the cloud corresponding to the ranges of the fixed targets. This limitation can be removed by holding the transmitted frequency in a memory until all the echoes are returned from the maximum range and then mixing this with the Doppler-shifted signals from the target. Modern Doppler radars are built on this principle, and are said to be 'coherent' because the reference frequency is coherent with the transmitter frequency. Such a radar was used in England in 1958 by Boyenval (1960) to measure the falling speeds of raindrops, and vertical air currents in convective clouds. In recent years several similar radars have been used with good effect in the United States.

In practice, the following information can be readily extracted from the signals from a vertically-pointing Doppler radar:

the minimum velocity of the Doppler spectrum corresponding to the vertical velocity of the slowest-falling particles;

the maximum velocity of the spectrum corresponding to the vertical velocity of the fastest-falling particles;

the vertical velocity corresponding to the peak intensity of the Doppler spectrum. Since the latter is usually Gaussian in form, the peak will correspond closely to the mean vertical velocity of the particles;

the width of the vertical velocity spectrum defined at a signal level equivalent to a particular value of the equivalent reflectivity Z_e; and

the equivalent radar reflectivity integrated over the entire Doppler spectrum as calculated from measurements of the back-scattered power and the characteristics of the radar.

In interpreting the signal, it is important to recognize that the Doppler spectrum is influenced by

(i) the mean vertical air motion within the pulse volume,

(ii) the spectrum of particle fall-speeds within the pulse volume,

(iii) gradients of vertical air motion across the pulse volume,

(iv) the spectrum of turbulence on scales smaller than the pulse volume, and

(v) radial components of the horizontal wind along the edges of the beam.

Factor (i) is revealed as a displacement of the entire spectrum, while factors (ii)–(v) produce a broadening of the spectrum. Some important results of investigations with Doppler radar are summarized in § 8.6 below, but first we shall briefly mention the type of information that may be obtained by analyzing the spectrum of intensity fluctuations provided by an incoherent radar.

8.5.3. *The Doppler spectrum and spectrum of intensity fluctuations*

Following earlier analyses by Siegert (1943), Fleisher (1953), Rogers (1963), and Lhermitte (1963), Atlas (1964) demonstrates that the power spectrum of the Doppler frequency, in which the frequency components are weighted according to the square of the amplitude of the back-scattered field and hence to the radar cross-sections of the scatterers, provides a weighted image of the actual velocity distribution within the pulse volume. On the other hand, the power spectrum of the intensity fluctuations may be interpreted as a *relative* velocity distribution. The two are uniquely related and the one may be derived from the other provided the actual velocity distribution (i.e. the Doppler spectrum) is symmetrical. In other words, measurements with an incoherent radar can be used to deduce the significant meteorological parameters that determine the width and shape of a symmetrical Doppler spectrum. However, relatively little has been accomplished in this direction, probably because of the difficulties of separating the effects of turbulence, wind shear, vertical air currents, and the breadth of the particle-velocity spectrum. Perhaps the most useful technique, capable of giving limited information, is incorporated in the R-meter designed by Rutkowski and Fleisher (1955) to measure the number of times in a sampling interval that the signal level crosses a particular preset threshold. Obviously this is related to the signal fluctuation rate. Lhermitte (1963) showed that the number of times per second, N_A, that a signal obeying the Rayleigh distribution of eqn (8.14) crosses a given amplitude level A with positive slope is

$$N_A = A/(\bar{A}^2)^{\frac{1}{2}}(2\pi\bar{F}^2)^{\frac{1}{2}}\exp(A^2/\bar{A}^2). \tag{8.25}$$

If $\qquad A = \bar{A} = 0.79(\bar{A}^2)^{\frac{1}{2}},$

then $\qquad N_{\bar{A}} = (\bar{F}^2)^{\frac{1}{2}} = \sqrt{2}\sigma_f = 2\sqrt{2}\sigma_v/\lambda, \tag{8.26}$

where $(\bar{F}^2)^{\frac{1}{2}}$ is the r.m.s. frequency of the fluctuation spectrum and σ_f is the standard deviation of the Doppler spectrum, and σ_v is the r.m.s. of the radial velocity of the particles. Lhermitte has confirmed these relations experimentally.

8.6. Meteorological information from Doppler radar

8.6.1. *Determination of drop-size distributions from the Doppler spectrum.*

When a Doppler radar is pointed vertically in rainfall from a stratiform cloud, in which both the mean vertical air velocity and the turbulence are negligible, and when the effective broadening of the beam width due to the horizontal wind is also negligible or easily calculated, the Doppler spectrum is determined completely by the falling speeds of the particles. Since the falling speed of the drop is completely defined by its diameter if proper allowance is made for the air density at the altitude concerned, the velocity scale of the spectrum becomes a diameter scale. If $N(v)$ dv denotes the number concentration of drops falling with velocities between v and $v+\mathrm{d}v$ and diameters between D and $D+\mathrm{d}D$, $N(v) = N(D)$ dD/dv. If the drops are Rayleigh scatterers, then the fraction of the received power returned by drops whose velocity components in the direction of the radar lie between v and $v+\mathrm{d}v$, is $S(v)$ dv, where

$$S(v) = kN(D)D^6 \, \mathrm{d}D/\mathrm{d}v, \qquad (8.27)$$

k being a constant for the radar. If $S(v)$ is measured and corrected for range attenuation for each v (or D) interval, relative values of $N(D)$ can be determined for the complete spectrum of drop diameters. This method has been used successfully by Boyenval (1960), and Rogers and Pilié (1962), who point out that a check on the resultant drop-size distribution can be obtained from the average power

$$P_\mathrm{r} = \int_{-\infty}^{+\infty} S(v) \, \mathrm{d}v = k \int_{-\infty}^{+\infty} N(D)D^6 \, \mathrm{d}D,$$

this being necessary since the spectrum may be disturbed by an unknown steady vertical motion of the air. Boyenval, using a 3·2-cm pulsed Doppler radar of peak power 10 kW with a pulse length of 0·8 μs, was able to measure line-of-sight velocities with a range resolution of 500 ft (150 m) and a velocity resolution of 1 m s^{-1}. The information was displayed in the form of a rectangular raster having 20 columns representing velocity intervals of 1 m s^{-1}, and 80 rows representing range (or height) intervals of 150 m. Fig. 8.23 is a photograph of a typical display obtained from a warm front with the radar beam pointing vertically. It shows snowflakes falling with velocities of about 1 m s^{-1} between the 4-km and 2-km levels, the acceleration of these particles during melting and, finally, the velocity distribution of the raindrops at lower levels. The echo from the hydrometeors appears as an intensity-modulated spot on the cathode-ray tube at a point corresponding to the range and radial velocity of the scatterer, the intensity indicating the power returned from scatterers falling within each element of the array. The relative intensities of the echoes in each

§ 8.6   RADAR STUDIES OF CLOUDS AND PRECIPITATION   443

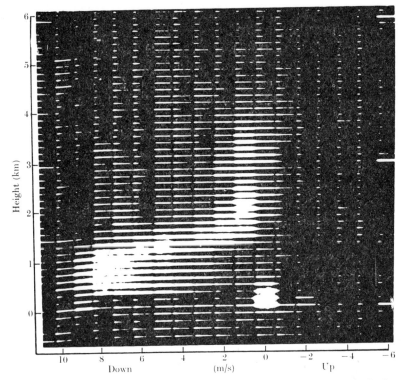

FIG. 8.23. Velocity-height display of a Doppler radar pointing vertically into steady precipitation. The signals show snow falling at about 1 m s^{-1} between 4 and 2 km, the acceleration of the snowflakes during melting, and raindrops falling at velocities of up to 9 m s^{-1}.

element were determined by measuring the attenuation required to make the signals disappear into the background noise. In a later version of the equipment, the signals from a number of range gates are recorded on multi-track magnetic tape and subsequently fed through a frequency analyser, the final output being a digital print-out of the returned power as a function of velocity, obtained within each 150-m range element once every 15 s.

Using the same Doppler radar, Caton (1966) made a series of 83 observations in fairly uniform steady rain. He derived an absolute scale of raindrop concentration from a comparison of the computed total flux of water at the lowest level of analysis (750 m) with that read from an open time-scale rain recorder 0·5 km from the radar. The total probable error in the determination of drop concentrations, averaged over three height intervals, each of 450 m, was estimated to be +40

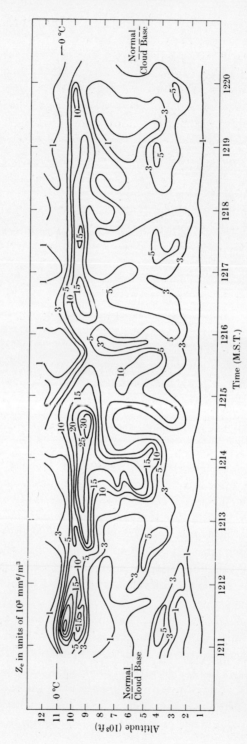

FIG. 8.24. Height-time contours of radar reflectivity from a layer cloud showing high reflectivity from a melting layer between 9000 and 11 000 ft. (From Du Toit (1967).)

per cent or −30 per cent. The 83 drop-size spectra showed substantial variations even from one 5-min period to another and, even when grouped according to the rainfall rate, still showed considerable variability and departures from, say, a Marshall–Palmer distribution. This accords with the filter-paper and raindrop-spectrometer observations of Mason and Andrews (1960). The observed median distributions at 525–975 m altitude contained fewer small drops ($D = 0.5$–0.75 mm) and fewer large drops ($D > 2$ mm) than the corresponding Marshall–Palmer distribution. The fact that the latter distribution overestimates the numbers of small drops from light and moderate rain is well known; the excess number of large drops at the lower levels may perhaps be attributed to coalescence.

Du Toit (1967) also used a vertically-pointing Doppler radar to measure the vertical velocities of the drops from continuous rain. Simultaneous measurements of drop size, and hence rate-of-rainfall, were made by exposing dyed filter papers. The height-time contours of radar reflectivity are shown in Fig. 8.24 in which the band of relatively high reflectivity between 9000 and 11 000 ft represents the melting layer. Within this layer, the height of the zone of maximum reflectivity shows considerable variations. The Doppler spectra and power distribution indicated that the occurrence of maximum reflectivity at relatively high levels was associated with particles of low falling speed, while the appearance of maximum reflectivity at low levels was associated with more rapidly falling particles. At one extreme there were particles in the melting layer falling at <3 m s^{-1}; these were probably of low density and melted completely over fall paths of only 200 m. At the other extreme, there were particles falling at 3–6 m s^{-1}, probably of high density, and having to descend about 1 km below the 0° C level before melting completely. These latter particles were thought to be soft hail pellets of density 0.3–0.4 g cm^{-3} and diameter 5–6 mm, falling from precipitation streamers and melting to produce raindrops of diameter 4 mm. The presence of such drops just below the melting layer was indicated by strong echoes in the 9–11 m s^{-1} velocity channels, in circumstances in which detailed analysis indicated the down-draughts to be negligible. Outside the regions of high reflectivity, and just below the melting zone, the drop-size spectra peaked at diameters of about 1 mm, but showed a marked deficiency of drops of $D \simeq 0.35$ mm. Between cloud base and ground, the distributions were very similar to those obtained on the filter papers, and to the ground measurements of Mason and Ramanadham (1953), except that there appeared to be high concentrations of small drops of $D < 0.5$ mm that were difficult to explain.

Rogers (1967) used a 3·2-cm pulsed Doppler radar to investigate warm orographic rains in Hawaii. Using a 0·5-μs range gate to isolate a small portion at a time of the weather echo for analysis, he obtained its average intensity within ± 4 dB, the mean Doppler velocity within $\pm\frac{1}{2}$ m s^{-1}, and the spread (standard deviation) of the Doppler spectrum using an R-meter. Height–time profiles of the various quantities were obtained with a vertically-pointing beam, but were not easy to interpret because they reflected both time and space variations. Nevertheless, the echo intensity profiles showed several features that are rather characteristic of maritime clouds. Shower echoes were always shallow and rarely extended more than 3·5 km above ground. The strongest echoes, with reflectivities approaching $Z = 10^5$ mm^6 m^{-3}, were associated with the greatest vertical development and strongest updraughts, this being consistent with raindrop growth by coalescence. The presence of considerable updraughts in most of the clouds made it difficult to determine the size distributions of raindrops very accurately because the deduced number density was very sensitive to the magnitude assumed for the updraught. However, as well as could be determined, the drop-size spectra were consistent with the coalescence process in showing a progressive increase in the proportion of larger drops at lower levels.

8.6.2. *Measurements of air motions and precipitation growth in showers and thunderstorms*

We have seen that in widespread 'steady' rain from stratiform clouds, where the vertical air motions are small compared with the falling speeds of the particles, pulsed Doppler radar can determine the raindrop size spectra and their variation with height. Conversely, if limits can be set to the terminal velocities of the particles, the Doppler velocities can be corrected to give bounds for the magnitudes of the vertical motions. For example, if the radar echo from a cloud shows a well-marked melting band, this indicates the presence of snowflakes above the 0° C level, with terminal velocities of about 1 m s^{-1}; in such a case the absence in the Doppler spectrum of velocities greater than 1 m s^{-1} would indicate the vertical motions to be weak. In the presence of steady updraughts or downdraughts, the Doppler spectrum would be displaced; knowledge of the largest or smallest falling speeds of the contributing particles in the sampling volume would then enable the air motions to be deduced.

This technique was applied successfully by Probert-Jones and Harper (1961), who noticed that, in modest shower clouds, the particles consistently accelerated in falling through the melting level, and took this as evidence for the existence of large ice crystals or snowflakes above that level. They therefore corrected the largest recorded Doppler velocities by 1 m s^{-1} to give the vertical air motions at all levels above the 0° C level. In extending the analysis below the melting level, continuity of vertical motion across the melting zone was assumed in order to calculate the terminal velocities of the drops just below it.

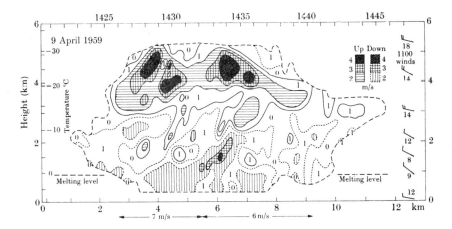

FIG. 8.25. The pattern of vertical motion in a shower cloud. The left-hand side is the leading edge of the shower. (From Probert-Jones and Harper (1961).)

The vertical air motion was then deduced from the maximum downward velocities of the Doppler spectrum on the assumption that the *largest* drops did not accelerate appreciably during their fall from just below the melting zone to the ground.

Fig. 8.25 shows a height-time analysis of the vertical motions so derived for a shower that produced 0·7 mm of rain at the ground, with a maximum rate of 8 mm h^{-1} for a short period near the start. The terminal velocities of the largest drops, as inferred from the Doppler record, were 7 m s^{-1} falling later to 6 m s^{-1}, the corresponding diameters being 2·5 and 2·0 mm. Fig. 8.25 shows the upward air motion to be concentrated in the top forward half of the cloud, while the downward motion is mainly below. The strongest upward motion of 4 m s^{-1} is close to the top of the echo, which reached 5·4 km (temperature −31° C). The downward motion is shown as mainly 2 m s^{-1}, but there is one cell of 3–4 m s^{-1} in the middle of the shower, just above the melting level. The initial echo appeared well above the 0° C level, between 2 and 2½ km, and it was 5 min before the shower reached the ground. During the decay stage of the shower, beyond 1440 h, the echo above the melting level was almost entirely associated with downward velocities of 1–2 m s⁻¹, and therefore almost certainly due to the fallout of snow. The pattern of vertical motion, though quite complex,

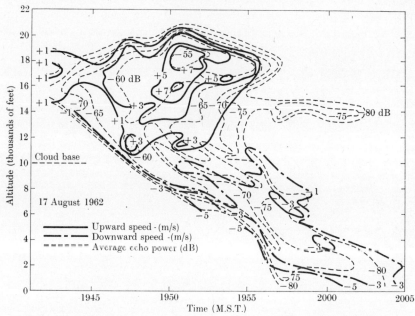

FIG. 8.26. Height–time contour map of range-normalized signal intensity, updraught speeds, and downdraught speeds, in a cumulonimbus passing over a vertically pointing Doppler radar. (After Battan (1964).)

was not complicated on this occasion by variation of wind with height and shear of precipitation across the plane of the height–time section. A detailed study of seven additional showers is described by Caton (1967).

A rather different technique has been used to investigate thunderstorms in Arizona by Battan and his colleagues. The results of Battan (1964) are shown in Fig. 8.26, where he has superimposed isopleths of updraughts and downdraughts on range-normalized contours of echo intensity. Prominent features of the diagram are the coincidence of the updraught and reflectivity maxima towards the top of the storm echo, the general similarity of the updraught and reflectivity patterns above cloud base, and the development of a moderately strong downdraught

§ 8.6   RADAR STUDIES OF CLOUDS AND PRECIPITATION   449

in the rain streamer below cloud base, especially near its leading edge where reflectivity decreases sharply downward. The decrease in reflectivity indicates strong evaporation of the hydrometeors in the dry Arizona environment, while the downdraught itself may be attributed to the evaporative cooling. When the first radar echo appeared, the cloud containing only weak updraughts, but these attained a maximum speed of 7 m s^{-1} and, within 3 min, a downdraught appeared in the lower part of the cloud. The updraught and downdraught intensified together, and the former is seen to consist of several maxima appearing at successively greater heights, suggesting a series of rising thermals. In the later stages of the storm, at about 1958 hours, a melting band appeared at about 14 000 ft (4·2 km). Having inferred the vertical velocities of the air, Battan took the difference between these and the maximum downward Doppler velocities to give the terminal velocities of the largest hydrometeors. He estimated that during the early life of the echo, the largest drops were 1–2 mm in diameter but, later on, particles with terminal velocities of 15 m s^{-1} appeared at about 11 000 ft and were therefore likely to be hail pellets, ranging from 5 to 9 mm in diameter, depending upon their density. Lightning and thunder occurred at 1955 hours, 10 min after hydrometeors had started to fall towards the ground.

A similar analysis of a growing thunderstorm by Battan and Theiss (1966) revealed many similar features, including the presence of hail pellets about 1 cm in diameter. Battan, Theiss, and Kassander (1964) made observations on a decaying thunderstorm and noted that, although the updraughts up to the altitude of 28 000 ft did not exceed 0·5 m s^{-1}, the cloud nevertheless continued to produce lightning strokes at 10-min intervals. There were, however, extensive regions between 15 000 and 25 000 ft with downward Doppler velocities of 5 m s^{-1} that indicated the presence of hail pellets.

Atlas (1964) has criticized Battan's method of computing the downdraughts below cloud base in Fig. 8.26. Battan assumed the presence of detectable quantities of 0·3-mm drops with terminal velocities of $\sim$1 m s^{-1} and corrected the lower bound of his Doppler spectra accordingly. But Atlas points out that, in the precipitation streamer shown in Fig. 8.26, the drops may have been sorted by the action of wind shear and so, in the absence of an independent estimate of the falling speed of the smallest drops actually present, the computed downdraughts may be in error. Of course, once the rain has reached the ground, the observer can determine the maximum sizes and fall speeds of the particles and estimate the corresponding fall speeds aloft. These should correspond to the maximum downward (particle) velocity in the Doppler spectrum; any excess can be attributed to the downdraught. However, in the presence of strong turbulence, wind shear, and hailstones, the Doppler spectrum is often ambiguous and difficult to interpret. The complications introduced by windshear underline the fact that a complete picture of the evolution of a storm, especially a convective storm, can come only from a four-dimensional space and time representation of the air and particle motions. The degree to which this can be achieved by the simultaneous use of two or three Doppler radars is discussed later.

An important step in this direction has been made by Browning, Harrold, Whyman, and Beimers (1968), who have used two Doppler

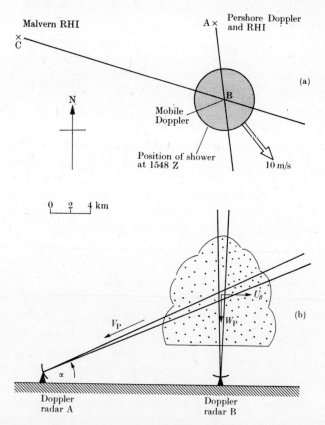

Fig. 8.27. (a) Location of radars and shower in plan view. (b) Schematic diagram showing two Doppler radars and the shower in vertical section. (From Browning *et al.* (1968).)

§ 8.6   RADAR STUDIES OF CLOUDS AND PRECIPITATION   451

radars and two conventional radars with RHI displays, to reveal two-dimensional patterns of both horizontal and vertical air motions and the growth of precipitation within a shower. The location of the radars relative to the position of the shower as it passed over the vertically-pointed Doppler radar, B, is shown in Fig. 8.27. Doppler A was operated with its aerial pointing along the azimuth of B at a sequence of elevations α between $1°$ and $45°$, each sequence taking about 1 min to complete. A velocity corresponding to the peak of the Doppler spectrum, V_p, was estimated for each range gate from the itennsity-modulated display of radar A.

Now, $$V_p = -U_\beta \cos \alpha + W_p \sin \alpha, \quad (8.28)$$

where U_β is the component of the horizontal wind along the azimuth of the beam (positive away from the radar) and W_p the vertical velocity corresponding to the peak of the spectrum (positive downwards). Since W_p was known vertically above B, it was possible to calculate the horizontal wind component U_β above B, and data from a sequence of elevation scans permitted the construction of height-time sections of U_β above B. The component of the wind along AB relative to the shower is
$$U_r = U_\beta - U_s, \quad (8.29)$$

U_s being the component of the horizontal motion of the shower along AB as measured by the conventional radars.

Fig. 8.28(a)–(c) show height-time sections in which the time scale may be interpreted as horizontal distance on the assumption that the shower maintains a steady state during its passage over the vertically-pointing Doppler radar. The diagrams, on which the time scales have been reversed to put the front of the storm on the right, show

(a) contours of the vertical velocity, W_p, and areas shaded to show where the minimum Doppler velocity, W_{min}, was $<0·5$ m s^{-1} and $>1·5$ m s^{-1};

(b) contours of W_p, and of the component of the horizontal wind, U_r, along AB relative to the shower;

(c) the distribution of the Doppler spectrum width (in this particular case almost synonymous with the intrinsic fall-speed spectrum of the hydrometeors), and contours of total radar reflectivity Z_e.

Fig. 8.28(c) shows that both Z_e and the spectrum width were much reduced above the $-30°$ C level where the particles were probably ice crystals having terminal velocities of about $0·75$ m s^{-1}. Estimates of the

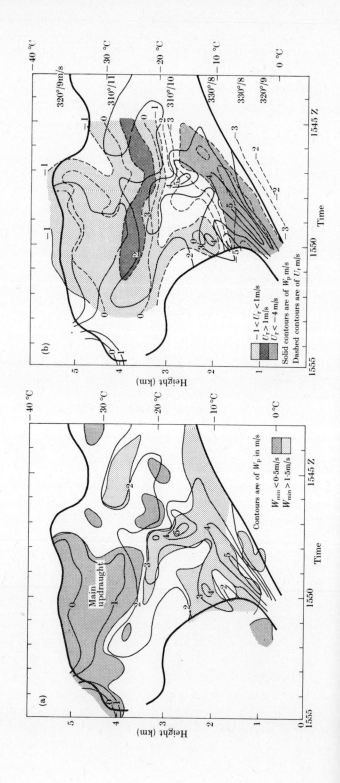

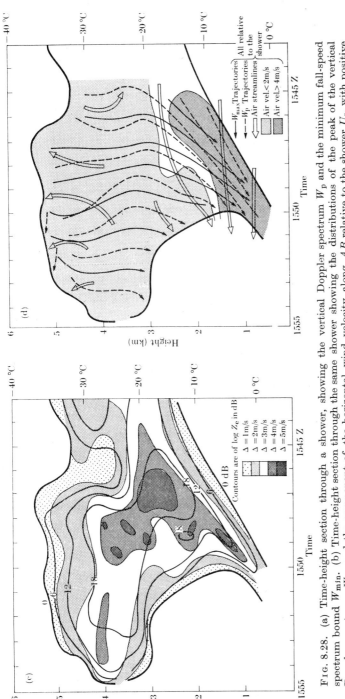

FIG. 8.28. (a) Time-height section through a shower, showing the vertical Doppler spectrum W_p and the minimum fall-speed spectrum bound W_{min}. (b) Time-height section through the same shower showing the distributions of the peak of the vertical Doppler spectrum, W_p, and the component of the horizontal wind velocity along AB relative to the shower U_r, with positive values representing winds in excess of the shower velocity. Environmental winds plotted to the right of the diagram are expressed in degrees and m s^{-1} relative to the ground. (c) Time-height section through the shower, showing the distribution of spectrum width Δ, and the total equivalent radar reflectivity factor Z_e, expressed in terms of $10 \ln Z_e$. (d) Time-height section through the shower showing streamlines of air motion, and trajectories of the largest particles (of fallspeed W_{max}) and of the particles corresponding to the peak of the vertical Doppler spectrum (W_p). All streamlines and particle trajectories are drawn relative to the shower. (From Browning et al. op cit.)

vertical motions were therefore made by subtracting 0·75 m s⁻¹ from
the values of W_{min}, and revealed the existence of only weak vertical
motions, a main updraught of only about 1 m s⁻¹ above 4 km, and a
fairly extensive downdraught of 1–2 m s⁻¹ below 2·5 km. The light
shading of Fig. 8.28(b) denotes areas where U_r was within ± 1 m s⁻¹
of the horizontal velocity of the shower, most of the updraught being
located in this area. In a small region near the 3·5-km level, the relative
wind was towards the front (right) of the shower, but at lower levels the
relative winds blew towards the rear, particularly along the axis of
the streamer (depicted in Fig. 8.28(b) by the W_p contours), where they
reached 4 m s⁻¹. This resulted in the streamer being strongly inclined.
Fig. 8.28(c) shows that the distribution of spectrum width was similar
to that of the reflectivity, Z_e, which reached maximum values of
$\sim 10^2$ mm⁶ m⁻³ in the middle parts of the shower, and shows that
particle growth, as indicated by the gradients of Z_e, occurred principally
within the updraught area between the 4- and 5-km levels. The contours
of W_p in Fig. 8.28(a) show that the particles contributing most to
Z_e were descending in most places, with fall-speeds increasing from 2
to 5 m s⁻¹; Browning *et al.* demonstrate that this behaviour is consistent with soft hail pellets growing to a maximum diameter of about
6 mm in a liquid-water concentration of 1 g m⁻³—about half the
adiabatic value. The predominance of downdraughts below 2·5 km
may have been due to chilling of the air by partial evaporation of the
precipitation.

The vertical and horizontal motions were combined to give the
streamlines of air motion and particle trajectories shown in Fig. 8.28(d).
This draws attention to the air-flow relative to the shower, and to the
trajectories of both the largest and 'average size' particles feeding the
precipitation streamer that produced a shower of hail and rain at
the ground.

8.6.3. *Determination of horizontal winds, wind shear and convergence*

As the beam of a Doppler radar is tilted down from the vertical, the
radial motion of the particles has an increasingly large component due
to the horizontal wind. In a horizontally uniform wind field of velocity
V_h at the sampling height, as indicated in Fig. 8.29(a), the mean radial
velocity measured by the Doppler radar is

$$V_d = V_t \sin \alpha + V_h \cos \beta \cos \alpha, \qquad (8.30)$$

where α is the elevation angle of the beam, β the azimuth angle of the

§ 8.6 RADAR STUDIES OF CLOUDS AND PRECIPITATION 455

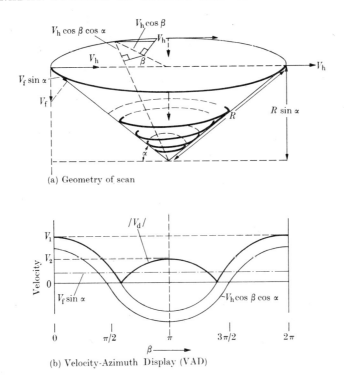

FIG. 8.29. (a) Geometry of scan for wind measurements by VAD techniques. (b) Wind and particle fall speed make up VAD pattern as indicated. (After Lhermitte and Atlas (1961).)

beam with respect to the upwind direction, and V_f is the mean fall speed of the particles relative to the ground. The contributions of both V_f and V_h to the Doppler velocity V_d as a function of β are illustrated in Fig. 8.29(b). When the sign of V_d is not presented, the radar displays only its magnitude, and the dashed negative portion of the curve is folded over as shown. The method, first developed by Lhermitte and Atlas (1961), has been called VAD for velocity-azimuth display. As will be seen from Fig. 8.29(b), the wind speed is obtained from the average of the two maxima and its direction is that of the major maximum. The particle fall speed is obtained from the difference between the two maxima. So
$$V_h = (V_1+V_2)/2 \cos \alpha, \tag{8.31}$$
and
$$V_f = (V_1-V_2)/2 \sin \alpha. \tag{8.32}$$

The method has been employed by Lhermitte and Atlas (1961), who estimated that the wind speed can be measured to better than $\pm \frac{1}{2}$ m s^{-1}, and its direction to within a few degrees. However, small residual errors are introduced if the particle fall speeds vary around the sampling circle, which may be several kilometres in diameter. But it may not be necessary for the entire circle to be filled with precipitation provided that enough of the VAD pattern is obtained to allow the fitting of a family of curves for computation of the wind.

Lhermitte (1962) improved the method by allowing a sampling gate to move out slowly in range as the beam rotates, thus following the conical spiral path indicated in Fig. 8.29(a). He then used a frequency meter to measure the r.m.s. Doppler frequency and thus the r.m.s. velocity which, for typical spectra obtained at elevations below 45°, is very close to the mean velocity. The r.m.s. velocity is recorded on a pen recorder, each circle of the spiral corresponding essentially to a single altitude, and so produces a major and minor peak. A curve running half-way between two curves joining the major and minor peaks respectively, then gives a profile of the wind speed versus height. A complete wind sounding can be obtained in a few minutes.

In England, Caton (1963) and Harrold (1966) have extended the VAD technique to obtain three-dimensional information on the winds and to measure their horizontal convergence/divergence. The normal method of operation is to determine, for a given elevation of the beam, components of the particle velocities towards the radar at various ranges and at 10-deg intervals of azimuth. Having measured the vertical motion of the particles with the beam pointing vertically, it is possible to deduce the horizontal components of the wind at these thirty-six different azimuths, and hence obtain the horizontal convergence/divergence of the wind over the scanning circle of radius r from the equation

$$-\text{div } V_\text{h} = \frac{1}{\pi r \cos^2 \alpha} \int_0^{2\pi} V_\text{d} \, d\beta - \frac{2 V_\text{f} \tan \alpha}{r}, \qquad (8.33)$$

Such measurements, when repeated at other ranges and elevations (heights), give the convergence at different levels in a vertical cone, and hence the vertical air motion at a given level, z, can be calculated from the equation

$$w(z) = \frac{1}{\rho_\text{a}(z)} \int_0^z \rho_\text{a} \, \text{div } V_\text{h} \, dz, \qquad (8.34)$$

where ρ_a is the air density. We note, however, that $w(z)$ involves integration from the lowest level, so errors at low levels are compounded at higher levels. The method also assumes that the vertical velocity of the precipitation remains constant over the circle of observation, and that the horizontal wind does not change appreciably during the 4 min taken to complete an azimuth rotation. It cannot therefore be used with convective clouds.

The various errors have been analysed by Harrold (1966), who demonstrates that the divergence can be measured more accurately in snow than in rain because the former has a narrower spectrum of fall velocities, and that the error decreases for lower aerial elevations and increasing height, i.e. as the area of the scanning circle increases. Thus with $\alpha = 30°$, the probable error in the measured divergence at a height of 4 km ($r = 7$ km) in snow was estimated to be 3×10^{-5} s^{-1}; at the same height, but with elevation 15° ($r = 15$ km), the corresponding error is only 1×10^{-5} s^{-1}. A more rigorous analysis of the errors involved in determining the properties of the wind field from the VAD mode is given by Browning and Wexler (1968). They suggest an optimum scanning procedure by which the errors imposed by inhomogeneities in the horizontal distribution of particle fall speed, the presence of strong vertical wind shear, inhomogeneities in reflectivity, and errors in the elevation setting, can be kept acceptably small.

Harrold (1966) identified layers of convergence and divergence in association with a frontal inversion, marked convergence below the front being largely compensated for by pronounced divergence of magnitude $\sim 5 \times 10^{-4}$ s^{-1} immediately above the front. Although these divergences were an order of magnitude greater than those measured in widespread precipitation by Caton (1963), Browning and Wexler (1968), and Wexler, Chmela, and Armstrong (1967), Harrold's computed vertical velocities, showing ascents of a $\sim$ 10 cm s^{-1} during rain, seem very reasonable.

Useful information on the structure of the horizontal wind field in the lower levels of convective storms may be obtained by the use of a quasi-horizontal Doppler beam scanning in the azimuth plane, the contribution to the Doppler shift due to the vertical velocities of the particles than being minimized. Some preliminary work on these lines has been reported by Donaldson (1967) and Easterbrook (1967) but, with a single radar, information is obtained only on the radial component of the velocity. The simultaneous use of two such radars, spaced 20–60 km apart, and scanning from two different directions, would give both quadratic components of the horizontal wind, allowing derivation of convergence and vorticity and, if observations were made at several different low elevations, also the vertical velocity from the continuity equation. These wind data could, of course, be related to the pattern of radar reflectivity.

Complete three-dimensional information on the wind field and determination of particle falling-speeds at several levels would require the simultaneous use of three Doppler radars (see Lhermitte (1968)).

8.7. The structure of precipitating layer clouds as revealed by radar
8.7.1. *General features*

The echoes received from widespread, steady precipitation normally associated with warm fronts and depressions lend themselves more readily to quantitative study than those from shower clouds, where conditions are changing very rapidly and interpretation becomes much more difficult. Typical photographs of the RHI and A-scope displays of warm-front precipitation are shown in Figs. 8.17 and 8.20. We have already mentioned the 'melting band' in the vicinity of the 0° C level, which separates a relatively steady echo obtained from the snow above the melting region and a rapidly fluctuating echo of often rather greater intensity from the rain below. The appearance of the melting band, which was accounted for by the theoretical calculations of Ryde (1946), can be explained qualitatively as follows. Above the 0° C level the scattering particles are snow crystals and snowflakes, growth and aggregation of which cause the echo intensity to increase with decreasing height. As the snowflakes fall through the 0° C level, they start to melt, and since the reflectively of a water particle is greater than that of an ice particle of equal volume, this melting will be accompanied by an increase in the echo intensity, which may be enhanced further by aggregation and preferred orientation of the wet snowflakes. It is usually assumed that, during the melting process, the particles retain the relatively slow fall velocities characteristic of snow, but once completely melted, they collapse to form raindrops having considerably larger fall speeds. As the precipitation rate is usually assumed to remain constant at all levels, this means that the spatial concentration of raindrops will be reduced relative to that of the parent snowflakes in the ratio of their respective terminal velocities. Thus just below the level where complete melting occurs, the echo intensity will show a sharp decrease, with the result that a narrow band of enhanced echo shows up between this level and the 0° C isotherm. This interpretation of the melting band has been confirmed by combined aircraft and radar observations (e.g. Bowen 1951).

We shall now discuss in greater detail the echo intensity and its changes with height in the region above the 0° C level, in the melting region, and also in the rain below the melting zone.

8.7.2. *The region above the 0° C level*

Well above the 0° C isotherm, the scattering particles are mainly snow crystals. In stratiform clouds, where the liquid-water content is generally small, it is likely that the crystals will grow more rapidly by deposition of water vapour than by accretion of cloud droplets. Their growth rate by both processes can, in principle, be computed in terms of the temperature, pressure, humidity, and liquid-water content of the cloud, and hence the back-scattered power, $P_r \propto \sum nm^2$, can be calculated as a function of height in the cloud, if a concentration and relative size-distribution function are assumed. In practice, such computations involve making rather bold assumptions concerning the density and shape of the growing crystals, the magnitude of the vertical air currents in the cloud, the effect of the cloud droplets on the crystal growth-rate, and whether or not a steady-state may be assumed; thus the answers obtained depend on what assumptions one regards as reasonable. Some calculations by Browne, Palmer, and Wormell (1954) indicated that Browne's measurements of echo intensity as a function of height could not be interpreted in terms of ice crystals growing by sublimation in a cloud in which the water budget is balanced.

Such computations of relative echo intensities at different levels, in which a steady state was not assumed, were made by Wexler (1952) for a model cloud following the 10° C saturated adiabatic, with base at 0° C and 805 mb, and sustained by a constant updraught of 10 cm s⁻¹. He first calculated the growth of ice crystals originating at the 3-km level and growing solely by diffusion of vapour in a water-saturated atmosphere, using equations analogous to (5.1)–(5.3). They were assumed to grow as plates with constant thicknesses of $x = 15$ and 40 μm respectively. The results are shown in Table 8.4.

At a given level, all crystals were assumed to have the same mass, and the crystal concentration was taken to be constant with height, so that the ratio of echo intensities at two levels is then $R = m_1^2/m_2^2$, where m is the crystal mass. Calculated values of R relative to the 2-km level are compared with the measured values obtained by Browne (1952a) and by Hooper and Kippax (1950a) in Table 8.5.

Because the calculated values of R in Table 8.5 increase much more rapidly than the observed values, and because the computed crystal

TABLE 8.4

Masses and terminal velocities of crystals originating at 3 km above base of model cloud (Wexler 1952)

Height above 0° C (km)	15 μm crystal		40 μm crystal	
	Mass (mg)	Velocity (cm s^{-1})	Mass (mg)	Velocity (cm s^{-1})
2·0	0·65	41	0·12	72
1·5	1·35	41	0·20	72
1·0	2·08	40	0·29	70
0·5	2·68	40	0·36	70
0	2·96	40	0·39	70

masses are much larger than those normally observed, Wexler concluded that, in clouds with small updraughts, ice crystals do not grow at water saturation except, perhaps, in the region below 1 km where there is quite good agreement between the observed and calculated ratios of echo intensities at 1 km and 0·5 km. In consequence, he considered the case where all the water created by the updraught within each height interval is deposited directly on the growing crystals within that interval, so that there is no storage of liquid water in the form of cloud droplets. Taking the upper boundary of the cloud at 4 km and $\rho_a U$ to be constant with height, where ρ_a is the air density and U the updraught speed, the value of R is now given by $R = (q-q_4)^2/(q_2-q_4)^2$, where q is the specific humidity. Values of R thus calculated for Wexler's cloud model are shown in the last column of Table 8.5; these are in much better agreement with the observations, suggesting that the crystals grow in conditions intermediate between ice- and water-saturation. However, it must be said that the calculated values of m

TABLE 8.5

Mean observed and theoretical ratios of echo intensity relative to 2-km level (Wexler 1952)

Height (km)	Browne	Hooper and Kippax	$R = m_1^2/m_2^2$ †		$R = \left(\dfrac{q-q_4}{q_2-q_4}\right)^2$
			$x = 15$ μm	$x = 40$ μm	
2·0	1·0	1·0	1·0	1·0	1·0
1·5	2·2	1·9	4·3	2·8	1·9
1·0	3·1	3·0	10·2	5·8	3·1
0·5	4·7	4·6	17·0	9·0	4·7
0		8·3	20·8	10·6	7·2

† Corrected values

§ 8.7   RADAR STUDIES OF CLOUDS AND PRECIPITATION   461

and R in Tables 8.4 and 8.5 depend a good deal on the assumptions made concerning the geometry of the growing crystal. Thus Browne (1952a), who assumed that instead of remaining practically constant, the thickness of a crystal was always one-fifth of its radius, concluded that the crystals would not grow fast enough to account for the observed echo intensity, even at water saturation, unless additional growth by accretion of supercooled droplets was postulated—a deduction directly opposed to that made by Wexler.

Calculations by Wexler and Atlas (1958) of a steady-state water budget for a stratiform cloud, which follows a saturated adiabatic, has a parabolic distribution of vertical velocity, and contains particles with an exponential size distribution, predicted that, while production of liquid water is likely to exceed consumption just above the 0° C level, at temperatures below $-2°$ C, the rate-of-growth of ice crystals by deposition is such that the cloud liquid water vanishes above the $-4°$ C level for a surface precipitation rate of 2 mm h^{-1}, and above the $-10°$ C level with 10 mm h^{-1}. The radar observations of Biswas, Murty, and Roy (1960), indicating only a very small increase of echo intensity between the 2 km and 1 km levels above the 0° C isotherm, suggested that the snow crystals were growing slowly in air only slightly supersaturated with respect to ice in the majority of their stratiform clouds.

Although aggregation of snowflakes probably occurs much more readily in the melting region, snow reaching the ground with temperatures at, or a few degrees below, freezing, generally contains aggregates consisting of several individual crystals, and suggests that aggregation may also occur above the 0° C level in the cloud.

If n individual crystals join together to form a cluster having a fall velocity f times that of an individual crystal, the echo intensity will be increased n/f times. Lhermitte and Atlas (1963), using a Doppler radar, deduced that the reflectivity of dry snowflakes increased sixfold, and their average terminal velocity doubled, in falling through a vertical distance of 750 m just above the melting zone. Allowing for some growth by deposition of vapour and droplets on the crystals, the authors interpreted these results in terms of an initial population of uniform-sized crystals forming aggregates of nine crystals during the interval. But it is generally difficult to locate the 0° C level with sufficient accuracy to delineate the freezing from the melting region, and so separate the effects of aggregation that may occur at temperatures just above and below 0° C. This should also be borne in mind when we come to discuss measurements of signal intensity in the melting zone.

8.7.3. *The melting region*

Measurements designed to locate the height of the melting band relative to the 0° C level, to determine its vertical thickness, and the radar reflectivity in the band relative to that in the snow above and the rain below it, have been made by Austin and Bemis (1950), Hooper and Kippax (1950b), and Browne (1952a). Difficulties arise because the measured range and width of the melting band depend on the pulse duration and the band-width of the receiver. Hooper and Kippax investigated the relationship between the heights of the melting band and the 0° C level using a vertically directed 3·2-cm radar based at a radio-sonde station, radar observations being made simultaneously with radio-sonde ascents from which the heights of the 0° C level were obtained. The results of observations made on ten occasions showed that, on average, the maximum of the melting band lay 330 ± 150 ft below the 0° C level. The melting-band echo was observed to be about 2·5 μs in duration using a set with pulse duration 1 μs, so that the vertical thickness of the band appeared to be about 750 ft. Austin and Bemis (1950), using a RHI display and working at ranges of more than 10 miles, give an average figure of 830 ft for the distance of the melting band below the 0° C level, but they admit of errors, of perhaps 500 ft, in the location of this level and of the melting band. The resolving power of their equipment did not allow them to make good measurements on the thickness of the band, but they separated their observations into occasions with stable lapse rates when the band was less than 1000 ft thick, and occasions with steep lapse rates when the band was more than 1000 ft thick. These authors quote one observation made with a vertically pointing radar with a 0·5 μs pulse, when the apparent thickness of the melting band was about 500 ft. Browne (1952a), using a similar arrangement, found the peak of the melting band to be 600 ± 200 ft below the 0° C level (but his observations included some cases of unsteady rain) and that, in warm fronts, the band was generally 300 to 400 ft thick. An analysis by Mason (1955c) of records obtained by Browne for twelve further occasions of steady rain reveals values of between 50 and 500 ft for the depth of the melting zone, the average value being 225 ft. A progressive lowering of the melting band may occur due to the cooling of the air by the melting snow (Wexler, Reed, and Honig 1954); it is also frequently observed to rise due to advection of warm air.

The depth of the melting zone, measured from the 0° C level to where it merges into the rain echo below, corresponds to the distance through which the majority of the snowflakes will fall before melting, and will therefore depend upon the lapse rate and on the sizes and fall velocities of the melting flakes. The author calculates that a flake of mass 1 mg composed of dendritic crystals and falling through an atmosphere with lapse rate 6° C/km, would melt completely after a fall of about 150 m. The results of similar but more extensive calculations have been published by Wexler (1955).

8.7.4. *The region below the melting band*

The echo from the rain below the melting band exhibits marked fluctuations in intensity from pulse to pulse, but the mean intensity changes only very slowly with height. On some occasions, an increase of signal strength with decreasing height is to be expected as the result of coalescence between raindrops and growth of raindrops by accretion of cloud droplets. These processes have been investigated theoretically by Mason and Ramanadham (1954) and Hardy (1963); the former calculated that the raindrops associated with precipitation of intensity 3 mm h^{-1}, in falling a distance of 1 km in a cloud of liquid-water content 0·2 g m^{-3}, should grow sufficiently to increase the radar echo intensity by a factor of about 2. Such increases were observed on a few occasions by Harper (1957) but, in the majority of warm-front situations, he observed no such increase.

8.7.5. *Polarization of the radiation scattered by hydrometeors*

Some information on the shape and orientation of the precipitation particles might be expected from a study of the parallel- and cross-polarized components of the back-scattered radiation. In § 8.5.2. it was argued that melting snowflakes, which may be treated approximately as oblate spheroids, will, if randomly oriented, produce a cross-polarized component, which is independent of the direction of polarization and the angle of incidence of the radiation. If they are oriented with their figure axes vertical, no cross-polarization will result, but if the aerial is vertically polarized, the intensity of the back-scattered radiation will increase as the beam moves towards the zenith. Browne and Robinson (1952) found that the particles in the melting region produced back-scattered radiation with a greater cross-polarized component than that given by the snow above and the rain below. The observations were made by using separate aerials for the transmitter and receiver and by comparing the echo received when the aerials had their planes of polarization parallel with that received when the plane of polarization for reception was at right angles to that for transmission. Wavelengths of both 3·2 cm and 8 mm were used with the results shown in Table 8.6.

The values for snow above the 0° C level are uncertain because of the very weak echoes, while the values given for rain may have been a measure of the cross-polarization of the aerial systems.

Further data were obtained by Hunter (1954) using a 3·2-cm transmitted beam that was circularly polarized. He measured the ratio of the orthogonal circular components, a quantity that is about twice the depolarization ratio for a plane-polarized beam. The most comprehensive measurements have been made by Newell (1956) using a common

TABLE 8.6
Ratio of parallel- to cross-polarized components of back-scattered radiation from 'steady' precipitation, i.e. (depolarization ratio) in dB

Author		Rain	Melting zone	Snow
Browne and Robinson (1952)	8 mm	15±1	7±1	12±2
	3·2 cm	21·8±0·4	16·9±0·3	19·9±0·6
Hunter (1954)	3·2 cm	26–38	16–23	29
Newell (1956)	3·2 cm	16–20	10–22	16–20

transmitting and receiving antenna that was first plane-polarized and then circularly polarized, the ratio of the intensities of circularly to plane-polarized signals being twice the depolarization ratio if the particles are randomly oriented. Inspection of Table 8.6 reveals considerable discrepancies between the results of Hunter and those of the other workers. The latter indicate that raindrops and dry snowflakes may show considerable departures from sphericity, the distortion being even greater for melting snowflakes. Hunter finds considerable smaller departures from sphericity in all cases, the axis/diameter ratios for raindrops rarely being less than 0·85. Newell found that rain and snow particles were randomly oriented as far as the radar could detect, but that, on occasions, particles in the melting zone showed some tendency towards horizontal orientation.

8.7.6. *The echo intensity as a function of height*

The intensity of the echo from the centre of the melting zone relative to that from the rain below and that from the snow crystals above has

§ 8.7 RADAR STUDIES OF CLOUDS AND PRECIPITATION

been measured by Hooper and Kippax (1950b), Austin and Bemis (1950), Browne (1952a), and Mason (1955c) using conventional radar, and by Lhermitte and Atlas (1963) with Doppler radar, the range of experimental values being given in Table 8.7.

The values of Austin and Bemis, obtained with a slanting radar working at 10 miles range, are probably not as accurate as the others obtained with vertically pointing sets; because of the width of their beam, the measured reflectivity above the melting band was an average value for several thousand feet and would lead to an overestimate of P_m/P_i and an underestimate of P_i/P_w.

TABLE 8.7
Relative echo intensities from melting region, snow, and rain

Author	P_m/P_w	P_m/P_i	P_i/P_w†
Hooper and Kippax	5–9		
Austin and Bemis	2–12	Mostly 2–25	0·1–1·0
		Occasionally > 100	(Average 0·42)
Browne	4–8	5–15	0·2–1·0
Mason	4–14	3–34	0·2–1·6
	(Average 7)	(Average 16)	(Average 0·8)
Lhermitte and Atlas	12	29	0·45
Biswas et al.	1·4–7·0		0·15–0·5
	(Average 3·7)		(Average 0·25)

† Subscripts m, w, i, refer to melt, water, ice, respectively.

8.7.7. *Theory of the melting band*

In the simple theory of the melting band, as first presented by Ryde (1946), it was either stated or implicitly assumed that the flux of particles was the same at all levels, that there were no vertical air currents, and that the particles suffered no change of mass after reaching the 0° C level.

Thus, the ratio of the reflected signal from snowflakes when almost completely melted to that from dry flakes, assuming no changes of fall speed or state of aggregation during the process, was given by

$$P_m/P_i = \frac{\rho_i^2}{\rho_w^2}\left(\frac{\epsilon_w-1}{\epsilon_w+2}\right)^2 \bigg/ \left(\frac{\epsilon_i-1}{\epsilon_i+2}\right)^2 \simeq 5 \quad \text{(for } \lambda = 3, 10 \text{ cm)}, \quad (8.35)$$

where ρ is the density of the scattering particle. Thus, if the flakes were to melt almost completely without any appreciable increase in their fall speed, the echo intensity would increase fivefold corresponding to the peak of the melting band. At any intermediate stage where the particle contains a fraction α of water, Ryde writes for the mixture:

$$K^2/\rho^2 = \{\alpha K_w^2/\rho_w^2 + (1-\alpha)K_i^2/\rho_i^2\} = 0{\cdot}94\alpha + 0{\cdot}19(1-\alpha),$$

where $K = (\epsilon-1)/(\epsilon+2)$. However, this formula may very well underestimate the reflectivity of the partially melted particle, since on acquiring a water skin, it will probably scatter to all intents and purposes as a water particle of the same cross-section.

When the snowflakes melt completely and collapse to form raindrops of fall-speed v_w, considerably greater than that v_i of the flakes, Ryde assumes that the spatial concentration of the scatterers, and hence the echo intensity, will decrease by a factor of v_w/v_i to produce a bright band rather than a step. If we assume a steady state, the particles to remain of constant mass and to be all of the same size,

$$n_i m_i v_i = n_w m_w v_w, \quad m_i = m_w,$$

and therefore the ratio of the signals from just above and just below the melting zone will be

$$P_i/P_w = \frac{\rho_w^2}{\rho_i^2} \frac{K_i^2}{K_w^2} \frac{\sum_{n_i} n_i m_i^2}{\sum_{n_w} n_w m_w^2} = \frac{\rho_w^2}{\rho_i^2} \frac{K_i^2}{K_w^2} \frac{v_w}{v_i} \simeq \frac{1}{5} \frac{v_w}{v_i}. \tag{8.36}$$

As raindrops will generally have fall velocities at least 5 times that of the parent snowflakes, eqn (8.36) predicts that the echo intensity above the melting zone should be at least as great as that below it. In fact, measurements show that the ratio P_i/P_w is generally less than unity and that the intensity of the signal from the melting zone is often much more than 5 times greater than that returned from the snow above or the rain below. This suggests that Ryde's formulation of the problem, though giving a satisfactory explanation of the main features, is not complete. Austin and Bemis (1950) pointed out that two further factors may cause enhancement of the melting band: aggregation of the flakes in the melting zone and their scattering cross-sections being increased because of their non-spherical shape. The following treatment is based on that of Mason (1955c).

If we assume that a uniform updraught u prevails throughout the cloud, that at any one level all particles are of the same mass, and that each of the snowflakes at the base of the melting zone is composed of ν initially dry flakes, we may write $n_i m_i(v_i - u) = n_w m_w(v_w - u)$, and $m_w = \nu m_1$, so that

$$P_i/P_w = \frac{1}{5} g_i \frac{1}{\nu} \frac{v_w - u}{v_i - u}, \tag{8.37}$$

where g_i is a shape factor expressing the scattering cross-section of a dry snowflake relative to that of a sphere of equal volume; v_i, v_w are respectively the fall velocities of a dry flake and a raindrop relative to the air.

A relation between the terminal velocities of dry snowflakes composed of dendrites and their melted diameter d (cm) has been given by Langleben (1954) as $v_i = 160d^{0\cdot 3}$. For raindrops of diameter D (cm), Spilhaus (1948) gives the approximate relation $v_w = 1420D^{0\cdot 5}$. Now if ν flakes aggregate to produce ultimately a raindrop of diameter D, then $D = \nu^{\frac{1}{3}}d$ and $v_i = 160\nu^{-0\cdot 1}D^{0\cdot 3}$. Substituting these relations in eqn (8.37), putting $g_i \simeq 1$, and neglecting u (a few centimetres per second for warm-frontal clouds) compared with v_i and v_w, we have

$$P_i/P_w = 1\cdot 8\nu^{-0\cdot 9}D^{0\cdot 2}. \tag{8.37a}$$

If no aggregation of the flakes were to occur during melting, i.e. $\nu = 1$, then according to eqn (8.37), the ratio P_i/P_w would be greater than unity for all values of $D > 0\cdot 5$ mm. This result is, generally speaking, contrary to observation; the echo from the snow above the melting zone is generally weaker than that from the rain below, as shown in Table 8.7.

Since most of the measurements were made in steady rain of intensity 0·5 to 3 mm/h, we may assume a median-volume drop diameter $D = 1\cdot 0$ mm. For values of $P_i/P_w = 1\cdot 0$, 0·4, 0·2, eqn (8.37) gives $\nu = 1$, 3, 6 respectively. This suggests that on those occasions when values of $P_i/P_w > 1$ are observed, aggregation of melting flakes occurs only rarely; but more usually a flake of mass about 0·2 mg may, on average, coagulate with two or more others during its fall through the melting zone, to produce a raindrop of 1 mm diameter. Putting $P_i/P_w = 0\cdot 4$, $\nu = 3$, $v_w = 425$ cm s^{-1} in eqn (8.37) gives $v_i = 70$ cm s^{-1}—a very reasonable value for dry flakes of mass 0·2 mg.

The ratio of the echo intensity from the melting band to that from the snow above, and to that from the rain beneath, are

$$\frac{P_m}{P_i} = 5g_m\nu\frac{(v_i-u)}{(v_m-u)} \simeq 5g_m\nu\frac{v_i}{v_m} \tag{8.38}$$

and

$$\frac{P_m}{P_w} = g_ig_m\frac{(v_w-u)}{(v_m-u)}b \simeq g_mb\frac{v_w}{v_m}, \tag{8.39}$$

g_m being the shape factor for the melting snowflakes, v_m their terminal velocity, and b a factor to allow for the break-up of a melting flake into an average of b drops. If the terminal velocity of the nearly-melted flakes is assumed to be 2 m s^{-1} (even larger values was suggested by the observations of Langleben (1954)), then $v_w/v_m = 2$, and a typical value of $P_m/P_w = 7$ implies a value of $g_m b = 3·5$, while an extreme value of $P_m/P_w = 14$ gives $g_m b = 7$. Since values of $g_m > 3·5$, corresponding to randomly-oriented oblate spheroids of diameter/axis ratio 5, are unlikely, the high observed values of P_m/P_w would appear to be evidence for the break-up of melting snowflakes to give mean values of $b \geqslant 2$. Substitution of the above values of $g_m = 3·5$, $\nu = 3$, $v_i/v_m = \frac{1}{3}$ in eqn (8.38) leads to $P_m/P_i = 17·5$, again a result that accords well with the mean of the observed values. On the other hand, a coalescence factor ν as high as 6 would be compatible with P_i/P_w values as low as 0·2 and P_m/P_i values as high as 35 that are occasionally observed.

8.7.8. *Radar upper bands*

The early radar pictures of snow, which were generally lacking in pattern, may have been due to the rather poor resolving power of the radar. In 1950, Swingle, using a 10-cm vertically-pointing radar, found considerable pattern in the snow above the melting band and concluded that 'with few exceptions, continuous surface precipitation is the result of non-uniform precipitation aloft'. The appearance of strong echoes above the melting band, which descended and merged with the melting band, thereby momentarily increasing its intensity, was reported about the same time by Hooper and Kippax (1950a). They were discussed in more detail by Bowen (1951) who termed them radar 'upper bands'. They show up in a very striking manner on a RHI display (Fig. 8.30) and can often be seen in warm-frontal conditions and less frequently in cold-front precipitation. They first appear several thousand feet above the 0° C level, occasionally remain almost stationary, but generally fall towards the melting band. Bowen described five occasions on which the bands appeared consistently around the −16° C level; in four cases they fell with speeds of between 5·5 and 7·5 ft s^{-1} towards the melting region and the process was repeated at intervals of about 20 min. Browne (1952b) described some observations made with a vertically pointing radar and A-scope display, in which the apparent velocity of fall of the bands was generally between 10 ft s^{-1} and 30 ft s^{-1}, the velocity generally decreasing as the melting band was approached. The merging of the upper band with the melting band was accompanied by an increase in intensity of the latter, and of the rain below, and after a minute or two the rainfall rate at the ground showed a temporary increase. Browne suggested that the high apparent

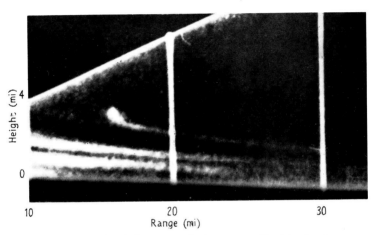

Fig. 8.30. An upper band shown as a sloping snow trail originating in a compact generating cell at a height of 16 000 ft. (By courtesy of Dr. J. S. Marshall, McGill University, Montreal.)

falling speeds of the upper bands on the radar screen, which were often considerably greater than the terminal velocities of snow crystals or snowflakes, could be produced by a slanting stream of precipitation particles, distorted in a vertical wind shear, drifting horizontally through the radar beam, the observed rate of fall of the band now corresponding to the rate of descent of the line of intersection of the curved streak with the radar beam (see Fig. 8.31). The same suggestion was made independently by Lhermitte (1952) and Marshall (1953).

The problem can be simplified by supposing that the streak is being generated at a certain point in the cloud and that the wind is blowing

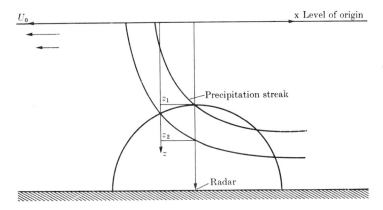

Fig. 8.31. Diagram to illustrate the interception of a precipitation streak by a vertically scanning radar. Vertical scale is greatly exaggerated.

in the same direction at all levels. Taking this point as origin of the coordinates z measured vertically downwards, and x measured horizontally into the wind, we can find the equation of the streak relative to these coordinates. If the wind speed u is assumed to increase uniformly with height, so that $u = u_0 - kz$, where k is the vertical wind shear and u_0 the wind speed at the origin, the equation of the streak referred to the coordinates, which are assumed to move with velocity u_0, is $x = \frac{1}{2}(k/v)z^2$, where v is the vertical velocity of the precipitation particles. Thus the streak is parabolic in shape. To find the apparent fall velocity of the upper band produced by the streak as it passes through the radar beam, we consider the streak to move with horizontal velocity u_0. Then at a time t after the origin of the streak passes through the beam, the radar is looking at that part of the streak which has its x coordinate equal to $u_0 t$ and $z = (2vu_0 t/k)^{\frac{1}{2}}$. The apparent velocity of fall is $W = (2vu_0/kt)^{\frac{1}{2}}$, so that W should decrease at $t^{-\frac{1}{2}}$.

Browne gave some examples of precipitation streaks in which reasonable agreement was obtained between the observed mean values of W as a function of height, and values calculated from radio-sonde data for u_0 and k. Very clear evidence for the existence of such streaks was obtained by Marshall (1953): his photographs of a RHI display suggested that the snow forming the streak was generated in compact cells having linear dimensions of order 1 km. An analysis by Gunn, Langleben, Dennis, and Power (1954) of twenty-two occurrences showed that the precipitation streaks were most frequent and well-defined when the air aloft was stable, so that instability does not appear to be the initiating mechanism. The radar echoes from the tops of the streaks were located on average about 500 m above the frontal surface, while the tops of layers of stratiform medium cloud were found (from upper air analysis) at an average height of 400 m above the frontal surface. The bases of the generating cells were in the mixing zone some 100 m below the frontal surface. A more extensive analysis by Douglas, Gunn, and Marshall (1957) confirmed many of the foregoing features, and demonstrated that the latent heat of sublimation released by the crystals growing in a moist stable environment was sufficient to produce vertical development comparable to the observed cell heights and also updraughts comparable with the terminal velocities of the snow particles themselves. Direct evidence as to the shape and extent of the cells in horizontal section was provided by Langleben (1956). From photographic records with a CPS–9 radar, he synchronized constant altitude pictures at the generating level. The horizontal extent of the individual cells was about 1 km in all directions, but the cells tended to form in line arrays. In plan view, there was a clear-cut pattern in the snow at the generating level, but the pattern was sometimes very diffuse near the ground.

Langleben (1954) measured the slope of the pattern, $dz/dx = v/(u_0-u)$, on the cathode-ray tube, and using the relevant wind data, calculated v, the fall velocity of the particles composing the trail. Having obtained average fall speeds in excess of 1 m s^{-1}, he concluded that the particles were snowflakes rather than single ice crystals, and that aggregation of crystals must have occurred at temperatures well below 0° C (perhaps as low as $-20°$ C) in the generating elements producing the streaks.

Usually, analysis of these precipitation streamers, which may seed lower clouds and may be important in releasing precipitation from stratiform cloud systems, is complicated by the wind direction varying with height and causing the trails to curve in three dimensions.

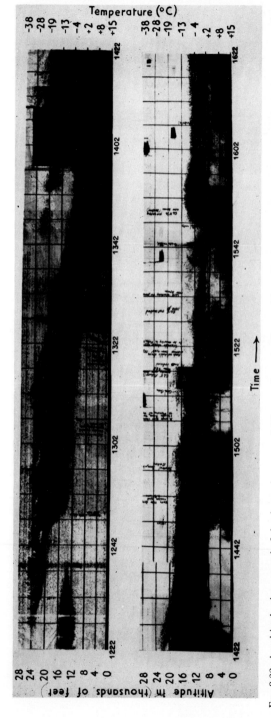

Fig. 8.32. An altitude–time record of clouds passing over a 1·25-cm radar near Boston, Mass. Sloping ice-crystal streamers are being generated at about 24 000 ft (between 1230 and 1240 hours) and falling to form an altostratus cloud deck. The low vertical echo beginning at 1324 hours represents the beginning of light rain. A very sharply defined band is noticeable just below 8000 ft after 1450 hours. (By courtesy of Dr. D. Atlas. Air Force Research Laboratories, Cambridge, Mass.)

8.8. Detection of non-precipitating clouds with millimetric radar

We have seen in § 8.2.5 that the radar reflectivity of non-precipitating clouds is so small that they cannot be detected with normal 3-cm and 10-cm equipments. However, since the reflectivity is proportional to $1/\lambda^4$, the possibility exists of detecting clouds with sets working in the millimetre waveband. Equipments working on wavelengths of either 1·25 cm or 8·6 mm have been used in a number of countries, but rather few systematic and detailed investigations have been reported. Usually the beam is directed vertically to give an A-scope display of echo-amplitude versus height. To obtain a continuous record of the cloud pattern, the signal output is fed to a chart recorder, the paper of which moves at constant speed to give an altitude-time record, i.e. a vertical cross-section of the clouds as they pass over the set, as shown in Fig. 8.32. Such radars, with a peak power output of, say, 15 kW and a beam width of $\frac{1}{3}°$, are able to penetrate low cloud and detect a reasonably high proportion (>50 per cent) of medium- and high-level cloud, but do not generally locate their bases and tops reliably. They are able, however, to detect precipitation in the form of streamers, virga, etc., which may not reach the ground, and also the early stages of precipitation in thermals and cumulus clouds. Used in conjunction with centimetric radars in both conventional and Doppler modes, the potential of this millimetric tool is far from exhausted. More detailed accounts of its uses and limitations are given by Plank, Atlas, and Paulson (1955) and Harper (1964).

8.9. The cumulonimbus and thunderstorm as revealed by radar

The most intensive study of large cumulonimbus clouds and thunderstorms was made during the Thunderstorm Project in 1946–7 and reported by Byers and Braham (1949). Radars with RHI and PPI facilities were used in conjunction with aircraft equipped to measure temperatures, updraught and downdraught speeds, gusts, etc., and with a network of surface observing stations to investigate clouds in Ohio and Florida. These and more recent investigations of modest thunderstorms show them to consist of one or more discrete units or cells, each of which goes through a fairly well-defined life cycle comprising a growth stage, a mature stage, and a decaying stage. During the growth stage, lasting for perhaps 10–15 min, updraughts of typically

5–10 m s^{-1}, but occasionally higher, exist throughout the cloud, and precipitation in the form of rain or hail develops rapidly in the middle troposphere. The onset of the mature stage coincides with the spread of precipitation towards the ground and the appearance of both updraughts and downdraughts at least in the lower part of the cell. As the concentration of rain-water builds up until it can no longer be supported by the updraught, it begins to fall towards the ground and initiates a downcurrent that is maintained partly by the weight of the precipitation, but mainly by chilling due to the melting, and especially the partial evaporation of the precipitation particles. A downdraught colder than its surroundings by only 1° C, or containing 4 g of rain-water per kg of air, will accelerate from zero velocity to 20 m s^{-1} over a height interval of 5 km. Thus in the mature stage of the storm, during which most of the precipitation and lightning discharges are released within 15–30 min, updraughts and downdraughts of 10–30 m s^{-1} may coexist, but gradually downdraughts spread throughout the cell and the storm enters the dissipating stage. During the next 30 min the cold downdraught occupies the whole cell and spreads out below it, but eventually it weakens and dies. The whole life of a cell may therefore occupy about 1 hr, but the spreading out of the cold downdraught near the ground may trigger new cells on the flanks of the storm and thereby ensure its regeneration and propagation over a period of a few hours.

The horizontal dimensions of a cell, as first detected by radar, may be only about 1 km with the top either below, or just above, the 0° C level. Thus Battan (1953) reported the mean temperatures of the bases and tops of the initial radar echoes appearing in the Ohio storms of the Thunderstorm Project to be +0·4° C and +10° C respectively. Ackerman (1960) indicated a modal temperature level of −10° C for the height of first echoes in convective showers in Arizona, but with a wide day-to-day variability, with two-thirds of the initial echoes falling between −4 and −16° C. Battan (1963), however, found the mean temperatures of the bases and tops of the initial echoes in a different sample of Arizona storms to be −4 and −9·5° C, the mean temperature of the cloud base being 11·7° C. Since, in addition, the altitude of the initial echo was positively correlated with that of the cloud base, Battan deduced that precipitation was initiated by the coalescence process.

Once a radar echo appears, it may grow rapidly towards the −30° C level or higher, and extend to several kilometres in each direction. Over 60 per cent of the storms investigated in the Thunderstorm

Project gave radar echoes extending above 35 000 ft (10·5 km), the mean value being nearly 38 000 ft (11·4 km), and the maximum height recorded, 56 000 ft (16·8 km). But, in estimating the maximum heights reached by the tops of storms, measurements made by radar alone must be regarded with caution, because spurious readings may arise from the finite width of the beam and the presence of side lobes, so that very strong echoes may produce a signal even when the main axis of the beam is several degrees above the storm top. The deceleration of updraughts might sometimes be sufficiently slow for there to be a considerable difference between the maximum height attained by the top of the radar echo and the top of the visible cloud. However, in England, Harper, Ludlam, and Saunders (1956) found that the difference between the height of the echo and that of the visible top, as measured by the double theodolite method, was usually <1000 ft (300 m), while in Miami, Saunders and Ronne (1962) found that the difference was never more than 3000 ft (900 m). Occasionally thunderstorms penetrate several thousands of feet into the stratosphere and experience rapid deceleration; in this case, the heights of the visible top and the radar echo should be practically the same.

Byers and Braham reported that the horizontal dimensions of the radar echoes usually decreased with increasing altitude above 3 km where, in more than half of the fully developed storms, they exceeded 8 km. The width of an echo was generally comparable with its height. The echo top often ascended in a series of steps, separated by intervals of about 15 min during which the echo top either remained nearly stationary or subsided by several thousand feet. The aircraft reported that within the updraught of a single cell there was more than one region of strong vertical ascent, and that these were associated with the development of individual cloud turrets at the summit of the visible cloud. These may be identified with a succession of thermals which would confer on the updraught something of the pulsating nature reported by several observers. Very similar behaviour was reported in the case of twelve thunderstorms in New Mexico by Workman and Reynolds (1949), where the appearance of the radar echo was quite sudden and was associated with a rapid increase in the expansion of the top of the visible cloud. The echoes first appeared at about the $-10°$ C isotherm and proceeded to grow upwards and outwards, the vertical speed varying from 2 m s^{-1} to 9 m s^{-1}, with an average speed of about 4 m s^{-1}. When the echo top approached the $-30°$ C level, the top of the visible cloud showed evidence of glaciation and a growing electric field was detected on the ground near the cloud. Then, the top of the radar echo began to descend at an average speed of about 4 m s^{-1}. The average period of the chain of events just described was about 25 min. The beginning of the downward motion of the echo top coincided with the first lightning strokes within the cloud and the appearance of precipitation at the cloud base. Cloud-to-ground lightning strokes sometimes followed within a few minutes. The top of the radar echo continued to move down but, after reaching, say, the $-10°$ C level, it often became obscured by newly arising cells. The electrical discharges and rain, however, continued for some minutes.

So far we have described the more important characteristics of rather modest thunderstorms but, in recent years, much attention has been given to severe storms of the type that frequently produces tornadoes and large hail in the mid-west of the United States. Important case studies have been published by Atlas and Ludlam (1961), Browning and Ludlam (1962), Browning and Donaldson (1963), and Browning (1964, 1965), while detailed review articles appear in Atlas *et al.* (1963). The most intensive study of an individual storm was made by Ludlam, Atlas, and Browning, who used five radars on wavelengths of 3·3, 4·7, and 10 cm to survey a storm that travelled from the south to the east coast of England and produced hailstones of golf-ball size for nearly ¾ h near the town of Wokingham. Throughout this intense phase, the radar echo, whose top reached a maximum height of 13 km, moved uniformly without any marked variation in character, and maintained the following important features which are illustrated in Fig. 8.33.

The 'wall'. In the central and right-hand parts of the echo mass, the leading precipitation near the ground, determined from ground observations to consist of large hail, formed a 'wall' with a sharply defined upright front face lying perpendicular to the direction of the storm motion. The wall was practically vertically beneath the highest echo top, and rather more than 1 km behind the position of *the most intense echo* centered at a height of about 25 000 ft (7·5 km). Atlas and Ludlam (1961) argued that the magnitude and the variation with wavelength of the intensity of this strong echo pointed to the presence of dry hailstones with an almost uniform diameter of about 5 cm. (Stones of this size reached the ground over areas about 1 km across.)

The 'anvil' echo. A diffuse echo extended forward of the storm, evidently corresponding to an anvil cloud. Its top was at about 35 000

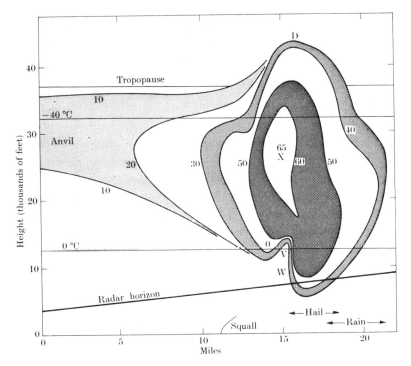

FIG. 8.33. Features of radar echo structure in the Wokingham storm during its intense phase, shown on a range–height section through the approaching storm. Isopleths are drawn of 4·7 cm reflectivity expressed as $10 \log Z$. (Z in mm^6 m^{-3}). X: position of maximum radar echo intensity; D: highest echo top, almost vertically above W, the echo 'wall', associated with large hail. O: the 'forward overhang', separated from the wall by V, the 'echo-free vault'. (After Browning and Ludlam (1962).)

ft (10·5 km) and its base descended gradually from about 7·5 km, 40 km ahead of the storm, to form a *forward overhang* of echo down to about 4 km, 2–5 km ahead of the wall and along its length. Between the forward overhang and the wall, the base of the echo rose again to 5 km or even more, enclosing the *echo-free vault*, which Browning and Ludlam (1962) regard as marking the zone in which a strong updraught enters the front of the storm and prevents the descent of the particles composing the base of the forward overhang. From the low echo intensity there, it appeared that the particles consisted of small (millimetric) hail whose fall speeds did not exceed about 10 m s^{-1}.

If pronounced wind shear exists through a deep layer of the troposphere, the storm as a whole will tend to move with the winds at middle levels, but the top will be pushed forward relative to the base and cause

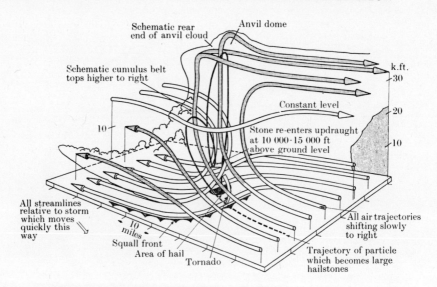

Fig. 8.34. A three-dimensional model of the air flow in a severe storm. (After Browning and Ludlam (1962).)

the updraught to become tilted as in Fig. 8.34. In these circumstances, precipitation may fall out of the updraught into relatively dry air having a steep lapse rate and so trigger a cool downdraught. Heavy rain, composed of large drops, may not be able to evaporate fast enough to keep the downdraught saturated so it may remain rather dry.

A tilted updraught and a strong downdraught may therefore co-exist and work continuously side by side to produce an almost steady three-dimensional circulation that may persist for several hours. The air-flow relative to such a storm is depicted in Fig. 8.34. Potentially warm air is fed into the front right quadrant of the storm at low levels, ascends to form the updraught, and leaves within the anvil cloud ahead of the storm at high levels and to the left of the mid-level steering wind. Potentially dry and cold air originating at middle levels descends in the downdraught, leaves the left rear flank of the storm, and spreads out near the ground towards the right flank to form a miniature cold front (or squall front) over which the warmer unstable air ahead of the storm is lifted to form the updraught.

The air in the sustained updraught of such a 'super cell' is likely to experience much less mixing and dilution than if it were composed of intermittent thermals, and is likely therefore to be more strongly accelerated. In the especially unstable situations that frequently occur during the summer in the Great Plains of the United States, updraughts in the upper troposphere may attain speeds of 40 m s^{-1} and more, and the cloud tops penetrate some kilometres into the stratosphere.

8.10. Measurement of rainfall by radar

The possibilities of measuring the intensity and distribution of rainfall over areas of hundreds or thousands of square kilometres with a single radar have been investigated by many workers because the inherent advantages of such a method compared with a network of rain-gauges recording at only a series of widely-spaced points are obvious. A very detailed discussion of the problem is given by Harrold (1965). Whether an accurate measurement of the rainfall in a region distant from the radar can be made by measuring only the intensity of the radar echo, must depend on whether there is a unique relationship between the echo intensity and the rate of rainfall, whether the absolute intensity of the signal can be measured with sufficient accuracy, and also on the attenuation suffered by the signal and the stability of the radar transmitter and receiver.

The existence of a unique relationship between the echo intensity which, for Rayleigh scatterers, is proportional to $Z = \sum ND^6$, where N is the number of raindrops per unit volume of diameter D, and the rate of rainfall, p, would imply that p is uniquely related to the drop-size distribution, whatever the type of rainfall and the physical process by which the drops are formed. Experimental methods for determining the size distribution of raindrops and various empirical relations between the drop-size spectra and the precipitation intensity are described in Appendix B. Although these formulae may represent quite well the average size distribution of a number of samples, or the characteristics of rain averaged over a considerable period of time, a sample taken over a period of several minutes on any one occasion may show considerable deviations from the smoothed distributions for reasons that will appear shortly.

The Marshall–Palmer (1948) formula

$$N_\mathrm{D} = 0{\cdot}08\,\mathrm{e}^{-\Lambda D}\ \mathrm{cm}^{-4}; \qquad \Lambda = 41 p^{-0{\cdot}21}\ \mathrm{cm}^{-1}, \qquad (8.40)$$

where $N_\mathrm{D}\,\delta D$ is the number of drops per unit volume of diameter between D and $D+\delta D$, best represents the average size distributions of raindrops originating as snowflakes in stratiform clouds, except that it tends to overestimate the numbers of small drops for which $\Lambda D < 4{\cdot}5$. Substitution of eqn (8.40) into $Z = \int N_\mathrm{D} D^6\,\mathrm{d}D$ and integrating from 0 to infinity gives $Z = 296\,p^{1\cdot47}$. Marshall and Palmer found that computations of Z and p from actual drop-size distributions showed a scatter about the latter curve and, in fact, found that $Z = 200\,p^{1\cdot60}$ gave the best fit to the observations, where Z is in $mm^6\,m^{-3}$ and p in $mm\,h^{-1}$. Many authors have proposed other relations of the form $Z = \sum ND^6 = Ap^b$ based on measured drop-size distributions; they have been reviewed by Battan (1959), Sivaramakrishnan (1961), Atlas (1964), and Stout and Mueller (1968), a representative selection being shown in Table 8.8. These formulae show considerable differences due partly to differences in methods and accuracy of measurement of drop size and of computing the precipitation rate, p, to the fact that they are based on drop-sampling volumes of only about 1 m^3,† and that some authors have obtained different values of A and b by using p instead of Z as the independent variable in analyzing their data, but the larger differences probably reflect the fact that rains of different origin and history have different drop-size spectra. The formulae of Andrews (and others) in Table 8.8 indicate marked differences of Z for a given value of p as between continuous, showery, and thundery rains in temperate latitudes; if the exponent b were taken as 1·6 for all three types, the corresponding values of A required to provide a reasonable representation of Andrews' data would be 200, 280, and 350, respectively. This suggests that if rains are classified according to locality and character, and this is not easy to do from the radar pattern alone, the average error in estimating the rainfall intensity at a point for a given value of Z, due to variations in drop-size spectra, will be about 25 per cent. It is also clear from Table 8.8 that orographic rains in Hawaii, which are composed of high concentrations of small drops, are associated with low values of A, while the opposite is true of heavy rains composed of large drops, and of rain that has fallen through a dry atmosphere (e.g. in Arizona) and lost its smaller drops by evaporation. Another cause of variability in the Z–p relations arises from the fact that locally-measured drop-size distributions may be influenced by drifting and size-sorting of the drops in the wind, and may therefore vary in both space and time within a particular storm.

But even if the measured drop-size distributions gave unique Z–p relations for all types of rain, errors in deducing the rainfall intensity in the volume occupied by the radar beam could arise through difficulties

† Joss and Waldvogel (1969) show, for example, that in widespread rain falling at 1 mm h⁻¹, a filter paper of area 1 m² must be exposed for 1 s to obtain, with a probability of 68 per cent, a Z value that deviates by <20 per cent from the mean.

TABLE 8.8
Z–p Relationships, $Z = Ap^b$

Author	A	b	Location	Rainfall type	Rainfall rate mm h^{-1}
Wexler (1948)	214	1·58	Washington, D.C.		0·37–114
Marshall and Palmer (1948)	200	1·60	Ottawa		0·15–35
Best (1950)	505	1·44	Shoeburyness		0·18–4·2
	257	1·45	Ynyslas		0·36–8·9
	436	1·64	East Hill		0·41–25·1
Blanchard (1953)	31	1·71	Hawaii	Orographic rain	1·00–20·8
Twomey (1953b)	127	2·29	Sydney		0·2–9·0
Jones (1956)	358	1·36	Illinois	Heavy shower	
	486	1·37	Illinois	Thunderstorm	
Andrews (1961)	204	1·52	London	Continuous	0·07–6·3
	280	1·46	London	Showers	0·05–35·3
	280	1·71	London	Thunderstorm	0·05–74·7
	215	1·60	London	all types	
Imai (1960)	300	1·60	Japan	Continuous	
	200	1·50	Japan	Showers	
Sivaramakrishnan (1961)	219	1·41	India	Thunderstorm	
Diem (1968)	278	1·30	Entebbe	Tropical showers	
	178	1·25	Germany	Spring and Autumn showers	
Foote (1966)	520	1·81	Arizona	Showers	
Mueller and Sims (1966)	286	1·43	Florida		
Stout and Mueller (1968)	301	1.64	Oregon		
	311	1·44	Indonesia		
	267	1·54	Alaska		
	593	1·61	Arizona		
	221	1·32	Marshall Islands		
Fujiwara (1967)	80	1·38	Hawaii	Orographic rains	
Imai et al. (1955)	500	1·60	Japan	Snow—small crystals	
Gunn and Marshall (1958)	2000	2·0	Canada	Snow—crystal aggregates and snowflakes	
Austin (1963)	1000	1·60	New England	Heavy snow	4–30

of maintaining the stability and calibration of the radar, of measuring the absolute intensity of the rapidly fluctuating signal, and of allowing for attenuation of the signal by intervening precipitation. The latter is usually serious in all but light precipitation at wavelengths below about 5 cm. Serious errors may also arise in the deduced value of p if the radar-echo intensity varies across the sampling volume, for example if the radar beam straddles the melting band through which the signal intensity changes rapidly with height. This may well cause errors of up to ±50 per cent in the deduced value of p depending on the height of the melting band and the range (see Harrold 1965). Additional uncertainty arises in relating the precipitation intensity aloft, as measured by the radar, to that obtaining at the ground, because the Z–p relationship may vary with height due to size-sorting of drops in the wind shear, and to evaporation or growth of drops during their fall, while the presence of strong vertical motions in the cloud will cause the value of p deduced aloft to be considerably different from that at the ground where the vertical motions will be negligible.

In principle, the difficulties that arise from the absence of a unique Z–p relationship, and from problems associated with the stability and calibration of the radar, can be overcome by using a network of rain-gauges to provide a direct calibration of the radar echo intensity in terms of a measured precipitation rate. This technique would require automatic rain-gauges capable of transmitting the readings immediately to the radar site but, even so, the rain-gauges may well not provide data that are representative of the much larger area covered by the radar. Moreover, the radar measurements of rainfall intensity aloft may not correlate closely with those measured at the ground because of the space and time lags introduced by the drops in falling through a horizontal wind.

Even with a well designed and calibrated radar working on a non-attenuating (e.g. 10-cm) wavelength, it is difficult to make individual measurements of received power to better than ±2 dB, while the discrepancy between the measured signals from rain and those calculated theoretically from the radar equation is at least −1·4 dB (Probert-Jones 1962). These errors, together with those that arise from uncertainties in the Z–p relation and other causes discussed above, produce standard errors of at least ±50 per cent for point samples of rainfall rate.

Since much of the error and uncertainty is due to wind sorting of the particles and similar time- or space-dependent phenomena, the accuracy of rainfall measurements may be greatly improved by averaging over space or time. Thus averaging over a period of time allows the total rainfall at a point during a known interval to be determined much more accurately than instantaneous rates of rainfall. For similar reasons, space averaging yields better estimates of rainfall over an area than at a point. An unusually good example of this was reported by Leber, Merritt, and Robertson (1961) whose integrated 10-cm radar rainfall measurements, based on the relation $Z = 200\, p^{1.60}$, over a 24-h period of very heavy rain in Indiana, agreed with those obtained from a network of 88 rain-gauges out to a range of 115 miles to within 2 per cent. Atlas (1964) attributes this remarkable high accuracy to the fact that, in these large, intense showers that filled the radar beam, the reflectivity probably remained fairly constant to considerable altitudes. In small showers, or in stratiform rain showing a melting band, considerably larger errors must be expected. Huff (1966), dealing with summer convective rains in the mid-west of the U.S.A., used data from the rain-gauge network to help choose the appropriate Z–p relation for individual storms and, in so doing, was able to reduce the error in determining the mean rainfall over an area of 200 miles^2, with 10-cm radar, from a median value of 44 per cent to 30 per cent. In order to produce the same improvement in accuracy, the average spacing of rain-gauges would have had to be reduced to 1 in 150 miles^2. By using the rain-gauges in the northern and southern parts of the networks to adjust the radar equation for the central part, the median radar error was reduced from 30 to 12 per cent and the equivalent gauge density lowered from 150 to 25 miles^2 per gauge. Wilson (1966) demonstrated that during heavy thunderstorm situations in Oklahoma, 25–50 rain-gauges would be required to determine the average rainfall rate over a 1000 miles^2 area with the same accuracy as may be obtained with a WSR-57 10-cm radar; the *actual* average density of rain-gauges in that area is only 1 per 1000 miles^2. Thus, it appears that, in general, a well-designed and operated 10-cm radar, equipped with an efficient signal integrating device, can determine the rainfall over an area more reliably than is likely with the rain-gauge networks that exist in most parts of the world.

Returning to point measurements, Austin (1963) correlated the output of a radar signal integrator with hourly rain-gauge readings and found that, in 30 of the 40 storms investigated, the equivalent rainfall deduced from the standard $Z = 200\, p^{1.6}$ relation differed from the actual amount by less than a factor of 2; in 6 storms the radar amounts appeared to be a half or less of the gauge values; and 4 storms showed amounts twice as large. In short, the radar could usually estimate hourly rainfall at a point to better than a factor of 2 by using a standard Z–p relation. Unusual deviations from this relation may have accounted for the occasional larger errors.

In 6 out of 9 snowstorms, Austin found excellent agreement between the radar and gauge estimates using the relation $Z = 1000\,p^{1.6}$, which is a compromise between those for single-crystal and aggregate snow (see Table 8.8). Five of the storms showing this agreement were major ones with precipitation rates ranging from 4 to 30 mm h^{-1}. In 3 storms of light snowfall (0.5–4 mm h^{-1}), precipitation rates were underestimated perhaps because the particles were predominantly single crystals with smaller Z than given by $1000\,p^{1.6}$.

9

The Electrification of Clouds

IN the past there has perhaps been a tendency to regard the electrical activity of clouds as a subject rather distinct from the condensation and precipitation processes that have been discussed in previous chapters. It is difficult to find logical justification for this since the electrical and precipitation processes are indissolubly linked and should be regarded only as different aspects of the field of study we call cloud physics. It is hoped, therefore, that this chapter will be regarded as an integral part of the subject-matter of this book.

Here we shall be concerned not so much with the electrical state of the atmosphere during fine weather, but rather with deviations from this 'normal' state during the passage of precipitating clouds, in particular, thunderstorms. It will, however, be useful to summarize the main features of the fine-weather electric field, which represents a reference state, and relative to which the complex changes that occur in disturbed weather may be measured.

9.1. The vertical electric field and current in fine weather

The fundamental electrical phenomenon of the lower atmosphere is the vertical electric field which, in clear-sky conditions, is directed downwards. This implies that the earth's surface carries a negative charge and the atmosphere a net positive charge. The surface density of charge on the conducting earth is about -3 e.s.u./m², the total fine-weather charge amounting to about 5×10^5 C. The intensity of the vertical electric field has a maximum value at the ground where its magnitude is 120 V/m when averaged over the whole earth, and 130 V/m over the oceans; in industrial regions, where the air is highly polluted, the field strength may be considerably enhanced, the mean value at Kew being 363 V/m. The surface-field strength F_0 is related to the surface density of charge σ by the equation

$$F_0 = \left(\frac{\partial V}{\partial z}\right)_{z=0} = -4\pi\sigma \quad (9.1)$$

and has the same sign and order of magnitude, in fine weather, over the whole earth. The field intensity (or potential gradient) decreases at greater heights, at 10 km falling to only 3 per cent of its surface value.

Thus, the lower atmosphere normally contains a positive space charge of density ρ, such that

$$\partial^2 V/\partial z^2 = -4\pi\rho, \tag{9.2}$$

where V is the potential difference between the earth and the atmosphere at height z. On integration (9.2) becomes

$$\partial V/\partial z - (\partial V/\partial z)_{z=0} = -4\pi \int_0^z \rho \, dz \tag{9.3}$$

and the surface value of the potential gradient is

$$(\partial V/\partial z)_{z=0} = 4\pi \int_0^\infty \rho \, dz = -4\pi\sigma. \tag{9.4}$$

Thus

$$\sigma = -\int_0^\infty \rho \, dz. \tag{9.5}$$

The potential V of the atmosphere with respect to the earth increases with altitude up to about 20 km, above which it remains sensibly constant at about 3×10^5 V. The very small potential gradients that exist in the atmosphere above 20 km indicate that the air at these levels is highly conducting.

Besides short-term fluctuations due to local changes in the space-charge density and conductivity of the air, the intensity of the fine-weather field undergoes diurnal and seasonal variations. Over the oceans, in polar regions, and in a few continental regions well removed from sources of atmospheric pollution, the potential gradient shows a maximum at about 1900 G.M.T. and a minimum at about 0400 G.M.T. independent of the local time. Wilson (1922), who suggested that the origin of the fine-weather field lay in thunderstorms, pointed out that, in consequence, its maximum value should occur when the most thundery regions of the earth attain their maximum activity. Whipple (1929) and Whipple and Scrase (1936) did find a very close correspondence between the variation in thunderstorm activity over the whole globe during the Greenwich day and the variation of potential gradient measured over the oceans and in the polar regions. At most land stations, the times of maxima and minima in potential gradient depend on local time, there usually being a double oscillation with minima at 0400–0600 hours and 1200–1600 hours, and maxima at 0700–1000 hours and 1900–2100 hours. The amplitude of this diurnal variation, which may amount to 50 per cent of the mean value, is closely correlated with the degree of atmospheric pollution, which controls the small-ion content and hence the conductivity of the air. There appears to be no marked annual variation of the potential gradient over the oceans, but land stations in both hemispheres show a maximum in the local winter and a minimum in the summer.

The lower atmosphere is slightly conducting due to the existence of ions produced by radioactive matter in the air and in the earth, and by cosmic rays. Over land, the ionization of the lowest layers is due mainly to radioactive substances in the earth's crust which produces ion-pairs at the rate of about 8 cm^{-3} s^{-1} compared with 2 cm^{-3} s^{-1} by cosmic rays but, at higher levels, cosmic radiation makes an ever-increasing contribution. Over the sea, cosmic radiation is the prime agency. The rate of ion production at first increases with increasing height, reaching a maximum value of about 45 cm^{-3} s^{-1} at 12 km in middle latitudes, but only 20 cm^{-3} s^{-1} near the equator. Above about 12 km, the rate of production falls off largely as the result of decreasing air density. Lower ion production at low latitudes may be attributed to deflexion of the charged primary cosmic rays from these regions by the earth's magnetic field. The space-charge density, ρ, decreases with increasing height in conformity with eqn (9.3) and the variation of the potential gradient with height. The positive space charge results from an excess of positive ions which, on average, balances the negative charge on the earth's surface.

The ions may be divided roughly into two groups: small ions that consist generally of a singly-charged molecule surrounded by a cluster of a few neutral molecules and possess a mobility of about 1·5 cm^2 s^{-1} V^{-1}, and large (Langevin) ions formed when small ions attach themselves to much larger neutral aerosol particles, with mobilities of 10^{-4} to 10^{-2} cm^2 s^{-1} V^{-1}. The concentration, n, of small ions, when equilibrium is established between their rate of production, q, and rate of destruction by recombination and capture by larger nuclei, is given, approximately, by the relation

$$q = \alpha n^2 + \beta nN, \qquad (9.6)$$

where n is the concentration of small ions of each sign, N the concentration of large particles (charged and neutral), α the recombination coefficient for small ions ($= 1\cdot 6 \times 10^{-6}$ cm^3 s^{-1} at s.t.p.), and β is a constant of magnitude about 2×10^{-6} cm^3 s^{-1}. In the lower layers of the atmosphere, the βnN term is the more important in determining the equilibrium concentration of ions, but above about 1 km, the αn^2 term becomes dominant. The average concentration of small ions is greater over the sea than over the land; their greater production over the land is more than offset by their high rate of capture by larger nuclei. Some average values for the rates of production and the concentrations of ions in different localities are shown in Table 9.1.

The concentrations of positive and negative ions are not very different, but there is a small excess of positive ions representing the positive space charge; the ratio of the concentrations of the small positive and negative ions is about 1·2.

The electrical conductivity of the air, which is proportional to the product of the concentration and mobility of the ions, increases quite rapidly with height up to about 20 km, an overall result of the varying intensity of cosmic radiation with height, the larger ionic mobilities at lower air densities, the lower rates of ion destruction, and the smaller

TABLE 9.1
Numbers of ions of each sign (after Wormell 1953)

	Rate of production ($cm^{-3}\,s^{-1}$)		Concentration (cm^{-3})	
	By radio-activity	By cosmic rays	Small ions	Large ions
Oceanic air	0	2	700	200
Country air	8	2	600	2000
City air	8	2	100	20 000

concentrations of radioactive material at greater heights. The conductivity of the lower atmosphere is slightly greater for positive than for negative ions, but the polar conductivities become nearly equal above about 7 km.

The drift of the positive ions downwards and of the negative ions upwards under the fine-weather field, constitute a downwardly directed conduction current that tends to neutralize the earth's negative charge. The current flowing through unit area normal to the field is

$$i = F(\lambda_+ + \lambda_-) = Fe(n_+ k_+ + n_- k_- + N_+ K_+ + N_- K_-), \qquad (9.7)$$

where λ_+ and λ_- are the polar conductivities, k, K the ionic mobilities of small and large ions respectively, n, N the ionic concentrations, and e the ionic charge. In country air and over the sea, the conductivity is due mainly to the small ions; in city air, the large ions may make a contribution of 10 or 20 per cent. Aircraft measurements made by Kraakevik (1958) over Greenland, where pollution and convection currents were not complicating factors, indicated the fair-weather conduction current to be $3\cdot 7 \times 10^{-16}$ A cm^{-2}, almost independent of height. Cobb (1968), from measurements made on Mauna Loa, Hawaii, at an altitude of 3·4 km and in clean air above the trade-wind inversion, obtained an average value of $5\cdot 4\times 10^{-16}$ A cm^{-2} for the year 1961. Measurements of the variation of conductivity with height allow the resistance of a unit column of the atmosphere to be calculated. Gish and Sherman (1936) used their balloon measurements to deduce a value of 10^{21} Ω cm^{-2}, implying a value of 200 Ω for the resistance of the whole atmosphere. Kraakevik's measurements over Greenland gave values of 8×10^{20} Ω cm^{-2} and 160 Ω, respectively. His measurements of conduction current, if typical of the whole atmosphere, would imply a

TABLE 9.2

Average values of parameters of the fine-weather field

Field intensity F (potential gradient)	120 V/m
Surface density of charge	-3 e.s.u./m^2 (10^{-9} C/m^2)
Total charge on earth's surface	$-1\cdot 5\times 10^{15}$ e.s.u. (-5×10^5 C)
Space-charge density	3×10^{-2} e.s.u./m^3 (10^{-11} C/m^3)
Conductivity of air (at sea level)	2×10^{-16} Ω^{-1} cm^{-1} ($1\cdot 8\times 10^{-4}$ e.s.u.)
Air–earth current density	2–4×10^{-16} A cm^{-2}
Total air–earth current	1800 A
Columnar resistance	10^{21} Ω
Total resistance of atmosphere	160 Ω
Potential difference between earth and high atmosphere	3×10^5 V

total air–earth current of about 1800 A and a value of $2\cdot 9\times 10^5$ V for the potential difference between the highly-conducting upper atmosphere and the earth. The average values of parameters of the fine-weather field are given in Table 9.2.

The earth can thus be regarded as a spherical condenser carrying a charge of -5×10^5 C with a leakage current of 1800 A. If it were not replenished, the negative charge on the earth would quickly disappear, the time constant of the system being about 4 min. The fact that the earth's charge remains sensibly constant speaks for the activity of a regenerative mechanism, the nature of which will appear later in this chapter.

9.2. The electrical fields produced at the earth's surface by thunderstorms and lightning discharges

9.2.1. *Field changes due to lightning flashes*

The first reliable determinations of the charges involved in thunderstorm processes were made by Wilson (1916, 1920), who measured the changes in the vertical electric field produced by lightning discharges. His experimental technique consisted in recording continuously the electric charge on a conductor of simple geometrical form, for example a flat metal plate of area 1 m² replacing a portion of the earth's surface and maintained automatically at earth potential. This was achieved by use of the capillary electrometer which is illustrated in Fig. 9.1. Any charge induced on the plate by a change in the earth's electric field, ($\Delta F = -4\pi\Delta\sigma$), passes to earth through the electrometer and causes the acid bubble to move. The film of acid surrounding the

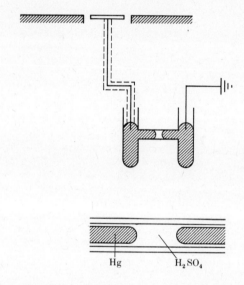

Fig. 9.1. The Wilson flat plate with capillary electrometer.

mercury, together with the mercury itself, forms a condenser by virtue of an electric double-layer existing at the mercury-acid interface. This interface is associated with a definite surface density of charge and any additional charge can be accommodated only by increasing the interfacial area. Thus when the plate receives an additional charge ΔQ, the acid bubble moves to the right in order to increase the area of the lefthand condenser which can now store the charge ΔQ; simultaneously the capacity of the right-hand limb is decreased by the same amount, which results in the passage of the charge ΔQ to earth through the electrometer. The charge passed is thus proportional to the displacement of the bubble which is recorded photographically on a moving film with the bubble acting as a cylindrical lens. The time response of the whole system to a sudden change in the vertical field is less than 0·1 s. In addition to the charge induced by the field, the instrument records that communicated to the plate by the ionization current and charged precipitation. In the absence of rain, the conduction current can be separated from the true induced charge by screening the plate from the field at regular intervals of a few minutes. To overcome the difficulties associated with charged rain, Wormell (1939) has used an inverted form of the test plate.

The flat-plate apparatus is not sufficiently sensitive to detect field changes due to storms more than a few kilometres distant; for these, Wilson and Wormell used an elevated sphere. A copper sphere of 30-cm diameter is mounted on insulators on an earthed metal pipe about 5 m above the ground; the pipe can be lowered so that the sphere becomes enclosed in an earthed metal case and shielded from the earth's field. In being raised to a height h where, in the absence of the sphere, the potential would be V, the sphere, which is maintained at earth potential, must acquire a charge Q of such a value as to make its potential zero. Thus the sum of the potentials V due to its position, Q/r due to the charge Q on a sphere of radius r, and $Q/2h$ the contribution of the image charge must be equated to zero:

$$V + Q/r - Q/2h = 0. \qquad (9.8)$$

The charge Q passing to the sphere when it is raised gives a measure of the field V/h and is measured by the capillary electrometer in the same manner as for the flat plate. Changes in the field are accompanied by changes in the charge Q and these are greater than the corresponding changes in the bound charge on the surface of the earth itself.

These methods yield absolute values of the vertical field over level ground and of the sudden changes that accompany lightning discharges, and delineate the form of the subsequent recovery of the field. A typical record in Fig. 9.2 shows the pre-discharge positive field of about 100 V/m, a sudden large positive field-change produced by a distant lightning flash and a relatively slow, approximately exponential recovery

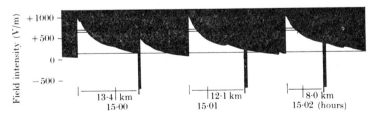

FIG. 9.2. Field-changes associated with lightning flashes as recorded by a capillary electrometer.

of the field. The vertical black line indicates the commencement of thunder, the time interval multiplied by the velocity of sound giving the range of the storm.

In order to obtain a continuous measurement of the potential gradient from the bound charge on an earthed conductor, various forms of instruments called field mills have been developed. The fixed test plate in the form of a sectored disc is alternately exposed to, and shielded from, the field by a motor-driven rotating earthed cover. The plate is connected to earth through a resistance R, and the alternating potential developed across R which is proportional to the field, is amplified, rectified, and may be displayed on a galvanometer or pen-recorder. There is usually some method of indicating the sign of the potential gradient and the instrument is calibrated in artificial fields of known magnitude. A typical instrument having a plate with four sectors is capable of following field changes of duration about 0·01 s, but models capable of responding to changes of <1 m s have been developed. Fig. 9.3 shows such an instrument, described by Malan and Schonland (1950), in which a multi-vaned earthed disc rotates over fixed electrodes in the form of studs mounted on an insulated disc and connected in parallel across the earthed resistance, R. The alternating potential generated across R is amplified and displayed on an oscilloscope; with N vanes on the disc rotating at n rev/s, the frequency of the output voltage is Nn/s. The instrument of Malan and Schonland produced a deflexion of 1 cm for a field change of 20 V m^{-1}, and had a response time of 0·4 ms. The variation of output with time is shown at the bottom of Fig. 9.3; the sudden increase in signal was produced by a lightning stroke. To indicate the direction of the field, two diametrically opposed slots in the disc were cut deeper than the rest and then, by mounting an extra pair of studs nearer the axis, it was possible to produce a temporary increase in the signal once every half revolution when these studs were uncovered. Thus, as shown at P in Fig. 9.3, an indication appears on one side of the trace when the field is positive, and on the opposite side when the field is negative.

Let us consider the thundercloud as a bipolar electrostatic generator with positive and negative centres of charge situated approximately vertically one above the other. Lightning flashes can be classified generally into two groups, cloud-to-ground discharges which generally lower a charge from the lower part of the cloud to earth, and internal flashes which tend to neutralize the bipolar charge distribution.

If a charge δQ, originally at height h_2, is lowered to earth, the vertical

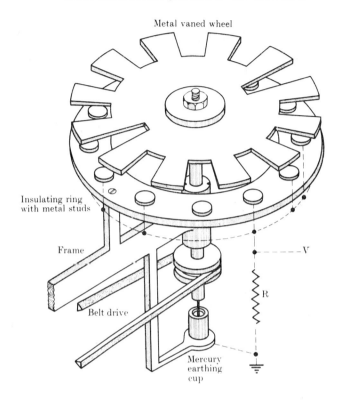

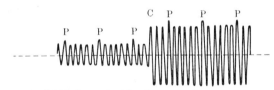

Record of field change C with polarity pips P

FIG. 9.3. The electrostatic fluxmeter. (From Malan, *Physics of lightning*, English Universities Press, London. (1963).)

field-change produced at an observing station at distance L is

$$\delta F = \frac{2\delta Q h_2}{(h_2^2 + L^2)^{\frac{3}{2}}}. \tag{9.9}$$

For a distant storm $(L \gg h)$ this reduces to

$$\delta F \simeq \frac{2\delta Q h_2}{L^3} = \frac{\delta M}{L^3}, \tag{9.9a}$$

where δM is the change in electric moment.

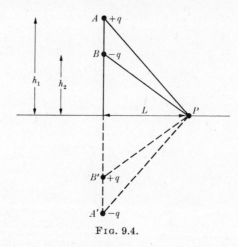

Fig. 9.4.

The field-change produced by an internal flash in which a charge δQ passes from A to B (Fig. 9.4) is

$$\delta F = 2\delta Q \left\{ \frac{h_2}{(h_2^2+L^2)^{\frac{3}{2}}} - \frac{h_1}{(h_1^2+L^2)^{\frac{3}{2}}} \right\}. \tag{9.10}$$

Again, if $L \gg h_1$ and h_2, (9.10) reduces to

$$\delta F = \frac{2\delta Q(h_2-h_1)}{L^3} = \frac{\delta M}{L^3}. \tag{9.10a}$$

We see from (9.10), that for a particular value of $L = (h_1 h_2)^{\frac{1}{3}}(h_1^{\frac{2}{3}}+h_2^{\frac{2}{3}})^{\frac{1}{2}}$, the field-change has value zero and at this point it reverses in sign. By studying the sign of the field-change as a function of distance, and in particular its reversal, Wilson was able to deduce the polarity of the thundercloud. He found that the field-change produced by a cloud-to-ground discharge was positive at all distances from the storm, i.e. it tended to augment the normal fine-weather positive field, whereas the field-change due to an internal flash was positive for near flashes and negative for distant storms. This information pointed to the main positive charge of the cloud being situated above the negative centre, the cloud then being classified as having positive polarity.

The reversal effect is particularly clear-cut in South African thunderstorms where the proportion of internal flashes is higher than in England. Schonland (1928a) reported, as a result of observations on some fifty storms, that for discharges within 7 km, positive field-changes were

21 times as frequent as negative; for discharges beyond 15 km, negative changes were nearly 10 times as frequent as positive. About 10 per cent of the flashes were cloud-to-ground discharges bringing negative charge to earth, most of the remainder being internal discharges in clouds of positive polarity. A prolonged series of observations at Cambridge by Wormell (1939) and Pierce (1955) has shown that, in English storms, the ratio of the numbers of positive and negative field-changes falls steadily from about 4·4 for flashes within 5 km to about 1·1 at 30 km, and remains constant beyond that distance. While the proportions of cloud-to-ground and internal flashes are thus quite different in the two series of observations, in both cases the typical flash to earth brings down negative charge and the typical cloud discharge destroys a bipolar distribution of positive polarity.

The evaluation of the field-change δF and the distance of the storm L from the electrometer records of Wilson and Wormell allows the magnitude of the electric moment destroyed by a flash to be determined, and if the heights of the charge centres are assumed, or independently determined, the quantity of charge neutralized can be estimated. Eqns (9.9a) and (9.10a) have been verified for the electrostatic effect of a lightning discharge for values of L from 20 to 200 km by Pierce (1955). Wormell and Pierce give 110 C km as a median value for δM. Wilson (1920) gave a mean value of 100 C km, while Schonland and Craib (1927) quoted a value of 94 C km for South African storms. Taking a mean value of 100 C km, values of h varying from 1·5 to 3·5 km for cloud-to-ground discharges and up to 5 km for internal discharges, the charges neutralized are calculated to range from 10 to 30 C.

Appleton, Watson-Watt, and Herd (1926) used a receiving aerial and cathode-ray tube recording to study the radiation component of the electric field-changes due to lightning discharges from distant storms, generally farther than 50 km. They found that in the majority of cases the field-change was negative, as would be expected from the passage of internal flashes within distant clouds of positive polarity.

Field-changes due to lightning discharges have been examined by Wormell (1930) and Whipple and Scrase (1936), by measuring the discharge current flowing through an elevated, insulated point conductor. When the electric-field intensity exceeds a certain minimum value, the air in the immediate neighbourhood of the point becomes ionized by collision, and a current passes through the point and a galvanometer connected in series to earth. Whipple and Scrase found empirically that the point-discharge current I is related to the field intensity F by the equation

$$I = a(F^2 - F_m^2),†$$

where F_m is the minimum field at which point discharge occurs (about 8 V/cm at tree-top height), and a is a constant. The field, whose sign is indicated by the direction of the current (downward current, positive field), was recorded photographically on a rotating drum, the time constant of the system being about 1 s. These authors, considering storms within a range of 5 km, found the ratio of positive to negative field-changes to be 3·5, a clear indication of the prevalence of clouds of positive polarity.

Workman and Holzer (1942), and Reynolds and Neill (1955), working in New Mexico, have attempted to locate the positions of the positive and negative charge centres involved in a lightning flash as well as measuring the electric moment destroyed, by recording the field-changes simultaneously on a network of field mills. The vertical field-change δF at a point (x_1, y_1) on the earth due to a flash to earth is

$$\delta F = \frac{2Qz}{\{(x-x_1)^2 + (y-y_1)^2 + z^2\}^{\frac{3}{2}}}, \qquad (9.11)$$

where x, y, and z are the space coordinates of the neutralized charge centre and Q is the charge dissipated. Thus, four independent measurements of δF are sufficient to fix the space coordinates of the charge centre; for an internal flash involving the neutralization of two equal and opposite charge centres, at least seven independent measurements are required. Unfortunately, even in a single isolated storm, the disposition of charges is usually too complex for them to be located unambiguously.

9.2.2. *Recovery of the field after a discharge—generation of the electric moment*

The pre-discharge fields beneath thunderstorms are predominantly negative, but periods of equally intense positive field also occur, often associated with very heavy rainfall. When the storm is, say, 20 km distant, there is on the average a definite, though not large, enhancement of the normal positive field. The pre-discharge field beneath a storm rarely exceeds 100 V cm^{-1}, whereas the field-change due to a lightning flash may exceed 500 V cm^{-1}. The field-change due to a

† This formula holds in calm conditions but needs correction in the presence of wind— see Chalmers (1967, p. 251).

§ 9.2 THE ELECTRIFICATION OF CLOUDS

discharge is followed by recovery of the field. Recovery curves associated with close flashes are so irregular and probably so much influenced by local point discharge following large field changes that they are difficult to interpret. Usually the discharge produces a positive field-change so that an intense positive field will exist at the ground due to the positive space charge (Fig. 9.5(b)). Most of the latter will disappear very rapidly, either by being driven into the ground by the intense field, or by being neutralized by a rush of negative ions from point discharge at the surface; this together with the regeneration of the electric moment in the cloud, will cause a rapid initial recovery of the field (Fig. 9.5(c)).

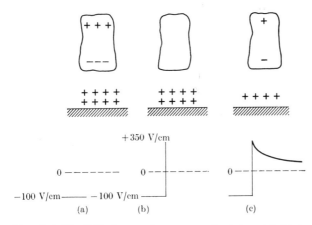

FIG. 9.5. The field changes due to a nearby internal lightning flash.

At greater distances from the storm, well beyond the reversal distance, the recovery curves tend to be more regular and often approximately exponential in shape, and for these Wilson (1929) and Wormell (1939, 1953) offered the following simple interpretation.

Just before the flash, the field is proportional to the electric moment of the cloud and of the space charge beneath it, the contribution of the latter being rather small because of its being restricted mainly to the lowest layers. The internal discharge causes a negative field-change at the distant station, and the subsequent rapid disappearance of the space charge a further small change in the same direction which, on a time-scale of 0·1 s, can hardly be distinguished from the effects of the discharge itself. Thereafter, the exponential recovery of the field effectively represents the regeneration of electric moment in the cloud. The initial rate of recovery is the rate at which the cloud can regenerate electric moment when the internal field is small; the subsequent slowing down is attributed to the increasing difficulty of separating the charged elements in the growing internal field, and to the increased rate of dissipation by continuous currents of the growing free charges as they develop. In both cases, the magnitude of the leakage effect is likely to be nearly proportional to the prevailing electric moment and so cause the recovery to approximate to the exponential form. The charge neutralized in an average flash is of the order 20 C and the initial rate of regeneration about $\frac{1}{7}$ s^{-1}; the vertical current inside the cloud is thus about 3 A immediately after the discharge.

However Illingworth (to be published), having found that the recovery rate of the field for distant flashes tends to increase with increasing distance from the storm, to decrease for larger field changes, but to show no correlation with the intervals between lightning flashes, deduces that the changes of field represented by the recovery curves are caused by rearrangement of the space charge existing around and above the storm as a consequence of the conductivity gradients in the atmosphere, and so give no useful information about the recharging of the thunderstorm itself.

The generation of electric charges and fields within a thunderstorm, their dissipation during lightning flashes, and the recovery of the electric field during the intervals between discharges are considered quantitatively in § 9.6.3.

9.3. The structure of the lightning flash

9.3.1. *Photography of the lightning flash*

The structure of the lightning flash has been elucidated by high-speed photography and by recording the rapid changes of electric field that accompany it.

Walter (1903) was probably the first to use a moving camera to investigate the structure of lightning and found a single flash to be composed of a number of separate discharges. Most of our information has been obtained with an ingenious camera designed by Boys (1926) and used with great success by Schonland and his collaborators in South Africa. In its original form, the camera had two lenses arranged to rotate at opposite arms of a diameter of a revolving disc. Images of the lightning flash were recorded on a stationary film behind the lenses. Each lens distorted the image of a non-instantaneous flash, but the two images were distorted in opposite directions so that, from a comparison

§ 9.3   THE ELECTRIFICATION OF CLOUDS   497

of the two pictures and a knowledge of the velocities of the lenses, it was possible to deduce the direction and speed of the visible discharge processes. The principle of the original Boys camera is shown in Fig. 9.6. Suppose that when one lens is vertically above the other, there occurs a downward moving discharge such as would produce a record like Fig. 9.6(a) on a fixed-lens camera, then the two moving lenses will give a double picture as shown in Fig. 9.6(b). The later, bottom, portion of the luminous channel will be displaced to the right by the upper lens and to the left by the lower one. To obtain an accurate measure of these

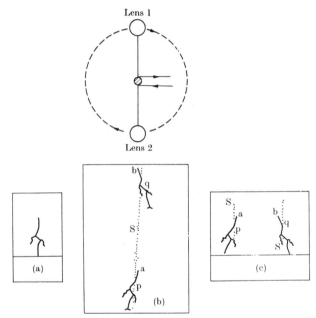

Fig. 9.6. Principle of operation of the Boys camera. (After Schonland, *The Flight of Thunderbolts*, Clarendon Press, Oxford (1964).)

shifts, a line psq is drawn on the print to connect corresponding points p and q, and the two portions of the photograph are mounted side by side with their sections of psq parallel to one another as shown in Fig. 9.6(c). The fact that the luminous process travelled downwards is revealed by the greater separation of the lower as compared with the upper portions of the images. The difference between the two separations pq and ab divided by twice the lens velocity gives the time interval during which the original process travelled from a to p. The same procedure can be used for any position of the two lenses, and by measuring the separations at various points along the channel, it is possible to draw up a time-table of the progress of the lightning stroke and to measure time intervals to an accuracy of a few microseconds.

Photographs taken with such a camera reveal that what appears to the eye to be a single flash of lightning, may sometimes consist of a number of successive strokes following along the same track in space at intervals of a few hundredths of a second. In very active South African storms, the most frequent number of recorded strokes per flash is four, but as many as fourteen have been recorded. The time interval between strokes ranges from 20 to 700 ms but is most frequently 40–50 ms. The average duration of a complete flash is about 250 ms.

The photographs show that a cloud-to-ground discharge is initiated by a streamer that develops downwards in a series of steps, each of which is revealed as a sudden increase in luminosity of the freshly ionized air at the tip of the streamer. This *stepped-leader* stroke approaches the ground at an average speed estimated most frequently at 1–2×10^7 cm s^{-1} from photographic records, and 2–4×10^6 cm s^{-1} from the duration of electric field-changes. It often travels along a branched path between irregularly distributed pockets of positive space charge to give forked lightning. The fact that the average overall velocity of the stepped leader agrees fairly closely with the calculated minimum velocity of an electron cloud that is able to perpetuate itself by ionization, that the velocity of a particular stepped leader remains remarkably constant during most of its journey to the ground, and because the attendant electrostatic field-changes also appear to vary in a continuous fashion, led Schonland (1938) to suggest that the insulation of the air is broken down by a faint *pilot leader* that advances continuously but, having progressed some tens of metres, causes the current in the lightning channel to increase and produce a luminous step. Further support for the existence of a pilot leader comes from the fact that long laboratory sparks in air are initiated by streamers of much the same velocity proceeding from the negative electrode, and such a streamer from a stepped discharge through oil has been photographed in ultraviolet light.

Stepped leaders are classified into two types. In the α-type leaders, the length of an individual step is usually between 10 and 200 m and is traversed with a velocity of about 5×10^9 cm s^{-1} in about 1 μs. There is then an apparent pause of 30–100 μs before the next step. In the β-type leaders, the velocity of propagation is about 5×10^7 cm s^{-1} at first, but later drops to about 10^7 cm s^{-1}, and sometimes the stroke never reaches the earth. The upper parts of the β-leaders are heavily branched and highly luminous, with long steps; in the lower portions, the streamer becomes fainter with shorter steps. The diameter of the stepped-leader channel is estimated from photographs to be between 1 and 10 m.

When the leader approaches to within 5 to 50 m of the ground, a streamer from some point on the earth comes up to meet it, and then there commences the upward *main or return stroke*, which travels up the ionized channel established by the leader. The luminosity of this return stroke is much greater than that of the leader. It travels at between 2×10^9 and 1.5×10^{10} cm s^{-1} and lasts for about 100 μs. It carries the main current of the discharge, which is typically of order 10 000 A, though currents of 100 000 A are occasionally exceeded. Although the luminosity is propagated upwards, the return stroke effectively lowers negative charge from the cloud to the earth. There are usually considerable charges in the branches of the leader channel; when the return stroke arrives at a junction it drains the negative charge from the branch and appears to move *down* the branch although moving *up* the main channel. Drained of charge, the leader-branch can no longer proceed and does not reach the ground. The return-stroke channel averages only a few centimetres in diameter, but the core of the channel, that carries most of the current, may be only a few millimetres across.

After the passage of the stepped leader and the first return stroke, there may be an interval of a few hundredths of a second, followed by a second leader and return stroke. The leaders to the second and subsequent strokes usually consist of streamers that develop in a single flight from cloud to ground and are called *dart leaders*. Since these travel along a pre-ionized path, there is no need for a pilot leader, there are no steps, the average velocity is higher at about 2×10^8 cm s^{-1}, and the duration shorter at about 1 ms, than for a stepped leader. However, when there has been an exceptionally long interval between the successive strokes of a flash, the ionization in the channel may fall to a low level by recombination and diffusion of ions, particularly at the bottom of the path. There may then be a need for a new pilot streamer to ionize the air afresh, which explains the occasional appearance of steps at the lower ends of dart leaders.

The luminous structure of intra-cloud discharges is usually obscured by the intervening cloud droplets so that only a general diffuse illumination (sheet lightning) is seen. Photographs show that the cloud remains luminous throughout the duration of a flash but that there are intermittent bursts of brighter luminosity that cause the flickering appearance of the intra-cloud discharges. Their duration varies between 75 and 700 ms, with a median value of 175 ms. The electric field-change records indicate that their structure is very similar to that of the long air discharges that begin inside the cloud and end in the clear air outside. Photographs show these air discharges to have continuously-luminous leaders but, since the discharge ends in a positive space charge in the air, there is no intense rapid return stroke. The negative charge on the channel is neutralized more slowly than when the leader reaches the conducting earth.

9.3.2. *Electrical field-changes during a lightning flash*

Recordings of the rapid changes of potential gradient produced by lightning flashes show three different components—electrostatic, inductive, and radiative fields. If the distance, L, to an observing station is large compared with the dimensions and height of the thundercloud, and if ionospheric effects may be neglected, the vertical electric field $F(t)$ at the earth's surface at any instant t is given by

$$F(t) = \frac{M}{L^3} + \frac{1}{cL^2}\frac{\mathrm{d}M}{\mathrm{d}t} + \frac{1}{c^2L}\frac{\mathrm{d}^2M}{\mathrm{d}t^2}, \qquad (9.12)$$

where all the electrical quantities are measured in e.s.u. and L in cm, M is the electric moment of the cloud, and c the velocity of light in cm s^{-1}. As we have seen, the electrostatic component is proportional to the electric moment of the discharge and varies as $1/L^3$; the electromagnetic induction term depends upon the current in the discharge and varies as $1/L^2$; and the electromagnetic radiation term, which depends upon the acceleration of the charges involved, varies as $1/L$. Thus, for close discharges, the electrostatic field predominates, while at large distances the radiative field is predominant.

We have discussed already measurement of the slow electrostatic field-changes, but rapid changes of <1 ms duration are best studied with the amplifier-oscilloscope method used by Chapman (1939), Pierce (1955), and Kitagawa and Brook (1960). An outdoor aerial is connected to earth through a condenser C shunted by a resistance R. Movement of charge along the lightning channel will cause a rapid change, ΔF, in the electric field at the aerial and a change in the actual potential of the aerial, $\Delta V = \Delta F \cdot h$, where h is the effective height of the aerial above ground. The charge induced on the aerial will change by $q = \Delta V \cdot C_A = \Delta F \cdot h \cdot C_A$, where C_A is the capacity of the aerial. The unbound charge q will now distribute itself between C_A and C, so that the resultant change of potential across the condenser becomes

$$\Delta v = q/(C_A+C) = \Delta F \cdot h \cdot C_A/(C_A+C), \qquad (9.13)$$

and is therefore proportional to the change of potential gradient, ΔF, produced by the flash. If the output voltage is to follow faithfully a rapid step-change in the field, the time constant CR should be long compared with the time of rise of ΔF. The sensitivity can be changed by varying C and R, keeping the product CR constant. The output signal may be fed into an electrometer valve acting as a cathode follower to match the high impedance of the aerial circuit to the low input impedance of a sensitive galvanometer, or amplified and fed to a cathode-ray tube recorder. A continuous time record of the field-changes can be obtained by photographing the cathode ray tube on moving film or, much more conveniently, by using a galvanometer recording on a light-sensitive paper chart with variable speed to give variable time resolution.

Appleton and Chapman (1937) found that the electric field-changes associated with a cloud-to-ground lightning discharge contained three components corresponding to the stepped leader, the return stroke, and a continuing discharge, while Schonland, Hodges, and Collens (1938) found good correlation between the rapid electrical field-changes and photographic records of the same flashes.

A positive field-change produced by a flash carrying negative charge to earth and consisting of only one stroke is shown in Fig. 9.7(a). It consists of a slow portion of duration about 0·1 s, the latter part of which is known as the L portion, followed by a rapid R step of duration <1 ms, and finally a slow S portion lasting for about 0·1 s. A discharge to earth consisting of two or more strokes causes field-changes that may again start and end with slow L and S portions, possess an R step corresponding to each stroke and, joining the R steps, slow J portions during which the field-change may, or may not, be considerable. Comparison of photographic and electrical records shows that the L portion is associated with the downwardly-developing stepped leader, and the R step with the return stroke. Pierce (1955) found that about half of his records finished with slow S portions, these being particularly frequent for flashes of only one stroke. The total duration of his flashes varied from 150 ms when there was only one stroke, to 515 ms for five strokes, the overall average for 373 flashes being 245 ms.

Negative field-changes, containing fast R steps, are also occasionally

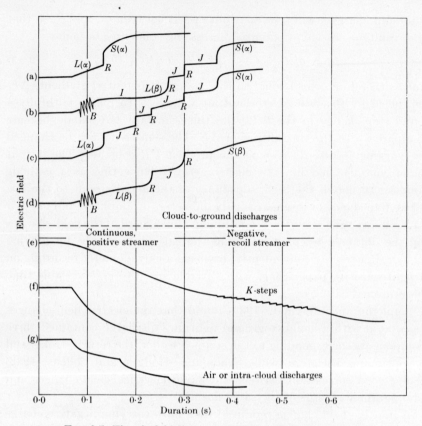

Fig. 9.7. Electric field changes produced by lightning flashes.

recorded. Their structure is very similar (but in reverse) to that of the corresponding positive variety, and they are probably associated with the passage of positive charge from cloud to earth.

Slow negative field-changes *without visible R steps* are shown in Figs. 9.7(e)–(g). Such changes, usually most rapid at the start, are due to air or cloud discharges that either lower positive charge or, more likely, raise negative charge. They may be continuous, or take place in stages interrupted by quiescent intervals as in Fig. 9.7(g).

Wormell (1953) has used the presence of an R step, which is an unambiguous indication of a flash to earth, to deduce the proportion of flashes of various types in storms recorded at Cambridge, England. The results of an analysis of some 1500 field-changes are shown in Table 9.3. The commonest types of discharge were, first, an internal discharge moving negative charge upwards and presumably these occur between

Table 9.3

Proportion of flashes of various types in thunderstorms at Cambridge
(after Wormell 1953)

Flashes to earth	(%)	Discharges not reaching the earth	(%)
Bringing down negative charge	31	Moving negative charge upwards	45
Bringing down positive charge	4	Moving negative charge downwards	10
Bringing down charges of both signs	2	Moving negative charge in both directions	8
Total	37	Total	63

the two main charges of the thundercloud, and, secondly, a discharge to earth from the negatively charged region in the lower part of the cloud.

We shall now discuss the form and interpretation of these field-changes in more detail.

The L field-changes fall into two distinct categories; in the α-type leaders of Fig. 9.7(a), (c), the field increases continually until the onset of the R steps, while in the β-leaders of Fig. 9.7(b), (d), an initial rapid field-change is succeeded by a steady (intermediate) portion, often followed by a further rise in field before the R step. According to Pierce (1955), the average durations of the $L(\alpha)$ and $L(\beta)$ field-changes are 50 and 175 ms respectively. An irregular series of pulses due to the radiation field usually appears at the beginning of the $L(\beta)$ record where the rate-of-change of the electrostatic field is greatest. These occur at longer intervals, of 80–200 μs, than later pulses, which coincide with the bright steps of the leader on the photographs. This, and the fact that the field-changes associated with the stepped leader are of much longer duration than the 10–20 ms indicated by photography, suggest that it is preceded by an initial breakdown process taking place within the cloud where it cannot be photographed.

Of the R changes associated with the first and second strokes of a flash, Pierce found that 40 per cent lasted for <1 ms, but later strokes were generally of longer duration. Again these times are considerably longer than the duration of the visual return stroke and suggest that much of the stroke is concealed within the cloud. There is evidence to suggest that the R elements often consist of a very rapid section followed by a slower field-change, and R elements are usually accompanied by radiative fields. The field-changes accompanying second and subsequent strokes show no portions that may be especially identified with the dart leaders, so presumably the charge transferred by these is very small.

The variations of field in the intervals between the strokes of multiple flashes to earth have been studied in detail by Malan and Schonland (1951a, b). They found these *interstroke changes* to be negative for storms nearer than 5 km and predominantly positive for storms between 12 and 20 km, and therefore concluded that the responsible processes connect a lower to a higher region of the cloud. These authors

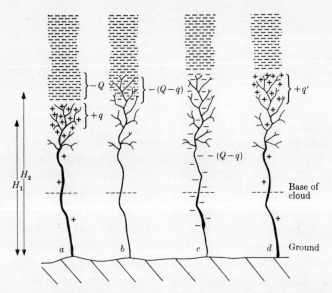

FIG. 9.8. Diagram illustrating the electrical processes taking place in the cloud during the interval between two successive strokes of a lightning flash to earth. (From Malan and Schonland (1951a).)

postulate that, during these intervals, a positively-charged junction (J) streamer travels from the top of the channel blazed by the previous return stroke towards the next volume of negative charge to be tapped. The development of the J process is illustrated in Fig. 9.8. Fig. 9.8(a) shows the suggested distribution of charges at the completion of a return stroke which has brought a positive charge $+q$ to the highly-conducting branched streamers at the top of channel at mean height H_1. In the strong field existing between $+q$ and the main negative charge of the thundercloud, $-Q$, positive streamers penetrate into $-Q$ until $+q$ has been neutralized, leaving a residual negative charge $-(Q-q)$ in a region that has been made conducting by the streamers now at a mean height H_2—see Fig. 9.8(b). The negative charge now flows down the channel as a dart leader as shown in (c). The return stroke that follows, again places a positive charge at the top of the channel as shown in (d). Conditions are now similar to those in (a), and the whole J process may be repeated, draining higher and higher regions of the negative column until the charge is exhausted.

Malan and Schonland (1951b) described five different methods of determining the heights of the charges involved in the separate strokes of a flash. The results agreed in showing that, in South Africa, the average height of the first stroke was 3·7 km above ground with later strokes rising in mean intervals of 0·7 km to reach a maximum height of 9 km. It was concluded that the main negative charge of the thundercloud was contained in a nearly vertical column, the base of which lay between the 0° C and −5° C levels, and the top reaching up to, but not beyond, the −40° C level, some 6 km higher.† Malan and Schonland find the average duration of the J process to be 30 ms; Pierce gives an average value of 93 ms, and a median value of 65 ms corresponding to a velocity of about 10^6 cm s^{-1}.

From observations on storms beyond 20 km a more complicated picture emerges. Thus Malan (1965) found that, for storms beyond 25 km, 19 per cent of the interstroke field-changes were positive, 37 per cent were zero, and 44 per cent were negative, in fair accord with Pierce (1955) who, working at Cambridge, England, found that only 25 per cent of the interstroke intervals showed positive field-changes, the rest showing no detectable change. Malan now suggests that the positive field-changes produced by these distant storms are caused by a continuous discharge bringing negative charge to earth, a conclusion supported by observations in New Mexico by Brook, Kitagawa, and Workman (1962) that a continuing current flows to ground in about 25 per cent of interstroke intervals accompanied by a continuous luminosity that may last for 40 ms or more. Brook et al. call the slow electric field-change that accompanies this continuing luminosity a C-(continuing) change to distinguish it from the junction or J-change that occurs in intervals between strokes when the channel is not luminous. Superimposed on the slow C field-changes are observed, at average intervals of 6 ms, small rapid changes, designated M changes, which are associated with sudden increases in intensity of the luminous channel. In

† The interpretation of these observations has recently been challenged by Ogawa and Brook (1969), who suggest that the 'nearly vertical' aspect of the charge distribution is over-emphasized.

TABLE 9.4
Characteristics of lightning discharges

		Length	Duration	Av. velocity (cm s^{-1})	Av. current (A)	Av. moment (C km)
cloud-to-ground discharge	Stepped leader					
	individual steps	1–4 km	10–30 ms	1–2×10^7	100	30
		10–200 m (av. 50 m)	~1 μs	5×10^9	500–2500	
	intervals between steps		30–100 μs			
	$L(\alpha)$ field change		Av. 50 ms			
	$L(\beta)$ field change		Av. 175 ms			
	Return stroke visible	1–4 km	10–100 μs	5×10^9	10^4–10^5	30
	field-change	1–9 km	up to 1 ms			
	Interstroke processes					
	J streamer (non-luminous)	500–1000 m	30–90 ms	2×10^6		<1
	with K-changes		1 ms			
	C streamer (luminous)		40–300 ms		~100	50
	with M-changes					
	Dart leader (visible)	1–3 km	1 ms	2×10^8	10^3	—
	S-changes					
	$S(\alpha)$		85 ms			20
	$S(\beta)$		145 ms			
	Complete C/G flash		150–500 ms Av. 250 ms		10^4–10^5	110
intra-cloud discharge	Positive, downward continuous leader		100–300 ms			
	Negative upward return stroke	1–3 km	50–200 ms	10^8	200	100
	individual K steps		1–3 ms		1–4×10^3	8
	Complete internal discharge		150–500 ms			100

non-luminous intervals, Brook *et al.* observe other rapid changes, K-changes, of duration <1 ms, which are actually components of the J-change. Both M and K changes are believed to be of similar origin and to be associated with streamer processes inside the cloud that tap local concentrations of charge.

Brook *et al.* deduce that the electric moments associated with the J-changes are very small, only about 1 C km, and show up only during the non-luminous intervals between strokes. On the other hand, they state that the C-changes associated with luminous intervals involve moments of order 100 C km, so perhaps the occasional large interstroke values of 50–100 C km quoted by Pierce were associated with the C rather than the J processes. Moreover, Kitagawa (1965) states that the charge lowered during a continuing current (C-change) may be as much as 30 C over a period of 300 ms, while that lowered in a stroke preceded by a dart leader may be only 1 C. The continuing current is therefore an efficient agent in carrying the cloud charge to earth. Characteristics of lightning discharges are listed in Table 9.4.

Malan (1955) believes that the negative interstroke field-changes may be attributed to the readjustment of the positive space charge in the top of the thundercloud following the disappearance of the lower negative charge and, in exceptional cases, this may involve discharge to the ionosphere.

According to Pierce (1955), the slow S field-changes that follow the last R element are of two types. In the first variety, called $S(\alpha)$, the slow change follows immediately after the R step and continues smoothly until the steady field is reached, as shown in Fig. 9.7(a), the average duration of the process being about 85 ms. In the second kind, termed $S(\beta)$, the gradual field-change is interspersed by periods of no change as in Fig. 9.7(d). The average duration of the $S(\beta)$-changes on Pierce's records was 145 ms. Pierce found that $S(\beta)$-changes tended to be more frequent in multiple-stroke flashes but, according to Malan (1954), the slow S-change occurs most frequently after flashes composed of fewer than four strokes, and is caused by the continuous discharge to ground of part of the negatively-charged column situated above the level reached by the last stroke. If Malan is correct, the S-change may be regarded as a special case of the C-change described above.

Turning now to the structure of intra-cloud discharges, by analysing the shape of the field-changes in the neighbourhood of the reversal distance, Smith (1957) concluded that an internal discharge could be described as the discharge of a vertical dipole in which, effectively, a negative charge was raised in the cloud. On the other hand, Takagi (1961) concluded that the cloud discharge involved a slowly descending positive streamer followed by a more rapidly ascending negative streamer, a view that is supported by a study of field-change records from intra-cloud discharges and photographs of air discharges by Ogawa and Brook (1964).

The field-change of an intra-cloud discharge may be divided into an initial part, an active, rapidly-changing part, and a final J-type part interrupted at about 10-ms intervals by short-period step changes. After studying how the field-changes varied with distance from the storm, Ogawa and Brook associated the initial and very active parts with a descending, positive, continuing-current streamer of duration 100–300 ms. This breakdown process is likely to be a positive rather than a negative streamer because a positive streamer produces a convergent field, which accelerates the photoelectrons into the tip, thus increasing the ionization ahead of the streamer and facilitating its steady advance. On the other hand, a negative streamer does not propagate as readily because the field at the tip repels the electrons ahead of it.

The sudden field-changes that occur in the later part of the cloud discharge are very similar to the K-changes that appear during the J process of the cloud-to-ground flash. Ogawa and Brook interpret these K-changes in a cloud discharge as a recoil or return stroke of opposite polarity to the breakdown streamer, i.e. as a negative return stroke that travels backwards from the tip of the channel already formed by the preceding positive streamer. The average change in electric moment associated with a K-change is about one-twelfth the total moment-change of the complete discharge i.e. about 8 C km. The duration of a single K-change is 1–3 ms, the length of the channel is 1–3 km, the average velocity of the process is about 10^8 cm s^{-1}, and the average current 1000–4000 A. A single discharge may produce as many as twenty detectable K-changes over an interval of 50–200 ms. The duration of the whole discharge appears to increase as the storm develops and the charge distribution becomes more complex and extensive, and may last for 500 ms or more.

On photographs of air discharges, Ogawa and Brook identify the short, bright events that follow a continuously-luminous streamer with the K field-changes, so that these may be regarded as constituting an intra-cloud return stroke, but of greater duration and less brilliance than the single return strokes of cloud-to-ground flashes.

9.3.3. *The mechanisms of electrical breakdown*

As pointed out on p. 503, the fact that the stepped leader, travelling at its minimum velocity of 10^7 cm s^{-1}, would travel from cloud base at 2 km to ground in 20 ms, while the associated electric field-changes frequently last for 50–150 ms, and the appearance on the early part of the record of radiation pulses at much longer intervals than those associated with the bright steps of the leader, suggest that the stepped leader is preceded by an initial breakdown process of very different nature. Malan and Schonland (1951a, b) have obtained evidence to suggest that a high proportion of the β-leaders originate near the 0° C level, in the region of strong field existing between the base of the main negative column and the subsidiary positive charge located near the cloud base that is generally associated with heavy rain. Now Macky (1931) found that, under the influence of strong electric fields, water drops became elongated in the direction of the field, and when this exceeds a value $F_c = 3875/\sqrt{R}$ Vcm^{-1}, where R is the drop radius in cm, they become unstable, and streamers develop at the ends with the onset of corona discharge. Thus, in the presence of drops of $R = 1$ mm, streamers would develop in fields exceeding 12 000 V cm^{-1}, the critical field for 2 mm drops being 8700 V cm^{-1}—much less than the minimum value of 30 kV cm^{-1} required to initiate the breakdown of dry air. The high field need only be very localized because a streamer starting from one drop may propagate itself from drop to drop under a much weaker field. The breakdown (B) streamers now establish a conducting path between the main negative column and the base of the cloud, the lower positive charge gets neutralized, and the streamer channels become progressively more negatively charged. This is indicated by a slow steady change (intermediate- or I-change) on the electric-field record. The large radiation pulses accompanying the B process are probably caused by separate streamers starting in succession in the region just above cloud base.

The *stepped leader* is a negative streamer advancing in step-wise fashion at an overall velocity similar to that of a negative streamer of a laboratory spark in air. This similarity suggests that each bright step-like advance is preceded by a pilot streamer rather like that of the laboratory spark. Laboratory experiments have shown that a minimum field of 30 kV cm^{-1} is required to propagate a negative streamer by an electron avalanche in which free electrons ionize the air molecules by collision, the drift velocity of electrons in this field being 10^7 cm s^{-1}. In addition, it appears that, for the development of the leader process, the pilot channel must be thermally ionized, which requires a minimum field of 60 kV cm^{-1}. The large-scale field below a thunderstorm is much less than this, perhaps only a few thousand V/cm, but the non-uniform field near the tip of the channel may well exceed 60 kV cm^{-1} for some distance ahead. The pilot streamer may therefore advance a limited distance to a point at which the field drops below the critical value. A sudden change from glow to arc conditions is now believed to take place in the newly-formed channel, which becomes heavily ionized and appears as a bright step. The whole channel from the cloud to the tip of the step is now highly conducting, and with the tip at much the same potential as the cloud, the strong field between it and the surrounding air is re-established, lateral corona currents develop, and the pilot leader now advances a further step. Different theories of the mechanism of the sudden transformation from glow to arc conditions in the channel have been proposed by various authors, notably Schonland (1938) and Bruce (1944) but, being largely based on the behaviour of discharges between solid electrodes, it is not certain which of these most nearly describes what happens in a lightning flash.

The *return stroke*, illustrated in Fig. 9.9(a), is a positive streamer, AB, advancing from the ground upwards into the ionized and negatively-charged channel of the leader stroke BC. In the vicinity of B, at the junction of the positive streamer and the highly-charged negative channel, the intense electric field will drive free electrons at great speed into the advancing positive tip. As they move downwards, these electrons will create electron avalanches, ionize and heat the air in their path, and leave behind a highly conducting channel. For these reasons, the positive streamer moves upwards at extremely high velocity, even though the positive ions are themselves relatively immobile. As it moves upwards, it passes to ground the electrons in both the stem and branches of the leader, and finally discharges a negative region of the cloud.

The *dart leader* is a negative streamer advancing along a previously ionized channel, so that the region in front of the advancing tip contains a high concentration of free electrons. In Fig. 9.9(b), AB is the negative dart leader, BCD the previously ionized channel in front of the tip B, and BC a region in which the electric field is strong enough to produce ionization by collision. The free electrons originally present in BC move forward away from B and produce electron avalanches, make the air highly conducting, and effectively extend the streamer to C. The streamer may be regarded as jumping forward from one avalanche to another at a velocity which may be much higher than the drift velocity

§ 9.4   THE ELECTRIFICATION OF CLOUDS   511

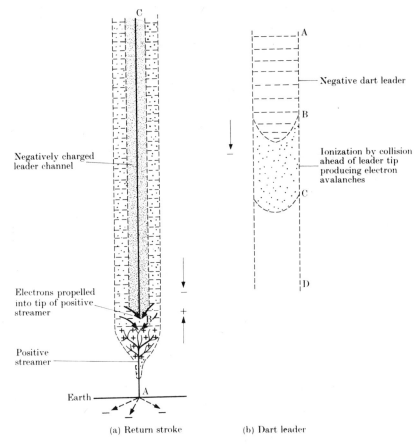

FIG 9.9. The mechanism of (a) the return stroke, (b) the dart leader. (After Schonland.)

of electrons in the critical field. This explains why dart leaders have velocities as high as 10^8–10^9 cm s^{-1}, the higher values occurring when the time interval since the previous return stroke is small, so that the density of free electrons in the channel has not appreciably decreased by recombination or attachment to air molecules.

9.4. The electrical structure of the thunderstorm

We have already seen how Wilson deduced the bipolar structure of a thunderstorm, with the main positive charge situated above the negative charge, from his measurements of field-changes at the ground. Wilson's conclusions have been amply confirmed by similar measurements made in South Africa by Schonland and his collaborators, and in New Mexico by Workman and Holzer (1942), who used a network of eight synchronized, continuous, potential-gradient recorders over an area of 80 miles². The latter obtained some evidence that pointed to the upper positive charge centre being displaced in the direction of motion relative to the lower negative charge, an observation that is probably to be attributed to the effects of wind shear. In the New Mexico storms, the negative and positive charge centres were found to be, on average, 2·5 and 4 km respectively above cloud base, where the temperature was about 8° C; the magnitudes of the charges involved in flashes were between 10 and 190 C, being most frequently between 20 and 50 C. Lightning activity generally continued for some minutes inside the cloud before the first cloud-to-ground strokes occurred. Workman and Reynolds (1950b) reported that, in the early stages of the storm, the positive centre appeared to be relatively close to the negative centre, but as the storm progressed, it moved upwards, whereas the negative centre appeared to remain fixed. We may recall here that Malan and Schonland (1951a, b), from a study of the field-changes between the component strokes of a lightning flash, concluded that in South African storms the negative charge is distributed throughout a nearly vertical column which, in the limit, may be bounded by the 0° C and −40° C levels. These authors, by studying how the field-changes due to the B (breakdown) process varied with distance from the storm, concluded that the base of the negative column was never more than 3·6 km, and the lower positive charge never less than 1·4 km, above ground. The existence of a pocket of positive charge in the base of some thunderclouds was indicated on some of Wilson's records by positive fields of about 100 V cm^{-1} beneath thunderclouds and has since been detected by other methods to be described later. It appears likely that the discharge of the negative column to earth may, in many cases, be triggered by the strong field existing between the positive pocket and the base of the negative column.

Measurement of the electric field *inside* thunderclouds was pioneered by Simpson and his collaborators at Kew, using a point-discharge apparatus (altielectrograph) on a balloon. Simpson and Scrase (1937) and Simpson and Robinson (1941) obtained continuous records of the direction of the point-discharge currents flowing between points D attached to the apparatus and a point A on a wire trailing below it, and hence the sign of the vertical component of the electric field. The principle of the apparatus is shown in Fig. 9.10. The direction of the current was indicated by a deposit of Prussian blue occurring at one or the

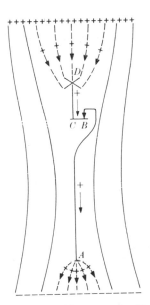

Fig. 9.10. The principle of the altielectrograph. (From Simpson and Scrase (1937).)

other of two iron electrodes, BC, pressing against pole-finding paper impregnated with a solution of potassium ferrocyanide and ammonium nitrate. The paper disk was rotated by a clockwork motor, the time scale of the trace being 3 mm min^{-1}, or 1 mm/100 m of ascent. The balloon carried an aneroid barometer to give the pressure, and hence the height of the balloon up to an altitude of about 8 km, at which the altielectrograph was released and returned by parachute. In addition to recording the sign of the field, it was found that the width of the trace on the pole-finding paper varied with the current passing through the instrument, and from this it was possible to estimate the magnitude of the field. The minimum field necessary to produce a legible trace was about 10 V cm^{-1}.

The vertical profiles of electric field obtained from fifty-eight legible records showed considerable variations, indicating the complexity of the electrical structure of a thundercloud, but complicated further by the movement of the clouds during the ascent and the fact that the balloons did not rise vertically. The records did not, then, give an instantaneous vertical section of the cloud, but a composite picture in both space and time. Nevertheless, the majority of the records clearly indicated that the upper portion of a thundercloud was positively charged, the compensating negative charge being situated in the lower part of the cloud. In the majority of storms there was also evidence of one or more localized regions of positive charge in the base of the cloud, these being usually associated with heavy rain. Positive charges predominated above 7 km, negative charges between 2 and 7 km, the lower positive charges being located below 2 km. Simpson and Robinson inferred that the upper positive charge was usually associated with temperatures below −20° C, and always below −10° C. The temperature at the negative charge centre was below 0° C in 13 out of 15 cases, while the lower positive charge, in 5 cases out of 7, was at temperatures above 0° C. These authors then constructed a theoretical model of a thundercloud to be consistent with the observed heights of the various charge centres and with the magnitude of observed fields at the ground. They found that, on average, these data could be represented by a model with an upper charge of +24 C distributed in a spherical volume of 2-km radius centred at a height of 6 km (about −30° C), a negative charge of −20° C in a sphere of radius 1 km centred at 3 km (−8° C), and a charge of +4 C in a sphere of radius 0·5 km centred at a height of 1·5 km (+1·5° C). To investigate whether the results of individual soundings could be explained in terms of the simple model, the fields along a number of tracks representing possible paths of the balloon in the cloud were calculated: the results are represented in Fig. 9.11 in which

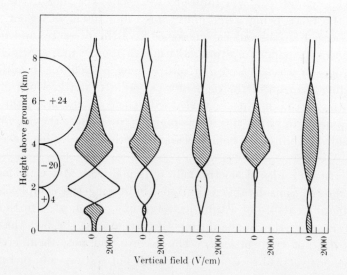

FIG. 9.11. The intensity of the electric field as a function of height in thunderstorms as deduced from altielectrograph records. (From Simpson and Robinson (1941).)

positive fields are indicated by cross-hatching. There was a very strong resemblance between the theoretical field distributions and those obtained experimentally, and only a very few of the actual records could not be classified immediately as one of the five types shown. The agreement was only qualitative, however, in that the altielectrograph records rarely showed fields exceeding 100 V cm^{-1}, although fields exceeding 1000 V cm^{-1} were predicted by the model. It must be remembered, however, that the measurement of the field was very crude and that on a number of occasions sparking occurred at the electrodes, so that the magnitudes of the fields were probably greater than indicated by the width of the trace. Also, probably very few of the balloons passed through the electrical centre of the storm where the strongest fields existed.

Balloon observations have been made more recently in the United States by Chapman (1950), who used a radio-sonde device, the observations being telemetered to the ground. His results on the charge distribution in a thundercloud broadly confirmed those of Simpson *et al*. The centres of negative charge for seven storms were found to lie within 0·5° C of the 0° C isotherm. The strongest field reported at the time was 210 V cm^{-1}, but Chapman later discovered an error in the calibration which suggests that the true value was about 2000 V cm^{-1}.

Some interesting data on the structure of thunderstorms were obtained by Kuettner (1950) working in the observatory on the Zugspitze (3 km a.s.l.), where, on 80 per cent of occasions, the apparatus was actually inside the thunderclouds. He attempted to construct the electrical pattern of a typical thunderstorm by studying the records of point-discharge current measured during the passage of over one hundred storms in different seasons, during which the temperature at the observatory varied over 25° C, corresponding to a height range of more than 10 000 ft. The mean picture which emerged from the electrical field measurements indicated the presence of a local positive charge situated near the cloud base, close to the 0° C level. This charge was less than 1 km in horizontal extent and was situated in the centre of the precipitation zone. The main negative charge, in the lower part of the cloud, was concentrated near the $-8°$ C level and was larger in both horizontal and vertical dimensions than the lower positive charge. The main positive charge appeared to extend over a rather large volume in the upper part of the cloud.

Important information on the electric-field intensity inside clouds has been obtained by Gunn (1948) from an aircraft, using field-meters situated above and below the fuselage in order to eliminate the effects of any net charge acquired by the aircraft. The mean of the maximum field intensities encountered in each of nine storms was 1300 V cm^{-1}, and on one occasion a field of 3400 V cm^{-1} was measured just before the aircraft was struck by lightning. The figures given represent the vertical field actually observed at the surface of the aircraft; the true field undistorted by the presence of the aircraft may have been smaller than the quoted value if its direction was nearly vertical, but may have been larger if it had a large horizontal component as was sometimes indicated by a current flowing into one wing-tip and out of the other. Gunn encountered the most intense fields near the 0° C level.

This technique has been developed by Fitzgerald (1965), who uses an analogue computer to add and subtract the signals from four field mills mounted on the wing-tips and fuselage to give outputs that are proportional to the aircraft charge and the three components of the external field. Fitzgerald reports that fields in and around swelling cumuli are generally weak, but that fields of several hundred V/cm are detected when precipitation develops. In and around the rain-sheets below thunderstorms vertical fields of up to 860 V cm^{-1} were recorded. The strong vertical field appeared to be associated with a negative charge centre in the cloud and a positive space charge between cloud base and ground, since the field on the ground did not exceed 300 V cm^{-1}.

Vonnegut, Moore, and Mallahan (1961) mounted radio-active ionizing probes on an aircraft and coupled them to a rotating capacitor system that compensated for asymmetry in the exposure of the probes. But although this arrangement compensated for the effects of charge on the aeroplane, it did not compensate for the space charge produced by the aeroplane, especially by the wing-tip and propellors and, consequently, the authors concluded that the fields measured in weakly electrified clouds were probably too large, and the large fields measured near thunderstorms were probably too small.

The magnitudes of the fields recorded by Gunn and Fitzgerald are of the order we should expect to exist near the electrical centre of a simple bipolar cloud, in which charges of about 20 C are separated by distances of some 3 km, and indicate that the total potential difference between the main centres may lie between 10^8 and 10^9 V. Taking the charge involved to be 20 C, the total energy dissipated in a lightning flash is 10^9–10^{10} J. Much of this is spent in heating up a narrow air column surrounding the discharge, the temperature rising in a few microseconds to about 30 000° K. The air in the channel expands explosively to create intense sound waves which are heard as thunder.

9.5. Correlation between lightning and precipitation

Lightning is usually accompanied by heavy precipitation although, in warm, dry climates, this may not reach the ground. Most theories of thunderstorm electrification have assumed that precipitation plays an important role in the generation and separation of the electric charge, and as the main charge centres appear at levels where the temperature is below 0° C, there has been a tendency to associate that generation with the presence of supercooled water and/or the ice phase. The most convincing evidence for this view comes from the following sources.

Kuettner (1950), from observations made inside thunderstorms capping the Zugspitze in Germany, affirmed that a high rate of precipitation was an indispensable requirement for electrical activity in the cumulonimbus, the central lightning area usually being coincident with the area of intense precipitation. He reported that solid precipitation elements were dominant in the greater part of the thundercloud and were present on 93 per cent of occasions. Snow pellets and pellets of soft hail were the most frequent form of hydrometeor, being present on 75 per cent of occasions, and were always accompanied by strong electric fields, but large hail was relatively rare.

By correlating measured intensities of radar echoes from cumulonimbus with reports of lightning and hail made by a large number of co-operative observers in New England, Shackford (1960) found that, not only was frequent lightning strongly correlated with the appearance of hail at the surface, but that the frequency of lightning strokes tended to increase as the height and reflectivity of the upper part of the radar echo increased. We have seen on p. 424 that very intense echoes can be produced by modest concentrations of hailstones if they are of fairly uniform size.

Fitzgerald and Byers (1962), using aircraft fitted with electric-field meters, reported that actively building regions of thunderstorms were regions of excess negative charge and that the electric fields, which increased rapidly with the onset of precipitation, indicated that the initial precipitation streamers carried a negative charge towards the earth, leaving a positive space charge in the upper regions of the cloud. The strongest fields, of up to 2300 V cm^{-1}, were associated with regions of heavy precipitation. In particular, a large hail shaft produced a strong, smoothly-increasing field that indicated a negative charge on the hail. Malan and Schonland (1951a, b) find that, in South African storms, the negative charge is often distributed in a nearly vertical column which may extend up to, but not beyond, the $-40°$ C level. This is consistent with the charge being generated by growing hail pellets because supercooled droplets exist at temperatures down to, but not below, −40° C.

Reynolds and Brook (1956), making simultaneous observations with a 3-cm RHI radar and measurements of the electric field on Mt Withington (3·7 km a.s.l.) in New Mexico, reported that the onset of precipitation in shower clouds was not a sufficient condition for their electrification because the electric field increased sharply only when the radar echo and the visual cloud showed rapid vertical development. Intensification of the electric field and the radar echo then went hand-in-hand. In agreement with earlier observations by Workman and Reynolds (1949), the initial radar echo usually appeared at about the −10° C level and the time interval between this and the first lightning flash was about 12 min. On many days, when clouds of considerable depth and vigour developed without precipitation, the electric field showed no significant departure from the fine-weather value.

On the other hand, Vonnegut, Moore, and Botka (1959)—see also Vonnegut (1965) and Moore (1965)—using the same radar equipment in the same location, claim that in some of the New Mexico storms electrification of the clouds and even lightning begins before the arrival of precipitation at the mountain top, and when the precipitation rate inside the cloud, as deduced by radar, is usually only a few mm/h and occasionally 1 mm/h. However, heavy gushes of rain or hail often reach the ground 2–3 min after a lightning flash, and Vonnegut and Moore take this as evidence that the lightning is the cause rather than the result of the precipitation. They further speculate that the rapid intensification of the precipitation from about 1 mm/h to 50 mm/h in this 2- to 3-min period is brought about by a greatly accelerated rate of coalescence of water drops under the influence of electrical forces by a mechanism that is obscure and has no convincing experimental or theoretical basis.

It is, in fact, often difficult to correlate, in both space and time, electrical activity and lightning with the development of precipitation as seen by a single radar. Lightning flashes often take a rather long and tortuous path, and it is almost impossible to locate them accurately in space, especially in the daytime, and to identify them with a particular precipitation echo on the radar screen. The evolution of the radar echo itself is also open to misinterpretation, because what may appear on the screen as a rapid vertical development of the echo may be caused by the advection of a vertically-inclined column of precipitation through the radar beam. Also, a radar having a poor minimum range because of receiver paralysis may fail to detect a low-level zone of precipitation until it has moved far enough away, and this may make any deductions about the order of events within intervals of a minute or two highly unreliable. Until much more precise observations have been made, and bearing in mind that it will usually take 2–3 min for the rain to fall from cloud base to ground, the available evidence may be interpreted as showing that the onset of lightning and heavy precipitation *within* the cloud are practically simultaneous and, if we think of the heavy precipitation as triggering the lightning flash in the manner suggested on p. 509, we avoid the embarrassment of explaining how the lightning might greatly enhance the growth rate of raindrops or hail pellets. However, the fact that the radar echo and the electric field may intensify simultaneously must be interpreted with caution. Because the radar echo intensity is $\propto \Sigma ND^6$ while the growth rate of the electric field is likely to be $\propto \Sigma ND^2 V \propto \Sigma ND^{2.5-3.0}$ (see eqn. (9.34)), the large hydrometeors will contribute more to the growth of the radar echo than to the growth of the electric field. Thus while production of small- and medium-sized particles in sufficient quantity to generate a detectable electric field may also produce a detectable radar echo and, initially, the two may grow in step, the radar echo may later be enhanced by the appearance of much larger hydrometeors that do not contribute greatly to the electrification. Vonnegut and Moore challenge the generally accepted thesis, that precipitation is the main source of electrification, on the additional grounds that the charge on thunderstorm rain reaching the ground is usually smaller in magnitude, and often of opposite polarity, to what one would expect if the principal negative charge of the storm were carried down by the precipitation. However, as Vonnegut and Moore agree, the charge on the rain reaching the ground has probably been acquired largely by the capture of positive ions produced by point-discharge at the earth's surface by the thunderstorm field. Also, it seems very likely that any original negative charge carried by hail pellets may be neutralized, and even reversed, during their melting and fall through the positive space-charge blanket towards the ground, since their time constant for discharge is probably <40 s.

Although the main charge-centres of large thunderstorms appear always to be located in the sub-freezing part of the cloud, there are on record reports by Foster (1950), Pietrowski (1960), and Moore, Vonnegut, Stein, and Survilas (1960) of lightning being observed from clouds whose tops were warmer than 0° C and could therefore have contained no ice. These observations were made from aircraft flying over the Caribbean Sea in twilight or darkness, and are not entirely convincing because, under such conditions, it could not have been easy to keep track of a cloud first spotted 100 miles away and be certain that it was not being illuminated by lightning originating in nearby larger storms.

Certainly lightning from such warmer clouds appears to be a rare phenomenon although rain falls frequently from such clouds in warm maritime climates. However, should detailed and well-documented observations firmly establish that lightning may actually originate inside isolated clouds whose tops do not, at any stage, extend well above the 0° C level, it will be necessary to look for a process of electrification not involving the ice phase.

9.6. Mechanisms of charge generation and separation in thunderstorms

9.6.1. *Basic requirements of a satisfactory theory*

The main features of the thunderstorm, which have been described in the foregoing paragraphs, and with which any satisfactory theory of charge generation and separation must be consistent, are as follows (see Mason (1953c)):

(i) The average duration of precipitation and electrical activity from a single thunderstorm cell is about 30 min.

(ii) The average electric moment destroyed in a lightning flash is about 100 C km, the corresponding charge being 20–30 C.

(iii) In a large, extensive cumulonimbus, this charge is generated and separated in a volume bounded by the $-5°$ C and the $-40°$ C levels and having an average radius of perhaps 2 km.

(iv) The negative charge is centred near the $-5°$ C isotherm, while the main positive charge is situated some kilometres higher up; a subsidiary positive charge may also exist near the cloud base, being centred at or below the 0° C level.

(v) The charge generation and separation processes are closely associated with the development of precipitation, probably in the form of soft hail.

(vi) Sufficient charge must be generated and separated to supply the first lightning flash within 12–20 min of the appearance of precipitation particles of radar-detectable size.

9.6.2. Basic mechanisms of cloud electrification

9.6.2.1. Influence mechanisms

(a) *Cloud particles rebounding from hydrometeors in polarizing electric fields.* Elster and Geitel (1913) considered how raindrops might become charged while falling in a vertical electric field and colliding with cloud droplets. A downwardly-directed field, such as exists in fine weather, would polarize the raindrop with its lower half positively charged and its upper half negatively charged, and cloud droplets rebounding from the underside of the raindrop would carry away some of the positive charge and leave the raindrop with a net negative charge. The falling drops, in carrying their negative charges towards the cloud base, would enhance the original field. This idea has been revived in a more general and quantitative form by Müller-Hillebrande (1954, 1955), Sartor (1961), Latham and Mason (1962), and Mason (1968) to include the case of ice crystals rebounding from hail pellets. The rate of charging of a hydrometeor of radius R, falling at velocity V relative to much smaller cloud particles of radius r and number concentration n in a field F is given by

$$dq/dt = -\pi R^2 V n \alpha \left(\tfrac{1}{2}\pi^2 F \cos\theta + \tfrac{1}{6}\pi^2 \frac{q}{R^2}\right) r^2, \qquad (9.14)$$

where α is the fraction of cloud particles lying in the cylinder swept out by the hydrometeor *which actually rebound from it.* Putting the average angle between the field and the line of approach of the colliding particles to be $\bar{\theta} = 45°$, eqn (9.14) may be integrated to give

$$q = -2\cdot 12 F R^2 \{1 - \exp(-\tfrac{1}{6}\pi^3 V n \alpha r^2)t\}, \qquad (9.15)$$

so that the maximum charge that may be acquired by the hydrometeor, and its relaxation time, are respectively,

$$q_{max} = -2\cdot 12 F R^2, \quad \text{and} \quad \tau = (\tfrac{1}{6}\pi^3 V n \alpha r^2)^{-1}.$$

The potential importance of this mechanism in thunderstorm electrification is considered in § 9.6.3.

(b) *Wilson's process of charging by selective ion capture.* Wilson (1929) pointed out that, under certain conditions, an electrically polarized raindrop in falling through a cloud of ions or charged cloud droplets could, by a process of selective ion capture, acquire a net charge. If the drop falls more rapidly through a downwardly directed field than the downwardly-moving positive ions, the latter are repelled from the lower half of the drop and deviated to one side, while the negative ions are attracted to it. Hence the drop acquires a net negative charge that tends to augment the pre-existing field. The fundamental condition for selective capture to occur may be written $V > k_+ F$, where V is the fall velocity of the drop relative to the air, k_+ the mobility of the positive ions, and F the field intensity. For a raindrop, V will not usually exceed 8 m s^{-1}, so that with ordinary small ions ($k \simeq 1.5$ cm^2 s^{-1} V^{-1}), selective capture is impossible if the field exceeds about 500 V cm^{-1}. Selective capture of large ions and charged cloud droplets is possible, however, in any field of less than breakdown strength. The mathematical theory of the process has been worked out by Whipple and Chalmers (1944), who showed that, if the fall velocity of the raindrop is large compared with the drift velocity of the ions under the existing vertical field (this will generally be the case if the ions are attached to cloud droplets):

(1) the final charge acquired by a drop of radius R, independent of its initial charge, in a vertical field of intensity F, is

$$q_{max} = -3(3 - 2\sqrt{2})FR^2 = -0.52 FR^2; \qquad (9.16)$$

(2) if its initial charge is zero, the initial (i.e. maximum) rate of charging of a drop is

$$\left(\frac{dq}{dt}\right)_{max} = -3\pi F R^2 \lambda_-, \qquad (9.17)$$

where λ_- is the polar conductivity for the slow negative ions;

(3) the time required for the drop to acquire half of its final charge is

$$\tau = 0.04/\lambda_- \text{ s.} \qquad (9.18)$$

As the above mechanism proceeds, the field increases, so that the equilibrium charge towards which the drops are tending is increasing with time. The quantitative aspects of this are considered in § 9.6.3.

(c) *Frenkel's theory.* Frenkel (1944, 1946, 1947) argues that a polarized raindrop will acquire a net *positive* charge by virtue of the polar conductivity, λ_+, for positive ions being greater than that, λ_-, for negative ions, the equilibrium charge being

$$q_{max} = \frac{(\lambda_+ - \lambda_-)}{(\lambda_+ + \lambda_-)} FR^2. \qquad (9.19)$$

The ratio of the polar conductivities, λ_+/λ_-, in clouds is not well known,

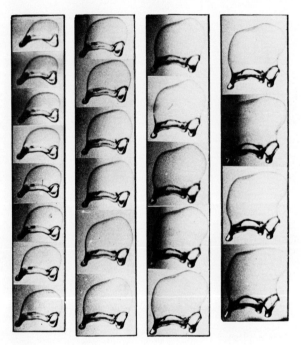

FIG. 9.12. A large falling water drop becomes unstable and forms an expanding bag supported on a toroidal ring of liquid. The bag eventually bursts producing large numbers of small droplets and the toroid breaks up into several large drops. The photographs are taken at intervals of 1 m s. (From Matthews and Mason (1964).)

but if we follow Frenkel and assume $\lambda_+/\lambda_- = 1\cdot3$,

$$q_{max} = 0\cdot13 F R^2. \tag{9.19a}$$

This mechanism would produce fields of opposite sign to those prevailing in thunderstorms and, in any case, would be outweighed by the Wilson mechanism if the two were to act simultaneously. It will not be considered further.

9.6.2.2. Electrification produced by the rupture of large drops

Electrification associated with the disruption of water drops was first noticed in 1890 by Elster and Geitel and was investigated further by Lenard (1892), who found that mists of small droplets carried upwards from the base of a waterfall were predominantly negatively charged, while the spray near the base carried a positive charge. Simpson (1909), in a series of careful experiments in which he largely eliminated this 'Lenard' effect, established that breaking of drops in a strong vertical air jet could also produce a considerable electrification. The large fragments of the broken drops carried positive charges, while the surrounding air carried ions of both signs, with an excess of negative charge. Ruptured drops of distilled water of diameter about 8 mm produced, on average, a charge of $5\cdot5 \times 10^{-3}$ e.s.u., i.e. about $2\cdot3 \times 10^{-2}$ e.s.u. cm^{-3}; with less violent shattering the charge produced was slightly less, about $1\cdot5 \times 10^{-2}$ e.s.u. cm^{-3}.

Lenard (1921) confirmed these results and quoted Hochschwender (1919) as having measured an average positive charge of $4\cdot2 \times 10^{-2}$ e.s.u. cm^{-3} on the heavier fragments. Hochschwender had also obtained some remarkable photographs showing the break-up of the drop. It first became flattened, and then the lower surface was blown concave by the air stream. The drop then expanded rapidly as a bubble or bag attached to an annular ring that contained the bulk of the water. Eventually the bubble burst and the annular ring broke up into a circlet of fairly large drops—see Fig. 9.12.

Zeleny (1933), using highly-purified water, found that the rupture of a drop in a 20 m s^{-1} horizontal air jet produced charges of about 2×10^{-2} e.s.u. cm^{-3}. Chapman (1952) allowed drops of distilled water of diameter 4 mm to fall into a vertical jet of speed 17·3 m s^{-1}. The violent disruptions were accompanied by charges of about 0·3 e.s.u. per drop, i.e. 10 e.s.u. cm^{-3}. However, when he allowed two drops to coalesce and the resultant unstable mass to break up into large fragments in a steady upcurrent of 8 m s^{-1}, the charge generated was of the same order as that found by Simpson and by Zeleny. It appears, then, that the charge separated by breaking drops depends markedly on the violence with which they are shattered.

Falling raindrops rarely exceed an equivalent spherical diameter of 6 mm. Laboratory experiments suggest that at this size they become distorted, unstable, and break up into several large, and many more small fragments. Simpson (1909, 1927) proposed that the separation of charge associated with the repeated rupture of large raindrops might account for the electrification of thunderstorms. It is clear, however, that if the large fragments acquire positive charges, and the negative charge is communicated to the air and to small cloud droplets, gravitational separation will tend to charge the cloud in the opposite sense to that which is observed. Furthermore, Mason (1953c), taking the values of 2×10^{-2} e.s.u. cm^{-3} given by Simpson and Zeleny, calculated that, even if the bulk of the rain-water were disrupted three times, the separated space charge would amount to only 9×10^{-2} C km^{-3}, which is two orders of magnitude too small to be the major charging mechanism of the thunderstorm, and too small even to account for the subsidiary positive charge found in the bases of many thunderstorms. Mason pointed out, though, that the break-up and charging of drops had not been studied in the presence of a strong electric field such as may be provided by the primary charging mechanism of the thunderstorm. The polarization charge, $Q = \frac{3}{4}FR^2$, on a raindrop of radius $R = 3$ mm in a field $F = 500$ V cm^{-1}, amounts to 1 e.s.u. cm^{-3}, and clearly provides the possibility of producing much larger charges during drop-breaking that those measured by Simpson and Zeleny.

A laboratory study of the break-up and electrification of large, freely-falling water drops in still air, and in electric fields of up to 1500 V cm^{-1}, was made by Matthews and Mason (1964). Drops of $d \approx 12$ mm, after falling 12 m, broke up between a pair of horizontal electrodes producing a vertical field. The charges on the larger fragments were measured as they fell through Faraday cylinders without splashing, and the volume of the fragments determined by collecting and weighing them. The mode of break-up was studied by high-speed photographs taken at the rate of 2000/s—see Fig. 9.12. In the absence of an applied field, the breaking drops produced charges of about 10^{-2} e.s.u. cm^{-3} in agreement with Simpson and Zeleny. The average charge per unit volume of water increased with increasing field strength and reached 5·5 e.s.u. cm^{-3} in fields of 1500 V cm^{-1}. These charges are two orders of magnitude greater than those quoted by Simpson and suggest that large raindrops breaking in the presence of the primary field of the thunderstorm may contribute significantly to the charge configuration in the lower part of the cloud; but other mechanisms, such as the melting of ice pellets and the capture by raindrops of positive ions from an upward-moving point-discharge stream, may also contribute to production of the subsidiary positive charge in the base of the thunderstorm.

9.6.2.3. Convective theories of thunderstorm electrification

Developing an earlier idea of Grenet (1947), Vonnegut (1955) proposed a convective theory of charge separation according to which the charges responsible for lightning, instead of being produced by the electrification and gravitational settling of hydrometeors, are drawn from the ionosphere by conduction and from the earth by point discharge, and are separated by vertical air motions in the cloud. It is envisaged that the convective updraughts of growing cumulus carry up positive space charge that normally resides in the lower atmosphere in fine weather, but that as the cloud grows to higher altitudes, where the air is more highly conducting, the positive charge in the cloud causes a counter-current of small negative ions to flow from the ionosphere to the surface of the cloud. Instead of neutralizing the positively-charged updraught, Vonnegut argues that these negative ions are carried away from the top of the cloud to lower levels by strong downdraughts existing on the flanks of the cloud. Accumulation of negative charge in the lower outer regions of the cloud will ultimately induce point discharge at the earth's surface; the positive ions so produced are thought to be carried up in the updraught to reinforce the positive charge in the centre of the cloud rather than neutralize the negative charges on its flanks. The growing positive charge in the cloud in turn increases the current of negative ions from the ionosphere, so that both the positive and negative charges build up cumulatively.

Vonnegut's hypothesis is unconvincing in that his picture of a strong central updraught surrounded by extensive downdraughts of comparable strength in the building stages of the cloud is unrealistic. While localized sinking motions are produced on the flanks of a cloud by evaporation and chilling of its edges, extensive and organized downdraughts can be triggered and sustained only by precipitation, and so appear in the decaying stages of the storm. Indeed the internal circulation within a growing cumulus would be more likely to transfer positive charge from the core to the flanks than to support the downward transport of negative charge from the ionosphere. Aside from the fact that it is conventional to think of point-discharge currents being a consequence rather than a cause of thunderstorms, electric-field changes accompanying lightning are more readily interpreted in terms of the point-discharge currents opposing, rather than promoting, the growth of the electric moment of the cloud.

9.6.2.4. Electrification associated with the freezing and melting of water

(a) *The Workman–Reynolds effect.* Workman and Reynolds (1948, 1950c) discovered that, during the freezing of water and dilute aqueous solutions, a potential difference developed across the ice–liquid interface, the sign and magnitude of which depended on the nature and concentration of the ionic species. For the majority of solutions tested, the ice became negative with respect to the liquid, with important exceptions in the case of ammonium salts. The largest potential difference, -232 V (potential of liquid with respect to ice), was obtained with a 5×10^{-5} N solution of ammonium hydroxide. For sodium chloride solutions, maximum charge separation and a potential of $+30$ V was obtained with a 10^{-4} N solution, but no effect was observed when the concentration was increased to 5×10^{-4} N. The electrical effects obtained with solutions of $CaCO_3$ and $Ca(OH)_2$ were very sensitive to the amounts of dissolved CO_2, while the freezing of doubly-distilled water, carefully freed of ammonia, produced no detectable potential.

These results have been broadly confirmed by more recent work. Lodge, Baker, and Pierrard (1956) found that the transient potentials produced during the freezing of NaCl solutions increased with increasing concentration to reach a maximum value of about 20 V at a concentration of 5×10^{-4} M, and thereafter decreased because of leakage of charge through the more highly conducting ice. The concentrations of both Na^+ and Cl^- ions in the ice were lower than in the original solution, the reduction being greater for Na^+ than Cl^-; this suggests a selective rejection of Na^+ ions by the ice, rather than a selective incorporation of Cl^- ions as suggested by Workman and Reynolds. Heinmets (1962) obtained rather similar results with solutions of HCl, and found that slow rates of freezing and the absence of convective stirring of the liquid were essential to the production of high transient freezing potentials which, in his experiments, attained values of up to 60 V with concentrations of 10^{-6} M. These high potentials disappeared after the completion of freezing to leave small steady potentials of order 100 mV. Gross (1965), using dilute solutions of KF, CsF, and LiI, deduced that the anions (even the large I anion), were always retained in greater concentration in the ice than the cations, and inferred that electronegativity was a more important factor for incorporation into the ice lattice than ionic size. When both types of ionic impurity were equally acceptable, as with HF, no freezing potential was detected. CsF solutions of concentration $2 \cdot 5 \times 10^{-4}$ M, freezing at linear rates of 3–10 μs^{-1}, developed potentials of about 5 V. Levi and Milman (1966), who tried to produce nearly constant freezing rates by gradually lowering the specimen into a cooling bath, obtained potentials of about 5 V for NaCl solutions, and confirmed that NH_3 solutions produced potentials of opposite polarity, with maximum values of about 100 V with concentrations of 10^{-6} M. This result was attributed to the selective incorporation of NH_4^+ ions into the ice lattice.

Workman and Reynolds suggested that their experiments pointed to a possible powerful mechanism for the generation and separation of charge in thunderstorms. They assumed that rapidly growing hailstones, being able to freeze only a fraction of the impinging cloud water, would acquire a liquid coat which would be shed in the form of small drops; they further assumed that, during freezing, the ice would acquire a negative charge as in their experiments on NaCl and $CaCO_3$ solutions, that the water drops flung off would thus carry away a positive charge, leaving the hailstone with a net negative charge. Gravitational separation of the negative hailstones and the small positive drops would then lead to a charge distribution in the cloud of the observed polarity.

These arguments may be challenged on the grounds that recent work on the growth and structure of wet hailstones (see § 6.6.7) shows that the unfrozen water tends to be retained in a skeletal framework of ice rather than shed into the airstream. Charge separation might occur if impinging droplets were to splash on impact, but laboratory tests indicate that this is an unlikely event even for large cloud droplets impinging at velocities of 10 m s^{-1}. Moreover, the fact that the Workman–Reynolds effect appears to be so sensitive to the concentrations of salts and carbon dioxide, and especially to traces of ammonia in the water, makes it unattractive as a proposed universal mechanism of charge generation in thunderstorms.

(b) *The Dinger–Gunn effect.* The first observations of electrical charging during the melting of ice were reported by Dinger and Gunn

(1946). A light current of air was passed over ice melting in a nickel dish, and the charge carried away in the air-stream measured by an electrometer connected to a metal tube containing close-packed copper turnings. Alternatively, the charge residing on the melt water was measured by connecting the dish directly to the electrometer. In all cases, the air acquired either zero or negative charge, and the melt water either zero or positive charge. The magnitude of the charge, which varied from a few tenths to about 2 e.s.u. per gramme of melt water for distilled water, was reduced in the presence of CO_2 and by the addition of small concentrations of ionic salts. It was also found to depend upon the rate of freezing of the ice specimens and upon their rate of melting. These results were not confirmed by Matthews and Mason (1963) who, in a series of measurements with three different experimental arrangements, including one in which snow crystals were grown and melted under clean conditions in a diffusion cloud chamber, failed to detect a separation of charge on melting above their detectable limit of 10^{-2} e.s.u./g. On the other hand, MacCready and Proudfit (1965) found that 1 cm^3 samples of ice made from distilled water acquired positive charges of about 0·1 e.s.u. when melted in an airstream of 8 m s^{-1}, Kikuchi (1965) found charges of up to 0·3 e.s.u./g depending on the air-bubble content of the ice, and Dinger (1965) reported charges as high as 6·6 e.s.u./g for triply-distilled water when extreme care was taken to avoid contamination of the water surface. These experimental results suggest that the charging effect is sensitive to the presence of CO_2 and other impurities in the air and the water and make it difficult to assess its importance in the real atmosphere where impurities are normally present.

It is now established that the separation of charge is a result of the bursting of air bubbles released from the ice during melting. Experiments by Blanchard (1963) and Iribarne and Mason (1967) have shown how the charge carried on droplets ejected by bubbles bursting at a water surface depends on the bubble radius, the ionic content of the water, and the lifetime of the bubble before bursting. Consistent results could be obtained only by taking the most scrupulous care to clean all the apparatus before use, avoiding the intrusion of surface-active impurities on the liquid/gas interface, using a continuously-renewed liquid surface, and by confining measurements to bubbles that burst immediately on arriving at the surface. Iribarne and Mason found that drops ejected from bubbles of radius 50–200 μm, bursting in dilute aqueous solutions, carried a negative charge, the magnitude of which decreased rapidly from about 10^{-3} e.s.u. per bubble for highly-purified water to become vanishingly small at concentrations greater than about 10^{-4} M. For still higher concentrations of salts or CO_2, the droplets carried away small *positive* charges of order 10^{-6} e.s.u. per bubble. A quantitative theory of the charging mechanisms, supported by the experiments of Jonas and Mason (1968), explains the negative charging of the drops in terms of a thin film of water rising from the inner surface of the bubble cavity to form a small jet that breaks up to produce drops, their charge resulting from the rupture of the electrical double layer at the air/water interface. The charge Q carried on the drops ejected from a bubble of radius R is calculated to be

$$Q = -5 \times 10^4 R^2 C^{\frac{1}{2}} \exp(-10^5 C^{\frac{1}{2}} R), \qquad (9.20)$$

where C is the concentration of the solution in mol/litre. The decrease of charge with increasing concentration is a result of the decreasing depth of the diffuse electrical double layer. The positive charging of the drops at high concentrations is attributed to the separation of charge during the break-up of an initially uncharged varicose jet, when water from the inner parts of the diffuse double layer containing an excess of positive ions is forced into the swelling regions that form the drops from the constricting necks between them. This positive current is calculated to be of order 1 e.s.u., almost independent of concentration, so that jets breaking up in times of 1–10 μs would produce charges of 10^{-6} to 10^{-5} e.s.u.

The analysis by Junge (1963) of rain-water collected at the ground shows that the concentrations of salts lie typically in the range 10^{-6} to 10^{-4} mol/litre, but in coastal regions the concentration of NaCl may be as high as 10^{-3} mol/litre. The equilibrium concentration of CO_2 in rain-water lies typically in the range 10^{-6} to 10^{-4} mol/litre. The sizes of air bubbles found in small hail pellets lie in the diameter range 1–500 μm. Hence the results of Iribarne and Mason (1967) anticipate a considerable variation in both magnitude and sign of the net charge separation during the melting of hail and snow particles in the atmosphere.

Drake and Mason (1966) observed that, during the melting of small ice particles supported in a vertical wind tunnel, strong convection currents developed in the melt water. This convective activity might well influence the production of charge by the Dinger–Gunn mechanism, since it produces a rapidly-moving and continuously-renewed water surface at which the bubbles burst. In many of the experiments just described, rather large specimens of ice were melted in solid containers and the charges separated by the slow passage of air over the container. These conditions both inhibit the development of convection within the melt water and allow the accumulation of impurity on the water surface. In the experiments of Dinger (1965), in which ice specimens were melted in a pan resting on a hot metal block, in a wind of 0·5 m s^{-1}, the onset of convection in the melt water was probably responsible for the high values of charge obtained with triply-distilled water.

Considerations such as these led Drake (1968) to investigate the electrical effects of melting, under a wide range of wind speed, temperature, and humidity conditions, ice specimens prepared from highly purified water and dilute aqueous solutions of the more common salts found in rain-water. A drop of a few millimetres diameter was placed on a small insulated wire loop, allowed to freeze, and then to melt in a stream of warm gas of controlled velocity, temperature, and humidity. The charge separated during melting was measured by a vibrating-reed electrometer. No charge was recorded until the specimen began to melt, but strong positive charging coincided with the onset of convective currents in the melt water and continued until the last traces of ice disappeared. The bubbles released during melting ranged from 5 to 200 μm in radius. When ice made from highly purified, de-ionized water was melted in a stream of nitrogen, the charging increased as the mean rate of heat transfer to the specimen was raised, and reached a maximum value of about 4 e.s.u./g. Similar experiments with ice specimens formed from solutions of $NaCl$, NH_4Cl, or $(NH_4)_2SO_4$ showed that the charging decreased as the concentration of salt was increased and became negligible for concentrations $>10^{-3}$ M. These results were consistent with Iribarne and Mason's explanation involving rupture of the electrical double layer, but the results obtained in the presence of CO_2 do not fit into this scheme.

Although previous workers had reported that the presence of CO_2 in the air had had an inhibiting effect on charge separation, Drake found this to be the case only for poorly ventilated ice specimens. If the melt water was agitated by convection, the presence of dissolved CO_2 caused a 50 per cent *increase* in the quantity of separated charge, while the replacement of the nitrogen stream by laboratory air containing CO_2 led to a doubling of the charge to about 8 e.s.u./g. This latter result is not easily explained but, even if we leave it aside, it does seem likely that the melting of falling hail and snow pellets, in conditions that would lead to convective motions in the melt water and ensure the removal of the droplets ejected from the bursting bubbles, may separate charge on the scale of 1–5 e.s.u./g.

Small and soft hail pellets, in a concentration of 5 g m^{-3}, and releasing 4 e.s.u./g during melting, would contribute a spatial concentration of charge of 6·7 C km^{-3}. This might be sufficient to account for the subsidiary pockets of positive charge located in the bases of many thunderstorms below the 0° C level and in the zone of heavy precipitation. However, the break-up of large raindrops and the capture of positive ions from the point-discharge stream may also contribute.

Continuous rain, which is predominantly positively charged, usually carries charges of $\sim$1 e.s.u./g, and these might well be produced during the melting of snowflakes by the Dinger–Gunn effect. Evidence that snowflakes do acquire a positive charge during melting, has been presented by Chalmers (1956) from observations that the total vertical current flowed in opposite directions in rain and snow, and by Magono and Kikuchi (1963, 1965) from measurements of the charges acquired by natural snow crystals as they fell down a heated tube, the charges being determined from deflection of the particles in a known horizontal alternating electric field. Snow crystals of melted diameter ranging from 30 to 400 μm acquired positive charges ranging from 10^{-6} e.s.u. to 10^{-2} e.s.u., with average values of 2×10^{-4} e.s.u. for a melted diameter of 150 μm. These charges, equivalent to about 100 e.s.u./g, are much higher than those normally associated with continuous rain, but then the crystals were melted much more rapidly in the experiments than would be the case in the atmosphere.

9.6.2.5. Thermoelectric effects in ice

Since many aspects of electrification associated with collisions and temporary contacts between ice particles—the fracture and evaporation of ice crystals, the freezing of supercooled water droplets and the formation of rime—appear to find at least partial explanations in terms of charge transfer in ice under the influence of transient temperature gradients, it seems appropriate to preface a description of these phenomena with a discussion of this basic process. Brook (1958) suggested that two pieces of ice in contact, with different temperatures, would produce an e.m.f. by virtue of the warmer ice transferring protons to the colder ice. A quantitative theory of this thermoelectric effect in ice, by which the hydrogen and hydroxyl ions, formed by the dissociation of a small fraction of the ice molecules, become separated under the influence of a temperature gradient, was formulated by Mason (Latham and Mason

1961a). The process depends essentially on two facts: that the concentrations of positive and negative ions increase quite rapidly with increasing temperature; and that the hydrogen ion (proton) diffuses much more rapidly through the ice lattice than does the hydroxyl ion. Thus if a steady temperature difference is maintained across a piece of ice, the warmer end will initially possess higher concentrations of both positive and negative ions, but the more rapid diffusion of H$^+$ ions down this concentration gradient will lead to a separation of charge, with a net excess of positive charge in the colder part of the ice. The space charge set up in the ice by this differential diffusion will produce an internal electric field tending to accelerate the OH$^-$ ions, decelerate the H$^+$ ions, and so oppose further separation of charge. There may also be an additional contribution to the separated charge by the differential migration of orientational (Bjerrum) defects in the lattice. These arise through a small proportion of hydrogen bonds being either doubly-occupied by two protons (D-defects), or left vacant (L-defects), but since there is no evidence to indicate that L- and D-defects have different mobilities in pure ice, the charge separation may be ascribed almost entirely to the differential migration of ion states. On these assumptions, Mason calculated the potential gradient developed across a piece of pure ice by a steady temperature gradient, and also the charge transfer between two pieces of ice of different temperatures during a momentary contact.

(i) *Steady temperature gradient.* When a steady temperature gradient is maintained across a uniform ice specimen, a steady state will be reached in which no net flow of current occurs, but a steady potential difference is established with the cold end positive with respect to the warm end. The fluxes of positive and negative ions in the direction of $+x$ are

$$\left. \begin{aligned} j^+ &= -\frac{d}{dx}(D^+ n^+) - \tfrac{1}{2} D^+ \frac{n^+}{T} \frac{dT}{dx} - u^+ n^+ \frac{dV}{dx}, \\ j^- &= -\frac{d}{dx}(D^- n^-) + \tfrac{1}{2} D^- \frac{n^-}{T} \frac{dT}{dx} - u^- n^- \frac{dV}{dx}, \end{aligned} \right\} \quad (9.21)$$

where the second term on the right-hand side represents the contribution of thermal diffusion, D^+, D^- are the diffusion coefficients for the H$^+$ and OH$^-$ ions, u^+, u^- the ionic mobilities, $-dT/dx$ and dV/dx are the temperature and potential gradients. In the steady state, with no net current flowing, $j^+ + j^- = 0$, and if we assume $n^+ \simeq n^- = n$, and that $D = ukT/e$ does not vary over a narrow range of temperature, we

have
$$-\frac{dV}{dx} = \frac{kT}{e}\left(\frac{u^+ - u^-}{u^+ + u^-}\right)\left[\frac{1}{n}\frac{dn}{dx} + \frac{1}{2T}\frac{dT}{dx}\right], \quad (9.22)$$

or, the thermoelectric power

$$-\frac{dV}{dT} = \frac{kT}{e}\frac{(u_+/u_- - 1)}{(u_+/u_- + 1)}\left(\frac{1}{n}\frac{dn}{dT} + \frac{1}{2T}\right). \quad (9.23\text{a})$$

At any point in the crystal, the ion concentration has the equilibrium value given by the law of mass action, so

$$n^+n^- \simeq n^2 = \alpha \exp(-\phi/kT),$$

where ϕ is the activation energy for dissociation of the molecule, and therefore $1/n\, dn/dT = \phi/2kT^2$, and

$$-\frac{dV}{dT} = \frac{k}{2e}\frac{(u^+/u^- - 1)}{(u^+/u^- + 1)}\left(\frac{\phi}{kT} + 1\right). \quad (9.23\text{b})$$

Putting $u^+/u^- = 10$ (Eigen and de Maeyer 1958), $\phi = 1\cdot 2$ eV, and $T = 260$ K, gives

$$-\frac{dV}{dT} = 1\cdot 9 \text{ mV}/^\circ\text{C}. \quad (9.24)$$

For the more general case of impure ice containing ionized foreign molecules, in which not only the ion states but the attendant thermally activated Bjerrum defects have to be taken into account, Jaccard (1963) derives a more complex expression for the thermoelectric power which reduces to eqn (9.23b) for pure ice.

If the separated space charge in the ice specimen be regarded as equivalent to a surface density of charge on the ends of the specimen, then the latter may be calculated from (9.23b) as

$$\sigma = +\frac{\epsilon}{4\pi}\frac{dV}{dx} = 5 \times 10^{-5}\frac{dT}{dx} \text{ e.s.u./cm}^2, \quad (9.25)$$

ϵ being the static permittivity of ice.

The thermoelectric potentials developed across an ice specimen by the application of a steady temperature gradient were first measured by Latham and Mason (1961a). The temperature difference across the specimen was measured with fine thermocouples, and the potential difference by attaching copper electrodes to the ends of the ice and connecting them to a vibrating-reed electrometer. Careful checks were made to see that spurious potentials did not arise at the copper–ice junctions, and some experiments with brass and aluminium electrodes produced very similar results. The steady potentials produced by various temperature differences across polycrystalline ice specimens, formed from distilled water passed through an ion-exchange column to achieve a conductivity of $10^{-6}\ \Omega^{-1}\ \text{cm}^{-1}$, are plotted in Fig. 9.13. The measurements show a linear relationship between V and ΔT, and excellent agreement with eqn (9.24) provided the temperature of the warm end of the ice was below about $-7°$ C. For higher temperatures,

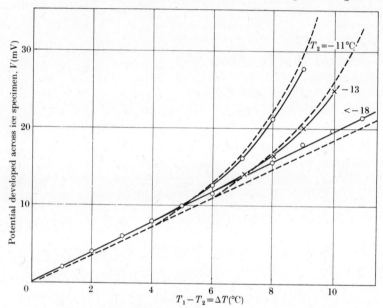

Fig. 9.13. The potentials developed across an ice specimen as a function of the applied steady temperature difference. ———, experimental values; ------, calculated values. (From Latham and Mason (1961a).)

however, the measured potentials were higher than those predicted by eqn (9.23b) on the assumption that ϕ remains constant. However, measurements on the electrical conductivity of ice indicate that, above $-7°$ C, the quantity $1/n\ dn/dT$ increases rather rapidly with increasing temperature in a manner that is consistent with the observed changes in thermoelectric power. Latham and Mason found that saturating the ice with CO_2, or adding HF in concentrations of up to 10^{-3} M, *increased* the thermoelectric power by up to 50 per cent, in contradiction to the predictions of Jaccard's theory, while the addition of NaCl produced the opposite effect.

Bryant and Fletcher (1965) measured the thermoelectric power of polycrystalline ice specimens between either brass or palladium electrodes. The specimens were heated from below, and the potentials across the electrodes measured by applying a balancing potential with a potentiometer, using an electrometer as a null detector. The ice samples were formed from water whose pH could be varied from 2 to 10 by the addition of HF or NH_4OH. The measured potentials included contact potentials at the electrodes, which were often of comparable magnitude to the thermoelectric potentials developed across the ice, but the authors believe that they were able to distinguish between the two potentials. With the top electrode held at constant temperature, the potential difference across the specimen was measured as a function of the temperature of the lower electrode. Despite variations in the standing (contact) potential, the slopes of the curves obtained with specimens of the same purity were consistent within a few per cent. When the warm end of the specimen was colder than $-10°$ C, the V–ΔT plots were practically linear, but they departed increasingly from linearity if the warm end was raised to higher temperatures, in agreement with the findings of Latham and Mason. Ice containing HF, with pH of the melt water between 2 and 6, had a thermoelectric power of about -2 mV/$°$ C, but that of fairly pure ice (pH = 6·0–7·5) reached a maximum value of $-3·5$ mV/$°$ C, while for ice doped with NH_3, the thermoelectric power was of opposite, i.e. positive, sign. The actual values for pure ice are in some doubt because of possible contamination of the specimen by CO_2, but probably more serious, are the systematic errors that may arise in all such experiments because of the existence of contact potentials at the two electrodes that cannot be disentangled from the thermoelectric potential if they have unequal temperature coefficients. It seems likely that the surprising result reported by Takahashi (1966), that the thermoelectric power of fairly pure ice changed from being negative to positive when the temperature of the specimen was raised above $-10°$ C, may have been caused by temperature-dependent contact potentials. It is therefore highly desirable to measure the thermoelectric power of single crystals without using electrodes in contact with the ice. To meet this requirement, Latham (1964) devised an ingenious experiment in which horizontal ice needles were grown from the vapour on a fine vertical fibre in a diffusion cloud chamber, and then heated differentially by radiation to produce a temperature gradient along the length of the needle. The charges produced at the ends of the needle were determined by measuring the rotation of the dipole in an applied electric field but, to avoid the thermoelectric effect being masked by polarization of the crystal in the electric field, the latter had to be very weak (a few V/m), and the quartz supporting fibre very thin ($d \sim 1$ μm). However, when properly set up, a crystal always rotated in the sense that its warmer end indicated a negative charge, the deflexion being roughly proportional to the temperature gradient along the crystal which reached maximum values of 4° C/cm. Moreover, the thermoelectric potential across the ice needle, as deduced from these measurements, agreed to within a factor of 2 with the value given in eqn (9.24).

Recently Brownscombe and Mason (1966) have measured the thermoelectric power of quite large single crystals of rather pure ice by an induction method involving no electrodes. The crystal was exposed to a vertical temperature gradient of about 2° C/cm along its length, and the potential difference developed across it was determined by observing the changes in potential induced on an insulated metal disc when the crystal was rotated to bring each end in turn close to it. At temperatures near $-20°$ C, a value of $-2 \cdot 3 \pm 0 \cdot 3$ mV/°C was obtained that may be compared with the theoretical value of $-1 \cdot 9$ mV/°C.

The same method was used to measure potential differences developed between the two halves of a polycrystalline ice specimen doped with different concentrations of HF and NH_3. Potential differences of 30–40 mV were produced by tenfold changes in the concentration of impurity. These results were in fair agreement with calculations made on the basis of Jaccard's theory, and even better agreement was obtained by Bryant (1967), who used single crystalline ice with the electrode method in the hope of eliminating the segregation effects that may occur in polycrystalline specimens.

(ii) *Transient thermoelectric effects in ice.* It follows from the above arguments, that when two pieces of ice of different temperatures are brought into temporary contact, the warmer piece should acquire a negative charge and the colder piece an equal positive charge. If two semi-infinite pieces of ice with initial temperatures T_1 and T_2, thermal conductivity K and electrical conductivity λ, are brought into contact for a time t, the charge transfer per unit area of contacting surface is given by Mason's theory as

$$\sigma = \frac{nkT}{2(\pi K)^{\frac{1}{2}}} (u^+ - u^-) \left\{ \frac{\phi}{2kT^2} + \frac{1}{(T_1 + T_2)} \right\} (T_1 - T_2) e^{-4\pi \lambda t/\epsilon} \int_0^t \frac{e^{4\pi \lambda t/\epsilon}}{t^{\frac{1}{2}}} \, dt,$$

(9.26)

and this reaches a maximum value of

$$\sigma_{\max} = 3 \times 10^{-3}(T_1 - T_2) \text{ e.s.u. cm}^{-2} \qquad (9.27)$$

when the surfaces are separated after about 0·01 s. If they are left in contact for longer times, the charge separation will be decreased as the two pieces of ice become more nearly equal in temperature.

The charge separation that results when two pieces of ice of different temperatures are brought into momentary contact, and separated under conditions in which rubbing or frictional contact is minimized, has been measured by Brook (1958) and Latham and Mason (1961a). The latter found good quantitative agreement with eqn (9.26) and showed that the charge transfer did decline for contact times in excess of about 0·01 s and became small after 0·5 s. They also confirmed Brook's result that if one piece of ice was contaminated with either CO_2 or NaCl, the charge transfer for a given temperature difference and area of contact, was increased if the contaminated ice was warmer, and reduced if it was colder, than the pure ice. Latham and Stow (1965) repeated these experiments, but with more attention to smaller times of contact in the range 1–50 ms, and confirmed Latham and Mason's result that the charge transfer decreased for contact times longer than about 1/100 s, but more rapidly than indicated by the theory, probably because of the finite thickness of the ice specimens. When the two cylindrical ice specimens were allowed to impact at 7·5 cm s^{-1}, the magnitude of the charge transfer was similar to that observed by Latham and Mason, but when the impact velocity was increased to 17·5 cm s^{-1}, the charge transfer was increased three fold without any apparent increase in the area of contact as indicated by measurements of current flowing between the contacting specimens. This result has never been satisfactorily explained. If one of the specimens was tapered, so that the areas of the two approaching surfaces were unequal, the charge transfer was markedly increased—as much as twelve fold when the ratio of the areas was 40 to 1. This result may be explained by the existence of enhanced gradients of both temperatures and excess charge carriers near the interface in this arrangement—see Latham and Stow (1967a).

(*iii*) *Electrification produced by asymmetric rubbing of ice on ice.* Reynolds, Brook, and Gourley (1957) showed that if two rods of 'pure' ice were rubbed together asymmetrically, so that frictional heating was confined to a small area on one specimen (the rubber), but spread over a larger area on the other, the former acquired a negative charge and the latter a positive charge. A more quantitative investigation of this effect was carried out by Latham (1963a), who allowed an ice cube to slide down a smooth inclined plane of ice and fall into an induction can connected to a sensitive electrometer. The undersurface of the cube was warmed relative to the plane by friction, and acquired a negative charge whose magnitude was roughly proportional to the distance travelled by the cube, and to the temperature difference between the two sliding surfaces as measured by a thermocouple. Although the temperature gradients across the two surfaces were neither very steady nor uniform, the measured charge transfer per unit area across the interface agreed with the steady-state value calculated from eqn (9.25), on the assumption that the actual area of contact was one-twentieth the geometric area. Variations of the environmental temperature over the range -10 to $-20°$ C produced no significant variations in the charge transfer but, above $-10°$ C, the ice surfaces tended to adhere and caused very erratic charging. Rather similar results were obtained by Magono and Shio (1967) in that the separation of charge produced by the asymmetric rubbing of two ice rods was consistent, both with the earlier experiments and the thermoelectric theory, when the rods were colder than $-10°$ C but, after prolonged rubbing, or when the specimens were warmer than $-5°$ C, the electrification tended to change sign, apparently as a result of surface melting. The negative charging of the colder surface under these circumstances might be explained by the formation of a thin layer of water on the warmer surface and the removal of the negative outer parts of the electrical double layer on the colder sliding surface. Certainly one expects the processes to become much more complex if a solid/liquid interface is formed.

Magono and Shiotsuki (1964) found that the presence of air bubbles in the ice had a marked effect. If two ice rods containing approximately equal concentrations of air bubbles were rubbed together asymmetrically, the rod with the smaller contact area, the 'rubber', acquired negative charge, leaving a positive charge on the other rod. However, if the rubber contained many fewer air bubbles than the other rod, it acquired a positive charge. Conversely, if the rubber contained a much higher concentration of air bubbles than the other specimen, it acquired a negative charge. Hobbs (1964) suggested that these results could be explained by the thermoelectric theory, if allowance were made for the fact that the surface of the specimen containing the more air bubbles will tend to become warmer by virtue of its lower thermal conductivity. This view is supported by Latham (1965) who repeated his sliding cube experiments with specimens containing various concentrations of air bubbles. If both the contacting bodies were of transparent ice, the magnitude and sign of the charge transfer were accounted for quite well by the simple steady-state thermoelectric effect (eqn (9.25)), but the measured charges were considerably greater than predicted if either or both the specimens contained high concentrations of air bubbles. However, close agreement cannot be expected between the theory that assumes steady, uniform temperature gradients and experiments in which steep and transient gradients probably existed over localized areas of contact, which could not be accurately measured.

The separation of electric charge in ice under the influence of temperature gradients is now an established fact; the extent to which it may be responsible for the electrification of ice particles in the atmosphere will be assessed in the next three sections.

9.6.2.6. Electrification associated with the collision and fracture of ice crystals

The electrical effects associated with blizzards were first studied in detail by Simpson (1919), who found that they were accompanied by a large increase in the normal positive potential gradient near the ground, indicating that the air contained excess positive charge, a compensating negative charge residing partly on the heavier settling snowflakes and on the underlying snow-cover. Simpson proposed that collisions between snowflakes caused them to acquire a negative charge, a compensating positive charge being communicated to the air by ions or on tiny fragments of ice broken off the large snowflakes. Later Simpson and Scrase (1937) and Simpson (1942) suggested that this might also be the main mechanism of thunderstorm electrification.

Pearce and Currie (1949) observed a large increase in the value of the positive space-charge density in a snowstorm, and showed that when a block of snow was eroded with an air blast, the visible fragments carried away a negative charge, leaving the air with a positive charge. When they allowed snow to blow against a block of snow they found that the snow became negatively charged and the air acquired a positive charge. A similar result was obtained by Norinder and Siksna (1953), who poured snow through an ice-coated funnel, collected it in a metal can connected to an electrometer, and found it to have acquired a negative charge.

There is considerable experimental evidence from the work of Findeisen (1940, 1943), Lange (1943), Kramer (1948), and Kumm

(1951) to show that when air currents of a few cm/s are allowed to flow past a deposit of frost grown by sublimation, small splinters are broken off the fragile dendritic crystals and carry away charges of predominantly one sign, leaving those of opposite sign on the parent crystals. Findeisen found that the deposit acquired a positive charge at the rate of 10^{-6} e.s.u. cm^{-2} s^{-1} in an air-stream of 35 cm s^{-1}. Kramer found a charging rate of the same magnitude and sign in the early stages of a growing frost layer, but found that, later, the polarity became reversed. Kumm found positive splinters to be about seven times as numerous as negative ones, the average charge per splinter being about $+3 \times 10^{-6}$ e.s.u. In a more recent investigation, Latham (1963b) allowed air currents to flow past a frost deposit whose temperature could be made higher or lower than that of the air-stream. The deposit became negatively charged if it were warmer than the air-stream and vice versa. The ejected splinters carried charges of opposite sign to that of the deposit, the average charge on splinters of linear dimensions about 150 μm $\times$ 40 μm being 6×10^{-7} e.s.u. for a temperature difference of 10° C. In an air-stream of 2 m s^{-1}, splinters were released at the rate of about 16 cm^{-2} s^{-1}, to produce charge at a rate of about 10^{-5} e.s.u. cm^{-2} s^{-1}.

Latham showed that these results could be accounted for by Mason's theory of the thermoelectric effect, and argued that this also produces a satisfactory qualitative explanation for the charging effects produced in blizzards by rubbing contact between snow surfaces. However, in a later experiment in which snow crystals were blown again a packed snow surface in a cold room, Latham and Stow (1967b) found that, although the signs of the charges on both the target and the rebounding crystals varied with their temperature difference according to the theory, the magnitude of the charges increased rapidly with increasing impact velocity, and was two orders of magnitude higher than predicted by eqn (9.27). The authors' attempts to explain this discrepancy in terms of empirical velocity and shape factors are unconvincing, and one suspects that inductive, and perhaps other spurious charging mechanisms were at work.

A variant of this type of charging process, and one which seems more relevant to the electrification of thunderstorms, was proposed by Reynolds, Brook, and Gourley (1957), who suggested that pellets of soft hail might become charged by collisions with much smaller ice crystals. They observed that when an ice sphere was rotated at a peripheral speed of 8 m s^{-1} in a mixed cloud of ice crystals and supercooled droplets, it acquired a negative charge, but if the cloud was composed entirely of droplets or entirely of crystals, there was negligible charging. Reynolds et al., following a suggestion by Henry (1953), attributed the charging to local temperature differences produced by asymmetric rubbing between the large ice sphere and the small crystals, and felt that the sole function of the cloud droplets was to produce a rimed surface on the sphere that would be warmer than the colliding ice crystals. They estimated that, with a temperature difference of a few degrees, the average charge carried away by a crystal of radius 50 μm was 5×10^{-4} e.s.u.

Latham and Mason (1961b) performed a similar experiment in which a smooth stationary ice probe was bombarded with small ice crystals, supercooled droplets being rigidly excluded in order to eliminate the possibility of charging by riming. The surface temperature of the probe was controlled by electrical or radiant heating and measured by small thermocouples. The magnitude of the charging was directly proportional to the temperature difference between the crystals and the ice target, but was rather insensitive to the crystal diameter in the range 20–50 μm and their impact velocity in the range 1–30 m s^{-1}. With a temperature difference of 5° C, a rebounding crystal of diameter 50 μm produced, on average, a charge of 5×10^{-9} e.s.u., i.e. five orders of magnitude less than the value of 5×10^{-4} e.s.u. reported by Reynolds et al. The ice target acquired a negative charge if it was warmer than the crystals and a positive charge if it was colder. The results were consistent with the predictions of the thermoelectric theory expressed by eqn (9.27) if the actual area of contact between an impacting crystal and the probe was one-tenth the geometric area of the crystal. It is tempting to attribute the much larger charges measured by Reynolds et al., which appeared only when their cloud contained a mixture of supercooled droplets and ice crystals, to the freezing and splintering of the droplets (see § 9.6.2.7) but, against this, is the fact that a cloud composed almost entirely of droplets produced little charging and that the droplets were probably too small to have splintered on freezing. Nevertheless, the high rates of charging reported by Reynolds et al. are difficult to account for and have not been confirmed by more recent work.

Evans and Hutchinson (1963) brought ice crystals, grown in different parts of a diffusion cloud chamber and differing in temperature by as much as 14° C, into contact for periods of 0·2–0·5 s, and then measured the charge on one of them by raising it into a small Faraday cage connected to a sensitive electrometer. No charges were detected with a system of sensitivity 8×10^{-5} e.s.u. Since the areas of contact were estimated to be between 0·2 and 2 mm², this result is not inconsistent with eqn (9.27), but fails to support the much higher charges reported by Reynolds *et al.* More recent attempts to repeat Reynolds' experiment in a cloud of natural snow crystals have produced conflicting results.

Magono and Takahashi (1963*a, b*), who performed their experiments both in a laboratory cold room and in a natural supercooled cloud on a mountain top, also detected no significant charging of an ice probe bombarded by small supercooled droplets, but impacting snow particles caused it to acquire a positive charge if its temperature was above −10° C, and a negative charge at lower temperatures. If the probe was coated with rapidly accreted, soft-rime ice, impacting crystals conferred upon it a negative charge at temperatures below −20° C and this increased if the rime surface was heated. The authors hint that the positive charging at the higher temperatures may have been associated with the formation of a liquid-water film on the rime surface, and that the negative charging at the lower temperatures may have been caused, at least in part, by the breaking off of delicate rime structures.

Latham and Miller (1965) whirled an ice sphere through a stream of falling millimetre-size snow crystals. The sphere was found to acquire a negative charge, the magnitude of which increased as the velocity of impact was increased and the surface of the sphere was made more irregular. When a sphere with a rough ice surface collided with ice crystals at 8 m s⁻¹, the average charge was estimated at about 10^{-3} e.s.u. per crystal collision—three orders of magnitude larger than predicted by the thermoelectric effect. But, in a similar outdoor experiment, Burrows, Hobbs, and Scott (1967) found the charging of the sphere to depend very little on its roughness or the velocity of impact, the most important factors being the air temperature and the direction of the atmospheric electric field. They considered that the sphere acquired charge largely from the initial charges on the impacting ice crystals rather than as a result of asymmetric rubbing.

Since the results of these various experiments are in conflict, and show little agreement as to how the charging depends on the nature and structure of the ice target, its temperature and purity, on the air temperature and the size and impact velocity of the crystals, there is an urgent need for much more careful experiments designed to eliminate all extraneous and unwanted electrical effects and elucidate the basic charging mechanisms. In this type of experiment, confusing results may arise from initial charges on the colliding particles, from charging of, and leakage across, insulating supports, and from the presence of unsuspected electric fields. The latter may polarize the ice target, which may then receive a net charge as the rebounding crystals carry away some of the induced charge (see p. 551). Provided that the times of contact are greater than the relaxation time for conduction in ice (about 0·01 s at $-10°$ C for pure ice), a single crystal of radius 100 μm could carry away a charge of 10^{-4} e.s.u. in a field of about 100 V cm^{-1}, and a crystal of radius 1 mm a charge of 10^{-3} e.s.u. in a field of only 10 V cm^{-1}. Certainly this induction charging of simulated hailstones by colliding ice crystals would seem capable of overwhelming any charging that might be produced (thermoelectrically) through the action of asymmetric rubbing or other locally induced temperature gradients. Its potential importance as a mechanism of charge generation in thunderstorms is examined in § 9.6.3.2.

9.6.2.7. Electrification associated with the freezing and splintering of water drops and the formation of rime

The fragmentation and splintering of freezing water drops is discussed in § 4.5.2. Realizing that this phenomenon was likely to be accompanied by electrification, Mason and Maybank (1960) suspended drops of distilled water of radius 0·3–1·0 mm from fine polythene fibres between vertical parallel-plate electrodes in a cold cell and observed the deflexions of the drop residues in an applied electric field. On the occasions when more than half of the drop remained on the fibre after fragmentation, about 80 per cent of the residues were found to carry a negative charge, whereas nearly half of the minor drop residues carried a positive charge. Drops of radius 1 mm nucleated at $0°$ C and frozen in air at $-10°$ C, produced charges ranging from $-1·1 \times 10^{-4}$ e.s.u. to $-7·2 \times 10^{-3}$ e.s.u., with an average value of $-0·86 \times 10^{-3}$ e.s.u. Drops of radius 0·35 mm behaved in much the same way but produced charges smaller by a factor of 2. Drops of NaCl solution of concentration 10^{-2} M failed to shatter and showed no detectable charge.

Kachurin and Bekryaev (1960) reported that the freezing of drops of $d = 0·2$–2·0 mm on a wire support left charges on the residues ranging from $+45 \times 10^{-3}$ e.s.u. to -90×10^{-3} e.s.u., with an average value of -3×10^{-3} e.s.u. The positive charges appeared to be associated with the ejection of large ice particles and the negative charges with the ejection of streams of minute water droplets. Evans and Hutchinson (1963) nucleated drops of $d = 1 \cdot 3$–$1 \cdot 5$ mm at about $-2°$ C, allowed them to freeze at $-15°$ C, and then drew them into a small Faraday cylinder connected to a sensitive electrometer to measure the charge on the residual drop. The results were rather similar to those of Mason and Maybank in that two-thirds of the residues resulting from the splitting of the drops or the ejection of a large spicule were negatively charged (highest -25×10^{-3} e.s.u., average $-2 \cdot 2 \times 10^{-3}$ e.s.u.), the remainder carrying positive charges averaging $1 \cdot 5 \times 10^{-3}$ e.s.u. Using the same technique, Stott and Hutchinson (1965) found the charges on the residues of about 100 drops of diameter $1 \cdot 0$–$1 \cdot 5$ mm that shattered on freezing, ranged from -17×10^{-3} e.s.u. to $+25 \times 10^{-3}$ e.s.u. Fractures across a central diametral plane tended to produce equal numbers of positive and negative residues. If spicules broke off in the early stages of freezing, when they contained some liquid water, the drop residues were usually negatively charged (average value $-1 \cdot 4 \times 10^{-3}$ e.s.u.), but if the breaking spicules were of solid ice, they usually left behind positive residues (average value $3 \cdot 4 \times 10^{-3}$ e.s.u.).

Johnson (1968) suspended drops of about 1 mm diameter from an insulating fibre in a cold cell, removed their initial charges by exposure to a radioactive source, nucleated them with ice crystals, allowed them to freeze at about $-20°$ C, and measured their charges, both before and after freezing, by withdrawing them from the centre of a small induction ring connected to a sensitive vibrating-reed electrometer and a galvanometer recorder. The smallest charge detectable with this system was 5×10^{-6} e.s.u., the time constant being 1 s. Photographs and careful microscope observations of the freezing process allowed shattering, the formation of spikes, splinters, etc. to be correlated with changes on the electrical record. Drops that were rapidly cooled after nucleation in air or nitrogen, and those that attained thermal equilibrium before nucleation in CO_2 or hydrogen, often shattered violently. The charges on the residues usually ranged from 10^{-5} to 10^{-3} e.s.u. with roughly equal numbers of positives and negatives. Occasionally charges of up to 10^{-2} e.s.u. were recorded but were attributed to movement of the drop on the fibre. (Drops that failed to shatter, but cracked and ejected splinters, also acquired charges of either sign up to 3×10^{-3} e.s.u.). But drops allowed to reach thermal equilibrium in air or nitrogen before nucleation rarely shattered and produced only small charges. Very similar results were obtained with drops of de-ionized water, 10^{-4} M NaCl solution, and 7×10^{-5} M NH_4Cl solution.

In summary, the results obtained by different workers indicate that the charges produced by similar drops freezing under identical conditions vary considerably in both magnitude and sign, as one might expect with such random and uncontrolled events as fragmentation and the ejection of spikes and splinters. The charges produced by millimetre drops rarely exceed 5×10^{-3} e.s.u. and are not much influenced by the presence of dissolved ionic salts in the water.

The mechanism of the charge separation has not yet been elucidated—indeed there may be more than one process at work. Mason and Maybank (1960) realized that charge separation under the influence of outwardly-directed temperature gradients in the ice shell might account for the ejection of positive splinters from the surface of the freezing drops and the predominantly negative charges on their major drop residues, but were not persuaded that this thermoelectric effect could explain the magnitudes of the observed charges. The maximum temperature gradients are likely to occur near the end of the freezing process, when the centre of the drop is at $0°$ C and its surface is approaching the air temperature. If the latter were $-T°$ C, the average temperature gradient would be of order T/R and, according to eqn (9.25), the separated charge would be of order 5×10^{-5} T/R e.s.u. cm^{-2}. For a drop of radius $R = 1$ mm supercooled to $-10°$ C, this implies a charge density of 10^{-2} e.s.u. cm^{-2}. A splinter of size $100\,\mu \times 100\,\mu$m would therefore be unlikely to carry away a charge in excess of 10^{-6} e.s.u., and the drop unlikely to acquire more than 10^{-5} e.s.u. It seems, then, that another mechanism is required to explain the observed charges of 10^{-3} e.s.u. or more. The Workman–Reynolds effect, which provides for the separation of large quantities of charge at an ice-liquid interface during freezing, would be an obvious candidate except that Johnson (1968) finds that the presence of even ammonium salts in the water has no detectable effect on the charges acquired by freezing drops. Other possibilities are the ejection of charged droplets by the bursting of air bubbles or during the formation of spikes and cracks on freezing drops. We have seen on p. 528 that 100 μm bubbles can eject droplets of size 10 μm carrying negative charges in excess of 10^{-3} e.s.u. The ejection of a liquid thread from a millimetre spike might take as long as 10^{-4} s, in which case, it might acquire a *positive* charge of order 10^{-4} e.s.u. by the Jonas–Mason mechanism described on p. 529. The ejection of several drops, as reported, for example, by Kachurin and Bekryaev (1960), might produce *negative* charges of order 10^{-3} e.s.u. on the parent drop. But, whatever the origin of the charges, the above experiments suggest that freezing raindrops are unlikely to produce space charges of more than about 0.1 C km^{-3}, which would not be of great importance in thunderstorm electrification. There remains, however, the evidence that strong electrification coincides with the appearance of hail pellets, which grow by the impaction and freezing of supercooled cloud droplets, and that growing rime deposits acquire a substantial charge.

Findeisen (1940, 1943) formed a rimed layer by spraying water droplets on to a cold metal surface and found that it acquired a *positive* charge. The charging ceased if the surface became smooth and glassy, as was the case if the drops froze slowly, or if it became wet. Rather stronger charging was obtained with a natural supercooled cloud than with an artificial spray, the difference being ascribed to more rapid freezing of the smaller cloud droplets. The rate of charging in the cloud was 10^{-3} e.s.u. $cm^{-2} s^{-1}$. Kramer (1948) found that a rime deposit acquired a *negative* charge that increased in proportion to the impact velocity of the droplets. With a velocity of $0·5$ m s^{-1}, the charging rate was 6×10^{-5} e.s.u. $cm^{-2} s^{-1}$ and, for a velocity of 5 m s^{-1}, ten times larger. Lueder (1951*a, b*) made experiments in natural supercooled clouds on a mountain top and stated that the growing rime deposit acquired a negative charge, an equal positive charge being communicated to the air, probably on the parts of the drops which was flung off without freezing. Unfortunately it is difficult to interpret his experiments and deduce the actual rate of charging.

Meinhold (1951) measured the electric field strength at the surface of the fuselage of an aircraft flying at 80 m s^{-1} through a supercooled cumulus congestus cloud. The deposition of rime was accompanied by a rapid rise in the field strength in a sense which indicated that the aircraft was acquiring a negative charge. The rate of charging was calculated to be $1·5 \times 10^{-2}$ e.s.u. $cm^{-2} s^{-1}$, but in view of the difficulty of interpreting the significance of the measured field in this case, not much weight can be attached to the actual magnitude of the result

The charging of a rime deposit on a cold metal surface was also studied by Weickmann and Aufm Kampe (1950). Water droplets in the diameter range 5–100 μm were sprayed at velocities from 5 to 15 m s^{-1} on to a metal rod of 5-mm diameter in a cold room kept either at -5 or $-12°$ C; they were therefore slightly supercooled on reaching the rod. The rate of charging, which was not sensitive to the presence of dissolved salts, increased with increasing velocity of the droplet stream and, for a velocity of 15 m s^{-1}, attained a value of $1·5 \times 10^{-2}$ e.s.u. $cm^{-2} s^{-1}$. When water at temperatures slightly above $0°$ C was sprayed onto the rod, the latter acquired a slight positive charge. Later the authors indicated that these results may have been seriously affected by electrification associated with the production of the spray.

The balance of the evidence from all these experiments points to the acquisition of a negative charge by a growing layer of rime, Findeisen's result being an outstanding contradiction. Using the rates of charging recorded in these experiments, and assuming that similar rates might apply during the growth of hail pellets by riming in a supercooled cloud, Mason (1953d) demonstrated that this could provide a powerful mechanism of charge generation and separation in a thunderstorm.

However, Reynolds (1954) and Reynolds, Brook, and Gourley (1957) reported that simulated hailstones, made by whirling an ice-coated metal sphere in a cloud of supercooled droplets, showed no detectable charging unless the cloud also contained some ice crystals. A similar result has been reported by Magono and Takahashi (1963a, b), but in light of the experiments now to be described, it now appears that the rate of charging depends on the size of the drops, their temperature, impact velocity, and the mode of freezing. Moreover, spurious results may arise from initial charging of the incident droplets and from electrification produced by splashing of droplets on the ice surface.

The most extensive study of charging produced by the impact and freezing of supercooled droplets on an ice surface was made by Latham and Mason (1961b). The experiments were conducted in a cold room, at air temperatures ranging from $0°$ C to $-17°$ C. The hailstone, simulated by a 5 mm-diameter, electrically-insulated copper sphere coated with a 0·25-mm layer of ice, was suspended in the centre of an earthed vertical brass tube through which the air-stream carrying the droplets could be drawn at velocities ranging from 0 to 30 m s^{-1}. Water drops of uniform diameter in the range 20–90 μm, produced by a spinning top, or larger drops from an atomizer, were allowed to fall several feet in the cold room and become supercooled before striking the hailstone target. For a given droplet size and air-stream velocity, the flux of droplets hitting the target was determined by allowing them to strike, for a given time, a Formvar-coated glass sphere of the same dimensions, and counting the droplet impressions under the microscope. Detection and counting of the ice splinters ejected from the target surface was attempted by inserting Formvar-coated slides just beneath the hailstone and later counting the plastic replicas of the crystals. The electric charge accumulating on the target was measured at 10-s intervals by a vibrating-reed electrometer able to detect 5×10^{-4} e.s.u.

The freezing of droplets of distilled water on the surface of the hail

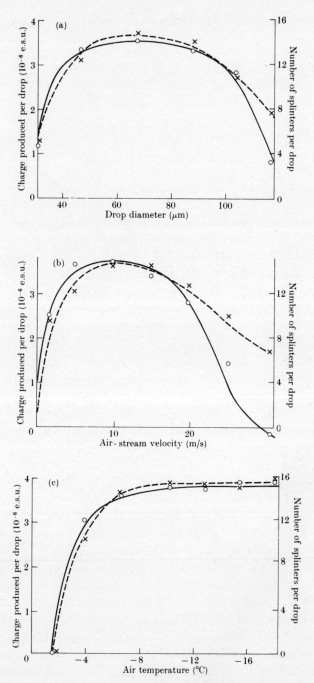

FIG. 9.14. The production of ice splinters (×) and electric charge (○) by the freezing of droplets on a rimed surface. (a) the influence of drop diameter: air temperature $-15°$ C; air velocity 10 m s^{-1}. (b) the influence of impact velocity: air temperature $-15°$ C, drop diameter 70 μm. (c) the influence of air temperature: air velocity 10 m s^{-1}, drop diameter 70 μm. (From Latham and Mason (1961b).)

stone caused it to become negatively charged and was accompanied by the ejection of small ice particles. The manner in which the average charge and number of splinters produced per drop varied with the drop diameter, impact velocity, and air temperature, is shown in Fig. 9.14 (a–c). In a typical experiment, with the air temperature at $-15°$ C and the air-stream moving at 10 m s^{-1}, 10^4 droplets of diameter 80 μm struck the hailstone within 10 s and produced a total charge of 4×10^{-2} e.s.u., i.e. an average of 4×10^{-6} e.s.u. per drop.

The crystal counts on the Formvar slides indicated that, on average, each droplet produced twelve splinters, but it is now thought that many of these particles may have been ejected as small supercooled droplets. Droplets of $d < 30$ μm produced few splinters and little charging, thus confirming the earlier result of Mason and Maybank (1960). Fig. 9.14(a) shows that the production of both charge and particles increased as the droplet diameter was increased to about 50 μm, remained fairly constant for diameters between 50 and 80 μm, and fell again for larger drops, which tended to splash on impact and communicate a positive charge to the ice target as observed originally by Faraday, and more recently by Gill and Alfrey (1952). Positive charging at high impact velocities is shown in Fig. 9.14(c). The rates of charging were almost independent of the air temperature in the range -6 to $-17°$ C, but fell off rapidly at higher temperatures because the hailstone surface became wet and the drops splashed on impact. No detectable charging occurred when the air-stream carried no droplets, and when droplets, impacting at very low velocity, froze slowly on the surface without producing splinters. The parallelism between the curves of charge and particle production is impressive and strongly suggests that the one is a consequence of the other.

Altogether, these experiments appear conclusive in showing that a hail pellet can become negatively charged by the impaction and freezing of large supercooled cloud droplets. However, more recent experiments in the author's laboratory indicate that the charging may depend on the mode of freezing of the droplets—in particular on the formation of a complete ice shell in the early stages which subsequently ruptures as the remainder of the liquid expands on freezing. This is more likely to occur when the large droplets impinge on small protuberances on a low-density ice pellet, and become cooled symmetrically by heat loss to the air, rather than by conduction to the underlying substrate. Charge separation may therefore be weak if the droplets are too small ($r < 20$ μm), or freeze progressively upwards from the substrate, and if there is no air-stream to enhance the cooling and separate the ejected fragments from the rimed deposit.

9.6.3. *The generation of electric charges and fields in precipitating clouds*

The rate of growth of an electric field, F, created by the separation of positive and negative charges at velocity V, is given by

$$\frac{\mathrm{d}F}{\mathrm{d}t}+4\pi\lambda'F = 4\pi QV, \tag{9.28}$$

where $\lambda'F$ represents the density of the dissipation current due to conduction, convection, and point discharge, λ' being the 'effective' conductivity of the air and Q the spatial concentration of separated charge. Whatever the exact nature of the charge-generation mechanism, Q will generally increase with time during the growth phase of the storm, and if Q, V, and λ' can be specified as functions of time, the growth of the electric field can be calculated from eqn (9.28). We have to consider two periods of field generation: an incubation period of 10–20 min during which the charges are generated and the field builds up to give the first lightning flash, and the much shorter periods of 10–20 s between flashes. We shall now apply eqn (9.28) to calculate the electric fields that could be produced by a number of possible charging mechanisms.

9.6.3.1. *The Wilson mechanism of selective ion capture*

According to the Wilson process, which invokes the selective capture of negative ions by hydrometeors polarized in the vertical electric field, the maximum charge that a particle can attain is very nearly $\tfrac{1}{2}FR^2$, and the *maximum* growth rate of the field is given by

$$\frac{\mathrm{d}F}{\mathrm{d}t}+4\pi\lambda'F = 4\pi . \tfrac{1}{2}F \sum^{R} NVR^2 = \tfrac{3}{2}F\frac{p}{\bar{R}\rho}, \tag{9.29}$$

where p is the precipitation intensity in cm s^{-1}, $\bar{R}$ the 'mean' radius of the hydrometeors, $\bar{\rho}$ their mean density, and N their spatial concentration. The field can grow only if

$$\frac{4\pi\Omega Vt}{F} > \frac{3p}{2\bar{R}\rho} > 4\pi\lambda',$$

THE ELECTRIFICATION OF CLOUDS

Ω being the rate of production of ions. Hence the maximum field that can be sustained by a precipitation rate p is

$$F_{max} = \frac{8\pi}{3} \Omega V t \frac{\bar{R}\rho}{p}, \qquad (9.30)$$

even if *all* the ions are segregated and captured by the falling particles. If $p = 5$ cm h^{-1}, $\bar{R} = 1$ mm, $V = 700$ cm s^{-1}, and $\Omega = 10$ ion-pairs/cm^3/s $= 4 \cdot 8 \times 10^{-9}$ e.s.u. cm^{-3} s^{-1}, the rate of ion production in the lower atmosphere, F_{max} would attain 600 V cm^{-1} after 1000 s. However, such a field could not actually be achieved because the drops would capture only a fraction of the ions, whose concentration would, in any case, be reduced by recombination. Moreover, selective capture of small, fast ions is impossible in fields greater than about 500 V cm^{-1}. Thus it appears that the Wilson mechanism cannot of itself produce electric fields of thunderstorm magnitude, but fields of up to 100 V cm^{-1} may be generated in this way.

9.6.3.2. *Electrification produced by the rebound of cloud particles from hydrometeors in polarizing electric fields*

A hydrometeor falling in a downwardly-directed electric field may be considered as a conducting sphere that becomes polarized with its upper half negatively charged and its lower half positively charged. Cloud particles colliding with, and rebounding from, the lower half, will carry away positive charge and leave a net negative charge on the hydrometeor, provided that the time of contact exceeds the time required for charge transfer by conduction between the two particles. The rate of charging of a hydrometeor of radius R, falling at velocity V relative to much smaller cloud particles of radius r and number concentration n, in a field F, is given by Latham and Mason (1962) as

$$\frac{dq}{dt} = -\pi R^2 V n \alpha \left(\tfrac{1}{2}\pi^2 F \cos\theta + \tfrac{1}{6}\pi^2 \frac{q}{R^2} \right) r^2, \qquad (9.31)$$

where α is the fraction of cloud particles lying in the cylinder swept out by the hydrometeor *which actually rebound from it*. Putting the average angle between the field and the line of approach of the colliding particles as $\bar{\theta} = 45°$, eqn (9.31) may be integrated:

$$q = -2 \cdot 12 F R^2 \{ 1 - \exp(-\tfrac{1}{6}\pi^3 V n \alpha r^2)t \}, \qquad (9.32)$$

so that the maximum charge that may be acquired by the hydrometeor and its time constant are respectively

$$q_{max} = -2 \cdot 12 F R^2, \quad \text{and} \quad \tau = (\tfrac{1}{6}\pi^3 V n \alpha r^2)^{-1}.$$

(a) *Ice crystals rebounding from hail pellets.* We first consider ice crystals rebounding from hail pellets, which thereby acquire a negative charge and, by gravitational separation, intensify the electric field at a rate given by

$$\frac{dF}{dt} + 4\pi i = -4\pi \sum NVq = 2 \cdot 12.4\pi \sum FNR^2V(1-e^{-t/\tau}), \quad (9.33)$$

where i represents the total leakage current. There is little evidence as to the magnitude of i or its dependence on the field strength. With very weak fields of < 100 V cm^{-1} in the cloud, (corresponding to < 10 V cm^{-1} near the ground), i may be represented by λF, where λ is the conductivity of the cloudy air. Stronger fields will produce point-discharge currents of which we have few representative measurements. Typical values of $0\cdot02$ A km^{-2}, and upper limits of $0\cdot16$ A km^{-2}, are quoted in the literature, so we shall represent the leakage current by $i = 10^{-3} (e^{0\cdot 2F} - 1)$, where F is the field in e.s.u. For weak fields, this gives $i = 2 \times 10^{-4} F$, an equivalent conductivity of 2×10^{-4} e.s.u. or about one-third the dry-air value at the 3-km level: large-scale fields of 10 e.s.u. inside the cloud would be associated with currents of $0\cdot02$ A km^{-2}, and fields of 20 e.s.u. with currents of $0\cdot18$ A km^{-2}. Eqn (9.33) thus becomes†

$$\frac{dF}{dt} + 4\pi \cdot 10^{-3}(e^{0\cdot 2F} - 1) = 2 \cdot 12.4\pi F \sum NR^2V(1-e^{-t/\tau})$$

$$= 6 \cdot 36 \frac{p}{\bar{R}\bar{\rho}} F(1-e^{-t/\tau}), \quad (9.34)$$

so that the growth rate of the field is determined largely by the precipitation rate p, the relaxation time τ, and the leakage current. Fig. 9.15 shows the growth of the field as calculated from eqn (9.34), for hail pellets of average radius $\bar{R} = 2$ mm and density $\bar{\rho} = 0\cdot5$ g cm^{-3}, falling at $V = 800$ cm s^{-1} through a cloud of ice crystals of radius $r = 50$ μm and concentration $n_c = 10^5$ m^{-3}, with $\alpha = 1$. In this case, $\tau = 100$ s, and Fig. 9.15 shows that the field inside the cloud would grow from an initial value of 5 V cm^{-1} to 3000 V cm^{-1} within 4·5 min if the precipitation intensity were 2·5 cm h^{-1}, and within 10 min if it were only 1 cm h^{-1}. The effect of increasing τ to 1000 s, for example by reducing n_c to 10^4 m^{-3}, is to increase the time taken for the field to reach

† This assumes that the vertical velocity of the crystals relative to the air is always $\ll V$. This is the case for the calculations represented in Fig. 9.15(a),(b) where the force exerted on a rebounding charged crystal by the field, $(\pi^2 r^2/2\sqrt{2})F^2 e^{-t/\tau}$, never greatly exceeds the gravitational force. This electrical force has even less influence on larger crystals of $r > 50$ μm.

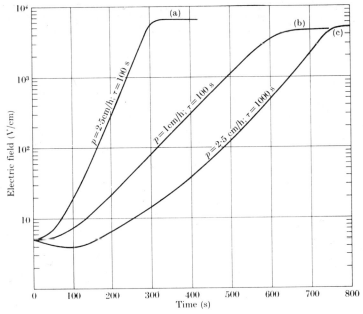

FIG. 9.15. Growth of electric field due to collisions of ice crystals with polarized hail pellets. (From Mason (1968).)

3000 V cm^{-1} from 4·5 min to 12 min, when $p = 2·5$ cm h^{-1}. As a result of the rapidly increasing point-discharge currents, the field does not grow indefinitely but saturates at realistic maximum values of about 6000 V cm^{-1}.

The charge actually separated during the build-up of these fields is given by $Q = AF/4\pi$, where A is the horizontal area of the charging region. If this is taken to be a circular region of radius 2 km, a field of 3000 V cm^{-1} would be associated with 33 C of separated charge available for the first lightning flash. The total charge carried by the precipitation elements is

$$Q_p = A \int_0^{z=(U+V)t} \Sigma Nq \, dz = \frac{6·36}{4\pi} \frac{A \cdot p}{\overline{R}\rho} \frac{(U+V)}{V} \int_0^t F(1-e^{-t/\tau}) \, dt, \quad (9.35)$$

and is contained in a column of height $z = \int (U+V) \, dt$, where U is the vertical velocity of the air. For the conditions represented by Fig. 9.15(c), with $p = 2·5$ cm h^{-1}, $\tau = 1000$ s, $t = 720$ s, $(U+V)/V = 1$, $Q_p = 48$ C.

If we suppose that the first flash neutralizes 20 C of the 33 C of separated charge, the field would drop to 1200 V cm^{-1} but, if p were to remain steady at 2·5 cm h^{-1}, it would recover to 3000 V cm^{-1} in 60 s. In similar circumstances, but with $\tau = 100$ s instead of 1000 s, the recovery time would be only 25 s, and lightning discharges could occur at such intervals until the precipitation intensity declines. Larger storms, containing greater quantities of separated charge and more intense precipitation, could, of course, produce more frequent lightning.

These calculations suggest that ice crystals rebounding from the undersides of polarized hail pellets may generate and separate electric charge at a rate that may be important in thunderstorm electrification, provided that the times of contact exceed the relaxation time, $\tau = \epsilon/4\pi\lambda_i$ (ϵ = static permittivity, λ_i = electrical conductivity), for the conduction of charge, and which is $\sim \frac{1}{100}$ s at $-10°$ C and $\sim \frac{1}{10}$ s at $-30°$ C. Latham and Mason (1962), who allowed ice crystals of diameter about 50 μm to impact on a *smooth* ice cylinder polarized in fields of up to 1000 V cm^{-1}, were unable to detect significant charging of the target, probably because the contact times in this case were less than the relaxation times. However, larger crystals colliding with rimed hail pellets having rough surfaces, may have longer times of contact and lead to appreciable charge separation in natural clouds. Indeed Latham and Miller (1965) observed that the charging produced by the collision of 1-mm snow crystals with a rotating ice-coated sphere, increased as the roughness of the sphere was increased, and it may be that their observed charges of about 10^{-3} e.s.u. per crystal collision were produced by the crystals bouncing preferentially from one half of the sphere polarized in a weak field of only $\sim$100 V cm^{-1}. Charges may also be separated in similar fashion and at similar rates by a small fraction, say 1 in 10^2 or 10^3, of cloud droplets rebounding after (grazing) collision with the undersurfaces of polarized ice pellets. In this case, the relaxation time for the conduction of charge is likely to be considerably shorter than for ice–ice contacts and is therefore unlikely to be a limiting factor.

(b) *Cloud droplets rebounding from polarized raindrops.* Calculations very similar to those just described can be made for cloud droplets bouncing off polarized raindrops. Could such a mechanism produce appreciable fields and perhaps lightning in an all-water cloud? Using eqn (9.34), and assuming the rainfall intensity to reach 2·5 cm h^{-1} and the average drop radius to be 1 mm, curves (a) and (c) of Fig. 9.15 also apply in this case. Let us consider a tropical maritime shower cloud containing 70 cloud droplets per cm^{-3}, of mean-volume radius 20 μm. If only 1 in 10^3 of these droplets that collide with raindrops actually rebound ($\alpha = 10^{-3}$), then $\tau = 1000$ s, and a field of 3000 V cm^{-1} could be established within 12 min.

In judging the probability of this actually happening in a cloud, much turns on the likelihood that perhaps 1 in 10^3 of cloud droplets can approach a polarized raindrop sufficiently closely for charge to pass between them, but without coalescence taking place. Sartor (1967) and Sartor and Abbott (1968) have obtained beautiful photographs of discharges between large, heavily- and oppositely-charged drops without coalescence, but these are probably not relevant to what may happen when a small, weakly-charged droplet approaches a raindrop. The important questions are whether, in this case, charge transfer can occur between the drops without physical rupture of the air film between them and, if a narrow liquid bridge is formed, can the drops rebound? Laboratory experiments suggest that this is unlikely, but one cannot rule out the possibility that only 1 in 10^3 droplets might rebound after temporary contact and charge transfer, and therefore that lightning might occasionally be produced in warm clouds by such a process.

9.6.3.3. *Electrification associated with the rupture of droplets impacting and freezing on hail pellets*

The evidence that freezing droplets of $r > 20$ μm tend to splinter and become electrically charged has been reviewed on pp. 543–50. Much of this evidence is conflicting, and the conditions under which hail pellets may acquire a systematic charge by the impaction and freezing of large cloud droplets have not been fully elucidated. Nevertheless, let us assume that droplets of $r > 20$ μm impacting on pellets of soft hail will, on average, impart a negative charge of 5×10^{-6} e.s.u., as reported by Latham and Mason (1961b), and enquire whether this might lead to the generation of thunderstorm charges and fields.

The rate of charging of an individual hail pellet is given by

$$\frac{dq}{dt} = E\pi R^2 V n_d q_d, \qquad (9.36)$$

where E, the collection efficiency, is taken to be unity, and n_d is the number concentration of droplets that contribute an average charge q_d. The rate-of-growth of the electric field is given by

$$\frac{dF}{dt} + 4\pi \times 10^{-3}(e^{0.2F}-1) = -4\pi \sum NqV = 3\pi \overline{(V/R\rho)} n_d q_d \int p\, dt. \qquad (9.37)$$

Taking $\overline{(V/R\rho)} = 5000$ g^{-1} cm^3 s^{-1}, $n_d q_d = 5 \times 10^{-6}$ e.s.u. cm^{-3} (i.e. 1 droplet/cm^3 contributing an average 5×10^{-6} e.s.u.), and the precipitation intensity to build up to a maximum value of $p = 5$ cm h^{-1}

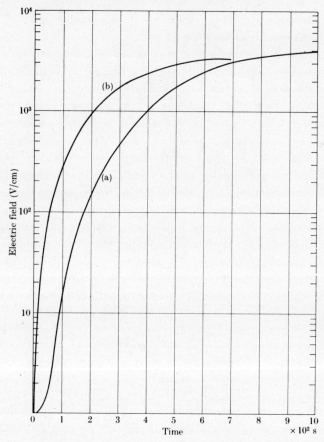

Fig. 9.16. Growth of electric field produced by water droplets freezing on hail pellets. (From Mason (1968).)

according to $p = p_m (1 - e^{-\beta t})$, with the relaxation time $1/\beta = 600$ s, integration of (9.37) shows that a field of 3000 V cm^{-1} would be achieved after 700 s as shown in curve (a) of Fig. 9.16. If p were to remain steady at 2·5 cm h^{-1}, the same field would be achieved in 500 s, as shown in curve (b). Fields of this magnitude would, in a cell of radius 2 km, be associated with separated charges of 33 C, and might therefore initiate the first lightning discharge within these time intervals.

However, should the field decay to 1200 V cm^{-1} after the first flash, it would take 100 sec to recover to 3000 V cm^{-1} under the conditions postulated for curve (b). A series of flashes at, say, 20-s intervals could be produced if larger quantities of charge are generated but partly masked by point-discharge currents during the incubation period, gravitational separation of these charges being sufficient to build up the necessary fields during the short intervals between flashes. Thus

Mason (1965) showed that, with a charging rate nearly four times greater than represented in curve (a), and a higher (but constant) effective conductivity of 2×10^{-3} e.s.u., a field of 3000 V cm^{-1} could be generated within 10 min of the first appearance of soft hail and that, after the first discharge, the field would recover to 3000 V cm^{-1} within 20 s.

It appears, then, that the fragmentation of freezing droplets on, and the rebound of ice crystals or cloud drops from, the surfaces of soft hail pellets are possible and, so far, the most likely, major mechanisms of charge generation and separation in thunderstorms. They would appear capable of reinforcing one another, perhaps with the induction process predominating once the field and the concentration of ice crystals have achieved threshold values by the splintering mechanism.

However, a good deal more work—laboratory work to establish the basic physics of charging mechanisms, and actual measurements of electric fields and hydrometeor charges in clouds—will be necessary before the origin of lightning is finally settled.

9.7. The transfer of electricity between the atmosphere and earth—the maintenance of the earth's charge

The observations described in previous paragraphs have established that the electrical structure of the thunderstorm is such as to maintain a vertical current from the earth upwards into the cloud base and also to drive a current upward from the top of the cloud. In view of Wilson's suggestion that the origin of the fine-weather field lay in thunderstorms, and of Whipple's discovery of a very close correlation between the variation of world-wide thunderstorm activity and potential gradient during the Greenwich day, it is now necessary to consider the various processes that convey the thunderstorm current and whether its magnitude is sufficient to balance the charge which leaks away from the upper atmosphere as a downwardly directed current of about 1800 A in fine-weather areas.

Electric charge is transported between the earth and the base of a thundercloud by lightning discharges, by ionization (point-discharge) currents, and by precipitation elements. We shall consider each of these three processes in turn.

9.7.1. *The charge transferred by lightning discharges*

The great majority of lightning discharges to earth bring down negative charge, the average magnitude being 20 C. In England, flashes to earth comprise about 40 per cent of all discharges, but in low latitudes where thunderstorms are more frequent, the proportion is only about

10 per cent. An active cell may produce several flashes per minute (an average value being perhaps three) over a period of 20 min or so, so that the effective current from a cell may be of the order 1 A. The total thunderstorm activity over the earth has been estimated roughly by Brooks (1925) to be equivalent, on average, to 100 flashes/s. If 10 per cent of these are flashes to earth, the total current will be of order 200 A, i.e. only one-ninth of the fine-weather current.

9.7.2. *Charge transfer by point-discharge currents*

The ionization currents in the intense fields beneath thunderstorms are very considerable, since a plentiful supply of ions originates from the brush discharges from prominent objects on the earth's surface, such as trees, grass, and other vegetation. Wilson (1923) showed that appreciable discharge currents occur from a grass-covered surface in fields not greatly exceeding those found beneath thunderstorms, and suggested that they probably play a major role in the vertical transport of charge from the earth in stormy conditions.

Wormell (1927, 1930) made a systematic investigation of the discharge from an elevated metal point, 12·3 m high, in the centre of a field fringed by fairly high trees, the nearest of which was 70 m from the point. The discharge current, during a thunderstorm or period of rain, was integrated by means of a specially designed gas micro-voltameter, which indicated the charge of each sign which had been lost from the point, as well as the net effect. Observations were also made of the actual behaviour of the current during a storm together with measurements of the vertical field. This investigation, which extended over about four years, demonstrated that, during periods when the field was sufficiently intense to cause appreciable discharge from the point (greater than about 8 V cm^{-1}), there was a marked predominance of negative fields. Over a lengthy period, the total loss of positive charge from the point was about twice the negative charge dissipated. The currents reached maximum values of about 15 μA and the net positive charge lost from the point during the passage of a storm might be 15 mC. The average net annual loss of positive charge from the point was 0·12 C. Experiments with various pointed objects, in an artificial field, showed that the current from a point of the type employed was not markedly different in magnitude from that from a twig, leaf, blade of grass, etc. at the same height. Wormell inferred, therefore, that the observed currents might be of the same magnitude as those from a tree of the same height as the artificial point, although recent work, notably by Ette (1966), casts doubt on this assumption. After making a census of trees in the surrounding countryside, Wormell estimated that the point-discharge current per km² should be at least as great as that from 800 points of the type he had used, which would imply currents of order 0·01 A km⁻² beneath the central parts of thunderstorms, and a flow of negative charge to the earth of 100 C/km²/year.†

This latter figure was also arrived at by Whipple and Scrase (1936) who recorded the currents from artificial points at Kew with a galvanometer. The ratio of the average annual positive current leaving the point to that entering it was 1·7 compared with Wormell's value of 2·0. Observations made by Chiplonkar (1940) and Sivaramakrishnan (1957) in India, and by Perry, Webster, and Baguley (1942) in Nigeria, yield ratios of 2·9, 2·0, and 2·9 respectively, in these thundery tropical regions.

Schonland (1928b) made observations on the discharge current from a small tree, 4 m high, which he uprooted and insulated. The maximum current measured was 4 μA in a field of 160 V cm⁻¹. Taking an average value of 0·8 μA, corresponding to an average steady field of 100 V cm⁻¹, for the current contributed per tree within a radius of 4·5 km of the centre of a storm, Schonland estimated a current of 0·16 A km⁻² due to point discharge. Simpson (1949), from a study of simultaneous records of the charge on rain, the point-discharge current, and the field, estimated that, in the neighbourhood of Kew, the point-discharge current per km² from the surrounding country was 2000 times that observed from the Kew artificial point; this gave a current of about 0·02 A km⁻² under the central part of a storm. Again, Smith (see Wormell (1953)) estimated the magnitude of this current at Cambridge from a study of the manner in which the very intense fields that follow close lightning discharges die away; he deduced a value of 0·018 A km⁻².

To sum up, it seems very probable that Schonland's figure is a considerable over-estimate and that a value of 0·02 A km⁻² in the regions of strongest field now seems about the best estimate that can be made.

9.7.3. *Charge transported by precipitation*

Many workers have measured the electric charge carried by precipitation reaching the ground. The general procedure has been to measure, at regular time or volume intervals, the net charge communicated to a receiver of known area by the precipitation, or to measure the charges on individual raindrops. The experimental results are summarized in Tables 9.5 and 9.6.

† More recent estimates tend to be a good deal higher, and Wormell (1953) himself suggests a revised figure of 170 C/km²/year, but these estimates may have to be revised downwards again if the report by Maund and Chalmers (1960), that trees in leaf produce smaller than expected point-discharge currents, is confirmed.

TABLE 9.5

The charges transported by rain in bulk

Observer	Recording interval	Charge per cm³ (e.s.u.)	Continuous steady rain	Thunderstorm rain	Ratio of positive to negative charge brought down	Current (A km^{-2})
Simpson (1909)	2 min				3·2 R	Gen. $<5\times10^{-4}$ R Max. 10^{-2} R
Baldit (1911)	15 s	+ −	2·49 1·07	4·0 3·46	1·36 T	Mean $3-5\times10^{-4}$ R
Schindelhauer (1913)	1 min	+ −	0·52 1·09	1·51 3·19	0·98 R	Max. $>10^{-3}$ R
Herath (1914)	Continuous				1·4 T 15·0 T	Mean 10^{-5} Max. 10^{-4}
McClelland and Nolan (1912)	30 cm³	+ −	0·21 0·08	0·72 0·84	4·5 T	Gen. $<5\times10^{-5}$
McClelland and Gilmour (1920)	30 cm³	+ −	0·21 0·08	0·72 0·84	4·8 T	Max. $6\cdot6\times10^{-3}$ NR $+1\cdot6\times10^{-5}$ -5×10^{-6} } NR
Schonland (1928b)					30 R	Mean 10^{-3} Max. 0·2 R

§9.7 THE ELECTRIFICATION OF CLOUDS

Author	Sample	Sign				
Marwick (1930)	30 cm³	+ / −	0·47 / 0·66	0·77 / 0·28	1·9 T	
Banerji and Lele (1932)	2 min	+ / −		0·11 / 0·12	0·69 R	
Banerji (1938)	2 min	+ / −		0·08 / 0·11	0·40 R	
Scrase (1938)	3 cm³	+ / −	0·43 / 1·24	1·23 / 1·42	1·1 T	$+5 \times 10^{-4}$ / -2×10^{-3} T
Chalmers and Little (1940)	10 min		0·20			Mean $+2 \times 10^{-5}$ NR
Chalmers (1956)	4½ min					$3·8 \times 10^{-6}$ NR
Ramsay and Chalmers (1960)	1 min					$3·5 \times 10^{-6}$ NR

T = total rainfall of all types. R = thunderstorm rain. NR = no thunderstorms included.

TABLE 9.6

The average charges carried by individual raindrops

Observer	Altitude (ft)	Charge per drop (e.s.u. ×10³)	Steady rain	Shower rain	Thunder-storm rain	Ratio of positive to negative charge brought down	Ratio of positive to negative drops
Gschwend (1922)	Surface	+	0.24	1.75	8.11	1.5	1.77
		−	0.53	5.43	5.88		
Banerji and Lele (1932)	Surface	+		6.4	6.9		0.8
		−		6.7	7.3		
Chalmers and Pasquill (1938)	Surface	+	2.2	1.3	3.7†	1.3	1.71
		−	3.0	2.3	9.2†		
Gunn (1947)	4000	+		24			
		−					
	12 000	+		41			
		−		100			
	20 000	+		63			
		−					

§ 9.7   THE ELECTRIFICATION OF CLOUDS

Reference	Height	Sign	N		
Gunn (1949)	Surface	+	15	1.6 R	0.77
		−	19		
Gunn (1950)	5000	+	81		
		−	63		
	10 000	+	148		
		−	112		
	15 000	+	123		
		−	76		
	20 000	+	52		
		−	62		
Hutchinson and Chalmers (1951)	Surface				0.95
Gunn and Devin (1953)				1.2 R	1.00 R
Arabadji (1959)				0.5	0.83
Jolivet (1959)	Surface				2.5
Krasnogorskaya (1960)	Surface				1.01

† Occurrence of lightning doubtful.

It will be seen that there are considerable differences between the results of different observers in regard to the magnitude of the charge carried by unit volume of rain or by individual drops, and the ratio of the positive to negative charge transported to the earth. These variations may, in part, be ascribed to the different sampling volumes (or periods) adopted, differences in intensity and character of the precipitation, in geographical location, and in the electrical conditions, but, in any case, it seems improbable that the charges carried by the drops on reaching the ground would bear much relation to those possessed by the drops on leaving the cloud, because they generally fall through a space charge in the intervening layer. Moreover, the observations, being made over only short periods of time and in only a few places, may not be representative, and spurious charging may have been produced by splashing of the drops and by induction on unshielded receivers. The balance of the experimental evidence shows, however, that: (a) precipitation of all kinds is sometimes positively and sometimes negatively charged; (b) the precipitation elements carrying positive and negative charges are mixed so that, during a short interval, they are seldom all of the same sign; (c) in all types of rain the amount of water that is positively charged is greater than the amount that is negatively charged, so that a net positive charge is communicated to the earth; (d) the quiet, steady rain of a depression is predominantly positively charged, the charge per cm^3 being rather small; (e) thunderstorm rain is much more highly charged, that falling from the centre of the storm often being predominantly positive.

It is significant that the *net* charge brought down is positive despite the predominance of negative charge near the cloud base and negative fields at the ground beneath thunderstorms. Indeed, Simpson (1949) found, from measurements made at Kew, that in fairly strong steady fields (greater than 20 V cm^{-1} at the ground), the sign of the charge brought down was opposite to that of the field. Moreover, the two often changed sign in opposite directions at the same time—the so-called 'mirror image' effect. Simpson deduced the following empirical relations between the rain current i (e.s.u. cm^{-2} s^{-1}), the charge Q (e.s.u. cm^{-3}), and the point discharge current I through a single point (e.s.u. sec^{-1}):

$$i = -2\times 10^{-8} I(p)^{0.57}, \tag{9.38}$$

$$Q = 7.22 \times 10^{-4} I(p)^{-0.43}, \tag{9.39}$$

and
$$i = 2.76 \times 10^{-5} Qp, \tag{9.40}$$

where p is the rate of rainfall in mm h^{-1}. Eqn (9.39) implies that in fields strong enough to cause brush discharge, the drops derive their charge largely by capture of the ions from the point-discharge stream. The charge found on the rain in bulk frequently exceeded that which could be acquired by Wilson capture, with ions of one sign only present in the field existing at the ground. This is strong evidence that the field increases as the cloud base is approached, as is required by the presence of the space charge derived from the point-discharge ions.

One might expect that more light would be shed on these problems from a study of the charges carried by individual precipitation elements. A summary of such observations made at the ground, and by Gunn (1947, 1950) in an aircraft, is given in Table 9.6. Gschwend (1922) and Gunn (1949) also measured the size of the individual drops by catching them on dyed filter paper and measuring the stains. Their observations pointed to the great complexity of the phenomena, showing frequent reversals of sign on successive particles and no obvious correlation with the size of the particles. Hutchinson and Chalmers (1951) measured the drop-charge and size, together with the point-discharge current and the field strength. They found a statistical correlation between Q and I and also between the ratio Q/I and the drop-size, the majority of the drops carrying charges of opposite sign to the point-discharge current and the potential gradient.

Smith (1955), in an extensive series of observations, measured the charge by allowing the drop to fall in succession between two insulated metal rings, the amplitude of the induced pulses being a measure of the charge, and the time interval between the two pulses giving the rate of fall of the drop, which was a sensitive indication of its size for small drops only. A similar arrangement was used by Gunn and Kinzer (1949) to determine the terminal velocity of freely falling drops. Smith also made an independent determination of drop size by allowing the drop to fall between the vertical plates of a condenser and determining the temporary change in capacity, as indicated by the change in frequency of a high-frequency oscillator of which the condenser formed a part. Pulses indicating the charge, fall velocity, and mass of the drop, were displayed on an oscilloscope and photographed continuously. Simultaneous measurements were made of the electric field and the point-discharge current. Smith found, during a sampling interval of 2 min, a large range of charges of both signs on drops of similar size. However, the mean charge on drops of a given size showed a systematic behaviour, being opposite in sign to the potential gradient for small drops, and of the same sign for large drops. This is interpreted in terms of the smaller drops being able to approach their equilibrium charge ($3Fr^2$), by Wilson capture of ions, more closely at all stages of descent than can the larger drops. Again, it is necessary to assume that the field is considerably greater at higher levels than at the ground in order to explain the magnitude of the observed charges. The charge transported to the ground by thunderstorm rain is generally between 0·1 and 5 e.s.u./cm³; taking a typical value of 1·0 e.s.u./cm³, (9.40) indicates that a rainfall of 25 mm h⁻¹ would produce a typical current of 2×10^{-3} A km⁻², and a maximum current of about 10^{-2} A km⁻². The highest value recorded is given by Schonland as doubtful at 0·2 A km⁻², while Chalmers and Little (1939) measured a current of $-7\cdot3 \times 10^{-2}$ A km⁻² associated with soft hail.

9.7.4. *Electrical balance-sheet for the earth's surface*

Wormell (1930), from measurements of the fine-weather current and those associated with precipitation, lightning discharges, and point discharge, drew up a rough electrical balance-sheet for a specified small area of the earth's surface at Cambridge, without claiming that the figures were representative of the whole earth. The items are listed in Table 9.7. Item 1, which was based on an assumed fine-weather

	Wormell (1930)	Later estimates
1. Fine-weather current	+60	+120
2. Precipitation of all types	+30	+30
3. Lightning discharges	−20	−20
4. Point-discharge current	−100	−170
Total	−30	−40

current of 2×10^{-16} A cm⁻², has been criticized by Whipple and Scrase (1936) as being about twice the true value, and by Gish and Wait (1950) as being about one-half the true value. Kraakevik's (1958) value of $3\cdot7 \times 10^{-16}$ A cm⁻² would imply a charge transfer of 120 C/km²/year. The value for item 3 would indicate that, on average, about one cloud-to-ground discharge occurs per km² per year. Later estimates by Wormell (1939) from field-change records, and by Golde (1945) from a survey of near-misses from lightning strokes, give values of 0·4 and 2 respectively, so there is no good reason for changing the estimate in Table 9.7. Later estimates of the contribution from point-discharge currents tend to be higher than Wormell's original estimate, for example -170 C/km^2/year at Cambridge and -180 C/km^2/year at Durham (see Chalmers 1967). Thus whether we take the original or the revised estimates of the items in Table 9.7, it seems plausible that the four processes probably balance out approximately, or even that, over land, the earth on the whole gains a negative charge. It must be noted also, that a very considerable part of the point-discharge contribution comes from occasions when there are no thunderstorms. Thus, the currents from showers and heavy frontal rain may make an appreciable contribution to the total interchange of electricity between the earth and atmosphere. It appears necessary that the earth should receive a net negative charge over land, since the contribution of point-discharge currents is likely to be smaller over the oceans.

Because of the difficulties which arise in making representative estimates of the items in Table 9.7, Gish and Wait (1950) sought an alternative approach. They made surveys of the electric current density i over the tops of twenty-four thunderclouds, using an aircraft equipped with rotating-vane field-meters and apparatus to measure the conductivity of the air $(i = F(\lambda_+ + \lambda_-))$. Owing to the absence of precipitation and the rarity of lightning in the clear air above a thundercloud, the electrical conditions are simpler than beneath the storm. Gish and Wait therefore expected that transfer of electricity in the air above the storm would occur mainly by conduction, and that measurements of the vertical component of the current density, made at short intervals (2 s) on a number of traverses over the cloud, would constitute an adequate basis for estimating the magnitude and direction of the total current from a storm. In the storms surveyed, the current was directed upwards, its magnitude varying from 0 to 1·4 A (except for one large value of 6·5 A), the average for all storms except this largest being 0·5 A. The fields were generally negative with occasional small positive excursions; the magnitudes were generally of order -100 V cm^{-1} but, on occasions, reached values of between 500 V cm^{-1} and 1000 V cm^{-1}. Measurements on 20 storms over Central Florida by Stergis, Rein, and Kangas (1957), using balloon-borne equipment, gave currents ranging from $+0·6$ A to 4·3 A, with an average value of 1·3 A.

If Gish and Wait's average value of 0·5 A is representative for the net current through a thunderstorm cell, then in order to supply a total current of 1800 A to balance the fine-weather current, 3600 cells or centres of electrical activity would have to be active on the earth at any one time. Brook's estimate of 1800, based on reports of the number of thunderstorm days at different stations, is probably an under-estimate, because he did not take into account that, at some places, two or more thunderstorms could be detected at a station on the same day, and that a single thunderstorm may contain several centres of lightning activity. On the other hand, if the average current of 1·3 A reported by Stergis *et al.* were typical of all thunderstorms, an average global population of only 1400 active storms would be sufficient to counterbalance the fine-weather current. More data are required, however, to settle the matter beyond all reasonable doubt.

APPENDIX A

The Collision and Coalescence of Water Drops Falling in Air

THE growth of incipient raindrops by collision and coalescence with smaller drops is critically dependent on two parameters: (i) the *collision cross-section* or *collision efficiency*, which may be defined as the probability that a larger drop will collide with a smaller one in its direct path, and (ii) the *coalescence efficiency*, defined as the fraction of colliding drops that coalesce. The product of these two parameters, both of which are likely to be sensitive functions of drop size, determines the growth rate of the drops and is termed the *collection efficiency*. It is this latter quantity that is determined in laboratory experiments which measure, in effect, the number of droplets captured by a collector drop; it has not been possible to determine collision and coalescence efficiencies separately. Collision efficiencies can, however, be calculated theoretically, the problem being to compute relative motions or trajectories of two approaching drops under the combined action of gravitational, hydrodynamical, and electrical forces, to determine grazing trajectories, and thereby the collision cross-sections of a drop for all smaller droplets. Ignoring the electrical forces for the present, we shall first consider a small droplet being overtaken by a much larger drop and then the more difficult case when the two droplets are of comparable size.

A.1. Theoretical computations of collision efficiencies

A droplet being overtaken by a large drop falling in still air will, when the vertical separation is large, approach the latter at a velocity equal to their differential terminal velocity along straight-line paths but, as they become closer together, the droplet will tend to follow the air flow round the larger drop and so be deflected from its initial path. A droplet possessing no inertia would follow a streamline, be carried round the drop and never collide with it. In practice, because the droplets have finite size and inertia, they do not follow the streamlines exactly and some, depending on their size and initial trajectories, will collide with the drop to give collision cross-sections that are greater than zero but generally smaller than the geometrical cross-section of the drop.

APPENDIX A

This situation is illustrated in Fig. A.1, where AB represents the trajectory of the centre of a droplet making just grazing contact with the drop. The effective collision cross-section of the drop of radius R for a droplet of radius r is πy_c^2, where the critical impact parameter, y_c, is the initial horizontal separation of the drop and droplet centres on a grazing trajectory. The ratio of πy_c^2 to the geometrical cross-section, πR^2, of the drop, viz.

$$E' = \pi y_c^2/\pi R^2 = y_c^2/R^2,$$

is sometimes called the collision cross-section and sometimes the collision efficiency. These terms are also applied to the parameter

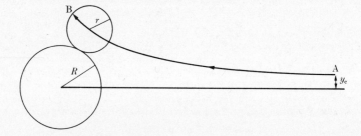

FIG. A.1. Diagrammatic representation of the trajectory of a small drop relative to a larger one.

$E = \pi y_c^2/\pi(R+r)^2 = y_c^2/(R+r)^2$ leading to some confusion in the literature. Numerical values of E', which depend on the radii R, r and on the viscosity and density of the air, may exceed unity when $R \simeq r$. We shall therefore define $E' = y_c^2/R^2$ as the *collision cross-section* of a drop of radius R for a droplet of radius r, and

$$E = y_c^2/(R+r)^2$$

as the *collision efficiency*. The latter quantity, being the ratio of the effective collision cross-section to the geometrical cross-section of the drop-droplet pair, is always <1. Of course, when $R \gg r$, the two parameters E' and E are practically identical. In what follows we shall mainly use the collision efficiency and convert all numerical values quoted from the literature to conform to the above definitions.

Calculation of the relative trajectory of two droplets, treated as rigid spheres, is carried out in a step-wise fashion. At each step the flow field round the two spheres is computed and the hydrodynamic force on each deduced. The drops are then allowed to move in response to the combined hydrodynamic and gravitational forces for a short interval of time after which their new positions are calculated.

The flow pattern round the drops in their new positions is then computed and the process repeated until the drops make their closest approach. Each trajectory is followed from a large initial vertical separation of the drops where the flow round each is not perturbed by the other. Trajectories for a series of impact parameters (initial horizontal separations) are computed until the grazing trajectory is defined to the required degree of accuracy, for example, collision may be deemed to occur when the gap between the drop surfaces becomes $<10^{-4} R$. The critical impact parameter y_c is thus determined and the collision efficiency $y_c^2/(R+r)^2$ follows directly.

The main difficulty lies in computing the air flow round two drops, or even one isolated drop, when both viscous and inertial terms in the equations of motion are taken fully into account. When the Reynolds number of the sphere is less than unity, Stokes's solution gives a good approximation to the flow very close to its surface, but is invalid at appreciable distances away. If the effect of the viscous forces is ignored, i.e. the fluid is regarded as an ideal fluid, we obtain a solution which corresponds to potential (ideal or aerodynamic) flow, which is valid at large distances from the surface of the sphere. No accurate solution has been found for intermediate distances. When the droplets are very much smaller than the collecting drop, i.e. $r \ll R$, it is legitimate to ignore their velocities relative to the air stream and to assume that they do not appreciably disturb the flow round the large drop. This is not permissible, however, when drop and droplet are of comparable size because their flow patterns will mutually interfere and more strongly as they move closer together. No complete and general solution of this complex two-body problem has yet been achieved, but approximate solutions for the limited case when both drop and droplet are small enough to obey Stokes's law are described later.

Langmuir (1948) was the first to make reliable calculations of the collision efficiencies for drops of radius R falling under gravity through a cloud of smaller droplets of radius r. Assuming the flow round the drop to be given by the potential flow solution, and the motion of the small droplets to be controlled by viscous forces, he found the collision efficiency E_p to be a function only of a dimensionless parameter

$$K = \frac{2\rho_L}{9\mu} \frac{r^2}{R} (V-v), \qquad (A.1)$$

where ρ_L is the density of water, μ the viscosity of air, and V, v the velocities of the drop and droplets respectively. K is a measure of the inertia of the smaller droplet, since for small values of K the trajectories show a large deviation, while for large values of K the deviation is small. Langmuir found that

$$E_p = 0 \quad \text{for} \quad K < 0.0833, \; E_p = K^2/(K+\tfrac{1}{2})^2 \quad \text{for} \quad K > 0.2. \quad (A.2)$$

If, however, a viscous flow regime was assumed for the drop, the collision efficiency became

$$E_v = 0 \quad \text{for} \quad K < 1.214, \quad \text{otherwise}$$
$$E_v = \{1 + (\tfrac{3}{4} \ln 2K)/(K - 1.214)\}^{-2} \quad (A.3)$$

The magnitude of E_p is always greater than that of E_v for the same value of K, as shown in Table A.1.

Langmuir suggested that the collision efficiencies for large raindrops should be computed from (A.2) but that the values obtained from (A.3) are more appropriate to droplets in the Stokes's law regime. For intermediate sizes he gave an interpolation formula

$$E_L = (E_v + E_p \, Re/60)/(1 + Re/60), \quad (A.4)$$

where Re is the Reynolds number of the large drop. Thus, for values of $Re \leqslant 60$, $E_L \simeq E_v$, and for $Re \geqslant 60$, $E_L \simeq E_p$. We shall call E_L the Langmuir collision efficiency, this being calculated on the assumption that the droplet could be treated as a point. But since the droplet has a finite radius, the true collision cross-section will depend not only on the parameter K, but also on the ratio r/R, and will be larger than the Langmuir value as indicated in Fig. A.1.

For the case of potential flow, Das (1950) computed the droplet trajectories taking the size of the droplet into account, and gave new values of collision efficiency for values of $r/R = 0.1, 0.2, 0.3$, and for values of $K \leqslant 2$, some of which are shown in Table A.1. More accurate calculations of droplet trajectories were made by Fonda and Herne (1957) with the aid of an electronic computer. Their values for E_p when $r/R = 0$ are in close agreement with those of Langmuir for values of $K > 0.5$. Their values of E_p with $r/R = 0.1$ are shown in Table A.1 and indicate that those of Das are a little too high. Fonda and Herne's values for viscous flow and $r/R = 0$ are in good agreement with those of Langmuir for values of $K > 5$. Values of E_v for finite values of r/R and $K < 1.214$ have been computed by Mason (unpublished) using the Fonda–Herne procedures, and specimen values are given in Table A.1.

The first attempt to calculate the collision efficiency for two drops of comparable size was made by Pearcy and Hill (1957), who used Oseen's approximate solution of the hydrodynamic equations to compute the velocity field around a water drop falling at terminal velocity. This solution, which possesses the correct asymptotic form at large distances from the sphere but fails close to it, predicts a

TABLE A.1

Values of collision efficiencies E† for potential and viscous flow as functions of the parameters K and r/R

	Potential flow E_p					Viscous flow E_v			
	$r/R = 0$		$r/R = 0.1$		$r/R = 0.2$	$r/R = 0$		$r/R = 0.1$	
K	L	F & H	D	F & H	D	L	F & H	F & H	M
0.0833	0	0	0.20		0.41	0	0		
0.1	0.01		0.22		0.42	0	0		
0.25	0.11	0.07	0.33	0.29	0.49	0	0		0.02
0.50	0.25	0.24	0.48	0.41	0.58	0	0		0.02
1.0	0.45	0.46	0.65	0.57	0.70	0	0		0.04
2.0	0.64	0.65	0.77	0.72	0.81	0.19	0.16	0.19	
3.5	0.76	0.76		0.80		0.38	0.35	0.36	
5.0	0.83	0.83		0.86		0.47	0.46	0.47	
10	0.91	0.91		0.93		0.63	0.64	0.65	
100	0.990	0.99		0.98		0.92	0.93	0.93	
200	0.995	0.995		0.98		0.96	0.96	0.96	

L = Langmuir; D = Das; F & H = Fonda and Herne; M = Mason.
† Note that the collision cross-section $E' = E(1+r/R)^2$.

parabolic wake for spheres of $Re > 1$ in which the velocities increase rapidly with increasing Re, the width of the wake being proportional to $(Re)^{-\frac{1}{2}}$. In estimating the relative motions of the two drops, Pearcy and Hill assumed that their individual flow fields could be superposed linearly to give the total flow. Collisions between droplets of $Re > 1$ and of nearly equal size could then conceivably occur in two different ways, both depending upon the flows within the wakes. Collision could occur *directly*, the large droplet (drop) falling and colliding with the smaller (droplet) as it becomes accelerated in the wake of the latter. Alternatively, the drop may pass the droplet so closely as to engage the latter within its wake and eventually make an indirect collision with it. In the case of direct collisions, Pearcy and

Hill deduced that the forces of attraction between the two drops increase rapidly with increasing radius of the droplet. Also, if drop and droplet are of nearly equal size, their velocity of approach is small, the drop spends a long time in the wake of the droplet, its trajectory is strongly affected, and consequently its collision cross-section may be many times the geometrical cross-section. Indeed their computations indicated that collision cross-sections greater than unity obtained for nearly equal-sized drops of $R > 13$ μm as a result of the attractive force exerted by the wake of the lower droplet.

The results of later computations and of laboratory measurements indicated that Pearcy and Hill's formulation of the problem, in which both the Oseen approximation and the linear superposition of the two flow fields would be least accurate when the drops are close together, is probably unrealistic, especially for small droplets of comparable size, but it represented an important step forward at the time and directed attention to the possibility of wake capture to which we shall return later.

Hocking (1959) was the first worker to allow fully for the mutual inference of the flows around two approaching drops. In the restricted case of droplets small enough ($R < 30$ μm) to obey the linear Stokes equations for steady viscous flow, he was able to obtain an exact solution by superimposing the flows for two rigid spheres moving along, and perpendicular to, their line of centres. By computing the drag forces on the two spheres resolved in these two directions, he followed trajectories from an initial vertical separation of 50 radii and obtained the grazing impact parameter, y_c, by trial and error. Hocking's results for collecting drops of $R = 19$, 20, 25, and 30 μm are shown in Fig. A.2. where the collision efficiency is plotted against the radius ratio r/R. The collision efficiencies are seen to be very sensitive to variations in R and r/R. Realizing that the accuracy of his computed forces decreased when the spheres are of very different sizes and close together, Hocking quoted no results for values of $r/R \leqslant 0 \cdot 2$ and intimated that the collision efficiencies given by the left-hand extrema of his curves should be regarded as tentative. His calculations also indicated that spheres of nearly equal size ($r/R \rightarrow 1$) tend to repel each other and so fail to collide. However, the most important deduction of Hocking's paper is that drops of $R < 19$ μm have zero collision efficiency for all smaller droplets. As we have seen in Chapter 3, this has important implications for the evolution of cloud-droplet spectra towards the precipitation stage.

APPENDIX A

Although Hocking's computations seem broadly correct and have received some experimental corroboration—see the next section—recent calculations by Davis and Sartor (1967) and Hocking and Jonas (1970) suggest that Hocking's original calculations contained

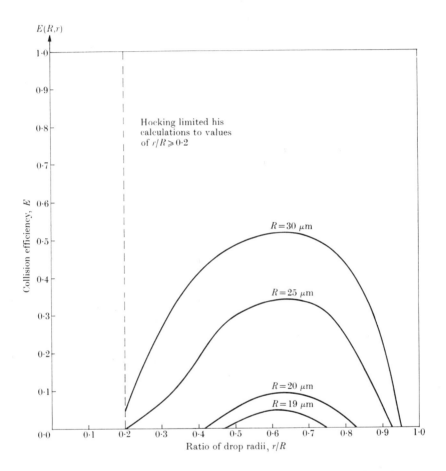

FIG. A.2. Hocking's computed values of collision efficiencies for small droplets of radius r and drops of radius R.

some inaccuracies and, in particular, that droplets of radius <19 μm do have finite but small collision efficiencies. Using a large computer, Davis and Sartor were able to retain more terms than Hocking in the expansions for the drag forces in inverse powers of the distance between droplet centres, to allow for the effects of droplet rotation, and so make more accurate calculations for dissimilar drops and close

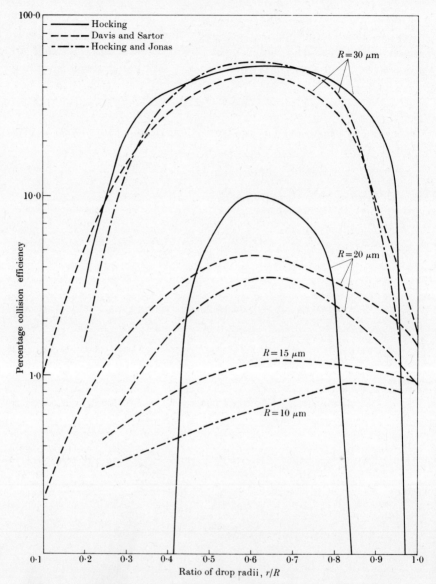

FIG. A.3. Computed values of collision efficiencies for small droplets of radius r and drops of radius R according to Hocking (1959), Davis and Sartor (1967), and Hocking and Jonas (1970).

separations. Their results are plotted in Fig. A.3. A method for determining the solution of the Stokes equations for two spheres moving arbitrarily was discovered by Dean and O'Neill (1963) and the values of the forces and couples for the transverse motion of unequal spheres have been found by O'Neill and Majumdar (1970) and Davis (1969).

These results, which enable the forces and couples on the droplets to be determined accurately for arbitrary speeds and sizes, and for arbitrary gaps even when these are small, have been combined with computations of the forces along the line of centres using the method of Stimson and Jeffrey (1926) to compute droplet trajectories and collision efficiencies by Hocking and Jonas (1970). An arbitrary assumption has to be made concerning the minimum separation of the droplet surfaces at which collision will occur because, when the gap is of the same size as the mean free path, λ, of the air molecules, the Stokes equations will no longer apply. Fig. A.3 shows the results obtained by Hocking and Jonas on the assumption that collision occurs when the gap narrows to 10^{-4} of the drop radius R. (The effect of increasing this to $10^{-3} R$ is shown in Fig. A.4; since these separations are $\ll \lambda \simeq 0.1\ \mu\text{m}$, the calculations probably underestimate the true collision efficiency.) These results are in good agreement with those of Davis and Sartor (1967) and, instead of predicting the sharp cut-off in collision efficiency for drops of $R < 19\ \mu\text{m}$ shown in Hocking's original paper, show that drops of radius as small as 10 μm have finite collision efficiencies although these are <1 per cent. It also appears that Hocking (1959) *overestimated* the collision efficiencies for drops of $R = 20\ \mu\text{m}$ and droplets of $r = 9$–$16\ \mu\text{m}$, the new values being only 2–3 per cent. The Hocking–Jonas calculations also indicate that the collision efficiency tends to a limiting value of ~ 0.01 independent of R as $r/R \to 1$. These modifications of the original Hocking calculations are interesting but lead to only very small changes in computed growth rates of droplets of $R = 10$–$30\ \mu\text{m}$.

Some valuable computations of collision efficiency for drop sizes intermediate between those represented by the work of Hocking and by Fonda and Herne have been made by Shafrir and Neiburger (1963). They used Jensen's (1959) method of computing, step by step, from the complete Navier–Stokes equations, the values for the stream function and vorticity around each drop treated separately and for values of $Re \leqslant 17$ ($R \leqslant 136\ \mu\text{m}$) above which a closed eddy appears in the wake. They then proceeded to calculate the drag force on each drop due to the isolated flow pattern of the other and hence the trajectories in the usual way. Their solution, which fails to satisfy the boundary condition at the drop surfaces, especially when they are close together, may well be inaccurate for drops of similar size but agrees well with the corresponding calculations of Langmuir and of Fonda and Herne for drops of $R = 60$–$150\ \mu\text{m}$ and values of $r/R \leqslant 0.2$. The results are plotted in Fig. A.5. In summary, the computations

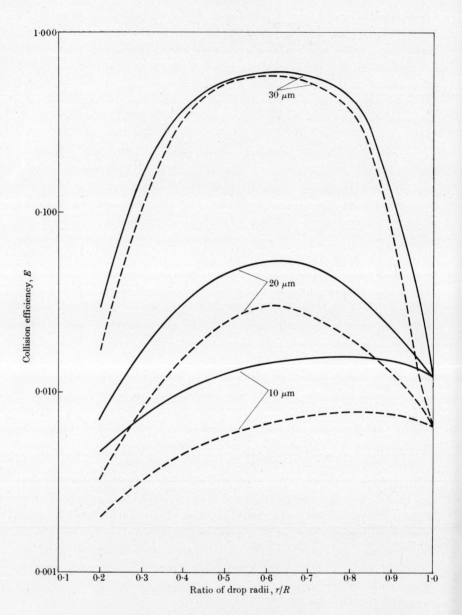

FIG. A.4. The effect on the calculated collision efficiency of varying the critical separation, δ, at which collision is assumed to occur. ----, $\delta = 10^{-4} R$; ———, $\delta = 10^{-3} R$. (From Hocking and Jonas (1970).)

APPENDIX A

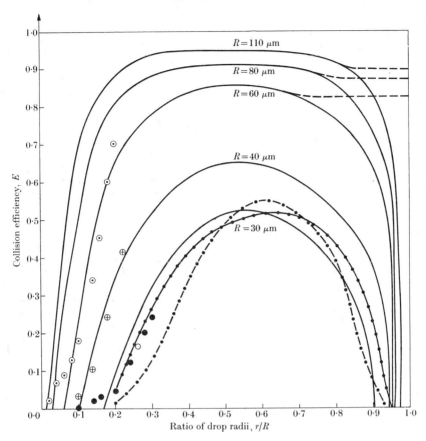

Fig. A.5. Computed and experimental values of collision efficiency for drops of radius $R = 30\text{–}110$ μm colliding with smaller droplets of radius r. Computed values: ———, Shafrir and Neiburger (1963), – – – –, values for $R = 60\text{–}110$ μm modified to allow for wake capture; —·—·—, Hocking (1959), —·—·—, Hocking and Jonas (1970). Experimental values: ●, Picknett (1968), Woods and Mason (1964), $R = 30\text{–}33$ μm; ⊕, Picknett, $R = 40$ μm; ○, Woods and Mason (1964), $R = 55$ μm.

described in this section provide reliable, consistent, and quite accurate values for the collision efficiencies of drops of radius greater than about 20 μm for smaller droplets. The specimen values shown in Table A.2, which is based on the results of Hocking and Jonas for $R \leqslant 30$ μm, of Shafrir and Neiburger for $30\ \mu\text{m} \leqslant R \leqslant 100\ \mu\text{m}$, and of Fonda and Herne for $R > 100$ μm, may be regarded as the best estimates available at present.

Table A.2

Collision efficiencies for drops of radius R colliding with droplets of radius r at 0° C and 900 mb

$R(\mu m)$	$r(\mu m)$							
	2	3	4	6	8	10	15	20
15	—	0.003	0.004	0.006	0.010	0.012	0.007	—
20	0.002	0.002	0.004	0.007	0.015	0.023	0.026	—
25	—	—	—	0.010	0.026	0.054	0.130	0.06
30	†	†	†	0.016	0.058	0.17	0.485	0.54
40	†	†	—	0.19	0.35	0.45	0.60	0.65
60	†	†	0.05	0.22	0.42	0.56	0.73	0.80
80	—	—	0.18	0.35	0.50	0.62	0.78	0.85
100	0.03	0.07	0.17	0.41	0.58	0.69	0.82	0.88
150	0.07	0.13	0.27	0.48	0.65	0.73	0.84	0.91
200	0.10	0.20	0.34	0.58	0.70	0.78	0.88	0.92
300	0.15	0.31	0.44	0.65	0.75	0.83	0.96	0.91
400	0.17	0.37	0.50	0.70	0.81	0.87	0.93	0.96
600	0.17	0.40	0.54	0.72	0.83	0.88	0.94	0.98
1000	0.15	0.37	0.52	0.74	0.82	0.88	0.94	0.98
1400	0.11	0.34	0.49	0.71	0.83	0.88	0.94	0.95
1800	0.08	0.29	0.45	0.68	0.80	0.86	0.96	0.94
2400	0.04	0.22	0.39	0.62	0.75	0.83	0.92	0.96
3000	0.02	0.16	0.33	0.55	0.71	0.81	0.90	0.94

†E takes values of order 1 per cent in this range but no accurate values have been computed.

A.2. The effect of electric fields and charges on collision efficiency

The collision efficiencies of cloud droplets will be influenced appreciably by the presence of electric fields or charges when the electrostatic forces between a pair of interacting droplets become comparable with the hydrodynamical and gravitational forces. The electrostatic forces between charged drops situated in an electric field have been calculated as a function of droplet size, separation, and orientation relative to the field by several workers using the equations derived by Davies (1964) for conducting spheres. The most important computations are those of Semonin and Plumlee (1966) and Hocking and Jonas (1970) who have taken into account the combined effects of hydrodynamical, gravitational, and electrostatic forces to compute trajectories and collision efficiencies of droplet pairs. Semonin and Plumlee showed that the collision efficiency of drops of $R > 30$ μm for smaller droplets was increased by the presence of electric fields or charges on the drops but that the effect was small for fields of <500 V cm^{-1}. It was thought that these effects would be greater for pairs of smaller, and more equal-sized, droplets. This is confirmed by the more accurate computations of Hocking and Jonas (1970) who extended their treatment described above to include the electrostatic forces. Their results, shown in Fig. A.6, indicate that fields in

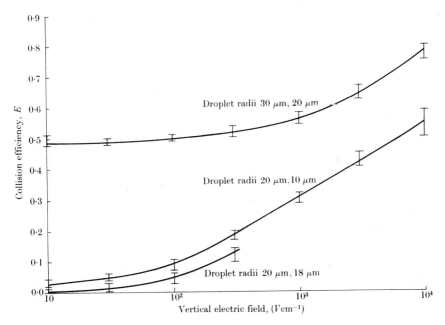

FIG. A.6. The effect of vertical electric fields on the collision efficiencies of freely-falling drops. (From Hocking and Jonas (1970).)

excess of 1000 V cm^{-1} are required to increase the collision efficiency of a 30–20 μm droplet pair by only 20 per cent, while that of a 20–10 μm pair attains a value of only 0·1 in a field of 100 V cm^{-1} and 0·3 in a field of 1000 V cm^{-1}.

Accordingly, we conclude that fields of 10–100 V cm^{-1} found in cumulus and shower clouds are likely to have very little effect on droplet growth by coalescence. Fields of order 1000 V cm^{-1}, capable of enhancing collision efficiencies, are found only in thunderstorms where the precipitation processes are already well under way.

A.3. Experimental studies of the collision and coalescence of water drops

Presently available laboratory techniques permit measurement of the collection efficiencies of water droplets, this parameter being the product of the collision efficiency and the coalescence efficiency.

Several workers, notably Sartor (1954), Schotland (1957), and Neiburger and Pruppacher (1965), have made some interesting model experiments in which the fall of cloud droplets in air was simulated by the fall of pairs of solid spheres or liquid drops through a viscous liquid. The requirements for dynamical similarity are that the Reynolds numbers, the radius ratio, and the ratio of the densities be matched, but the third condition is impossible to achieve in such experiments. The occurrence of a collision was determined from visual observation, changes in momentum, or charge transfer between the spheres. The results of these three studies show considerable differences which are probably a consequence of the difficulty of determining whether the interacting simulated droplets actually came into contact. For this reason, together with the inability to attain complete dynamical similarity and the fact that oil and air films may act differently as barriers to contact, these model experiments, although useful in revealing many aspects of the relative motion of interacting cloud droplets, cannot be regarded as providing independent reliable values of collision efficiencies.

Several experiments have been designed to measure the collection efficiencies of water droplets in air. Gunn and Hitschfeld (1951) allowed water drops of 1·6 mm radius to fall through a 3-metre column of cloud droplets produced by either condensation ($r = 2$–20 μm) or by atomization ($r = 7$–50 μm) and then into a receiving cup. The dropper and cup were weighed before and after the transit of 500 drops through the cloud. The liquid-water content of the clouds, which varied from 8 g m^{-3} to 20 g m^{-3}, was determined by sealing off part of the column, evaporating the water, determining the vapour density with a dew-point hygrometer, and subtracting the saturation vapour density at the initial temperature of the cloud. The measured mass increase per drop was, within the limits of experimental error, in agreement with Langmuir's calculated values of the collision efficiency and with the assumption that every collision resulted in coalescence. The presence of charges up to 0·2 e.s.u. per drop had no noticeable effect on the collection efficiency.

Kinzer and Cobb (1958) supported small water drops in a vertical air stream carrying a cloud of smaller droplets in a narrow tube of 8 mm diameter, and measured the growth rate of an individual drop by the changes in air velocity required to keep it stationary in the tube. From the growth rate and the liquid-water content of the cloud, the authors derived the collection efficiency of the drop at various stages of its growth. Unfortunately, droplet collisions were caused not only by gravitational settling but also by turbulence, and the drop growth was also affected by evaporation–condensation processes. Because of these factors, and undesirable velocity gradients and wall effects in the tube, these results cannot be regarded as reliable.

A direct experimental determination of the collection efficiencies of falling water drops for much smaller particles was performed by Starr and Mason (1966) by allowing water drops ranging from $\frac{1}{10}$ to 1 mm in radius to fall through airborne clouds of spores of *Lycoperdon* (mean radius $\bar{r} = 2\cdot25$ μm), spores of black rust ($\bar{r} = 2\cdot6$ μm), and grains of Paper Mulberry pollen ($\bar{r} = 6\cdot4$ μm), all of which are nearly spherical and uniform in size. After falling through the cloud contained in a long, vertical glass tube, the drops were allowed to fall without splashing on a specially-prepared surface and the number of captured particles in each counted under the microscope. For drops of $R > 400$ μm, the experimental results were in quite good agreement with calculations based on the assumption of potential flow around the drops. With both kinds of small spore the collection efficiencies reached maximum values for drops of $R = 400$ μm, and fell sharply for smaller drops. In the latter regime, where the potential flow solution is inappropriate, and the viscous flow solution predicts zero values, a sharp drop in collection efficiency is to be expected.

Carefully controlled experiments which permitted the measurement of collection efficiencies in the size range covered by the computations of Hocking have been performed by Picknett (1960) and by Woods and Mason (1964), using similar techniques. In a particular experiment, several thousand equal-sized water drops produced by a spinning disk or the vibrating-needle device of Mason, Jayaratne, and Woods (1963), were allowed to fall through a 1-m depth of cloud composed of saline droplets of radius 1–12 μm maintained at their equilibrium radii in an ambient humidity of 84 per cent, and then to fall onto a grease-coated slide. A drop landing on this slide evaporated, leaving behind a small salt particle if it had previously collected one of the cloud droplets. Recondensation upon the salt particle in a chamber which reproduced the temperature and humidity existing within the cloud produced a hemispherical saline droplet of volume equal to that originally captured. Since the probability that the collector drops had undergone more than one collision was negligible, it was possible to determine, by microscopic examination of the slides, the proportion of collector drops that had undergone coalescence and the fraction of these that had coalesced with each size of salt droplet. Having determined the total number and size of the collector drops introduced into the cloud, and also the sizes and number concentrations of the saline droplets, the collection efficiencies of drops of this one particular size for smaller droplets of different sizes could be evaluated.

These experiments confirmed Hocking's prediction that drops of $R < 18$ μm are unable to capture droplets of $r = 1$–12 μm or, at any rate, that the collection efficiencies were less than 5 per cent. This result is in accord with the computations of Hocking and Jonas (1970). Fig. A.5 shows that for drops of $R = 30$–60 μm, the experimental collection efficiencies agree, within the limits of experimental error, with the theoretical values of collision efficiency calculated by Hocking and by Shafrir and Neiburger, except for values of $r/R < 0.1$, where the experimental values of collection efficiency are consistently higher than the calculated ones and fail to show the sharp cut-off predicted by the earlier theoretical work but not by the more recent calculations of Davis and Sartor and Hocking and Jonas.

The prediction by Pearcey and Hill (1957) that collision cross-sections of pairs of similar sized drops may be many times the geometrical cross-section because of wake capture appeared to be confirmed by the experimental observations of Telford, Thorndike, and Bowen (1955), who took streak photographs of groups of interacting, nearly equal-sized drops of radius about 75 μm while they moved slowly upwards in a vertical wind tunnel. Coalescences produced drops of double mass which fell downwards. The number concentration, n, of original drops moving upwards and that, N, of the coalesced drops moving downwards were determined by photographing a known volume of the air stream. The relative velocities of pairs of neighbouring but not interacting drops were measured from streak photographs and the average relative velocity, Δv, for approaching pairs was calculated. This average value of Δv, together with values of N and n, were then substituted in the equation

$$NV \tfrac{1}{2} = n^2 E' \pi R^2 \, \Delta v H,$$

where V is the settling velocity of the coalesced drops, R the radius of the original drops, and H the height of the settling column, to give a computed value of the collision cross-section $E' = 12 \cdot 6 \pm 3 \cdot 4$ or a collision efficiency of $E = 3 \cdot 15 \pm 0 \cdot 85$ which would require the upper drop to be sucked into the wake of the lower one. This value of E is surprisingly high for drops of $Re < 6$ and the experiment has been criticized on the ground that collisions may not have resulted from gravitational settling but may have been influenced either by turbulence or horizontal velocity gradients in the wind tunnel. If this result were correct, it would provide strong support for the theory of Pearcey and Hill, but it has not been confirmed and is in direct contradiction to the experiments of Woods and Mason (1965) described below. Actually Telford and Thorndike (1961) used the same technique to study collisions between drops of $R = 10$–30 μm. Coalescence was occasionally observed to occur between nearly equal drops of radius about 22 μm but no values were given for the collision efficiency. When smaller drops, of $R = 15$–18 μm, came close together, they fell with increased velocity but later separated and coalescence was never observed. These results were interpreted by their authors as being in accord with the calculations of Hocking (1959) but contrary to those of Pearcey and Hill; they are also consistent with the recent experiments of Woods and Mason (1965), who developed a technique of determining the collision efficiencies entirely from streak photographs such as that shown in Fig. A.7.

Many thousands of photographs of streams of drops of $R < 35$ μm failed to show a single coalescence event. With drops of $R = 35$–40 μm coalescence was occasionally observed, which may have been a result of small-scale eddying in the air stream. Drops of radius 40–100 μm had well-developed wakes, and coalescences between drops of equal size were relatively frequent. The corresponding collection efficiencies increased from 0·5 to 0·9 with increasing drop radius in this range, but there was no evidence of one drop being attracted laterally into the wake of the other. Because the initial vertical separation of their drops was only about ten radii, Woods and Mason conceded that larger collision efficiencies such as those reported by Telford *et al.* might result from weak interactions over larger distances for longer times, but thought it more likely that the relative motion of the interacting drops would have been influenced at large separations by random eddies in the air stream. Although these experimental contradictions are not fully resolved, the excellent agreement in Table A.3 between

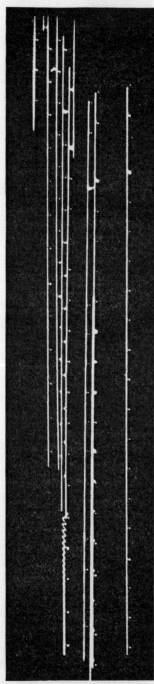

FIG. A.7. A streak photograph showing the coalescence of two drops of 62-μm radius. (From Woods and Mason (1965).)

Table A.3

Comparison of computed and experimentally determined collision efficiencies for pairs of nearly-equal drops falling in air

$R(\mu m)$	$r(\mu m)$	E†(expt)	E‡(computed)
113	90	0·9	0·89
70	70	0·9	0·90
68	47	0·9	0·90
64	64	0·85	0·92
60	47	0·8	0·81
50	30	0·5	0·47
40	40	0·5	0·41

† Woods and Mason (1965). ‡ Neiburger (1967).

the calculations of Neiburger (1967) and the measurements of Woods and Mason for drops of $R > 50$ μm suggest that the collision efficiencies of drops of nearly equal size are close to unity for radii greater than about 60 μm, as indicated by the dashed lines in Fig. A.5, and tend towards zero for radii <40 μm. The exact numerical values are of little practical importance in cloud physics because collisions between nearly equal-sized drops of $R > 40$ μm are likely to occur only very rarely in non-precipitating clouds and therefore contribute very little to the development of precipitation elements.

A.4. The mechanism of coalescence

It is very difficult to study the final stage of the approach of two colliding droplets, culminating in their coalescence, because of the minute and inaccessible volume and short time intervals in which the crucial processes occur. If the droplets come into close proximity the barrier to coalescence is presented by the air film between the droplet surfaces, which has to be expelled by the action of dynamical, gravitational, or electrical forces working against the viscous resistance of the air. Schotland (1960) studied the coalescence of electrically neutral water drops, of diameter ranging from 200 to 800 μm, falling on large hemispherical water surfaces. He found that the minimum height from which the drops had to fall in order to coalesce with the target was greater when the drops impinged at smaller angles of incidence and when the density of the surrounding gas was increased. A much more comprehensive investigation of this problem was carried out by Jayaratne and Mason (1964) who studied the impaction, bouncing, and coalescence of small water drops of radius ranging from 60 to 200 μm on plane and curved air–water interfaces. The influence of drop radius r, impact velocity V, angle of impact θ, drop charge q, and vertical field strength F upon the probability of coalescence was measured over a wide range of values. For instance, Fig. A.8 presents the measured relationships between drop radius and the

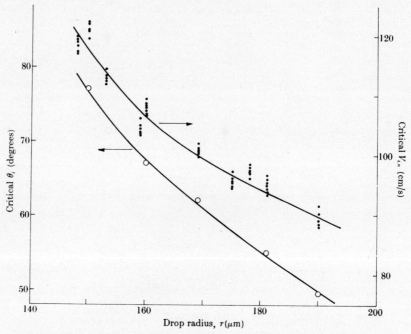

FIG. A.8. Variation with the drop radius of the critical values of the normal component of incident velocity $V_{i,n}$ and the impact angle θ_i at which an uncharged drop coalesces with a plane water surface. (From Jayaratne and Mason (1964).)

critical values of V and θ at which coalescence occurs between uncharged droplets and a plane water surface. Uncharged drops impacting at nearly normal incidence were found to remain in contact with the surface for about 1 ms, to lose about 95 per cent of their kinetic energy during impact, and to rebound with an effective coefficient of restitution of about 0·2. Drops carrying a net charge and drops polarized in an applied electric field were found to coalesce more readily than uncharged drops of the same size and impact velocity. Accurate measurements were made of the magnitudes of the critical values of q and F required to cause coalescence as a function of

V, θ, and r. Typically, droplets of radius 150 μm impacting at 100 cm s^{-1} coalesce if the charge exceeds about 10^{-4} e.s.u. or if the field exceeds about 100 v cm^{-1}.

If an impinging drop is to coalesce with a water surface it must first expel and rupture the intervening air film. By treating the undersurface of the drop as a flattened circular disk, Jayaratne and Mason calculated that the minimum thickness, δ, achieved by the film during the period of contact is related to V, θ, and r by the equation

$$\delta = \left(\frac{3\mu}{8\sigma}\frac{\phi}{1+\epsilon}\right)^{\frac{1}{2}}(V\sin\theta)^{\frac{1}{2}}r, \qquad (A.5)$$

where ϵ is the coefficient of restitution, $\phi = \frac{15}{32}$, σ is the surface tension, and μ is the viscosity of air. This equation predicts values of $\delta \sim 0\cdot 1$ μm, below which fusion may well take place under the influence of van der Waals forces. Several features of the observed relations between V, θ, and r are accounted for by this simplified theory, but the behaviour of drops impacting at nearly glancing incidence, and of relatively large energetic drops impacting normally, is not. Jayaratne and Mason showed that the electrical forces produced between the liquid surfaces by fields or charges at the moment of coalescence were much smaller than the dynamical forces produced by deceleration of the impinging drop. They concluded, therefore, that it is difficult to understand why charging or polarization should promote coalescence unless the electrical forces distort the liquid surfaces in such a way that drainage of the air film and the formation of a liquid bridge are facilitated. They pointed out that such a distortion has been observed by Allan and Mason (1962) with two water drops of radius 750 μm approaching each other in silicone oil under an external field of 1000 V cm^{-1}, (see also Fig. A.9(b)), and they suggested that distortion may not only cause more rapid thinning of the air film but may also induce its rupture by means of a micro-discharge.

Subsequent work has confirmed these suggestions concerning the role of electrical forces in inducing coalescence, and it has provided a more detailed physical explanation of this phenomenon. Photographic evidence for the deformation of the adjacent surfaces of colliding water drops carrying high charges of opposite sign, has been obtained by Sartor (1967). In addition, the calculations of Latham and Roxburgh (1966) of the localized fields near the surfaces of closely separated drops situated in an external field demonstrate that distortion followed by disintegration will occur at close separations even if the external forces are extremely weak, owing to the mutual interaction of the polarization charges. As has been demonstrated by Miller, Shelden, and Atkinson (1965), the onset of instability of the droplets in this situation is generally accompanied by the ejection of a fine jet of water, which travels along the field lines to strike the second droplet. The penetration of the air film by the water filament may therefore promote coalescence of the droplets. This argument is consistent with the observations of several workers that the delay time between apparent collision and coalescence of two droplets decreases with increasing external field strength.

Telford and Thorndike (1961) observed in a wind-tunnel experiment that while the collection efficiency of nearly equal-sized uncharged droplets of $R = 16$ μm was zero in the absence of an electric field, the number of coalescences was observed to increase markedly in the presence of applied horizontal fields exceeding 1000 V cm^{-1}. Woods (1965), by counting the number of coalescence events in thousands of streak photographs, found that the coalescence rate of identically-sized and oppositely-charged droplets of $R < 40$ μm increased linearly with drop charge above a threshold value of 5×10^{-5} e.s.u. For drops $R \geqslant 40$ μm, which coalesce spontaneously by wake attraction even when uncharged, the coalescence rate increased linearly with the magnitude of the (opposite) charges. When the drops all carried charges of the same sign, no coalescence events were recorded for drops of $R < 40$ μm, and for $R = 40$ μm the coalescence rate was decreased by a factor of 5 (compared with the uncharged rate) when they carried charges of 2×10^{-4} e.s.u.

The fact that the number of coalescences between drops of $R < 40$ μm remained constant until the charge exceeded the threshold value suggests that a further increase in charge increased the collision rather than the coalescence efficiency. This was also indicated by the curvature of the trajectories on some of the photographs, and by the fact that the vertical electric field required to produce electrostatic forces equivalent to those of the threshold charge of 5×10^{-5} e.s.u. on drops of radius 30 μm would be 1000 V cm^{-1}, and just such fields are required to produce a significant increase in collision efficiency according to Fig. A.6.

Since, in non-precipitating cumulus clouds, the charges on droplets are typically only 10^{-8} e.s.u. and the fields only about 10 V cm^{-1}, these cannot have a significant effect on collection efficiencies and

Fig. A.9(a). Two streams of droplets AA' and BB' interacting to produce larger drops by coalescence at C and C'. The smaller drops are about 50 μm radius. (From Mason (1964).)

Fig. A.9(b). The distensions produced by the enhanced field between the near surfaces of two 374-μm radius drops in an ambient electric field parallel to the line of centres of 6400 V cm^{-1}. (By courtesy of Mr. J. D. Sartor.)

hence the early stages of droplet growth. The fact that experimentally-determined *collection* efficiencies of freely-falling drops of $R > 30$ μm for smaller droplets agree quite well with theoretically derived *collision* efficiencies suggests that the great majority of collisions are followed by coalescence to give coalescence efficiencies of practically unity. Further evidence of this is provided by photographs such as Fig. A. 9(a), which record the impaction of droplets of $R = 50$–150 μm at velocities of <1 m s^{-1} with only very rare instances of the drops bouncing apart after collision.

APPENDIX B

The Physical Properties of Freely Falling Raindrops

B.1. The terminal velocities of water drops falling in still air

B.1.1. *Computations of terminal velocity*

A SPHERE of radius r, mass m, moving with constant velocity v through a fluid of density ρ' and viscosity μ experiences a resistance (drag force)

$$F_D = \pi r^2 \tfrac{1}{2} C_D \rho' v^2, \tag{B.1}$$

where C_D is known as the drag coefficient and is a function of the Reynolds number

$$Re = 2vr\rho'/\mu \tag{B.2}$$

and is defined by the equation

$$C_D Re/24 = F_D/6\pi\mu rv, \tag{B.3}$$

where the term $6\pi\mu rv$ represents the viscous forces acting on the sphere. For a sphere falling at its terminal velocity under gravity, the drag force F_D is equal to the weight of the sphere (actually its weight minus the upthrust of the fluid). In the case of a water drop falling through air, the upthrust of the air may be neglected and we have

$$\frac{C_D Re}{24} = \frac{\tfrac{4}{3}\pi r^3 \rho g}{6\pi\mu rv} \tag{B.4a}$$

or

$$\frac{C_D Re^2}{24} = \frac{4}{9\mu^2} r^3 \rho' \rho g, \tag{B.4b}$$

where ρ is the density of the drop.

In the special case where the drag force on the sphere is balanced by the viscous forces we have

$$F_D = 6\pi\mu rv = \tfrac{4}{3}\pi r^3 \rho g$$

whence

$$C_D = 24/Re$$

and

$$v = 2\rho g r^2/9\mu = 1\cdot 2 \times 10^6 r^2 \text{ (at 1013 mb, 20°C)}. \tag{B.5}$$

Equation (B.5) represents Stokes's law and is accurate to within about 2 per cent for water droplets of $Re \leqslant 0\cdot 1$ (radius $\leqslant 20$ μm at s.t.p.). For drops of $Re = 1$ (radius 40 μm) the actual velocity is already 10 per cent smaller than the corresponding Stokes velocity.

Drag coefficients of rigid spheres and hence relations between $C_D Re$ and Re have been determined experimentally by a number of workers but, as Pruppacher and Steinberger (1968) point out, the relationships quoted in the literature show considerable variation, and therefore uncertainty, especially at low Reynolds numbers. This induced Pruppacher and Steinberger to measure the terminal velocities of solid spheres falling in a tank of oil with an accuracy of better than $\pm 0 \cdot 1$ per cent and hence to determine the drag on solid spheres of $0 \cdot 2 < Re < 200$. Beard and Pruppacher (1969) used a well-designed and controlled wind tunnel to determine the drag on small water drops falling at terminal velocity in saturated air. Both sets of results are in close agreement, showing that water drops behave aerodynamically as rigid spheres for $Re < 200$, and are well represented by the following empirical formulae:

$$C_D Re/24 = 1 + 0 \cdot 10 \, Re^{0 \cdot 955}, \, 0 \cdot 01 \leqslant Re \leqslant 2, \qquad \text{(B.6a)}$$

$$= 1 + 0 \cdot 11 \, Re^{0 \cdot 81}, \, 2 \leqslant Re \leqslant 21, \qquad \text{(B.6b)}$$

$$= 1 + 0 \cdot 189 \, Re^{0 \cdot 632}, \, 21 \leqslant R \leqslant 200. \qquad \text{(B.6c)}$$

Moreover these empirical results agree within a few per cent with numerical computations of the drag on spheres of $0 \cdot 01 < Re < 400$ made by Le Clair, Hamielec, and Pruppacher (1970) using the steady-state Navier–Stokes equations.

A graph of $C_D Re^2/24$ against Re plotted on the basis of (B.6) may be used together with eqn (B.2) to compute the terminal velocity of a sphere (drop) of given radius falling in a specified environment. Thus once ρ, ρ', μ, and r are specified, the value of $C_D Re^2/24$ may be calculated from eqn (B.4b), the corresponding value of Re read off from the graph and then the terminal velocity appropriate to that value of Re calculated from eqn (B.2). Beard and Pruppacher used this procedure to calculate the terminal velocities of water drops of $r = 10\text{–}475$ μm in water-saturated air at pressures of 400, 500, 700, and 1013 mb and corresponding temperatures of -16, -8, 14, and $20°$ C. Specimen values for 1013 mb, $20°$ C are given in Table B.1. These are probably the most reliable data on the terminal velocities of drops in this range of Reynolds numbers; they are a little lower than the corresponding velocities measured by Gunn and Kinzer (1949), probably because the latter's small drops suffered some evaporation

TABLE B.1

Terminal velocities of water drops in still air, pressure 1013 *mb, temperature* 20° C

Drop diameter (mm)	Terminal velocity (cm s^{-1})	Re†	Drop diameter (mm)	Terminal velocity (cm s^{-1})	Re†
0·01	0·3		1·80	609	731
0·02	1·2	0·015	2·0	649	866
0·03	2·6		2·2	690	1013
0·04	4·7	0·12	2·4	727	1164
0·05	7·2	0·24	2·6	757	1313
0·06	10·3	0·41	2·8	782	1461
0·08	17·5	0·93	3·0	806	1613
0·10	25·6	1·69	3·2	826	1764
0·12	34·5	2·74	3·4	844	1915
0·16	52·5	5·55	3·6	860	2066
0·20	71	9·4	3·8	872	2211
0·30	115	22·8	4·0	883	2357
0·40	160	42·3	4·2	892	2500
0·50	204	67·5	4·4	898	2636
0·60	246	97·5	4·6	903	2772
0·70	286	132	4·8	907	2905
0·80	325	172	5·0	909	3033
0·90	366	218	5·2	912	3164
1·00	403	267	5·4	914	3293
1·20	464	372	5·6	916	3423
1·40	517	483	5·8	917	3549
1·60	565	603			

† Calculated values. For $D < 1·0$ mm, the terminal velocities are based on the measurements of Beard and Pruppacher (1969); for larger drops, the terminal velocities are those of Gunn and Kinzer (1949).

in falling through air of only 50 per cent relative humidity before they were collected and measured for size. For drops of $Re > 400$, the measurements of Gunn and Kinzer are probably still the best; certainly drops larger than this become distorted from the spherical and their drag coefficients become appreciably larger than those of the corresponding rigid sphere and their terminal velocities become correspondingly less.

B.1.2. *Measurements of the terminal velocities of raindrops*

The falling velocities of water drops of raindrop size were first measured by Lenard (1904) by suspending them in a vertical air stream and measuring the air velocity at the point of suspension.

The size of the drops was determined by catching them on a sheet of absorbent paper as they slipped out of the air stream and measuring the diameters of the spots.

Schmidt (1909) measured the fall velocities of raindrops varying in diameter from 0·4 to 3·5 mm with an ingenious apparatus consisting of two disks mounted on an axle and rotated at a known rate. A raindrop, which by chance fell through a small sector cut in the upper disk, would land upon a piece of absorbent paper fixed on the lower disk and turning with it. The location of the spot relative to the projection of the sector on the paper gave a measure of the velocity, while the diameter of the spot gave a measure of the drop size.

But only in more recent years, with the aid of modern electronic and optical techniques, have really accurate measurements become possible. The results of Laws (1941) and of Gunn and Kinzer (1949), which are in quite close agreement, show that the fall velocities given by both Lenard and Schmidt are about 15 per cent too low.

Laws measured the velocities of water drops with diameters between 1 and 6 mm falling in still air from heights of 0·5 to 20 m, and also made some measurements on natural raindrops. The drops, formed at the tip of a capillary tube, were allowed to fall through an optical system employing dark-field illumination, the light scattered by each drop being collected by a lens and focused on to the plate of a still camera in front of which rotated a 'chopper disk' at constant speed. The image of the falling drop therefore appeared on the plate as a line broken at regular intervals during which the light was interrupted by the disk. The duration of these intervals was determined from the geometry of the disk and its speed of rotation, and the actual distance travelled by the drop during an interval was found by photographing a calibrated scale. Thus the fall velocity of the drop could be calculated, its mass being ascertained by collecting a large number of them for weighing. By varying the height of the capillary tube above the camera, it was found that drops of up to 6 mm diameter reached 95 per cent of their terminal velocities after falling less than 8 m, and for all practical purposes could be said to reach their full terminal velocities after falling 20 m. Having made a number of separate determinations for each size of drop, Laws plotted the average values of fall velocity against drop diameter; the values given in Table B.1 were read off from the curves of best fit and are appropriate to a pressure of 1013 mb and a temperature of 20° C. Law's few measurements on raindrops indicated that their terminal velocities were slightly lower than drops

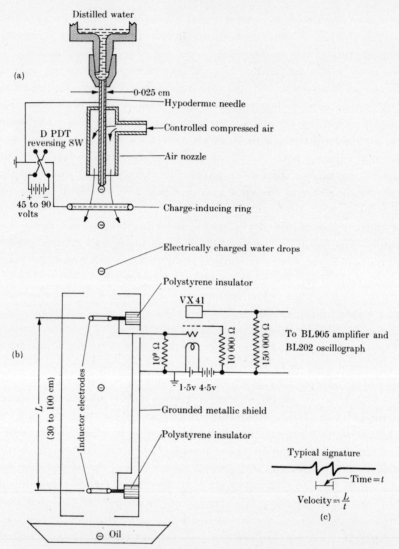

Fig. B.1.(a) Apparatus for producing electrically charged water drops of selected size; (b) arrangement of induction rings and circuits for measuring fall velocities of drops; (c) the signal produced by a falling drop. (From Gunn and Kinzer (1949).)

of the same size falling from 20 m in the still air of the laboratory; he attributed this to the presence of small-scale turbulent motions in the atmosphere.

The most extensive, and probably the most accurate, set of measurements have been made by Gunn and Kinzer (1949) covering a range of drop diameters from 8 μm to 5·8 mm. The apparatus is shown in Fig.

B.1. The drops were detached from a carefully ground hypodermic needle by a downwardly directed axial flow of air, the size of the drops being controlled by varying the velocity of the air stream. They were then allowed to fall through a metal-ring electrode and so acquired an electrostatic charge. Thereafter they fell into an earthed, vertical metal tube and passed successively through two insulated induction rings spaced about 1 m apart and connected to the grid of a valve electrometer. On passing through each ring, the charged drop produced a voltage pulse on the grid of the valve, the time interval between the two events being measured upon a moving-tape oscillograph. The average velocity of fall between the two electrodes was then calculated from the spacing of the rings and this measured time. The separation of the rings was reduced for the smaller droplets in order to reduce the evaporation to negligible proportions. The largest drops were allowed to fall about 20 m before their velocity was measured. The masses (and hence the diameters) of the larger drops were determined by weighing, but drops less than 2 mm across were caught in a shallow dish of oil and their diameters measured under the microscope. The results of these experiments, which were carried out at 760 torr pressure, temperature 20° C, and relative humidity 50 per cent, are given in Table B.1. Each value represents the average of at least fifty separate determinations, an error of less than 1 per cent being claimed for drops of mass between 10 and 10^5 μg. For smaller droplets, the errors were larger but were not fully evaluated.

It will be seen that Gunn and Kinzer's values are slightly lower than those of Laws at both ends of the latter's curve, the discrepancy being about 1·5 per cent for the largest drops.

It would be convenient to have a formula for the terminal velocity of a drop of any size falling through air of any density, but there appears to be little hope of deriving a reasonably simple expression which would take account of all the factors involved, for example shape, circulation effects, etc. However Davies (1942), on the basis of his measurements of terminal velocities of drops of diameter 3·38–7·25 mm in air of different temperatures and pressures, has given the following empirical equation to predict (to within 3 per cent) the terminal velocities of drops of any liquid in air at pressures down to half an atmosphere within the limits $0·4 < 4r^2\rho g/\sigma < 1·4$, when circulation effects are negligible:

$$\log Re = 2·655 \sqrt{\left(\frac{\log C_D Re^2 - f_1}{f_2}\right)} - f_3, \qquad (B.7)$$

where
$$f_1 = 0.460 + 1.012 \log X + 0.225 (\log X)^2,$$
$$f_2 = 0.933 - 0.167 \log X,$$
$$f_3 = 3.306 + 0.118 \log X + 0.0765 (\log X)^2,$$
$$X = 4r^2 \rho g/\sigma,$$
$r = $ drop radius,

ρ, σ are respectively the density and surface tension of the liquid.

Foote and du Toit (1969) have used Davies's original data to obtain the following empirical formula for the terminal velocity of water drops in air of density ρ' and temperature T in terms of the value v_0 in air at pressure 1013 mb and temperature 20° C:
$$v = v_0 10^\gamma \{1 + 0.0023(1.1 - \rho'/\rho_0')(T_0 - T)\}, \tag{B.8}$$
where
$$\gamma = 0.43 \log (\rho_0'/\rho') - 0.4 \{\log (\rho_0'/\rho)\}^{2.5}.$$

The predictions of eqn (B.8) using the Gunn–Kinzer data for v_0 are always with 2·5 per cent of Davies's data over the size range of his observations (radius 1·69–2·98 mm), most of the discrepancy being explained by the 2 per cent difference in the two sets of data for v_0.

B.2. The shape and disintegration of large raindrops

Water drops exist as mechanically stable systems because the surface forces at the water–air interface continually try to minimize the interfacial area. When the surface tension forces are predominant, as in the case of cloud droplets, drizzle, and even small raindrops, the shape is one of minimum surface-to-volume ratio, i.e. a sphere. When however, factors other than surface energy contribute significantly to the total energy of the drop, minimum total energy may become inconsistent with a spherical shape.

Lenard (1904), while studying the fall velocities, deformation, and break-up of water drops suspended in a vertical air stream, was the first to notice that large drops were deformed considerably from the spherical, and suggested that the deformation might be due to centrifugal distortion set up by internal circulations within the drop, these being established by the frictional drag of the air at the drop surface. The existence of these internal circulations has been definitely established for drops falling through another liquid, but until recently, evidence on the internal circulation of drops falling in a gas was contradictory. Blanchard (1949a) reported no evidence of internal circulation in water drops suspended in a vertical wind tunnel, while Garner and Lane (1959) observed circulation velocities of 7 to 18 cm s^{-1} in water drops of 2–3 mm radius.

McDonald (1954), convinced that internal circulations were likely to be of negligible importance in controlling drop shape, argued that the aerodynamic pressures and the surface-tension pressure increments combine to produce an internal pressure pattern which satisfies the hydrostatic equation. As the surface-tension and hydrostatic pressure differences at any point on the surface of the drop may be computed if the surface geometry is known, it remains only to specify the fall velocity of the drop in order to calculate the aerodynamic pressure profile along a meridian. McDonald's computed pressure profiles were, at comparable Reynolds numbers, in good agreement with the pressure profiles around spheres experimentally determined by Fage (1937), and the drag forces computed from the aerodynamic pressures compared well with the weight of the drops. Despite the convincing arguments of McDonald, it remains to be checked whether the four drop photographs taken from Magono (1954) on which he based his arguments are representative for the drop shape, and whether his assumption that the internal circulation does not affect the drop shape is correct. In the latter connexion, Foote (1969) demonstrates that, contrary to the conclusions of McDonald, internal circulations of order 10 cm s^{-1} near the equator of the drop should significantly affect its shape in the sense that the curvature in the vicinity of the drop's equator is decreased.

Very recently, Pruppacher and Beard (1970) have studied the internal circulation and shape of water drops falling at terminal velocity by means of a wind tunnel. Drops of equivalent radius $R < 140$ μm showed no detectable deformation from the spherical shape. Drops of 140 μm $< R < 500$ μm were slightly deformed into oblate spheroids. The deformation of drops of 0.5 mm $< R < 4.5$ mm was linearly related to the drop-size as predicted by the semi-empirical calculations of Savic (1953). By means of a tracer technique it was established that water drops at terminal velocity in air have a well-developed internal circulation, the surface velocity at the equator of a drop being about 1 per cent of the terminal velocity. Nevertheless, because of the good agreement between their experimentally determined drop deformations and those predicted by Savic's theory,

Pruppacher and Beard conclude that the internal circulation has no significant effect on the shape of a drop falling in air.

There is little information on the shape of natural, freely-falling raindrops. Jones (1959), using two 35-mm cameras synchronized to take simultaneous flash photographs from two different angles, recorded about 2000 drops and deduced that large drops oscillate about a preferred oblate shape, the mean axial ratios ranging from 0·98 for drops of 2 mm diameter to as small as 0·60 for drops of diameter 5–6 mm.

Since drops of diameter greater than 6 mm are rarely observed in natural rain, it may be assumed that drops larger than this usually break up during their journey towards the ground. Lenard (1904) made the first observations on the disruption of large water drops supported by a vertical air stream. While drops of diameter 4·5 mm sometimes floated for several seconds without breaking up, drops exceeding 5·4 mm diameter were almost invariably disrupted on entering the upward current. The break-up was irregular, the number and size of the fragments varying greatly from drop to drop, but mostly there was one large drop ($d = 1\cdot5$–$3\cdot5$ mm) together with a number of smaller ones of diameter 1 mm and much less. Occasionally, a large drop, having floated for some seconds in a particularly smooth air stream, was seen to break up into a circular ring of 7 to 9 equally-spaced smaller drops of uniform size. This phenomenon has been explained in terms of the drop taking the form of a vortex ring and disintegrating when it develops oscillations of amplitude sufficient to overcome the surface-tension restoring forces.

The break-up of freely-falling water drops has been studied by C. N. Davies (unpublished), who carried out a few experiments with drops of distilled water released from heights of 30 and 40 ft in still air. He found that, up to a critical size, all the drops reached the ground intact but, quite suddenly, at a critical size, some of the drops disintegrated in the last ten feet or so of their fall. Merrington and Richardson (1947), who extended this work to release heights of 125 ft, reported that the drops either broke into 2 or 3 large fragments or into a drop of moderate size surrounded by a chaplet of much smaller drops. Fournier d'Albe and Hidayetulla (1955) reported on the break-up of drops of distilled water released from heights of 20 m in the open air when the wind speed was generally less than 5 knots. Under these conditions the smallest drop observed to disintegrate had a diameter of about 8 mm. Drops exceeding 9 mm diameter usually broke into a number of fragments which tended to increase as the size of the parent drop was increased from 9 to 12·5 mm, and which varied between 3 and 97. The larger drops tended to give, on break-up, a few drops of 6–8 mm diameter accompanied by about 30 small drops, mainly of diameter 1–3 mm.

The deformation and disintegration of large water drops have been studied by means of high-speed photography by Lane (1951) with drops exposed to both steady and transient air streams, and by Matthews and Mason (1964) who photographed drops in free fall. They confirmed the observations of earlier workers that such a drop becomes distorted and flattened, and, if its velocity relative to the air exceeds a critical value, eventually blows up to form a large bag supported on a toroidal ring of liquid. Finally, the bag bursts to produce a fine spray of droplets and the toroid breaks up into several large drops as shown in Fig. 9.12. The critical conditions for break-up were calculated by Matthews and Mason on the assumption that on the point of instability the drop possessed a hemispherical profile with its lower surface plane, and that, momentarily, a balance existed between the aerodynamic pressure and the total pressure difference set up across the vertical extremities of the drop by surface tension and hydrostatic forces. It was demonstrated that for drops of equivalent spherical diameter $d \leqslant 8$ mm, the surface tension forces dominated the hydrostatic forces and that the critical velocity, U_c, for break-up is given by

$$U_c^2 d = 8\sigma/n\rho' C_D, \qquad (B.9)$$

where σ is the surface tension of the liquid, ρ' the air density, C_D the drag coefficient of the drop, and n a numerical factor of between 1 and 2 expressing the ratio of the curvature of the drop at the upper pole to its equivalent spherical radius. This accords quite well with the experimental result that the critical velocity for break-up is given by

$$U_c^2 d = \text{const.} = 6 \times 10^5 (\text{c.g.s. units}) \qquad (B.9a)$$

so that drops of $d = 6$ mm disintegrate if their velocity relative to the air exceeds 10 m s^{-1}.

The critical drop size for break-up, and possibly also the number and size distribution of the fragments, appear to depend also on the turbulence regime of the air through which the drop falls. The exact manner in which natural raindrops disintegrate may therefore be rather different from those observed in the laboratory and, indeed, may vary under different conditions, but there is little reason to doubt that raindrops, soon after attaining a diameter of about 5 mm, tend to break up into a few large and a greater number of small fragments.

B.3. The size distribution of raindrops

B.3.1. *Experimental techniques*

A good deal of effort has been devoted in the last half-century to the measurement of raindrop sizes. At first sight this might appear to be a fairly simple experimental problem, but the development of a reliable, accurate, and convenient method, capable of continuously analysing a true, representative sample of the rain has proved difficult.

A very simple method was developed by Wiesner (1895) and, with minor modifications, has been used subsequently by many workers. An absorbent paper, stretched across a frame, is dusted with a water-soluble dye (e.g. eosin, rhodamine, methylene blue), the size of the impinging raindrops being determined from the diameters of the coloured stains they produce. In order to obtain consistent results, it is important to use paper of standard thickness, texture, and absorptive properties, and to control the humidity of the paper during storage. Whatman No. 1 filter paper appears to be very satisfactory and is now generally used for this purpose. For drops impacting at terminal velocity, Best (1950) and Andrews (1961) find that the stain diameter increases almost as the $\frac{5}{4}$ power of the drop diameter, drops of diameter 1 mm and 5 mm producing stains of 4 mm and 35 mm respectively. The method has the merit of simplicity and cheapness, but has the disadvantages that large drops hitting the paper at high velocity may splatter, while drops arriving with a tangential component of velocity (as in a strong wind), produce non-circular stains. Moreover, measurement of large numbers of stains for classification into size groups is a very laborious and tedious procedure. The method has been made continuously recording by Blanchard (1953) and Sivaramakrishnan (1961) using a moving tape of filter paper.

To minimize the splashing which occurs when a drop hits a solid, hard surface, Blanchard (1949b) experimented with screens of fine wire mesh. He found that drops pass quite easily through 50–100 gauge wire mesh without flattening or spreading out laterally to the degree that they do when hitting a hard impenetrable surface. When such screens were coated with soot, the drops passed through to leave impressions of diameter very nearly proportional to the drop diameter. Experiments have also been made with screens of nylon mesh treated with a benzine-lanolin solution and then covered with powdered sugar.

A rather ingenious, though at first sight unattractive, technique was devised by Bentley (1904) and used with good effect by Laws and Parsons (1943). The raindrops were allowed to fall into a layer, about 1 in. deep, of fine, uncompacted flour with a smooth surface, contained in shallow trays. A tray was exposed for only a few seconds when the rain was heavy, but for longer periods in lighter rains. The raindrops were allowed to remain in the flour until the dough pellets they produced were dry and hard. Laws and Parsons dried the pellets in an oven and then separated them by means of a set of standard sieves into groups of several sizes. The average mass of pellet in a particular size group was determined from the weight of the catch in each sieve and a count of the number of pellets. Because some of the larger drops were flattened by impact with the flour and so produced flattened pellets, the limits of drop size were not given accurately by the dimensions of the sieve openings. The calibration, which took the form of determining the ratio mass of drop to mass of pellet, was performed with the aid of drops of known size falling from glass tubes with carefully ground tips. This mass ratio was not constant for drops of all sizes, but was greater for larger pellets. Blanchard (1949c), who mixed a little methylene blue with the flour to make the pellets more easily visible, found that in order to prevent splattering of drops larger than 4 mm in diameter, it was advisable to use a fairly deep layer of flour having an irregular surface. A careful calibration revealed that the drop diameter D (mm) and the mass of the pellet M (mg) were related by the formula $D = 1 \cdot 29 M^{\frac{1}{3}}$.

A number of rather more sophisticated instruments for measuring raindrop size have been developed in recent years with a view to reducing the labour of analysis and to obtaining ultimately a completely automatic record of the drop-size distribution.

A method which has attracted considerable effort in recent years was first used by Schindelhauer (1925). The drops were allowed to impinge on a membrane and the sounds produced were amplified electrically and operated recording apparatus which indicated the total number of drops falling on the receiver in a given time. The technique has been developed by Maulard (1951), who arranged for the drops to strike a leather disk attached to the diaphragm of a telephone receiver. Each impulse, having been converted into a voltage signal, was amplified and used to trigger an electro-mechanical counter activated by a thyratron. The apparatus was capable of recording the number of raindrops of diameter greater than 0·5 mm striking the receiver, provided that the time interval between two successive drops exceeded 0·2 s.

A balloon-borne version of such an instrument, designed to measure raindrop size and rain-water content at different altitudes, has been described by Cooper (1951). The drops impinged on a specially designed microphone diaphragm, 2 inches in diameter, and each produced a transient frequency modulation of the transmitter carrier to an extent which depended upon drop size. The frequency deviation was roughly proportional to the momentum of the impinging drop. The frequency-modulated transmission from the balloon was received at the ground and converted into voltage pulses of amplitude proportional to raindrop momentum. Because the latter increases rapidly with drop diameter, the range of pulse voltages was compressed by means of a logarithmic network before they were fed to the counting circuit, thus enabling the whole range of drop sizes to be covered more conveniently. Finally, the pulses were sorted electronically into six amplitude groups and the number in each was registered by an electro-mechanical counter. Discrimination between drops of diameter greater than 3 mm was not possible.

There are several inherent difficulties in this method. It is difficult to obtain uniform response over a microphone diaphragm which is large enough for edge errors to be unimportant; the noise level increases rapidly with the size of the diaphragm so that the detection of small drops is prejudiced. Errors arise from shattering of the larger drops on impact and also from the fact that, as the response is proportional to the momentum of the drops, this will be a function of the rate of ascent of the balloon and of the angle of incidence of the drops on the diaphragm. Cooper states that, apart from counting losses due to the finite time of the counting circuits (about 30 ms), an independent check of the instrument on the ground showed that the number of drops in each group could be registered with a probable error of ± 10 per cent, except in the smallest size group ($d = 0.5$ to 0.7 mm) where the error may reach ± 20 per cent. But considerably larger errors occur, no doubt, when the instrument is airborne.

Several of these limitations have been overcome in the instrument described by Joss and Waldvogel (1967) in which the force exerted by a drop impinging on a rigid circular diaphragm of radius 10 cm is compensated by an equal and opposite force applied by the vertical movement of an electrical solenoid in a magnetic field. The impacting drop induces an e.m.f. in a primary solenoid, the signal from which is fed to a second solenoid causing it to move and communicate a force to the receiver to counteract the impact of the drop. The pulses of current in the compensating solenoid are separated into twenty size groups by a pulse-height analyser and recorded on counters. The instrument is calibrated with drops of known size. Drops greater than 0·3 mm diameter can be detected above the noise level and can be recorded at rates of up to 200 per second. Because the compensating device ensures that the receiver is subjected to only a small fraction of the kinetic energy of the impacting drop, the instrument is ready to record the next drop within a few milliseconds.

An instrument having the merits of simplicity and the ability to give a continuous record of raindrop sizes has been devised by Bowen and Davidson (1951). In principle, the apparatus is a mass spectrograph in which falling raindrops are deflected by a horizontal air current. The drops fall through an orifice in a low-velocity wind tunnel and are deflected horizontally through a distance which is approximately inversely proportional to their mass. They then strike a moving dyed filter paper, their sizes being determined from the diameters of the stains. The stains are thus sorted into size groups but the numbers in each group have to be counted. The instrument has three main disadvantages; it has a sampling orifice of only 1 in^2 so that edge errors are large, it can deal with only vertically-falling drops and is suitable only for drops of diameter between 0·3 and 1·5 mm.

In an attempt to overcome many of the inherent disadvantages of previous methods and to obtain a continuous, fully automatic record of raindrop-size distribution, Mason and Ramanadham (1953) developed a photo-electric spectrometer. The raindrops are allowed to fall through a parallel rectangular beam of light, only those falling through the sampling area of 72 cm^2 being recorded. The light scattered at an angle of about 20° by the falling drops is focused on to a photo-multiplier cell, each drop producing a voltage pulse of amplitude roughly proportional to the cross-sectional area of the drop. The voltage pulses, corresponding to various drop sizes, are sorted by electronic circuits into eight different size groups, the number in each group being recorded continuously by an electromechanical counter. The bank of counters, together with the recorder of a very sensitive rain-gauge and a clock, are photographed every half-minute. Hence a histogram of the drop-size distribution can be plotted for half-minute intervals and the variations of the spectrum with time can be studied. The instrument is calibrated with artificially produced drops of known size and, under conditions of low and steady background intensity, can detect drops of radius greater than 100 μm. Some of the advantages of the system are: (a) the sampling area is optically defined so that the results are not vitiated by splashing, which occurs in mechanically-limited systems; (b) the sampling area is well defined and chosen so that a drop of given size produces the same response at all points in the sensitive area except, of course, at the very edges where corrections have to be applied; (c) the sampling area is large enough to give a representative sample of drops in a 10-minute shower but not so large that coincident counts will be numerous; (d) use of the instrument is not restricted to vertically falling drops; (e) it is entirely automatic. The first model, working with visible light, suffered from the disadvantage that it could not be used satisfactorily during showery weather in the daytime because of the variability in background illumination, but this limitation was largely removed in a later version of the instrument working with ultraviolet light.

A spectrometer working on the same principle is described by Dingle and Schulte (1962) but in this case the sensitive field is a vertical ribbon of light 10 cm long, $\frac{1}{2}$ cm wide and 4 cm deep, and both the light source and photometer rotate about a vertical axis to scan a volume at the rate of 9330 cm^3 s^{-1}. A schematic diagram of the optical arrangement is shown in Fig. B.2. The electrical pulses from the photometer are displayed on a multi-channel oscillograph and analyzed manually.

A number of devices have been developed to measure raindrop sizes from an aircraft but few have been satisfactory and few data have been published. Perhaps the most successful instrument is the simple foil sampler described by Garrod (1957). An aluminium foil tape, 2·5 cm wide and 2·5 μm thick, is fed at the rate of 7·5 cm s^{-1} between two spools past an aperture of area 3 cm^2. The impacting drops impinge on the foil, which rests against a fine wire mesh screen soldered to a brass plate, and make firm impressions showing the pattern of the mesh. The foil records quite clearly drops of diameter $\geqslant$ 100 μm and also snow and ice crystals of these sizes. The instrument is calibrated on a whirling arm in a wind tunnel, the diameter of the impressions being 1·4 times the drop diameter for air speeds of 100

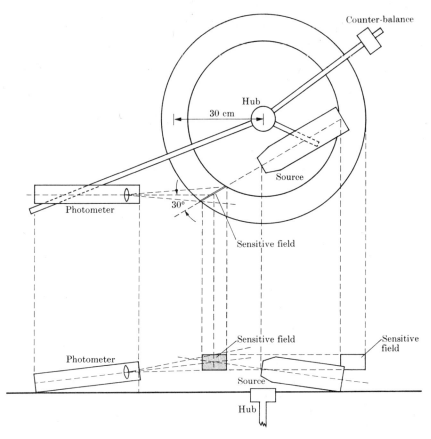

Fig. B.2. Diagrammatic representation of the photoelectric raindrop spectrometer described by Dingle and Schulte (1962).

m s^{-1}. Counting and measurement of the impressions are laborious and errors may well arise due to splashing of large drops at high impact velocities. Moreover, the sampling volume, being only of order 0·1 m³ extending over a flight path of 400 m, is too small for accurate determination of drop concentrations at diameters greater than about $1\frac{1}{4}$ mm.

By far the most promising technique for determining the size distribution of raindrops in the free atmosphere is Doppler radar which, in principle, allows drop-size spectra to be obtained at many levels simultaneously and changes from level to level to be followed. This method, together with some specimen results, is described in § 8.6.1.

B.3.2. *Summary of observational data*

The size distribution of raindrops might be expected to vary considerably with the character of the rain (e.g. continuous, steady rain, thunderstorm, shower, and orographic rain), with the type of cloud from which they fall, and also with the rainfall intensity. The most casual observer will not fail to notice that the largest raindrops are generally associated with thunderstorms. Rather more careful observation will reveal that in nearly all rains there is a considerable spread of drop sizes and that, in general, the number of drops tends to increase rather rapidly as the drop size decreases. Furthermore, the drop-size distribution may change with time—this may often be seen to occur during the life of a thunderstorm. And although there is little in the way of direct observational evidence, there is good reason to believe that the drop-size spectrum must change continually as the rain falls towards the ground because of coalescences between raindrops, the separating effects of wind shear and gravitational settling, and evaporation of the drops in the sub-cloud layer. Unfortunately, most of the available data refer to intermittent samples taken at single stations on the ground, so that we know very little about how the number concentration and size distribution of raindrops vary in space and time during a storm. Moreover, the samples are often very small and the errors correspondingly large. For example, a filter paper of diameter 25 cm exposed for 5s in thunderstorm rain will collect only $\sim$100 drops from a spatial volume of only $\sim$ 1 m³, whereas Mueller and Sims (1966) state that a sample of about 50 m³ is required to estimate the rainfall intensity to within 10 per cent with 95 per cent confidence. There has also been an unfortunate tendency to smooth the data so that they may be fitted to simple empirical formulae, thus suppressing features which may be of physical significance.

Lenard (1904), using the filter-paper method, was the first to publish data on the frequency of drops of different sizes in several rains which varied widely in intensity. While both Defant (1905) and Niederdorfer (1932), who grouped their measurements according to the character of the rainfall, claimed that the sizes of their drops tended to be grouped around masses m, 2m, 4m, 8m, the existence of these modal sizes does not appear to be confirmed by other workers.

Most of the data on raindrop-size distributions available at the time was analyzed by Best (1950), who found that most of the observations could be represented by the formula

$$1 - F = \exp\{-(x/a)^n\}, \tag{B.10}$$

where $F(x)$ is the fraction of liquid water comprised of drops of diameter $<x$; a and n are constants for any particular rainfall, and $a = \alpha p^\beta$, where α and β are numerical factors. If W is the mass of rainwater in unit volume of air, $W = cp^f$, where c and f are constants. Best deduced that the median-volume drop diameter could be represented by $D_n = (0 \cdot 69)^{1/n} a$, and gave the values shown in Table B.2 for the parameters characterizing the drop-size data of

TABLE B.2

Parameters characterizing raindrop-size distributions

$$1 - F = \exp\{-(x/\alpha p^\beta)^n\}; \quad D_n = bp^\beta; \quad W = cp^f$$

where D_n is measured in mm, W in mg m^{-3}, and p in mm h^{-1}.

Location/rain type	Author	n	α	b	β	c	f
	(a) Values quoted by Best (1950)						
Germany	Lenard (1904)	2·59	1·42	—	0·272	61	0·840
U.S.A.	Laws and Parsons (1943)	2·29	1·25	1·06	0·199	72	0·867
Canada	Marshall et al. (1947)	1·85	1·00	—	0·240	72	0·880
Hawaii	Anderson (1948)	4·48	0·88	—	0·283	82	0·838
Ynyslas		2·49	1·88	—	0·203	74	0·845
Shoeburyness	Best (1950)	2·29	1·56	—	0·209	59	0·816
East Hill	Gt. Britain	1·99	1·38	—	0·269	65	0·829
	(b) Some more recent values						
Canada continuous rain	Marshall & Palmer (1948)			0·91	0·21	72	0·88
Hawaii							
warm orographic rain							
in cloud	Blanchard (1953)			0·30	0·40	235	0·58
at cloud base				0·40	0·37	150	0·70
non-orographic rain				1·18	0·19	61	0·89
Illinois, U.S.A. heavy showers	Jones (1956)			1·48	0·05	52	0·97
Massachussetts, U.S.A. continuous rain with melting band	Atlas & Chmela (1957)			0·95 to 1·22	0·17 to 0·29	62 to 80	0·88 to 0·94
India orographic monsoon	Ramana Murty & Gupta (1959)					76	0·84
thunderstorms				0·82	0·29	62	0·90
continuous rain with melting band	Sivaramakhrishnan (1961)			0·71	0·29	70	0·83
						86	0·77
warm layer cloud				0·49	0·50	101	0·69
monsoon thunderstorms	Kelkar (1968)					70	0·68

several workers. The value of n showed no systematic variation with the character or intensity of the rain but the parameter a and hence D_n took higher values in heavier rains. The drop-size spectra obtained by Anderson in Hawaii, and which were associated with non-freezing, orographic clouds, were very different from those obtained in temperate latitudes and were characterized by small values of D_n. Similar differences are apparent in Blanchard's (1953) data which show warm orographic rains to have much narrower spectra than those associated with thunderstorms or cyclonic storms and to contain very few drops exceeding 2 mm in diameter.

Several other workers have deduced empirical relations between D_n, W, and p of the form $D_n = bp^\beta$, $W = cp^\gamma$ for different types of rain falling in different geographical locations. Some representative values of the empirical parameters are given in Table B.2 and illustrate that neither the median-volume drop diameter nor the concentration of rain-water in the air are uniquely related to the precipitation intensity, but that they also vary with the type and character of the rainfall. For fairly steady continuous rain produced by the Wegener–Bergeron process, the following relations are fairly representative of average drop-size spectra, p being measured in mm h^{-1}.

Mean raindrop mass $\qquad\qquad \bar{m}(\mu g) = 180 p^{\frac{3}{4}}$

Mean-volume drop diameter $\qquad D_v(\text{mm}) = 0.7 p^{\frac{1}{4}}$

Median volume drop diameter $\qquad D_n(\text{mm}) = p^{0.23}$

Rain-water concentration $\qquad\qquad W(\text{mg m}^{-3}) = 70 p^{0.85}$

Hence $\qquad\qquad\qquad\qquad\qquad\qquad p = \tfrac{1}{70} W D_n^{\frac{1}{2}}$

Radar reflectivity $\qquad\qquad\qquad Z\ (\text{mm}^6\ \text{m}^{-3}) = 200 p^{1.6}$

Marshall and Palmer (1948), found that the expression

$$N_D = 0.08 e^{-x'D}\ \text{cm}^{-4};\ x' = 41 p^{-0.21}\ \text{cm}^{-1}. \qquad (\text{B.11})$$

where $N_D\,\delta D$ is the number of drops per unit volume having diameters between D and $D+\delta D$, and p is the rainfall intensity in mm h^{-1}, provided a good average fit to the experimental data of Laws and Parsons, apart from a tendency to overestimate the numbers of small drops for which $x'D < 4.5$.

More recent investigations, for example, those of Mason and Andrews (1960) in England, Sivaramakhrishnan (1961) in India, and Dingle and Hardy (1962) in the United States, confirm the earlier

APPENDIX B 611

findings of Mason and Ramanadham (1954) that while the Marshall–Palmer distribution, (B.11), represents quite well the average size distribution of a large number of samples from continuous cyclonic or warm-frontal rain, or the characteristics of the rain averaged over a considerable interval of time, samples taken over periods of only a few minutes may show considerable deviations from these average spectra. Thus Mason and Andrews (1960) found that while in continuous warm frontal rain, where the radar signals from both above

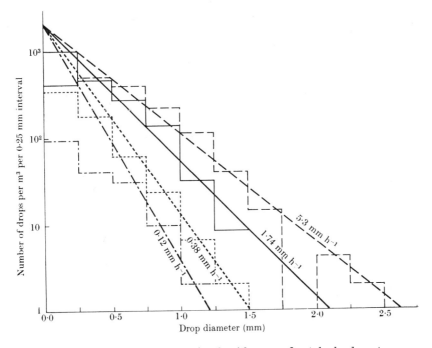

Fig. B.3. Drop-size spectra associated with warm-frontal cloud systems exhibiting marked radar melting-bands but otherwise little structure. (From Mason and Andrews (1960).)

and below the melting level remained nearly uniform and steady over periods of several minutes, the drop-size distribution followed the Marshall–Palmer formula quite well except that the latter, as usual, overestimated the numbers of small drops—see Fig. B.3. However, frontal rain was often far from steady and uniform and then the fluctuations in rainfall intensity and drop-size distribution were often correlated with the presence of streaks, cells, or patches of heavier precipitation. The drop-size spectra associated with showery-type radar echoes, which did not extend above the 0° C level and were

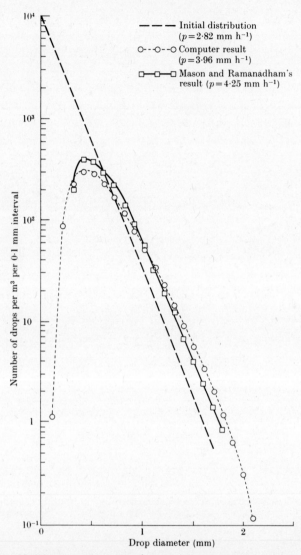

FIG. B.4. Modification of raindrop-size distribution due to coalescence and accretion for a fall of 1 km through a cloud containing 0·2 g m^{-3} of liquid water. ——— initial distribution ($p = 2·82$ mm h^{-1}), -○-○ Hardy's result ($p = 3·96$ mm h^{-1}) ▫—▫—▫ Mason and Ramanadham's result ($p = 4·25$ h^{-1}). (From Hardy (1963).)

therefore presumably produced by the coalescence process, bore little resemblance to Marshall–Palmer distributions but were narrow and sharply cut-off at both ends. The smallest drops were probably removed by coalescence, evaporation, and sorting in wind-shear, while the size of the largest drops was limited by the restricted depth of the non-freezing cloud. Thunderstorm rains usually contained considerably higher concentrations of both very small and very large drops than predicted by the Marshall–Palmer formula, the excess of small drops being attributed to the splashing and break-up of very large drops near the ground. Several of these features are confirmed by the observations of Sivaramakhrishnan (1961) and Dingle and Hardy (1962).

The size distribution of raindrops will change during their fall from the level of origin to the ground due to (a) coalescences between raindrops of different sizes; (b) the growth of raindrops by accretion with cloud droplets; (c) differential rates of evaporation of drops of different sizes when falling between cloud base and ground. There will be corresponding changes in the rainfall intensity. These matters were first examined by Rigby and Marshall (1952) who concluded that these processes cannot be of great importance in determining or modifying the drop-size spectrum. Rather larger effects were predicted by the more realistic and detailed calculations of Mason and Ramanadham (1954) and Hardy (1963) who agree in showing that a flux of raindrops having an initial exponential size distribution of the Marshall–Palmer type may undergo considerable modification after falling through only 1 km in a layer cloud of modest liquid-water content, the smaller drops being seriously depleted and the larger ones increased. The results of calculations to show the magnitude of the effects are depicted in Fig. B.4. The drop-size spectrum may undergo considerable further modification by evaporation of the drops in the sub-cloud layer, the smaller drops being preferentially depleted. It is important to be aware of such changes when attempting to correlate the intensity of precipitation or radar echoes aloft with measurements of rain on the ground. They merit further study, for example, by precision Doppler radar techniques.

Some Useful Physical Constants

Universal constants

Avogadro's number $\mathcal{N} = 6 \cdot 023 \times 10^{23}$ mol^{-1}
Universal gas constant $\mathbf{R} = 8 \cdot 314 \times 10^7$ erg mol^{-1} K^{-1} = $8 \cdot 314$ J mol^{-1} K^{-1}
Boltzmann's constant $k = \mathbf{R}/\mathcal{N} = 1 \cdot 381 \times 10^{-16}$ erg K^{-1} = $1 \cdot 381 \times 10^{-23}$ J K^{-1}

Physical properties of dry air

Molecular weight = 28·964
Gas constant for dry air = $2 \cdot 87 \times 10^6$ erg g^{-1} K^{-1} = 287 J kg^{-1} K^{-1}
Specific heats: $c_p = 0 \cdot 240$ cal g^{-1} K^{-1} = $1 \cdot 005$ J kg^{-1} K^{-1} $c_v = 0 \cdot 172$ cal g^{-1} K^{-1} = $0 \cdot 718$ J kg^{-1} K^{-1}. $\gamma = c_p/c_v = 1 \cdot 40$
Density of air at 0° C and 1000 mb pressure = $1 \cdot 2754$ g cm^{-3} or kg m^{-3}
(for other temperatures and pressures multiply by $0 \cdot 273\, p/T$)

The density, viscosity, thermal conductivity of air and the diffusion coefficient of water vapour in air as functions of temperature at 1000 mb pressure

T (°C)	40	20	10	0	−10	−20	−40
Density ρ (g cm^{-3} or kg m^{-3})	1·1125	1·1884	1·2303	1·2754	1·3238	1·3761	1·4942
Dynamic viscosity μ (poise × 10^4), (N.s.m^{-2} × 10^5)	1·908	1·815	1·766	1·717	1·667	1·616	1·512
Thermal conductivity K (J m^{-1} s^{-1} K^{-1}) × 10^2	2·71	2·55	2·48	2·40	2·32	2·23	2·07
(cal cm^{-1} s^{-1} K^{-1}) × 10^5	6·47	6·09	5·92	5·73	5·54	5·33	4·94
Diffusion coefficient D† (cm^2 s^{-1} × 10), (m^2 s^{-1} × 10^5)	2·84	2·53	2·39	2·25	2·11	1·98	1·72
Kinematic viscosity ν (cm^2 s^{-1} × 10), (m^2 s^{-1} × 10^5)	1·72	1·53	1·44	1·35	1·26	1·17	1·01

Note: μ and K are independent of pressure, $\rho \propto p$, and $D \propto 1/p$.

† There is some evidence (see Thorpe and Mason (1966)), that these values are about 10 per cent too high.

Physical properties of water substance

Molecular weight = 18·015
Gas constant for water vapour treated as ideal gas
  = $4 \cdot 615 \times 10^6$ erg g^{-1} K^{-1} = $461 \cdot 5$ J kg^{-1} K^{-1}

Latent and specific heats

T (°C)	40	20	10	0	−10	−20	−40
L_f (cal g^{-1})				79·7	74·5	69·0	56·3
(J.kg^{-1}) × 10^{-5}				3·34	3·12	2·89	2·36
L_v (cal g^{-1})	547·7	586·0	591·7	597·3	603·0	608·9	621·7
(J.kg^{-1}) × 10^{-6}	2·406	2·454	2·477	2·500	2·525	2·549	2·603
L_s (cal g^{-1})				677	677·5	677·9	678
(J.kg^{-1}) × 10^{-6}				2·835	2·837	2·838	2·839
C_w (cal g^{-1} K^{-1})	0·998	0·999	1·001	1·007	1·020	1·040	1·140
(J.kg^{-1} K^{-1}) × 10^{-3}	4·179	4·182	4·192	4·218	4·27	4·35	4·77
C_i (cal g^{-1} K^{-1})				0·503	0·485	0·468	0·433
(J.kg^{-1} K^{-1}) × 10^{-3}				2·106	2·031	1·959	1·813

L_f, L_v, L_s are respectively the latent heats of fusion, vaporization, and sublimation.
C_w, C_i are the specific heats of liquid water and ice respectively.

SOME USEFUL PHYSICAL CONSTANTS

Surface tension of pure water

T(°C)	30	20	10	0	−10	−20	−40
σ(dyn/cm), (Nm$^{-1}\times 10^3$)	71·18	72·75	74·22	75·70	77·29	79·14	83·9†

† Extrapolated.

Density of saturated water vapour

	Over water					Over ice			
T (K)	300	290	280	270	260	270	260	250	240
ρ_v (g m^{-3})	25·8	14·5	7·74	3·93	1·87	3·82	1·65	0·67	0·25

Saturation vapour pressures over pure liquid water (e_w) and over pure ice (e_i) as functions of temperature

T (°C)	e_w (mb)	e_i (mb)	T	e_w	e_i	T	e_w	T	e_w
−50	0·0635	0·0393	−24	0·8826	0·6983	1	6·565	26	33·606
−49	0·0712	0·0445	−23	0·9647	0·7708	2	7·054	27	35·646
−48	0·0797	0·0502	−22	1·0536	0·8501	3	7·574	28	37·793
−47	0·0892	0·0567	−21	1·1498	0·9366	4	8·128	29	40·052
−46	0·0996	0·0639	−20	1·2538	1·032	5	8·718	30	42·427
−45	0·1111	0·0720	−19	1·3661	1·135	6	9·345	13	44·924
−44	0·1230	0·0810	−18	1·4874	1·248	7	10·012	32	47·548
−43	0·1379	0·0910	−17	1·6183	1·371	8	10·720	33	50·303
−42	0·1533	0·1021	−16	1·7594	1·505	9	11·473	34	53·197
−41	0·1704	0·1145	−15	1·9114	1·651	10	12·271	35	56·233
−40	0·1891	0·1283	−14	2·0751	1·810	11	13·118	36	59·418
−39	0·2097	0·1436	−13	2·2512	1·983	12	14·016	37	62·759
−38	0·2322	0·1606	−12	2·4405	2·171	13	14·967	38	66·260
−37	0·2570	0·1794	−11	2·6438	2·375	14	15·975	39	69·930
−36	0·2841	0·2002	−10	2·8622	2·597	15	17·042	40	73·773
−35	0·3138	0·2232	−9	3·0965	2·837	16	18·171	41	77·798
−34	0·3463	0·2487	−8	3·3478	3·097	17	19·365	42	82·011
−33	0·3817	0·2768	−7	3·6171	3·379	18	20·628	43	86·419
−32	0·4204	0·3078	−6	3·9055	3·684	19	21·962	44	91·029
−31	0·4627	0·3420	−5	4·2142	4·014	20	23·371	45	95·850
−30	0·5087	0·3797	−4	4·5444	4·371	21	24·858	46	100·89
−29	0·5588	0·4212	−3	4·8974	4·756	22	26·428	47	106·15
−28	0·6133	0·4668	−2	5·2745	5·173	23	28·083	48	111·65
−27	0·6726	0·5169	−1	5·6772	5·622	24	29·829	49	117·40
−26	0·7369	0·5719	0	6·1070	6·106	25	31·668	50	123·39
−25	0·8068	0·6322	—	—	—	—	—	—	—

Bibliography and Author Index

ACKERMAN, B. (1960) Orographic–convective precipitation as revealed by radar. *Physics of precipitation*, Geophys. Monogr. No. 5, p. 79. American Geophysical Union. (472)

ADDERLEY, E. E. and BOWEN, E. G. (1962) Lunar component in precipitation data. *Science, N.Y.* **137,** 749. (205)

ADEN, A. L. (1951) Microwave reflection from water spheres, *Am. J. Phys.* **19,** 163. (417)

—— (1888–92) Related papers on the development and use of the dust counter. *Collected scientific papers* (1923) (ed. C. G. Knott) pp. 187, 207, 236, 284. Cambridge University Press. (31, 32–4)

AITKEN, J. (1880–1) On dusts, fogs and clouds. *Trans. R. Soc. Edinb.* **30,** 337. (31)

—— (1912) The sun as a fog producer. *Proc. R. Soc. Edinb.* **32,** 183. (53)

ALLAN, R. S. and MASON, S. G. (1962) Particle motions in sheared suspensions; coalescence of liquid drops in electric and shear fields. *J. Colloid Sci.* **17,** 383. (589)

ALPERT, L. (1955) Notes on warm-cloud rainfall. *Bull. Am. met. Soc.* **36,** 64. (295)

ANDERSON, L. J. (1948) Drop-size distribution measurements in orographic rains. *Bull. Am. met. Soc.* **29,** 362. (609)

—— DAY, J. P., FRERES, C. H., and STOKES, A. P. D. (1947) Attenuation of 1·25 cm radiation through rain. *Proc. Inst. Radio Engrs* **35,** 351. (428)

ANDERSON, V. G. (1915) The influence of weather conditions upon the amounts of nitric acid and of nitrous acid in the rainfall at and near Melbourne, Australia. *Q. Jl R. met. Soc.* **41,** 99. (90)

ANDRÉN, L. (1917) Zählung und Messung der Komplexen Moleküle einiger Dämpfe nach der neuer Kondensationstheorie. *Annln Phys* **52,** 1. (18)

ANDREWS, J. B. (1961) Size distribution of precipitation elements. Ph.D. Thesis, London University. (479, 602)

ANON. (1948) Artificial stimulation of precipitation. *South African CSIR Report.* (372, 377)

APPLETON, E. V. and CHAPMAN, F. W. (1937) On the nature of atmospherics, IV. *Proc. R. Soc.* A**158,** 1. (501)

—— WATSON-WATT, R. A. and HERD, J. F. (1926) On the nature of atmospherics, II. *Proc. R. Soc.* A**111,** 615. (493)

ARABADJI, V. I. (1959) On the electrical attributes of storm precipitation (in Russian). *Dokl. Akad. Nauk. SSSR* **127,** 298. (563)

ATLAS, D. (1953) Device to permit radar contour mapping of rain intensity in rainstorms. U.S. Patent No. 2656531, U.S. Govt. Printing Office, Washington, D.C. (431)

—— (1964) Advances in radar meteorology. *Adv. Geophys.* **10,** 317. (399, 434, 441, 449, 478, 481)

—— and CHMELA, A. C. (1957) Physical-synoptic variations of raindrop size parameters. *Proc. 6th Weather Radar Conf., Cambridge,* p. 21. (609)

—— HARPER, W. G., LUDLAM, F. H., and MACKLIN, W. C. (1960) Radar scatter by large hail. *Q. Jl R. met. Soc.* **86,** 468. (419–20)

—— KERKER, M. and HITSCHFELD, W. (1953) Scattering and attenuation by non-spherical atmospheric particles. *J. atmos. terr. Phys.* **3,** 108. (406–13)

—— and LUDLAM, F. H. (1961). (1962) Multi-wavelength radar reflectivity of hailstorms. *Q. Jl R. met. Soc.* **87,** 523; **88,** 207. (334, 416, 425, 474)

ATLAS, D. and WEXLER, R. (1963) Back-scatter by oblate ice spheroids. *J. atmos. Sci.* **20**, 48. (422–3)
—— et al., (1963) Severe local storms. *Met. Monogr.* **5**, No. 27. American Meteorological Society, Boston. (474)
AUFDERMAUR, A. N., LIST, R., MAYES, W. C., and DE QUERVAIN, M. R. (1963) Kristallachsenlagen in Hagel Körnern. *Z. angew. Math. Phys.* **14**, 574. (335, 342)
—— and MAYES, W. C. (1965) Correlations between hailstone structures and growth conditions. *Proc. int. Conf. on Cloud Physics, Tokyo*, p. 281. (342)
AUFM KAMPE, H. J. (1950) Visibility and liquid water content in clouds in the free atmosphere. *J. Met.* **7**, 54. (118)
—— KELLY, J. J. and WEICKMANN, H. K. (1957) Seeding experiments in subcooled stratus clouds. *Met. Monogr.* No. 2, p. 86. American Meteorological Society, Boston. (381–2)
—— WEICKMANN, H. K. and KEDESDY, H. H. (1952) Remarks on electron-microscope study of snow-crystal nuclei—Kumai, M. *J. Met.* **9**, 374. (197)
—— —— and KELLY, J. J. (1951) The influence of temperature on the shape of ice crystals growing at water saturation. *J. Met.* **8**, 168. (251–2)
—— —— —— (1956) A continuously recording water-content meter. *J. Met.* **13**, 64. (116)
AUSTIN, P. M. (1963) Radar measurements of the distribution of precipitation in New England storms. *Proc. 10th Weather Radar Conf., Washington*, p. 247. (479, 481–482)
—— and BEMIS, A. C. (1950) A quantitative study of the 'bright band' in radar precipitation echoes. *J. Met.* **7**, 145. (462, 465–466)
—— and WILLIAMS, E. L. (1951) Comparison of radar signal intensity with precipitation rate. *M.I.T. Weather Radar Research Tech. Rep.* No. 14. (414)

BAILEY, I. H. and MACKLIN, W. C. (1968a) The surface configuration and internal structure of artificial hailstones. *Q. Jl R. met. Soc.* **94**, 1. (287, 342, 345, 348)
—— —— (1968b) Heat transfer from artificial hailstones. *Q. Jl R. met. Soc.* **94**, 93. (287, 346, 357–358, 368)
BALABANOVA, V. N. (1961) Determination of the water content of clouds by the filtration method (in Russian). *Izv. Akad. Nauk SSSR Ser. geofiz.* No. 1, p. 100. (118)
—— MALEEV, M. N. and ZHIGALOVSKAYA, T. N. (1960) The extent of destruction of the silver iodide particles under the thermal methods of dispersion. *Izv. Akad. Nauk. SSSR Ser. geofiz.* No. 9, p. 941. (231)
—— and ZHIGALOVSKAYA, T. N. (1962) Dispersed state of a silver iodide aerosol (in Russian). *Izv. Akad. Nauk SSSR Ser. geofiz.* No. 3, p. 293. (231)
BALDIT, A. (1911) Observations sur l'électricité de la pluie pendant l'été au Puy-en-Velay. *Annls Soc. mét. Fr.* **59**, 105. (560)
BANERJI, S. K. (1938) Does thunderstorm rain play any part in the replenishment of the earth's negative charge? *Q. Jl R. met. Soc.* **64**, 293. (561)
—— and LELE, S. R. (1932) Electrical charges on raindrops. *Nature, Lond.* **130**, 998. (561–2)
BARNARD, A. J. (1954) A study of condensation and the operation of diffusion cloud chambers. Ph.D. Thesis, Glasgow University. (14, 16)
BARRETT, P. and BROWNE, I. C. (1953) A new method of measuring vertical air currents. *Q. Jl R. met. Soc.* **79**, 550. (439)

BARTHAKUR, N. and MAYBANK, J. (1963) Anomalous behaviour of some amino-acids as ice nucleators. *Nature, Lond.* **200**, 866. (226)
—— and MAYBANK, J. (1965) The growth of ice crystals on amino-acid substrates. *J. Rech. Atmos.* **2**, 475. (225, 226)
BARTLETT, J. T. (1966) The growth of cloud droplets by coalescence. *Q. Jl R. met. Soc.* **92**, 93. (147–9)
—— (1968) Condensation in a turbulent updraught. *Proc. int. Conf. on Cloud Physics, Toronto*, p. 515. (144)
—— VAN DEN HEUVEL, A. P. and MASON, B. J. (1963) The growth of ice crystals in an electric field. *Z. angew. Math. Phys.* **14**, 599. (266)
BARTNOFF, S., ATLAS, D., and PAULSEN, W. H. (1952) Experimental statistics in cloud and rain echoes. *Proc. 3rd Radar Weather Conf., Montreal*, p. G1. (435)
BASHKIROV, G. M. and KRASIKOV, P. N. (1957) Experiments with certain substances as crystallization agents for super-cooled fog. (in Russian) *Trudy glav. geofiz. Obs. A. I. Voeikova* **72**, 118. (223, 225)
BATTAN, L. J. (1953) Observations on the formation and spread of precipitation in convective clouds. *J. Met.* **10**, 311. (292, 472)
—— (1958) Influence of the environment on the initiation of precipitation in tropical cumuli over the ocean. *Tellus* **10**, 466. (296)
—— (1959) *Radar meteorology*. University of Chicago Press, Chicago. (399, 478)
—— (1963) Relationship between cloud base and initial radar echo. *J. appl. Met.* **2**, 333. (292, 472)
—— (1964) Some observations of vertical velocities and precipitation sizes in a thunderstorm. *J. appl. Met.* **3**, 415. (448–9)
—— (1965) Some factors governing precipitation and lightning from convective clouds. *J. atmos. Sci.* **22**, 79. (293)
—— (1966) Silver iodide seeding and rainfall from convective clouds. *J. appl. Met.* **5**, 669. (387)
—— and BRAHAM, R. R. (1956) A study of convective precipitation based on cloud and radar observations. *J. Met.* **13**, 587. (291)
—— and HERMAN, B. M. (1962) The radar cross sections of 'spongy' ice spheres. *J. geophys. Res.* **67**, 5139. (421–2)
—— and KASSANDER, A. R. (1967) Summary of results of a randomized cloud seeding project in Arizona. *Proc. 5th Berkeley Symposium on Mathematical Statistics and Probability*, Vol. 5, p. 29. University of California Press. (387)
—— and REITAN, C. H. (1957) Droplet size measurements in convective clouds. In *Artificial stimulation of rain* (eds. H. Weickmann and W. Smith), p. 184. Pergamon Press, New York. (105)
—— THEISS, J. B. and KASSANDER, A. R. (1964) Some doppler radar observations of a decaying thunderstorm. *Proc. 11th Weather Radar Conf., Boulder*, p. 362. (449)
—— —— (1966) Observations of vertical motions and particle sizes in a thunderstorm. *J. atmos. Sci.* **23**, 78. (449)
BAYARDELLE, MAUD (1954) Sur le mécanisme de la congélation de l'eau dans les nuages. *C. r hebd. Séanc. Acad. Sci., Paris* **239**, 988. (160, 168)
BEARD, K. V. and PRUPPACHER, H. R. (1969) A determination of the terminal velocity and drag of small water drops by means of a wind tunnel. *J. atmos. Sci.* **26**, 1066. (593)
BECKER, R. and DÖRING, W. (1935) Kinetische Behandlung der Keimbildung in übersättigten Dämpfen (Kinetic treatment of embryo formation in supersaturated vapours). *Annln Phys.* **24**, 719. (11)

BECKWITH, W. B. (1956) Hail observations in the Denver area. *United Air Lines Met. Circ.* No. 40. (334)

—— (1960) Analysis of hailstorms in the Denver network, 1949–1958. *Physics of precipitation, Geophys. Monogr.* No. 5, p. 348. American Geophysical Union, Boston. (332)

BELYAEV, V. I. (1961) Drop-size distribution in a cloud during the condensation stage of development. *Izv. Akad. Nauk SSSR* Ser. geofiz., p. 1209. (143)

BENTLEY, W. A. (1904) Studies of raindrops and raindrop phenomena. *Mon. Weath. Rev. U.S. Dep. Agric.* **32**, 450. (603)

—— and HUMPHREYS, W. J. (1931) *Snow crystals.* McGraw-Hill, London and New York. (236)

BENWELL, G. R. R. and TIMPSON, M. S. (1968) Further work with the Bushby-Timpson 10-level model. *Q. Jl R. met. Soc.* **94**, 12. (281)

BERGERON, T. (1935) On the physics of cloud and precipitation. *Proc. 5th Assembly U.G.G.I. Lisbon*, Vol. 2, p. 156. (284, 306)

—— (1949) The problem of artificial control of rainfall on the globe. I. General effects of ice nuclei in clouds. *Tellus* **1**, 32. (285)

BERRY, E. X. (1967) Cloud droplet growth by collection. *J. atmos. Sci.* **24**, 688. (146, 149–51)

—— (1968) A parameterization of the collection of cloud drops. *Proc. int. Conf. on Cloud Physics, Toronto*, p. 111. (314)

BEST, A. C. (1950) The size distribution of raindrops. *Q. Jl R. met. Soc.* **76**, 16. (479, 602, 608–9)

—— (1951a) Drop-size distribution in cloud and fog. *Q. Jl R. met. Soc.* **77**, 418. (111)

—— (1951b) The size of cloud droplets in layer-type cloud. *Q. Jl R. met. Soc.* **77**, 241. (124, 140)

—— (1952) Effect of turbulence and condensation on drop-size distribution in cloud. *Q. Jl R. met. Soc.* **78**, 28. (140)

BETHWAITE, F. D., SMITH, E. J., WARBURTON, J. A., and HEFFERNAN, K. J. (1966) Effects of seeding isolated cumulus clouds with silver iodide. *J. appl. Met.* **5**, 513. (375)

BIGG, E. K. (1953) The supercooling of water. *Proc. phys. Soc.* B**66**, 688. (157–162)

—— (1956) Counts of atmospheric freezing nuclei at Carnavon, Western Australia. *Aust. J. Phys.* **9**, 561. (199, 204)

—— (1957) A new technique for counting ice-forming nuclei in aerosols. *Tellus* **9**, 394. (175, 178)

—— (1961) Natural atmospheric ice nuclei. *Science Prog., Lond.* **49**, 458. (187)

—— (1963) A lunar influence on ice nucleus concentrations. *Nature, Lond.* **197**, 172. (205)

—— (1965) Problems in the distribution of ice nuclei. *Proc. int. Conf. on Cloud Physics, Tokyo*, p. 137. (189, 200)

—— and GIUTRONICH, J. (1967) Ice nucleating properties of meteoritic material. *J. atmos. Sci.* **24**, 46. (205)

—— and HOPWOOD, S. C. (1963) Ice nuclei in the Antarctic. *J. atmos. Sci.* **20**, 185. (200)

—— and MEADE, R. T. (1959) Continuous automatic recording of ice nuclei. *Bull. Obs. Puy de Dôme*, No. 4, p. 125. (178, 182, 187)

—— and MILES, G. T. (1963) Stratospheric ice nucleus measurements from balloons. *Tellus* **15**, 162. (201)

—— —— (1964) The results of large-scale measurements of natural ice nuclei. *J. atmos. Sci.* **21**, 396. (201, 205)

BIGG, E. K., MILES, G. T., and HEFFERNAN, K. J. (1961) Stratospheric ice nuclei. *J. Met.* **18**, 804. (178, 201)

—— MOSSOP, S. C., MEADE, R. T., and THORNDIKE, N. S. C. (1963) The measurement of ice nucleus concentrations by means of Millipore filters. *J. appl. Met.* **2**, 266. (178)

BILHAM, E. G. and RELF, E. F. (1937) The dynamics of large hailstones. *Q. Jl R. met. Soc.* **63**, 149. (349)

BIRSTEIN, S. J. (1952) The effect of relative humidity on the nucleating properties of photolyzed silver iodide. *Bull. Am. met. Soc.* **33**, 431. (231–2, 233)

—— (1955) The role of adsorption in heterogeneous nucleation. I, Adsorption of water vapor on silver iodide and lead iodide. *J. Met.* **12**, 324. (234)

—— (1957) Studies on the effects of certain chemicals on the inhibition of nucleation. In *Artificial stimulation of rain* (eds. H. Weickmann and W. Smith), p. 376. Pergamon Press, London, New York. (194)

—— (1960) Studies on the effect of chemisorbed impurities on heterogeneous nucleation. *Physics of precipitation, Geophys. Monogr.*, No. 5, p. 247. American Geophysical Union, Boston. (194)

BISWAS, K. R., RAMANA MURTY, B. V., and ROY, A. K. (1960) Freezing rain at Delhi and associated melting band characteristics. *Ind. J. Met. Geophys.* **13** (Suppl), 137. (461)

BLANCHARD, D. C. (1949a) Experiments with water drops and the interaction between them at terminal velocity in air. *G. E. Res. Lab. Project Cirrus Occ. Rep. No. 17.* (599)

—— (1949b) The use of sooted screens for determining raindrop size and distribution. *G. E. Res. Lab. Project Cirrus Occ. Rep. No. 16.* (602)

—— (1949c) The size distribution of raindrops in natural rain. *G. E. Res. Lab. Project Cirrus, Occ. Rep. No. 15.* (603)

—— (1953) Raindrop size-distribution in Hawaiian rains. *J. Met.* **10**, 457. (479, 602, 609–10)

—— (1963) The electrification of the atmosphere by particles from bubbles in the sea. *Progr. Oceanogr.* **1**, 71. Pergamon Press, London, (76, 528)

—— (1969) The oceanic production rate of cloud nuclei. *J. Rech. Atmos.* **4**, 1. (78)

—— and WOODCOCK, A. H. (1957) Bubble formation and modification in the sea and its meteorological significance. *Tellus* **9**, 145. (75, 77)

BOROVIKOV, A. M. (1953) Some results on an investigation of the structure of crystal clouds. (in Russian) *Trudy Tsentral. Aerolog. Obs.* No. 12. (247)

—— et al. (1961) Fizika Oblakov (Cloud Physics). Gidrometeor. Izdatel'stvo, Leningrad (English translation by Israel Program for Scientific Translations, Jerusalem. 1963). (110, 121, 382)

BOWEN, E. G. (1950) The formation of rain by coalescence. *Aust. J. scient. Res.* **A3**, 193. (286, 318–20, 379)

—— (1951) Radar observations on rain and their relation to the mechanisms of rain formation. *J. atmos. terr. Phys.* **1**, 125. (293, 458, 468)

—— (1952a) Australian experiments on artificial stimulation of rainfall. *Weather, Lond.* **7**, 204. (374)

—— (1952b) A new method of stimulating convective clouds to produce rain and hail. *Q. Jl R. met. Soc.* **78**, 37. (379)

—— (1953) The influence of meteoritic dust on rainfall. *Aust. J. Phys.* **6**, 490. (194, 201)

—— (1956a) The relation between rainfall and meteor showers. *J. Met.* **13**, 142. (202)

—— (1956b) An unorthodox view of the weather. *Nature, Lond.* **177**, 1121. (202)

BOWEN, E. G. (1956c) A relation between meteor showers and the rainfall of November and December. *Tellus* **8**, 394. (202)
—— (1956d) January freezing nucleus measurements. *Aust. J. Phys.* **9**, 552. (204)
—— (1957) Relation between meteor showers and the rainfall of August, September, and October. *Aust. J. Phys.* **10**, 412. (202, 204)
—— (1961) Freezing nuclei—methods of measurement and some of their characteristics. *Nubila* **4**, 7. (188, 199)
—— (1966) The effect of persistence in cloud-seeding experiments. *J. appl. Met.* **5**, 156. (387)
—— and DAVIDSON, K. A. (1951) A raindrop spectrograph. *Q. Jl R. met. Soc.* **77**, 445. (605)
BOYCE, S. G. (1954) The salt spray community. *Ecol. Monogr.* **24**, 29. (75)
BOYENVAL, E. H. (1959) Echoes from precipitation using pulsed doppler radar. *Royal Radar Establishment Men.* No. 1606. (440, 442)
—— (1960) Echoes from precipitation using pulsed doppler radar. *Proc. 8th Weather Radar Conf., San Francisco*, p. 57. (440, 442)
BOYLAN, R. K. (1926) Atmospheric dust and condensation nuclei. *Proc. R. Ir. Acad.* **37**(A), 58. (53)
BOYS, C. V. (1926) Progressive lightning. *Nature, Lond.* **118**, 749. (496)
BRADLEY, D. A., WOODBURY, M. A., and BRIER, G. W. (1962) Lunar synodical period and widespread precipitation. *Science, N.Y.* **137**, 748. (205)
BRAHAM, R. R. (1964) What is the role of ice in summer rain showers? *J. atmos. Sci.* **21**, 640. (211)
—— BATTAN, L. J., and BYERS, H. R. (1957) Artificial nucleation of cumulus clouds. *Met. Monogr.* **2**, 47. American Meteorological Society, Boston. (374, 378)
—— and FLUECK, J. A. (1967) Contribution to the discussion of Professor Neyman's paper. *Jl R. statist. Soc.* **A130**, 316. (389)
—— REYNOLDS, S. E. and HARRELL, J. H. (1951) Possibilities for cloud seeding as determined by a study of cloud height versus precipitation. *J. Met.* **8**, 416. (291)
—— and SPYERS-DURAN, P. (1967) Survival of cirrus crystals in clear air. *J. appl. Met.* **6**, 1053. (193)
BREWER, A. W. and PALMER, H. P. (1951) Freezing of supercooled water. *Proc. phys. Soc.* **B64**, 765. (173)
BROOK, M. (1958) *Recent advances in thunderstorm electricity* (ed. L. G. Smith), p. 383. Pergamon Press, Oxford. (531, 537)
—— KITAGAWA, N., and WORKMAN, E. J. (1962) Quantitative study of strokes and continuing currents in lightning discharges to ground. *J. geophys. Res.* **67**, 649. (505–7)
BROOKS, C. E. P. (1925) The distribution of thunderstorms over the globe. *Geophys. Mem., Met. Office Lond.* **3**, No. 24. (558)
BROWN, E. N. and WILLETT, J. H. (1955) A three-slide cloud droplet sampler. *Bull. Am. met. Soc.* **36**, 123. (95)
BROWNE, I. C. (1952a) Radar studies of clouds. Ph.D. Thesis, Cambridge University. (307, 459, 462, 465)
—— (1952b) Precipitation streaks as a cause of radar upper bands. *Q. Jl R. met. Soc.* **78**, 590. (468)
—— PALMER, H. P., and WORMELL, T. W. (1954) The physics of rainclouds. *Q. Jl R. met. Soc.* **80**, 291. (307, 459)
—— and ROBINSON, N. P. (1952) Cross polarisation of the radar melting band. *Nature, Lond.* **170**, 1078. (463–4)

BROWNING, K. A. (1964) Airflow and precipitation trajectories within severe local storms which travel to the right of the winds. *J. atmos. Sci.* **21**, 634. (474)
—— (1965) The evolution of tornadic storms. *J. atmos. Sci.* **22**, 664. (474)
—— (1966) The lobe structure of giant hailstones. *Q. Jl R. met. Soc.* **92**, 1. (287, 345, 474)
—— and BEIMERS, J. G. D. (1967) The oblateness of large hailstones. *J. appl. Met.* **6**, 1075. (333)
—— and DONALDSON, R. J. (1963) Airflow and structure of a tornadic storm. *J. atmos. Sci.* **20**, 533. (474)
—— HALLETT, J., HARROLD, T. W., and JOHNSON, D. (1968) The collection and analysis of freshly fallen hailstones. *J. appl. Met.* **7**, 603. (359)
—— and HARROLD, T. W. (1969) Air motion and precipitation growth in a wave depression. *Q. Jl R. met. Soc.* **95**, 288. (302–6)
—— —— WHYMAN, A. J. and BEIMERS, J. G. D. (1968) Horizontal and vertical air motion and precipitation growth within a shower. *Q. Jl R. met. Soc.* **94**, 498. (450–4)
—— and LUDLAM, F. H. (1962) Airflow in convective storms. *Q. Jl R. met. Soc.* **88**, 117. (287, 359–61, 474, 475–6)
BROWNING, K. A., LUDLAM, F. H., and MACKLIN, W. C. (1963) The density and structure of hailstones. *Q. Jl R. met. Soc.* **89**, 75. (287, 355, 361–2)
—— and WEXLER, R. (1968) A determination of kinematic properties of a wind field using Doppler radar. *J. appl. Met.* **7**, 105. (457)
BROWNSCOMBE, J. L. and HALLETT, J. (1967) Experimental and field studies of precipitation particles formed by the freezing of supercooled water. *Q. Jl R. met. Soc.* **93**, 455. (338)
—— and MASON, B. J. (1966) Measurement of the thermoelectric power of ice by an induction method. *Phil. Mag.* **14**, 1037. (536)
BROWNSCOMBE, J. L. and THORNDIKE, N. S. C. (1968) Freezing and shattering of water drops in free fall. *Nature*, London **220**, 687. (211)
BRUCE, C. E. R. (1944) The initiation of long electrical discharges. *Proc. R. Soc.* **A183**, 228. (510)
BRYANT, G. W. (1967) Thermoelectric power of single crystals of ice containing HF or NH_3. *Phil. Mag.* **16**, 495. (536)
—— and FLETCHER, N. H. (1965) Thermoelectric power of ice containing HF or NH_3. *Phil. Mag.* **12**, 165. (534)
—— HALLETT, J. and MASON, B. J. (1959) The epitaxial growth of ice on single-crystalline substances. *Physics Chem. Solids* **12**, 189. (214, 220–1, 267)
—— and MASON, B. J. (1960a) Photolytic de-activation of silver iodide as an ice-forming nucleus. *Q. Jl R. met. Soc.* **86**, 354. (233)
—— —— (1960b) Etch pits and dislocations in ice crystals. *Phil. Mag.* **5**, 1221. (273)
BURLEY, G. and HERRIN, D. W. (1962) Effect of additives on silver iodide particles exposed to light. *J. appl. Met.* **1**, 355. (235)
BURROWS, D. A., HOBBS, P. V., and SCOTT, W. D. (1967) Factors affecting the electric charge acquired by an ice sphere moving through natural snowfall. *Mon. Weath. Rev. U.S. Dep. Commerce.* **95**, 878. (542)
BUSHBY, F. H. and TIMPSON, M. S. (1967) A 10-level atmospheric model and frontal rain. *Q. Jl R. met. Soc.* **93**, 1. (281)
BYERS, H. R. and BRAHAM, R. R. (1949) *The thunderstorm: report of the thunderstorm project.* U.S. Govt. Printing Office, Washington. (471–3)

—— and HALL, R. K. (1955) A census of cumulus-cloud height versus precipitation in the vicinity of Puerto Rico during the winter and spring of 1953–54. *J. Met.* **12**, 176. (295)

—— SIEVERS, J. R. and TUFTS, B. J. (1957) Distribution in the atmosphere of certain particles capable of serving as condensation nuclei. *Artificial stimulation of rain*, p. 47. Pergamon Press, New York. (62, 63, 87)

CADLE, R. D., FISCHER, W. H., FRANK, E. R., and LODGE, J. P. (1968) Particles in the Antarctic atmosphere. *J. atmos. Sci.* **25**, 100. (72)

—— and ROBBINS, R. C. (1960) Kinetics of atmospheric chemical reactions involving aerosols. *Discuss. Faraday. Soc.* No. 30, p. 155. (71)

CARTE, A. E. (1956) The freezing of water droplets. *Proc. phys. Soc.* **B69**, 1028. (160–1, 167–8)

—— (1961) Air bubbles in ice. *Proc. phys. Soc.* **77**, 757. (339, 345)

—— (1963) Hail studies in the Pretoria–Witwatersrand area. *C.S.I.R. (South Africa) Newsletter* No. 145. (334)

—— (1966) Features of Transvaal hailstorms. *Q. Jl R. met. Soc.* **92**, 290 (334)

—— and KIDDER, R. E. (1966) Transvaal hailstones. *Q. Jl R. met Soc.* **92**, 382. (334, 343)

CARTWRIGHT, J., NAGELSCHMIDT, G., and SKIDMORE, J. W. (1956) The study of air pollution with the electron microscope. *Q. Jl R. met Soc.* **82**, 82. (58)

CATON, P. G. F. (1963) The measurement of wind and convergence by Doppler radar. *Proc. 10th Weather Radar Conf., Washington*, p. 290. (456)

—— (1966) A study of raindrop-size distributions in the free atmosphere. *Q. Jl R. met. Soc.* **92**, 15. (443)

—— (1967) A study of vertical air motion and particle size in showers using a Doppler radar. *Sci, Pap., Met. Office, Lond.* No. 26. (448)

CHALMERS, J. A. (1956) The vertical electric current during continuous rain and snow. *J. atmos. terr. Phys.* **9**, 311. (531, 561)

CHALMERS, J. A. (1967) *Atmospheric electricity*, 2nd edn, p. 251. Pergamon Press, Oxford, London. (494)

—— and LITTLE, E. W. R. (1939) Electric charge on soft hail. *Nature, Lond.* **143**, 3615. (566)

—— —— (1940) The electricity of continuous rain. *Terr. Magn. atmos. Elect.* **45**, 451. (561)

—— and PASQUILL, F. (1938) The electric charges on single raindrops and snowflakes. *Proc. phys. Soc.*, **50**, 1. (562)

CHAPMAN, F. W. (1939) Atmospheric disturbances due to thundercloud discharge, I. *Proc. phys. Soc.* **51**, 876. (562)

CHAPMAN, S. (1950) Hydrometeors and thunderstorm electricity. *Proc. Conf. on Thunderstorm Electricity, Chicago*, p. 149. (515)

—— (1952) Thundercloud electrification studies, II. *Cornell Aero. Lab. Rep.* No. VC 603-P-1. (523)

CHATERJEE, R. N., BISWAS, K. R., and RAMANA MURTY, B. V. (1968) Result of cloud seeding experiment at Delhi as assessed by radar. *Indian J. Met. Geophys.* **20**, 11. (380)

CHIPLONKAR, M. W. (1940) Measurement of point discharge current during disturbed weather at Colaba. *Proc. Indian Acad. Sci.* **12**, 50. (559)

CLAGUE, L. F. (1965) An improved device for obtaining cloud droplet samples. *J. appl. Met.* **4**, 549. (95)

CLARK, W. E. and WHITBY, K. T. (1967) Concentration and size distribution measurements of atmospheric aerosols and a test of the theory of self-preserving size distributions. *J. atmos. Sci.* **24**, 677. (59)

COBB, W. E. (1968) The atmospheric electric climate at Mauna Loa Observatory, Hawaii. *J. atmos. Sci.*, **25**, 470. (486)

COONS, R. D., JONES, E. L., and GUNN, R. (1948) Second partial report on the artificial production of precipitation: cumuliform clouds, Ohio, 1948.

—— —— —— (1949) Gulf States 1949. *Bull. Am. met. Soc.* **29**, 544; **30**, 289. (372, 377)

COOPER, B. F. (1951) A balloon-borne instrument for telemetering raindrop-size distribution and rainwater content of cloud. *Aust. J. appl. Sci.* **2**, 43. (604)

CORNFORD, S. G. (1967) Sampling errors in measurements of raindrop and cloud droplet concentrations. *Met. Mag., Lond.* **96**, 271. (98)

CORRIN, M. L., EDWARDS, H. W., and NELSON, J. A. (1964) The preparation of silver iodide free of hygroscopic impurities and its interaction with water vapor. *J. atmos. Sci.* **21**, 565. (234)

COULIER, P. J. (1875) Note sur une nouvelle propriété de l'air. *J. Pharm. Chim., Paris* **22**, 165. (31)

COURTNEY, W. G. (1963) Kinetics of condensation of water vapor—experimental. *J. chem. Phys.* **38**, 1448. (12)

CRADDOCK, J. M. (1949) The development of cumulus cloud: results of observations made in Malaya. *Q. Jl R. met. Soc.* **75**, 147. (294)

CUNNINGHAM, R. M. (1951) Some observations of natural precipitation processes. *Bull. Am. met. Soc.* **32**, 334. (297)

—— (1952) Distribution and growth of hydrometeors around a deep cyclone. *M.I.T. Weather Radar Res. Rep.* No. 18. (297–302)

CWILONG, B. M. (1947a) Sublimation in a Wilson chamber. *Proc. R. Soc.* A**190**, 137. (164, 175–6)

—— (1947b) Observations on incidence of supercooled water in expansion chambers and on cooled solid surfaces. *J. Glaciol.* **2**, 53. (173)

—— (1947c) Sublimation in outdoor air. *Nature, Lond.* **160**, 198. (176, 200)

DAS, P. K. (1950) The growth of cloud droplets by coalescence. *Indian J. Met. Geophys.* **1**, 137. (572)

DAVIES, C. N. (1942) Investigations on falling drops. (Quoted by O. G. Sutton) *Met. Res. Pap., Met. Office, Lond.* No. 40. (597)

DAVIES, D. A. (1950) Tropical rainfall from cloud which did not extend to the freezing level. *Met. Mag., Lond.* **79**, 354. (294)

—— HEPBURN, D. and SANSOM, H. W. (1952) Report on experiments at Kongwa on artificial control of rainfall, Jan.–April 1952. *Mem. E. Afr. met. Dep.*, **2**, No. 10. (380)

DAVIS, B. L. (1969) Chemical complexing of AgI–NaI aerosols. *J. atmos. Sci.* **26**, 1042. (229)

DAVIS, M. H. (1964) Two charged spherical conductors in a uniform electric field; forces and field strength. *Q. Jl Mech. appl. Math.* **17**, 499. (152, 580)

—— (1969) The slow translation and rotation of two unequal spheres in a viscous fluid. *Chem. Engng Sci.* **24**, 1769. (576)

—— and SARTOR, J. D. (1967) Theoretical collision efficiencies for small cloud droplets in Stokes flow. *Nature, Lond.* **215**, 1371. (152, 575–7)

DAY, G. A. (1953) Radar observations of rain at Sydney, N.S.W. *Aust. J. Phys.* **6**, 229. (293)

—— (1958) Sublimation nuclei. *Proc. phys. Soc.* **72**, 296. (192)

DAY, J. A. (1964) Production of droplets and salt nuclei by the bursting of air-bubble films. *Q. Jl R. met. Soc.* **90**, 72. (76)

DEAN, W. R. and O'NEILL, M. E. (1963) A slow motion of viscous liquid caused by the rotation of a solid sphere. *Mathematika* **10**, 13. (576)

DECKER, W. L. and SCHICKEDANZ, P. T. (1967) The evaluation of rainfall records from a five year cloud seeding experiment in Missouri. *Proc. 5th Berkeley Symposium on Mathematical Statistics and Probability*, **5**, 55. (389)

DEFANT, A. (1905) Gesetzmässigkeiten in der Verteilung der verschiedenen Tropfengrössen bei Regenfällen. *Sber. Akad. wiss. Wien* **114**, 585. (608)

DENNIS, W. L. (1960) The growth of hygroscopic drops in a humid air stream. *Discuss. Farad. Soc.* No. 30, p. 78. (125)

DESSENS, H. (1946a) La brume et le brouillard étudiés à l'aide des fils d'arraignées. *Annls Géophys.* **2**, 276. (45, 56)

—— (1946b) Étude d'une particule de brume. *Annls Géophys.* **2**, 343. (45, 56)

—— (1947a) Brume et noyaux de condensation. *Annls Géophys.* **3**, 68. (45, 56)

—— (1947b) Les noyaux de condensation de l'atmosphère. *Météorologie*, p. 321. (45, 56)

—— (1949) The use of spiders' threads in the study of condensation nuclei. *Q. Jl R. met. Soc.* **75**, 23. (28, 45, 56)

—— (1952) Sur la microstructure et la précipitation artificielle d'un brouillard surfondu. *C. r. hebd. Séanc. Acad. Sci., Paris* **235**, 1675. (381)

—— LAFARGUE, C. and STAHL, P. (1952) Nouvelles réchercher sur les noyaux de condensation. *Annls Géophys.* **8**, 21. (56)

DESSENS, J. (1968) Experience de suppression de la grêle dans le sud-ouest de la France. *Proc. int. Conf. on Cloud Physics, Toronto*, p. 773. (391)

DIEM, M. (1942) Messungen der Grösse von Wolkenelementen, I. *Annln Hydrogr. Berl.* **70**, 142. II. (1948) *Met. Rdsch.* **1**, 261. (96, 98, 109, 114)

—— (1968) Zur Struktur der Niederschläge III. *Arch. Met. Geophys. Bioklim.* B**16**, 347. (479)

DINGER, J. E. (1965) Electrification associated with the melting of snow and ice. *J. atmos. Sci.* **22**, 162. (528, 530)

—— and GUNN, R. (1946) Electrical effects associated with a change of state of water. *Terr. Magn. atmos. Elect.* **51**, 477. (527)

DINGLE, A. N. and HARDY, K. R. (1962) The description of rain by means of sequential raindrop-size distributions. *Q. Jl R. met. Soc.* **88**, 301. (610, 613)

—— and SCHULTE, H. (1962) A research instrument for the study of raindrop-size spectra. *J. appl. Met.* **1**, 48. (606–7)

DONALDSON, R. J. (1958) Vertical profiles of radar echo reflectivity in thunderstorms. *Proc. 7th Weather Radar Conf., Miami*, p. B8. (423)

—— (1959) Analysis of severe convective storms observed by radar, Pt. 2. *J. Met.* **16**, 281. (423)

—— (1961) Radar reflectivity profiles in thunderstorms. *J. Met.* **18**, 292. (423)

—— (1967) Horizontal wind measurement by Doppler radar in a severe squall line. *Proc. Conf. on Severe Local Storms, St Louis*, p. 89 (457)

DORSCH, R. G. and HACKER, P. T. (1950) Photomicrographic investigation of spontaneous freezing temperatures of supercooled water droplets. *Tech. Notes natn. advis. Comm. Aeronaut., Wash.* No. 2142. (160)

DOUGLAS, R. H., GUNN, K. L. S., and MARSHALL, J. S. (1957). Pattern in the vertical of snow generation. *J. Met.* **14**, 95. (470)

—— and HITSCHFELD, W. (1961) Radar reflectivities of hail samples. *Proc. 9th Weather Radar Conf., Kansas City*, p. 147. (425)

DRAGINIS, M. (1958) Liquid water within convective clouds. *J. Met.* **15**, 481. (112, 120)

DRAKE, J. C. (1968) Electrification accompanying the melting of ice particles. *Q. Jl R. met. Soc.* **94**, 176. (530)
—— and MASON, B. J. (1966) Melting of small ice spheres and cones. *Q. Jl R. met. Soc.* **92**, 500. (368, 529)
DRIVING, A. Ia., MIRONOV, A. V., MOROZOV, V. M., and KHVOSTIKNOV, I. A. (1943) The study of optical and physical properties of natural fogs. *Izv. Akad. Nauk SSSR, Ser. geograf. geofiz.* **2**, 70. (97)
DURBIN, W. G. (1959) Droplet sampling in cumulus clouds. *Tellus* **11**, 202. (138–9)
DU TOIT, P. S. (1967) Doppler radar observations of drop sizes in continuous rain. *J. appl. Met.* **6**, 1082. (445)
DYE, J. E. and HOBBS, P. V. (1966) Effect of carbon dioxide on the shattering of freezing water drops. *Nature, Lond.* **209**, 464. (209)
—— —— (1968) The influence of environmental parameters on the freezing and fragmentation of suspended water drops. *J. atmos. Sci.* **25**, 82. (210)

EAST, T. W. R. and MARSHALL, J. S. (1954) Turbulence in clouds as a factor in precipitation. *Q. Jl R. met. Soc.* **80**, 26. (153)
EASTERBROOK, C. C. (1967) Some Doppler radar measurements of circulation patterns in convective storms. *J. appl. Met.* **6**, 882. (457)
EDWARDS, G. R. and EVANS, L. F. (1960) Ice nucleation by silver iodide. I, Freezing vs. sublimation. *J. Met.* **17**, 627. (190, 220)
—— —— (1968) Ice nucleation by silver iodide. III, The nature of the nucleating site. *J. atmos. Sci.* **25**, 249. (190, 220)
EIGEN, M. and DE MAEYER, L. (1958) Self-dissociation and protonic charge transport in water and ice. *Proc. R. Soc.* A**247**, 505. (533)
ELDRIDGE, R. G. (1957) Measurement of cloud drop-size distributions. *J. Met.* **14**, 55. (97)
ELSTER, J. and GEITEL, H. (1913) Zur Influenztheorie der Niederschlagselektrizität. *Phys. Z.* **14**, 1287. (521)
ELTON, G. A. H., MASON, B. J., and PICKNETT, R. G. (1958) The relative importance of condensation and coalescence processes on the stability of a water fog. *Trans. Faraday Soc.* **54**, 1724. (138, 146)
ERIKSSON, E. (1952) Composition of atmospheric precipitation. I, Nitrogen compounds. *Tellus* **4**, 215. (72)
—— (1955) Airborne salts and the chemical composition of river waters. *Tellus* **7**, 243. (87)
ETTE, A. I. I. (1966) Measurement of electrode by-passing efficiency in living trees. *J. atmos. terr. Phys.* **28**, 295. (558)
EVANS, D. G. and HUTCHINSON, W. C. A. (1963) The electrification of freezing water droplets and of colliding ice particles. *Q. Jl R. met. Soc.* **89**, 370. (208, 541, 544)
EVANS, L. F. (1966) Ice nucleation by amino acids. *J. atmos. Sci.* **23**, 751. (191, 225)

FAGE, A. (1937) Experiments on a sphere at critical Reynolds numbers. *Aeronaut. Res. Comn. Rep. and Mem.* No. 1766. (599)
FARKAS, L. (1927) Keimbildungsgeschwindigkeit in übersättigten Dämpfen. *Z. phys. Chem.* A**125**, 236. (11)
FARLEY, F. J. M. (1952) The theory of condensation of supersaturated ion-free vapour. *Proc. R. Soc.* A**212**, 530. (6–10)
FENN, R. W., GERBER, H. E., and WASSHAUSEN, D. (1963) Measurements of the

sulphur and ammonium component of the arctic aerosol of the Greenland ice cap. *J. atmos. Sci.* **20**, 466. (72)
—— and WEICKMANN, H. K. (1959) Some results of aerosol measurements. *Geofis. pura appl.* **42**, 53. (178, 200)
FETERIS, P. J. and MASON, B. J. (1956) Radar observations of showers suggesting a coalescence mechanism. *Q. Jl R. met. Soc.* **82**, 446. (294)
FINDEISEN, W. (1938) Die kolloidmeteorologischen Vorgänge bei der Niederschlagsbildung (Colloidal meteorological processes in the formation of precipitation). *Met. Z.* **55**, 121. (370)
—— (1939a) Das Verdampfen der Wolken-und-Regentropfen (The evaporation of cloud-and raindrops). *Met. Z.* **56**, 453. (282)
—— (1939b) Zur Frage der Regentropfenbildung in reinen Wasserwolken (On the question of rain drop formation in pure water clouds). *Met. Z.* **56**, 365. (285, 308)
—— (1940) Über die Entstehung der Gewitterelektrizität. *Met. Z.* **57**, 201. (539, 546)
—— (1942a) Experimentelle Untersuchungen über die Eisteilchenbildung. *Met. Z.* **59**, 349. (175, 189)
—— (1942b) Ergebnisse von Wolken- und Niederschlagsbeobachtungen bei Wettererkundungsflügen über See. *Forsch.-u. Erfahr Ber. Reichsomt Wetter-Dienst.* Ser. B, No. 8. (290)
—— and FINDEISEN, E. (1943) Untersuchungen über die Eissplitterbildung an Reifschichten. *Met. Z.* **60**, 145. (539, 546)
—— and SCHULZ, G. (1944) Experimentelle untersuchungen zur atmosphärischen eisteilchenbildung, I. *Forsch-u. Erfahr Ber. Reichsamt WetterDienst.* Ser. A, No. 27. (175, 183–6)
FIRST, M. W. and SILVERMAN, L. (1953) Air sampling with membrane filters. *Archs ind. Hyg.* **7**, 1. (48)
FITZGERALD, D. (1965) Measurement techniques in clouds. *Problems of atmospheric and space electricity*, p. 199. Elsevier, Amsterdam. (516)
—— and BYERS, H. R. (1962) Aircraft electrostatic measurement instrumentation and observations of cloud electrification. *Contract Rep. AF* 19(604) 2189. University of Chicago. (517)
FLEISHER, A. (1953) Information contained in weather noise. *M.I.T. Weather Radar Res. Rep.* No. 22. (441)
FLETCHER, N. H. (1958) Size effect in heterogeneous nucleation. *J. chem. Phys.* **29**, 572; **31**, 1136. (22, 217–20)
—— (1962a) *The physics of rainclouds*, p. 52. Cambridge University Press. (21)
—— (1962b) *The physics of rainclouds*, p. 241. Cambridge University Press. (187)
—— (1970) The chemical physics of ice, chapter 4. Cambridge University Press. (172)
FLOOD, H. (1933) Doctoral dissertation, Berlin, quoted in Volmer, *Kinetik der Phasenbildung*, p. 132. Steinkopff, Dresden and Leipzig, 1939. (18)
FONDA, A. and HERNE, H. (1957) The aerodynamic capture of particles by spheres. *National Coal Board Min. Res. Est. Rep* No. 2068. (572–3)
FOOTE, G. B. (1966) A Z–R relation for mountain thunderstorms. *J. appl. Met.* **2**, 229. (479)
—— (1969) On the internal circulation and shape of large raindrops. *J. atmos. Sci.* **26**, 179. (599)
—— and DU TOIT, P. S. (1969) Terminal velocity of raindrops aloft. *J. appl. Met.* **8**, 249. (598)
FOSTER, H. (1950) An unusual observation of lightning. *Bull. Am. met. Soc.* **31**, 40. (519)

FOURNIER, d'ALBE, E. M. (1949) Some experiments on the condensation of water vapour at temperatures below 0°C. *Q. Jl R. met. Soc.* **75**, 1. (164, 189, 191)
—— (1951) Sur les embruns marins. *Bull. Inst. Océanogr. Monaco* **48**, No. 995. (46)
—— and HIDAYETULLA, M. S. (1955) The break-up of large water drops falling at terminal velocity in free air. *Q. Jl R. met. Soc.* **81**, 610. (600)
—— LATEEF, A. M. A., RASOOL, S. I., and ZAIDI, I. H. (1955) The cloud-seeding trials in the central Punjab, July–September 1954. *Q. Jl R. met. Soc.* **81**, 574. (380)
FRANK, F. C. (1949) The influence of dislocations on crystal growth. *Discuss. Faraday Soc.* No. 5, p. 48. (273)
FRANK, H. S. and WEN, W. Y. (1957) Structural aspects of ion solvent interaction in aqueous solutions: a suggested picture of water structure. *Discuss. Faraday Soc.* No. 24, p. 133. (171)
FRASER, D., RUSH, C. K., and BAXTER, D. (1952) Thermodynamic limitations of ice accretion instruments. *Natn. Aero. Est. Canada Lab. Rep. LR*-32. (115)
FRENKEL, J. (1946) *Kinetic theory of liquids*, pp. 382, 374. Clarendon Press, Oxford. (5)
FRENKEL, Y. I. (1944) A theory of the fundamental phenomena of atmospheric electricity (in Russian). *Fiz. Zh.* **8**, 285. (522)
—— (1946) Influence of water drops on the ionization and electrification of air (in Russian). *Fiz. Zh.* **10**, 151. (522)
—— (1947) Atmospheric electricity and lightning. *J. Franklin Inst.* **243**, 287. (522)
FREY, F. E. (1941) Über die Kondensation von Dämpfen in einem Trägergas. *Z. phys. Chem.* **B49**, 83. (14, 15)
FRIEDLANDER, S. K. (1960) Similarity considerations for the particle size spectrum of a coagulating sedimenting aerosol. *J. Met.* **17**, 479. (59)
—— (1961) Theoretical considerations for the particle size spectrum of the stratospheric aerosol. *J. Met.* **18**, 753. (59)
—— and PASCERI, R. E. (1965a, b) Measurements of particle size distribution of the atmospheric aerosol. I, Introduction and Experimental methods. II, Experimental results and discussion. *J. atmos. Sci.* **22**, 571, 577. (58–9)
FRITH, R. (1951) The size of cloud particles in stratocumulus cloud. *Q. Jl R. met. Soc.* **77**, 441. (94, 109, 121, 308)
FRÖSSLING, N. (1938) Über die Verdunstung Fallender Tropfen. *Beitr. Geophys.* **52**, 170. (351)
FUCHS, N. and PETRYANOFF, I. (1937) Microscopic examination of fog-, cloud- and rain-droplets. *Nature, Lond.* **139**, 111. (93)
FUJIWARA, M. (1967) Preliminary report on Hawaii rain mechanism. *Tellus* **3**, 392. (479)
FUKUTA, N. (1958) Experimental investigations on the ice-forming ability of various chemical substances. *J. Met.* **15**, 17. (214, 216, 218)
—— (1963) Ice nucleation by metaldehyde. *Nature, Lond.* **199**, 475. (225)
—— and MASON, B. J. (1963) Epitaxial growth of ice on organic crystals. *Physics Chem. Solids* **24**, 715. (223, 225)
FUQUAY, D. M. and WELLES, H. J. (1957) The Project Skyfire cloud seeding generator. *U.S. Advisory Committee on Weather Control Final Report*, Vol. 2, p. 273. (230)

GABRIEL, K. R. (1967) The Israeli artificial rain stimulation experiment. Statistical evaluation for 1961–65. *Proc. Berkeley Symposium on Mathematical Statistics and Probability*, Vol. 5, p. 91. (390)

—— AVICHAI, Y. and STEINBERG, R. (1967) A statistical investigation of persistence in the Israeli artificial rainfall stimulation experiment. *J. appl. Met.* **6**, 323. (390)

GAIVORONSKII, I. I., SEREGIN, J. A., and VORONOV, G. S. (1968) Investigations of hail processes and their artificial modification in flat regions of the USSR. *Proc. int. Conf. on Cloud Physics, Toronto*, p. 760. (392)

GANS, R. (1912) Uber die Form Ultramikroskopischer Goldteilchen. *Annln Phys.* **37**, 881. (406)

GARNER, F. H. and LANE, J. J. (1959) Mass transfer to drops of liquid suspended in a gas stream. *Trans. Instn chem. Engrs* **37**, 167. (599)

GARROD, M. P. (1957) Recent developments in the measurement of precipitation elements from aircraft. *Met. Res. Pap., Met. Office, Lond.* No. 1050. (606)

GARTEN, V. A. and HEAD, R. B. (1964) Hydrogen-bonding patterns and ice nucleation. *Nature, Lond.* **204**, 573. (225)

—— and HEAD, R. B. (1965) A theoretical basis of ice nucleation by organic crystals. *Nature, Lond.* **205**, 160. (225)

GEORGII, H. W. (1959) Neue Untersuchungen über den Zusammenhang zwischen atmosphärischen Gefrierkernen und Kondensationskernen (Recent studies on the relationship between atmospheric freezing nuclei and condensation nuclei). *Geofis. pura appl.* **42**, 62. (177)

—— (1963) Investigations on the deactivation of inorganic and organic freezing-nucleii. *Z. angew. Math. Phys.* **14**, 503. (194)

—— and METNIEKS, A. L. (1958) An investigation into the properties of atmospheric freezing nuclei and sea-salt nuclei under maritime conditions at the west coast of Ireland. *Geofis. pura appl.* **41**, 159. (199)

GEOTIS, S. G. (1963) Some radar measurements of hailstorms. *J. appl. Met.* **2**, 270. (423–5)

GERHARD, E. R. and JOHNSTONE, H. F. (1955) Photochemical oxidation of sulphur dioxide in air. *Ind. Engng Chem. ind. Edn* **47**, 972. (70)

GERHARDT, J. R., TOLBERT, C. W., BRUNSTEIN, S. A., and BAHN, W. W. (1961a) Experimental determinations of the back-scattering cross-sections of water drops and of wet and dry ice spheres at 3·2 centimeters. *J. Met.* **18**, 340. (418)

—— —— (1961b) Further studies of the back-scattering cross-sections of water drops and wet and dry-ice spheres. *J. Met.* **18**, 688. (420)

GIBBS, J. W. (1875) On the equilibrium of heterogeneous substances. *Trans. Conn. Acad. Arts Sci.* **3**, 108. See also *Collected papers (thermodynamics)*, p. 55. Longmans, London, 1906. (2)

GILL, E. W. B. and ALFREY, G. F. (1952) Production of electric charge on water drops. *Nature, Lond.* **169**, 203. (549)

GISH, O. H. and SHERMAN, K. L. (1936) Electrical conductivity of air to an altitude of 22 km. *Natn. geogr. Soc. Pap.* Stratosphere Ser. 2, p. 94. (487)

—— and WAIT, G. R. (1950) Thunderstorms and the earth's general electrification. *J. geophys. Res.* **55**, 473. (566)

GOETZ, A. and PREINING, O. (1960) The aerosol spectrometer and its application to nuclear condensation studies. *Physics of precipitation, Geophys. Mem.* No. 5, p. 164. American Geophysical Union. (45)

GOLD, L. W. and POWER, B. A. (1952) Correlation of snow-crystal type with estimated temperature of formation. *J. Met.* **9**, 447. (246–7)

—— —— (1954) Dependence of the forms of natural snow crystals on meteorological conditions. *J. Met.* **11**, 35. (247)

GOLDE, R. H. (1945) Frequency of occurrence of lightning flashes to earth. *Q. Jl R. met. Soc.* **71**, 89. (566)

GOLDSMITH, P., DELAFIELD, H. J. and COX, L. C. (1963) The role of diffusiophoresis in the scavenging of radioactive particles from the atmosphere. *Q. Jl R. met. Soc.* **89**, 43. (89)

GOLOVIN, A. M. (1963) The solution of the coagulation equation for cloud droplets in a rising air current. *Izv. Akad. Nauk SSSR Ser. geofiz.* No. 5, p. 783. (146–7)

GRANT, L. O. and STEELE, R. L. (1966) The calibration of silver iodide generators. *Bull. Am. met Soc.* **47**, 713. (231)

GRENET, G. (1947) Essai d'explication de la charge éléctrique des nuages d'orages. *Annls Géophys.* **3**, 306. (525)

GROSS, G. W. (1965) The Workman–Reynolds effect and ionic transfer processes at the ice-solution interface. *J. geophys. Res.* **70**, 2291. (527)

GRUNOW, J. (1960) Snow crystal analysis as a method of indirect aerology. *Physics of precipitation*, Geophys. Monogr. No. 5, p. 130. American Geophysical Union. (247)

GSCHWEND, P. (1922) Beobachtungen über die electrischen Ladungen einzelner Regentropfen und Schneeflocken. *Jb. Radioakt. Electronik.* **17**, 62. (562, 565)

GUCKER, F. T., O'KONSKI, C. T., PICKARD, H. B., and PITTS, J. N. (1947) A photoelectronic counter for colloidal particles. *J. Am. chem. Soc.* **69**, 2422. (48)

—— and ROSE, D. G. (1954) A photoelectronic instrument for counting and sizing aerosol particles. *Br. J. appl. Phys.* Suppl. 3, p. S.138. (48)

GUNN, K. L. S. and EAST, T. W. R. (1954) The microwave properties of precipitation particles. *Q. Jl R. met. Soc.* **80**, 522. (427–8)

—— and HITSCHFELD, W. (1951) A laboratory investigation of the coalescence between large and small water drops. *J. Met.* **8**, 7. (582)

—— LANGLEBEN, M. P., DENNIS, A. S. and POWER, B. A. (1954) Radar evidence of a generating level for snow. *J. Met.* **11**, 20. (470)

—— and MARSHALL, J. S. (1958) The distribution with size of aggregate snowflakes. *J. Met.* **15**, 452. (479)

GUNN, R. (1947) The electrical charge on precipitation at various altitudes and its relation to thunderstorms. *Phys. Rev.* **71**, 181. (562, 565)

—— (1948) Electric field intensity inside natural clouds. *J. appl. Phys.* **19**, 481. (515)

—— (1949) The free electrical charge on thunderstorm rain and its relation to droplet size. *J. geophys. Res.* **54**, 57. (563, 565)

—— (1950) Free electrical charge on precipitation inside an active thunderstorm. *J. geophys. Res.* **55**, 171. (563, 565)

—— and DEVIN, C. (1953) Raindrop charge and electric field in active thunderstorms. *J. Met.* **10**, 279. (563)

—— and KINZER, G. D. (1949) The terminal velocity of fall for water droplets in stagnant air. *J. Met.* **6**, 243. (565, 594, 596–7)

HAGEMANN, V. (1935) Eine methode zur Bestimmung der Grösse der Nebel- und Wolkenelemente. *Beitr. Geophys.* **46**, 261. (110)

HALL, F. (1957) The weather Bureau ACN project. *Met. Monogr.* **2**, 24. American Meteorological Society, Boston. (395)

HALL, P. G. and TOMPKINS, F. C. (1962) Adsorption of water vapour on insoluble metal halides. *Trans. Faraday Soc.* **58**, 1734. (234)

HALLETT, J. (1961) The growth of ice crystals on freshly-cleaved covellite surfaces. *Phil Mag.* **6**, 1073. (268)

—— (1964) Experimental studies of the crystallization of supercooled water. *J. atmos. Sci.* **21**, 671. (248, 335, 338, 342)

HALLETT, J. and MASON, B. J. (1958a) The influence of temperature and supersaturation on the habit of ice crystals grown from the vapour. *Proc. R. Soc.* A**247**, 440. (258–61, 264)

—— —— (1958b) Influence of organic vapours on the crystal habit of ice. *Nature, Lond.* **181**, 467. (265)

HAMA, K. and ITOO, K. (1956) Freezing of supercooled water droplets. *Pap. Met. Geophys., Tokyo* **7**, 99. (195)

HANNAN, E. J. (1955) A test for singularities in Sydney rainfall. *Aust. J. Phys.* **8**, 289. (203)

HARDY, K. R. (1963) The development of raindrop-size distributions and implications related to the physics of precipitation. *J. atmos. Sci.* **20**, 299. (463, 612–3)

HARPER, W. G. (1957) Variation with height of rainfall below the melting level. *Q. Jl R. met. Soc.* **83**, 368. (463)

—— (1962) Radar back-scattering from oblate spheroids. *Nubila* **5**, 66. (420)

—— (1964) Cloud detection with 8·6-millimetre wavelength radar. *Met. Mag., Lond.* **93**, 337. (471)

—— LUDLAM, F. H. and SAUNDERS, P. M. (1956) Preliminary report on cumulus investigations. *Met. Res. Pap., Met. Office, Lond.* No. 1019. (473)

HARROLD, T. W. (1965) Estimation of rainfall using radar—a critical review. *Sci. Pap., Met. Office, Lond.* No. 21. (477, 480)

—— (1966) Measurement of horizontal convergence in precipitation using a Doppler radar. *Q. Jl R. met. Soc.* **92**, 31. (456–7)

HEAD, R. B. (1961) Steroids as ice nucleators. *Nature, Lond.* **191**, 1058. (223, 225)

—— (1962a) Ice nucleation by some cyclic compounds. *Physics Chem. Solids* **23**, 1371. (223, 225)

—— (1962b) Ice nucleation by alpha-phenazine. *Nature, Lond.* **196**, 736. (225)

HEIM, F. (1914) Diamantstaub und Schneekristalle in der Antarktis. *Met. Z.* **31**, 232. (244)

HEINMETS, F. (1962) Measurement of ice-liquid interphase potentials in protonated and hydroxylated electrolytes. *Trans. Faraday Soc.* **58**, 788. (526)

HENRY, P. S. H. (1953) The role of asymmetric rubbing in the generation of static electricity. *Br. J. appl. Phys.* Suppl. **2**, 31. (541)

HERATH, F. (1914) Die Messung der Niederschlagselektrizitat durch das Galvonometer. *Phys. Z.* **15**, 155. (560)

HERMAN, B. M. and BATTAN, L. J. (1961a) Calculations of the Mie back-scattering of microwaves from ice spheres. *Q. Jl R. met. Soc.* **87**, 223. (417, 419)

—— —— (1961b) Calculations of Mie back-scattering from melting ice spheres. *J. Met.* **18**, 468. (417, 420–1)

HESS, V. F. On the concentration of condensation nuclei in the air over the North Atlantic. (1948) *Terr. Magn. atmos. Elect.* **53**, 399. (1951) *J. geophys. Res.* **56**, 553. (54, 55)

HEVERLY, J. R. (1949) Supercooling and crystallization. *Trans. Am. geophys. Un.* **30**, 205. (157)

HEYWOOD, G. S. P. (1940) Rain formation in the tropics. *Q. Jl R. met. Soc.* **66**, 46. (294)

HIGUCHI, K. (1956) A new method for the simultaneous observation of shape and size of a large number of falling snow particles. *J. Met.* **13**, 274. (238)

—— (1958) The etching of ice crystals. *Acta metall.* **6**, 636. (273)

—— (1962a) A case study of snowfall from clouds under a subsidence inversion. *J. met. Soc. Japan* **40**, 65. (247)

—— (1962b) Horizontal distribution of snow crystals during the snowfall, II and III. *J. met. Soc. Japan* **40**, 73, 266. (247)

HIGUCHI, K. and FUKUTA, N. (1966) Ice in the capillaries of solid particles and its effect on their nucleating ability. *J. atmos. Sci.* **23**, 187. (192)

HOBBS, P. V. (1964) The effect of air bubbles in ice on charge transfer produced by asymmetrical rubbing. *J. atmos. Sci.* **21**, 706. (538)

—— (1965) The aggregation of ice particles in clouds and fogs at low temperatures. *J. atmos. Sci.* **22**, 296. (250)

—— and ALKEZWEENY, A. J. (1968) The fragmentation of freezing water droplets in free fall. *J. atmos. Sci.* **25**, 881. (210)

—— and MASON, B. J. (1964) The sintering and adhesion of ice. *Phil. Mag.* **9**, 181. (250)

HOCHSCHWENDER, E. (1919) Ph.D. Dissertation, Heidelberg. (523)

HOCKING, L. M. (1959) The collision efficiency of small drops. *Q. Jl R. met. Soc.* **85**, 44. (145, 152, 574–7)

—— and JONAS, P. R. (1970) The collision efficiency of small drops. *Q. Jl R. met. Soc.* **96**, 722. (152–3, 575–81, 584)

HOFFER, T. E. (1961) A laboratory investigation of droplet freezing. *J. Met.* **18**, 766. (160, 162, 165, 167, 181, 190–1)

—— WEBER, K. E. and FRITZEN, J. S. (1964) Note on preparation of sodium silicate solutions for ice nuclei detection. *J. appl. Met.* **3**, 489. (182)

HOOPER, J. E. N. and KIPPAX, A. A. (1950a) Radar echoes from meteorological precipitation. *Proc. Instn. elect. Engrs* **97**, 89. (413–5, 459, 468)

—— —— (1950b) The bright band—a phenomenon associated with radar echoes from falling rain. *Q. Jl R. met. Soc.* **76**, 125. (462, 465)

HOSLER, C. L. (1950) Preliminary investigation of condensation nuclei under the electron microscope. *Trans. Am. geophys. Un.* **31**, 707. (35)

—— (1951) On the crystallization of supercooled clouds. *J. Met.* **8**, 326. (214)

—— (1954) Factors governing the temperature of ice-crystal formation in clouds. *Proc. Toronto Met. Conf.*, 1953, p. 253. *R. met. Soc.*, London. (157)

—— and HALLGREN, R. E. (1961) Ice crystal aggregation. *Nubila* **4**, 13. (249)

—— JENSEN, D. C. and GOLDSHLAK, L. (1957) On the aggregation of ice crystals to form snow. *J. Met.* **14**, 415. (250)

HOUGHTON, H. G. (1950) A preliminary quantitative analysis of precipitation mechanisms. *J. Met.* **7**, 363. (275, 286)

—— and RADFORD, W. H. (1938) On the measurement of drop size and liquid water content in fogs and clouds. *M.I.T. Pap. Phys. Oceanogr. Met.* **6**, No. 4. (110)

HOWELL, W. E. (1949) The growth of cloud drops in uniformly cooled air. *J. Met.* **6**, 134. (125, 127–132)

HUFF, F. A. (1966) The adjustment of radar estimates of storm mean rainfall with raingage data. *Proc. 12th Weather Radar Conf.*, Norman, Oklahoma. p. 198. (481)

HUMPREYS, W. J. (1929) *Physics of the air*, p. 516. McGraw-Hill, London and New York. (245)

HUNT, T. L. (1949) Formation of rain. *Met. Mag., Lond.* **78**, 26. (294)

HUNTER, I. M. (1954) Polarisation of radar echoes from meteorological precipitation. *Nature, Lond.* **173**, 165. (464)

HUTCHINSON, W. C. A. and CHALMERS, J. A. (1951) The electric charges and masses of single raindrops. *Q. Jl R. met. Soc.* **77**, 85. (563, 565)

INN, E. C. Y. (1951) Photolytic inactivation of ice-forming silver iodide nuclei. *Bull. Am. met. Soc.* **32**, 132. (231–2)

IMAI, I. (1960) Raindrop size distributions and Z–R relationships. *Proc. 8th Weather Radar Conf.*, San Francisco, p. 211. (479)
—— FUJIWARA, M., ICHIMURA, I., and TOYAMA, Y. (1955) Radar reflectivity of falling snow. *Pap. Met. Geophys.*, Tokyo **6**, 130. (479)
IRIBARNE, J. V. and MASON, B. J. (1967) Electrification accompanying the bursting of bubbles in water and dilute aqueous solutions. *Trans. Faraday Soc.* **63**, 2234. (528–30)
ISONO, K. (1957) On sea-salt nuclei in the atmosphere. *Geofis pura appl.* **36**, 156. (65)
—— (1958) Mode of growth of ice crystals in air and other gases. *Nature, Lond.* **182**, 1221. (264)
—— and IKEBE, Y. (1960) On the ice-nucleating ability of rock-forming minerals and soil particles. *J. met. Soc. Japan* **38**, 213. (195)
—— KOMABAYASI, M. and ONO, A. (1957) On the habit of ice-crystals grown in the atmospheres of hydrogen and carbon dioxide. *J. met. Soc. Japan* **35**, 327. (264)
—— —— —— (1959) The nature and the origin of ice nuclei in the atmosphere. *J. met. Soc. Japan* **37**, 211. (195, 199)
—— —— YAMANAKA, Y. and FUJITA, H. (1956) An experimental investigation on the growth of ice crystals in a super-cooled fog. *J. met. Soc. Japan* **34**, 158. (278)
ISRAËL, H. (1930) Untersuchungen über schwere Ionen in der Atmosphäre. *Beitr. Geophys.* **26**, 283. (54)
—— (1931) Zur Theorie und Methodik der Grossen bestimmung von Luftionen. *Beitr. Geophys.* **31**, 173. (38)
—— (1932) Zum Problem der Randstörungen bei Ionenmessungen. *Beitr. Geophys.* **35**, 341. (38)

JACCARD, C. (1963) Thermoelectric effects in ice crystals. *Phys. Kondens. Mater.* **2**, 143. (533)
JACOBI, W. (1955) Homogeneous nucleation in supercooled water. *J. Met.* **12**, 408. (160, 168)
JACOBS, M. B., BRAVERMAN, M. M., and HOCHHEISER, S. (1957) Ultramicro determination of sulfides in air. *Analyt. Chem.* **29**, 1349. (67)
JACOBS, W. C. (1937) Preliminary report on the study of atmospheric chlorides. *Mon. Weath. Rev. U.S. Dep. Agric.* **65**, 147. (75)
JAFFRAY, J. and MONTMORY, R. (1956) Congélation orientée de l'eau surfondue sur des surfaces cristallines: 1° cas du mica muscovite. *C. r. hebd. Séanc. Acad. Sci.*, Paris **243**, 126.—— and MONTMORY, R. (1957) Épitaxies de la glace sur l'iodure d'argent. *C. r. hebd. Séanc. Acad. Sci.*, Paris **244**, 2221. (220)
JAYARATNE, O. W. and MASON, B. J. (1964) The coalescence and bouncing of water drops at an air/water interface. *Proc. R. Soc.* A**280**, 545. (587–9)
JAYAWEERA, K. O. L. F. and MASON, B. J. (1965) The behaviour of freely-falling cylinders and cones in a viscous fluid. *J. Fluid Mech.* **22**, 709. (248)
JAYAWEERA, K. O. L. F. and MASON, B. J. (1966) The falling motions of loaded cylinders and discs simulating snow crystals. *Q. Jl R. met. Soc.* **92**, 151. (248)
JEFFREYS, H. (1918) Some problems of evaporation. *Phil. Mag.* **35**, 270. (275)
JELLINEK, H. H. G. (1961) Liquid-like layers on ice. *J. appl. Phys.* **32**, 1793. (250)
—— (1962) Ice adhesion. *Can. J. Phys.* **40**, 1294. (250)
JENSEN, V. G. (1959) Viscous flow round a sphere at low Reynolds numbers. *Proc. R. Soc.* A**249**, 346. (577)

JIUSTO, J. E. (1966) Maritime concentrations of condensation nuclei. *J. Rech. Atmos.* **2**, 245. (86)

JOHNSON, D. A. (1968) An experimental investigation of charge separation due to the fracture of freezing water drops. *Proc. int. Conf. on Cloud Physics, Toronto*, p. 624. (554–5)

—— and HALLETT, J. (1968) Freezing and shattering of supercooled water drops. *Q. Jl R. met. Soc.* **94**, 468. (209)

JOLIVET, J. (1959) La charge électrique des gouttes de pluie dans les Antilles françaises. *Annls Géophys.* **15**, 153. (563)

JONAS, P. R. and MASON, B. J. (1968) Systematic charging of water droplets produced by the break-up of liquid jets and filaments. *Trans. Faraday Soc.* **64**, 1971. (529)

JONES, D. M. A. (1956) Rainfall drop-size distribution and radar reflectivity. *Illinois State Water Survey. Met. Lab. Rep.* No. 6. (479, 609)

—— (1959) The shape of raindrops. *J. Met.* **16**, 504. (600)

JONES, R. F. (1950) Radar weather echoes. Pts. I–IV. *Met. Mag., Lond.* **79**, 109, 143, 170, 198. (432–3)

JOSS, J. and AUFDERMAUR, A. N. (1965) Experimental determination of the radar cross sections of artificial hailstones containing water. *J. atmos. Sci.* **4**, 723. (424)

—— and WALDVOGEL, A. (1967) Ein Spektrograph für Niederschlagstropfen mit automatischer Auswertung. *Geofis. pura appl.* **68**, 240. (604)

—— and WALDVOGEL, A. (1969) Raindrop size distribution and sampling size errors. *J. atmos. Sci.* **26**, 566. (478)

JUNGE, C. (1936) Ubersattigungsmessungen an atmosphärischen Kondensationskernen. *Beitr. Geophys.* **46**, 108. (29)

—— (1950) Das Wachstum der Kondensationskerne mit der relativen Feuchtigkeit. *Annln Met., Hamburg* **3**, 128. (29)

—— (1952a) Die Konstitution des atmosphärischen Aerosols. *Annln Met. (Beiheft)* 1952, p. 1. (28, 30, 42, 46)

—— (1952b) Gesetzmässigkeiten in der Grössenverteilung atmosphärischer Aerosole über dem Kontinent. *Ber. dt. Wetterd. U.S. Zone* **35**, 264. (51, 62)

—— (1953) Die Rolle der Aerosole und der gasförmigen Beimengungen der Luft im Spurenstoffhaushalt der Troposphäre. *Tellus* **5**, 1. (42, 56, 64, 70, 89)

—— (1954) The chemical composition of atmospheric aerosols. I, Measurements at Round Hill Field Station, June–July 1953. *J. Met.* **11**, 323. (64, 70)

—— (1956) Recent investigations in air chemistry. *Tellus* **8**, 127. (64)

—— (1957a) Chemical analysis of aerosol particles and of gas traces on the island of Hawaii. *Tellus* **9**, 528. (64)

—— (1957b) Some facts about meteoritic dust. In *Artificial stimulation of rain* (eds. H. Weickmann and W. Smith), p. 24. Pergamon Press, London and New York. (204)

—— (1958) Atmospheric chemistry. *Adv. Geophys.* **4**, 1. (57, 60, 82)

—— (1960) Sulphur in the atmosphere. *J. geophys. Res.* **65**, 227. (67)

—— (1963) *Air chemistry and radioactivity*, chapter 2. Academic Press, New York and London. (61, 82, 87, 529)

—— and MANSON, J. E. (1961) Stratospheric aerosol studies. *J. geophys. Res.* **66**, 2163. (201)

—— and RYAN, T. G. (1958) Study of the SO_2 oxidation in solution and its role in atmospheric chemistry. *Q. Jl R. met. Soc.* **84**, 46. (70, 71)

KACHURIN, L. G. and BEKRYAEV, V. I. (1960) Investigation of the electrification of crystallizing water (in Russian). *Dokl. Akad. Nauk SSSR* **130**, 57. (543, 545)

KANTROWITZ, A. (1951) Nucleation in very rapid vapour expansions. *J. chem. Phys.* **19**, 1097. (12)

KARASZ, F. E., CHAMPION, W. M., and HALSEY, G. D. (1956) The growth of ice layers on the surfaces of anatase and silver iodide. *J. phys. Chem., Ithaca* **60**, 376. (234)

KARTSIVADZE, A. I. (1968) Modification of hail processes. *Proc. int. Conf. on Cloud Physics, Toronto*, p. 778. (392)

KASSANDER, A. R., SIMS, L. L., and MCDONALD, J. E. (1957) Observations of freezing nuclei over the south-western U.S. *Artificial stimulation of rain*, p. 392. Pergamon Press, London and New York. (200)

KATZ, R. E. and CUNNINGHAM, R. M. (1948) Aircraft icing instruments. *M.I.T. Met. Dep. De-icing Res. Lab. Report.* (114)

KATZ, U. (1960) Zur Eiskeimbildungsfähigkeit von Kupferoxyden und Kupfersulfiden. *Z. angew. Math. Phys.* **11**, 237. (214, 216, 218)

—— (1962) Wolkenkammeruntersuchungen der Eiskeimbildungsaktivität einiger ausgewählter Stoffe. *Z. angew. Math. Phys.* **13**, 333. (219)

KAZAS, V. I. (1963) The use of a continuous photoelectric device for studying the microstructure of clouds from an aircraft. *Izv. Akad. Nauk SSSR Ser. geofiz.* (in Russian) No. 5, p. 494. (97)

KEILY, D. P. and MILLEN, S. G. (1960) An airborne cloud-droplet-size-distribution meter. *J. Met.* **17**, 349. (95)

KEITH, C. H. and ARONS, A. B. (1954) The growth of sea-salt particles by condensation of atmospheric water vapour. *J. Met.* **11**, 173. (125)

KELKAR, V. N. (1968) Size distribution of raindrops, Pt. VI. *Indian J. Met. Geophys.* **19**, 143. (609)

KELVIN, LORD (1870) On the equilibrium of vapour at a curved surface of a liquid. *Proc. R. Soc. Edinb.* **7**, 63. (2)

KERKER, M., LANGLEBEN, P., and GUNN, K. L. S. (1951) Scattering of microwaves by a melting, spherical ice particle. *J. Met.* **8**, 424. (406)

KERR, D. E. (1951) *Propagation of short radio waves.* M.I.T. Radar Lab. Ser. 13, McGraw-Hill, New York. (417, 434)

KIENTZLER, C. F., ARONS, A. B., BLANCHARD, D. C., and WOODCOCK, A. H. (1954) Photographic investigation of the projection of droplets by bubbles bursting at a water surface. *Tellus* **6**, 1. (75)

KIKUCHI, K. (1965) On the positive electrification of snow crystals in the process of their melting. *J. met. Soc. Japan* Ser. 2, **43**, 343, 351. (528)

KINGERY, W. D. (1960) Regelation, surface diffusion, and ice sintering. *J. appl. Phys.* **31**, 833. (250)

KINZER, G. D. and COBB, W. E. (1958) Laboratory measurements and analysis of the growth and collection efficiency of cloud droplets. *J. Met.* **15**, 138. (582)

—— and GUNN, R. (1951) The evaporation, temperature and thermal relaxation-time of freely-falling water drops. *J. Met.* **8**, 71. (125)

KIRKWOOD, J. G. and BUFF, F. P. (1949) The statistical mechanical theory of surface tension. *J. chem. Phys.* **17**, 338. (4)

KITAGAWA, N. (1965) Types of lightning. *Problems of atmospheric and space electricity*, p. 337. Elsevier, Amsterdam. (507)

—— and BROOK, M. (1960) A comparison of intracloud and cloud-to-ground lightning discharges. *J. geophys. Res.* **65**, 1189. (500)

KLEBER, W. and WEIS, J. (1958) Keimbildung und Epitaxie von Eis(I). *Z. Kristallogr. Kristallgeom.* **110**, 30. (220)

KLINE, D. B. and BRIER, G. W. (1961) Some experiments on the measurement of natural ice nuclei. *Mon. Weath. Rev. U.S. Dep. Commerce.* **89**, 263. (180)

—— and WALKER, J. A. (1951) Meteorological analysis of icing conditions encountered in low-altitude stratiform clouds. *Tech. Notes natn. advis. Comm. Aeronaut., Wash.* No. 2306. (96, 121)

KNELMAN, F., DOMBROWSKI, N., NEWITT, D. M. (1954) Mechanism of the bursting of bubbles. *Nature, Lond.* **173**, 261. (75)

KNIGHT, C. A. (1968) On the mechanism of spongy hailstone growth. *J. atmos. Sci.* **25**, 440. (342)

—— and KNIGHT, N. C. (1968) Spongy hailstone growth criteria. I Orientation fabrics. II Microstructures. *J. atmos. Sci.* **25**, 445, 453. (342)

KOBAYASHI, T. (1957) Experimental researches on the snow crystal habit and growth by means of a diffusion cloud chamber. *J. met. Soc. Japan* 75th Anniv. Vol., p. 38. (261)

—— (1958) On the habit of snow crystals artificially produced at low pressures. *J. met. Soc. Japan* **36**, 193. (264)

—— (1960) Experimental researches on the snow crystal habit and growth using a convection-mixing chamber. *J. met. Soc. Japan* **38**, 231. (262)

—— (1961) The growth of snow-crystals at low supersaturations. *Phil. Mag.* **6**, 1363, (262)

—— (1965) Vapour growth of ice crystals between -40 and $-90°C$. *J. met. Soc. Japan* **43**, 359. (262)

KODAIRA, N. (1957) An iso-echo contouring device. *Proc. 6th Weather Radar Conf., Cambridge, Mass.*, p. 307. (432)

KOENIG, L. R. (1962a) Ice in the summer atmosphere. *University of Chicago Cloud Physics Lab. Tech. Note* No. 24. (189)

—— (1962b) A note on a method to determine the orientation of crystals within hailstones. *Z. angew. Math. Phys.* **13**, 165. (335)

—— (1963) The glaciating behaviour of small cumulonimbus clouds. *J. atmos. Sci.* **20**, 29. (199, 211)

—— (1968) Some observations suggesting ice multiplication in the atmosphere. *J. atmos. Sci.* **25**, 460. (211)

KÖHLER, H. (1921) Zur Kondensation des Wasserdampfes in der Atmosphäre. *Geofysiske Publikationer*, Vol. 2, No. 1, p. 3. No. 3, p. 6. (24)

—— (1926) Zur Thermodynamik der Kondensation an hygroskopischen Kernen und Bemerkungen über das Zusammenfliessen der Tropfen. *Meddn St. met.-hydrogr. Anst.* **3**, No. 8. (24)

KOMABAYASI, M. and IKEBE, Y. (1961) Organic ice nuclei; ice-forming properties of some aromatic compounds. *J. met. Soc. Japan* **39**, 82. (223, 225)

KORNFELD, P., SHAFRIR, U., and DAVIS, M. H. (1968) A direct numerical simulation experiment of cloud droplet growth by accretion. *Proc. int. Conf. on Cloud Physics, Toronto* p. 107. (151)

KOTSCH, W. J. (1947) An example of colloidal instability of clouds in tropical latitudes. *Bull. Am. met. Soc.* **28**, 87. (294)

KOZIMA, K., ONO, T., and YAMAJI, K. (1953) The size distribution of fog particles. *Studies on fogs*, p. 303. Hokkaido University. (110)

KRAAKEVIK, J. H. (1958) The airborne measurement of atmospheric conductivity. *J. geophys. Res.* **63**, 161. (486–7, 566)

KRAMER, C. (1948) Electrische ladingen aan berijpte oppervlakken (Electric charges on rime-covered surfaces). *Neth. Met. Inst. Med. en Verhard* A54. (539, 546)

KRAMERS, H. (1946) Heat transfer from spheres to flowing media. *Physica, s'Grav.* **12**, 61. (351)

KRASNOGORSKAYA, N. V. (1960) Investigation of electrification processes of cloud particles and of precipitation (in Russian). *Izv. Akad. Nauk SSSR* Ser. geofiz., No. 1, p. 54. (563)

KRAUS, E. B. and SQUIRES, P. (1947) Experiments on the stimulation of clouds to produce rain. *Nature, Lond.* **159**, 489. (370–1, 379)

KUETTNER, J. (1950) The electrical and meteorological conditions inside thunderclouds. *J. Met.* **7**, 322. (515, 517)

—— and BOUCHER, R. J. (1958) A study of precipitation systems by means of snow crystals, synoptic and radar analysis. *Final Report to U.S. Army Signal Corps. Eng. Lab. Contract* No. DA-36-039. (247)

KUHNS, I. E. (1966) The supercooling and nucleation of water. Ph.D. Thesis, University of London. (210)

—— and MASON, B. J. (1968) The supercooling and freezing of small water droplets in air. *Proc. R. Soc.* **A302**, 437. (165–7, 170–2)

KUMAI, M. (1951) Electron-microscope study of snow-crystal nuclei. *J. Met.* **8**, 151. (197–8)

—— (1961) Snow crystals and the identification of the nuclei in the northern United States of America. *J. Met.* **18**, 139. (198)

—— and FRANCIS, K. E. (1962) Nuclei in snow and ice crystals on the Greenland Ice Cap under natural and artificially stimulated conditions. *J. atmos. Sci.* **19**, 474. (198)

KUMM, A. (1951) Über die Entstehung von elektrischen Ladungen bei Vorgängen in der kristallinen Eisphase. *Arch. Met. Geophys. Bioklim.* **A3**, 382. (539)

KUROIWA, D. (1951) Electron-microscope study of fog nuclei. *J. Met.* **8**, 157. (65)

—— (1953) Electron-microscope study of atmospheric condensation nuclei. *Studies on fogs*, p. 349. Hokkaido University. (65)

—— (1956) The composition of sea-fog nuclei as identified by electron microscope. *J. Met.* **13**, 408. (65)

—— (1961) A study of ice sintering. *Tellus* **13**, 252. (250)

—— and KINOSITA, S. (1953) A balloon fog meter and the vertical distribution of liquid water contents in the lower atmosphere. *Studies on fogs*, p. 187. Hokkaido University. (121)

LABRUM, N. R. (1952a) Some experiments on cm-wavelength scattering by small obstacles. *J. appl. Phys.* **23**, 1320. (413)

—— (1952b) The scattering of radio waves by meteorological particles. *J. appl. Phys.* **23**, 1324. (413)

LABY, T. H. (1908) The supersaturation and nuclear condensation of certain organic vapours. *Phil. Trans. R. Soc.* **A208**, 445. (18)

LAKTIONOV, A. G. (1959) Automatic flow device for investigation of natural aerosols. *Izv. Akad. Nauk SSSR* Ser. geofiz., No. 11, p. 1165. (97)

LA MER, V. K. and GRUEN, R. (1952) A direct test of Kelvin's equation connecting vapour pressure and radius of curvature. *Trans. Faraday Soc.* **48**, 410. (23)

LANDSBERG, H. E. (1934) Zählungen von Kondensationskernen auf dem Taunus Observatorium. *Bioklim. Beibl.* **1**, 125. (54)

—— (1938) Atmospheric condensation nuclei. *Ergebn. kosm. Phys.* **3**, 155. (52)

LANE, W. R. (1951) Shatter of drops in streams of air. *Ind. Engng Chem. ind. Edn* **43**, 1312. (601)

LANGE, E. (1943) Polarisationseinflüsse auf voltapotentiale von reifschichten. *Met. Z.* **60**, 154, 303. (539)

LANGER, G. ROSINKI, J., and EDWARDS, C. P. (1967) A continuous ice nucleus counter and its application to tracking in the troposphere. *J. appl. Met.* **6**, 114. (182)

LANGHAM, E. J. and MASON, B. J. (1958) The heterogeneous and homogeneous nucleation of supercooled water. *Proc. R. Soc.* **A247**, 493. (157, 159–62, 165–8, 207)

LANGILLE, R. C. and THAIN, R. S. (1951) Some quantitative measurements of 3 cm radar echoes from falling snow. *Can. J. Phys.* **29**, 482. (415)

LANGLEBEN, M.P. (1954) The terminal velocity of snowflakes. *Q. Jl R. met. Soc* **80**, 174. (239, 242, 467, 470)

—— (1956) The plan pattern of snow echoes at the generating level. *J. Met.* **13**, 554. (470)

LANGMUIR, I. (1944) Supercooled water droplets in rising currents of cold saturated air. *G. E. Res. Lab. Rep.* W-33-106-SC-65. (138)

—— (1948) The production of rain by a chain-reaction in cumulus clouds at temperatures above freezing. *J. Met.* **5**, 175. (286, 318, 328, 571–3)

—— and BLODGETT, K. B. (1946) A mathematical investigation of water droplet trajectories. *U.S. Army Air Forces Tech. Rep.* No. 5418. (40)

—— et al. (1947) Summary of results thus far obtained in artificial nucleation of clouds. *G. E. Res. Lab. First Quarterly Prog. Rep., Met. Res.* 1947. (370)

LANGSDORF, A. (1936) A continuously sensitive cloud chamber. *Phys. Rev.* **49**, 422. (35)

LATHAM, J. (1963a) Electrification produced by the asymmetric rubbing of ice on ice. *Br. J. appl. Phys.* **14**, 488. (538)

—— (1963b) The electrification of frost deposits. *Q. Jl R. met. Soc.* **89**, 265. (540)

—— (1964) Charge transfer associated with temperature gradients in ice crystals grown in a diffusion chamber. *Q. Jl R. met. Soc.* **90**, 266. (535)

—— (1965) The effect of air bubbles in ice on charge transfer produced by asymmetric rubbing. *J. atmos. Sci.* **22**, 325. (538)

—— and MASON, B. J. (1961a) Electric charge transfer associated with temperature gradients in ice. *Proc. R. Soc.* A**260**, 523. (531–7)

—— —— (1961b) Generation of electric charge associated with the formation of soft hail in thunderclouds. *Proc. R. Soc.* A**260**, 537. (211, 541 547–9)

—— —— (1962) Electrical charging of hail pellets in a polarizing electric field. *Proc. R. Soc.* A**266**, 387. (521, 551–2, 554)

—— and MILLER, A. H. (1965) The role of ice specimen geometry and impact velocity in the Reynolds-Brook theory of thunderstorm electrification. *J. atmos. Sci.* **22**, 505. (542, 554)

—— and ROXBURGH, I. W. (1966) Disintegration of pairs of water drops in an electric field. *Proc. R. Soc.* A**295**, 84. (589)

—— and SAUNDERS, C. P. R. (1964) Aggregation of ice crystals in strong electric fields. *Nature, Lond.* **204**, 1293. (249)

—— and STOW, C. D. (1965) The influence of impact velocity and ice specimen geometry on the charge transfer associated with temperature gradients in ice. *Q. Jl R. met. Soc.* **91**, 462. (537)

—— —— (1967a) The distribution of charge within ice specimens subjected to linear and non-linear temperature gradients. *Q. Jl R. met. Soc.* **93**, 121. (537)

—— —— (1967b) A laboratory investigation of the electrification of snowstorms. *Q. Jl R. met. Soc.* **93**, 55. (540)

LAWS, J. O. (1941) Measurements of the fall-velocity of water-drops and raindrops. *Trans. Am. geophys. Un.* **22**, 709. (595–7)

—— and PARSONS, D. A. (1943) The relation of raindrop size and intensity. *Trans. Am. geophys. Un.* **24**, 452. (603, 609)

LAWSON, J. A. and UHLENBECK, G. E. (1950) *Threshold signals*. M.I.T. Radar Lab. Ser. 24. McGraw-Hill, New York. (434)

LEBER, G. W., MERRITT, C. J., and ROBERTSON, J. P. (1961) WSR-57 analysis of heavy rains. *Proc. 9th Weather Radar Conf., Kansas City*, p. 102. (481)

LE CLAIR, B. P., HAMIELEC, A. E., and PRUPPACHER, H. R. (1970) A numerical

study of the drag on a sphere at low and intermediate Reynolds numbers. *J. atmos. Sci.* **27**, 308. (593)
LEE, C. W. and MAGONO, C. (1967) On the vertical distribution of snow crystals in relation with conditions revealed by two point radio-sonde soundings. *J. met. Soc. Japan* **45**, 343. (247)
LENARD, P. (1892) Über die Elektrizität der Wasserfälle. *Annln Phys.* **46**, 584. (523)
—— (1904) Ueber Regen. *Met. Z.* **21**, 248. (594, 598, 600, 608–9)
—— (1921) Zur Wasserfall-Theorie der Gewitter. *Annln Phys.* **65**, 659. (523)
LEOPOLD, L. B. and HALSTEAD, M. H. (1948) First trials of the Schaefer–Langmuir dry-ice cloud seeding technique in Hawaii. *Bull. Am. met. Soc.* **29**, 525. (294)
LEVI, L. and MILMAN, O. (1966) Freezing potentials of electrolytic solutions. *J. atmos. Sci.* **23**, 182. (527)
LEVIN, L. M. (1954) Distribution function of cloud and raindrops by sizes. *Dokl. Akad. Nauk SSSR* **94**, 1045. (113)
—— and SEDUNOV, Y. S. (1966) Stochastic condensation of drops and kinetics of cloud spectrum formation. *J. Rech. Atmos.* **2**, 425.(143)
—— and SEDUNOV, Y. S. (1968) The theoretical model of the drop spectrum formation process in clouds. *Geofis. pura appl.* **69**, 320. (143)
LEWIS, W. and HOECKER, W. H. (1949) Observations of icing conditions encountered in flight during 1948. *Tech. Notes natn. advis. Comm. Aeronaut., Wash.* No. 1904. (96, 114)
LHERMITTE, R. (1952) Les "bandes supérieures" dans la structure verticale des échos de pluie. *C. r. hebd. Séanc. Acad. Sci., Paris* **235**, 1414. (469)
—— (1960a) New developments of the echo fluctuation theory and measurements.
—— (1960b) The use of a special pulse Doppler radar in measurements of particle fall velocities. *Proc. 8th Weather Radar Conf., San Francisco*, pp. 263, 269. (439)
—— (1962) Note on wind variability with doppler radar. *J. atmos. Sci.* **19**, 342. (456)
—— (1963) Weather echoes in doppler and conventional radars. *Proc. 10th Weather Radar Conf., Washington*, D.C., p. 323. (435, 441)
—— (1968) New developments in Doppler radar methods. *Proc. 13th Weather Radar Conf., Montreal*, p. 14. (458)
—— and ATLAS, D. (1961) Precipitation motion by pulse Doppler radar. *Proc. 9th Weather Radar Conf., Kansas*, p. 218. (455)
—— —— (1963) Doppler fall speed and particle growth in stratiform precipitation. *Proc. 10th Weather Radar Conf., Washington*, p. 297. (461, 465)
LIDDELL, H. F. and WOOTTEN, N. W. (1957) The detection and measurement of water droplets. *Q. Jl R. met. Soc.* **83**, 1263. (94)
LIST, R. (1958) Kennzeichen atmosphärischer Eispartikeln. 1, Teil. Graupeln als Wachstumszentren von Hagelkörnen. *Z. angew. Math. Phys.* **9A**, 180. (334, 340)
—— (1959a) Wachstum von Eis-Wassergemischen im Hagelversuchskanal. *Helv. phys. Acta* **32**, 293. (336, 341)
—— (1959b) Zur Aerodynamik von Hagelkörnen. *Z. angew. Math. Phys.* **10**, 143. (347)
—— (1960) Zur Thermodynamik teilweise wässriger Hagelkörner. *Z. angew. Math. Phys.* **11**, 273.(287, 336, 342, 351)
—— (1961) Physical methods and instruments for characterizing hailstones. *Bull. Am. met. Soc.* **42**, 452. (335, 340)

LIST, R. (1963) General heat and mass exchange of spherical hailstones. *J. atmos. Sci.* **20**, 189. (287, 351)

—— CHARLTON, R. B. and BUTTULS, P. I. (1968) A numerical experiment on the growth and feedback mechanisms of hailstones in a one-dimensional steady-state model cloud. *J. atmos. Sci.* **25**, 1061. (362)

—— and DUSSAULT, J. G. (1967) Quasi steady-state icing and melting conditions and heat and mass transfer of spherical and spheroidal hailstones. *J. atmos. Sci.* **24**, 522. (351)

—— SCHUEPP, P. H. and METHOT, R. G. J. (1965) Heat exchange rates of hailstones in a model cloud and their simulation in the laboratory. *J. atmos. Sci.* **22**, 710. (351)

LITVINOV, I. V. (1956) Determination of the terminal velocity of snowflakes (in Russian). *Izv. Akad. Nauk SSSR* Ser. geofiz. No. 7, p. 853. (242)

LODGE, J. P. (1954) Analysis of micron-sized particles. *Analyt. Chem.* **26**, 1829. (50)

—— BAKER, M. L. and PIERRARD, J. M. (1956) Observations on ion separation in dilute solutions by freezing. *J. chem. Phys.* **24**, 716. (526)

LOEB, L. B., KIP, A. F., and EINARSSON, A. W. (1938) On the nature of ionic sign preference in C.T.R. Wilson cloud chamber condensation experiments. *J. chem. Phys.* **6**, 264. (18, 20)

LOTHE, J. and POUND, G. M. (1962) Reconsiderations of nucleation theory. *J. chem. Phys.* **36**, 2604. (12)

LOW, R. D. H. (1969) A generalized equation for the solution effect in droplet growth. *J. atmos. Sci.* **26**, 608, 1345. (25)

LUDLAM, F. H. (1950) The composition of coagulation elements in cumulonimbus. *Q. Jl R. met. Soc.* **76**, 52. (287, 341, 351)

—— (1951a) The production of showers by the coalescence of cloud droplets. *Q. Jl R. met. Soc.* **77**, 402. (286, 287, 317, 321–326, 379)

—— (1951b) The heat economy of a rimed cylinder. *Q. Jl R. met. Soc.* **77**, 663. (115)

—— (1952) The production of showers by the growth of ice particles. *Q. Jl R. met. Soc.* **78**, 543. (287, 316–7, 329–32)

—— (1955) Artificial snowfall from mountain clouds. *Tellus* **7**, 277. (396)

—— (1958) The hail problem. *Nubila* **1**, 12. (351–7, 366–8)

—— and MACKLIN, W. C. (1959) Some aspects of a severe storm in S.E. England. *Nubila* **2**, 38. (334)

—— and SAUNDERS, P. M. (1956) Shower formation in large cumulus. *Tellus* **8**, 424. (326)

LUEDER, H. (1951a) Ein neuer elektrischer Effekt bei der Eisbildung durch Vergraupelung in naturlichen unterkühlten Nebeln. *Z. angew. Phys.* **3**, 247. (546)

—— (1951b) Vergraupelungs-Elektrisierung als eine Ursache der Gewitterelektrizität. *Z. angew. Phys.* **3**, 288. (546)

McCLELLAND, J. A. and NOLAN, J. J. (1912) The electric charge on rain. *Proc. R. Ir. Acad.* **A29**, 81; **A30**, 61. (560)

—— and GILMOUR, A. (1920) Further observations of the electric charge on rain. *Proc. R. Ir. Acad.* **A35**, 13. (560)

MACCREADY, P. B. and PROUDFIT, A. (1965) Self-charging of melting ice. *Q. Jl R. met. Soc.* **91**, 54. (528)

—— and TAKEUCHI, D. M. (1968) Precipitation initiation mechanisms and droplet characteristics of some convective cloud cores. *J. appl. Met.* **7,** 591. (140)

—— and TODD, T. C. (1964) Continuous particle sampler. *J. appl. Met.* **3,** 450. (95)

McCULLOUGH, S. and PERKINS, P. J. (1951) Flight camera for photographing cloud droplets in natural suspension in the atmosphere. *NACA Res. Memor.* E50K01A. (95)

McDONALD, J. E. (1953) Erroneous cloud physics applications of Raoult's law. *J. Met.* **10,** 68. (25)

—— (1954) The shape and aerodynamics of large raindrops. *J. Met.* **11,** 478. (599)

—— (1963) Use of the electrostatic analogy in studies of ice crystal growth. *Z. angew. Math. Phys.* **14,** 610. (276)

MACKLIN, W. C. (1961) Accretion in mixed clouds. *Q. Jl R. met. Soc.* **87,** 413. (336, 341)

—— (1962) The density and structure of ice formed by accretion. *Q. Jl R. met. Soc.* **88,** 30. (336)

—— (1963) Heat transfer from hailstones. *Q. Jl R. met. Soc.* **89,** 360. (351, 357, 366–8)

—— and BAILEY, I. H. (1966) On the critical liquid water concentrations of large hailstones. *Q. Jl R. met. Soc.* **92,** 297. (345)

—— —— (1968) The collection efficiencies of hailstones. *Q. Jl R. met. Soc.* **94,** 393. (359)

—— and LUDLAM, F. H. (1961) The fallspeeds of hailstones. *Q. Jl R. met. Soc.* **87,** 72. (347)

—— MERLIVAT, L. and STEVENSON, C. M. (1970) The analysis of a hailstone. *Q. Jl R. met. Soc.* **96,** 472. (362–5)

—— and PAYNE, G. S. (1968) Some aspects of the accretion process. *Q. Jl R. met. Soc.* **94,** 167. (337)

—— and RYAN, B. F. (1962) On the formation of spongy ice. *Q. Jl R. met. Soc.* **88,** 548. (338)

—— —— (1965) The structure of ice grown in bulk supercooled water. *J. atmos. Sci.* **22,** 452. (338, 342)

—— STRAUCH, E. and LUDLAM, F. H. (1960) The density of hailstones collected from a summer storm. *Nubila* **3,** 12. (346)

MACKY, W. A. (1931) Some investigations on the deformation and breaking of water drops in strong electric fields. *Proc. R. Soc.* **A133,** 565. (509)

MADONNA, L. A., SCIULLI, C. M., CANJAR, L. N., and POUND, G. M. (1961) Low-temperature cloud-chamber studies on water vapour. *Proc. phys. Soc.* **78,** 1218. (14, 16, 174)

MAGONO, C. (1953) On the growth of snowflakes and graupel. *Scient. Rep. Yokohama Univ.* Ser. 1, No. 2, p. 18. (239, 241, 242)

—— (1954*a*) On the falling velocity of solid precipitation elements. *Scient. Rep. Yokohama Univ.* Ser. 1, No. 3, p. 33. (340)

—— (1954*b*) On the shape of water drops falling in stagnant air. *J. Met.* **11,** 77. (599)

—— (1960) Structure of snowfall revealed by geographic distribution of snow crystals. *Physics of precipitation, Geophys. Monogr.* No. 5, p. 142. American Geophysical Union. (247)

—— (1968) Problems on physical understanding of snowfall phenomena. *Proc. int. Conf. on Cloud Physics,* Toronto, p. 243. (248)

—— and KIKUCHI, K. (1963, 1965) On the positive electrification of snow

crystals in the process of their melting. *J. met. Soc. Japan* **41**, 270; **43**, 331. (531)
—— and SHIO, H. (1967) Frictional electrification of ice, and change in its contact surface. *Physics of snow and ice*, Vol. 1, Pt. 1, p. 137. Hokkaido University. (538)
—— and SHIOTSUKI, Y. (1964) The effect of air bubbles in ice on frictional charge separation. *J. atmos. Sci.* **21**, 666. (538)
—— and TAKAHASHI, T. (1963a) On the electrical phenomena during riming and glazing in natural supercooled clouds. (1963b) Experimental studies on the mechanism of electrification of graupel pellets. *J. met. Soc. Japan* **41**, 71, 197. (542, 547)
—— and TAZAWA, S. (1966) Design of 'snow crystal sondes.' *J. atmos. Sci.* **23**, 618 (247)
MALAN, D. J. (1954) Les décharges orageuses intermittentes et continués de la colonne de charge négative. *Annls Géophys.* **10**, 271. (507)
—— (1955) La distribution verticale de la charge négative orageuse *Annls Géophys.* **11**, 420. (507)
—— (1965) The theory of lightning. *Problems of atmospheric and space electricity*, p. 323. Elsevier, Amsterdam. (505)
—— and SCHONLAND, B. F. J. (1950) An electrostatic fluxmeter of short response time for use in studies of transient field changes. *Proc. phys. Soc.* **B63**, 402. (490)
—— —— (1951a) The electrical processes in the intervals between the strokes of a lightning discharge. (1951b) The distribution of electricity in thunderclouds. *Proc. R. Soc.* **A209**, 145, 158. (504–5, 509, 512, 517)
MANN, G. (1940) Untersuchungen über die aerologischen Bedingungen für die Niederschlagsbildung in der Atmosphäre an Hand des Aufstiegsmaterials der Wetterflugstelle zu Königsberg. *Beitr. Phys. frei Atmos.* **26**, 121. (289)
MANSON, J. E. (1955) X-ray diffraction study of silver iodide aerosols. *J. appl. Phys.* **26**, 423. (228)
MARSHALL, J. S. (1953) Precipitation trajectories and patterns. *J. Met.* **10**, 25. (469–70)
—— (1957) The constant altitude presentation of radar weather patterns. *Proc. 6th Weather Radar Conf., Cambridge*, p. 321. (430)
—— and GORDON, W. E. (1957) Radiometeorology, *Met. Monogr.* No. 3, pp. 73–113. American Meteorological Society, Boston. (399)
—— and GUNN, K. L. S. (1961) Wide dynamic range for weather radar. *Beitr. Phys. Atmos.* **34**, 69. (432)
—— and HITSCHFELD, W. (1953) The interpretation of the fluctuating echo for randomly distributed scatterers, Part 1, *Can. J. Phys.* **31**, 962. (434–5)
—— HITSCHFELD, W. and GUNN, K. L. S. (1955) Advances in radar weather. *Adv. Geophys.* **2**, 1–56. (399)
—— and LANGLEBEN, M. P. (1954) A theory of snow-crystal habit and growth. *J. Met.* **11**, 104. (254–6)
—— and PALMER, W. McK. (1948) The distribution of raindrops with size. *J. Met.* **5**, 165. (405, 477, 609, 610–3)
MARWICK, T. C. (1930) The electric charge on rain. *Q. Jl R. met. Soc.* **56**, 39. (561)
MASON, B. J. (1950) The nature of ice-forming nuclei in the atmosphere. *Q. Jl R. met. Soc.* **76**, 59. (192, 194)
—— (1951) Spontaneous condensation of water vapour in expansion chamber experiments. *Proc. phys. Soc.* **B64**, 773. (4, 15)
—— (1952a) The production of rain and drizzle by coalescence in stratiform clouds. *Q. Jl R. met. Soc.* **78**, 377. (140–2, 288, 308–13)

—— (1952b) The spontaneous crystallization of supercooled water. *Q. Jl R. met. Soc.* **78**, 22. (166–70, 174)

—— (1953a) Progress in cloud physics research. A progress report on recent investigations at Imperial College, London. *Arch. Met. Geophys. Bioklim.* A**6**, 1. (162)

—— (1953b) The growth of ice crystals in a supercooled water cloud. *Q. Jl R. met. Soc.* **79**, 104. (251–2, 275–8)

—— (1953c) A critical examination of theories of charge generation in thunderstorms. *Tellus* **5**, 446. (520, 524)

—— (1953d) On the generation of charge associated with graupel formation in thunderstorms. *Q. Jl. R. Met. Soc.* **79**, 501. (547)

—— (1954) Bursting of air bubbles at the surface of sea water. *Nature, Lond.* **174**, 470. (76, 78)

—— (1955a) The appearance of ice crystals in expansion and mixing-cloud chambers. *Bull. Obs. Puy de Dôme*, p. 65. (186)

—— (1955b) The physics of natural precipitation processes. *Arch. Met. Geophys. Bioklim.* A**8**, 159. (207)

—— (1955c) Radar evidence for aggregation and orientation of melting snowflakes. *Q. Jl R. met. Soc.* **81**, 262. (462, 465–6)

—— (1956a) The nucleation of supercooled water clouds. *Sci. Prog. Lond.* No. 175, p. 479. (160, 164, 165)

—— (1956b) On the melting of hailstones. *Q. Jl R. met. Soc.* **82**, 209. (368)

—— (1957) The oceans as a source of cloud-forming nuclei. *Geofis. pura appl.* **36**, 148. (76, 78)

—— (1958) The growth of ice crystals from the vapour and the melt. *Adv. Phys.* **7**, 235. (220)

—— (1960a) The evolution of droplet spectra in stratus cloud. *J. Met.* **17**, 459. (140–2)

—— (1960b) Nucleation of water aerosols. *Discuss. Faraday Soc.* No. 30, p. 20. (165–6)

—— (1960c) Ice-nucleating ability of clay minerals and stony meteorites. *Q. Jl R. met. Soc.* **86**, 552. (195, 205)

—— (1965) Charge generation in thunderstorms. *Problems of atmospheric and space electricity*, p. 239. Elsevier, Amsterdam. (557)

—— (1968) The generation of electric charges and fields in precipitating clouds. *Proc. int. Conf. on Cloud Physics, Toronto*, p. 657. (521, 553, 555–6)

—— and ANDREWS, J. B. (1960) Drop-size distributions from various types of rain. *Q. Jl R. met. Soc.* **86**, 346. (445, 610–11)

—— BRYANT, G. W. and VAN DEN HEUVEL, A. P. (1963) The growth habits and surface structure of ice crystals. *Phil. Mag.* **8**, 505. (268–72)

—— and CHIEN, C. W. (1962) Cloud-droplet growth by condensation in cumulus. *Q. Jl R. met. Soc.* **88**, 136. (133–8)

—— and EMIG, R. (1961) Calculations of the ascent of a saturated buoyant parcel with mixing. *Q. Jl R. met. Soc.* **87**, 212. (133)

—— and GHOSH, D. K. (1957) The formation of large droplets in small cumulus. *Q. Jl R. met. Soc.* **83**, 501. (125–7)

—— and HALLETT, J. (1956) Artificial ice-forming nuclei. *Nature, Lond.* **177**, 681. (214–5, 228)

—— —— (1957) Ice-forming nuclei. *Nature, Lond.* **179**, 357. (214, 215)

—— and HOWORTH, B. P. (1952) Some characteristics of stratiform clouds over N. Ireland in relation to their precipitation. *Q. Jl R. met. Soc.* **78**, 226. (289, 312)

JAYARATNE, O. W. and WOODS, J. D. (1963) An improved vibrating

capillary device for producing uniform water droplets of 15 to 500 μm radius. *J. scient. Instrum.* **40,** 247. (583)
—— and MAYBANK, J. (1958) Ice-nucleating properties of some natural mineral dusts. *Q. Jl R. met. Soc.* **84,** 235. (181, 191, 193, 195–7)
—— —— (1960) The fragmentation and electrification of freezing water drops. *Q. Jl R. met. Soc.* **86,** 176. (207, 211, 543–5, 549)
—— and RAMANADHAM, R. (1953) A photoelectric raindrop spectrometer. *Q. Jl R. met. Soc.* **79,** 490. (445, 605)
—— —— (1954) Modification of the size distribution of falling raindrops by coalescence. *Q. Jl R. met. Soc.* **80,** 388. (463, 611–3)
—— and VAN DEN HEUVEL, A. P. (1959) The properties and behaviour of some artificial ice nuclei. *Proc. phys. Soc.* **74,** 744. (181, 190–1, 215–6, 218, 222)
MATTHEWS, J. B. and MASON, B. J. (1963) Electrification accompanying melting of ice and snow. *Q. Jl R. met. Soc.* **89,** 376. (528)
—— —— (1964) Electrification produced by the rupture of large water drops in an electric field. *Q. Jl R. met. Soc.* **90,** 275. (524, 601)
MAULARD, J. (1951) Mesure du nombre de gouttes de pluie. *J. Sci. Mét.* **3,** 69. (603)
MAUND, J. E. and CHALMERS, J. A. (1960) Point discharge from natural and artificial points. *Q. Jl R. met. Soc.* **86,** 85. (559)
MAY, K. R. (1945) The cascade impactor: an instrument for sampling coarse aerosols. *J. scient. Instrum.* **22,** 187. (42, 93)
—— (1950) The measurement of airborne droplets by the magnesium oxide method. *J. scient. Instrum.* **27,** 128. (94)
—— (1961) Fog-droplet sampling using a modified impactor technique. *Q. Jl R. met. Soc.* **87,** 535. (43)
MAYBANK, J. and MASON, B. J. (1959) The production of ice crystals by large adiabatic expansions of water vapour. *Proc. phys. Soc.* **74,** 11. (174)
MAZIN, I. P. (1965) On the theory of formation of the size spectrum of particles in clouds and precipitation. *Moscow Central Aerolog. Obs. Trudy vyp.* **64,** 57: **65,** 106. (143)
MAZUR, J. (1943) The number and size distribution of water particles in natural clouds. *Met. Res. Pap., Met. Office Lond.* No. 109. (93, 94, 98, 114)
—— (1952) On the sampling of water droplets in natural clouds and in radiation fogs. *Proc. phys. Soc.* **B65,** 457. (93)
MEINHOLD, H. (1951) Die elektrische Ladung eines Flugzeuges bei Vereisung in Quellwolken. *Geofis. pura appl.* **19,** 176. (546)
MERLIVAT, L., NIEF, G., and ROTH, E. (1965) Formation de la grêle et fractionnement isotopique du deuterium. *Abhand. Akad. Wiss. Berl.* **7,** 839. (363)
MERRINGTON, A. C. and RICHARDSON, E. G. (1947) The break-up of liquid jets. *Proc. phys. Soc.* **59,** 1. (600)
METNIEKS, A. L. (1958) The size spectrum of large and giant sea-salt nuclei under maritime conditions. *Geophys. Bull. Dubl.* **15,** 1. (61)
MEYER, J. and PFAFF, W. (1935) Zur Kenntnis der Kristallisation von Schmelzen, III. *Z. anorg. allg. Chem.* **224,** 305. (160, 168)
MILLER, A. H., SHELDEN, C. E., and ATKINSON, W. R. (1965) Spectral study of the luminosity produced during coalescence of oppositely charged falling water drops. *Physics Fluids* **8,** 1921. (590)
MONTMORY, R. (1956) La congélation de l'eau surfondue sur une surface crystalline. *Bull. Obs. Puy de Dôme*, No. 4, p. 126. (220)

MOORE, C. B. (1965) Charge generation in thunderstorms. *Problems of atmospheric and space electricity*, p. 255. Elsevier, Amsterdam. (518)
—— VONNEGUT, B., STEIN, B. A., and SURVILAS, H. J. (1960) Observations of electrification and lightning in warm clouds. *J. geophys. Res.* **65**, 1907. (519)
MOORE, D. J. (1952) Measurements of condensation nuclei over the North Atlantic. *Q. Jl R. met. Soc.* **78**, 596. (55, 62)
—— (1955) The role of sea salt nuclei in atmospheric condensation. Ph.D. Thesis, London University. (55)
—— and MASON, B. J. (1954) The concentration, size distribution and production rate of large salt nuclei over the oceans. *Q. Jl R. met. Soc.* **80**, 583. (61, 78, 86)
MORDY, W. A. (1959) Computations of the growth by condensation of a population of cloud droplets. *Tellus* **11**, 16. (125, 129, 137)
—— and EBER, L. E. (1954) Observations of rainfall from warm cloud. *Q. Jl R. met. Soc.* **80**, 48. (296)
MOSSOP, S. C. (1955) The freezing of supercooled water. *Proc. phys. Soc.* B**68**, 193. (157, 161, 164–5, 168)
—— (1956a) Sublimation nuclei. *Proc. phys. Soc.* B**69**, 161. (191)
—— (1956b) The nucleation of supercooled water by various chemicals. *Proc. phys. Soc.* B**69**, 165. (214, 216)
—— (1963a) Atmospheric ice nuclei. *Z. angew. Math. Phys.* **14**, 456. (198, 201)
—— (1963b) Stratospheric particles at 20 km. *Nature, Lond.* **199**, 325. (201)
MOSSOP, S. C. (1968) Comparison between concentration of ice crystals in cloud with the concentration of ice nuclei. *J. Recherches. Atmos.* **3**, 119. (211)
—— CARTE, A. E. and HEFFERNAN, K. J. (1956) Counts of atmospheric freezing nuclei at Pretoria. *Aust. J. Phys.* **9**, 556. (200)
—— and JAYAWEERA, K. O. L. F. (1969) AgI–NaI aerosols as ice nuclei. *J. appl. Met.* **8**, 241. (229)
—— and KIDDER, R. E. (1961) Hailstorm at Johannesburg on 9th November 1959. Part 2, Structure of hailstones. *Nubila* **4**, 74. (342, 344, 347)
—— and ONO, A. (1969) Measurement of ice crystal concentration in clouds. *J. atmos. Sci.* **26**, 130. (211)
—— ONO, A. and WISHART, E. R. (1970) Ice particles in maritime clouds near Tasmania. *Q. Jl R. met. Soc.* **96**, 487. (211)
—— RUSKIN, R. E. and HEFFERNAN, K. J. (1968) Glaciation of a cumulus at approximately $-4°C$. *J. atmos. Sci.* **25**, 889. (211)
—— and THORNDIKE, N. S. C. (1966) The use of membrane filters in measurements of ice nucleus concentration. *J. appl. Met.* **5**, 474. (179)
—— and TUCK-LEE, C. (1968) The composition and size distribution of aerosols produced by burning solutions of AgI and NaI in acetone. *J. appl. Met.* **7**, 234. (229)
MUCHNIK, V. M. and RUDKO, Y. S. (1961) Peculiarities of freezing supercooled water drops. *Trudy ukr. nauchno-issled. gidromet Inst.* **26**, 64.
MUELLER, E. A. and SIMS, A. L. (1966) The influence of sampling volume on raindrop size spectra. *Proc. 12th Weather Radar Conf., Norman, Oklahoma*, p. 135. (479, 608)
MUGURUMA, J. and HIGUCHI, K. (1959) On the etch pits of snow crystals. *J. met. Soc. Japan* **37**, 71. (273)
MÜLLER-HILLEBRAND, D. (1954) Charge generation in thunderstorms by collision of ice crystals with graupel falling through a vertical electrical field. *Tellus* **6**, 367. (521)
—— (1955) Zur Frage des Ursprunges der Gewitterelektrizität. *Ark. Geofys.* **2**, 395. (521)

MURAI, G. (1956) On the relation between natural snow crystal forms and the upper-air conditions. *Low Temp. Sci.* Ser. A, **15**, 14. (247)

MURGATROYD, R. J. and GARROD, M. P. (1957) Some recent airborne measurements of freezing nuclei over southern England. *Q. Jl R. met. Soc.* **83**, 528. (185, 200)

—— and GARROD, M. P. (1960) Observations of precipitation elements in cumulus clouds. *Q. Jl R. met. Soc.* **86**, 167. (211)

NAKAYA, U. (1951) The formation of ice crystals. *Compendium of meteorology*, p. 207. American Meteorological Society. (252–5)

—— (1954) *Snow crystals.* Harvard University Press. (236, 243, 252)

—— (1955) Snow crystals and aerosols. *J. Fac. Sci. Hokkaido Univ.* **4**, 341. (264)

—— HANAJIMA, M. and MUGURUMA, J. (1958) Physical investigations on the growth of snow crystals. *J. Fac. Sci. Hokkaido Univ.* Ser. II, **5**, 87. (265)

—— and HIGUCHI, K. (1960) Horizontal distribution of snow crystals during snowfall. *Physics of precipitation*, Geophys. Monogr. No. 5, p. 118. American Geophysical Union. (247)

—— and MATSUMOTO, A. (1954) Simple experiment showing the existence of 'liquid water' film on the ice surface. *J. Colloid Sci.* **9**, 41. (249)

NAKAYA, U. and TERADA, T. (1935) Simultaneous observations of the mass, falling velocity and form of individual snow crystals. *J. Fac. Sci. Hokkaido Univ.* Ser. II, **1**, 191. (237–40, 340)

NATHAN, A. M. and HILL, D. (1957) Freezing nuclei meter. *New York Univ. Res. Div. Tech. Rep.* 389.02. (178)

NEEL, C. B. (1955) A heated-wire liquid-water-content instrument and results of initial flight tests in icing conditions. *NACA Res. Memor.* RMA54123. (115)

—— and STEINMETZ, C. P. (1952) The calculated and measured performance characteristics of a heated-wire liquid-water-content meter for measuring icing severity. *Tech. Notes natn. advis. Comm. Aeronaut. Wash.* No. 2615. (115)

NEIBURGER, M. (1949) Reflection, absorption, and transmission of insolation by stratus cloud. *J. Met.* **6**, 98. (110, 121)

—— (1967) Collision efficiency of nearly equal cloud drops. *Mon. Weath. Rev. U.S. Dep. Commerce* **95**, 917. (587)

—— and CHIEN, C. W. (1960) Computations of the growth of cloud drops by condensation using an electronic digital computer. *Physics of precipitation*, Geophys. Monogr. No. 5, p. 191. American Geophysical Union. (125, 129, 132)

—— and PRUPPACHER, H. R. (1965) Experimental tests of a method of computing collision efficiencies of spheres falling in a viscous medium. *Proc. int. Conf. on Cloud Physics, Tokyo*, p. 97. (582)

NÉMETHY, G. and SCHERAGA, H. A. (1962) Structure of water and hydrophobic bonding in proteins. 1. A model for the thermodynamic properties of liquid water. *J. chem. Phys.* **36**, 3382. (171–2)

NEWELL, R. E. (1956) MS Thesis, Massachusetts Institute of Technology. (464)

NEYMAN, J. and SCOTT, E. L. (1967) Some outstanding problems relating to rain modification. *Proc. 5th Berkeley Symposium on Mathematical Statistics and Probability*, Vol. 5, p. 293. (389, 391)

—— —— and WELLS, M. A. (1969) Statistics in meteorology. *Rev. int. statist. Inst.* **37**, 119. (389)

NIEDERDORFER, E. (1932) Messungen der Grösse der Regentropfen. *Met. Z.* **49**, 1. (608)

NIESSEN, C. W. and GEOTIS, S. G. (1963) A signal level quantizer for weather radar. *Proc. 10th Weather Radar Conf., Washington*, p. 370. (432)

NOLAN, P. J. and DOHERTY, D. J. (1950) Size and charge distribution of atmospheric condensation nuclei. *Proc. R. Ir. Acad.* **A53**, 163.(38)
—— and POLLAK, L. W. (1946) The calibration of a photoelectric nucleus counter. *Proc. R. Ir. Acad.* **A51**, 9. (34)
NORINDER, H. and SIKSNA, R. (1953) On the electrification of snow. *Tellus* **5**, 260. (539)
ODDIE, B. C. V. (1960) The variation in composition of sea-salt nuclei with mode of formation. *Q. Jl R. met. Soc.* **86**, 549. (75)
OGAWA, T. and BROOK, M. (1964) The mechanism of the intracloud lightning discharge. *J. geophys. Res.* **69**, 5141. (508)
—— —— (1969) Charge distribution in thunderstorm clouds. *Q. Jl R. met. Soc.* **95**, 513. (505)
OGIWARA, S. and OKITA, T. (1952) Electron-microscope study of cloud and fog nuclei. *Tellus* **4**, 233. (65)
OHTA, S. (1951) On the contents of condensation nuclei and uncharged nuclei on the Pacific Ocean and the Japan Sea. *Bull. Am. met. Soc.* **82**, 30. (55, 62)
OKITA, T. and KIMURA, K. (1954) Ice crystal growth in the atmosphere. *J. met. Soc. Japan* **32**, 129. (278)
O'MAHONY, G. (1962) Singularities in daily rainfall. *Aust. J. Phys.* **15**, 301. (203)
—— (1965) Rainfall and moon phase. *Q. Jl R. met. Soc.* **91**, 196. (205)
O'NEILL, M. E. and MAJUMDAR, S. R. (1970) Asymmetrical slow viscous fluid motions caused by the translation or rotation of two spheres. In press (576)
ORIANI, R. A. and SUNDQUIST, B. E. (1963) Emendations to nucleation theory and the homogeneous nucleation of water from the vapour. *J. chem. Phys.* **38**, 2082. (12)
ORR, J. L., FRASER, D., and PETTIT, K. G. (1949) Canadian experiments on artificially inducing precipitation. *U.N. Conf. on Conservation*, Vol. 4, *Water resources*. p. 27. (373)
ORR, C., HURD, F. K., HENDRIX, W. P., and JUNGE, C. (1958) The behaviour of condensation nuclei under changing humidities. *J. Met.* **15**, 240. (28)
OURA, H. and HORI, J. (1953) Optical method of measuring drop-size distribution of fog. *Studies in fogs*, p. 327. Hokkaido University. (96, 110)
OWENS, J. S. (1922) Suspended impurity in the air. *Proc. R. Soc.* **A101**, 18. (41)
—— (1926) Condensation of water from the air upon hygroscopic crystals. *Proc. R. Soc.* **A110**, 738. (28)
—— (1940) Sea-salt and condensation nuclei. *Q. Jl R. met. Soc.* **66**, 2. (75)

PALMER, H. P. (1949) Natural ice-particle nuclei. *Q. Jl R. met. Soc.* **75**, 15. (186, 200)
PAINTER, R. and SCHAEFER, V. J. (1960) Permanent replicas of the crystalline structure of hailstones. *Z. angew. Math. Phys.* **11**, 318. (335)
PARKINSON, W. C. (1952) Note on the concentration of condensation nuclei over W. Atlantic. *J. geophys. Res.* **57**, 314. (55)
PARUNGO, F. P. and LODGE, J. P. (1965) Molecular structure and ice nucleation of some organics. *J. atmos. Sci.* **22**, 309. (225)
—— —— (1967) Amino-acids as ice nucleators. *J. atmos. Sci.* **24**, 274. (225)
PATERSON, M. P. and SPILLANE, K. T. (1967) A study of Australian soils as ice nuclei. *J. atmos. Sci.* **24**, 50. (199)
PEARCE, D. C. and CURRIE, B. W. (1949) Some qualitative results on the electrification of snow. *Can. J. Res.* **A27**, 1. (539)
PEARCEY, T. and HILL, G. W. (1957) A theoretical estimate of the collection efficiencies of small droplets. *Q. Jl R. met. Soc.* **83**, 77. (573–4, 584)

PEPPLER, W. (1940a) Unterkühlte Wasserwolken und Eiswolken. *Forsch.-u. Erfahr Ber. Reichsamt Wetter Dienst.* **B1**. (289)

—— (1940b) Beiträge zum Kumulus und Kumulonimbus. *Z. angew. Met.* **57**, 341. (290)

PERKINS, P. J. (1951) Flight instrument for measurement of liquid-water content in clouds at temperatures above and below freezing. *NACA Res. Memor.* E50J12A. (115)

PERRY, F. R., WEBSTER, G. H., and BAGULEY, P. W. (1942) The measurement of lightning voltages and currents in Nigeria. *J. Instn elect. Engrs.* **89**, 185. (559)

PETTERSSON, H. (1958) Rate of accretion of cosmic dust on the earth. *Nature, Lond.* **181**, 330. (204)

PICKNETT, R. G. (1960) Collection efficiencies for water drops in air. *Aerodynamic capture of particles*, p. 160. Pergamon Press, London. (579, 583)

PIERCE, E. T. (1955) Electrostatic field-changes due to lightning discharges. *Q. Jl R. met. Soc.* **81**, 211.(493, 500, 503, 505, 507)

PIETROWSKI, E. L. (1960) An observation of lightning in warm clouds. *J. Met.* **17**, 562. (519)

PLANK, V. G., ATLAS, D., and PAULSEN, W. H. (1955) The nature and detectability of clouds and precipitation as determined by 1·25 cm. radar. *J. Met.* **12**, 358 (471)

PLUMLEE, H. R. and SEMONIN, R. G. (1964) Cloud droplet collision efficiency in electric fields. *Tellus* **17**, 357. (153)

PODZIMEK, J. (1959) Measurement of the concentration of large and giant chloride condensation nuclei during flight. *Studia geophys. geod.* **3**, 256. (74)

PODZIMEK, J. and ČERNOCH, I. (1961) Höhenverteilung der Konzentrationen von Riesen-kernen aus Chloriden und Sulphaten. *Geofis. pura appl.* **50**, 96. (74)

POLLAK, L. W. (1952) A condensation nuclei counter with photographic recording. *Geofis. pura appl.* **22**, 75. (34)

—— and METNIEKS, A. L. (1957) Photo-electric condensation nucleus counters of high precision for measuring low and very low concentrations of nuclei. *Geofis. pura appl.* **37**, 174. (34)

—— and O'CONNOR, T. C. (1955) A photo-electric condensation nucleus counter of high precision. *Geofis. pura appl.* **32**, 139. (34)

POUND, G. M., MADONNA, L. A., and PEAKE, S. L. (1953) Critical supercooling of pure water droplets by a new microscopic technique. *J. Colloid Sci.* **8**, 187. (160, 168)

POWELL, C. F. (1928) Condensation phenomena at different temperatures. *Proc. R. Soc.* **A119**, 553. (14, 18)

POWER, B. A. and POWER, R. F. (1962) Some amino-acids as ice nucleators. *Nature, Lond.* **194**, 1170. (223, 225)

PRIESTLEY, C. H. B. (1953) Buoyant motion in a turbulent environment. *Aust. J. Phys.* **6**, 279. (134)

—— (1954) Buoyant motions and the open parcel. *Met. Mag., Lond.* **83**, 107. (134)

PROBERT-JONES, J. R. (1962) The radar equation in meteorology. *Q. Jl R. met. Soc.* **88**, 485. (402)

—— and HARPER, W. G. (1961) Vertical air motion in showers as revealed by Doppler radar. *Proc. 9th Weather Radar Conf., Kansas City*, p. 225. (447)

PROBSTEIN, R. F. (1951) Time lag in the self-nucleation of a supersaturated vapor. *J. chem. Phys.* **19**, 619. (12)

PRUPPACHER, H. R. and BEARD, K. V. (1970) A wind tunnel investigation of the

internal circulation and shape of water drops falling at terminal velocity in air. *Q. Jl R. met. Soc.* **96,** 247. (599)
—— and NEIBURGER, M. (1963) The effect of water soluble substances on the supercooling of water drops. *J. atmos. Sci.* **20,** 376. (162)
—— and SÄNGER, R. (1955) Mechanismus der Vereisung unterkühlter Wassertropfen durch disperse Keimsubstanzen. *Z. angew. Math. Phys.* **6,** 407. (195, 214, 216, 218)
—— and STEINBERGER, E. H. (1968) An experimental determination of the drag on a sphere at low Reynolds numbers. *J. appl. Phys.* **39,** 4129. (593)
PRZIBRAM, K. (1906) Über die Kondensation von Dämpfen in ionisierter Luft. *Sber. Akad. Wiss. Wien* **115,** 33. (18)

RADFORD, W. H. (1938) An instrument for sampling and measuring liquid fog water. *M.I.T. Pap. phys. Oceanogr. Met.* **6,** 19. (121)
RADKE, L. F. and HOBBS, P. V. (1969a) An automatic cloud condensation nuclei counter. *J. appl. Met.* **8,** 105. (37)
—— —— (1969b) Measurement of cloud condensation nuclei in the Olympic Mountains of Washington. *J. atmos. Sci.* **26,** 281. (72)
RAMANA MURTY, B. V. and GUPTA, S. C. (1959) The size distribution of raindrops. *Indian J. scient. ind. Res.* **18A,** 352. (609)
RAMSAY, M. W. and CHALMERS, J. A. (1960) Measurements on the electricity of precipitation. *Q. Jl R. met. Soc.* **86,** 530. (561)
RANZ, W. E. and MARSHALL, W. R. (1952) Evaporation from drops, Pts. I and II. *Chem. Engng. Prog.* **48,** 141, 173. (125, 352)
—— and WONG, J. B. (1952) Inspection of dust and smoke particles on surface and body collectors. *Ind. Engng Chem. ind. Edn* **44,** 1371. (40, 41)
RAU, W. (1944) Gefriervorgänge des Wassers bei tiefen temperaturen. *Schr. dt. Akad. Luft Forsch.* **8,** 65. (173)
—— (1953a) Nachweis der allgemeinen Gultigkeit des Gefrierkernspektrums. *Geofis. pura appl.* **26,** 75. (157)
—— (1953b) Uber der Einfluss des Tropfenvolumens auf die Unterkühlbarkeit von Wassertropfen und die Bedeutung des Gefrierkernspektrums. *Z. Naturf.* **8,** 197. (157)
—— (1954) Die Gefrierkerngehalte der verschiedenen Luftmassen. *Met. Rdsch.* **7,** 205. (187)
—— (1955) Grösse und Häufigkeit der Chloridteilchen in kontinentalen Aerosol und ihre Beziehungen zum Gefrierkerngehalt. *Met. Rdsch.* **8,** 169. (74)
REYNOLDS, S. E. (1952) Ice-crystal growth. *J. Met.* **9,** 36. (278)
—— (1954) Compendium of thunderstorm electricity, p. 77. New Mexico Institute of Mining and Technology. (547)
—— and BROOK, M. (1956) Correlation of the initial electric field and the radar echo in thunderstorms. *J. Met.* **13,** 376. (518)
—— —— and GOURLEY, M. F. (1957) Thunderstorm charge separation. *J. Met.* **14,** 426. (537, 540–1, 547)
—— HUME, W., VONNEGUT, B., and SCHAEFER, V. J. (1951) Effect of sunlight on the action of silver iodide particles as sublimation nuclei. *Bull. Am. met. Soc.* **32,** 47. (231, 232)
—— —— and MCWHIRTER, M. (1952) Effects of sunlight and ammonia on the action of silver-iodide particles as sublimation nuclei. *Bull. Am. met. Soc.* **33,** 26. (193)
—— and NEILL, H. W. (1955) The distribution and discharge of thunderstorm charge centres. *J. Met.* **12,** 1. (494)

RICH, T. A. (1955) A photo-electric nucleus counter with size discrimination. *Geofis. pura appl.* **31**, 60. (34)

RIGBY, E. C. and MARSHALL, J. S. (1952) Modification of rain with distance fallen. *McGill University Stormy Weather Group, Sci. Rep. MW-3*. (613)

ROBBINS, R. C., CADLE, R. D., and ECKHARDT, D. L. (1959) The conversion of sodium chloride to hydrogen chloride in the atmosphere. *J. Met.* **16**, 53. (73–4)

ROBERTS, D. E. (1959) Melting bands and precipitation rates. *Decca Radar. Radar Res. Lab. Rep. RL* 1902. (415)

ROBERTS, P. and HALLETT, J. (1968) A laboratory study of the ice nucleating properties of some mineral particulates. *Q. Jl R. met. Soc.* **94**, 25. (190, 192)

ROBERTSON, S. D. and KING, A. P. (1946) The effect of rain upon the propagation of waves in the 1–3 cm region. *Proc. Inst. Radio Engrs* **34**, 178. (428)

ROGERS, R. R. (1963) Radar measurement of velocities of meteorological scatterers. *J. atmos. Sci.* **20**, 170. (441)

—— (1967) Doppler radar investigation of Hawaiian rain. *Tellus* **19**, 432. (446)

—— and PILIÉ, R. J. (1962) Radar measurements of drop-size distribution. *J. atmos. Sci.* **19**, 503. (442)

RUCKLIDGE, J. (1965) The examination by electron microscope of ice crystal nuclei from cloud chamber experiments. *J. atmos. Sci.* **22**, 301. (198)

RUPE, J. H. (1950) Critical impact velocities of water droplets as a problem in injector-spray samplings. *Jet Prop. Lab. Calif. Inst. Tech. Rep.* 4–80. (93)

RUTKOWSKI, W. and FLEISHER, A. (1955) R meter; an instrument for measuring gustiness. *M.I.T. Weath. Radar Res. Rep.* No. 24. (441)

RYDE, J. W. (1946) The attenuation and radar echoes produced at centimetre wavelengths by various meteorological phenomena. *Meteorological factors in radio-wave propagation*, p. 169. Physical Society and Royal Meteorological Society. (404, 428, 548)

SAFFMAN, P. G. and TURNER, J. S. (1956) On the collision of drops in turbulent clouds. *J. Fluid Mech.* **1**, 16. (154)

SANDER, A. and DAMKÖHLER, G. (1943) Übersättigung bei der spontanen Keimbildung in Wasserdampf. *Naturwissenschaften* **31**, 460. (14, 16, 18, 173)

SANO, I., FUJITANI, Y., and MAENA, Y. (1960) The ice-nucleating property of some substances and its dependence on particle size. *Mem. mar. Obs. Kobe* **14**, 1. (214, 216 218, 219)

SANSOM, H. W. (1968) A four year hail suppression experiment using explosive rockets. *Proc. int. Conf. on Cloud Physics, Toronto*, p. 768. (391)

—— BARGMAN, D. J. and ENGLAND, G. (1955) Report on experiments on artificial stimulation of rainfall at Mityana, Uganda, Sept.–Dec. 1954. *E. Afr. Met. Dep. Mem.* Vol. 3, No. 4. (380)

SARRICA, O. (1965) Observational results on hail formation and structure. *Ricerca scient.* **35**, 345. (345)

SARTOR, J. D. (1954) A laboratory investigation of collision efficiencies, coalescence and electrical charging of simulated cloud droplets. *J. Met.* **11**, 91. (582)

—— (1961) Calculations of cloud electrification based on a general charge separation mechanism. *J. geophys. Res.* **66**, 831. (521)

—— (1967) The role of particle interactions in the distribution of electricity in thunderstorms. *J. atmos. Sci.* **24**, 601. (555, 559)

—— and ABBOTT, C. E. (1968) Charge transfer between uncharged water drops in free fall in an electric field. *J. geophys. Res.* **73**, 6415. (555)

SAUNDERS, C. P. R. (1968) The influence of cloud electrification on ice crystal aggregation. *Proc. int. Conf. on Cloud Physics, Toronto*, p. 619. (249)

SAUNDERS, P. M. and RONNE, F. C. (1962) A comparison between the height of cumulus clouds and the height of radar echoes received from them. *J. appl. Met.* **1,** 296. (473)

SAVIC, P. (1953) Circulation and distortion of liquid drops falling through viscous medium. *Natn. Res. Council of Canada, Div. Mech. Engng, Rep. No. NRC-MT-22.* (599)

SAWYER, K. F. and WALTON, W. H. (1950) The conifuge—a size-separating sampling device for airborne particles. *J. scient. Instrum.* **27,** 272.(43)

SAX, R. I. (1970) Drop freezing by Brownian contact nucleation. Ph.D. Thesis, University of London. (180, 212)

SCHAEFER, V. J. (1946) The production of ice crystals in a cloud of supercooled water droplets. *Science, N.Y.* **104,** 457. (240, 370)

—— (1949a) The detection of ice nuclei in the free atmosphere. *J. Met.* **6,** 283. (175)

—— (1949b) The formation of ice crystals in the laboratory and the atmosphere. *Chem. Rev.* **44,** 291. (194, 265)

——(1951) Report on cloud studies in Puerto Rico. Project Cirrus Report, p. 11. (295)

—— (1953) Flight experiments under Project Cirrus. *Project Cirrus, Final Report,* Chap. 4. (295)

—— (1954) The concentration of ice nuclei in air passing the summit of Mt. Washington. *Bull. Am. met. Soc.* **35,** 310. (187, 199)

—— (1957) The question of meteoritic dust in the atmosphere. In *Artificial stimulation of rain* (eds. H. Weickmann and W. Smith), p. 18. Pergamon Press. (205)

SCHARRER, L. (1939) Kondensation von übersättigten Dämpfen an Ionen. *Annln Phys.* **35,** 619. (18)

SCHINDELHAUER, F. (1913) Über die Elektrizität der Niederschläge. *Phys. Z.* **14,** 1292. (560)

—— (1925) Versuch einer Registrierung der Tropfenzahl bei Regenfällen. *Met. Z.* **42,** 25. (603)

SCHMID, P. (1967) On 'Grossversuch III' a randomized hail suppression experiment in Switzerland. *Proc. 5th Berkeley Symposium on Mathematical Statistics and Probability,* Vol. 5, p. 141. (391)

SCHMIDT, W. (1909) Eine unmittelbare Bestimmung der Fall-geschwindigkeit von Regentropfen. *Sber. Akad. Wiss. Wien* **118,** 71. (595)

SCHOLZ, J. (1932) Vereinfachter Bau eines Kernzählers. *Met. Z.* **49,** 381.(34)

SCHONLAND, B. F. J. (1928a) The polarity of thunderstorms. *Proc. R. Soc.* **A118,** 233. (492)

—— (1928b) The interchange of electricity between thunderclouds and the Earth. *Proc. R. Soc.* **A118,** 252. (559, 560)

—— (1938) Progressive lightning. IV The discharge mechanism. *Proc. R. Soc.* **A164,** 132. (498, 510)

—— and CRAIB, J. (1927) The electric field of South African thunderstorms. *Proc. R. Soc.* **A114,** 229. (493)

—— HODGES, D. B. and COLLENS, H. (1938) Progressive lightning. V. A comparison of photographic and electrical studies of the discharge process. *Proc. R. Soc.* **A166,** 56. (501)

SCHOTLAND, R. M. (1957) The collision efficiency of cloud droplets. *New York Univ. Coll. Engng Final Report No. AF*19(604)-993. (582)

—— (1960) Experimental results relating to the coalescence of water drops with water surfaces. *Discuss. Faraday Soc.* No. 30, p. 72. (587)

SCHUMANN, T. E. W. (1938) The theory of hail formation. *Q. Jl R. met. Soc.* **64,** 3. (227, 341, 349-51)

SCHWERDTFEGER, W. (1948) Über die Bildung von Regenschaüern über See. *Met. Rdsch.* **1**, 453. (291)

SCORER, R. S. and LUDLAM, F. H. (1953) The bubble theory of penetrative convection. *Q. Jl R. met. Soc.* **79**, 94. (315)

SCOTT, W. D. and HOBBS, P. V. (1967) The formation of sulfate in water droplets. *J. atmos. Sci.* **24**, 54. (70)

SCOTT, W. T. (1968) Analytic studies of cloud drop coalescence. *J. atmos. Sci.* **25**, 54. (146–7)

SCRASE, F. J. (1938) Electricity on rain. A discussion of records obtained at Kew Observatory. 1935–6. *Geophys. Mem., Met. Office Lond.*, No. 75. (561)

SEDUNOV, Y. S. (1965) Fine cloud structure and its role in the formation of the cloud particle spectrum. *Izv. Akad. Nauk SSSR* Ser. Faio, **7**, 722. (English edn, p. 416). (143)

SEELY, B. K. (1952) Detection of micron and submicron chloride particles. *Analyt. Chem.* **24**, 576. (50)

SEMONIN, R. G. and PLUMLEE, H. R. (1966) Collision efficiency of charged cloud droplets in electric fields. *J. geophys. Res.* **71**, 4271. (580)

SERPOLAY, R. (1958) L'activité glaçogène des aérosols d'oxydes métalliques. *Bull. Obs. Puy de Dôme*, p. 81. (197, 218)

—— (1959) L'activité glaçogène des aérosols d'oxydes métalliques. II Les oxydes catalyseurs. *Bull. Obs. Puy de Dôme*, p. 81. (191, 197, 217, 218)

SHACKFORD, C. R. (1960) Radar indications of a precipitation-lightning relationship in New England thunderstorms. *J. Met.* **17**, 15. (517)

SHAFRIR, U. and NEIBURGER, M. (1963) Collision efficiencies of two spheres falling in a viscous medium. *J. geophys. Res.* **68**, 4141.(147, 149, 577–9)

SHAW, D. and MASON, B. J. (1955) The growth of ice crystals from the vapour. *Phil. Mag.* **46**, 249. (256-8, 279)

SHIMADA, T., SEKIHARA, K., and KAWAMURA, K. (1955) An effect of ultraviolet radiation on the action of silver iodide particles as sublimation nuclei. *J. met. Soc. Japan* **33**, 276. (231–2)

SHIRATORI, K. (1934) Ionic balance in air and nuclei over ocean. *Mem. Fac. Sci. Agric. Taihoku imp. Univ.* **10**, 175. (54)

SIEGERT, A. J. F. (1943) The fluctuations in signal from many independently moving scatterers. *M.I.T. Radiation Lab. Rep.* No. 465. (441)

SIMPSON, G. C. (1909) On the electricity of rain and its origin in thunderstorms. *Phil. Trans. R. Soc.* **A209**, 379. (523–4, 560)

—— (1919) Atmospheric electricity. British Antarctic Expedition 1910–13. *Meteorology*, Vol. 1, p. 302. Thacker, Spark & Co., Calcutta. (539)

—— (1927) The mechanism of a thunderstorm. *Proc. R. Soc.* **A114**, 376. (524)

—— (1941) On the formation of cloud and rain. *Q. Jl R. met. Soc.* **67**, 99. (285)

—— (1942) The electricity of cloud and rain. *Q. Jl R. met Soc.* **68**, 1. (539)

—— (1949) Atmospheric electricity during disturbed weather. *Geophys. Mem., Met. Office Lond.*, No. 84. (559, 564–5)

—— and ROBINSON, G. D. (1941) The distribution of electricity in thunderclouds, II. *Proc. R. Soc.* **A177**, 281. (512–5)

—— and SCRASE, F. J. (1937) Distribution of electricity in thunderclouds. *Proc. R. Soc.* **A161**, 309.(512–5, 539)

SIMPSON, J., BRIER, G. W., and SIMPSON, R. H. (1967) Stormfury cumulus seeding experiment 1965 statistical analysis and main results. *J. atmos. Sci.* **24**, 508. (375–7, 397)

SIVARAMAKRISHNAN, M. V. (1957) Point discharge current, the earth's electric field and rain charges during disturbed weather at Poona. *Indian J. Met. Geophys.* **8**, 379. (559)

―― (1961) Studies of raindrop size characteristics in different types of tropical rain using a simple raindrop recorder. *Indian J. Met. Geophys.* **12,** 189. (478–9 602, 609, 613)
SKATSKIĬ, V. I. (1963) An airborne water-content meter for clouds with sufficient liquid to form droplets. *Izv. Akad. Nauk SSSR* Ser. geofiz. No. 9 (English edn p. 885). (116)
SMITH, E. J. (1949) Experiments in seeding cumuliform cloud layers with dry ice. *Aust. J. scient. Res.* **A2,** 78. (373)
―― (1950) Observation of precipitation with an airborne radar. *Aust. J. scient. Res.* **A3,** 214.(293)
―― (1951) Observations of rain from non-freezing clouds. *Q. Jl R. met. Soc.* **77,** 33. (293)
―― (1967) Cloud seeding experiments in Australia. *Proc. 5th Berkeley Symposium on Mathematical Statistics and Probability*, Vol. 5, p. 161. (386)
―― ADDERLEY, E. E. and WALSH, D. T. (1963) Cloud seeding experiment in Snowy mountains, Australia. *J. appl. Met.* **2,** 324. (385)
―― and HEFFERNAN, K. J. (1954) Airborne measurements of the concentration of natural and artificial freezing nuclei. *Q. Jl R. met. Soc.* **80,** 182. (177, 184–5, 200, 230)
―― ―― (1956) The decay of the ice-nucleating properties of silver iodide released from a mountain top. *Q. Jl R. met. Soc.* **82,** 301. (231–3)
―― ―― and THOMPSON, W. J. (1958) The decay of the ice-nucleating properties of silver iodide released from an aircraft. *Q. Jl R. met. Soc.* **84,** 162. (232)
―― SEELY, B. K., and HEFFERNAN, K. J. (1955) The decay of ice-nucleating properties of silver iodide in the atmosphere. *J. Met.* **12,** 379. (231–2, 380)
SMITH, L. G. (1955) The electric charge of raindrops. *Q. Jl R. met. Soc.* **81,** 23. (565)
―― (1957) Intracloud lightning discharges. *Q. Jl R. met. Soc.* **83,** 103. (507)
SOUDAIN, G. (1951) Réalisation d'un compteur automatique de noyaux de chlorure de sodium. *J. scient. Mét.* **3,** 137. (48)
SOULAGE, G. (1955) Étude de générateurs de fumées d'iodure d'argent. *Bull. Obs. Puy de Dôme*, No. 1, p. 1.(230)
―― (1958) Contribution des fumées industrielles à l'enrichissement de l'atmosphère en noyaux glaçogènes. *Bull. Obs. Puy de Dôme*, No. 4, p. 121. (197)
SPAR, J. (1957) Project Scud. *Met. Monogr.* **2,** 5. American Meteorological Society, Boston. (395)
SPILHAUS, A. F. (1948) Drop size, intensity, and radar echo of rain. *J. Met.* **5,** 161. (467)
SPYERS-DURAN, P. A. and BRAHAM, R. R. (1967) An airborne continuous cloud-particle replicator. *J. appl. Met.* **6,** 1108. (95)
SQUIRES, P. (1952) The growth of cloud drops by condensation. *Aust. J. scient. Res.* **A5,** 59. (125)
―― (1956) The microstructure of cumuli in maritime and continental air. *Tellus* **8,** 443. (105)
―― (1958a) The microstructure and colloidal stability of warm clouds. Pt. I. The relation between structure and stability. (1958b) Pt. II. The causes of the variations in microstructure. *Tellus* **10,** 256, 262. (83, 94, 105–7, 111, 112).
―― (1958c) The spatial variation of liquid water and droplet concentration in cumuli. *Tellus* **10,** 372. (120)
―― (1966) An estimate of the anthropogenic production of cloud nuclei. *J. Rech. Atmos.* **2,** 297. (79)
―― and GILLESPIE, C. A. (1952) A cloud-droplet sampler for use on aircraft. *Q. Jl R. met. Soc.* **78,** 387. (94, 98)

SQUIRES, P. and SMITH, E. J. (1949) The artificial stimulation of precipitation by means of dry ice. *Aust. J. scient. Res.* A**2**, 232. (373)
—— and TWOMEY, S. (1966) A comparison of cloud nucleus measurements over central north America and the Caribbean Sea. *J. atmos. Sci.* **23**, 401. (83)
—— —— (1958) Some observations relating to the stability of warm cumuli. *Tellus* **10**, 272. (108)
—— and WARNER, J. (1957) Some measurements in the orographic cloud of the island of Hawaii and in trade wind cumuli. *Tellus* **9**, 475. (105)
STARR, J. R. and MASON, B. J. (1966) The capture of airborne particles by water drops and simulated snow flakes. *Q. Jl R. met. Soc.* **92**, 490. (583)
STEPHENS, J. J. (1961) Radar cross-sections for water and ice spheres. *J. Met.* **18**, 348. (417–8)
STERGIS, C. G., REIN, G. C., and KANGAS, T. (1957) Electric field measurements above thunderstorms. *J. atmos. terr. Phys.* **11**, 83. (567)
STEVENSON, C. M. (1968) An improved millipore filter technique for measuring the concentrations of freezing nuclei in the atmosphere. *Q. Jl R. met. Soc.* **94**, 35. (179)
STEWART, J. B. (1964) Precipitation from layer cloud. *Q. Jl R. met. Soc.* **90**, 287. (290)
STEYN, K. (1950) The Pretoria hailstorm. *Publ. Wks S. Afr.* **10**, No. 75. (346)
STICKLEY, A. R. (1940) An evaluation of the Bergeron–Findeisen precipitation theory. *Mon. Weath. Rev. U.S. Dep. Agric.* **68**, 272. (289)
STIMSON, M. and JEFFREY, G. B. (1926) The motion of two spheres in a viscous fluid. *Proc. R. Soc.* A**111**, 110. (577)
ST. LOUIS, P. T. and STEELE, R. L. (1968) Certain environmental effects on silver iodide nuclei. *Proc. int. Conf. on Cloud Physics*, Toronto, p. 178. (234)
STOTT, D. and HUTCHINSON, W. C. A. (1965) The electrification of freezing water drops. *Q. Jl R. met. Soc.* **91**, 80. (208, 544)
STOUT, G. E. and MUELLER, E. A. (1968) Survey of relationships between rainfall rate and radar reflectivity in the measurement of precipitation. *J. appl. Met.* **7**, 465. (478–9)
STYLES, R. S. and CAMPBELL, F. W. (1953) Radar observations of rain from non-freezing clouds. *Aust. J. Phys.* **6**, 73. (293)
SULAKVELIDZE, G. K. (1968) On the principles of hail control method applied in the USSR. *Proc. int. Conf. on Cloud Physics*, Toronto, p. 796. (392–3)
SWIFT, D. L. and FRIEDLANDER, S. K. (1964) The coagulation of hydrosols by Brownian motion and laminar shear flow. *J. Colloid Sci.* **19**, 621. (59)

TAKAGI, M. (1961) The mechanism of discharges in a thundercloud. *Proc. Res. Inst. Atmospherics. Nagoya Univ.* **8B**, 1. (508)
TAKAHASHI, T. (1966) Thermoelectric effect in ice. *J. atmos. Sci.* **23**, 74. (535)
TELFORD, J. (1955) A new aspect of coalescence theory. *J. Met.* **12**, 436. (146)
—— (1960a) Freezing nuclei from industrial processes. *J. Met.* **17**, 676. (197)
—— (1960b) Freezing nuclei above the tropopause. *J. Met.* **17**, 86. (201)
—— and THORNDIKE, N. S. C. (1961) Observations of small drop collisions. *J. Met.* **18**, 382. (585, 590)
—— —— and BOWEN, E. G. (1955) The coalescence between small water droplets. *Q. Jl R. met. Soc.* **81**, 241. (584)
THAMS, J. C. et al. (1966) Die Ergebnisse des Grossversuches III zur Bekampfung des Hagels in Tessin in den Jahren 1957–63. *Met. Zentralanstalt Veroff.*, No. 3. Zurich. (390)
THORPE, A. D. and MASON, B. J. (1966) The evaporation of ice spheres and ice crystals. *Br. J. appl. Phys.* **17**, 541. (125, 277, 352, 614)

TOBA, Y. (1966) Critical examination of the isopiestic method for the measurement of sea-salt nuclei masses. *Tokyo Geophys. Inst. Spec. Contrib.* **6**, 59. (46)
TOHMFOR, G. and VOLMER, M. (1938) Die Keimbildung unter den Einfluss elektrischer Ladungen. *Annln Phys.* **33**, 109. (19)
TOLBERT, C. W. and GERHARDT, J. R. (1956) Measured rain attenuation of 4.3 mm wavelength radio signals. *Univ. Texas elect. Engng Res. Lab. Rep.* No. 83. (429)
—— and BAHN, W. W. (1959) Rainfall attenuation of 2.15 mm radio wavelengths. *Univ. Texas elect. Engng Res. Lab. Rep.* No. 109. (429)
TOLMAN, R. C. (1949) The effect of droplet size on surface tension. *J. chem. Phys.* **17**, 333. (4)
TRABERT, W. (1901) Die Extinction des Lichtes in einem trüber Medium (Schweite in Wolken). *Met. Z.* **18**, 518. (118)
TRUBY, F. K. (1955) Hexagonal microstructures of ice crystals grown from the melt. *J. appl. Phys.* **26**, 1416. (273)
TURNBULL, D. and FISHER, J. C. (1949) Rate of nucleation in condensed systems. *J. chem. Phys.* **17**, 71. (165, 167)
TURNER, J. S. (1955) The salinity of rainfall as a function of drop size. *Q. Jl R. met. Soc.* **81**, 418. (90)
TWOMEY, S. (1953a) The identification of individual hygroscopic particles in the atmosphere by a phase-transition method. *J. appl. Phys.* **24**, 1099. (28)
—— (1953b) On the measurement of precipitation intensity by radar. *J. Met.* **10**, 66. (479)
—— (1954) The composition of hygroscopic particles in the atmosphere. *J. Met.* **11**, 334. (28)
—— (1955) The distribution of sea-salt nuclei in air over land. *J. Met.* **12**, 81. (63)
—— (1959a, b) (a) The nuclei of natural cloud formation. Pt. 1. The chemical diffusion method and its application to atmospheric nuclei. (b) Pt. 2. The supersaturation in natural clouds and the variation of cloud droplet concentration. *Geofis. pura appl.* **43**, 227, 243. (37, 83)
—— (1960) On the nature and origin of natural cloud nuclei. *Bull. Obs. Puy de Dôme*, No. 1, p. 1. (76, 80)
—— (1964) Statistical effects in the evolution of a distribution of cloud droplets by coalescence. *J. Met.* **12**, 436. (146–9)
—— (1966) Computations of rain formation by coalescence. *J. atmos. Sci.* **23**, 404. (146–9)
—— and SEVERYNSE, G. T. (1963) Measurements of size distributions of natural aerosols. *J. atmos. Sci.* **20**, 392. (58)
—— —— (1964) Size distributions of natural aerosols below 0.1 micron. *J. atmos. Sci.* **21**, 558. (58)
—— and SQUIRES, P. (1959) The influence of cloud nucleus population on the microstructure and stability of convective clouds. *Tellus* **11**, 408. (85)
—— and WARNER, J. (1967) Comparison of measurements of cloud droplets and cloud nuclei. *J. atmos. Sci.* **24**, 702. (86, 108)

VALI, G. (1968) Ice nucleation relevant to the formation of hail. *McGill Univ. Stormy Weather Group Sci. Rep.* MW-58. (163–4)
VAN DEN HEUVEL, A. P. and MASON, B. J. (1959) Habit of ice crystals grown in hydrogen, carbon dioxide, and air at reduced pressure. *Nature, Lond.* **184**, 519. (264)
—— —— (1963) The formation of ammonium sulphate in water droplets exposed to gaseous sulphur dioxide and ammonia. *Q. Jl R. met. Soc.* **89**, 271. (71)

VERAART, A. W. (1931) Meer Zonneschijn in het Nevelig Noorden; meer regen in de Tropen. *Seyffardt's Boek en Muziekhandel* Amsterdam. (370)

VERZÁR, F. (1953) Kondensationskernzähler mit automatischer Registrierung. *Arch. Met. Geophys. Bioklim.* **A5**, 372. (34)

VIRGO, S. E. (1950) Tropical rainfall from cloud which did not extend to the freezing level. *Met. Mag., Lond.* **79**, 237. (294)

VITTORI, O. (1955) Détérmination de la nature chimique des aérosols. *Arch. Met. Geophys. Bioklim.* **A8**, 204. (50)

—— and DI CAPORIACCO, G. (1959) The density of hailstones. *Nubila* **2**, 51. (346)

VOLMER, M. (1939) Kinetik der Phasenbildung. Steinkopff, Dresden and Leipzig. (20)

—— and FLOOD, H. (1934) Tröpfchenbildung in Dämpfen. *Z. phys. Chem.* **A170**, 273. (14, 15)

—— and WEBER, A. (1926) Keimbildung in übersättigten Gebilden. *Z. phys. Chem.* **A119**, 277. (11)

VONNEGUT, B. (1947) The nucleation of ice formation by silver iodide. *J. appl. Phys.* **18**, 593. (227–8)

—— (1948a) Variation with temperature of the nucleation rate of supercooled liquid tin and water drops. *J. Colloid Sci.* **3**, 563. (156)

—— (1948b) Influence of butyl alcohol on shape of snow crystals formed in the laboratory. *Science, N.Y.* **107**, 621. (265)

—— (1949a) A capillary collector for measuring the deposition of water drops on a surface moving through clouds. *Rev. scient. Instrum.* **20**, 110. (116)

—— (1949b) Nucleation of supercooled water clouds by silver iodide smoke. *Chem. Rev.* **44**, 277. (230)

—— (1955) Possible mechanism for the formation of thunderstorm electricity. *Proc. int. Conf. on Atmospheric Electricity Portsmouth, New Hampshire, Geophys. Res. Pap.* No. 42, p. 169. (525)

—— (1965) Thunderstorm theory. *Problems of atmospheric and space electricity*, p. 285. Elsevier, Amsterdam. (518)

—— and MAYNARD, K. (1952) Spray-nozzle type silver-iodide generator for airplane use. *Bull. Am. met. Soc.* **33**, 420. (372)

—— MOORE, C. B. and BOTKA, A. T. (1959) Preliminary results of an experiment to determine initial precedence of organized electrification and precipitation in thunderstorms. *J. geophys. Res.* **64**, 347. (518)

—— —— and MALLAHAN, F. J. (1961) Adjustable potential-gradient-measuring apparatus for airplane use. *J. geophys. Res.* **66**, 2393. (516)

—— and NEUBAUER, R. (1951) Recent experiments on the effect of ultraviolet light on silver iodide nuclei. *Bull. Am. met. Soc.* **32**, 356. (231–2)

—— (1953) Counting sodium-containing particles in the atmosphere by their spectral emission in a hydrogen flame. *Bull. Am. met. Soc.* **34**, 163. (48)

WAKESHIMA, H. (1954) Time lag in self-nucleation. *J. chem. Phys.* **22**, 1614. (12)

WALDMANN, L. (1959) The force of a non-homogeneous gas on small suspended spheres. *Z. Naturf.* **14a**, 589. (89)

WALL, E. (1947) Uber die Entstehung der Schneekristalle. *Wiss. Arb. dt. Dienst. franz. zone.* **1**, 151. (246-7)

WALTER, B. (1903) Über die Entstehungweise des Blitzes. *Annln Phys.* **10**, 393. (496)

WALTON, W. H. (1950) Practice of electron microscopy—airborne particles. *J. R. microscop. Soc.* **70**, 45. (47)

—— and PREWETT, W. C. (1949) The production of sprays and mists of uniform drop size by means of spinning-disc type sprayers. *Proc. phys. Soc.* **B62,** 341. (94)

WARNER, J. (1955) The water content of cumuliform cloud. *Tellus* **7,** 449. (120)

—— (1957) An instrument for the measurement of freezing nucleus concentration. *Bull. Obs. Puy de Dôme*, p. 33. (176,186)

—— (1968) Gust velocities and the droplet spectrum in cumulus clouds. *Proc. int. Conf. on Cloud Physics*, Toronto, p. 138. (108)

—— and NEWNHAM, T. D. (1952) A new method of measurement of cloud-water content. *Q. Jl R. met. Soc.* **78,** 46. (116,120)

—— and TWOMEY, S. (1956) The use of silver iodide for seeding individual clouds. *Tellus* **8,** 453. (375)

WARREN, D. R. and NESBITT, M. V. (1955) An airborne silver iodide dispensing burner. *Aust. Dep. Supply Aero. Res. Lab., Mech. Engng Note ARL/ME* 200. (230)

WARSHAW, M. (1967) Cloud droplet coalescence: statistical foundation and a one-dimensional sedimentation model. *J. atmos. Sci.* **24,** 278. (146,149-51)

—— (1968) Cloud-droplet coalescence: effects of the Davis-Sartor collision efficiency. *J. atmos. Sci.* **25,** 874. (149-51)

WATSON, H. H. (1936) A system for obtaining dust samples from mine air. *Trans. Instn. Min. Metall.* **46,** 155. (46)

WEGENER, A. (1911) Thermodynamik der Atmosphäre, pp. 81 and 289. Barth, Leipzig. (189, 284)

WEGENER, P. and SMELT, R. (1950) Summary of research on liquefaction phenomena in hypersonic wind tunnels. *U.S. Naval Ordinance Lab. Memo.* 10772. (12)

WEICKMANN, H. K. (1947) Die Eisphase in der Atmosphäre. *Reports and Translations* No. 716. Ministry of Supply (A) Völkenrode. (244-7)

—— (1950) 'Biologie' der Schneekristalle. *Umschau* **50,** 116. (255)

—— (1957a) Recent measurements of the vertical distribution of Aitken nuclei. In *Artificial stimulation of rain* (eds. H. Weickmann and W. Smith), p. 81. Pergamon Press, London and New York. (54)

—— (1957b) The snow crystal as aerological sonde. In *Artificial stimulation of rain* (eds. H. Weickmann and W. Smith), p. 315. Pergamon Press, London and New York. (247)

—— and AUFM KAMPE, H. J. (1950) Preliminary experimental results concerning charge generation in thunderstorms concurrent with the formation of hailstones. *J. Met.* **7,** 404. (546)

—— (1953) Physical properties of cumulus clouds. *J. Met.* **10,** 204. (98–103, 119)

WESTMANN, J. (1907) Form und Grösse der Schneekristalle. *Met. Z.* **24,** 333. (244)

WEXLER, H. (1945) The structure of the September 1944 hurricane when off Cape Henry, Virginia. *Bull. Am. met. Soc.* **26,** 156. (294)

WEXLER, R. (1948) Rain intensities by radar. *J. Met.* **5,** 171. (479)

—— (1952) Precipitation growth in stratiform clouds. *Q. Jl R. met. Soc.* **78,** 363. (306,459-61)

—— (1955) The melting layer. *Blue Hill Obs., Harvard Univ., Met. Radar Studies* No. 3. (463)

—— and ATLAS, D. (1958) Moisture supply and growth of stratiform precipitation. *J. Met.* **15,** 531. (307,461)

—— CHMELA, A. C. and ARMSTRONG, G. M. (1967) Wind field observations by doppler radar in a New England snowstorm. *Mon. Weath. Rev. U.S. Dep. Commerce* **95,** 929. (457)

—— REED, R. J. and HONIG, J. (1954) Atmospheric cooling by melting snow. *Bull. Am. met. Soc.* **35,** 48. (462)

WHIPPLE, F. J. W. (1929) On the association of the diurnal variation of electric potential gradient in fine weather with the distribution of thunderstorms over the globe. *Q. Jl R. met. Soc.* **55,** 1. (484)

—— and CHALMERS, J. A. (1944) On Wilson's theory of the collection of charge by falling drops. *Q. Jl R. met. Soc.* **70,** 103. (522)

—— and SCRASE, F. J. (1936) Point discharge in the electric field of the Earth. An analysis of continuous records obtained at Kew Observatory. *Geophys. Mem., Met. Office, Lond.* No. 68, p. 1. (484, 493, 559, 566)

WHIPPLE, F. L. and HAWKINS, G. S. (1956) On meteors and rainfall. *J. Met.* **13,** 236. (204)

WHYTLAW-GRAY, R. and PATTERSON, H. S. (1932) *Smoke: a study of aerial disperse systems,* Arnold, London. (80)

WIELAND, W. (1956) Die Wasserdampfkondensation an natürlichen Aerosol bei geringen Übersattigungen. *Z. angew. Math. Phys.* **7,** 428. (37, 83–4)

WIESNER, J. (1895) Beiträge zur Kenntnis des tropischen Regens. *Sber. Akad. Wiss. Wien* **104,** 1397. (602)

WILLIAMS, E. L. (1949) The pulse integrator. Part A; description of the instrument and its circuitry. *M.I.T. Weather Radar Res. Tech. Rep.* No. 8a. (437)

WILLIS, J. T., BROWNING, K. A., and ATLAS, D. (1964) Radar observations of ice spheres in free fall. *J. atmos. Sci.* **21,** 103. (348, 420)

WILSON, A. T. (1959) Surface of the ocean as a source of airborne nitrogenous material and other plant nutrients. *Nature, Lond.* **184,** 99. (75)

WILSON, C. T. R. (1897) Condensation of water vapour in the presence of dust-free air and other gases. *Phil. Trans. R. Soc.* **A189,** 265. (1)

—— (1899) On the comparative efficiency as condensation nuclei of positively and negatively charged ions. *Phil. Trans. R. Soc.* **A193,** 289. (14, 17, 18)

—— (1916) On some determinations of the sign and magnitude of electric discharges in lightning flashes. *Proc. R. Soc.* **A92,** 555. (487–9)

—— (1920) Investigations on lightning discharges and on the electric field of thunderstorms. *Phil. Trans. R. Soc.* **A221,** 73. (487–9, 493)

—— (1922) The maintenance of the earth's charge. *Observatory* **45,** 393. (484)

—— (1923) Atmospheric electricity. *Dictionary of applied physics,* Vol. 3, p. 101. Macmillan. (558)

—— (1929) Some thundercloud problems. *J. Franklin Inst.* **208,** 1. (521–2)

WILSON, J. W. (1966) Storm-to-storm variability in the radar reflectivity–rainfall relationship. *Proc. 12th Weather Radar Conf., Norman, Oklahoma,* p. 229. (481)

WOODCOCK, A. H. (1950a) Sea salt nuclei in a tropical storm. *J. Met.* **7,** 397. (61)

—— (1950b) Condensation nuclei and precipitation. *J. Met.* **7,** 161. (286)

—— (1952) Atmospheric salt particles and raindrops. *J. Met.* **9,** 200. (62)

—— (1953) Salt nuclei in marine air as a function of altitude and wind force. *J. Met.* **10,** 362. (61, 62)

—— and GIFFORD, M. M. (1949) Sampling atmospheric sea-salt nuclei over the ocean. *J. mar. Res.* **8,** 177. (46, 59)

—— KIENTZLER, C. F., ARONS, A. B., and BLANCHARD, D. C. (1953) Giant condensation nuclei from bursting bubbles. *Nature, Lond.* **172,** 1144. (75)

WOODS, J. D. (1965) The effect of electric charges upon collisions between equal-size water drops in air. *Q. Jl R. met. Soc.* **91,** 353. (590)

—— and MASON, B. J. (1964) Experimental determination of collection efficiencies for small water droplets in air. *Q. Jl R. met. Soc.* **90,** 373. (152, 583)

—— —— (1965) The wake capture of water drops in air. *Q. Jl R. met. Soc.* **91,** 35. (585–7)

WORKMAN, E. J. and HOLZER, R. E. (1942) The electrical structure of thunder

storms. *Tech. notes natn. advis. Comm. Aeronaut., Wash.* No. 850. (494, 511)
—— and REYNOLDS, S. E. (1948) A suggested mechanism for the generation of thunderstorm electricity. *Phys. Rev.* **74,** 709. (526–7)
—— —— (1949) Electrical activity as related to thunderstorm cell growth. *Bull Am. met. Soc.* **30,** 142. (473, 518)
—— —— (1950a) Thunderstorm electricity. *New Mexico Instn Min. Tech. Prog. Rep.* No. 6. (186)
—— —— (1950b) Thunderstorm research programme in the New Mexico School of Mines. *Proc. Conf. on Thunderstorm Electricity, Chicago,* p. 29. Chicago University Press. (512)
—— —— (1950c) Electrical phenomena occurring during the freezing of dilute aqueous solutions and their possible relationship to thunderstorm electricity. *Phys. Rev.* **78,** 254. (526)
WORMELL, T. W. (1927) Currents carried by point discharges beneath thunderclouds and showers. *Proc. R. Soc.* **A115,** 443. (558)
—— (1930) Vertical electrical currents below thunderstorms and showers. *Proc. R. Soc.* **A127,** 567. (493, 558, 566)
—— (1939) The effects of thunderstorms and lightning discharges on the earth's electric field. *Phil. Trans. R. Soc.* **A238,** 249. (489, 493)
—— (1953) Atmospheric electricity; some recent trends and problems. *Q. Jl R. met. Soc.* **79,** 3. (486, 502, 559)
WRIGHT, H. L. (1932) Observations of smoke particles and condensation nuclei at Kew Observatory. *Geophys. Mem., Met. Office Lond.* No. 57. (53)
—— (1936) The size of atmospheric nuclei: some deductions from measurements of the number of charged and uncharged nuclei at Kew Observatory. *Proc. phys. Soc.* **48,** 675. (24)
—— (1939) Atmospheric opacity: a study of visibility observations in the British Isles. *Q. Jl R. met. Soc.* **65,** 411. (39)
—— (1940) Atmospheric opacity at Valentia. *Q. Jl R. met. Soc.* **66,** 66. (39)
WYLIE, R. G. (1953) The freezing of supercooled water in glass. *Proc. phys. Soc.* **B66,** 241. (158, 160, 168)

YAMAMOTO, G., OGIWARA, S., YOSHIDA, K. *et al.* (1952) Observations of the rate of growth of ice crystals artificially produced in the atmosphere. *Sci. Rep. Tôhoku Univ.* Geophys. Ser. **4,** 83. (278)
—— and OHTAKE, T. (1953) Electron microscope study of cloud and fog nuclei. *Sci. Rep. Tôhoku Univ.* Geophys. Ser. **5,** 141. (65)
—— —— (1955) Electron microscope study of cloud and fog nuclei—II. *Sci. Rep. Tôhoku Univ.* Geophys. Ser. **7,** 10. (65)
YOUNG, R. E. G. and BROWNING, K. A. (1967) Wind tunnel tests of simulated spherical hailstones with variable roughness. *J. atmos. Sci.* **24,** 58. (348)

ZAITSEV, V. A. (1948) New method of determining the water content of cloud and fog (in Russian). *Trudy glav. geofiz. Obs.* **13,** 75. (104, 117)
—— (1950) Liquid water content and distribution of drops in cumulus cloud (in Russian). *Trudy glav. Geofiz. Obs.* **19,** 122. (119)
ZELDOVICH, J. (1942) Theory of the formation of a new phase. *Zh. eksp. teor. Fiz.* **12,** 525. (11)
ZELENY, J. (1933) Variation with temperature of the electrification produced in air by the disruption of water drops and its bearing on the prevalence of lightning. *Phys. Rev.* **44,** 837. (523)
ZETTLEMOYER, A. C., TCHEUREKDJIAN, N., and CHESSICK, J. J. (1961) Surface properties of silver iodide. *Nature, Lond.* **192,** 653. (234)

Subject Index

Acoustic ice-nucleus detector, 182.
Adsorption of water vapour
  on ice nuclei, 189.
  on silver iodide, 234–5.
Aerosols
  in the atmosphere (*see* Condensation nuclei and Ice nuclei).
  chemical composition of, 63–6.
  coagulation of, 80–2.
  production of, 66–80.
  removal of, 87–91.
  size distribution of, 50–1, 56–63.
  techniques for collection and measurement of, 32–50, 92–8.
Aerosol spectrometer, 45.
Air bubbles, bursting of, in sea-water, 75–9.
Aitken counter (*see* Nucleus counters).
Aitken nuclei (*see* Condensation nuclei).
Altielectrograph, 512–15.
Ammonia
  in the atmosphere, 72–3.
  effect of, on ice nuclei, 193–4.
Ammonium sulphate particles, in the atmosphere, 70–1.
  formation of, 70–1.
Aqueous solutions, freezing of, 162–3, 526–7.
Artificial ice nuclei (*see under* Ice nuclei).
Artificial stimulation of precipitation, 369–98 (*see also* Cloud seeding).
Attenuation of radar signals (*see* Radar).

Back-scattering cross-section (*see* Radar reflectivity).
Bergeron mechanism of rain formation, 284–5, 306–8 (*see also* Ice-crystal mechanism).
Boys camera, 497–8.
Brownian capture of aerosols, 80–2, 88, 212.
Bubbles, bursting of, in sea-water, 75–9.
  electrification of, 528–31.

Cadmium iodide, as ice nucleus, 191, 214–16, 220–2.
Capillary collector water-content meter, 116.
Capillary electrometer, 488.
Cascade impactor, 42.
Chain reaction of raindrop multiplication, 327–9.

Charge (*see* Electric charge).
Chemical constitution of condensation nuclei, 64–6.
Chemical detection of aerosol particles, 50, 64–6.
Chemical reactions in the atmosphere, 66–75.
Chlorides
  in the atmosphere, 74.
Clay minerals as ice nuclei, 190–3, 194–9.
Clouds
  constitution of (*see* Cloud droplets, Ice crystals, Precipitation).
  electrification of, 483–568.
  evaporation of, 315–17.
  formation of, by penetrative convection, 314–17.
  ice crystals in (*see* Ice crystals, Snow crystals).
  layer-cloud systems, 288–90, 297–306.
  liquid-water content of, 113–21 (*see also* Liquid-water content).
  mother-of-pearl clouds, 174.
  precipitating clouds (*see* Precipitation).
  radar studies of (*see* Radar).
  seeding of (*see* Cloud seeding).
  shower clouds (*see* Showers).
  supersaturations achieved in, 1, 31, 125–38.
  thunderclouds (*see* Thunderstorms).
  updraughts in, 133–6, 301–3, 314–17, 326–7, 446–58, 471–7.
  visibility in, 102, 118.
Cloud chambers
  diffusion type, 34–8.
  expansion chamber (*see* Expansion cloud chamber).
  mixing type, 176–8 (*see also* Mixing cloud chamber).
Cloud droplets
  charges upon, 153.
  coalescence of, 145–53.
  collision efficiency of, 152–3, 569–87.
  condensation nuclei in, 65
Cloud-droplet camera, 95.
Cloud-droplet growth
  growth of single droplet, 122–5.
  growth of droplet population
    by coalescence, 145–53.
    by condensation, 125–45.
  effect of mixing upon, 133–8.
  theoretical treatments of, 122–54.
  effect of turbulence upon 140–5, 153–4

SUBJECT INDEX

Cloud-droplet radius, 98.
  mean radius, 98.
  mean-volume radius, 98.
  median-volume radius, 98.
  mode radius, 98.
  predominant radius, 98.
Cloud-droplet samplers, 94–8.
Cloud-droplet size and size distribution, 92–113.
  in relation to condensation nuclei, 85–6, 140.
  in cumulus, 98–108.
  in fog, 110–11.
  in layer clouds, 106–7, 109–10.
  summary of results, 111–13.
  techniques for measurement of, 92–8.
  theoretical treatments of, 122–54.
  effect of turbulence upon, 140–5.
Cloud-seeding, 369–98.
  of cumulus with dry ice, 370, 371–4.
  of cumulus with hygroscopic nuclei, 380–1.
  of cumulus with silver iodide, 374–7, 385–91.
  of cumulus with water drops, 377–80.
  of cumulonimbus, to suppress hail, 391–3.
  dissipation of clouds by, 380–2.
  of layer clouds with dry ice, 270, 380–2.
  of layer clouds with silver iodide, 371.
  of orographic clouds, 387–8, 395–6.
  large-scale seeding operations, 382–93.
    design of, 385–93.
    evaluation of, 383–5.
    operation of, 385–93.
    problems arising from, 394–5.
  overseeding of clouds, 382, 397.
  potentialities of, 395–8.
  principles of, 369.
  randomized procedures for, 384–91.
Coagulation of nuclei, 80–2.
Coalescence of cloud drops, 145–53.
Coalescence of water drops, 587–91.
Coalescence efficiency of water droplets, 569, 591.
  effect of electric charge upon, 588–91.
  effect of electric field upon, 52–3, 588–91.
Coalescence mechanism of rain formation
  artificial stimulation of, 369, 377–80.
  description and theory of, 285–7, 308–14, 317–27.
  initiation of, by giant salt nuclei, 286–7, 292, 320.
  radar evidence for, 285, 291–7.
  theory of, in layer clouds, 288, 308–14.
  observational evidence for, in layer clouds, 289–90, 295.
  theory of, in shower clouds, 286–7, 317–27.
  observational evidence for, in shower clouds, 291–4, 294–7, 326–7.
Collection efficiency of ice particles, 249.
Collection efficiency of water drops, 145, 152, 569.
  experimental determination of, 582–7.
Collision efficiency of water drops, 152, 569–87.
  calculation of, 569–80.
  effect of electric charge upon, 152–3, 580–1.
  effect of electric field upon, 152–3, 580–1.
  effect of droplet wake upon, 573–4, 584–6.
  influence of turbulence upon, 140–5.
Condensation coefficient, 6, 12.
Condensation nuclei, 31–91.
  activation of, in clouds, 83–7, 125–38.
  Aitken nuclei, 31–9.
    concentrations of, 52–6, 73.
    counting and measurement of, 32–9.
    distribution with heght, 54.
    electron-microscope examination of, 38.
    identification of, 38.
    ionic mobility of, 38.
    size of, 35.
    variability in concentration over land, 52–4.
    variability in concentration over sea, 54–6.
  coagulation of, 80–2.
  critical radii of, 27.
  critical supersaturations for activation of, 27.
  equilibrium sizes as functions of relative humidity, 25–9.
    of insoluble nuclei, 27.
    of mixed nuclei, 29–30.
    of soluble nuclei, 25–9.
  removal of, from atmosphere, 87–91.
  large and giant nuclei, 32, 39–50, 56–66, 83–9.
    chemical analysis of, 50, 64–6.
    in cloud formation, 83–9.
    collection and measurement of, 39–50.
    concentrations of, 56–63, 83–6.
    constitution of, 64–6.
    identification of, 49–50, 65.
    mean size from visibility observations, 39.
    size determination by ultra- and electron-microscope, 50, 58.
    size distribution of, over land and sea, 51, 56–63.

Condensation Nuclei *contd.*
  mixed nuclei, 29
  multiplication of nuclei by crystallization of solution droplets, 28.
  nature of condensation nuclei, 63–6.
  oceans as a source of, 75–9.
  origin of condensation nuclei, 66–80.
  production of condensation nuclei, 66–80.
    by combustion, 67–74.
    by coagulation, 80–2.
    by mechanical disruption of matter, 66, 75–9, 80.
    from sea salt, 75–9.
    by smokes and gaseous reactions, 67–75.
    rate of production at ocean surface, 78.
    rate of production over land, 79–80.
  role in rain formation, 286–7, 292, 320.
  size and size distribution of, 51, 56–63.
  variations with height and wind-speed, 59–63.
  sources of, 66–80.
  transport of, 62–3, 97–8.
  washout by rain, 89–90.
Condensation nucleus counters (*see* Nucleus counters).
Condensation of water vapour, 122–5.
  upon atmospheric nuclei, 20–30.
  in expansion chambers, 1, 13–20.
  homogeneous (spontaneous) condensation, 1–17.
  upon insoluble nuclei, 20–4.
  upon mixed nuclei, 29–30.
  upon soluble nuclei, 24–9.
  preferential condensation upon negative ions, 20.
  role of, in precipitation release, 283.
  upon small ions, 17–20.
  in supersonic wind tunnels, 12.
Conductivity (electrical) of the dry atmosphere, 486–7.
  in clouds, 552.
  variation with height, 487.
Conifuge, 43–5.
Convection, role of, in formation of cumulus clouds, 314–17.
Cosmic rays, as source of ions, 485.
Crystal growth, theories of, 267–73.
Crystallization of solution droplets, 28.
Crystals (*see* Ice crystals, Snow crystals, etc.).
Cumulonimbus, (*see also* Showers and Thunderstorms).
  characteristics of, 290–7, 314–17, 471–7.
  growth and development of, 314–17, 471–7.
  precipitation release from, 314–32, 348–61.
  radar studies of, 291–7, 446–54, 471–7.
  structure of, 314–17, 446–54, 471–7.
  vertical velocities in, 446–54, 471–7.
Cupric sulphide as ice nuclus, 214–16, 221–2.
Currents (electrical) of fine-weather field, 486–7.
  point-discharge (*see under* Point-discharge current).
  in thunderstorms, 499, 506–8, 567–8.

Dielectric constants of ice and water at centimetre wavelengths, 404.
Diffusion cloud chambers for studies
  of condensation nuclei, 34–8.
  for ice nuclei, 213–17.
Diffusiophoresis, 88.
Dinger–Gunn effect, 527–31.
Dislocations in ice crystals, 273–4.
Doppler spectrum of radar signals from clouds, 438–58.
Drag coefficients for water drops 592–3.
Drizzle, 282.
  evaporation of, 282.
  formation of, in layer clouds, 288, 308–14.
Droplet formation
  in clean, supersaturated water vapour, 9–17.
  in clouds (*see* Cloud droplets).
  by condensation on nuclei (*see* Condensation nuclei).
  on ions, 17–20.
Droplet-size distribution (*see* Cloud and Raindrops).
Dry ice, as ice-nucleating agent,
  use of, in cloud seeding (*see* Cloud seeding).
Dust counter, 42.

Electric charge
  on cloud droplets, 153.
  on continuous rain, 531, 560–1, 564.
  density of, in the atmosphere, 485–7.
  density on earth's surface, 483.
  earth's charge and its maintenance, 487, 557–68.
  neutralization of, by lightning, 492–3.
  on precipitation, 559–66.
  on soft hail, 566.
Electrification produced by:
  air bubbles bursting in water, 528–31.
  breaking raindrops, 523–5.
  collision and fracture of ice crystals, 539–43.

Electrification *contd.*
  collision between ice crystals and hail pellets, 540–3.
  freezing of aqueous solutions, 526–7.
  freezing and splintering of drops, 543–6.
  melting of ice, 527–31.
  rime formation, 546–50.
Electric charge in thunderstorms
  distribution of, 490–4, 505, 511–15.
  generation and separation of, 520–57.
    basic requirements of theory of, 520.
    by breaking raindrops, 523–5.
    by convective transport of charge, 525–6.
    by collision and separation of droplets and polarized hydrometeors, 521, 554–5.
    by collision and separation of ice crystals and polarized hail pellets, 551–4.
    by droplet splintering during growth of hail pellets, 555–7.
    Frenkel's theory of, 522.
    by melting snow and hail, 527–31.
    Wilson's influence theory of, 521–2, 550–1.
    Workman–Reynolds theory of, 527.
  leakage of, due to point discharges and precipitation, 496, 552.
  lightning flashes, charges involved in, 493, 507, 514.
  transfer of charge between atmosphere and earth, 557–68.
    by lightning discharges, 557–8.
    by point-discharge currents, 558–9.
    by precipitation, 559–66.
Electric currents
  associated with fine-weather field, 486–7.
  point-discharge currents (*see under* Point discharge).
  in thunderstorms, 499, 506–508, 567–8.
Electric fields
  of the atmosphere, in fine weather, 483–4.
    in small cumulus, 581
    currents associated with, 486–7.
    diurnal and seasonal variations of, 484–5.
    intensity of, 483.
    potential difference between atmosphere and earth, 487.
    variation of field intensity with height, 484.
  produced by thunderstorms and lightning flashes, 487–96, 500–8.
    breakdown fields in thunderstorms, 509.
    fields measured inside thunderstorms, 512–17.
    calculation of fields in thunderstorms, 550–7.
    fields associated with component strokes of lightning flash, 500–8.
    interpretation of field-changes at the ground, 490–6.
    measurement of field-changes at ground, 487–94, 518.
    radiation fields produced by lightning, 493, 500.
    recovery of field after a flash, 494–6, 553, 556.
    reversal of field changes with distance, 492–3.
Electric moment of thunderclouds:
  destruction of, by lightning, 492–3.
  moments associated with components of lightning flash, 493, 506–8.
  regeneration of, 496, 553, 556.
Electrical balance-sheet for earth's surface, 566–8.
Electrical conductivity,
  of dry atmosphere, 486–7.
  in clouds, 552.
  variation with height, 487.
Electrical resistance of atmosphere, 487.
Electron diffraction studies
  of condensation nuclei, 64–5, 70.
  of snow-crystal nuclei, 197–9.
  of silver iodide, 228–9.
Electron microscopy
  of Aitken nuclei, 38.
  of larger condensation nuclei, 65.
  of snow-crystal nuclei, 197–9.
Electrostatic cloud-droplet sampler, 95.
Elevated sphere method of measuring electric field, 489.
Embryo droplets formed in supersaturated water vapour, 1–13.
  dynamical equilibrium between embryos of different sizes, 6–7.
  in statistical equilibrium with the vapour, 5–6.
  size distribution of, in supersaturated vapour, 7–9.
Evaporation
  of clouds, 315–17.
  of cloud drops, 283.
  of drizzle drops, 282, 312–13.
  of raindrops, 282, 312–13.
Expansion chambers
  condensation upon ions in, 17–20.
  droplet-concentration as function of supersaturation in, 13–17.
  ice-nucleation studies in, 164, 173–4, 175–6, 213–14.
  spontaneous condensation of water vapour in, 1, 13–20.

SUBJECT INDEX

Fallstreifen, 283.
Field-mill instruments for measuring electric fields, 490–1.
Filter-paper method for raindrop size determinations, 602.
Fine-weather electric field of the atmosphere, 483–7.
Flame counter for detection of salt particles, 48.
Flour method for raindrop sizes, 603.
Fog droplets, nature of nuclei in, 65.
  droplets, size distribution of, 110–11.
  dispersal of, 382.
Foil sampler for raindrops, 606.
Freezing (*see also* Supercooling)
  of cloud droplets, 155.
  of droplets in cloud chambers, 155, 164, 173–8.
    freezing nuclei, 157–64, 189–91.
      (*see also* Ice nuclei).
  and shattering of water drops, 207–12.
  spontaneous freezing, 155, 164–72.
  of water samples containing foreign particles, 156–64.

Giant nuclei (*see* Condensation nuclei).
Graupel particles (*see* Soft hail).
Grossversuch III, 390–1.

Hail, 283, 332–68.
  classification of, 332–3.
  duration of, 332.
  frequency of, 332.
  prevention of, 391–3.
  radar reflectivity of, 423–6.
  small hail, 333, 341.
  soft hail (graupel)
    charge upon, 566.
    definition of, 333.
    density of, 238, 340.
    dimensions of, 238–40.
    fallspeed of, 240.
    formation of, in shower clouds, 329–32.
    mass of, 355, 356.
    melting of, 368, 531.
    polarization of, in electric fields, 551–5.
    role in thunderstorm electrification, 517, 551–5.
    sizes of, 238–40.
    structure of, 340.
Hailstones
  aerodynamics of, 347–8.
  air-bubble structure of, 343–5.
  crystalline structure of, 342–4.
  density of, 346–7.
  drag coefficients of, 347–8.
  growth of, 348–61.
  heat balance of, 350–9.
  isotopic analysis of, 362–4.
  lobed structure of, 345–6, 359–61.
  melting of, 366–8.
  radar cross-sections of, 423–6.
  recycling of, in storm, 359–61.
  size and shape of, 333–5.
  structure of, 341–6.
  terminal velocities of, 347–9, 359.
  theories of hailstone growth, 348–66.
  wet hailstones, 341–6, 351–9.
Hailstorm
  effect of wind shear upon, 359–60.
  structure of, 359–61, 474–7.
  suppression of, by seeding, 391–3.
  updraughts in, 359–62, 365–6.
Homogeneous condensation (*see* Condensation).
Homogeneous nucleation (*see* Nucleation).
Hot-wire water content meter, 115–16.

Ice
  accreted ice
    density, structure, and air content of, 336–40.
    compact ice, 336.
    rime ice, 335.
    spongy ice, 336.
  thermoelectric effects in, 531–9.
Ice crystals (*see also* Snow crystals)
  adhesion of, 249–51.
  classification of, 236–7.
  charging of, by collision and fracture, 539–43.
    by splintering, 539–40.
  combination forms of, 246–7.
  concentrations of, in cumulus, 211–12.
  dendritic forms of, 237–40, 245–6.
    growth of, in the laboratory, 251–5, 260–1, 263–4.
    occurrence of, in clouds, 245–6, 288–90, 297–306.
  etch pits on, 270–74.
  growth of, in the laboratory, 251–67.
  growth layers on surface of, 267–8, 271–3.
  growth rate of, 274–9.
    calculation of, 274–8.
    experimental measurement of, 278–9.
    in layer clouds, 287, 306–8, 459–62.
  growth habit of, 251–70.
    experimental studies of, 251–67.
    in electric field, 265–7.
    mechanism of, 267–270.
  hopper crystals, 270–2.
  international classification of, 236–7.
  needle-like forms of, 237–40, 246.

Ice crystals contd.
  plate-like forms of
    dimensions of, 245.
    in medium-level clouds, 245.
    growth of, in the laboratory, 251-7, 260-4.
    occurrence at the ground, 244.
    trigonal plates, 245.
  prismatic forms of
    in cirrus, 244.
    clusters of, 245, 248.
    dimensions of, 245.
    growth of, in the laboratory, 251-7.
    hollow prisms, 245.
    occurrence at the ground, 244.
    with pyramidal ends, 247, 257, 262.
  secondary production of, 207-12.
  surface structure of, 270-4.
Ice crystal mechanism of precipitation release
  artificial stimulation of (see Cloud seeding).
  description and theory of, 287, 306-8, 329-32.
  radar evidence for, 288, 301-6, 464-8.
  theory of, in layer clouds, 306-8, 459-62.
  observational evidence for, in layer clouds, 288-90, 297-306.
  theory of, in shower clouds, 329-32.
  observational evidence for, in shower clouds, 290-3.
Ice-forming nuclei, 174-235.
  artificial ice nuclei, 212-35.
    experimental techniques for study of, 174-83, 213-17, 220-3.
    industrial sources of, 195, 197.
    inorganic compounds as, 213-23.
      ammonium fluoride, 214-16, 222
      cadmium iodide, 214-16, 220-2.
      cupric sulphide, 214-16, 221-2.
      cuprous iodide, 215.
      iodine, 214-16, 221-2.
      lead iodide, 214-16, 220-2.
      mercuric iodide, 214, 216, 221-2.
      metal oxides, 216-18, 222.
      silver iodide, 190, 214-16, 220-2, 227-35.
        (see also under Silver iodide)
      silver oxide, 214-16, 222.
      silver sulphide, 214-16, 222.
      vanadium pentoxide, 214-16, 221-2.
    organic compounds as, 223-7.
    in relation to crystal properties, 214, 222, 225-7.
  atmospheric ice nuclei
    abundance of, 183-9.
    activation at water saturation, 189-90.
    activation at sub-water saturation, 189-90.
    activity as function of cooling rate, 183.
    activity as function of temperature, 183-7.
    concentration of, in space and time, 187-9, 199-202.
    concentrations of, as function of altitude, 200-2.
    identification of, in snow crystals, 197-9.
    multiplication of, by bursting of frozen drops, 207-12.
    multiplication of, by splintering of ice crystals, 207
    nature of, 194-200.
    origin of, 194-9, 201-6.
    preactivation of, 190-3.
    soil, silicate particles as, 190-3, 194-9, 206.
    techniques for detection and counting of, 174-83.
  critical radius of, 169.
  deactivation of, 193-4.
  effect of size, 190.
Impaction methods for collecting aerosols, 40-6, 92-6.
Iodine, as ice-forming nucleus, 214-16, 221-2.
Ions
  capture of, by raindrops, 519, 521-3, 525, 550, 564-6.
  concentrations of, in the atmosphere, 485-6.
  condensation upon, 17-20.
  mobility of, 38.
  production of, 485-6.
  recombination of, 485-6.
Isopiestic method of measuring soluble nuclei, 46.

Junction streamers in lightning discharges, 501-7.

Kaolinite as ice nucleus, 190-3, 195-9.
Kelvin equation for equilibrium vapour pressure
  over a curved liquid surface, 2, 24.
  derivation of, 3-5.
Konimeter, 42.
Koschmieder's formula, 118.

Layer-cloud systems, 289-90, 297-306, 441-6, 454-7, 458-68.
Lead iodide as ice nucleus, 214-16, 220-2.

SUBJECT INDEX

Lightning discharges
  air discharges, 500, 502, 508.
  breakdown processes, 509–10, 512.
  correlation with hail, 517.
  correlation with precipitation, 517–20.
  continuous streamer, 505–7.
  currents in, 499, 506–8.
  dart leader strokes of, 499, 504–6.
  electric field-changes produced by, 500–8.
  energy in, 516.
  initiation of, 509–12.
  interstroke processes, 501–7.
  intra-cloud discharges, 500, 502, 507–8.
  junction streamers, 501–7.
  K-changes, 506–8.
  M-changes, 505–7.
  movement of electric charges by, 502–3
  occurrence in relation to precipitation and radar echoes, 517–20.
  photography of, 496–500.
  pilot leader, 498, 509–10.
  return strokes of, 499, 501–3, 506, 510–11.
  S-field changes, 501–2, 506–7.
  stepped leader strokes of, 498, 501–3, 506, 509–10.
  structure of, 496–511.
  from warm clouds, 519–20, 554–5.
  world-wide frequency of, 558, 567.
Liquid-water content of clouds, 102, 113–21.
  calculated values of, 133–6.
  of cumulus clouds, 102, 119–20.
  of fogs, 121.
  of layer clouds, 121.
  measurement of, 113–21.
    by capillary collector, 116
    by hot-wire instrument, 115–16.
    by paper-tape instrument, 116–17.
    by riming cylinders, 114–15.
    by rotating disk, 114–15.
    summary of results, 119–21.
    by transmissometer method, 118.

Magnesium-oxide method of droplet-size measurement, 93–5.
Metallic oxides as ice nuclei, 191, 197.
Meteoritic dust as ice nuclei, 194, 202–6.
Meteor showers and rainfall, 202–6.
Millipore filters, 47, 50, 178–80.
Mixing-cloud chamber
  used for study of ice nuclei, 176–8, 213–18, 223–5.
Mobility of ions, 38.
Mother-of-pearl clouds, 174.

Nitrates in the atmosphere, 64, 73.
Nitric acid, formation of, in the atmosphere, 73.
Nitrogen oxides in the atmosphere, 73–4.
Nucleation
  of a crystal face, 273.
  homogeneous nucleation in water vapour, 1–17.
    comparison of theory and experiment, 13–17.
    critical radius of nucleus in, 1, 3–5.
    rate of nucleus formation, 9–11.
    relaxation time of, 11–12.
    thermodynamic and kinetic theory of, 1–13.
  spontaneous nucleation of supercooled water, 155, 164–72.
    in cloud chambers, 161, 164, 173–4.
    rate of, 167–70.
    dependence on rate of cooling, 166–7, 170.
    theory of, 167–72.
  of supercooled water by foreign particles, 156–64.
Nuclei
  Aitken nuclei (see Aitken nuclei).
  condensation nuclei (see Condensation nuclei).
  ice-forming nuclei (see Ice nuclei).
  spontaneous formation of nuclei in clean water vapour, 1–17.
  sublimation nuclei, 189.
Nucleus counters
  Aitken nucleus counters, 32–5.
  diffusion cloud chambers, 35–8.
  flame counter for salt nuclei, 56.
  Nolan–Pollak photo-electric counter, 34.
  Owens dust counter, 42.
  Rich counter, 34.

Oiled-slide method for cloud-droplet size measurement, 93.
Opacity of atmosphere (see Visibility).
Optical methods, for cloud-droplet size determination, 96–9.
Organic compounds as ice nuclei, 223–7.
Overseeding of clouds, 307.
Owens dust counter, 42.

Paper-tape water-content meter, 116–17.
Point-discharge currents
  as measure of electric field-changes, 493–4.
  related to charges on precipitation, 525, 565.

role of, in limiting fields beneath thunderstorms, 495.
role of, in transfer of charge between atmosphere and ground, 558-9.
from trees, vegetation, etc. 558-9.
Polarity of thunderclouds, 492-4.
Polarization of radar signals from precipitation, 406-13, 422, 463-4.
Potential of atmosphere relative to earth, 487.
Potential gradient of the atmosphere (see Electric field).
Precipitating clouds
air motions in, 301-5.
characteristics of, 280-3, 288-97, 297-306, 314-17.
drizzle formation in, 288-90.
effect of dynamical factors upon, 280-1, 293, 296, 301-5, 394-7.
electric fields in (see under Electric fields).
geographical differences in behaviour of, 291-6.
in low latitudes, 294-7.
observations of cumuliform clouds in middle latitudes, 290-4.
observations of layer clouds in middle latitudes, 289-90, 297-306.
radar studies of, 285, 288, 291-7, 429-77.
rain formation in, 288-329.
snow formation in, 289-90, 299-306.
structure of, 297-306, 314-17.
vertical velocities in, 133-6, 301-4, 314-17, 326-7, 446-54, 456-8.
Precipitation
artificial stimulation of, 369-98 (see also Cloud seeding).
role of condensation in release of, 283
elements
electric charges on, 562-3, 565-6.
growth of, in layer clouds, 306-14.
growth of, in shower clouds, 316-32.
scattering of radar waves by (see under Radar).
factors, determining development of, 280-1.
forms of, 282-3.
release of, from layer-cloud systems, 297-314.
release of, from shower clouds, 314-32.
Precipitation mechanisms, 283-8, 306-14, 317-32.
coalescence mechanism (see Coalescence mechanism).
ice-crystal mechanism (see Ice-crystal mechanism).

natural precipitation processes, 280-368.
Pulse integrator, 437.

Radar
basic theory of, 400-3, 413-17.
detection of non-precipitating clouds by, 426, 471.
detection of precipitating clouds by, 429-77.
displays, A-scope, PPI, RHI, CAPPI, 429-32.
equation for power received from a target, 402.
verification of, 413-17.
Doppler radar signals from precipitation
determination of drop size distribution, 441-6.
determination of horizontal winds and convergence, 454-8.
determination of particle fallspeed, 441-6.
determination of vertical air motions, 446-54, 456-8.
theory of, 438-41.
VAD technique, 303, 454-7.
millimetric radar, 471.
Radar echoes from precipitation
amplitude fluctuations of, 434-6.
Doppler spectrum of, 438-58.
from cold fronts, 432.
from cumulonimbus, 433, 471-7.
from hurricanes, 433.
intensity measurements of, 413-14, 434-7.
iso-echo contours of, 431-2.
from layer-cloud systems, 297-306, 441-6, 454-7, 458-68.
polarization of, 406-13, 463-4.
presentation of, 429-32.
from severe storms, 474-7.
from showers, 291-4, 446-54.
from thunderstorms, 471-7.
from warm fronts, 433.
Radar equipment
angular resolution of, 403.
beam width of, 403-4.
parameters, choice of, 403-4.
pulse length, 403.
range resolution of, 403.
sensitivity of, 403.
siting of, 404.
wavelength, choice of, 403.
Radar melting band, 458, 462-8.
general explanation of, 458.
location of, 462.
polarization of signals from, 463-4.

SUBJECT INDEX

Radar melting band *contd.*
  shape and orientation of particles in, 463–4, 467–8.
  theory of, 465–8.
  thickness of, 463–5.
Radar reflectivity
  of hail, 423–6.
  of ice and snow particles, 408–13, 415.
  of ice-water particles, 420–2, 424–5.
  of non-precipitating clouds, 426, 471.
  of precipitation echo as function of height, 464, 465.
  of rain, 404–5.
  of snow, 409–10, 415–17, 458–70.
  of spherical scattering particles with $D \ll \lambda$ (Rayleigh regime), 400–5.
    with $D > \lambda/20$ (Mie regime), 416–26.
    measurements of, 419–422.
  of non-spherical particles,
    in Rayleigh regime, 405–13.
    of larger particles, 422–6.
    measurements of, 422–3.
  effect of polarization on, 406–13, 422, 463–4.
  effect of shape and orientation on, 405–13, 466–8.
Radar signals from precipitation
  amplitude fluctuations of, 434–6.
  analysis of, 434–41.
  attenuation of, 402–3, 426–9.
  average intensity of, 436–7.
  Doppler spectrum of, 438–51 (*See also* under Doppler radar).
  integration of, 432, 437.
  measurements of, 413–17, 435–7.
Radar upper bands, 468–70.
  description of, 468–9.
  generation of, 470.
  location of, 470.
  theory of, 468.
Rain (*see also* Precipitation).
  definition of, 282.
  from non-freezing clouds, 285–8, 290–4.
  electrification of, 559–61, 564–5.
Raindrops
  charges upon, 562–3, 565–6.
  disruption of, 327–9, 509, 523–5, 600–2.
  drag coefficients of, 592–3.
  evaporation of, 282, 312–14.
  growth of, by coalescence (*see* Coalescence).
  polarized radar signals from, 464.
  production by chain reaction, 327–9.
  shape of, 598–600.
  size distribution of
    from Doppler radar, 441–6.
    formula for, 608–10.

    related to radar-echo intensity, 405 477–80.
    techniques for measurement of, 441–6, 602–8.
    variation of, with height, 441–6, 613.
    variation of, with type of rain, 609–13.
  terminal velocities of, 441–6, 592–8.
    calculated values of, 592–4.
    formula for, 597.
    measurement of, 441–6, 594–7.
Raindrop microphone, 603–5.
Raindrop spectrographs, 605–8.
Rainfall rate, measured by radar, 477–82.
  accuracy of, 478–82.
  related to radar signal intensity, 405, 477–80.
Rainmaking (*see* Cloud seeding).
Recombination of ions, 485–6.
Replica techniques for
  registration of cloud droplets, 95.
  registration of ice crystals, 95, 243.
Resistance (electrical) of atmosphere, 487.
Rime
  charge associated with formation of, 546–50.
Riming-cylinder method for cloud-droplet size determination, 96.
Rotating-cylinder water-content meter, 114–15.
Rotating-disk water-content meter, 114–15.

Salt particles
  chemical detection of, 50, 64–6.
  collection of, on hydrophobic slides, 46
  used in cloud seeding, 380–1.
  detection of, in cloud and fog droplets, 65.
  detection by flame counter, 48.
  as source of condensation nuclei, 75–9.
  transport of, inland, 62.
Saturation ratio, 5, 13–19.
Showers, 282.
Shower clouds
  chain reaction of raindrop multiplication in, 327–9.
  characteristics of, 290–7, 314–17.
  radar studies of, 291–7, 446–54.
  release of precipitation from, 314–32.
  structure of, 314–17, 446–54.
Shower production
  by coalescence of water drops, 285–7, 317–27 (*see also* Coalescence).
  criteria for, 316–17, 322–4.
  by growth of ice particles, 287, 329–32 (*see also* Ice-crystal process).

Shower production *contd.*
  over the sea,
    290-1, 294-7.
  by seeding (*see* Cloud seeding).
Silicates, as ice nuclei, 190-3, 194-9.
Silver iodide, as ice nucleus, 190, 214-16, 220-2, 227-35.
  activity of, as function of temperature, 217, 221, 230
  contaminants, effects upon, 193-4.
  used in cloud seeding (*see* Cloud seeding)
  crystal structure of, 222, 228-9.
  decay of, in the atmosphere, 231-2.
  discovery of, as an ice nucleus, 227.
  dispersion of, from airborne and ground generators, 229-33.
  electron-diffraction patterns of smokes, 228-9.
  generators, 229-31.
  photolysis of, 231-5.
  production of, as a smoke, 229-31.
  reduction of, to metallic silver, 231.
  effect of u.v. radiation upon, 231-5.
  X-ray patterns of smokes, 228-9.
Snow
  electrification of, 531, 539.
Snow crystals, 236-79 (*see also* Ice crystals).
  aggregation of, 248-51, 461, 466-8.
  classification of, 236-7.
  in cirrostratus clouds, 245.
  in cirrus clouds, 245.
  combination forms of, 246-7.
  correlation of type with cloud temperature, 244-8.
  correlation of type with ground temperature, 243-4, 247
  dendritic forms of, 245-6, 237-40.
  dimensions of, 237-40.
  fall-speeds of, 239-42.
  growth of, in the laboratory, 251-67.
  growth rates of, 274-9.
  masses of, 237-9.
  needle forms, 237-40, 246.
  photography of, 242-3.
  plastic replicas of, 242-3.
  plate-like forms, 244-5.
  prism forms, 244-5.
  riming of, 238-40, 248.
  sampling of, 242-4.
  splintering of, 207.
Snow-crystal habit
  influence of crystalline structure upon, 261.
    of environmental gas upon, 264.
    of fall motion upon, 277-8.
    of impurities upon, 264-5.

  of surface structure upon, 267-73, 279,
  mechanisms of, 267-70.
  metamorphosis of, 261.
  dependence upon temperature and supersaturation, 251-64.
Snow-crystal sonde, 247.
Snowflakes 236, 239-42.
  aggregation of, 461, 466-8.
  dimensions of, 241.
  fall speeds of, 239, 241-2.
  formation of, 248-51.
  melting of, 442, 458, 462-4, 465-8.
  orientation of, 408-13, 463-8.
  polarization of radar signals from, 408-13, 463-4.
  riming of, 241.
  scattering of radar waves by, 409-10.
  structure of, 248,
Sodium compounds in the atmosphere, 74-5.
Soil particles
  detected in clouds and fog droplets, 65.
  as ice nuclei, 190-9, 206.
Solution droplets
  crystallization of, 28.
  equilibrium radii of, as function of humidity, 24-9.
Spontaneous nucleation (*see* Nucleation).
Spiders' threads for collection of aerosol, 45-6.
Sublimation nuclei, 189.
Sulphur compounds in atmosphere, 67-72.
  sulphuric acid, 68, 71.
  sulphur dioxide, 67-70.
  sulphates, 67, 70-2.
Supercooled solutions for detection of ice nuclei, 175, 182.
Supercooled droplets
  in clouds, 155.
  spontaneous freezing of, 155, 164-72.
Supercooling
  of alcohol-water mixtures, 173.
  of aqueous solutions, 162-3.
  of water, 155-74 (*see also* Freezing).
    containing foreign nuclei, 156-64.
    dependence upon volume of sample, 157-62, 164-70.
    dependence upon rate of cooling, 160-61, 166-7, 170.
    structure of supercooled water, 171-2.
Supersaturation
  achieved in Aitken counter, 35.
  achieved in clouds, 1, 31, 128-36.
  achieved in diffusion chambers, 35-8.
  for condensation on pure water droplets, 26-8.

SUBJECT INDEX

Supersaturation *contd.*
   for condensation on salt nuclei, 26–8.
   for condensation on ions, 17–19.
   for spontaneous condensation, 1, 13–16.
   relative to ice in mixed clouds, 283–4.
Surface migration of molecules on ice crystals, 267–70.
Surface tension (or free energy).
   effect of curvature of surface upon, 4.
   of a small droplet, 2.
   of water-ice interface, 169–70.

Terminal velocities
   of water drops, 592–8.
   of snow crystals, 239–42.
Thermals, 315–17.
Thermal precipitator, 46, 178.
Thermodynamic functions, 2–3.
Thermoelectric effects in ice, 531–9.
   calculation of, 523–33, 536.
   effect of impurities upon, 534–5.
   measurements of, 533–6.
   mechanism of, 531–2.
   role in charging of ice crystals by collision, fracture, and asymmetric rubbing, 537–42.
Thunderstorms (*see also* Cumulonimbus).
   breakdown fields in, 509
   charge distribution in, 490–4, 505, 511–15.
   charge generation in, 520–57 (*see also* Electric charge).
   electric currents in, 499, 506–8, 567–8.
   electric fields associated with (*see* Electric fields).
   electric moments of, 492–3, 496.
   electrical structure of, 511–16.
   lightning in (*see* Lightning).
   life cycle of, 471–4.
   meteorological structure of, 471–7, 517–20.
   pocket of positive charge in base of, 512–14.
   polarity of, 492–4, 511–15.
   precipitation elements in, 517.
   radar studies of, 471–7, 518–19.
Trabert's formula, 118.

Transmissiometer for measuring visibility in clouds, 118.
Turbulence, in clouds
   effect upon droplet growth, 140–5, 153–4, 309, 312–13.
   effect upon disruption of raindrops, 327.

Ultra-microscope, for particle size determination, 48.
Ultraviolet light, effect upon silver iodide, 231–5.

Van't Hoff factor, 25.
Vapour pressure (of water)
   equilibrium value over surface of charged droplet, 18–19.
   equilibrium value over surface of solution droplet, 24–8.
   saturation vapour pressures over water and ice, 615.
Virga, 283.
Visibility in clouds, 102, 118.
   dependence upon nucleus population and humidity, 39.

Water content of clouds (*see* Liquid-water content).
Water drops (*see also* Raindrops).
   deformation of, in an electric field, 509.
   evaporation of, 312–13.
   shape and disruption of, 598–602.
   terminal velocities of, 592–8.
Water, supercooling and freezing of, 155–74.
Wegener–Bergeron mechanism of precipitation release, 284–5, 306–8. (*See also* Ice crystal mechanism).
Whitetop project, 388–9.
Wilson's flat-plate instrument, for electric-field measurement, 488.
Workman–Reynolds effect, 526–7.

Zinc sulphide particles, used as tracers in medium-range diffusion experiments, 231.